WORLD RAILWAY ACCIDENTS 1931-1945

Francis Voisey

ISBN: 978-1-9162011-8-7

This book is dedicated to Signalman Brian Stone,
who first awakened my lifelong interest in railways way back in 1966.

Thanks to the staff of the Exeter Reference Library and the University of Exeter Library for their assistance while I was researching this book, to my friend June Nightingale for encouraging me to go ahead and get it finished when my enthusiasm was waning a little, and to my publishing team, without whose guidance and assistance it is doubtful if I would have been able to navigate the mysteries of getting books published.

CONTENTS

INTRODUCTION

This book is the fifth of what it is hoped will be a series detailing railway accidents worldwide. A good many books have been written over the years about railway accidents, some concentrating on single disasters, others covering accidents in a given country or geographical area. Accidents in countries such as Great Britain, Germany, the Netherlands, the United States, Australia, New Zealand and so on have been documented comprehensively in one or more books but no attempt has been made previously to detail every significant accident worldwide. This is clearly an ambitious project, but four volumes dealing with the earlier years having been completed. This book will attempt to detail the accidents which occurred between 1931 and 1945. Previous volumes have covered the periods 1811-80, 1881-1900, 1901-15 and 1916-30. Further volumes are planned to cover the remainder of the twentieth century and the opening years of the twenty-first century. As before, an attempt will be made to include all accidents involving five or more fatalities to train passengers. A number of less serious accidents which have been considered of interest have also been included. The accidents have been grouped into chapters according to their primary cause and are dealt with as far as possible chronologically within each chapter.

The gathering of information about accidents in distant parts of the world has not been easy, though it is much easier nowadays than when this project began in the days before the internet. Outside of Great Britain and Ireland few countries seem to have had a formal body charged with investigating all serious railway accidents until recent years, though one or two major disasters such as the Bussey Bridge failure in the United States and the Baptiste Creek collision in Canada were the subject of full formal investigations even in the early days. Otherwise reliance has largely had to be placed on contemporary newspaper reports and previously published works on the subject, though the author has tried as far as possible to keep to primary sources. If the book appears to be biased towards accidents which occurred in Great Britain and the United States this is probably partly because these are the countries about which information on accidents has been easiest to obtain, though these were also in any case the countries where many of the most serious accidents did occur.

Contemporary spellings of place names have been used where appropriate and for the many American accidents, terms such as railroad, engineer, caboose and so on have been used instead of railway, driver, brake van etc. The 24-hour clock was gradually coming into use, beginning with Italy from 1893 and the 12-hour clock and 24-hour clocks will both be used depending upon the custom in the country concerned.

By the 1930s railways had spread to almost all parts of the world, though some countries only had very small systems or perhaps just one line.

The names and the boundaries of countries have undergone many changes over the years. Considerable changes occurred following the end of the First World War and the Treaty of Versailles in 1919 and there have of course been many name changes since, plus the splitting up of the former Soviet Union and Yugoslavia, the separation of India into India, Pakistan, and Bangladesh and so on. The names of some other countries and of numerous places have changed. The names of locations which were in use at the time of an accident occurred have been given, together with present names where appropriate, as for example Ürsküb (now Skopje).

CHAPTER 1

DERAILMENTS: TRACK DEFECTS

The tracks over which trains run are obviously a vital part of the railway and any problems with the track are likely to lead to the train being derailed. By 1931 the technology involved in laying railway track had much advanced since the early days, but train speeds and weights had also much increased, while accidents occurred from time to time due to failures to keep the track in a good state of maintenance, to insufficient allowance being made for hot weather expanding the rails, to damage caused to the track from floodwaters and so on. Also included in this chapter, though they were not strictly due to defects in the track, are accidents due to trains having become derailed on open points. Broken rails were another hazard which caused mishaps and worse from time to time.

A train carrying soldiers of the First Artillery Regiment who were returning from manoeuvres at Morella was derailed at Apasco, Mexico, between Queretaro and Mexico City on Friday 6th March 1931. Seven soldiers died and six were seriously injured. The derailment was caused by the rails spreading.[1] Some sources stated that there had been 11 fatalities in all, the other deaths being those of civilians, and that the casualties had mostly resulted from the victims having been crushed by the shells of artillery pieces which had broken from their fastenings and rolled over when the cars had toppled from the track.[2]

An express travelling at 90 kilometres an hour was derailed as it arrived at Étampes station, France, at 20:20 on the evening of Thursday 19th March 1931. The last two vehicles of the train, one of them a restaurant car, left the rails on a set of points. The restaurant car struck the third-class carriage of an omnibus train which was just being made up, the carriage of which was fortunately empty. The restaurant car itself was completely smashed and flattened over a length of 10 to 15 metres. It was the hour when passengers were taking their meals and of those seated at two tables, three were killed

[1] *The Corsicana Daily Sun*, Corsicana, Texas, United States, 7th March 1931
[2] *The Circleville Herald*, Circleville, Ohio, United States, 7th March 1931

outright along with two restaurant car staff. A dozen others were seriously injured by fragments of broken glass. The remainder of the train continued along the line for 700 metres before what had now become the rear coach left the rails in turn. The force of the derailment ripped the roof and left-hand wall from this coach; fortunately this was the corridor side so the passengers in their compartments escaped serious injury. The train was shortly pulled up and screaming passengers descended to the track, but calm was soon restored.

Station staff and other railwaymen immediately began the rescue work after the accident and help soon came from airmen at the nearby Étampes-Montplaisir camp, who had ambulances and automobiles in which the injured could be taken to the hospital. Shortly afterwards M. Devise, procurer de la Republique (crown prosecutor), M. Lechevallier, the *juge d'instruction* (examining magistrate) and M. Auzeau, the police commissioner, were on the scene and M. Deligne, the Minister of Public Works, arrived at 01:00 the next morning.

A newspaper reporter managed to interview M. Doublet, a lawyer who had been a passenger on the train. "I was riding in a compartment at the front of the train and had been intending to go to the restaurant car for the second sitting. Then as we arrived at Étampes we felt a shock. We did not react to the first shock but when it was followed by a second, we wanted to know what was happening. I didn't have time to see though before the train stopped amidst a smashing of windows and breaking of timber. With my fellow travellers I descended to the track. Cries were coming from both ends of the station. Only then did I realise exactly what had happened. The rescue work was very quickly organised. One must praise the company's employees for the dedicated way in which they dealt with the injured."

Three of those injured and taken to Étampes hospital died in the hours after the accident. Thirty-five more seriously injured patients were admitted to the hospital, with others less badly hurt being taken to local hotels.[3] By 27th March, with the death of Mme Francois Carranzé, the number of passengers who had lost their lives in the accident had risen to ten.[4]

Excessive rain (13 inches had fallen in just two days) undermined the track of the Canton – Kowloon railway and caused the derailment of a passenger train from Shunchun to Kowloon at Taipo station, Hong Kong, at 5.15 p.m. on Monday 20th April 1931. The engine and four carriages fell wholly or partly down an embankment, the first two carriages going into a stream and being buried in mud and debris, while the other two telescoped one another. One report spoke of imprisoned women and children struggling to escape amid the agonised cries of victims trapped in the wreckage. A large force of firemen, police and railwaymen struggled in the darkness to rescue the passengers, with the ever present danger of further slips in the embankment. As usual early reports of the number of casualties were exaggerated, speaking of 30 and even 100 dead, plus 20 or 30 injured. Two days after the accident it was stated that there were in fact 12 dead, plus 20 injured. The latter included two schoolboys who had been pinned in the wreckage and had to undergo amputations of their legs to release them. Sir William Peel, the Governor

---

[3] *La Liberté*, Paris, France, 21st March 1931
[4] *La Charente*, Angoulême, France, 27th March 1931

of Hong Kong, visited the scene of the accident. A relief train from Kowloon was stopped by another washaway and road vehicles sent to the scene found a road blocked by a landslide, so rescuers had to make their way to the train on foot. The victims were all thought to be Chinese; had the accident happened the day before there would have been many European golfers in the first-class carriage of the train, while had the accident happened a little later a crowded express rather than a local train would have been involved.[5] [6] [7]

A train was derailed near Leushkovskaya, USSR, 95 miles south of Rostov, on Wednesday 27th May 1931. During track repairs a rail had been removed but a watchman at Leushkovskaya station allegedly failed to warn the driver of a Moscow – Sotchi express. The engine and three carriages of the train left the track, and six people were killed and 11 injured.[8]

A way freight (what in Great Britain would be called a pick-up goods – one calling to pick up and set down at intermediate stations and sidings along a line) of the Canadian Pacific Railway from Prince Albert to Regina was derailed at Govan, Saskatchewan, Canada, on Saturday 12th September 1931. Over a dozen trespassers were riding in two cars of the train which were wrecked and at least five of them were killed. Five cars of wood, five of hay and one of wheat were piled up on the track. The derailment was attributed to a broken rail.[9]

Chemins de fer de Paris à Lyon et à la Méditerranée special train No. 403, which had been made up the evening before, set off at 07:15 on Wednesday 14th September 1932 from Sidi-bel-Abbès, Algeria, bound for Oujda, in Morocco, with two officers, Lieutenants Pariscot and Pénicaud; 27 non-commissioned officers and 481 men of the French Foreign Legion, whose headquarters were at Sidi-bel-Abbès. The men were all volunteers who were intended to take part in the final Pacification of Morocco, a military operation which had been in progress from 1907 and would end in 1934.[10] Behind the engine the train consisted of a van, eight goods waggons, 11 third-class coaches of four compartments, one third-class coach with five compartments, one third-class coach with six compartments, a second-class coach, five goods waggons and a van: 29 vehicles in all, weighing 329 tonnes. The maximum permitted load over the section of line was 40 loaded or 65 empty waggons, or with a four-couple engine weighing 55 tonnes a train weighing 452 tonnes. The train involved was therefore nowhere near the maximum weight permitted. Nor was its speed excessive, it was travelling at 36 kilometres an hour and the maximum authorised speed was 45 kilometres an hour. More significant was that fact that work on track renewal was being carried out in the area. The work had progressed to a point four kilometres beyond the scene of the derailment, but the new

[5] *The Malaya Tribune*, Singapore, 21st and 22nd April 1931
[6] *The Times*, London, England, 21st and 22nd April 1931
[7] *The Singapore Free Press and Mercantile Advertiser*, Singapore, 29th April 1931
[8] *Olean Evening Times*, Olean, New York, United States, 27th May 1931
[9] *The Sunday Star*, Washington, D.C., United States, 13th September 1931
[10] Webpage: Foreign Legion Info: 1932 Turenne Rail Accident

track there was still lacking some of its sleepers and hence in a weakened condition.

Most of the soldiers had removed their jackets because the stormy weather had made the atmosphere more oppressive and the heat more unbearable. However, they were chatting and singing, 'because the legionnaire, as we know, does not get sad,' and in any case, they were not likely to be bothered by the heat. The train had passed Tabia, Lamoricière, Tlemcen, Mansoura, and Zelboun and was entering the charming Zittoun valley.

Suddenly, at km 163.750, 9 km 900 from Zelboun, 5 kilometres from Turenne (now Sabra) and close to the exit from Tunnel No. 4, there was a terrible crash, frightful cries, broken windows, crashing metal, broken wood, and the engine, tender and 31 coaches were precipitated into a 20 ft deep ravine, leaving just the rear van on the track. The accident happened at 14:53.

Having been saturated by the heavy rains of the previous few days the embankment gave way beneath the weight of the train and the engine had left the rails, gouged deep ruts in the ballast and then gone into the ravine, taking most of the train with it. The regiment was buried under the mass of wreckage. Those who were unhurt or only suffered slight injuries were soon able to free themselves, and without even waiting to recover from their fright, they began, under a leaden sky, to quickly get to work to rescue those still buried. A shepherd had witnessed the accident and realised that he could not do much to assist on his own and it would be better to make his way as quickly as possible to Turenne, 5 kilometres away, and raise the alarm. On his way he met a policeman and quickly told him what had happened. The two of them then continued to Turenne, getting there at about the same time as Conductor Valéro, who had been riding in the rear van.

The Turenne stationmaster telegraphed Tlemcen at 15:25 and also informed the local population, who proceeded to the accident scene, led by their mayor, M. Ducros. A rescue train was made up at Tlemcen with doctors and nurses from the garrison and set off from there at 16:30, while a second train, which included a crane, started from Sidi-bel-Abbès at 17:30, though it was found that the terrain at the accident site did not permit the crane to be used. The rescue work was supervised by General Rollet, Inspector-General of the Foreign Legion, and commandant of the Tlemcen subdivision. Some of the injured soldiers, a reporter wrote, refused to let themselves be evacuated but insisted on remaining at the scene to continue the work of rescuing their mates who were still buried in the wreckage. At the time that the correspondent telephoned in his report the rescue work was still in progress with the aid of acetylene lamps, darkness having fallen, while the first batch of badly hurt soldiers had been admitted to the military hospital in Tlemcen.

Among the first dead to be identified were two of the railwaymen operating the train, Conductor Beltra and Driver Prieto. Fireman Rigaud managed to jump as the engine left the rails and escaped unhurt.[11]

A week after the disaster it was stated that 56 legionnaires and five railwaymen had lost their lives and 220 people had been injured.[12]

---

[11] *L'Echo d'Alger*, Algeria, France, 15th September 1932

[12] *L'Echo d'Tlemcen*, Tlemcen, Algeria, 20th September 1932

A train accident in Romania late on Monday 17th October 1932 cost 24 lives. It occurred just outside of the station at Temisoara when a coach and a postal van of a passenger train left the rails. The bodies of 17 victims were removed from the wreckage and seven more passengers died shortly afterwards, while another seven were seriously injured. The derailment was blamed on a faulty set of points.[13]

A serious error of judgement led to an accident in which five people died near Raynes Park, England, on the Southern Railway, on the afternoon of Thursday 25th May 1933. Two steam-hauled passenger trains were involved: the 3.10 p.m. Down service from London Waterloo to Alton and the 12.11 p.m. Up train from Southampton Terminus to London Waterloo. The Alton train had consisted of 0-4-4 tank engine No. 107 and five bogie coaches and the Southampton train of 2-6-0 engine No. 1621, a luggage van and five bogie coaches. The Alton train was in charge of Guard Robert White with Driver W.J. Adams and Fireman W.H. Mitchell; the Southampton train was in charge of Guard G. Sanders with Driver Downs and Fireman James Kemble.

The Alton train, running on the Down through line, was completely derailed a quarter of a mile east of Raynes Park signal box, fouling the Up through line, and before it had come to rest its leading coach was struck by the Southampton train, which was running under clear signals on the Up through line. A series of further impacts followed as the Up train struck successive coaches of the Down one. The accident happened at about 3.25 p.m. The passengers who died were Colonel Harold Norman and his wife, of Farnham; Mrs W.R. Dykes, of Mayford, near Woking; Mr. A.A. Hitchcock, of Woking; and nine-year-old Roderick Macdonald Duthie, of Portland, Oregon, United States. All had been on the Alton train. Another 34 passengers were injured, and Driver Downs of the Up train was badly hurt.

Several passengers told their stories to reporters: 'Mrs Coffey, of King Alfred Avenue, Bellingham, who was travelling in the Alton train with three children, said: - "There was a terrible crash, and it seemed that the whole of the train was being splintered to pieces. I had my baby in my arms, and I was thrown forward violently, but somehow managed to protect my baby. My little boy and little girl were both buried under various articles. I was helpless. I could not help them and could only listen to their cries. Some men pulled out the two children, and I walked along by the side of the stretcher comforting my little boy, who broke his leg. The little girl had head injuries. We were all taken to hospital. One moment we were on our way to Aldershot to stay with friends, and the next moment everything seemed to collapse upon us."

'Mr T.A. Duthie, of Portland, Oregon, one of whose boys was killed, said: - "I was born in Aberdeen. My wife comes from Chester. We have been in America for some years, and I have been in business in Portland, Oregon. I have been over in this country with my family on business for the past month. We were on our way to Aldershot to stay with friends and both the boys were looking forward to seeing the British soldiers."'

Fireman Mitchell of the Alton train was also interviewed: "I think the driver and I are two of the luckiest men alive tonight. It is a miracle that we are not among the casualties. All I remember is a violent jolt, and I was thrown against the tender with terrific force. When I recovered the train had stopped, and our first duty was to prevent fire spreading

---

[13] *The Times*, London, England, 19th October 1932

from the firebox. We looked behind and saw the damage that had been done. It seemed that half the train was completely wrecked. We had scarcely climbed down from the engine when rescuers had come along and were rendering first-aid." [14]

Lieut.-Colonel E.P. Anderson, who conducted the Ministry of Transport inquiry into the accident, wrote in his report that he had visited the scene of the accident only three hours after it had occurred. He stated that at the point of derailment the line ran on a low clay bank. The rails were 95 lb and 45 ft long, laid on 48 lb London and South Western Railway chairs. The creosoted wooden sleepers had mostly been laid in 1925 but a few dated from 1912. Flange marks on the sleepers indicated the point of derailment, a track jack was lying undamaged nearby and a number of sleepers on either side of the jack were off their beds, leaving the rails unsupported.

Mr G. Ellison, the Chief Engineer of the Company, had explained that it had been decided to realign the curve at the accident site from 64 to 80 chains while preserving the six-foot space between the tracks. The work had been undertaken in three stages; firstly realigning the curve between the 8 and 8½ mileposts, secondly extending the track drainage system and finally giving all roads a three-inch lift on new stone ballast. The realignment of the curves had been completed and it was then intended to put the road in sufficiently good order to run on until the drainage work was completed, with a final lift on new ballast.

The first witnesses at the inquiry were Examining Fitter E.J. Collins and Locomotive Foreman E. Buck, who stated that there had been nothing wrong with engine No. 107 before the accident.

Driver Adams testified that he had been running under clear signals and travelling at between 35 and 40 miles an hour approaching Raynes Park. He had seen a gang of platelayers alongside the line and then his engine had lurched violently to the right and then to the left.

Fireman Mitchell and Guard White could not add anything to Driver Adams' evidence, while Driver F. Stichley of the 3 p.m. Down train from Waterloo said that he had noticed no problems with the track at Raynes Park.

Ganger G. Hendley said that he had been in charge of all five roads from mile 8¼ to mile 9¼. His normal gang of six men had been increased to 20 men for the work and Permanent Way Inspector W. Puttick had instructed him to lift the high rail by 1½ inches between Wimbledon and Raynes Park, then correct the super-elevation by lifting the low rail. The work had been begun at 1.15 p.m. on the day of the derailment and the lifting of the high rail had been completed by 3.25, while only two lengths of the low rail remained to be done. The job had been undertaken between trains, with no speed restriction and no flagman to caution trains; only a lookout man to warn the platelayers in time for them to clear the track. He had considered that the ten-minute interval between the 3 p.m. and 3.10 p.m. trains on the Down Through line was sufficient to lift the low rail and had begun to do so as soon as the first train had passed. A short time later the lookout man had whistled to warn of a train on the Up Through line, then immediately afterwards for the 3.10 train on the Down Through line. There had just been time for the men to remove the jack and he was sure that one sleeper on each side of the jack had been packed and had been under the impression that five sleepers on either side were packed.

[14] Ibid, 26th May 1933

Lengthman A.E. Holman, one of the men operating the jack, said that two sleepers on the Raynes Park side of the jack, and five or six sleepers on the Wimbledon side had not been packed at the time of the derailment. They had made two or three lifts between the two trains.

Lengthman T. Forrester generally confirmed Holman's evidence.

Permanent Way Inspector Puttick stated that he had told Ganger Hendley to lift the high rail a maximum of 1½ inches, and that he considered this "quite reasonable and proper" without imposing a speed restriction. Had he known that the low rail was to be lifted three inches he would have imposed a speed restriction or done it in two lifts, not in one lift between trains. He had been satisfied with the ballast, but some of it had been too large to pack with shovels.

A statement issued by the Company showed that there was only one interval of 11 minutes between trains on the Down Through line between 3.25 p.m. and 5 p.m.; all the other intervals were ten minutes or less.

Permanent Way Superintendent J. Cole testified that he had considered that the track work could be done during the week; he had not anticipated a lift of ½ or ¾ inch. He had not intended the track to be lifted as much as it was. He considered there was sufficient small ballast for shovel packing.

Divisional Engineer H.I. Bond considered that Ganger Hendley had been trying to do too much, perhaps because he had been anxious to finish the job before he went home at 5 p.m.

Concluding his report, Lieut. Colonel Anderson remarked that the derailment had not been caused by any defect of the engine or rolling stock but had clearly been due to the train passing over a length of temporarily defective track. Three or four lifts of the inner rail of the curve had been made in the ten-minute interval between the two Down trains, the last of 1¾ inches and perhaps more, but the time available had been insufficient to pack the ballast under all 10 or 12 sleepers. The sleepers immediately adjoining the removed jack had been packed but several adjoining ones had been unsupported at their inner ends, which had led to the rolling of the engine first to the right and then to the left.

Ganger Hendley, the inspecting officer considered, had committed a serious error in attempting to carry out the work in the way that he had, and that he was mainly responsible for the accident. A considerable amount of blame attached, however, to Permanent Way Inspector Puttick, whose instructions to the ganger had been somewhat vague and therefore unsatisfactory, while Permanent Way Supervisor Cole and Divisional Engineer Bond did not seem to have appreciated the risks involved in carrying out the work on an electrified line with such fast and dense traffic.

The collision which had followed the derailment must be regarded as entirely fortuitous.

Lieut. Colonel Anderson recommended that the railway companies should amend their rules so as to define more precisely what work could be done on the track during normal traffic. [15]

At the inquest at Wimbledon on the five passengers the jury took only ten minutes to

---

[15] Ministry of Transport report

reach a verdict of "accidental death", exonerating any particular individual from blame. [16]

At the beginning of June 1933 torrential rain fell in Turkey, leaving many parts of the country under water. Near Sazak on Friday 9th June 1933 the flooding led to the derailment of an Eskishehir – Angora passenger train. Seven passengers lost their lives, some by drowning, and 14 were injured. [17]

On Thursday 20th September 1934 No. 46 Up passenger train on the Bengal-Nagpur Railway was derailed between Kesinga and Rupra Road stations, India, on the Raipur – Vizianagram section of the railway. A flood had washed away an embankment, leaving the sleepers and rails unsupported. Thirteen people died. [18]

Chemins de fer de Paris à Lyon et à la Méditerranée train No. 1017, an overnight Paris – Saint Etienne express, was crowded with passengers on Thursday 29th July 1937, including pilgrims returning from the shrine at Lisieux, children bound for summer holiday camps, people who had been holidaying in the French capital going home and Paris people going on holiday. It was derailed just before 23:00 300 metres from the station at Villeneuve St Georges, France, about 25 kilometres south of Paris. The engine overturned, as did its tender, a wooden luggage van and a steel mail van. The third vehicle in the train was a steel second-class carriage which remained almost upright and still coupled to the mail van in front, with its middle part up against the heavy tender. This steel carriage, together with the engine, tender and steel postal van, formed an obstacle which, in the words of the newspaper report, was as heavy and hard as rock, against which the next, wooden, third-class carriage, which was filled with passengers, was crushed to matchwood, and it was here that most of the casualties occurred. Many of the passengers were impaled and chopped by wood fragments, or else crushed. The remaining vehicles of the train, which were either wholly or partly of steel construction, were little damaged. A fire also broke out in the wreckage, but it was extinguished by the fire brigade before it could get a serious hold. [19]

A major disaster resulted from the derailment; by the evening of the next day it was announced that there were 28 dead and 50 injured, three of whom were in a critical condition. Those who lost their lives included Conductor Lhuillier and Mail Clerk Desavis; the others were all passengers.

A newspaper reporter managed to interview a M. Copine, from Paris, who had been a passenger in the leading coach, who said, "I felt the carriage leave the rails, and saw the locomotive lying next to us. A burning coal briquette came through the window and buried itself in the seat. Our steel coach remained intact, but the two following wooden carriages were terribly damaged, as was the mail van. Jets of steam escaping from the engine, the darkness and the obstructed corridor combined to make it a frightening scene."

---

[16] *The Times*, London, England, 31st May 1933
[17] Ibid, 10th June 1933
[18] Report by the Railway Board on Indian Railways for 1934-35 Volume 1
[19] *The Times*, London, England, 31st July 1937

Cars on the parallel national highway stopped and used their headlights to illuminate the accident scene, while lighting was also provided by the acetylene lamps of the permanent way staff. The leading mail van was described as having been literally flattened like a pancake.

The preliminary investigation carried out by the railway company suggested that a set of facing points had been moved under the train. On duty in Poste (signal box) No. 5 at Villeneuve St Georges were Head Signalman Sontag and Signalman Dauvergne. This box controlled the junction for the lines to Montgeron and Corbeil. Trains from Paris were signalled from Poste 1 to a *poste* at kilometre 8 and from the latter to Poste 5 at Villeneuve St Georges. Trains for the Montgeron line were signalled by one stroke on a bell or similar device and those for the Corbeil line by seven strokes. Examination of the train register book showed that the evening of the accident had been a busy one, with 11 trains from Paris in the 90 minutes before the accident, as follows:

At 21:42 train No. 43

At 21:43 train No. 43 bis

At 21:46 train No. 43 ter

At 21:53 train No. 43 quater

At 22:28 train No. 667

At 22:33 train No. 667 bis

At 22:36 train No. L.33

At 22:43 train No. 613

At 22:51 train No. L.43

At 22:59 train No. 511 [20]

At 23:02 the Paris dispatcher had advised the Villeneuve St Georges train that the train following L.43 had been described in error as No. 511 and it was in fact train No. 1017, which had departed from Paris at 22.50, 15 minutes late. Train No. 511 was routed via the Montgeron line and train No. 1017 via the Corbeil line. Allegedly train No. 1017 was not shown on the illuminated diagram in the signal box as yet being in the proximity of the junction so the signalmen, anxious to avoid delay, had replaced the signals and then reversed the facing points. In fact the train had been close at hand and the road could not be reset in time before the engine had reached the points. Given that it was a modern installation with power signalling it was of course strange that there was no time delay device to prevent the signals being replaced in front of a train and the facing points immediately reversed. Such devices certainly existed by this date, but none were installed at this location. The two signalmen were questioned all night by the *juge d'instruction* (examining magistrate). The railway officials wanted to interrogate the driver and conductor of the train the same night as well, but both men were lying injured in hospital and the officials were told that the driver at any rate was too badly hurt to be subjected to an examination at that time. [21]

A union official pointed out that regardless of a possible mistake made by the

[20] *L'Humanité*, Paris, France, 31st July 1937

[21] *L'Homme Libre*, Paris, France, 31st July 1937

signalmen the accident would have been much less serious had there not been a wooden coach involved. Seizing upon this, a newspaper accused the railway company of acting criminally in refusing to ensure passengers' safety by eliminating wooden carriages, despite the lessons of previous accidents involving wooden carriages; listing four which had occurred in France in the recent past:

24th December 1933 Pomponne, near Lagny – 235 dead

22nd September 1936 Lourdes – 25 dead

13th March 1937 Châteauneuf-sur-Cher – 15 dead

10th July 1937 Le Mans – 10 dead

The newspaper referred to wooden coaches as 'coffin-waggons' which must be removed from the whole networks if further hecatombs were to be avoided. Reference was also made to a derailment of extreme violence at Dax on 30th March 1937 of a Bordeaux – Hendaye express in which, because the train had been composed wholly of steel or part steel carriages, there had been only one death. The article demanded that with the beginning of the holidays, when thousands of people would be travelling across France, the public authorities must insist that the railway companies use only steel carriages, and that the railway tycoons and their methods must be judged and condemned. The article conveniently ignored that fact that with the heavy holiday traffic the railways had had to use all the rolling stock that was available, and that the replacement of all of the wooden rolling stock with steel could not be accomplished overnight.

According to a follow-up newspaper piece on 7th August by the secretary-general of a railway union, it was Head Signalman Sontag who, upon receiving the telephone message that train No. 1017 was coming first, had operated levers Nos. 1 and 2 to replace the signals to danger and then gone to the window to see where the train was. As he could not see it coming Signalman Dauvergne had looked at the track diagram depicting the lines, signals, and points to determine the train's whereabouts. The 'Train Approaching' field on the diagram had been yellow but the 'Track Occupied' field was still green, and so he had moved the lever to set points No. 26 right for the Corbeil line. Almost at once had come the crash of the derailment. Theoretically, the union official continued, the accident should not have happened. But it had. It was 430 metres from the protecting signal to the junction points and this length was not completely track circuited. The train had passed a berth track circuit 36 metres in the rear of the signal, where there were also co-acting detonators, placed on the track when the *carré* was at danger. The head signalman had failed to see the train, presumably because the headlights of its locomotive were too feeble, and thus as soon as he had replaced the signal Dauvergne had operated the points. However, the train had already passed the track circuit and the signal, which had been replaced to danger behind it, this explaining why no detonators had exploded. The train had been travelling at 22 metres a second. It took six seconds to operate the points; add three or four seconds to make the decision to do so, thus it could be easily seen that the front of the train was on the points before their movement had been completed. The official asked if it was the signalman's fault and remarked that it would be good to know why the points were only locked when a train reached a track circuit 47 metres to their rear instead of the track circuit extending over the whole 430 metres from the protecting signal (this second track circuit presumably corresponding to the 'Track Occupied' field on the box diagram). It was asked if the company had not foreseen the possibility of such an accident, to which the company had simply replied that there were regulations to be followed (presumably relating to points not being altered in the face of an approaching train until the signalman had assured himself that the train was at a stand).

The same union representative also questioned why passenger trains composed of wooden carriages had still been leaving Paris Lyon two days after the accident, when steel carriages were apparently available. The company responded that the steel carriages were under repair.

Meanwhile with the death of one of the injured the number of fatalities had increased to 30. [22]

Eight transients died and eight others were seriously injured in the derailment of a transcontinental Canadian National Railway freight train 97 miles north-west of Sudbury, Ontario, Canada, early on Saturday 31st July 1937. The derailment occurred in a remote wooded area; the nearest flag station was Makwa. The transients or hobos had been riding in box cars surrounded by lumber. The fast freight train was reportedly travelling at 50 miles an hour when eight of its 53 cars left the rails and tumbled down a steep embankment, taking the transients with them. The train was in charge of Conductor Alexander Stewart with Engineer James McGee.

Wreckage was scattered over the tracks for a quarter of a mile. A relief train was sent from Capreol, 75 miles to the east, and work on clearing the track began before daybreak. Rails were torn and bent over hundreds of yards. The single line was completely blocked, of course, and east and westbound passenger trains were 'delayed many hours.' [23]

By the following day seven of those who lost their lives had been identified. They were John Choppick, of Bishop, Manitoba; Kenneth Gordon, of Whitoff, Ontario; John High, of Coldwater, Ontario; Alexander Lalone, of Montreal, Quebec; John McArthur, of Manitoba; Charles Riley, of Pembroke, Ontario; and Wilfred Rioux, of Montreal, Quebec. The remaining victim was later tentatively identified as a man named McDermott, possibly an Australian. [24] [25]

The cause of the derailment was unclear, but railwaymen suggested that it had been due to a broken rail. [26]

A special pilgrims' train was passing through the station at Holzheim, Germany, at 08:30 on Sunday 5th September 1937 when it became derailed. Eighteen passengers and the driver died in the accident and 33 people were seriously injured. Work on the track at the point of the derailment had been completed the day before. As it had been left on the Sunday morning, 70 metres of track which had been lifted and re-ballasted was connected by a ramp 20 metres long to the old track which had not been lifted. With a 15 centimetre difference in height between the new track and the old the line was not in a safe condition for the passage of trains at full speed.

---

[22] *L'Humanité*, Paris, France, 7th August 1937

[23] *Winnipeg Evening Tribune*, Winnipeg, Manitoba, Canada, 1st August 1937

[24] *Winnipeg Free Press*, Winnipeg, Manitoba, Canada, 2nd August 1937

[25] *The Lethbridge Herald*, Herald, Alberta, Canada, 4th August 1937

[26] *Sunday Journal and Star*, Lincoln, Nebraska, United States, 1st August 1937

In June 1939 the *Bahnmeister* (permanent way inspector) and *Rottenführer* (ganger) in charge of the work found themselves in court at Düsseldorf charged with having caused the accident through negligence. The ganger stated, "During the work so-called 'Slow Signals' were used. When the work was finished on the Saturday evening these signals were removed. I assumed that trains using the line would be travelling at the normal permitted speed. I never imagined that the special would be allowed to run as fast as 85 kilometres an hour." The dozen trains which had passed over the affected area between the end of the work on Saturday and the pilgrims' train on Sunday morning had all stopped at Holzheim and so the problem with the track had not been so apparent because of their lower speeds. This perhaps explained the ignorance of the ganger of the pilgrim train's likely higher speed, and his and his superior's carelessness.

The special had consisted of a class BR 38 steam locomotive and 13 Prussian coaches. Sixty metres from the end of the raised track the engine had swayed violently, and its driver had at once shut the regulator and applied the brakes. His fireman had cried, "*Donnerkeil* [thunderbolt], the track is bad here!" At almost the same moment the fireman had noticed that the engine had moved to the right and after another 60 metres it overturned to the right, continuing for another 200 metres, partly derailed, before it stopped. With the sudden stoppage of the engine and tender the following coaches were forced up against the obstruction and some were telescoped into one another.

What remained unclear and was not questioned in contemporary newspaper reports of the accident was why the drivers of previous trains had not reported the problems with the track which must have been evident, even at the lower speeds at which they were running.

The ganger and permanent way inspector were both found guilty of negligence by the court and were sentenced to respectively six and ten months' imprisonment, with their pre-trial detentions taken into account. The inspector received the heavier sentence in view of his higher position. [27]

A passenger train on the Canton – Hankow railway was derailed between Canton and Shaokwan, China, on Sunday 2nd January 1938, due to a subsidence of the roadbed. Over 100 of the 300 people on the train were reportedly killed or injured. [28]

Torrential rain accompanying a typhoon caused a bank to collapse between Wake and Kumayama, Japan. An Up night train from Shimonoseki to Kyoto on the San'yo main line was derailed and overturned when it reached the spot on Wednesday 15th June 1938. The accident was compounded when a Down train from Kyoto to Uno ran into the wreckage. Twenty-five people died and 108 were injured. Problems with track inspections and the suspension of traffic were blamed for the disaster. [29]The Up train was an excursion carrying schoolchildren, at least 16 of whom died. [30]

---

[27] H.J. Ritzau, *Katastrophen der deutschen Bahnen Teil II*, pgs 88-91 (in German)

[28] The Railway Gazette, London, England, 11th March 1938

[29] Webpage: Japanese Railway Safety and the Technology of the Day by Masao Saito

[30] The Malaya Tribune, Singapore, 16th June 1938

What were described as phenomenal floods due to a cloudburst washed away an embankment between Ayyalur and Vadamadura, India, on the South Indian Railway, and caused the derailment of train No. 11 from Madras to Shencottah early on Sunday 21st August 1938. The engine and four leading coaches of the train plunged into the gap and 33 people died and 81 were injured. The train had been crowded and the victims included members of a wedding party and a number of refugees from anti-Muslim riots in Rangoon, Burma. The train driver escaped with injuries; the fate of his fireman was not stated. [31]

Damage to rolling stock and the permanent way was estimated at 82,000 and 530 rupees respectively. [32]

A succession of storms in north-west Spain caused the Caudal River to overflow its banks and it washed away an adjoining railway embankment at Figaredo, near Mieres and south of Oviedo. In turn, this led to the derailment of a Mieres – Collenzo mail train on Saturday 10th September 1938. The engine and three of the four coaches of the train left the rails and plunged down the embankment into the river. Many of the victims were rescued by boats but about 10 died, mostly by drowning, and another 30 were injured. The brief reports of the accident did not state if the engine crew survived. [33] [34]

On Tuesday 7th June 1939, passenger train P 2418 was derailed at Göschwitz, Germany, on the line between Gera and Jena, due to distortion of the track (*Gleisverwerfung*), probably heat distortion (sun kinking). Six people died and 14 were injured. [35]

At 02:30 on Wednesday 28th June 1939, No. 33 Up Delhi – Dehra Dun Mixed Passenger train met with an accident at mile 26/2 of the Bijnor branch, between Chanpur Siau and Haldaur, United Provinces, India, on the East Indian Railway. Following heavy rain the engine, three waggons and three 'lower class bogies' fell into a breach in the embankment near culvert 25, the latter being described as having two spans two ft high in the four ft high embankment. Ten people died in the accident and 25 were injured. [36]

The report of the government inspector estimated the cost of the accident as 80,000 rupees in damage to rolling stock and 2,070 rupees damage to the permanent way. As to the cause of the accident the report stated that it had been due to a 'breach caused by the permeation of flood water through the bank either through ballast backing behind a bridge abutment or due to burrows and holes made by animals.' In any event the inspector considered that neither the railway administration nor any of its staff were responsible or had been negligent. [37]

---

[31] The Times, London, England, 22nd and 23rd August 1938
[32] Report by the Railway Board on Indian Railways for 1938-39 Volume 1
[33] Webpage: Federación Castellano Manchega de Amigos del Ferrocarril (in Spanish).
[34] The Glasgow Herald, Glasgow, Scotland, 12th September 1938
[35] H.J. Ritzau, *Katastrophen der deutschen Bahnen Teil II*, pg. 95 (in German)
[36] *The Bombay Chronicle*, Bombay, India, 29th June 1939
[37] Report by the Railway Board on Indian Railways for 1939-40

A passenger train from Bucharest was derailed in the station at Petroaïa, Romania, early on Sunday 13th August 1939. [38] Fifteen people were reportedly killed and 25 gravely injured. The accident was stated to have resulted from the points having been moved under the train. [39]

A nasty accident occurred on the Tehuantepec Railway in Mexico on Saturday 21st October 1939. It happened at Tolosa, between Santa Lucretia and Matias Romero. A train which included two tank waggons of petrol became derailed at a point where workmen had left cooking fires burning a short time earlier. As a result, fire broke out almost immediately after the accident and an explosion followed a little later, though not before the driver had uncoupled his engine and set off to Matias Romero for assistance. The blaze quickly spread to other vehicles of the train, including two coaches filled with track workers, their wives, and children, 40 of whom lost their lives. The derailment was blamed on the poor condition of the track. [40]

Early on Thursday 11th January 1940, torrential rain swelled the River Capivari in Brazil, causing it to overflow its banks and wash away an embankment of the Estrada de Ferro Sorocabana line. This led to the derailment of a passenger train from São Paulo to Bauru. The *London Times* wrote that 'it is reported that many persons were killed and injured.' [41]

An account of the accident in a Brazilian newspaper published the day after the accident stated, however, that only two people had lost their lives while a third was 'missing' and about 30 were injured. A statement by the railway company indicated that the accident had occurred at Kilometre 276 between Embaúba and Victoria stations at 05:22. Train N7 was drawn by engine No. 818 with Driver Virgolino Franco. As a result of the derailment the leading luggage van was completely broken up, second-class coach C-70 was partly overturned, and second-class coach C-83 was derailed. The following first-class and sleeping cars remained on the line. Porter José Bugno (name illegible in newspaper) de Barros and Train Helper Benedicto Camargo were killed, and the unnamed conductor had not been found but was presumed dead. The driver and fireman were unhurt. The injured passengers had been taken by special train No. 1008 to Piramboia as there were no facilities for their treatment at Embauba. Relief train No. 256 had left Laranjal for the accident scene at 07:18 with the senior district engineer and his staff and ballast train No. 811 had followed from Piramboia at 08:00. Meanwhile the line was completely blocked to traffic.

The newspaper added that it had been initially reported that three coaches had gone into the river and there were 100 dead and injured. [42]

---

[38] *L'Humanité*, Paris, France, 14th August 1939
[39] *The Railway Gazette*, London, England, 18th August 1939
[40] *The Dubois Courier*, Dubois, Pennsylvania, United States, 23rd October 1939
[41] *The Times*, London, England, 13th January 1940
[42] O Estado de São Paulo, São Paulo, Brazil, 12th January 1939

Japan suffered a major catastrophe on the morning of Monday 29th January 1940. The train involved in the 7 a.m. accident was a three-coach petrol railcar, which was crowded with passengers, mostly workers bound for the Sumitomo copper works. The rear coach of the train was derailed at Ajikawaguchi station, near Osaka, and overturned. Petrol from damaged fuel tanks immediately set the carriages alight and many passengers were burnt to death before they could escape. Estimates as to the number of dead varied from 180 to 220, with 80 or 100 more injured. Many of the survivors had suffered serious burns and only ten of the 300 passengers on the train escaped unhurt. Shelters were erected at the station to house the bodies of those who had died. [43] [44]

Initially it was stated wrongly that the train had collided with a row of oil tank waggons. Later it emerged that the derailment had been caused by the premature movement of a set of facing points after only two of the train's three coaches had passed over them. [45]

A disaster in France early on Saturday 4th May 1940 involved the collapse of a bridge but the initial event was the washing away of an embankment, so it has been included in this chapter. The train involved in the accident was Société nationale de chemins de fer français (SNCF) express No. 426 from Aurillac to Paris Austerlitz, due to arrive in the French capital at 05:35. The accident occurred at about 01:20, on the Montluçon – Bourges line, at a point where the railway ran parallel to the Canal de Berry and the River Cher and where it crossed the Queugne, a tributary of the River Cher, between Vallon-en-Sully and Urcay stations.

The express consisted of a Pacific class steam engine, two box cars, eight carriages and a mail van. When it reached the point of the accident the embankment, weakened by floodwater, gave way beneath the weight of the train, followed by the collapse of the span across the Queugne stream. The engine, tender and leading van overturned on the left-hand side, as did the following five carriages, of steel and timber construction, and these were wrecked and two of them were telescoped. Twenty-nine passengers were killed more or less outright and four of the 28 injured who were taken to Bourges hospital died of their injuries. M. Albertin, Under Secretary of State, and M. Claudon, Director General of the SNCF, were involved, while M. Albertin also visited the chapelle ardente at Saint-Amand-Montrond where the bodies of the dead had been taken.

Heavy rain had fallen all day on Friday 3rd. The investigation into the accident found that because of this the line had been inspected by permament way staff, the last inspection before the accident having been at 23:55. At that time it was found that the Queugne had overflowed its banks and flooded the adjoining meadows but had not reached as far as the railway embankment. By then it had stopped raining and the banks of the canal and the River Cher had not been breached. It was decided therefore that there was no danger. It appears, however, that there must have been another cloudburst between midnight and 00:30 which caused the Queugne and a tributary stream of it, the Bœuf, to rise abnormally, and their waters scoured the railway embankment just before the arrival of the train. The accident report produced by the SNCF blamed the derailment

---

[43] *Journal de Geneve*, Geneva, Switzerland, 30th January 1940

[44] *The Bakersfield Californian*, Bakersfield, California, United States, 29th January 1940

[45] *Galveston Daily News*, Galveston, Texas, United States, 30th January 1940

on 'a rapid, violent and absolutely abnormal flooding of the Queugne and the Bœuf.' [46]

A newspaper commented that there had been no similar flooding at the location since 1856. [47]

A train on the Aidin line was derailed somewhere in the Smyrna region of Turkey on about Tuesday 4th February 1941 due to flooding damaging the tracks. Twelve people died and 15 were injured, eight very seriously. [48]

A mistake by a signalman led to the route being reset prematurely and a set of facing points being moved before passenger train P 3936 had completely passed them at Forbach-Gaussbach, Germany, on the Rastatt – Freudenstadt line, on Saturday 30th August 1941. The train was derailed and eight people lost their lives and 16 were injured. [49]

A passenger train on the Circumvesuviana railway was entering Cercola station, Italy, on the Sarno – Napoli line, at 07:00 on Saturday 20th December 1941, when it became derailed on the points at the entrance to the station. Various carriages of the train overturned and 25 people died, about 50 were injured and others were bruised or shaken up. An immediate investigation into the cause of the accident was ordered. [50]

Severe flooding in the Drau Valley of Oberkärnten (Upper Carinthia), Austria, washed away all the bridges across the Drau River and resulted in the derailment of a whole passenger train on Monday 28th September 1942 near Nikolsdorf in the neighbouring province of Osttirol (East Tyrol). Twenty-two people died, 20 were seriously injured and 80 slightly hurt. Despite the accident being one of the most serious in Austrian history it was not mentioned in the local press due to wartime censorship. At the time of the disaster Austria was a part of Germany. [51]

Mail train No. 810 from Linares to Almeria was derailed between Pedro Martinez and Moreda, Spain, on Friday 15th October 1943, and three carriages went down an embankment. The train consisted of engine No. 441, a van, a first-class coach, three third-class coaches, an armoured van, a livestock waggon, and engine No. 410. It was fitted with vacuum brakes in addition to hand brakes. The derailment was apparently due to a broken rail. The train became divided into three, with divisions between the first-class and leading third-class carriages and between the rear third-class carriage and the armoured van. It was the three third-class carriages which went down the embankment, rolling over several times. The armoured and livestock vans and the rear engine were derailed and ended up leaning over on the embankment but did not overturn. Reports of

---

[46] Wikipedia – Déraillement du train Aurillac – Paris
[47] *L'Action Française*, Paris, France, 5th May 1940
[48] *The Times*, London, England, 7th February 1941
[49] *Katastrophen der deutschen Bahnen Teil II* by H.J. Ritzau, pg. 104 (in German)
[50] *La Stampa*, Torino, Italy, 21st December 1941
[51] Webpage – Hochwasserchronik Kärnten

the accident did not clearly state the number of casualties but there were probably about 20 dead and 100 injured. [52]

On the evening of Sunday 17th October 1943, passenger train No. 3393 which had left Berne at 19:23 for Lucerne had to cross express train No. 369 which had left Lucerne for Berne at 20:47 at Schüpfheim, Switzerland. The train from Berne was waiting in the station for the arrival of the express from Lucerne but as the latter train entered the station the facing points at the Lucerne end of the station were moved before the whole of it had cleared them. The last two carriages of the express were thrown off the track and its rear coach struck engine No. 3392 of the train from Berne while the penultimate coach was thrown against the leading van of the stationary train. Three of the passengers on the express were killed outright and 40 were injured, some seriously. The injured were taken to hospitals in Langnau and Lucerne. [53] The collision occurred at 21:20. The signalman who it appeared had operated the facing points prematurely was arrested. [54]

M. Sonvico Pino of Lugano died in Lucerne Hospital on Thursday 21st October, bringing the number of dead up to five. [55]

Two streamlined express passenger trains of the Atlantic Coast Line Railroad were involved in a major catastrophe in the early hours of Thursday 16th December 1943. The Interstate Commerce Commission gave the location of the accident as near Rennert, North Carolina, United States; the location is given by other sources as Buie, and as near Lumberton. The trains involved were the southbound and northbound *Tamiani East Coast Champions*. Train No. 91, southbound, consisted of diesel electric units (engines) Nos. 515, 753 and 514, two express cars, a mail car, a passenger baggage car, a coach, a dining car, a tavern car, four coaches, a Pullman sleeping car, a lounge car, three Pullman sleeping cars, a dining car and a Pullman sleeping car. Train No. 8, northbound, consisted of diesel electric units Nos. 506, 503 and 500, a passenger baggage car, two coaches, four Pullman sleeping cars, a dining car, and eight Pullman sleeping cars. William Myers was the engineer of the southbound train and Frank Belknap was the engineer of the northbound one; the names of the other members of the train crews do not seem to have been published.

Southbound train No. 91 left Fayetteville, 20 miles north of the accident scene, at 12.25 a.m., an hour, and five minutes late, passed Rennert, passed automatic block signal No. 3239 on the double track main line, and was travelling at about 85 miles an hour when its last three cars left the rails at a point 3,510 feet south of the aforementioned signal. The cars remained upright but the rear two came to a stand foul of the northbound line at 12.50 a.m. Northbound train No. 8 passed Pembroke, just over 11 miles south of the accident site, at 1.20 a.m., an hour and 40 minutes late, passed automatic signal No. 2338, which was showing ‘Proceed’, and travelling at about 80 miles an hour collided with the derailed cars of the southbound train at about 1.30 a.m. As a result of the accident 72 passengers died and 160 passengers, 15 Pullman staff, nine dining car staff, a

---

52 Webpage: Federación Castellano Manchega de Amigos del Ferrocarril (in Spanish)
53 *Journal de Geneve*, Geneva, Switzerland, 18th October 1942
54 *Gazette de Lausanne*, Lausanne, Switzerland, 19th October 1942
55 Ibid, 23rd October 1942

coach attendant, a train porter, and the flagman of train No. 8 were injured. All but one of those who lost their lives were in train No. 8, mostly in the second and third coaches. [56]

A newspaper reported 'Heroic acts abound in tragedy,' and continued with an account of a passenger's experiences. '"I was a passenger on the Tamiani West {sic} Coast Champion, southbound to Tampa, Fla., which derailed at about 1.10 a.m.," said William Wood, publisher of Small Homes Guide Magazine, unhurt in the wreck." Capt. P.W. Allen, of the Navy, Washington, D.C., and I took charge of helping the injured and trying to reach the dead. We got blankets from the Pullman cars and hot coffee and whiskey from the diner, then made fires in the snow-covered ground. Captain Allen did some heroic work in that smashed wreckage. Through State Highway Patrol radio we reached Robert Samet, at Lumberton, who brought acetylene torches used to cut out the dead. The A.C.L. wrecker arrived about 9 a.m. and ambulances had been streaming in and out from Pembroke and other places. You have never seen such a smash-up. Bodies were mangled rails twisted into hairpins and loops all over the place."

'Wood said he and others stood at the side of the railroad track and watched the speeding streamliner scream into the rear end of the southbound train. Wood said he and other passengers of the derailed southbound train frantically tried to build fires out of newspapers to warn the northbound engineer that the smoking car of his train had jumped across the northbound track. Two southbound freight trains had already been stopped by makeshift flares when they heard the streamliner roaring north. The speeding train ripped into the smoking car, he said, cutting it almost in half and five coaches of the northbound train leaped the track and folded together like an accordion.

'"It was like a bad dream, filled with screams and in the dark we couldn't see exactly what had happened. You could hear people shouting and crying and in the below freezing temperature their voices echoed and re-echoed in the night," he said. Wood said one woman expecting her baby had her legs and thighs crushed and mangled. "She kept saying, 'I won't lose my baby, I won't lose my baby. God help me, I won't. I must be calm. Please dear God, help me.' She had more spunk than I've ever seen in a woman before," Wood said. "A short time later a doctor came up and examined her, gave her a sedative, and told her that her baby would be born. "Thank God,' she said."

'Wood said another woman badly crushed about the hips and legs held a baby in her arms. Both were unconscious with the mother still clutching the baby close to her. Finally, he said, the baby regained consciousness and began to cry. A moment later the mother's eyes fluttered open. She patted the baby softly and cooed to it for a moment, and then lapsed back into unconsciousness.

'Wood said an Army officer he was talking to a short time before the southbound train was derailed was killed and the man who occupied a lower berth next to his was literally cut to ribbons but not killed. At least 30 were killed at the time of the accident, he said, and "I don't know how many more may have died later of injuries." Shortly after the wreck, he said, a police car with a radio transmitter arrived, and after a quick look at the scene, police summoned acetylene torches to cut the steel coaches. Ambulances and doctors arrived almost simultaneously with the police car.

'The cars were jammed with passengers, he said. "It was a terrible sight. A lot of the people were cut in half when the seats snapped together by the impact clipped them in

[56] Interstate Commerce Commission report

two like paper." Torsos and legs were pulled from the coaches, he said, and it was impossible to match them up. Wood said that he believed that many of the servicemen were going home on "last leave" to spend the holiday with friends and he estimated that at least half of those killed were servicemen.' [57]

The initial derailment was found to have been caused by a broken rail. The rail had broken into many pieces, of which 64 fragments were recovered. It was a 39 foot 100 lb open hearth rail manufactured by the Tennessee Coal, Iron and Railroad Company in 1937 and laid the same year. The rail had apparently broken under the train, due to a transverse fissure, as otherwise the break would have caused signal No. 3239 to go back to danger. The faulty rail had not been identified by a detector car which had passed through the area on 17th October or by the section foreman on his motor car on 13th December.

The Interstate Commerce Commission (ICC) report stated that as train No. 91 had approached Rennert the enginemen had been in the control compartment of the leading unit, the baggageman and train porter had been in the fourth car, the conductor had been in the thirteenth car and the flagman in the rear car. Their first warning of the derailment had been the brakes applying in emergency. The flagman was the only member of the train crew who could see that the rear cars had become derailed and that two of them were foul of the opposite running line. He had lit and placed a 10-minute fusee at the south end of the derailed cars, had signalled to the other members of a train crew with a white lantern that the train had parted and had then set of back to protect the rear of the train. As he was setting off a passenger had asked him about the protection of the obstructed northbound line, and he had assured them that the train crew at the front would take care of that matter. About 30 minutes later he had flagged a southbound freight train, told them what had happened, and the engine of that train was detached and going forward towards the derailed cars when the collision occurred.

The conductor told the ICC enquiry that he had seen the fusee lit by the flagman but not his white hand signal. He had not realised that any part of his train had derailed and went forward towards the front. He had found that the train had become separated between the second and third cars due to a broken coupling knuckle. A little later he had seen a railwayman going southwards with a red lantern, then had met the engineer who had confirmed that he had sent the fireman southwards to protect the northbound line. The conductor, engineer and baggageman had then occupied themselves with replacing the broken knuckle and recoupling the train, then with attempting to contact the dispatcher via a portable telephone.

The fireman said that he had proceeded southward with red and white lanterns and one fusee, but no torpedoes, having forgotten to provide himself with them. When he had seen the lights of the northbound train, he had attempted to light the fusee but at that moment had slipped on the wet snow and broken the fusee. He had then attempted to signal to the engineer with his red lantern but did not think that his warning had been seen.

Engineer Belknap of train No. 8 testified that he had been able to see the reflection of the headlight of train No. 91 from about 4 miles away. The last automatic signal that he had passed had been clear and although he had seen a stationary red light at the front of that train from a few hundred feet away no warning signals had been given. It was not

---

[57] *The Burlington Times-News*, Burlington, North Carolina, United States, 16th December 1943

until he had actually been passing the front of train No. 91 that he had seen stop signals from the rear of the train (given by a passenger using the rear marker lamp) and at the same moment had seen the obstruction. The speed of his train could be little reduced by this belated warning.

The ICC pointed out that the operating rules of the company provided that when a train was stopped suddenly from an unknown cause and there was a possibility that adjoining lines might be obstructed these lines must be at once protected. About 40 minutes had elapsed between the initial derailment and the subsequent collision so there had been ample time to provide protection. The rear had been protected and the fireman had gone forward, though probably reluctantly in the bitter cold, and without torpedoes or sufficient fusees, while his red hand signal had not been seen. Various suggestions, the ICC continued, had been made as to means of warning trains on adjoining lines, such as the use of radios, or rockets, or track circuit operating clips to replace the automatic signals to danger, but these all presupposed that it was known that the opposite line was obstructed. In this case only the flagman was aware that the opposite road was blocked, and he had been obliged to go back and protect the rear of his train. The train crew at the front had assumed that the train had stopped simply because it had become divided and would not therefore have used any of these alternative means of protection even had they been available. Had they provided adequate front end protection and then made a careful examination of the whole train the accident would not have occurred. [58]

No inquest on the victims of the accident was held, Roberson County Coroner W.D. Biggs stating, "It was absolutely an accident and there will be no inquest for that reason." [59] The ICC clearly did not agree.

Less than eight months after the Rennert disaster the Atlantic Coast Line Railroad suffered another accident which was almost as bad, again caused by a broken rail. On Friday 4th August 1944 westbound passenger train No. 57 from Waycross, Georgia, to Montgomery, Alabama, consisted of engines Nos. 1514 and 1513, a mail express car, an express car, a mail car, a baggage car, four coaches, a Pullman sleeping car, three coaches, a passenger baggage car, and a U.S. hospital car, in charge of Conductor J. B. Harris, Engineers A. C Darling and G.W. Patterson, and Firemen Aiken and A.D. Robinson. All of the cars were of steel construction. The train had left Waycross at 10.44 p.m. 29 minutes late, and Dupont, Georgia, the last open office, at 11.36 p.m., 32 minutes late. Just over 3¼ miles further west, when it was travelling at 65 miles an hour, it was derailed at about 11.45 p.m. at Stockton, Georgia, United States. The ninth to fourteenth cars left the rails and the tenth car struck the locomotive of second No. 214, an eastbound freight train which was standing in the siding at Stockton. The tenth car was sheared diagonally for almost its entire length, and almost all of the fatalities and injuries occurred to the occupants of this car, which chanced to be the Jim Crow car, reserved for black passengers. The eleventh and twelfth cars stopped up against the side of the freight engine and the last two cars came to a stand practically upright. Forty-seven passengers, all of them negro members of railroad work crews who had been on their way home for the weekend, were killed, while 40 of their mates were injured. None of the other

---

[58] Interstate Commerce Commission report

[59] *The Anniston Star*, Anniston, Alabama, United States, 17th December 1943

passengers or crew of train No. 57 was hurt but Fireman Knight of the freight train was scalded. One of the passengers who died reportedly alighted from the car, walked the few steps to the adjoining highway, asked for a cigarette, lit it, and then dropped dead. [60]

None of the crew of train No. 57 was aware of anything wrong before the derailment. The conductor and Baggagemaster W.P. Bethea felt an unusual jar and the conductor opened the emergency valve but the brakes had already been applied by the division of the train.

As at Rennert the rail which broke was a 39 ft 100 lb one manufactured by the Tennessee Coal, Iron and Railroad Company, though it was rather older, dating from January 1927 and having been laid in June of that year. It had broken through the head, web, and base in seven places and there was a transverse fissure at the site of the initial fracture. A Sperry rail detector car had passed over the rail on 7th August 1943 and the section foreman had examined the track 36 hours before the accident, but the break was not such as could be discovered by a casual visual inspection.

The Interstate Commerce Commission noted that between 1st January 1940 and 30th June 1944 the Atlantic Coast Line Railroad had reported 61 accidents due directly or indirectly to broken rails, and remarked, 'These accidents indicate that the operation of trains is such that excessive stresses are being exerted upon the track structure.'

A Kantara – Haifa passenger train was derailed 'north of the Sinai border,' near Chanyunis, Palestine, on Wednesday 8th November 1944, flooding following a cloudburst having washed a portion of the track away. Two of the seven coaches which left the rails were telescoped. Seven people died and 40 were injured. 'Most of the casualties were in the military part of the train.' [61] [62]

---

[60] *Thomasville Times-Enterprise*, Thomasville, Georgia, United States, 5th August 1944

[61] *The Times*, London, England, 9th November 1944

[62] *The Railways of Palestine and Israel* by Paul Cotterell, pg. 78

CHAPTER 2

DERAILMENTS - BROKEN TYRES, BROKEN AXLES AND OTHER VEHICLE DEFECTS

One of the frequent causes of derailments of trains other than problems with the track was defects in the trains themselves, such as broken tyres, wheels, and axles. Such accidents were not as frequent by the twentieth century as in the early days of railways, but they still happened quite often and were to a fair degree unavoidable.

Missouri Pacific Railroad freight train No. 66 from Pueblo, Colorado, to Kansas City, Kansas, via Hoisington, was derailed near the station at Pendennis, Kansas, United States, on Saturday 9th May 1931, due, it was thought, to a broken journal. Fifteen cars left the rails, and six trespassers who had been riding in one of the cars which was overturned were killed while one, named Theodore Johnson, was seriously injured, and had to have his right arm amputated. About 20 other trespassers on the train and the train crew escaped injury. Two of the dead hobos were identified from papers found in their pockets as Normal and Ora Rogers, of Crowley, Colorado, while the injured Johnson confessed that he was an army deserter from Fort Warren, Wyoming. He said that two of those killed had been fellow deserters named Ralph Hoover and Howard Wheeler. A fifth man had also apparently been identified, leaving one unknown victim. Johnson's condition was said to be serious three days after the accident; the Fort Warren authorities asked that he be arrested and sent back should he recover. [63] [64] [65]

Chemins de fer du Nord omnibus train No. 21 which had left Paris Nord at 17:37 for Abbeville and Boulogne-sur-Mer was derailed 300 metres short of the station at St-Just-en-Chaussé, France, at 19:10 on the evening of Sunday 17th January 1932. Three carriages and a van left the track. The third-class carriage, fourth in the train, which was

[63] *The Sedalia Democrat*, Sedalia, Missouri, United States, 10th May 1931
[64] *The Maryville Daily Forum*, Maryville, Missouri, United States, 9th May 1931
[65] *Evening News-Journal*, Clovis, New Mexico, United States, 13th May 1931

the first vehicle to leave the rails, was telescoped by the following first- and second-class composite carriage. The sixth vehicle, another third-class carriage, was not badly damaged but the rear van was half flattened.[66] Eleven passengers died and 20 people, three of them railwaymen, were injured and taken to hospital. Both main lines were blocked by the accident.

M. Deligne, the Minister of Public Works, and various senior officials went to the scene of the accident.

The work of releasing the victims and the bodies of those killed was quickly begun, and work on clearing the tracks went on through the night, with local inhabitants assisting firemen and soldiers. Also quickly begun was an investigation into the cause of the derailment. It was established that Driver Deschamps, of Caen depot, had entered the station at the normal speed of 50 kilometres an hour, and the speed of the train was thus ruled out as a cause of its having left the rails.[67] The initial investigation indicated that the derailment had been due to the breakage of one of the axles of the demolished third-class carriage. Even so, the accident would have been a good deal less serious had not the derailed vehicles struck a signal box which chanced to lie in their path. The three railwaymen hurt in the accident were Signalman Taupard and two other men named Bailleur and Vaillant, all of whom had been in the box at the time. Most of the passengers on the train were returning home after spending their weekends with relatives in the capital. Although the train was bound for Boulogne to connect with the Channel ferries there were no Britons or Americans amongst the victims. After the derailment Driver Deschamps had run down the track with a red lantern to stop a freight train due shortly from the opposite direction, while Conductor Blond had gone back to protect his train in the rear. [68]

M. Maurinvilliers, a Nord employee from Saint Just who had been a passenger on the train, told his story to a newspaper reporter. "I had been riding in the third-class carriage which, a short time later, was wrecked in the accident. Then when the train reached Saint-Remy-en-L'Eau station some colleagues suggested that I join them in their coach, which I did, and thus probably saved my life. I was sat with my mates when all of a sudden, I heard a sound like tearing metal. The train pulled up suddenly. I shouted to my companions, 'Our train has knocked down the signal box.' At the same moment I saw a great violet light and jumped down to the track. What I saw first was a man lying on the ground, his back torn off. Then I saw the terrible spectacle of the crushed carriages piled up one upon the other, from which cries of pain and distress were coming. I rushed to the station nearby and alerted the stationmaster, and he called Dr Quest, the company doctor. Then I returned to the scene and helped with the rescue work." [69]

A Baltimore and Ohio Railroad freight train was derailed Sir John's Run, West Virginia, United States, on Saturday 29th April 1933. About 50 train riders were reportedly riding on the fast eastbound train and five of them were killed or fatally injured and three or five others were hurt when 34 of the waggons of the train left the

---

66 *L'Intransigeant*, Paris, France, 19th January 1932
67 *Le Peuple*, Paris, France, 19th January 1932
68 *Chicago Daily Tribune* (European Edition), Chicago, Illinois, United States, 19th January 1932
69 *Excelsior*, Paris, France, 19th January 1932

track on a set of trailing points. One of the dead was identified by a card with the name 'Bud Williams, Plainfield, N.J.' The injured were taken to hospitals at Martinsburg and Berkeley Springs.[70]

Eastbound train Extra No. 6103 consisted of engine No. 6103, 120 loaded cars and a caboose, in charge of Conductor Tederick, with Engineer Staub and an unnamed fireman. It had left Cumberland, Maryland, almost 50 miles west of Sir John's Run, at 11.15 a.m., passed Sir John's Run at 12.40 p.m. and became derailed shortly afterwards while travelling at 35 miles an hour. The rear bogie of the 30th car was the first to leave the track, followed by all those up to the 63rd inclusive. Many of the cars were damaged beyond repair and all four tracks were blocked.

Engineer Staub told the Interstate Commerce Commission (ICC) inquiry that the brakes of the train had been tested before it left Cumberland. He had had no warning of the derailment until he had felt a sudden jerk and had seen the train parting.

Head Brakeman Bloss, who had been riding on the engine, said that he had been looking back at intervals but had seen no problem before the derailment. Afterwards he had found a broken arch bar near the first car to leave the track.

Conductor Tederick said that he, too, had noticed no problems before the derailment. The only defect which he could find afterwards was the broken arch bar on the 30th car, due to an old fracture.

Car Inspector Decker stated that he had examined the defective waggon (Buffalo and Susquehanna car No. 9056) at Cumberland at 10.55 a.m. on the 28th of April. This was a class "O" inspection for general defects and would not have detected the broken arch bar, which could only have been discovered by dismantling the car.

General Car Foreman Scott stated that the car had received a more rigid class "A" inspection on 26th April at Somerset, Pennsylvania, but agreed that this would not have detected the concealed fracture either and added that cars were only dismantled when undergoing heavy repairs.

Train Master Groves had accompanied the train from Cumberland. He put the speed at derailment at 35 or 40 miles an hour. Examining the track afterwards he found the first sign of derailment at Lineburg, 7¼ miles back. He considered that all the evidence pointed to the defective arch bar as the cause of the derailment.

The ICC concluded that the derailment was due to a broken arch bar, which had caused the truck of the car to sag such that it had fouled crossing planks at several points and then struck the frog of the trailing points, causing the truck to collapse and the derailment.[71]

An inquest into the deaths of the trespassers was not deemed necessary.[72]

The bodies of eight hobos were recovered from the wreckage of a Nashville Chattanooga and St Louis Railway freight train which was derailed near La Vergne, Tennessee, United States, on the afternoon of Monday 9th July 1934. About a dozen other

---

[70] *The Charleston Gazette*, Charleston, West Virginia, United States, 1st May 1933

[71] Interstate Commerce Commission report

[72] *Cumberland Evening Times*, Cumberland, Maryland, United States, 4th May 1933

hobos were injured. The engine and about half the train remained on the track and none of the train crew was hurt. Twenty-two cars in all were derailed. The derailment was blamed on a broken wheel.[73] [74] As often seems to have been the case because all those killed had been trespassers, the Interstate Commerce Commission did not bother to investigate the accident, despite its severity.

At 21:59 on the evening of Wednesday 6th March 1935, Chemins de fer de Paris à Orléans et du Midi express train No. 695 to Bordeaux, which was formed at Libourne from the omnibus train from Angoulême and a second train from Bergerac, became derailed 700 metres beyond the station at Saint-Loubes, France. Four of its 62 passengers were killed outright in the accident and a number were injured, at least one of whom succumbed to his or her injuries later. First-aid was given by a party of medical students who chanced to be on the train and most of the injured were taken to the Sabatié hospital in Libourne, with a few going to the Saint-Andre hospital in Bordeaux, 15 kilometres distant, including one victim whom it was thought was likely to require an amputation, and whose condition was causing some concern.

The cause of the accident was not immediately certain, but it was believed to have been due to the breakage of an axle of a third-class carriage. This vehicle had overturned onto its left-hand side and the following coaches were derailed and telescoped one another. The postal van was caught between two others and M. Dupart, the travelling sorter, was lucky to escape with his life.

Conductor Léonard was bruised in the accident but his first thought as a good railwayman was to protect the opposite running line, which was obstructed, and immediately after the accident he went forward towards Bordeaux and was able to stop a Bordeaux – Geneva fast train and thus prevent a worse accident.

According to the Flaman speed recorder the train had been travelling at 95 kilometres an hour at the moment of the derailment, a speed which a newspaper reporter considered somewhat high. However, a company spokesman stated that the maximum authorised speed over the line in question was 120 kilometres an hour. Driver Raymond, Fireman Lacoste, and Head Conductor Cavaillé, all from Bordeaux depot, were asked to be available for possible interrogation by the magistrates. [75]

Another report stated that the train, which must have already been derailed, had snapped off an iron column supporting the station roof as it passed through the platform.

The passengers who lost their lives at the scene were Mme Hélène Duparti, née Duhamel; Mme Carrier, M. Marcel Duthil and M. Pierre Boyeux. [76]

It was later determined that the derailment had been caused not by a broken axle but by the failure of a steel tyre, the result of a faulty casting. A further, less serious, accident was to occur on 14th April at Marchprime, France, from the same cause. [77]

---

[73] *The Times Recorder*, Zanesville, Ohio, United States, 10th July 1934
[74] *The Piqua Daily Call*, Piqua, Ohio, United States, 13th July 1934
[75] *La Charente*, Angoulême, France, 8th March 1935
[76] *Le Radical de Marseille*, Marseilles, France, 7th and 8th March 1935
[77] *The Times*, London, England, 16th April 1935

In contrast to the Pendennis and La Vergne derailments, detailed above, an accident on the Missouri-Kansas-Texas Railroad near Klondike, Missouri, United States, where again all those who lost their lives were hobos, was investigated by the Interstate Commerce Commission, which issued quite a lengthy report on it.

Northbound Missouri-Kansas-Texas third-class freight train No. 74 consisted of engine No. 864, 63 loaded cars, two empty cars and a caboose, in charge of Conductor Toler with Engineer Clayton and Fireman Bryan. It left Franklin, Missouri, at 4.30 p.m. and was derailed half a mile north of the station at Klondike at about 8.45 p.m. on Thursday 10th September 1936. The 32nd to 48th cars left the rails. Twelve of these cars, beginning with the leading one, were carrying oil, and a post-derailment fire resulted in a large blaze, with flames rising 200 ft into the air, attracting the attention of local farmers. After the accident, the engine was able to pull the un-derailed front cars clear of the fire and the farmers helped to manually push the un-derailed rear cars back, one by one, until they were clear of the fire as well. Almost two hours after the wreck screams were heard coming from near the train and a trespasser, later identified as Walter Bragg, of Columbus, Ohio, was discovered. He was very badly hurt and died while being taken to hospital in St Charles, Missouri. The fire continued to burn fiercely all night and was still burning at noon the next day. The morning after the accident the charred bodies of four further trespassers were found in the wreckage. [78]

Engineer Clayton told the Interstate Commerce Commission (ICC) inquiry that nothing unusual had happened short of Klondike, where he had reduced speed for Rock House curve, north of the station. He had looked back along the train at intervals and seen nothing until just short of the curve, when he had seen a reflection of light from the rear of the train. Shortly afterwards the brakes had come on in emergency.

Fireman Bryan and Head Brakeman Charles corroborated the statement of the engineer. The head brakeman said that he was certain that he had seen the flames of a fire 10 or 15 seconds before the brake application.

Conductor Toler, who was riding in the caboose with Flagman Lux, testified that they had inspected the train and there had been no sign of fire or dragging equipment until immediately before the derailment, when he had seen sparks flying from the right-hand side, followed at once by flames and the brakes applying in emergency.

Acting Car Foreman Donahue stated that he had personally inspected the train at Franklin and had carried out a brake test. He had gone to Klondike as wreck foreman early the following morning and examined the wreckage. In his opinion the derailment was due to a broken flange of one of the wheels of a car.

Section Foreman Koelling, in charge of the section of line at the site of the accident, stated that he had examined the track carefully after the accident accompanied by Roadmaster Humphrey. He had found some low spots in the track, but they were not bad. While he could not form any opinion as to the cause of the derailment, he believed that the train had been moving at high speed when it had occurred.

Roadmaster Humphreys said that he had been over the line a number of the times in the weeks preceding the accident, on track motor cars and trains, most recently on 8th September, and had noticed no track irregularities. He did not think that low joints found

[78] *Daily Cosmos-Monitor*, St Charles, Missouri, United States, 11th September 1936

after the accident would have caused the cars to sway to any appreciable extent and attributed the derailment to a broken wheel flange.

District Engineer Dunlay spoke of marks on the track which he attributed to a wheel with part of its flange missing.

Car Foreman Rochus testified that tank car S.H.P.X. 11577, the leading derailed car, had a trailing wheel with a broken flange. Through the wheel having been in the fire it was impossible to say if the fracture had occurred during the derailment or was the result of an old defect.

The ICC's inspectors also examined the track and noted some minor defects such as some evidence of rail creep and noted that the condition of the track might be criticised to some degree, but it did not appear that track defects were the primary cause of the derailment. Having summarised and reviewed the evidence it concluded that whether the derailment had been caused by a broken flange or by an explosion as a result of sparks from dragging equipment or from some other cause could not be determined. [79]

An inquest upon the five victims returned a verdict of 'accidental death.' [80]

The derailments of freight trains in the United States resulting in the deaths of trespassers travelling across the country looking for work during the Great Depression continued with an accident at Kerman, California, on Tuesday 23rd November 1937. The train involved was westbound Southern Pacific Railroad freight No. 401. It consisted of engine No. 2433, 21 loaded cars, three empty cars and a caboose, in charge of Conductor William Erwin with Engineer Leal and Fireman Randall. It left Fresno at 8.50 p.m. and was derailed at Kerman at about 9.20 p.m. The 10th to 17th cars left the track, the 10th and 13th cars were badly damaged, six trespassers were killed or fatally injured and another six less badly hurt.

Engineer Leal told the Interstate Commerce Commission (ICC) that the air brakes had been tested before leaving Fresno. Looking back along the train as they neared Kerman he had seen nothing amiss but then the brakeman had shouted that there was a fire on his (the engineer's) side and he had applied the brakes, thinking that some brake rigging had dropped. Speed had dropped and then the train had broken in two.

Fireman Randall said that he had felt a slight jerk before his engineer had applied the brakes.

Head Brakeman Paris spoke of seeing the fire and warning the engineer.

Conductor Erwin testified to having noticed a strange smell as his train approached Kerman, not that of a hot-box but perhaps from the acid which the train was carrying. As he had gone out onto the rear platform of his caboose the brakes had been applied.

Flagman Minton spoke of seeing a larger flash of sparks near Kerman, followed almost at once by the application of the brakes.

Car Foreman Charles Safford said that he had arrived at the scene of the accident about midnight with the wrecker. He had found some damage to the track but thought

[79] Interstate Commerce Commission report
[80] *Daily Cosmos-Monitor*, St Charles, Missouri, United States, 12th September 1936

that this was a result rather than the cause of the derailment. Picking up the wreckage, he had found two pieces of a broken arch bar belonging to S.P. 47873, a tank car carrying acid sludge, the first vehicle to become derailed, and it was his opinion that the broken arch bar had been the cause of the derailment.

Car Inspectors Mattox, McClure and Morris stated that they had examined the safety appliances, brake rigging, arch bar trucks wheels and so forth of train Extra No. 3253 when it had arrived at Fresno at 6 p.m. This train included S.P. tank car No. 47873, which was to forward in train No. 401. The inspection was made using an electric light, with particular attention being given to arch bars, and they did not think that the broken arch bar had existed at that time. In any case the defect would have been hard to discover by an ordinary inspection due to the location of the arch bar and it being covered with dirt, oil, and rust.

Assistant Superintendent Smith made a statement in which he also expressed the opinion that the derailment had been due to a broken arch bar, and that a chipped flange and broken spring plank found on the same car were a result of the derailment.

The ICC noted that S.P. tank car No. 47873 had been built by the American Car and Foundry Company in April 1909 and acquired by the Southern Pacific in 1924. The arch bar which had failed was not the original one; it contained an old flaw on its underside.

Concluding its report the ICC cautiously stated that the cause of the derailment could not definitely be determined but it was probably due to the breaking of an arch bar. [81]

Having heard the evidence of, amongst others, Car Foreman Safford, the inquest jury on 29th November were more categorical than the ICC in declaring that the derailment had been caused by a broken arch bar. [82]

On Sunday 9th November 1941, an unlucky accident occurred on the Pennsylvania Railroad at Dunkirk, Ohio, United States, when the failure of a part of the engine of a freight train led to the derailment of a passenger train on an adjoining line.

Westbound freight train Extra No. 8196, or FW-33, consisted of 2-10-2 engine No. 8196, 31 loaded and 17 empty cars and a caboose. It had left Crestline, 48 miles east of Dunkirk, at 8.18 p.m.; eight cars had been added at Colsan, 36½ miles east of Dunkirk, and a brake test had been conducted, and it passed Dunkirk at 10.19 p.m., travelling at 35 or 40 miles an hour. Shortly afterwards, 750 feet west of the station, and 680 feet west of Dunkirk tower, where the double-track Pennsylvania line was crossed by a single track New York Central Railroad line, the left crosshead of engine No. 8196 broke. The breaking of the crosshead resulted in the tearing loose of the cylinder head from the left cylinder. This fell between tracks 1 (eastbound) and 2 (westbound), then bounced and ended up in the middle of track 1.

Eastbound passenger train No. 78, *The Pennsylvanian,* consisted of 4-6-2 engine No. 3874, a passenger baggage car, a coach, a dining car and five Pullman sleeping cars, all of steel construction, in charge of Conductor Tom Scott, with Engineer R.S. Schuler and Fireman J.L. Gebhard. This train had left Fort Wayne, 83½ miles west of Dunkirk, at

[81] Interstate Commerce Commission report
[82] *Berkeley Daily Gazette*, Berkeley, California, United States, 30th November 1937

9.01 p.m., passed Sugar Street, 23¼ miles from Dunkirk, at 10.01 p.m., and was travelling at 70 to 75 miles an hour approaching Dunkirk when it struck the cylinder head at 10.22 p.m. The leading pair of wheels of the engine were derailed and it continued eastwards to a trailing crossover, where general derailment of the train occurred. The engine overturned here and slid along the track to Dunkirk tower, collided with the tower and ended up at right-angles across the track, where it was struck by the second car of the train, which was telescoped. Fireman Gebhard, ten passengers and a dining car employee were killed outright, Engineer Schuler was fatally injured and died several days later and 35 passengers, 11 dining car employees, the conductor and front brakeman of train No. 78 and either Operator Roy Schwartzkopf or his assistant Bucky Davis in Dunkirk tower were injured.

Operator Schwartzkopf said afterwards, "I noticed the headlight of the train swerving back and forth and swaying from side to side. It started to weave about 400 or 500 feet from our tower. As it came nearer, it turned sideways and started to tear up the track, as if its wheels were loose or even off."

Conductor Scott also spoke to a reporter. "I had just finished my book-keeping (making out my ticket reports, etc.) and had just seated myself in the front seat of the combination car, immediately behind the tender, when all of us in the car, about 15 in number, heard a jolting and heard a resounding crash to the rear. We were too stressed and shocked at first, all being thrown against the seats in front of us when our car came to a sudden halt about 100 feet past the passenger depot and 250 feet east of the pile-up, to know really what had happened.

"As soon as we gained our senses we climbed out and went back to give what assistance we could to the dying and injured. The right side of our car was dented nearly its entire length, evidently rubbing along the front of the overturned locomotive but not being sliced in two as was the coach immediately behind. The tender broke loose from our car and went 50 or 75 feet further up the track, both running on the ties."

Lynn Mahan, a passenger on the train, said, "I had retired to my stateroom just before the crash about 10.10 p.m. The first knowledge I had of anything unusual was when my Pullman started to bump and sway. I was thrown to the floor, but fortunately I instinctively grabbed the pillows, and they protected my head as I was knocked down." There was a moment of awful silence, grim and foreboding, before he realised what had happened. "I dressed and went toward the rear of the train but found the coach from the rear crushed against the one I had occupied. I then went to the forward end and found it badly wrecked but was able, by crawling thru the door, to drop about 10 feet to the ground." Mahon said that he had not at first appreciated the magnitude of the catastrophe. "Strangely enough there was little moaning or screaming."

Mahon then made his way along the track and came across Engineer Schuler. "He was bleeding profusely from the left hand which was badly lacerated and from which the thumb had been severed. The engineer's right arm was broken, also. I received a terrific shock as I went further ahead up the tracks and found the headless body of the fireman. In a few minutes all was confusion as many trapped in cars attempted to escape. By that time fire had started in the wreckage. I suppose this was from the blazing coals in the firebox of the overturned locomotive. I don't remember how much time elapsed, but I, in company with two or three other men from my car, attempted to lift the injured from the cars, and then we heard a siren. Soon a fire truck came and easily extinguished the flames. The firemen, in company with other townsfolk, helped in removing the dead and dying. Three or four doctors appeared on the scene quickly and went about giving first-aid and ordering the dead to one side. Quickly, then, it seemed to me, ambulances started to arrive on the scene, and we just went along carrying the injured to the vehicles. The spirit of the folks of Dunkirk in this tragedy was the kind that makes me glad that I live in America."[83]

The Interstate Commerce Commission (ICC) found that: 'The crosshead involved failed as the result of a progressive fracture in the outer web of the body. This fracture progressed from the inner surface of the web outward through the web and apparently broke through the outer surface at the time of the failure. This fracture was concealed from detection during ordinary inspection, as there was no crack in the outer surface at the time of the last inspection prior to the time of the accident, however, work reports filed after the completion of the last four trips prior to the one involved indicated that the left crosshead pin was working loose. The condition of the main-rod brasses indicated but little pounding; therefore, it is reasonable to assume that flexing of the metal surrounding the fracture was a factor to be considered concerning the loosening of the crosshead pin. Since the crosshead pin was found loose on successive trips, it appears that the pin should have been removed for examination. Had such action been taken, it is possible the defective condition of the crosshead might have been detected.'

In addition to finding fault with the failure of the company's employees to discover the defective crosshead before it gave way catastrophically, the ICC pointed out that the rules required that if there was a danger of adjacent tracks being obstructed measures must be taken to protect them. When the cylinder head had dropped train No. 78 had been a mile or a mile and a half away. Had a lighted fusee been displayed at once by the crew of the freight train the engineer of train No. 78 might have been warned in time to avert the accident. All the members of the crew of train No. 8196 maintained, however, that as they had been unaware that the cylinder head had been lost, as far as they had been concerned there had been no emergency, and hence no need to protect the opposite running line. Their view was supported by the general superintendent of the company.[84] It will be noted that this accident has features which are identical with the Rennert disaster detailed in the previous chapter.

Early on the evening of Monday 6th September 1943, the Pennsylvania Railroad suffered, on the outskirts of Philadelphia, what a newspaper described as 'the most disastrous wreck in recent railroading history.' By 10.30 p.m. the number of known dead was 43. Deputy Coroner Luke Barrett, at the scene of the accident, believed that over 100 had died while Chief Deputy Coroner Matthew Ross placed the number at 150. Their estimates were too high, but 79 passengers and a dining car employee lost their lives and 102 passengers, 22 dining car employees and five Pullman employees were injured.

The train involved was eastbound passenger Extra 4930, an advance section of *The Congressional Limited.* The train consisted of electric engine No. 4930, eight coaches, two dining cars and six Pullman parlor cars, all of steel construction, in charge of Conductor G.L. McPherson with Engineer Fred Schaller and Fireman G.R. Gierman. It left Washington, D.C. at 4 p.m. and ran non-stop over the 141½ miles to the scene of the

---

[83] *The Lima News*, Lima, Ohio, United States, 10th November 1941

[84] Interstate Commerce Commission report

accident. The last open office, at North Philadelphia, was passed at 6.03 p.m., and the train became derailed just under three miles further on at Shore, Pennsylvania, United States, at 6.06 p.m. (The location of the accident is usually given as Frankford Junction.) Travelling at 56 miles an hour, the trains left the track 151 feet east of the shore home signal. The engine and leading six cars remained on the track and stopped about half a mile down the line. The rear truck of the sixth car was detached, the seventh and fourteenth cars were completely derailed as was the front truck of the fifteenth car. What transformed a perhaps fairly minor derailment into a major catastrophe was the fact that the derailed seventh car came into contact with a signal bridge carrying the westbound Shore home signals, by which it was sheared diagonally for almost its entire length from floor to roof. In addition the eighth car was crushed, and it was in these two coaches that almost all the fatalities occurred.

As might be expected, given its magnitude, the accident received quite expensive coverage in the press, the war news notwithstanding. Thus, 'Robert Calvert, of Birmingham, Ala., a veteran of the war in North Africa, who was walking in the aisle of the third car, said something hit him in the stomach and – "It seemed a million people were on top of me. I couldn't get my breath. I thought, 'this is it; will die now.'" Roy Howard, president of the New York World-Telegram, and his wife, and Lin Yutang, the Chinese author, were on the train but escaped injury.

'Traffic on the main line was still blocked early today, but freight and through passenger trains were rerouted around the city on other lines. Armed guards with fixed bayonets and sub-machine guns headed a crowd away from the wreckage. Stretcher bearers had difficulty getting the dead and injured through the thousands who earlier had gathered at the scene three miles east of the North Philadelphia station. All available ambulances went to the wreck and hundreds of air raid wardens were mobilized to help police, firemen, soldiers, and sailors in rescue work. Hospitals pleaded for registered nurses and blood donors.

'The last person to be taken alive from the wreckage was Mrs Christmas Nix, 28, of Long Island City, N.Y., who lay for six hours with her head and torso outside a sheared coach and both legs held firmly inside. It was she who had said, "Take your time, I can stand it. There are dead people in here all around me," she moaned to a physician who administered a drug to relieve the pain. A Marine who had occupied the seat next to Mrs Nix lay dead nearby. For hours rescuers cut through the wreckage with acetylene torches before Mrs Nix could be removed and taken to a hospital, where her condition was reported critical. During that time she was given several transfusions but not once, witnesses said, did she complain.

'In the same car, rescue workers saw the bodies of a man and a woman, the man's arms protectively around the woman's shoulders.

'"Please take care of me, my mother would want you to," pleaded a soldier to Mrs Margaret Akers, a volunteer worker, before he died in her arms. His identification tag torn away by the wreck the soldier's body was placed with the unidentified dead.

'Bailey Beard, 54, the Congressional Limited's head waiter and a veteran of 26 years' service with the Pennsylvania on the Washington – New York run, was in the diner, which was midway in the train, when he was thrown to the floor by a severe jolt. There were about 50 people in the diner at that time, Beard said, about half of them women, {when} a terrific crash whipped the diner sideways and threw half a dozen passengers through the windows. "Coffee and hot water and dishes flew all over the galley," he related, and men working there were badly scalded. When the car finally stopped, Beard was lying on the floor beneath three tables. "There was panic everywhere," he said. "Women were screaming at the top of their lungs. The men in the galley were screaming from the pain of the scalding. Beard extricated himself and was helped outside by rescue workers. He was badly battered about the head and body and received leg injuries.

'At Frankford hospital, where hundreds of people went in response to a plea for blood donations, a woman, her clothing torn to shreds, her face covered with blood and dirt, helped carry the inert body of her husband into the emergency room. An intern made a quick examination and waved the body to an improvised morgue. The wife was taken to another room for treatment before her name was given.' [85]

Another newspaper reporter managed to interview Mr Roy Howard, the New York newspaper editor. 'Mr and Mrs Howard were in the 11th car from the locomotive. "I knew there was a wreck, and I pushed my wife to the floor, believing she would be safest there if the car turned over," Howard said, "I lay down on the floor too. Fortunately, the car did not overturn, and we were not hurt. I took my time, and as soon as I was able, I left the train and telephoned the story of the wreck to New York." [86]

The Interstate Commerce Commission (ICC) found that the derailment had been caused by the failure of the left front journal of the front truck (bogie) of Pennsylvania Railroad car No. 1860, the seventh vehicle of the train. It noted that during the 82 days preceding the accident the coach had been in regular use on the Washington – New York run, as well as on other services, and over this period had run about 30,000 miles. For the previous ten days it had run daily between Washington and New York, and on the day of the accident it had arrived in Washington at 11.20 a.m. on train No. 109. In the 4½ hours it had been at Washington it had been inspected and the journal boxes supplied with lubricating oil. On the return trip to New York it been regularly observed by the train crew, by the crews of six other trains which had passed it, by the operators at 15 stations and by railwaymen at Baltimore, Wilmington, Philadelphia, and North Philadelphia, none of whom had detected any problem.[87] Finally Harold McClintock, the engineer of a yard engine standing on track 5, a mile west of Shore, did see flames and smoke coming from the overheated journal. He shouted to Yard Clerk A. J. Carlin, and the latter ran to a telephone and advised Operator John Bower at Shore, but the train was in the process of leaving the track as the latter took the message. [88]

Continuing, the ICC stated that coach No. 1860 had been built in 1909 and modernised in 1935. Its journals had been repacked on 16th June 1943. Post-accident examination of the defective journal in the laboratory determined that the metal used for it had been within specifications. It was found that the journal had overheated very rapidly to a temperature of 1,400 degrees Fahrenheit, and then failed, but it could not be determined why it had done so. [89]

---

[85] *The Sandusky Register-Star News*, Sandusky, Ohio, United States, 7th September 1943

[86] *Miami Daily News-Record*, Miami, Oklahoma, United States, 7th September 1943

[87] Interstate Commerce Commission report

[88] *The Wisconsin State Journal*, Madison, Wisconsin, United States, 10th September 1943

[89] Interstate Commerce Commission report

Irving Ginsburg, 17, of New York, died in Frankford Hospital on Sunday 12th, the final fatality in the accident. [90]

An evening train on the Trem de Cantareira section of the Estrada de Ferro Sorocabana was derailed while travelling at speed between Tucuruvi and Jaçanã stations, Brazil, at about 5.30 p.m. on Saturday 25th March 1944. The train consisted of a *Baldwin* 4-6-0 steam locomotive and several wooden coaches and was crowded with homeward bound workers and shoppers, the railway being the only reliable form of public transport between central São Paulo and Guarlhos in the northern zone. The driver of the train had made an emergency brake application when he had heard a very loud metallic noise, which had resulted from the breakage of a coupling between two coaches, probably caused by the overloading of the train. Unfortunately the sudden brake application resulted in a traction shock which derailed the train. Thirty-three people died and perhaps 100 were injured. [91] [92]

The Pennsylvania Railroad suffered a third disaster on Friday 15th June 1945, near Milton, Pennsylvania, United States. Two trains were involved in the accident. Eastbound freight train Extra No. 4491 consisted of engine No. 4491, 102 cars of iron ore and a caboose. It had passed Allens, almost 25 miles west of Milton, at 11.48 p.m. passed MU tower, three quarters of a mile east of Milton, and was travelling at about 20 miles an hour when the rear truck of the 35th car, the 36th to 49th cars, and the 61st to 64th cars left the track about a mile and a half east of Milton.

Train No. 575, the *Dominion Express*, westbound from Washington D.C. to Buffalo, consisted of engine No. 3847, a storage mail car, an express car, a mail car, a passenger baggage car, three coaches and six Pullman sleeping cars, all of steel construction. It had passed Montandon, 3¾ miles east of Milton, at 1.02 a.m., five minutes late, and while travelling at 60 miles an hour it collided with the derailed freight cars at about 1.05 a.m., and the engine and seven cars were derailed. The passenger baggage car and the first coach, marshalled fourth and fifth in the train, were demolished. Sixteen passengers and the Engineer R.C. Stratton, Fireman N.E. Graeff and Assistant Baggageman A.B. Hoverter of the passenger train were killed and 42 passengers, three railwaymen, two mail clerks and an express messenger were injured.

Harrisburg ticket collector J.N. Gamble said, "There was one bump and I saw the glass of the door at the end of the car coming at me before I realised that I was flying through the air… The worst thing of all was the engine. It was still on the track, but the mechanical parts outside were stripped off and the tank had split away. Lying on the ground was the engineer. He was dead."

Passenger James Double, a toolmaker of Washington, D.C., described how the coach in which he and his wife Mildred were riding had careened against the freight train and "the flooring was shoved up until it almost met the ceiling, and one side was pushed in," while a soldier described the crash as "louder than anything I heard on European

---

[90] *The Bradford Era*, Bradford, Pennsylvania, United States, 13th September 1943
[91] Wikipedia: Tucuruvi Rail Accident
[92] *The Ogden Standard-Examiner*, Ogden, Utah, United States, 26th March 1944

battlefronts."[93]

As is often the case there was tragedy for some and relief for others. 'The family of a Buffalo soldier, homeward bound after two and a half years in a German prison camp, spent hours awaiting his arrival at Central Terminal yesterday, only to learn he had been killed in the Dominion Express train wreck at Milton, Pa. Pte. Herbert E. Swan had telephoned his mother he was coming home on the Washington-Buffalo flyer. She, accompanied by his brothers, Robert, and Carl, went to the station to greet him. Informed of the wreck, they went home and returned to meet a later relief train, but again went away disappointed. The family was informed last night Swan was trapped in the car in which most of the victims lost their lives and that his body had been identified in a Milton undertaking parlor.'[94]

For others there was better news. 'For 45 long minutes, F.J. Quirk thought his daughter had been a victim of the disastrous wreck at Milton, Pa., last week, then he heard her voice on the telephone. Miss Kathleen Quirk, a freshman at Drexel Institute, was a passenger on the wrecked train. Her father hurried to Milton when he heard of the crash, and through a window of a smashed coach saw a body of a girl he believed to be that of Kathleen. "Then they brought over a purse that had been found nearby," said Quirk, and I identified it as Kathleen's. The girl wore a ring similar to my daughter's and that seemed to clinch it." Then Quirk received a telegram advising him that his daughter was back home safely. "Even then," he said, "I didn't believe it until I heard her voice on the 'phone. It was as though she had come back from the dead."'[95]

The Interstate Commerce Commission noted that under the rules the crew of the freight train were required to protect adjoining lines in both directions when their train was stopped by an emergency application of the brakes, but in this case the collision had occurred within a few seconds of the derailment and so there had been no time to afford protection.

Post-accident examination showed that the left front journal of the rear truck of the 35th car of the freight train had broken off, having overheated. The car had been loaded at Erie, 275 miles west of Milton, on 13th June. The journal boxes had been lubricated at Renovo, 80 miles west of Milton, and the train had left there at 8.18 p.m. the evening before. Stops were made at Ferney and Linden, 63½ and 32½ miles west of Milton, for water, and no problems had been noticed by the crew, or at a stop west of Milton station at 12.50 a.m. On the other hand a crossing watchman 6¼ miles west of the point of derailment had seen the 'hot-box', as had three other crossing watchmen 6, 1½ and 1 mile short of the accident point and the assistant trainmaster and a flagman on westbound passenger train No. 581, which passed the freight train near Milton. The crossing watchmen and the flagman had given hand signals with a white lantern and whistled but none of these signals had been observed by the crew of train No. 4491.

As in the Shore derailment, laboratory examination of the journal found that it had heated to 1,400 degrees Fahrenheit and that its failure was due to overheating but was unable to determine why it had overheated.[96]

---

[93] *The Morning Herald*, Uniontown, Pennsylvania United States, 16th June 1945

[94] *The Daily Messenger*, Canandaigua, New York, United States, 16th June 1945

[95] *Indiana Evening Gazette*, Indiana, Pennsylvania United States, 18th June 1945

[96] Interstate Commerce Commission report

CHAPTER 3

OTHER DERAILMENTS

Other chapters in this book cover derailments caused by track defects, defects with the trains themselves and those caused deliberately by acts of vandalism or sabotage. This chapter will deal with derailments due to other causes such as excessive speed, collisions with animals and other accidental obstructions, traction shocks and so on, together with those for which no cause was given.

An Angora – Istanbul passenger train was derailed between Polatli and Beylik Keupru, Turkey, on Saturday 9th July 1932, its engine and carriages being 'completely wrecked.' Telegraph wires were brought down by the accident, making information on it harder to obtain, but it was believed that at least 19 people had been killed and 18 seriously injured. Most of the victims were soldiers going home on leave or on the completion of their service. [97]

Omnibus train No. 3965 of the Chemins de fer du Midi, which had left Quillan at 13:56 and was due at Perpignan at 16:29 on Friday 16th December 1932, met with an accident at Maury, France. The train included passengers from a Paris express who would normally have reached Perpignan via Narbonne but the line between Narbonne and Perpignan was blocked between Salsen and Fitou, and they had therefore had to travel from Carcasonne via Quillan to Perpignan.

At 15:30 the train had been approaching Maury and the crossing keeper there had lowered the barriers of the level crossing. Unfortunately, the driver of a motor vehicle, who was probably going too fast, was unable to stop at the crossing and smashed through the lowered barrier. With the train close at hand the motorist then drove off. Seeing what had happened and having no time to clear the remains of the barriers from the line the crossing keeper attempted to warn the drivers of the double-headed train. The driver of the leading engine saw his warning and immediately applied the brakes, but his colleague on the train engine apparently did not. Striking the obstruction, the two engines were

[97] *The Times*, London, England, United States, 11th July 1932

derailed. The pilot engine left the rails completely and overturned into a field, the second engine overturned on the line. The leading third-class coach was completely demolished, and the remaining coaches of the train were derailed, but less badly damaged. The noise of the collision alerted the residents of Maury, who immediately went to the aid of the victims. [98]

Seven people died, including Drivers Alary and Anglade and Fireman Laveran, and 26 persons were injured, including Fireman Surjus. [99]

The vehicle which had caused the accident was described as a commercial one of the torpedo type (that is streamlined with a foldable or detachable soft top). [100] It had had a blue body with a soft black canvas top. It should have been easy to trace as its front right mudguard and front right running board had been demolished in the accident. [101]

On the late evening of Boxing Day, Monday 26th December 1932, freight train No. 3727, consisting of an engine and 32 waggons, including one loaded with dynamite, ran into a landslide, and was derailed at the exit from No. 8 Tunnel, between El Vacar and Alhondiguillas, Spain, on the Córdoba – Bélmez line. The engine and 18 waggons went down an embankment. The bodies of Head Conductor Don Luis Guerra Delgado and Porter Luis Sánchez Nieto were recovered from the wreckage along with that of a third unidentified person, while Fireman Francisco Sanz Díaz and Brakeman Rafael Behmonte Alvarez were seriously injured. Driver Enrique López Navarro and two other railwaymen were only slightly hurt. The dynamite fortunately did not explode.

A follow-up report stated that the fireman and brakeman had both died in hospital of their injuries. The unknown body had been identified as one Manuel Rodriguez Barba, who had been riding hidden in a waggon, so must have been a trespasser. [102] [103]

The engine and four third-class carriages of a passenger train were derailed near the station at Ventaquemada, Colombia, late on Monday 10th April 1933. According to an initial report in a Colombia newspaper there were 21 dead and 43 injured. There were 144 passengers on the train, of whom six were riding in a first- and second-class composite carriage and the other 138 in third-class carriages. Most of the victims were peasants, returning from a fiesta in the capital, Bogotá. Eighteen of the injured were taken to the hospital at San José, 15 to the hospital at Tunja and the remainder were treated by doctors at the scene. Some of the other passengers suffered slight bruises. The accident occurred at 22:45 on the line of the Ferrocarril del Nordeste to the train which had left Bogotá at 16:00. The train left the track at the summit of the line and in what was described as the most desolate part of the journey. It was the coaches in the centre of the train which suffered most by a sudden brake application and were completely wrecked. The first and last coaches of the train were little damaged. The leading wheels of the

---

[98] Le Matin, Paris, France, 17th December 1932

[99] L'Homme Libre, Paris, France, 17th December 1932

[100] Wikipedia – Torpedo (car)

[101] Le Peuple, Paris, France, 18th December 1932

[102] La Vanguardia, Barcelona, Spain, 28th December 1932

[103] Webpage: Federación Castellano Manchega de Amigos del Ferrocarril (in Spanish)

engine remained on the track.

Notice of the accident reached Bogotá at 05:15 the following morning when telephone communication was restored between there and Ventaquemada. There was no telegraphic communication with the capital. News of the accident was, however, sent to Tunja soon after it had occurred by a Boyacá government official.

Doctor D'Hertague, the director general of the railway company, issued a statement regarding the disaster to the newspapers. He stated that the derailment had occurred between Albarracin and Ventaquemada stations. As soon as he had received notice of the accident he had communicated with Drs Petterson and Danay, the secretary general and transport chief of the company, and a special train had left Bogotá for the scene of the accident at 07:00. Regarding the cause of the accident, Dr D'Hertague stated that matters were made more difficult by the death of the two key witnesses to it, namely Driver Luis Ramirez and Fireman Misael Montaño. The driver's assistant, a lad of 15 named Valderrama, had had a lucky escape, falling on a pile of coal. He was not able to say what had happened and was still suffering severely from shock. Possibly when he had recovered somewhat he might be able to give some information.

The director general gave his theory as to how the accident had happened. He stated that the train had departed from Bogotá at 16:00, 20 minutes late, with engine No. 1006 in charge of Driver Ramirez, one of the most senior, best, and most careful of the company's men. Engines were changed at Villapinzón, and the journey had continued with engine No. 52. There had been heavy rain throughout the journey, causing the train to lose more time, and it was three quarters of an hour late when it reached Albarracin. The train had left there at 22:00 and because of the rain the engine wheels had slipped badly and sand had been put down. After three kilometres the supply of sand had run out. Driver Ramirez had then stopped the train to replenish the sand from one of the boxes alongside the line. Between ten and fourteen minutes more were thus lost.

The Ventaquemada signalman says that he had heard the whistle of the approaching train, then there had been a tremendous noise followed by confused shouting. He assured the director general that it was a brake whistle that he had heard, and that it had suddenly been interrupted, without his knowing why.

It was to him absolutely certain, Dr D'Hertague said, that Driver Ramirez had, in an instinct of self-preservation, made a sudden violent brake application, and had thus caused the disaster. He must have seen some obstruction on the line, very possibly some rocks which were found alongside of it afterwards. Optical effects due to a combination of ground mist and the light from the engine headlamp might have made the rocks appear to be much larger than they actually were. It was also possible that the engine wheels striking something had caused the brakes to apply. The brakes were of the modern, efficient, Westinghouse design, which pulled up a train very quickly.

Rumours had circulated that the driver had been drunk. This seemed impossible, or at least highly unlikely in such a conscientious and reliable employee as Driver Ramirez. No complaints as to his drinking had ever been made in all the years of his service with the company. Among the director general's acquaintances with drivers not only in Colombia but also in visits to Italy, France and Belgium he had never met a driver better than Ramirez. The latter had been a serious man, 32 or 35 years old, married with children. He knew how to do his duty and thoroughly understood the rules. Furthermore, and in contrast to the unfounded rumours of his intoxication, there was the testimony of those at Villapinzón, where he had changed engines, who said that he had not had anything to drink there, not even a lemonade.

There were also allegations that Driver Ramirez had driven the train at a fantastic speed, so that stations had passed 'like a kaleidoscope.' There was no evidence for this. He had stopped at all the booked stations and had lost three quarters of an hour because of the rain. He was clearly a very conscientious driver. He had been upset that his train was running so late. The doctor did not believe that a driver who was so dedicated and who knew the line so well would have caused such a serious derailment.

Some of the afternoon papers had stated, the director general continued, that the derailment had occurred on a curve and was due to excessive speed but this was not the case, the curve on which the derailment had occurred was a considerably easier one than many others on the line and it could therefore be taken at considerable speed.

Furthermore the high ground did not have the characteristics of other parts of the line where drainage measures had been necessary. Marks on the track provided confirmation that a sudden hard brake application had caused the derailment. It was the traction shock which had caused the coaches to sway from side to side and the destruction of the middle vehicles of the train. The fact that the middle carriages had formed a zig-zag whilst the front wheels of the engine were not derailed and the leading and rear coaches were little damaged confirmed the manner of the derailment. As already stated the middle coaches of the train had borne the full brunt of the disaster. Of the 21 who had lost their lives 18 had been killed outright, one had died while being treated at the scene, one at Bogotá and the other in hospital at San José.

The director general spoke movingly of having seen the corpse of Driver Ramirez, for whom, he said, he had felt a genuine affection. The poor driver had been almost decapitated by the brake lever. He said that what he had seen when he had arrived at the scene of the derailment, 350 or 400 metres from Ventaquemada station, had caused him the greatest anguish.

Continuing, the director general said that doctors had been called to the scene as soon as possible and Dr Pinto, the Ventaquemada parish priest, had also attended and given spiritual comfort to the dying. The mayor of Ventaquemada had organised the rescue operation and the local inhabitants had assisted with the care of the injured.

This was the first major accident to occur on the line since it had opened a year and a half earlier, though there had been a minor derailment. The company had chartered a bus and cars to tranship passengers past the site of the derailment until the line could be cleared. The work of removing the wrecked coaches and clearing the track would not be begun until the investigation of the accident had been completed. [104]

The following day it was reported that the number of dead had risen to 22. Government inspectors considered that the derailment had been caused by an obstruction deliberately placed on the line, a view contradicted by the railway engineers, who said that no traces of any such obstruction had been found. Meanwhile a government official stated that he was especially interested to know if the railway company intended to compensate the families of the victims. When the question was raised the railway

[104] *El Tiempo*, Bogotá, Colombia, 11th April 1933

directors asked for time to respond. It was considered that the railway company should be easily able to meet the cost.

The investigation into the disaster continued, and in contradiction of the statement of the managing director of the company the day before it was alleged that on the afternoon of the disaster Driver Ramirez had been in a state of complete intoxication in Villapinzón, and that he had caused a serious disturbance in a canteen so that the police had been obliged to remove him by force. Dr D'Hertague repeated, however, that he did not believe that Driver Ramirez had been drunk and pointed out that this would have been noticed by the train conductor and the stationmasters along the line. The director general said that he had spoken to the Villapinzón stationmaster, who had assured him that the driver had been in a normal condition, likewise the train conductor, who had spoken to the driver there, had noticed nothing wrong with him. A senior director of the company agreed that stories of the driver drinking beer in Villapinzón were hard to believe. [105]

The following day Dr Barrios Carbonell, the chief judge in charge of the investigation, declared that the derailment had been caused by poor track and the driver being intoxicated. Secretary General Peterson, meanwhile, declared that the telegram from Villapinzón reported in the newspaper the previous day which had alleged that Driver Ramirez had been drunk had been misinterpreted. The alleged disturbance had occurred in a store when the driver had attempted to apprehend two thieves, who were later found robbing the bodies of the victims of the disaster and were currently in prison at Tunja. Driver Ramirez had therefore not been involved in any scandal, nor had he been drunk, and unfair allegations had been made against him. The company was paying compensation to the widows of the driver and fireman.

Obstructions had been found on the line before. The previous week Driver Antonio Pérez had found a large log lying across the track and had had to stop the train to remove it. Two months ago as a train had emerged from the tunnel at Km 110, it had encountered a large rock on the line which had undoubtedly been placed there. These incidents the investigators should bear in mind, the newspaper wrote.

Traffic Chief Danay stated that the number of dead was only 21, not 22 as stated the day before. [106]

A Chemin de fer de Paris à Orléans freight train was derailed due to a broken axle on the evening of 3rd June 1933 at Les Blottereaux, France, three kilometres outside of Nantes. No one was hurt, but the track was damaged for a distance of some 800 metres. A temporary track had been installed to keep traffic running, the main line being slewed into a wartime siding constructed by the Americans, with a speed restriction of 30 kilometres an hour in force through the connection. It was here that train No. 141 *ter*, the last of three expresses to Croisic, which had left Paris at 21:45, was derailed at 04:50 the next morning, Sunday 4th June 1933. The number of passengers, heading for the beaches of the Côte D'Amour for the Pentecostal holiday, had resulted in the train being run in triplicate. The locomotive and seven coaches were overturned, and the tender fell down an embankment.

---

[105] Ibid, 12th April 1933
[106] Ibid, 13th April 1933

The noise of the accident awoke the residents of Grand Blottereaux who would have alerted the emergency services. As soon as he became aware of the accident M Mathivet, the *préfet* of Loire-Inférieure organised a rescue operation, notifying the fire service, health workers, civil and military doctors, and the police. The rescuers worked hard in distressing conditions, though some of the injured were trapped in a coach for three hours before they could be released. Notably at 09:00 a woman who was trapped between the bodies of her sister and her mother was finally released. The injured were taken either to the Hôtel Dieu or to the Broussais military hospital in Nantes. By 13:30 the rescuers had recovered 14 bodies, nine of whom could be identified. Another 124 people were more or less injured.

The initial investigation placed the blame on the driver of the express, who had been advised of the restriction but failed to observe it until he had braked at the last moment. The track recorder on the derailed engine indicated a speed at the moment it had left the track of 92 kilometres an hour. In addition to the notice of the restriction which the driver should have received signals indicating the restriction were displayed 1,200, 800 and 200 metres from the diversion.

Deputy Stationmaster Eugène Halgaud, who was in charge of *poste* (signal box) P, working the points at the location of the derailment, declared after the accident in a statement, "I saw the express approaching, and was very surprised, not to say apprehensive, to see that it was running at 90 or 100 kilometres an hour. As the engine reached the points the driver, no doubt realising his mistake, applied the brakes abruptly. I saw fire literally flying from the engine wheels. The engine and tender crossed the points but the leading first class carriage jumped the rails at the points and ploughed up the ballast for 15 metres or so. Then the coach overturned, pulling the engine and the other coaches after it. The engine crossed two or three tracks and then overturned as well, while the coaches piled up into one another."

After the accident, *le parquet* (the prosecutor's office) interrogated Driver Pierre Cuzon and Fireman Robert Mathis. The driver was then taken into custody, but the fireman was released. Driver Cuzon declared that he had not seen the signals to reduce speed, nor had he received notice of the speed restriction at Angers. He was an experienced driver, and in fact had been due to retire in October. He was prostrated by the accident and spent the day in an office at Nantes station, haggard and drained. He said, "I don't know what could have happened. I saw the last signal when I turned round. But it was then too late to brake. It's a catastrophe." [107]

A second accident happened the same morning at Saint-Pierre-des-Corps. A railbus providing a Nantes – Tours service was struck by a train from Nantes which had overrun a signal. A conductor and two mail clerks suffered bruises and a dozen passengers were slightly hurt but were able to continue their journeys after treatment. [108]

Driver Cuzon and Fireman Mathis were questioned by M. Delrieu, the *juge d'instruction* (examining magistrate) on 9th June.

The fireman stated that he had known nothing of any speed restriction and saw neither the notice at the depot nor the warning signals at Les Blottereaux.

---

[107] *L'Intransigeant*, Paris, France, 6th June 1933

[108] *Le Temps*, Paris, France, 5th June 1933

Driver Cuzon also denied having seen any notice regarding the temporary speed restriction at the depot. He stated that the green signal on the approach to the diversion had lacked any indication of a speed restriction immediately ahead and had led him to believe that the work of repairing the track was at a point further down the line. He added that the restriction should have been protected by a red hand signal and a 'crocodile' (automatic train control ramp).

Deputy Stationmaster Halgaud repeated his assertation that Driver Cuzon had braked hard, precipitating the derailment. Driver Cuzon denied it, stating that he had braked gradually. The prosecutor sent to the scene of the accident in the hope of determining from the Flaman track recorder whether or not a hard brake application had been made.

Driver Cuzon stood trial in May 1938. In the course of the evidence the three signals approaching the site of the diversion were described. The first was a green disc advising the driver to have his train under control. Then came a triangular signal indicating the 30-kilometres-an-hour restriction and finally a T signal indicating the point at which the restriction commenced. Driver Cozens stated that not having seen any notice regarding the speed restriction at Angers, he had assumed that the first signal applied to work further ahead. He had then failed to see the second signal altogether and the T signal had taken him by surprise. Various points such as the fact that the train still had wooden coaches and the lack of a 'crocodile' were raised in his defence [109]

The tribunal announced its findings on 1st June 1934. It found Driver Cuzon guilty of causing the accident and he was given a year's imprisonment, suspended, and fined 300 francs. The railway company was found to be civilly liable. [110]

Eight workmen died and seven were injured in the derailment of a train loaded with sugar cane on the Casa Grande estate in Trujillo, Peru, on Sunday 10th December 1933. [111]

Early on Sunday 7th January 1934, the night mail train from Maritzburg to Durban, carrying 200 passengers, became derailed on a curve near Donnybrook, Natal, South Africa. The engine and six carriages went down an embankment. It was regarded as very fortunate that there were only six fatalities – the driver and five coloured passengers. [112] The cause of the derailment was not immediately apparent. [113]

The wreck of a Pennsylvania Railroad passenger train on the evening of Monday 26th February 1934 on the outskirts of Pittsburgh, Pennsylvania, United States, was attributed initially by railwaymen to a 'frozen switch.'

Eastbound passenger train No. 1638 from Akron, Ohio, to Pittsburgh, consisted of engine No. 7274, a combination passenger and baggage car, a coach, a dining car, and two Pullman sleeping cars, all of steel construction, in charge of Conductor Walter Shank

---

[109] *L'Ouest Éclair*, Rennes, France, 19th May 1934
[110] Ibid, 2nd June 1934
[111] *The Evening Star*, Washington, D.C., United States, 11th December 1933
[112] *Bredasche Courant*, Breda, The Netherlands, 8th January 1934
[113] *The Evening Sun*, Hanover, Pennsylvania, United States, 8th January 1934

with Engineer W.M. Shaw and Fireman O.C. Douthlitt. It passed Homewood Junction 38¾ miles from Pittsburgh, at 8.40 p.m., three minutes late, but by the time that it left Rochester, 25¾ miles out, at 9.06 p.m., it was 17 minutes late, having lost time because the engine was not steaming well. Pennsylvania Avenue, three quarters of a mile from Federal Street station, was passed at 9.31 p.m., the train now being only 13 minutes late, and approaching Federal Street at an estimated speed of 60 miles an hour the train was derailed at about 9.32 p.m. The entire train left the rails and the engine collided with a steel boxcar on a siding and knocked it into a building. The engine then demolished a transformer house and fell into Merchant Avenue. The five vehicles behind the tender ended up in various positions, one demolishing a disused signal tower. The engine was badly damaged, the combination car and coach were almost destroyed, the dining car was badly damaged, and the Pullman cars were considerably damaged as well. Nine passengers, Engineer Shaw, and Fireman Douthlitt were killed and 28 passengers, Conductor Shank, Baggagemaster Charles Nee, Brakeman A. Weaver, Flagman Clyde Thorndell, eight off-duty railwaymen, four dining car employees and two Pullman employees were injured. [114]

A newspaper report described the tragedy of one family arising from the accident. 'When D.C. Mackey, of 515 Sheridan avenue {New Castle, Pennsylvania} took his brother-in-law, C. J. Mayer, of Ingram, Pa., to the train, Monday evening, neither of them had any idea it would be the last time they would be together in this life. Less than two hours later, Mayer was dead, buried in the wreckage of the Pennsylvania train that crashed through the bridge near the Federal Street station, Pittsburgh.

'Mr Mayer, who was about 48 years of age, was married to a sister of Mr Mackey. Some years ago he worked in the Shelby Tube plant in Ellwood City, and before that had resided near Portersville. Several years ago, he went into business with a relative in Pittsburgh and his trip to New Castle Monday was in connection with the settling of a business deal he made when he disposed of property he owned in Lawrence County. When he had finished his business, he went to the home of Mr and Mrs Mackey and had dinner with the family. Dinner over they sat around chatting and about 7.30 Mr Mackey took Mr Mayer to the Pennsylvania station where he boarded the bus for New Castle Junction.

'Mr and Mrs Mackey are grief stricken today as they recall the pleasant dinner scene but a few short hours ago and today their brother-in-law is dead. They left at noon for Pittsburgh to be with Mr Macey's sister in Ingram.' [115]

Conductor Shank was interviewed by Interstate Commerce Commission investigators in hospital and said that he had not thought that the train had been going too fast, though it had been making good time. Brakeman Weaver and Baggagemaster Nee likewise had not thought the speed of the train excessive.

Flagman Thorndell, on the other hand, testified that he had thought the speed of the train had been too high in view of the 25-mile-an-hour speed restriction ahead. He put the speed at 35 to 40 miles an hour.

Passenger Trainmaster Lowery put the speed of the train at Pennsylvania Avenue at 60 miles an hour. When the speed was not reduced on passing the tower there he had got

---

[114] Interstate Commerce Commission report

[115] *New Castle News*, New Castle, Pennsylvania, 27th February 1934

up and signalled to the enginemen to reduce speed. He had then sat down again, and the derailment had occurred before speed had been reduced.

The statements of off-duty railwaymen travelling on the train varied but most of them did not notice any brake application after passing Pennsylvania Avenue and some of them said that they had been alarmed at the speed of the train.

Pennsylvania Avenue Block Operator Leverman McLaughlin stated that the train had passed his tower at its usual speed of 45 to 50 miles an hour but that its brakes had not been applied there as they usually were. His colleague, Telegraph Operator Lively, put the speed of the train at 50 to 55 miles an hour, though he added that he was not a good judge of speed.

Mr J.W. Fellinger said that he had been driving his car along the north side of the tracks at 35 miles an hour and the train had passed him, going at at least 45 miles an hour. The speed of his train had attracted his attention and he had thought that if it did not slow down it would leave the track. Then he saw the engine overturn. He saw no sparks from a brake application.

Track Supervisor Callahan and Superintendent Adams spoke of finding and examining a damaged set of facing points known as Clark's Switch at the point of the derailment. They agreed that it was not open points that had caused the derailment but rather that the derailment had damaged the points mechanism.

Master Mechanic Chafin spoke of examining the engine at 3.40 a.m. the next morning and of failing to find anything which could have had a bearing on the accident.

Crew Dispatcher Lewis and others who had seen the driver and fireman at Akron on the day of the accident said that they had both appeared to be normal.

Assistant Medical Examiner Wilkinson stated that he had examined Engineer Shaw, who had given his age as 59, on 23rd February 1934, thus only three days earlier. The engineer's heart had appeared normal, with no murmurs or sclerosis, his blood pressure had been normal, his pulse regular and his urine had had no albumin, sugar, or casts.

The ICC concluded that the train had derailed due to excessive speed. All the witnesses but one agreed that the train had been moving at practically normal speed at Pennsylvania Avenue and there had been no subsequent brake application. The position and condition of the wreckage were also conclusive proof of the high speed of the train when it left the track. Trains passing over the curve on which the accident occurred were restricted to 25 miles an hour and there was a 40 miles an hour restriction on the preceding curve. In addition, the aspects of three signals, the first two showing *Approach restricting* and the third showing *Approach*, required the engineer to reduce speed. It could not be determined why Engineer Shaw had failed to reduce speed. The ICC concluded that he must have either been dead or incapacitated in the moments before the accident.

Had Passenger Trainmaster Lowery, as he said, realised that the train was going too fast it was his duty, the commissioners pointed out, to have applied the brakes himself.

Automatic cab signals were in use on the railroad between Washington and New York and between Baltimore and Indianapolis. Had they been in operation at Pittsburgh a whistle would have sounded if the train had failed to reduce speed in accordance with the signal indications, alerting either the engineer or the fireman. The ICC hoped that such protection would be considered on the Pittsburgh to Chicago line, which had frequent high-speed trains. [116]

On the day after the La Vergne derailment in Tennessee, Tuesday 10th July 1934, another freight train was derailed at Morrisville, North Carolina, United States. Westbound Southern Railway freight train No. 67, from Selma to Spencer, consisted of 61 cars, which had left Raleigh at 6.30 a.m., in charge of Conductor J.A. Fowler and Engineer Jenkins. [117] The train had only covered 12 miles and the engine and leading cars had just crossed Crabtree creek when cars were seen to sway and 31 of them then left the rails at 7.10 a.m. The derailed vehicles included eight gasoline tank cars and fire broke out in the wreckage, eventually involving both these tank cars and at least 15 cars of merchandise. The accident happened in the middle of the little town of Morrisville, several buildings of which were set alight. Fifteen or 20 hoboes were riding on the train and the bodies of four of them were quickly discovered, three of them blacks and one white man. They had been burned to death.

Two young Durham men named Larry Lynn and Bill Mitchell had a narrow escape. They were sat in a bakery truck at a level crossing waiting for the train to pass. When they saw the train derail they just had time to jump clear. Their truck was caught by the derailed cars, dragged down the line and burnt. A Mr R.S. Dodd, whose house and small store adjoined the railway, was less fortunate. The tank cars piled up against it and burnt it down. He lost everything, to a value of some $15,000, but was reassured that his loss was covered by insurance. [118]

By the next day, the discovery of further bodies had raised the number of dead to 10, all trespassers. Four injured negro hoboes were taken to Durham Hospital and one railwayman was slightly hurt. Total damage caused was estimated at $200,000. [119]

There had been a previous derailment and fire in the same place on 28th April 1933. Only one tank car was involved in this accident, in which one hobo was slightly bruised. [120]

Five passengers were killed and 24 were injured, nine seriously, when one coach of a passenger train was derailed in Verba station, Bulgaria, on the line between Sofia and Kustendil, on Thursday 9th August 1934. [121]

The Genève to Vintimille international express of the Chemins de fer de Paris à Lyon et à la Méditeranée, which had departed at 20:00 on Saturday 11th August, became derailed at Avignon, France, at about 03:20 on Sunday 12th August 1934. All the vehicles of the train left the rails, and the leading mail van was entirely destroyed when it struck a row of goods waggons standing on an adjoining line. The following vehicles of the train then piled up behind it, and an electric cable standard was demolished, throwing the

---

[116] Interstate Commerce Commission report
[117] *The Daily Times-News*, Burlington, North Carolina, United States, 10th July 1934
[118] *The Daily Times*, Wilson, North Carolina, United States, 10th July 1934
[119] *Gastonia Daily Gazette*, Gastonia, North Carolina, United States, 11th July 1934
[120] *The Daily Times*, Wilson, North Carolina, United States, 12th July 1934
[121] *The Times*, London, England, 10th August 1934

station into darkness. The engine allegedly caught fire. The darkness made the initial rescue work very difficult but the seven railwaymen who were on duty at that early hour did all they could, and more help and doctors arrived within half an hour.

'According to the initial information, which is backed up by the statement of the signalman in the box at the entrance to the station, the accident resulted from the train entering the station at too high a speed, trains being restricted to a speed of 20 kilometres an hour over the connections at the throat of the station. It appears that the express was travelling at twice that speed, 40 kilometres an hour, as it came into the station. However, the Flaman speed recorder fitted to the locomotive was damaged in the accident, so it was difficult to determine the exact speed of the train.' [122]

Possibly significant in this respect was the statement of a railwayman who had been a passenger on the train, whose son had lost his life in the accident and whose wife had been injured. He stated that the train had passed through Grenoble 23 minutes late and he had noticed an excessive amount of vibration there when the brakes had been applied, which he had attributed to undue play between the couplings. At Valence there had been a shock so violent that he had anticipated immediate disaster. Things had then returned to normal, though, until they were approaching Avignon, when he had been struck by the train's high speed. He had just risen from his seat to go to the corridor when the crash had come, and he had been thrown head over heels the full length of the carriage. [123]

Driver Emile Achard and Fireman Alfred Héritier survived the accident. Driver Achard stated, "We were 200 metres from the station, travelling at about 20 kilometres an hour, when the derailment began. I believe it was the first vehicle behind the engine which initially left the rails. I came into the station completely normally. The mail van got completely across the line; it obstructed track No. 2; a first-class coach, two second-class coaches and the van at the head of the train got off the rails."

At 15:00 it was reported that Conductor Auguste Merindol and seven passengers had lost their lives in the accident and 30 passengers and Fireman Héritier were injured.

Driver Achard and Fireman Héritier were interrogated by *Juge D'instruction* (examining magistrate) Rouquette. [124] They continued to maintain that the train had not been travelling too fast as it entered the station and that the brakes had been applied, but this was disputed by the statements of Signalmen Marius Bourges and Paul Carnot in No. 2 signal box, who told M. Rouquette, "After it had passed *carré* [home signal] No. 1 the Genève – Vintimille express was still travelling at about 60 kilometres an hour when it should not have been doing more than 20. We waved and shouted. A few seconds later there was a tremendous noise, and all the lights went out."

It seems clear that the derailment was due to excessive speed and perhaps also a traction shock when the engine brakes were violently applied at the last moment in an attempt to check the speed of the train. There were rumours that the Flaman tape had not been accidentally destroyed in a post-accident blowback on the engine but had been deliberately damaged to get rid of incriminating evidence.

[122] *L'Homme Libre*, Paris, France, 12th August 1934
[123] *The Times*, London, England, 13th August 1934
[124] *L'Homme Libre*, Paris, France, 12th August 1934

An eighth passenger, M. Clary Gérard, died of his injuries the day after the accident. [125]

A ballast train which was also carrying a party of workmen struck two horses which had wandered onto the line near Lulea, Sweden, on Thursday 27th June 1935. The engine was propelling its waggons and because of a curve its driver failed to see the animals in time to pull up. The collision overturned the waggons of the train and seven of the workmen were killed. [126]

The heaviest rain ever recorded in northern Japan caused a serious accident on Sunday 27th October 1935 near Kawamaye station on the Banyetsu – Higashi line, which had only opened the year before. At 5.54 p.m. that day a train which had just left the station ran into a landslide, though a train coming in the opposite direction had passed the spot without incident and without anyone on it noticing any problem. The train which met with the accident was a mixed one, consisting of a locomotive, four carriages and three vans. The engine and three leading coaches fell into a ravine, about 30 metres deep, and nine passengers were killed, nine passengers and three railwaymen were seriously injured, and 35 passengers were less badly hurt. [127]

Twenty-three people died and 63 were injured on Tuesday 28th January 1936 when a train ran into or was struck by a snowslide east with Yamagata in northern Japan. Three coaches in which engineers and workmen were riding were thrown over a cliff. [128]

An omnibus train was derailed on a curve on the line between Naples and Nola near Bufola, Italy, early on Wednesday 10th June 1936. Three coaches well filled with workmen bound for Naples overturned. Rescue workers were quickly sent from Naples and by the evening 15 bodies had reportedly been recovered and there were estimated to be 200 injured, though officially the figures were only six and 40 respectively. The cause of the derailment was not immediately known. [129] [130]

Ten people were reportedly killed and at least six seriously injured in the derailment of a Warsaw – Posen express near Posen, Poland, on the evening of Wednesday 1st July 1936. A worse accident was averted when a second express was stopped close to the scene of the accident. The cause of the derailment was not immediately known. [131]

The last carriage of a passenger train was derailed on Monday 26th October 1936 near the station at Foscani, in Moldova, Romania. Six people died and 60 were injured. All the

---

[125] *L'Ouest Éclair*, Rennes, France, 14th August 1934
[126] *The Railway Gazette*, London, England, 12th July 1935
[127] Ibid, 6th December 1935
[128] *Daily News-Record*, Harrisonburg, Virginia, 28th January 1936
[129] *The Times*, London, England, 11th June 1936
[130] *L'Homme Libre*, Paris, France, 11th June 1936
[131] *International Herald Tribune*, Paris, France, 2nd July 1936

victims were Romanians; most had been returning from holidays. [132] The dead were described as five peasants and a schoolboy. The cause of the derailment could not be determined. [133]

Briefly reported in an English newspaper was an accident to a motor train near Trnovce, Czechoslovakia, on Monday 23rd November 1936, in which eight people died. No further details of the accident were given other than that it had occurred in Western Slovakia but as only one train was mentioned in the report it was presumably a derailment. In another accident the same day four people were killed in a collision involving the *Balkan Express* and a goods train at Letovice in Moravia, and the report added that a total of 26 people had been injured in the two accidents. [134]

An accident befell a Port Elizabeth to Cape Town mail train at Vlakteplaats, near Oudtshoorn, South Africa, on Monday 21st December 1936. Its engine left the rails on a curve and the leading six carriages of the train followed and were overturned and smashed. At least seven passengers died, and many were seriously injured, about a dozen of whom were not expected to survive. The train was full of holidaymakers.

A breakdown gang, assisted by people from local farms, worked all day to extricate the dead and injured. The number of casualties overwhelmed the facilities at Oudtshoorn Hospital and several of its convalescent patients agreed to be discharged early to make room for the injured. [135]

On or about Friday 19th March 1937 a passenger train was derailed at an unstated location in Siberia, USSR. Fourteen people were killed and many injured. The train was carrying workers and their families to a new town at Komosol.

In a sequel to the accident the Irkutsk *East Siberia Pravda* newspaper of 11th August of the same year announced that no less than 72 railwaymen had been executed for causing this and other accidents. The alleged saboteurs were accused of being Trotskyist rightist terrorists who had wrecked trains at the behest of the Japanese intelligence services, in order to weaken Soviet transportation before a possible war with Japan. [136] [137]

A passenger train was reportedly derailed at Sestao, Spain, near Bilbao, on Tuesday 13th April 1937. It has not been possible to obtain any information on this accident other than a brief Wikipedia reference. Six passengers were reportedly killed and 30 injured. At the time the government forces in Bilbao were in the final months of a siege by Franco's Fascists. [138]

---

[132] *La Dépêche de Brest*, Brest, France, 27th October 1936
[133] *Le Peuple*, Paris, France, 27th October 1936
[134] *The Times*, London, England, 25th November 1936
[135] Ibid, 22nd December 1936
[136] *The Railway Gazette*, London, England, 13th August 1937
[137] *The Daily Herald*, Biloxi, Mississippi, United States, 11th August 1937
[138] Wikipedia – List of Rail Accidents in Spain

Two local trains were involved in an accident at Barón terminus of the Empresa de los Ferrocarriles del Estado in Valparaiso, Chile, on Monday 28th June 1937. Seven people lost their lives and 30 were injured, many seriously. [139] [140] The trains involved were an outbound one going to Calera and one entering the station from Peña Blanca. The incoming train left the rails at some points at the entrance to the station and the outgoing train was unable to pull up before colliding with it. The driver of one of the trains was amongst those killed. The damage resulting was estimated at 4,000,000 Chilean pesos. [141]

In the early hours of Saturday 17th July 1937, between 3.50 a.m. and 4 a.m., train No. 18 of the East Indian Railway, the Down Punjab Express, was derailed as it was crossing bridge No. 191, about 500 yards west of the signal cabin at Bihta, India. The engine ploughed up the track and then capsized, the leading coach was derailed but not badly damaged, and the next four carriages, in the words of the report on the accident, 'folded up concertina fashion' and were wrecked. The sixth and seventh carriages were derailed but remained intact and the rear three carriages remained on the line. The majority of the 450 passengers on the train were riding in the front coaches and 107 of them lost their lives, while 117 were badly enough hurt to require hospital treatment. The derailed carriages were all third-class ones and 'all the European passengers are safe.' Driver Blinkhurst and his two firemen also escaped. [142]

Initially it was suspected that the derailment had resulted from an act of sabotage, but this was quickly shown not to be the case when examination of the accident site by the railway authorities and police showed no evidence of tampering. The catastrophe aroused considerable interest worldwide and was the subject of a judicial inquiry, conducted by the Honourable Mr Justice Thom.

Mr H. A. Joscelyne, the Senior Government Inspector who carried out the initial investigation into the derailment, concluded that, 'I am confident that the following describes what took place. Engine XB 1916 has a strong proclivity for violent hunting at high speeds. The train topped the rise at the Sone bridge at a speed considerably higher than the permissible limit, ahead of time, and then increased in speed on the long steep down grade. The track being 88½ lb on D.O. plate sleepers was stiff enough to hold the side thrust of the engine in check and to prevent hunting. A few seconds after passing the Down warner signal of Bihta station the engine ran on to the wooden sleeper track on a new bank. This being weaker than the D.O. plate sleeper track immediately responded and gave way to the side thrust of the engine which at once sensed its freedom and started hunting violently from side to side. The distortion of the track is for several inches either side and this, with the normal play of the engine, must have allowed a side lashing of at least six inches. As long as the engine was on the wooden track, which yielded to its hunting motion, the engine kept the rails, but the time came when it had to run back onto the D.O. plate sleeper track. This, on an open bank, with the extra resistance of the D.O. plate sleepers, would have been bad enough, but the transition point is on the pucca

---

[139] *Lincoln State Journal*, Lincoln, Nebraska, United States, 29th June 1937
[140] *The Railway Gazette*, London, England, 2nd July 1937
[141] Ibid, 13th August 1937
[142] *The Times*, London, England, 19th and 21st July 1937

bridge where the D.O. sleeper was rigidly held both horizontally and vertically and unable either to depress or to give to either side. The engine was too much out of hand to steady into a straight course, something had to give way. The engine jumped the rails.

'I find that the accident was due to the derailment of XB class engine 1916 caused by hunting while running at an excessive speed over track that was not sufficiently strong to withstand the strain thrown on to it.'

The Railway Administration did not agree with the senior inspector's report and a series of test derailments of trains were then conducted at Jamalpur, with the Chief Engineer, the Chief Mechanical Engineer, and the Chief Operating Superintendent. The Senior Government Inspector then submitted a final report in which he stated that:

(a) The fact that the track became distorted was the primary cause of the derailment.

(b) The distortion was caused by an XB class engine hunting which started as a result of a lurch due to a low joint.

(c) If it had not been for the coincidence that the track at the point where the engine started to hunt was on new wooden sleepers on a bank which had been completed a few months previously, the hunting would not have had the effect of distorting the track in the manner which occurred.

(d) The driver apparently did not realise at the onset, possibly due to lack of experience of this class of engine, that the engine had started to hunt, and he only realised this when the hunting became violent and the engine was beyond control, when it was too late for him to take effective action.

(e) The serious consequences of the accident were mainly due to particular features in the design of the tender of the engine, with respect to the position of the spring hangers relative to the track, and the effect they must have on it whenever a tender of this design derails.

At Mr Thom's inquiry the Senior Inspector, Mr Joscelyne, repeated his conclusions that:

(1) The engine was being driven at an excessive rate of speed.

(2) The engine was hunting.

(3) The track, though sufficiently strong for normal traffic, was unable to withstand the strain imposed upon it by an XB engine running at a speed of over 45 miles an hour and hunting.

The Railway Company, on the other hand, maintained that:

(a) The speed of the engine was not excessive.

(b) That the engine was not hunting sufficiently violently to distort a track of normal strength.

(c) That not only was the track weak but prior to the arrival of the No. 18 Down it had been distorted by two other trains, the No. 6 Down and the Imperial Mail, and that the accident was due to this prior distortion which started the engine of the No. 18 Down lurching, and to the inherent weakness of the track.

Mr Thom's inquiry was lengthy, running to 55 pages, and included the statements of Driver Blinkhurst, Guard Carr, Bihta Switchman Sheopujan Ram and other officials, details of previous accidents and incidents, and much technical evidence. Concluding his report Mr Thom stated that he was satisfied that the Railway Administration had been guilty of negligence, that the Operating Department was guilty of negligence and that the officer in charge of Control at Dinapore was negligent in having failed to impose a speed restriction after the driver of No. 6 Down train had reported a 'lurch' at Bihta. Mr Thom noted with displeasure that Mr Robertson, Head of the Operating Department, had been permitted to go on leave prior to his requirement, due to 'certain commitments', such as visiting a winter house in Switzerland, and so had not attended the inquiry. The Dinapore control officer was less fortunate, he stood trial for his failure to impose a speed restriction and was found guilty, but released pending an appeal. [143]

Mr Thom went on, 'I consider it my duty in all the circumstances to make one recommendation. XB class engines have been a source of danger to the Public since they were introduced into this Country in the year 1928.

As Counsel for the Railway Company declared in his opening address his instructions were: -

"...That in the earlier stages of running this type of engine in 1928 their behaviour was so extraordinary and unforeseen that the engineers sat down to rebalance them."

'From the evidence it is abundantly clear that the engines have been, even as modified and improved, far from satisfactory. The Superintendent of Power, Dinapore, admitted in evidence that he wished they were "dead and buried." The Chief Operating Superintendent noted his view that their purchase was not justified. The Divisional Superintendent of Dinapore in March 1936, in his letter Exhibit 277, declared that they were obviously faulty in design. They have continued to hunt and to distort the track; they have been involved on the East Indian Railway in three accidents; they have been involved in accidents and given trouble on other railways. In these circumstances, I am of the opinion that in the interests of the public the design, the purchase, and the continued purchase of these engines should be the subject of a thorough, searching, and independent enquiry.'

Mr Thom's final conclusions were:

(1) 'That the derailment of the 18 Down near Bihta on July 17, 1937, resulted from distortion of the track which was caused by XB class engine running at excessive speed and hunting, and

(2) That the accident is directly attributable to negligence on the part of the Railway Company, and the Railway Company are accordingly liable to damages to those who were injured and to the dependants of those who were killed in the accident.' [144]

The Indian Government accepted the report and ordered the East Indian Railway Company, as a special case and without admitting responsibility, to meet claims for damages. It also set up an enquiry into the design, purchase, and continued purchase of the XB engines as recommended in the report and required the railway company to impose speed restrictions on the XB engines immediately, pending the findings of the report.

---

[143] *The Railway Gazette*, London, England, 17th September 1937

[144] Report by the Honourable Mr Justice Thom on the accident on the East Indian Railway on 17th July 1937

The report finally came out in 1939. It found that the class of engines was sensitive to poor maintenance and defective track and recommended the fitting of speed recorders to all express locomotives to ensure strict compliance with speed restrictions. [145]

A passenger train was derailed on the Ferrocarril de Antioquia, in Central Colombia, on a curve on 13:50 Sunday 30th January 1938. Three carriages left the rails and overturned. The train had left Puerto Berrio for Medellin at 12:30. It consisted of engine No. 39, two goods waggons, a luggage van, two third-class carriages, two second-class carriages and one first-class carriage, in charge of Conductor Carlos Mejia Maya with Driver Baudilio Gómez and Fireman Roberto Arango.

The train left the rails at kilometre 159 on the Cantarrana curve, between Zabaleta and Cabañas stations. The rear first-class coach was the first to leave the track and it dragged off in turn the two second-class coaches and one of the third-class ones. The other third-class coach was derailed, damaging the track, but did not overturn. Initially it was reported that eight people had been killed and there were 40 injured. Those who lost their lives included Fireman Arango and Brakesman Nicolás Velásquez. Conductor Maya was among the injured.

A train carrying the injured and the dead arrived at Medellin at 23:00 and was followed by a second with the uninjured passengers which arrived at 01:00. As soon as news of the accident had been announced there the station was besieged by people who had friends or relatives on the train, to the extent that a strong detachment of police had to be sent there to maintain order. A train pulled by engine No. 52 had been sent to the accident site with doctors from Cisneros station, and a diesel railcar was sent from Puerto Berrio with doctors, railwaymen, and Señor Juvenal Tobón, the mayor of that place. At 15:30 railcar No. 20 left Medellin for the accident scene with senior company officials and a doctor, followed by railcar No. 19 at 16:00 with more officials and another doctor, and finally railcar No. 10 left Medellin at 16:30 with another official, Dr Roberto Diaz.

The cause of the accident was not immediately apparent. Experts and laymen, a newspaper reporter wrote, found it strange (ironic?) that the first-class carriage, placed according to the regulations at the rear of the train, had been the first to overturn. This coach was one of the most badly damaged; its roof was detached and remained some distance behind where the train stopped. The scene of the accident was said to be indescribable. The train ran for some distance after the rear coach overturned, and during this time the coaches in front overturned as well. Confusion and panic seized the passengers. Mothers travelling with their little ones rather than worrying about themselves pressed their youngsters in close to try and shield them. Other passengers tried to shelter by windows and platforms because jumping from the train was felt to be more risky. Nothing was left intact in the restaurant car and the damage to the kitchenware involved a tremendous financial loss.

There was much talk of the courage and serenity of Padre Félix Ruiz. He was one of the most seriously hurt, with a shattered leg and his intestines protruding from an abdominal injury, but he remained serene and asked that other badly injured passengers be brought to him so that he could administer the absolution to them. He admonished rescuers who wanted to treat him first because of the gravity of his injuries and told them

[145] *The Railway Gazette*, London, England, 13th January 1939

to attend to others instead, declaring that he would die calm and satisfied, giving thanks to God for having allowed him until the last moments of his life to be able to help others and that it would be useful to him to fulfil the sacred duties of his sacred ministry. The padre and another passenger, Señor Esteban Vallejo, were taken to hospital in Medellin in a critical condition.

A reporter spoke warmly of the efforts of the rescuers, who were said to have worked tirelessly to care for the injured and prevent the number of dead from rising, while others strove to release the bodies of the dead which were trapped in the wreckage, some of whom were badly disfigured. At least some of the uninjured passengers helped with the rescue work.

Also praised by the reporter was the heroic behaviour of seven-year-old Octavio Correa, who after emerging from the wreckage with only scratches and bruises and becoming aware of the suffering of the seriously injured, took on the task of bringing water from the restaurant car to the injured who were crying out for it, and calmly and conscientiously cared for many people. He was said to have distinguished himself.

Another seriously injured woman passenger who was travelling with a little girl would not give her name and flatly refused to be helped from the wreckage until her three-year-old child had been found. Fortunately the child was unhurt and the woman then allowed herself to be released. She then did not complain at all of her injuries but just held her child, 'a pretty little doll,' in her embrace.

The first train from the accident site arrived at Medellin with three of the dead and most of the injured. (The details in the same issue of the newspaper vary at different places. Here it was stated that the first train arrived at Medellin at 22:00, not 23:00 as stated earlier, that the first name of one of the dead was Nicomedes rather than Nicolás, Velásquez, and that he was a driver rather than a brakesman.) The injured were immediately taken to San Juan de Dios Hospital; the bodies were taken to a room in the station. Five hundred people remained at the station to await the arrival of the other trains. [146]

On Tuesday 1st February it was announced that the number of people who had lost their lives had risen to 11 as Señor Vallejo, his son, 18-year-old Hernando Vallejo, and Padre Ruiz had all succumbed to their injuries, while several other passengers remained in a serious condition. [147]

By 2nd February the number of people who had lost their lives had risen again to 14. As to the cause of the derailment Superintendent Emilio Montoya Gaviria told the press that it was believed to be due to a defect in the first-class coach, and that Engineers Rafael de la Calle and Isaias Cuartas were conducting a scientific study of the case. Others blamed the derailment on an unfortunate alteration to the profile of the line. A reporter named Ximénez, who witnessed the derailment, told the *El Diario* newspaper that the excessive speed of the train, necessary to keep time, had caused it. It was also stated that the rails at the site of the derailment were very worn. A newspaper noted that all the different versions of how the accident had occurred made for an impatient wait for an expert opinion, while the general population demanded greater attention to the maintenance of the track and rolling stock and a stricter supervision of train speeds,

---

[146] *El Tiempo*, Bogotá, Colombia, 30th January 1938

[147] Ibid, 1st February 1938

because the frequency of bad accidents put the railway company in a bad light. [148]

As was often, though by no means always the case, the initial reports of a derailment of a troop train in Brazil on Saturday 5th February 1938 were exaggerated. They spoke of the loss of 15 lives and the injury of over 100 soldiers in the derailment of a train on the Estrada de Ferro Goyaz at Moro de Mesa, Brazil. A follow-up more detailed account stated that only four soldiers had died, together with Guard Manuel Miranda, though another 42 soldiers had been injured, some seriously. The train was carrying the 6 B.C. (Batalhão de Caçadores – Battalion of Hunters) back to Ipamery (now Ipameri) after they had been in São Paulo on a parade to commemorate the foundation of the city. The train, consisting of a locomotive and ten carriages, had left Araguary at 7 a.m. and had passed Ararpira station when the engine had left the track at kilometre 48, dragging the coaches with it. As soon as the accident became known in Araguary four doctors and several nurses went to the scene, and a military aircraft flew medical supplies from the capital to Ipamery. The injured, who included Major Amilcar Salgado dos Santos, the commander of the battalion, were taken to a military hospital. [149]

At 9.45 p.m. on Good Friday, 25th March 1938, an excursion train consisting of engine No. Ab 829, six carriages and a guards van left Wellington, New Zealand, for New Plymouth. Guard White was in charge of the train, with Driver Edward Percival and Fireman Ken Miller on the engine.

Just after 2 a.m. the following morning the train was wrecked on a sharp curve about three quarters of a mile beyond the station at Ratana. After leaving the rails the engine ploughed through soft sand and came to rest on its side. The leading carriage was little damaged, but the following vehicles swept past the engine and had their sides ripped open by it before they came to rest in a cutting. Four passengers in the front compartment of the second carriage were killed instantly and two others died of their injuries later. Over 40 passengers were injured, 13 badly enough to be admitted to hospital, and Driver Percival and Fireman Miller were also badly hurt, the fireman later succumbing to his injuries.

Dr Barmcoat of Wellington chanced to be a passenger on the train and although lacking and supplies did what he could to treat the injured pending the arrival of more doctors, ambulances, and police from Wanganui. He earned much praise from the passengers for his efforts. The location of the accident caused considerable difficulty for the rescuers and in one case a woman on a stretcher had to be lifted through the roof of a carriage and then be passed across the top of a second carriage before her stretcher could be laid on an embankment. Two ambulances came from Wanganui, plus one from Palmerston North, and every local doctor. Assistance was also given, a newspaper reporter added, by Maoris from the Ratana settlement.

An ambulance officer said, "There was no panic at any time. The passengers behaved splendidly and the injured bore their sufferings with fortitude. Many were still half asleep when they realised what had happened. The suddenness of the accident seemed to have

---

[148] Ibid, 2nd February 1938

[149] *O Estado de São Paulo*, São Paulo, Brazil, 6th and 8th February 1938

stunned the senses of the others."

Mrs Staveley, of Wellington, her head bandaged, agreed. "All that I can tell you is that there was a grinding of brakes, then a tremendous crash. There was no panic that I saw at all."

'Among heroes,' a newspaper reporter continued, 'was Arthur Hare, of Wellington, who gave the alarm by a telephone from a control box at Aromoho. He had made the journey three weeks previously and remembered seeing the telephone box at the side of the line. Directly he got clear of the wreckage, he located the box and succeeded in getting the alarm through early in the morning. "We were sitting in the last carriage," said Hare. "Suddenly there was a terrific bump, and I was shot through the door."'

Another passenger, a young man, said, "To see the rear of the carriage coming through on top of us was the most awful experience of all." He described the scene as like a battlefield, and said that there was no panic, although the darkness of the scene added to the horror.

Dr Picking, of Palmerston North Hospital, described the accident as one of the most ghastly scenes he had ever witnessed, despite his having gone through the whole war. He remarked, "Our work was made difficult because we could not reach the passengers, many of whom were blocked in the carriages, also because of the darkness. We were also handicapped by lack of medical equipment." [150]

The board of inquiry into the accident found that the track was in good condition and there was no evidence of malicious interference. The accident was due solely to the excessive speed of the train. Driver Percival was experienced but had not driven over the stretch on which the accident occurred for about eight years. He could have declined to operate the train on the grounds that he was not sufficiently familiar with the line or perhaps asked for a pilot, but he had not done so. Matters were complicated by the darkness and by patchy fog. The curve on which the train had derailed had a speed restriction of 20 miles an hour. Driver Percival said that he had been looking for the speed restriction sign and had made a slight brake application but in the darkness and the fog had failed to realise how far he had gone and was on the curve before he knew it. The board found that he had not been guilty of culpable negligence but had committed an honest error of judgement.

The board recommended the installation of speed recorders on express, passenger and mail trains. [151] [152]

Six people died and 18 were injured when three carriages of a Belgrade – Sarajevo passenger train ran into a large boulder which had fallen onto the track. Following the collision the coaches were said to have plunged down a mountain precipice near Sargan, Yugoslavia, on Sunday 3rd April 1938. [153]

---

[150] *The Telegraph*, Brisbane, Queensland, Australia, 26th March 1938
[151] *The Townsville Daily Bulletin*, Townsville, Queensland, Australia, 21st July 1938
[152] *Tragedy on the Track* by Geoff Conly and Graham Stewart, pgs. 99-102
[153] *Butte Montana Standard*, Butte, Montana, United States, 4th April 1938

As passenger train No. 39-39 was entering the station at Saint Trond, Belgium, at 08:14 on Sunday 24th July 1938 a third-class carriage left the rails on a set of points. The coach then struck a signal post, was overturned, and cut in two. The alarm was immediately raised and by 08:30 the bodies of five badly mutilated corpses had been recovered by rescuers together with six seriously injured victims. It was feared that number of the injured would die. Three of the dead were from the same family.

Breakdown cranes were ordered from Landen and Liége to clear the wreckage while railway officials and judicial authorities were struggling to ascertain why the derailment had occurred. [154] The train had departed from Landen at 08:00 for Hasselt.

The island of Jamaica suffered a disaster on Saturday 30th July 1938. A passenger train from Kingston to Montegro Bay consisted on its departure of an engine and tender, a baggage car and four coaches. The train was crowded, especially as Monday 1st August was Emancipation Day, a public holiday on the island which commemorates the freeing of 311,000 slaves in 1840. Some of the passengers were Kingston people going into the country and others were country people returning from Kingston. At a point about 50 miles from Kingston, probably Kendal, a second banking engine was coupled to the rear of the train to assist it up the severe gradients ahead. All went well as the journey continued until the train reached Oxford Siding, just short of Balaclava station, where the leading engine left the rails on a curve. Unaware of what had happened ahead the crew of the banking engine continued to push, though it has been suggested that had the regulator of the banker been closed it would have made little difference. At any rate the tender, baggage car and coaches were telescoped into one another and three of the coaches were described as having been 'reduced to matchwood,' though the rear carriage was less badly damaged. The leading coach was forced over an embankment and ended up in an adjoining river.

As usual the initial reports of the number of dead were exaggerated, with statements that perhaps 60 of the 300 passengers had lost their lives, and the known dead supposedly being 52, though it was later determined that only 32 had died, including the driver and fireman of the leading engine and two or three other railwaymen. About 70 others were injured.

Slightly injured passengers helped those who had escaped altogether to release victims from the twisted steel, wood, and glass. A corps of doctors and nurses went to the scene and a special train with medical supplies, civilian and army doctors was sent from Kingston at midnight. Rescuers used acetylene torches to cut through the wreckage in the search for further victims. The splintered first coach reportedly lay in the river bed for hours before breakdown cranes could reach it. [155] [156]

Initial talk of sabotage was quickly discounted. It was found that the train had been travelling at 35 miles an hour, rather faster than the authorised speed of 25, but the cause of the derailment could not quickly be definitely established. [157]

---

[154] *Ce Soir*, Paris, France, 25th July 1938
[155] *San Antonio Express*, San Antonio, Texas, United States, 1st August 1938
[156] *Nebraska State Journal*, Lincoln, Nebraska, United States, 1st August 1938
[157] *The Railway Gazette*, London, England, Vol 71, pgs. 110-1

Initially if was said that a party of 40 Girl Guides was missing, but it was later determined that they had not been on the train, and all were safe.

The acting governor of the island went to the scene of the disaster, visited the injured in hospital, and issued a statement expressing the colony's profound regret at what had happened. [158]

The report of a special commission set up to investigate the accident was published in July 1939. It stated that the train had consisted of 4-8-0 engine No. 30 and tender, two wooden baggage vans, a wooden coach, a steel coach, three wooden coaches, three waggons and engine No. 39 at its rear. The derailment had occurred on a five chain radius left-hand curve, preceding which were speed limit boards drawing attention to the 20-mile-an-hour limit over this section of the line, a limit which was also laid down in the regulations covering the running of trains over this section.

Following the derailment the engine stopped against the side of a cutting, the leading two vehicles shot past it, the wooden coach third in the train was completely telescoped by the following steel coach, the next coach ended up at right angles and was wrecked and the next coach fell to the left. The rear vehicles of the train and banking engine remained on the track.

The weight of evidence was that the train had been travelling at 35 miles an hour when it left the rails. Such a speed was not sufficient in itself to account for the derailment without other factors being involved. One of these was minor tracks defects, consisting of:

(a) Unevenness in two or three rail lengths beyond Oxford Siding

(b) An uneven rate of run-off of superelevation

(c) A soft spot near the point of derailment

In addition it appeared that the driver of engine No. 30 had braked as he had reached Oxford Siding and then, realising that he was still approaching the curve too fast, had given one short blast, calling a brake application from the banking engine, but instead the banker had accelerated, pushing the train engine off the curve. There was no evidence of either driver having been drunk or taken ill, or of a brake failure.

The commissioners made several recommendations. The use of banking engines should be restricted to 'well-defined up grades,' the use of unofficial hand signals should be banned and a more distinctive braking whistle was recommended. Facilities for weighing engines should be provided and they should be equipped with speed indicators. A closer medical examination of drivers were proposed and various recommendations were made regarding track maintenance.

No comment was apparently made about the mixing of steel and wooden coaches in the train.

The use of banking engines was altogether prohibited after the disaster. [159]

At 17:10 on Thursday 8th September 1938 a departmental train left Josselin, France,

[158] *The Times*, London, England, 1st, 2nd and 3rd August 1938
[159] *The Railway Gazette*, London, England, 21st July 1939

for Lochminé and Vannes. It was carrying pilgrims who were returning to the latter two places after visiting the fêtes at Notre-Dame-du-Roncier. It was only travelling at 20 kilometres an hour when it became derailed on a curve near Reguiny, France. The driver felt his engine slipping under his feet and it immediately left the rails and ran into an embankment. A violent shock ensued, and the two leading carriages struck the locomotive and were shattered. Three bodies were recovered from the wreckage: a girl of about 12, a young man of about 17 and a woman in Breton costume. They could not be immediately identified. About 20 injured passengers were taken to a clinic in Vannes. Two others, who were seriously injured, were taken by ambulance to Ploermel, where they died. [160] The driver and fireman were only slightly hurt. The cause of the derailment was not immediately obvious. [161]

A train from Mexico City to Vera Cruz was derailed on the Ferrocarril Mexicano near Xala, in the State of Mexico, Mexico, 48 miles from Mexico City, at 9.20 a.m. on Tuesday 20th December 1938. The train was crowded, with 1,034 passengers according to the number of tickets sold. Most initial reports spoke of eight dead and 52 injured, but an unofficial report spoke of about 40 dead. The passenger were largely government employees and their families bound for Vera Cruz for the Christmas holidays. The locomotive uncoupled from its train and five passenger carriages overturned. [162] [163]

The day after the accident the railway company stated that in fact only five passengers had died. [164]

On New Year's Day, Sunday 1st January 1939, a passenger train was reportedly derailed at Castro Urdiales, Spain, and one carriage fell seven metres down an embankment. Seven people died and 16 were injured. [165]

A passenger train was derailed on the Cape Town – Port Elizabeth line at Daljosephat station, South Africa,[166] between Paarl and Wellington, on Monday 13th February 1939. 'Four Europeans and a coloured woman were killed, and eight persons injured.' [167]

Accidents involving trains colliding with animals, even relatively large beasts such as cows, rarely prove serious except to the animal itself, but there have been exceptions, such as the disaster which occurred near Châteauroux, France, on Monday 13th March 1939.

At 23:02 produce train No. 1124, with Driver Berland, left Châteauroux for Vierzon. As it was passing signal box B, a kilometre from the station, Signalman Palleau there

---

[160] *Le Journal*, Paris, France, 9th September 1938
[161] *Le Matin*, Paris, France, 9th September 1938
[162] *The Wisconsin State Journal*, Madison, Wisconsin, United States, 21st December 1938
[163] *Berkeley Daily Gazette*, Berkeley, California, United States, 21st December 1938
[164] *Nevada State Journal*, Reno, Nevada, United States, 22nd December 1938
[165] Webpage: Federación Castellano Manchega de Amigos del Ferrocarril (in Spanish)
[166] *The Railway Gazette*, London, England, 17th February 1939
[167] *The Times*, London, England, 14th February 1939

glimpsed a white mass moving along the line in the same direction as the train. It was a cow. When the animal had gone about 80 metres it was struck by the train. The engine of the train was derailed and thrown to the left, while the two following waggons became uncoupled from it, ripped up the track and were thrown to the right, obstructing the opposite running line. Signalman Palleau said that on seeing this he had immediately thought of *rapide* (express) No. 67, which generally passed the food train in the vicinity of his box. He had thrown back the signal 1,183 metres from his cabin, then signal No. 25, 139 metres away.

The signalman could do no more, but as chance would have it the crowded express, which had left Paris at 20:25 for Toulouse and Cerbère, was too close. Driver Boisranier, the driver of the express, which was hauled by an electric locomotive, told the *juge d'instruction* what happened next. "I was travelling at my normal speed, about 110 kilometres an hour. I saw the fire in front of the produce train which we usually pass there, then, suddenly, the signal was thrown back in my face, and I saw men by the track giving stop signals with their lanterns. I jammed on the brakes, but only 70 metres further on I ran into some waggons fouling the line.

"Before the collision I had time to cut off the power, apply the brakes and sand and open the door into the engine room, into which I fell. As to the circumstances which led to the accident, I have no idea."

Assistant Driver Combes exactly confirmed his colleague's testimony.

The engine of the express was derailed in turn when it struck the derailed trucks. The following mail van was overturned but none of the postal workers in it were hurt. Next in the train came a third-class coach, which was telescoped between the engine and the first-class coach which was marshalled third in the train. It was in these two coaches that the casualties occurred. By the next morning the bodies of 17 victims had been removed from the wreckage and 20 injured passengers had been taken to hospital, where one had passed away. [168]

Driver Berland told the *juge d'instruction*, "I saw the animal, an enormous white beast, when it was only a few metres away. It was too close to be able to stop. My engine was thrown into the air when it hit the beast. Then we saw the express coming when it was perhaps four or five hundred metres away. We jumped down to the track and I ran towards it waving a red light."

The Flaman speed recorders from the locomotives showed that the produce train had been travelling at 50 kilometres an hour, and the express at 107, reduced to 75 at the moment of the collision.

Assistant Stationmaster Claverie said, "We had been transhipping cattle just after 22:30 that evening. We had to transfer the animals from one waggon to another so that the train could leave at 01:50. I only had three men to help me. The waggon door was opened and the first animal which came out, a large, white, bullock, escaped and ran onto the track. We tried in vain to recapture it, and a little later train No. 1124 passed."

According to a newspaper report the assistant stationmaster had therefore apparently failed to warn Driver Berland of the possible obstruction on the line. A technician was also asked by a reporter if it would have been possible to cut the current to stop the

[168] *Le Radical de Marseille*, Marseille, France, 14th March 1939

express, but he said that the electricity substation was 4 kilometres away. Even if the signalman, the only external witness to the initial collision, had telephoned the sub-station there would not have been time to have cut the current early enough to have stopped the express. [169]

Despite the newspaper implication that the negligence of the assistant stationmaster had led to the disaster, M. De Monzie, the Minister of Public Works, following the initial inquiry into the accident, stated that it not been due to any failure of either staff or equipment. He also praised the signalman and the driver of the express, whose vigilance and coolness, he said, had prevented a worse accident. [170]

On Monday 5th June 1939, a special train carrying officials to fight a forest fire was derailed and fell into a river in Kochi Prefecture, Japan. Thirteen people died and 33 were injured. [171]

The Roma – Wien – Warsaw international express due in the Polish capital at 12.22 p.m. was derailed at 12.15 ten kilometres from its destination at Pruszkow, Poland, on Wednesday 7th June 1939. The engine and five leading vehicles, including *Wagons-Lits* coaches, left the track and two wooden coaches, attached to the train in Bohemia, were telescoped. At least six people lost their lives, among them the driver and fireman. The others who lost their lives were the Polish controller of the *Wagons-Lits* coaches; a Polish doctor; Mr. Stefen Steinhhagen, a well-known Polish industrialist; and a young woman whose name and nationality could not be immediately established. About 30 others were seriously injured. The train was travelling at 80 kilometres an hour when it left the track at a set of points and the derailment was attributed to excessive speed. [172 173]

The Minister of Communications, Mr Jullusz Ultych, went to the accident scene, accompanied by Vice Minister Julian Piasecki. [174]

On Thursday 15th June 1939 the driver of train D 148, a Berlin – Praha – Bratislava international express, disregarded the signals at Mittelgrund, Germany*, on the Dresden – Bodenbach main line with the result that the train approached a set of facing points at excessive speed and became derailed, demolishing a signal box in the process. The *Einfahrsignal* was set for the loop at Mittelgrund because the main track was occupied by a freight train bound for Bodendach. Speed through the crossover was limited to 40 kilometres an hour. Thirteen or fourteen people died and 26 were injured. [175] Those who lost their lives included a Berlin driver who was learning the road and the fireman, who were killed more or less outright, and the driver actually in charge of the engine, from Dresden-Altstadt shed, who died in hospital. Two relief trains were sent to the scene from

[169] *Le Matin*, Paris, France, 15th March 1939
[170] *The Times*, London, England, 15th March 1939
[171] *Edwardsville Intelligencer*, Edwardsville, Illinois, United States, 5th June 1939
[172] *La Charente*, Angoulème, France, 9th June 1939
[173] *The Times*, London, England, 8th June 1939
[174] *La Croix du Nord*, Lille, France, 8th June 1939
[175] *Katastrophen der deutschen Bahnen Teil II* by H.J. Ritzau, pg. 95 (in German)

Dresden and a third from Aussig, while the Wehrmacht (army), German Red Cross, police and others assisted in the rescue work. Whether or not the signalman survived the destruction of his box was not reported.

With the death of the engine crew any hope of determining the cause of the accident disappeared. [176]

*Mittelgrund was in Czechoslovakia, at this date occupied by the Germans, and was in the Protectorate of Bohemia and Moravia, which was considered part of the German Reich. It is now Decin -Prostředni Žleb in the Czech Republic.

Two coaches of a train reportedly derailed and overturned between Puebla and Matamoros, Mexico, on or about Tuesday 8th August 1939. Seven people died and eight were injured. [177]

At least 17 people died and 108 were injured, 32 seriously, when a two-car electric train was derailed and plunged into 30 ft down an embankment into a stream near Nikko, Japan, in Tochigi Prefecture. The accident occurred on Thursday night, 12th October 1939. [178]

Wednesday 25th October 1939 saw a major disaster at St Valentin, Austria, Germany. As train D54 was approaching the station at 18:55 the *Einfahrsignal* (home signal) indicated that the points were set for a diverging route, but its driver failed to reduce speed. The engine, No. 12 006, managed to negotiate the first set of points but left the rails at the second set, and derailment of the coaches it was pulling followed. Seventeen passengers died in the accident, nine civilians, among them the renowned surgeon Prof Dr Anton Freiherr von Eiselsberg, and eight German soldiers. [179]

A derailment occurred between Eger and Pilsen, Germany, on Sunday 5th November 1939. These places were in Bohemia, before the Second World War part of Czechoslovakia, but as noted above with relation to the Mittelgrund accident they had now been incorporated into Germany. Five people died in the accident. [180]

Five people died and at least 20 were injured when a Roma – Genova – Torino express became derailed on a set of points near the station at Pisa, Italy, on Monday 12th February 1940. King Victor Emmanuel and Queen Elena went to the scene of their accident from their nearby villa at San Rossore, where they were holidaying, and visited the injured in hospital. [181] Officials blamed the derailment on excessive speed. [182]

---

[176] https://www.saechsische.de/plus/das-drama-am-bahnhof-von-mittelgrund-864418.html
[177] *Journal de Geneve*, Geneva, Switzerland, 9th August 1939
[178] *The Times*, London, England, 14th October 1939
[179] Wikipedia – Eisenbahnunfall von St Valentin 1939
[180] *The Railway Gazette*, London, England, 10th November 1939
[181] *The Times*, London, England, 13th February 1940

Fifteen passengers were killed and about 30 were injured, ten seriously, when a passenger train ran into a landslide near Zaluka station, Yugoslavia, on the Carlovatz-Ljubljana line, on Sunday 17th March 1940. The second carriage of the train was thrown into the River Kupa and passengers in it were drowned. Other passengers reportedly escaped by swimming to safety. The river was especially deep at this point as it adjoined the dam of an electricity works.[183] Soldiers riding in six carriages of the train performed the initial rescue work. While the rescue operation was in progress a second landslide struck the front of the train, pushing the engine and leading mail van into the river as well and injuring five of the soldiers. [184]

In the United States, the New York Central Railroad had what it promoted as the Water Level Route between the cities of New York and Chicago, in contrast to the considerably more heavily graded line of its rival, the Pennsylvania Railroad. The Water Level Route was achieved by following river valleys as much as possible, but at Little Falls, New York, this led to another difficulty. Here the river ran in a narrow valley between hills to the south and Adirondack Mountains to the north, and river, canal, highway and two railroads (the New York Central and the West Shore) were squeezed into a narrow gap. The river had a bend at this point and so the railroad had to bend correspondingly, resulting in the Gulf Curve, the sharpest on the route, with a permanent speed restriction of 45 miles an hour. It was this this spot that disaster struck late on Friday 19th April 1940.

The train involved in the disaster was train No. 19, the westbound *Lake Shore Limited,* which consisted of 4-6-4 steam engine No. 5615, an express car, a baggage car, two coaches, Pullman sleeping cars *Red Ash, Poplar Arch, Elkhart Valley* and *Poplar Drome*, a dining car, Pullman sleeping cars *Lake Brum, East Bernard* and three others, and a final coach, in charge of Conductor Grattan with Engineer Jesse Earl and Fireman J.Y. Smith. Also riding on the engine footplate was Road Foreman of Engines Andrew Bayreuther. As the train approached the Gulf Curve the engine and all but the last three coaches were derailed, and the two coaches marshalled third and fourth in the train and the *Red Ash* Pullman car were badly damaged. The impact also caused the engine boiler to explode. The accident happened at 11.33 p.m. [185]

The aftermath of the accident was the usual mixture of chaos and of survivors rising to the occasion. 'The roar of the boiler explosion, the screech of escaping steam, the crunching of metal against metal, awakened the town, and within five minutes townsmen and townswomen with overcoats drawn over their nightclothes were at the scene, trying to succour the injured.

'But the surviving trainmen had the situation in hand. A Pullman conductor, blood on his face, was screaming orders to the white coated porters and another conductor was marshalling the uninjured male passengers. Blankets and seat cushions were pulled out of the upright Pullmans and laid in a row on the side of the right of way. The porters and the

---

[182] *The Racine Journal-Times*, Racine, Wisconsin, United States, 12th February 1940
[183] *The Times*, London, England, 19th March 1940
[184] *Gazette de Lausanne*, Lausanne, Switzerland, 18th March 1940
[185] Interstate Commerce Commission report

passengers pulled out the injured and carried them to this row of patients where they were required to lie down.

'Little Falls doctors and nurses had been quickly assembled and they quickly and efficiently ministered to the injured. At first private cars were commandeered to take the seriously injured to the local hospital, but soon ambulances were on the scene from Herkimer, Utica, and other nearby towns. Pullman porters had produced sheets from their cars and were tearing them into long strips for bandages. Other porters had made coffee and poured it for the shocked but unhurt passengers.

'Meanwhile railroad men had arrived from the Little Falls yards and were laboring to force an entrance into the first two Pullman coaches and the locomotive cab. They knew there was life in the cab because they could hear groans. Within fifteen minutes acetylene torches had arrived and a plate was cut out of the cab so that rescuers could get in. The engineer, J. Earle, of Syracuse, his legs horribly mangled, was still alive. He died on the way to hospital. His fireman was dead.

'Other torches were at work on the first Pullman coach, smashed down on top of the telescoped mail coach. Rescuers finally crawled into the holes. In the crimped steel they cut and came out with a few men and women who seemed barely alive. A few minutes were enough to account for all of the living. Then they began carrying out the dead.

'To get into the second Pullman took almost two hours. Both ends had been smashed by the cars in front of and behind it and its sides had been smashed in. The workers had to cut off one end of the coach with acetylene torches and as they worked they heard shrill, piercing screams that ceased after a while, and groans and cries. State policemen, a priest and railroad men climbed up on top of the tangled ball of wreckage and consoled the trapped victims as best they could. At last the end of the coach fell off and the rescuers crawled in, crawling out with more injured – and more dead.

'Meanwhile a hospital train arrived from Utica with more doctors and nurses, and with medical supplies, food and clothing. It left after an hour with some of the injured and most of the uninjured passengers.'

Rescue work went on through the night with the aid of searchlights, in rain which later turned to snow. As to the cause of the accident it soon became apparent that it was likely due to excessive speed, a newspaper reporting two days after the accident that the speed recorder on the locomotive had stopped at 59 miles an hour at the moment of the derailment. [186]

Twenty-six passengers, Engineer Earl, Fireman Smith, two Pullman porters and a train porter died and 47 passengers, Foreman Bayreuther, a Pullman porter and two dining car employees were injured and almost all of the passenger baggage which had been taken from the wrecked as well as the undamaged coaches.

'The mail car had not been reached at dawn. To get to it, the wreck - workers had to clear the two Pullman coaches on top of it.

'The news spread rapidly and within an hour 1,000 persons had arrived by automobile and foot. Within two hours there were at least 3,000 persons and at dawn approximately 5,000 persons were as near the wreck as state policemen, assembled hastily from barracks

---

[186] *Ogdensburg-Advance-News*, Ogdensburg, New York, United States, 21st April 1940

within 100 miles, would permit them.' [187]

Carroll Wright, of Watertown, New York, President of North Country Security Corporation, a passenger on the train who was injured, spoke to a newspaper reporter from hospital. "I was in the small smoker, walking towards the mail car, the next car ahead, at the time of the wreck. My sensation, standing in the aisle, was of sailing through the air as the car lurched. I saw a porter ahead of me reach for an emergency cord, and then lost consciousness. When I came to I was lying in the Pullman section of the car, ahead of the smoking compartment. The car lay on its side. I was buried in cushions, which softened my fall and saved me from more serious injury. A state trooper stood over me.

"My first thought was of my travelling companion, Norman C. Hicks, 45, 415, Pawling Street, Watertown, vice president of the North Country Security Corporation. I remember the trooper asking me if I was hurt. I was too dazed to reply, although I was fully conscious. As soon as I was able to talk, I asked him if he would help me find Mr Hicks. The trooper helped me to my feet. I had trouble walking. I had severe cuts on the thigh, a bruised hip, and a back injury. Apparently I had been thrown out of the smoking department when my car went off the rails. It was a strange sensation, finding myself in another part of the car and not knowing how I got there. There were several other injured persons in the car, and I could faintly hear some of them calling for help, although the sound of escaping steam from the engine drowned out almost all other noises.

"I scarcely remember getting out of the car, with the trooper's help. Thinking back now over the few minutes preceding the wreck, I would say the train was late. I had planned to get off at Utica, 20 miles farther on, and the train was due there at 11.28 p.m., as I recall.

"It is plain now, that I was very close to death. The car through which I was walking was the third in the train, and a few steps one way or another might have been the end of me. The baggage car, next to the engine, miraculously escaped damage, but the mail car and the Pullman car behind it, toward which I was walking, were smashed to bits. In a few seconds I would have been in that car. A physician who examined me at the scene of the wreck suggested that I come to the hospital here {Syracuse}. I did so and was among the first to arrive. I soon had to give up my bed as a stream of more seriously injured began coming in. I don't know what happened to Mr Hicks." [188]

As the only survivor of the three men on the footplate the key witness at the subsequent enquiry into the accident by the Interstate Commerce Commission (ICC) was of course Road Foreman Bayreuther. He told the inquiry that he had examined the engine before leaving Albany and it had been in good condition. Approaching Little Falls the speed had been 74 miles an hour and the fireman had called out the indications of automatic signals 21421 and 21531, both of which had displayed green over yellow, indicating *Medium speed*, not to exceed 30 miles an hour. The signals, 9.046 and 3.086 feet from the point of derailment, were provided with automatic train control apparatus, which Engineer Earl had forestalled. Between the two signals the engineer had also checked the amount of water in the tender. After passing the second signal the engineer had made a brake application, but not one to adequately reduce the speed of the train for

---

[187] *Dunkirk Evening Observer*, Dunkirk, New York, United States, 20th April 1940

[188] *Syracuse Herald-Journal*, Syracuse, New York, United States, 20th April 1940

the curve. Foreman Bayreuther said that he had then become alarmed, had warned Earl that he was going too fast and had told him to make another brake application. As the train was entering the curve the engineer had then suddenly closed the throttle and mumbled something. He had not seemed to be ill but rather in despair and realising that something was wrong. The rear of the engine had been the first part to leave the rails and there had then been a jack-knife between the engine and the tender. Foreman Bayreuther thought that the maximum speed that a train could negotiate go safely around Gulf Curve was 52 miles an hour but that if Engineer Earl had not suddenly closed the throttle the train would not have become derailed.

Conductor Grattan agreed that the run to Little Falls had been normal, and he had not been alarmed at the speed. Albany had been left 21½ minutes late.

Various other witnesses gave evidence at the ICC investigation, and the commission noted that there was no problem with the engine brakes, evidence of any problems with the track, which was in excellent condition, or of any obstruction on the line. It concluded that the derailment must be attributed to the excessive speed of the train on the curve combined with the sudden run in of slack produced upon Engineer Earl's abrupt closing of the throttle. [189]

As he was killed it was of course impossible to obtain any explanation from Engineer Earl for his actions. Robert Shaw has theorised that possible factors were the engineer's age – he was almost 74 – and the presence of Foreman Bayreuther on the footplate.

The curve was later made less severe by a major engineering operation which involved rerouting the river. [190]

The London Midland and Scottish Railway had what might be justifiably described as an unlucky accident at Wembley, England, on Saturday 12th October 1940. Just as it was getting dark, at 7.10 p.m., the 11.50 a.m. Up express passenger train from Liverpool Lime Street to London Euston collided with a luggage trolley and was derailed. Nine passengers, the driver and fireman were killed, and four passengers were seriously injured.

The events which led up to the accident began with the arrival at Wembley station of a consignment of bleach ointment weighing about a ton, a regular wartime traffic from the place. The goods were due to be loaded onto the 6.46 p.m. parcels train running on the Down Slow line. This train was late that evening, but the consignment also arrived at the station late, and it had to be weighed and stamped before being sent, so time was short. A hoist was available to take the consignment of ointment, which was packed into 33 lb cartons and loaded onto a luggage trolley, from the parcels office to Platform 6, which served the up slow line, but the trolley had to be taken over a barrow crossing over the Slow lines to the island platform serving the Down Slow and Up Fast lines.

Acting Parcels Porter Glanmor Thomas loaded a standard four-wheeled luggage trolley with half the consignment – 10 cwt – at the parcels office and brought it down in the hoist to No. 6 platform just after 7 p.m.. He then told Station Foreman R.F. Beckett

[189] Interstate Commerce Commission report
[190] *A History of Railroad Accidents, Safety Precautions and Operating Practices*, by Robert B. Shaw, pgs. 299-301

that he would need some assistance to take the trolley over to No. 5 platform. The only staff available to help were Porter Signalman Edward Parker and Temporary Porter Cyril Bird. The three men took the trolley along No. 6 platform and down the ramp to the barrow crossing, with Parker in front and the other two behind. They brought the trolley across the Up and Down Slow lines without any problem and then had to pull it up the 1 in 8 ramp to the island platform. It was here that things began to go wrong. As they began to pull the trolley up the slope and had got about a third of the way up, Thomas and Bird lost their foothold. The trolley was then gently lowered to the foot of the ramp, Bird went to the front in place of Parker, who joined Thomas at the rear, and they tried again. This time they almost got to the top when Parker's foot slipped, and he fell on his back. He recovered himself, but at that point Thomas slipped in turn, and the barrow began to run back down the slope. Bird tried to divert the trolley away from the line but was thrown off by the handle and he and the trolley ended up on the Up Fast line. Thomas saw the train coming and shouted to Parker, "One on the Up Fast." Thomas then pulled Bird to safety while Parker rushed to a telephone to warn the signalman, but it was of course much too late.

The Liverpool express consisted of 4-6-0 steam engine No. 5529, 11 corridor bogie coaches and two four-wheeled goods waggons, with Driver Reginald Clynes and Fireman John Heath. It was travelling at about 55 miles an hour under clear signals when it struck the trolley, whereupon the engine and seven coaches were derailed. The engine overturned and its boiler was penetrated by a rail, while the leading three coaches were badly damaged. [191]

The rescue work took place during the blackout, with railwaymen assisted by Air Raid Precautions workers and an air raid actually took place at one point, with the sound of enemy aircraft and the anti-aircraft fire plainly audible, but the rescue operation was not interfered with. Doctors and nurses arrived within 25 minutes. Discretion was given in the circumstances to ignore the blackout restrictions but only hand lamps were used, though fortunately there was also moonlight. Jets of scalding steam prevented a porter reaching the engine cab and a signalman who tried to rescue injured passengers was driven back at times by flames and steam.

A Mrs Thorn, who lived close to the station, said, "My daughter Pat and I made tea for the passengers. Some were lying in the signal box seriously injured; others were trapped and moaning. When I tried to give a soldier a sip of tea he said, 'Tell my mother and sister I shall be all right.' He died soon afterwards." [192]

Great disruption to traffic was caused by the accident, with Euston station being closed to steam services until the evening of Thursday 17th, with trains diverted to Watford, diversions to St Pancras and extra services at Kings Cross and Paddington.

Major G.R.S. Wilson conducted the Ministry of Transport inquiry into the accident. He noted that Parker and Thomas were both small in stature and quite lightly built, while Bird, though taller, appeared less robust than the other two.

Station Foreman Beckett considered that the trolley was quite within the capability of three ordinary men, though he agreed that Bird was not a man of normal strength. He said that had he considered the trolley overloaded he would have ordered some of the load

---

[191] Ministry of Transport report

[192] *The Times*, London, England, 14th October 1940

removed, but that it was necessary to load trolleys to 8 cwt or more to deal with the volume of traffic.

Relief Stationmaster Charles Jackson on the other hand did not think that it was necessary to load the trolley so heavily to cope with the traffic and said that if the consignment could not have been brought to Platform 5 in time it could have gone by a later train. He agreed with Beckett that three men should have been able to handle the trolley.

Major Wilson said that the trolley was clearly overloaded and disagreed with the station foreman and stationmaster that it had been within the capacity of three men. He did not place too much responsibility on Thomas, Parker and Bird in view of their inexperience (Parker had 2½ years' service, Thomas only ten months and Bird only seven months). They had evidently been doing their best to deal with the traffic expeditiously. Nevertheless, a more experienced railwayman would have realised that the trolley was overloaded, especially after the first failed attempt to get up the ramp. Driver Clynes had clearly acted promptly when he saw the obstruction and was in no way responsible for the accident.

In conclusion the inspecting officer noted that this was the first serious accident involving a luggage trolley since that at Wellingborough in 1898, but the results of such collisions were fortuitous, and the increased wartime traffic and shortage of staff made the supervision of the loading and working of trolleys in the neighbourhood of running lines. He also suggested that the foothold of the ramps should be improved by roughening the surface or the provision of wood or metal cross bars. [193]

At the inquest on the victims of the accident Relief Stationmaster Jackson said that it was not normal for luggage trolleys to have brakes. Only one of the eight trolleys at Wembley had them and on ramps they were more of a nuisance than a help.

Mr F.O. Bates, representing the railway company, stated that as far as he was aware there was no Ministry of Transport requirement for barrows to have brakes.

The inquest returned a verdict of "Accidental death." [194]

Towards the end of 1940 Great Western Railway of England had a very good safety record given the extent of its system and had not suffered a major disaster since that at Norton Fitzwarren on 11th November 1890. Then by a strange coincidence, a week short of 50 years later, it suffered another disaster at exactly the same place, on Monday 4th November 1940.

Norton Fitzwarren is about two miles west of Taunton, and since the 1930s had been the end of the quadruple track section which began at Cogload Junction, where the Bristol and Westbury main lines diverge east of Taunton. Three important down trains ran through Taunton in the early hours of the morning, the Down Postal, the 12.50 a.m. Paddington to Penzance newspaper train and the 9.50 p.m. Paddington to Penzance Down passenger train (*the Owl*). The Postal and the *Owl* ran via Bristol, but the newspaper train took the more direct route via Westbury. The 9.50 p.m. passenger train (the *Owl*)

---

[193] Ministry of Transport report
[194] *The Times*, London, England, 17th October 1940

consisted of 4-6-0 King class engine No. 6028 *King George VI* and 13 bogies coaches, in charge of Guard J.W. Winyard with Driver Percy Stacey and Fireman V. Seabridge. It was crowded with 900 passengers. The newspaper train consisted of another King class engine, No. 6013 *King Henry VIII*, and five bogie vehicles in charge of Guard E. Baggett with Driver H. Hawkins and Fireman T. Lindsey.

On the morning of the accident the *Papers,* as the newspaper train was usually known, was three minutes early at Taunton, but the *Owl* was 68 minutes late. The two trains thus arrived at Taunton more or less at the same time on the Down Main and Down Relief lines respectively and Signalman Harry Wadham in Taunton West Station signal box decided to give the *Papers* priority. He cleared his Down Main line signals for it and it ran through Taunton station at 3.45 a.m., the *Owl* having departed from the Down Relief line platform on the Down Relief line a minute earlier. Both trains had clear signals at Taunton West Junction and Silkmill Crossing boxes, but at Norton Fitzwarren the Down Relief line signals were necessarily at danger, the *Owl* having to stop here at the end of the Relief line until the *Papers* had passed and cleared the next box at Victory Siding. Disaster occurred when the passenger train failed to stop and was still travelling at at least 45 miles an hour as it became derailed on the catch points at the end of the Relief line. The engine overturned 140 ft beyond the catch points and the leading five carriages were wrecked. Twenty-six passengers were killed or fatally injured, and the fireman also lost his life, while 56 passengers were badly enough hurt to need hospital treatment and 18 were slightly hurt or suffered from shock. Even in disaster the Great Western was lucky, for the newspaper train had overhauled the '*Owl*' at Silkmill and just managed to get clear of the derailing train at Norton, though a rivet-head from the derailed engine struck its fourth van and flying ballast peppered the fifth van. It was a narrow escape. The accident happened at about 3.48 a.m. [195]

'Rescue work,' a reporter wrote, 'was carried out under great difficulty. People from Norton Fitzwarren, a village near, gave valuable assistance, and the landlord of the village inn gave all his sheets to be used as bandages. The village hall was opened and less seriously hurt were given first-aid treatment there. Others were taken to Taunton hospital.

'After working throughout the morning, breakdown gangs had recovered 20 bodies from the wreckage, and they were taken to Taunton. Among them were eight sailors, apparently returning from leave, and two women. A sailor who escaped said: "It was a nightmare. The crash came with such startling suddenness that I thought we had been bombed. The pluck of the injured passengers was really wonderful. A pal of mine who had been sleeping in the corridor emerged from the debris with a baby in his arms, and a woman, pinned under the wreckage, exclaimed: 'Thank God I have got away with my life.' She could not be extricated at once."

'Unstinted praise for members of the Forces who were travelling in the train was given by an Exeter passenger, Mr. R.J. Woodley. "Those fellows worked as though they had been doing rescue jobs every day," he said. "Three was not the slightest sign of panic, and it was reassuring to hear the soldiers, sailors, and airmen calling out, 'It's all right.' The night was inky. In front I saw something which looked like aeroplane wings and then I realised it was the debris of a coach sticking at queer angles into the air. The

---

[195] Ministry of Transport report

engine appeared to be on its side with steam pouring from it. Fortunately there was no fire. So far as I could see the first coaches had been smashed, and there was one coach without its bogie ahead of the engine. People who had been passengers in that coach told me that when the crash came they simply shot ahead of the engine. Huge pieces of debris littered the Up line, and we helped to move them so that a relief train could come by.'" [196]

The Ministry of Transport inquiry into the accident was conducted by Lt Col Alan Mount.

Driver Stacey told the inquiry that a minute after his train had stopped at the Down Relief line at Taunton he had seen the Down Relief to Down Main home signal (No. 3) worked from Taunton West Station box change from red to green. A little later he had seen it at green again and he had seen a green light a third time just before he had started off, but this time was not sure which light it was. The driver's statement continued: -

"I proceeded with my train on what I thought was the Down Main line, and on running over the ramp at the end of Taunton platform I heard the A.T.C. siren. After leaving Taunton and having the ramp signal, of which I heard the siren, I did not touch the A.T.C. again, by which I mean I did not have another siren. On approaching Norton Fitzwarren a train passed on my right-hand side. As this train passed me on my right-hand side I immediately shut my regulator and applied the brake, being under the impression that we were getting nearer the verging point and I saw this train and the station on my right-hand side, which brought to my mind that I was on the Down Relief. At the time I turned to my mate and told him that we were on the Down Relief, and with that the train became derailed at the catchpoint."

Driver Stacey put the speed of his train through Norton Fitzwarren station at 40 miles an hour and stated that he had managed to reduce the speed of his train somewhat, which the inspecting officer doubted. Driver Stacey said that he had not noticed the Norton Fitzwarren starting or advanced starting signals or the explosion of any detonators. The engine had overturned to its left and Driver Stacey being on the right-hand side escaped, while his fireman was buried under tons of coal and died. Climbing out of the cab Stacey had waded through waist-deep floodwater, then made his way down the line to Victory box. Signalman W.W. Langford there said that the driver had looked very agitated and dazed, as well he might, of course, and had told him (Signalman Langford) that he was afraid that he (Driver Stacey) was responsible for the accident.

Guard Winyard could not help the inquiry much other than to say that the train was usually routed via the Down Main.

Travelling Ticket Collector J. Pearce spoke of reminding Fireman Seabridge at Taunton that they must stop at Teignmouth. He put the speed at Norton Fitzwarren at 35 or 40 miles an hour.

Travelling Parcel Porter J. Hedey said that Guard Winyard had sent him down the line to provide protection against Up trains and he had met Driver Stacey, who was returning from Victory box, but he would not let the driver talk about the accident.

Driver Hawkins and Fireman Lindsey of the newspaper train said that they had noticed nothing of the accident.

---

[196] *The Times*, London, England, 5th November 1940

Guard Baggett was also unaware of the passenger train but felt something strike his van and applied the brakes. The train came to a stand somewhere beyond Victory box. He examined his train, consulted with his driver and they continued to Wellington, where they were told of the accident.

Signalman Wadham stated that the Down Postal had left the Down Relief platform at 2.58 a.m. via the crossover (No. 49) to the Down Main. He had then reset the crossover in order to run a shunting engine via the Down sidings to the loco shed. He was positive that subsequently he had not reset the crossover from Down Relief to Down Main until after the departure of the *Owl*. Having been advised by Westbury between 3.30 and 3.35 a.m. that the *Papers* was running a little early, he had rung Athelney box, 8 miles east of Taunton, and had been told that the *Papers* was still running early. He had then decided to let it have a clear run on the Down Main and to send the passenger train down the Relief line to Norton Fitzwarren. He had asked clear for both trains to Signalman Tucker at Taunton West Junction and had cleared the signals for the Relief line for the passenger train and a minute later those for the Main line for the *Papers*. He was emphatic that none of the Relief line signals had been lowered prior to his pulling off down the Relief at 3.37 (but see below).

Signalmen Tucker at Taunton West Junction and Birch at Silkmill confirmed that their signals had been off for both trains. Signalman Birch said that the newspaper train had been quickly gaining on the *Owl* and was only 200 yards behind it at his box.

Signalman W.J. Coles at Norton Fitzwarren stated that he had accepted the passenger train at 3.36 and the *Papers* at 3.37. When he saw the berth track circuit indicator for the track in the rear of his Down Relief home signal go to "Occupied" he had cleared the signal, then been surprised to see that the train was still travelling at 40 miles an hour. The detonators in a 3-shot detonator placer located just in advance of the starting signal should have exploded but he had not heard them. He had sent the "Obstruction Danger" signal promptly on witnessing the derailment.

Concluding his report Lt-Col Mount noted that the accident had not been due to enemy action or sabotage and there were no problems with the permanent way or signalling. It was due solely to an unaccountable lapse on the part of Driver Stacey, who admitted full responsibility for the accident.

The normal route for the train was out onto the Down Main from the Down Relief via No. 49 crossover at Taunton West Station. Driver Stacey said that he had seen the signal (No. 3) for this movement clear a minute after his train had arrived at Taunton, but Signalman Wadham denied having cleared it and then changed his mind, and the inspecting officer believed the signalman and thought Driver Stacey mistaken.

Between Taunton and Norton Fitzwarren the majority of Down Relief signals were to the left of the relief line, while the majority of the Down Main signals were between the Down and Up Main lines. It was clear that Driver Stacey had been reading the Down Main signals, which were clear for the newspaper train, as being his, not noticing that they were a little further over to the right than normal.

As it was the Great Western, the distant signals were fitted with Automatic Train Control (ATC) apparatus. Driver Stacey denied having received any ATC or brake application at the Norton Fitzwarren Down Relief distant. There could be no doubt that the distant signal had been properly at caution, the clearing of the Main line signals would have locked the lever of the Down Relief distant, and the interlinking of the block instrument would have prevented it being placed at "Line Clear" unless the arm had been horizontal. The ATC apparatus, both on the engine and the ramp at the distant signal, was tested soon after the accident and shown to be in order. It could only be assumed that under the impression that he was running under clear signals Driver Stacey had automatically cancelled the brake application.

It seemed unlikely that there had been fresh detonators in the machine at Norton Fitzwarren (these were responsibility of the station staff rather than the signalman). In the circumstances their failure to explode was immaterial but the need for the conscientious replacement of detonators in the machine was obvious.

The inspecting officer remarked that the reason for the lapse on the part of such an experienced driver was difficult to understand, and the explanation was probably largely psychological. Quite apart from the difficulty of operating trains at night in the blackout the driver's house had been damaged in an air raid attack.

Lt-Col Mount rejected suggestions that drivers should be specially advised of unexpected diversions – that was the purpose of the signals – or that there was any problem with the driver's position on the right-hand side of the footplate.

The position of the Down Main signals to the right of that line was due to insufficient space between the Down Relief and Down Main lines. It would be well to consider re-siting them should it become possible to do so.

Installation of a sand drag beyond the catch points might have reduced the severity of the derailment. [197]

At the close of the inquest upon the victims of the accident at Taunton the jury, after a short retirement, returned a verdict that, "We find that the driver has been guilty of an error of judgement, but we cannot decide on the actual degree of error. It was certainly not criminal, especially in view of the abnormal weather and other conditions." [198]

Driver Stacey never drove on the main line again and died within the year. After the war it emerged that he had not been wholly to blame. His statement that when he had first arrived at Taunton station he had seen signal No. 3 cleared to bring his train from the Down Relief to the Down Main line, which the inspecting officer had rejected, turned out to be true. In 1946 Signalman Wadham, who incidentally was also a local magistrate, confessed to another engine driver that he *had* in fact cleared No. 3 signal for the *Owl* as Driver Stacey had said, before changing his mind after finding out from Athelney where the newspaper train was. Signalman Wadham was thus to some small degree responsible for the accident, something which never emerged from the official investigation. [199]

Twenty people were said to have died in the derailment of a northbound express train near Palos Quemados Tunnel, Chile, 32 kilometres from Calera, at 3.40 p.m. on Saturday 8th March 1941. The engine and leading five third-class carriages of the 14-carriage train jumped the rails and overturned. Sixty other people were more or less injured. The accident occurred on a falling gradient and a photograph shows that it happened on a

[197] Ministry of Transport report
[198] *The Times*, London, England, 21st November 1940
[199] *Great Western Echo, Journal of Great Western Society*, No. 226 pg. 25

sharp right-hand curve, implying that it was possibly due to excessive speed. [200] Driver Pedro Carroza was scalded to death. Two trains with doctors, nurses and medical supplies were sent to the scene from Valparaiso. [201] A later report stated that only nine people had died and 32 were injured. [202]

Twenty people died in the derailment of a passenger train near Kampala, Uganda, on Wednesday 23rd April 1941. [203]

On the morning of Monday 1st September 1941, a passenger train was derailed at Tablada, Spain. The leading three of the seven coaches of the train left the track and the first two overturned. Some of the passengers who had jumped when they had realised that the train was leaving the track were trapped beneath the capsized coaches. Eight passengers lost their lives and 16 were injured, three seriously. The train involved was a workman's special for employees of the Maestranza Aérea (an aircraft maintenance factory). It was a service which ran daily between Plaza de Armas station and the factory. The accident occurred at a level crossing adjoining the CAMPSA (State petroleum company) siding. The accident was possibly due to excessive speed through a crossover. [204]

Passenger train No. 404 from Ciudad Rodrigo to Medina del Campo was derailed near Tejares, Spain, at about 10:15 a.m. on Wednesday 8th October 1941. The engine left the rails on a curve and went into a ditch. The leading steel van and the following third-class carriage, which was well-filled, were telescoped into one another. The remaining two coaches, one first and the other third-class, were not damaged. Driver Octavio Morollón and 11 passengers were killed. The fireman, who had jumped from the engine, escaped with a slight chest injury and 36 passengers were hurt.

In the following days the Republican press in exile referred to another accident on the same line, for which no other reference has been found, due, the same source stated, to presumed sabotage, and which involved a military train bound for the Portuguese frontier. This may be a confused reference to the Tejares accident, or it may refer to a different accident, which may or may not have involved a military train and may or may not have been accidental. [205] [206]

On Saturday 14th February 1942 severe weather conditions with a heavy snowfall led to the pilot engine of passenger train P 2409 becoming derailed when it ran into a drift between Gaschwitz and Meuselwitz, Germany. A relief train was sent to its rescue, fitted with a snow plough, but apparently was either not cautioned to proceed with care, or perhaps the location of the failed train was misunderstood. A head-on collision ensued

---

[200] *Sunday Journal and Star*, Lincoln, Nebraska, United States, 9th March 1941

[201] *Syracuse Herald-American*, Syracuse, New York, United States, 9th March 1941

[202] *The Railway Gazette*, London, England, 30th May 1941

[203] Ibid, 2nd May 1941

[204] Webpage: Federación Castellano Manchega de Amigos del Ferrocarril (in Spanish)

[205] Ibid

[206] *La Vanguardia*, Barcelona, Spain, 9th October 1941

and six people died and 16 were injured. [207]

The city of New York, United States, saw a serious accident on the evening of Sunday 26th April 1942, involving one of its subway trains. At 10.48 p.m. (Eastern War Time) a well-filled Hudson and Manhattan Railroad train was derailed at the western end of Exchange Place station. The first three cars of the five car train left the track, the leading power car, MU coach 1972, striking a signal post and the wall of a tunnel, and the second and third cars were badly damaged.

A newspaper reporter described the aftermath of the accident. 'The derailment cut power lines, plunging the tunnel into blackness at the deepest point on the line, just before the train burrows under the Hudson river. Acrid fumes gushed from storage batteries in the wrecked car, which left the track with a blinding flash and burst into flames. Women shrieked; car windows were kicked out. One Negro girl, pinned under a seat, had to be extricated by acetylene torches.'

Five passengers died and 188 passengers and three railwaymen were injured. The passengers who lost their lives were:

Mrs Ann Stevens, of Woodhaven, Queens, New York

Morris Huttler, of Province, Rhode Island, and Newark, New Jersey

Christensen Tiempe, of Jersey City, New Jersey, a civilian machinist at the Brooklyn Navy Yard

Rube Greenberg, of Brooklyn, New York

An unidentified Negro aged about 30

The reporter continued, 'The passengers had graphic stories to relate. Sixteen-year-old Martin Stevens kicked out a window and carried his mother out of the third car, unaware that she was dead. Martin sobbed out his story, "We were riding in a forward car. Suddenly there was a terrible crash and our car turned over. I was knocked on my face. When I got up there was my mother beside me."' [208]

Colourlight signals were provided on this line, with automatic train stops. At the western end of Exchange Place station there were five parallel tunnels, designated E, G, L, H, and F from north to south. The tracks of Tunnel H and Tunnel F converged at a point 207 feet east of the west end of the station and those of Tunnel H and Tunnel L at a point 110 feet east of the west end of the station at switch 5. The derailment began at a point 30 feet west of switch 5. Approaching this switch there were two curves to the left followed by one to the right and in its immediate vicinity the line fell at about 1 in 29. The maximum authorised speed for trains at this point was 12½ miles an hour.

A terminal air brake test had been carried out on the eastbound train involved in the derailment at the start of its journey at Newark and the brakes had worked normally on the short journey to Exchange Place. The train had left Newark at 10.32 p.m., had left Journal Square at 10.42 p.m., and was travelling at a speed estimated at between 17 and 50 miles an hour when it left the track.

---

[207] *Eisenbahnunfälle in Europa* by Erich Preuß pg. 123 (in German)

[208] *The Frederick Post*, Frederick, Maryland, United States, 28th April 1942

Motorman Louis Vierbuchen told the Interstate Commerce Commission (ICC) investigators that the signals between Newark and the point of derailment had all been clear. Signal No. 399B, five feet east of the point of derailment, had been at caution, indicating that signal No. 229, 126 feet in advance of it, was at danger. The motorman stated that he had been travelling at 20 miles an hour when he had applied the electro-pneumatic brakes about 1,350 feet short of signal No. 399B, but they had not responded. Four hundred feet short of the signal, with the speed not reduced, he had placed the automatic brake valve in the emergency position. He could not explain why he had not done so earlier. The brakes were found to be working properly after the derailment. It was calculated that the speed of the train on derailment must have been between the maximum safe speed for the curve of 19.8 miles an hour and the overturning speed of 39 miles an hour, as the train had not overturned.

The ICC commented that the controls were such that if the train had exceeded an average speed of 6.6 miles an hour between signals Nos. 399B and 299 the automatic train stop system would have applied the brakes of the train. However, this provided no protection against excessive speed over the sharp curve in the rear of signal No. 399B. Had such protection existed the accident would not have occurred.

The ICC concluded that the derailment had been due to excessive speed on a sharp curve. [209]

Early on the day after the accident Public Safety Commissioner David Casey stated that Motorman Vierbuchen, aged 48, had been arrested on a manslaughter charge. The motorman admitted having drunk five glasses of beer, while having had nothing to eat, at his meal break shortly before taking the train out from Newark. [210]

Perhaps a little surprisingly at the close of his trial in December the motorman was acquitted of the manslaughter charge. Fourteen fellow employees had testified at the trial that he had not been intoxicated on the day of the accident. It took the jury just over six hours to reach their verdict. [211]

A brief news report from China stated that more than 300 Japanese had died in the derailment of five coaches of a troop train on Tuesday 2nd June 1942. The coaches fell into a river north of Hsuchowfu in Kiangsu Province. [212]

A Pennsylvania Railroad passenger train from Atlantic City to New York became derailed at Minson, near Delair, New Jersey, United States, on the evening of Sunday 23rd May 1943. The train was crowded with almost 1,300 passengers homeward bound after a weekend spent on the Atlantic coast. Twelve passengers, the fireman and conductor of the train died, and 80 passengers and the engineer were injured. The derailment happened on a sharp curve, and the following day Pennsylvania General Manager W.C. Higginbotham announced that, "The train was derailed on a 14 degree curve, and the speed limit on that section of track is 15 miles per hour. Early tracks show that the

[209] Interstate Commerce Commission report

[210] *Somerset Daily American*, Somerset, Pennsylvania, United States, 28th April 1942

[211] *The Evening Star*, Washington, D.C., United States, 19th December 1942

[212] *Joplin Globe*, Joplin, Missouri, United States, 12th June 1942

equipment and tracks were in good condition, but definite indications that the train was moving faster than the authorized speed when the derailment occurred." [213]

Westbound passenger train No. 1080 consisted of 4-6-2 engine No. 3806 and 15 coaches which had been converted from Pullman parlor cars, in charge of Conductor C.H. Bohr with Engineer Christopher Pittock and Fireman H.N. Becker. The train left Atlantic City at 9 p.m., according to the Interstate Commerce Commission (ICC) report, and passed Vernon, the last open office, at 10.1 p.m., eight minutes late, passed automatic signal No. 2, which indicated *Approach Medium*, passed signal No. 121-B, which indicated *Approach,* passed signal 26-l, which was showing *Slow-Approach*, ran through switch No. 25, worked from Jersey tower and controlling the junction with the single track connection to Minson on the Camden – Trenton line, and became derailed on the curve at a speed variously estimated as 15 and 30 miles an hour at 10.8 p.m. The engine and first seven coaches left the track. The cab of the engine was demolished, the second to fifth cars were badly damaged and the sixth and seventh somewhat less so. [214]

A reporter described the accident. 'Passengers and baggage were hurled brutally to the floor. Screams of women rose above the crash. One car perched above the engine. Another landed on its side near the tender, with its rear telescoped by the third. The fourth slid underneath the wreckage and two others remaining upright on the right-of-way, tilted rakishly on their crumpled trucks.

'Police, railroad crews and civilian defense volunteers struggled to aid the injured and rescue trapped passengers. For five hours workers using acetylene torches cut into the wrecked first car, seeking additional bodies. They were crushed so badly that attempts at identification were confined to a search of their possessions.

'A soldier, seated in one of the cars praying for his critically ill mother, whom he was rushing home to see, suddenly found himself thrown to his side, other passengers and their baggage piled on top of him, his foot squeezed so tightly against the side of the car he could not remove it. The soldier, Pvt. Felix A Shagin, 19, of Passaic, N.J., took his knife, cut away his shoe to free his foot, and in doing it badly cut the foot.

'Someone screamed "fire" as wisps of steam curled through the car from the shattered engine. Women shrieked in panic as everyone fought to get out. Then cool-headed soldiers restored order and began helping the safe and injured outside.

'Two priests, Fathers Raymond Cavanaugh and Cletus Moran, of nearby St Cecila's church, climbed into the wrecked coaches to administer to the dead and dying the last rites of the Catholic church.' [215]

Engineer Pittock told the ICC inquiry that approaching signal No. 2 the speed of his train had been 65 miles an hour. As they passed the signal he had closed the throttle and made an 8-pound brake reduction, followed by a 4- or 5-pound reduction a quarter mile west of the signal. He maintained that a quarter mile east of signal No. 121-B he had released the brakes and the speed of his train had fallen to 30 miles an hour. Half a mile east of signal No. 26-L his fireman had called that it was at *Slow-Approach*, whereupon he had made an 8-pound reduction, followed by a 4- or 5-pound reduction, At a point

---

[213] *The Coshocton Tribune*, Coshocton, Ohio, United States, 24th May 1943

[214] Interstate Commerce Commission report

[215] *Mason City Globe-Gazette*, Mason City, Iowa, United States, 23rd May 1943

2,200 feet east of signal No. 26-L, when the speed was down to 22 miles an hour, the brakes had been released. He said that the speed had fallen to about 15 miles an hour at switch No. 25, had entered the curve, lurched to the left and overturned.

No statements could be obtained from the fireman or conductor of course and the front brakeman and the swing brakeman had been busy collecting tickets and could make no estimate of the train's speed. The flagman, riding in the 14th car, put the speed of the train at 50 miles an hour when the brakes were applied. He said that 40 or 50 seconds later there were two sudden lurches, and the train was off the road.

The crew of the previous train to pass the point, Passenger Extra No. 3728, stated that they had noticed nothing unusual there.

A track foreman had examined the switch the previous day and observed nothing wrong.

A test train, similar in make-up to train No. 1080, was run past signal No. 2 at 53 miles an hour, thus 12 miles an hour less than train No. 1080, and the throttle was closed, and the brakes were applied as Engineer Pittock said he had done. Signal No. 121-B was passed at 38 miles an hour speed of the test train at switch No. 25 was still 34 miles an hour. The route at the switch was set for straight ahead rather than around the curve for obvious reasons.

The ICC concluded that the derailment had been caused by excessive speed on a sharp curve. [216]

The Camden County Grand Jury indicted Engineer Pittock on a technical charge of manslaughter on 9th July. [217] It is not clear if his case ever came to trial.

New Zealand was not a country which suffered a great number of serious railway accidents, but it suffered a major disaster midway through the Second World War. At 1.45 p.m. on Friday 4th June 1943 the engine of a Cromwell to Dunedin express became derailed on a sharp curve on the Central Otago line, the accident happening in a deep cutting south of Hyde. Twenty-one passengers were killed or fatally injured and 47 were hurt. There had been 113 passengers on the train, many going either to the Winter Show in Dunedin or to the races at Wingatui.

The accident apparently came as no surprise to at least some of the passengers. Philip Banon, a merchant seaman and one of the few passengers to escape injury, said afterwards that, "Minutes before the accident passengers were falling about the carriage because of the way in which it was swaying. Engine No. Ab 782 and all seven carriages of the train left the track as the engine entered a 10 chains radius curve and four of the carriages were telescoped." Banon said that everything had happened very quickly. The leading carriage was one of those telescoped, but the next two, in one of which he was riding, were not, although they were overturned. Their passengers got out by climbing through windows. A passenger in the penultimate carriage said that it appeared to go up in the air and then lurch forward. [218] Some passengers had lucky escapes. A male passenger was thrown out through a carriage window against the side of the cutting and

---

[216] Interstate Commerce Commission report

[217] *Doylestown Daily Intelligencer*, Doylestown, Pennsylvania, United States, 9th July 1943

[218] *Townsville Daily Bulletin*, Townsville, Queensland, Australia, 7th June 1943

back though another window into the train again and still escaped with cuts. A woman gave up her seat to a mother with a baby; they were both killed, she was unhurt. Others were not so lucky. An elderly woman passenger survived but lost her husband, daughter and two grandchildren. [219] A farmer who witnessed the derailment ran a mile to a telephone to call for assistance, then returned to the scene to find that his own son, who had been returning on leave from an Army camp, was one of those killed. Fortunately, there was no fire. Some of the bodies were badly mangled and jacks had to be used to free them from the wreckage. There were also problems with identification, as the passengers came from different places and were mostly unknown to one another. Four remained unidentified two days after the accident. Thirty-one of the more seriously injured were admitted to Ranfurly Hospital, while others with minor injuries were treated and sent home.

The board of inquiry which investigated the accident found that it had been clearly due to excessive speed, as evidenced by the finding that the underframe of one of the carriages had been bent into an 'S' shape, indicating that a tremendous force had been exerted in the crash, and the speed of the train must have been not less than 70 miles an hour. The maximum permitted speed over the curve in question was 30 miles an hour. As to the damage resulting from the accident the board commented that, 'It is safe to say that no previous railway accident has resulted in such a mass of distorted and wrecked rolling stock.'

The board concluded that the locomotive of the train had overturned through the action of centrifugal force after having entered the curve at excessive speed, and that the man in charge of the engine, Driver John Corcoran, had operated his engine in a 'reckless manner' and had been guilty of a serious dereliction of duty. Guard Robert Parata was also criticised for not taking action when he found that his driver was going too fast. It emerged at the inquiry that Driver Corcoran, Guard Parata and Fireman Stanley Hollow had all been drinking at the bar of the Ranfurly Hotel before they had taken charge of the train and that Driver Corcoran had taken more beer onto the engine. The board decided that he had consumed sufficient alcohol to affect his judgement and sense of responsibility to a marked degree.

Driver Corcoran was subsequently charged with manslaughter at Dunedin Supreme Court, found guilty and sentenced to three years' imprisonment, while railway regulations regarding the use of alcohol and drugs were tightened up. Neither Guard Parata nor Fireman Hollow was charged with any offence. [220] [221]

From time to time after accidents there have been allegations that train drivers have been intoxicated, as for example after the Elliot Junction disaster in Scotland in 1906, but only occasionally, as in the Exchange Place accident (see above) and the Hyde disaster, has this been proven to be the case.

An initial report stated that 18 people had died and 44 were injured in the derailment of No. 13 passenger train on the Madras and Southern Mahratta Railway between Bezwada and Guntakal, India, at 5.15 a.m. on Tuesday 6th July 1943. The cause of the

[219] *West Australian*, Perth, Western Australia, 7th June 1943

[220] *Tragedy on the Track* by Geoff Conly and Graham Stewart, pgs. 103-4

[221] Wikipedia: The Hyde Railway Disaster

derailment was not known but sabotage was not suspected.[222] The next day the number of dead had increased to 20 and of injured to 58. [223]

The Indo-Ceylon Boat Mail, bound for Madras, was derailed at 3.30 a.m. on Sunday 14th November 1943, near Sendanur station yard, India, on the South Indian Railway. The engine, mail van, a second-class and three third-class carriages were derailed. The engine crew and postal staff had miraculous escapes, but the passengers were less fortunate, especially those in the third-class carriages, while an early report stated that an underguard was missing. On news of the accident being received a relief train was sent from nearby Villupuram Junction and the injured were taken to hospitals in Villupuram and Cuddalore. A duplicate train took the uninjured passengers on to Madras. These included the occupants of the women's coach, which left the rails but remained upright. Initially it was reported that there were 30 dead and about 100 injured, 12 seriously. [224] A later report spoke of 39 dead and 88 injured and stated that the cause of the derailment had not been determined.

Southbound Louisville and Nashville Railroad second-class passenger train No. 47, was a troop train carrying 1,006 new recruits to Fort Benning, Georgia, on Thursday 6th July 1944. At 8.42 p.m. that evening it passed Williamsburg, Kentucky, running eight hours and two minutes late. Twenty-three minutes later, at 9.05 p.m., it became derailed near High Cliff, Tennessee, United States, not far from the town of Jellicoe. The results of the derailment were disastrous: 33 passengers (soldiers) and the engineer and fireman were killed and 93 passengers, three Pullman employees and two railwaymen were injured.

The train consisted of 4-8-2 steam locomotive No. 418, four Pullman tourist cars, a Pullman kitchen car, a Pullman troop sleeping car, two Pullman tourist cars, a baggage car, three Pullman troop sleeping cars, a Pullman kitchen car, two Pullman troop sleeping cars and a baggage car, all of steel construction, in charge of an unnamed conductor with Engineer John (Lyle) Rollins and Fireman John Tummins. It was negotiating a left-hand curve over which the speed was limited to 35 miles an hour when the engine and first eight cars left the rails. None of the surviving members of the train crew was aware of anything being wrong until the brakes were applied in emergency immediately before the derailment, which occurred when the train was travelling at an estimated speed of 45 miles an hour. The engine and tender fell over 30 ft to the bed of the Clear Fork River, the leading two cars stopped against the engine and the third and fourth cars ended up on top of the second cars. All four cars were described as having been 'practically demolished.' The fifth car went down the embankment and the sixth, seventh and part of the eighth car were derailed. Fire broke out soon after the derailment and the combustible portions of the derailed cars were burned. [225]

The first rescue work was undertaken by locals (described in a newspaper report as 'nearby mountainfolk'), who hoisted the injured up the shrubbery-covered gorge with

---

[222] *The Bombay Chronicle*, Bombay, India 7th July 1943
[223] Ibid, 8th July 1943
[224] *The Bombay Chronicle*, Bombay, India, 15th November 1943
[225] Interstate Commerce Commission report

block and tackle slings. Early on, Dr Ned Watts was the only medical man there but doctors from neighbouring towns joined him as soon as news of the accident became known. At least two soldiers were trapped in the wreckage and the doctors gave blood transfusions to one of them. The first casualties were transported from the wreck by a lorry driver who chanced to be passing, later ambulances took the injured to hospitals in Lake City, LaFollette, Jellico, Corbin, and Williamsburg. At the time of the accident most of the soldiers had either been in their berths or in the wash rooms preparing for bed, according to an Army public relations officer. Personal possessions had been thrown all together in the crash and some identification tags lost, making identification of some of the victims difficult.

An unnamed soldier said that the crash had happened just after they had finished chow and that he thought that the fire had begun in the kitchen car. He continued, "I was in an upper berth, and was almost thrown out when we went around a curve. Just a moment later she banged off the track."

Another soldier, Private Wallace Lewis, of Canton, Ohio, said, "I saw a big flash, and someone said, 'There's going to be a wreck.' There was. I crawled out of the car, fell into the shallow creek, then stumbled out." [226]

John Ruggles, from the Knoxville office of the Federal Bureau of Investigation, stated that the possibility of sabotage was being investigated, but the Interstate Commerce Commission (ICC) came to a different conclusion. It noted that at 45 miles an hour the speed of the train had been 20 miles an hour in excess of its equilibrium speed and 10 miles an hour above the maximum safe speed around the curve. It also found that there were deficiencies in the track. The curvature should have been 10 degrees instead of 11 degrees 15 minutes and the gauge 4ft 8½ inches instead of 4 ft 9⅛ inches. Had, the commission declared, the gauge and alignment of the track been in accordance with the specifications of the railroad the engine would probably not have rolled laterally as it had entered the curve, even at its excessive speed, and the left front engine truck (bogie) wheel would not have mounted the guard rail, leading to the derailment. [227]

According to one witness Engineer Rollins should have been relieved at Corbin but his relief not having turned up, he was forced to take the train on himself. "He was very mad and possibly under the influence of alcohol," the witness stated. There is, however, no suggestion in the ICC report of the engineer having been intoxicated.[228]

A Southern Pacific Company intercontinental train, *The Challenger*, en route from Chicago, Illinois, to Oakland, California, was derailed early on Wednesday 8th November 1944. The accident happened two and a half miles west of the station at Colfax, California, United States, in the foothills of the Sierra Nevada mountains.

The train involved was westbound first No. 87, which consisted of 4-8-8-2 steam locomotive No. 4176, two baggage cars, nine coaches, a dining car and six coaches, all of the vehicles being of steel construction, in charge of Conductor Carl Bohle with Engineer Fred Graham and Fireman Donald Hurley. It had left Sparks, Nevada, at 12.55 a.m., 22

---

[226] *Kingsport Times*, Kingsport, Tennessee, United States, 7th July 1944
[227] Interstate Commerce Commission report
[228] http://www.drwebman.com/trooptrainwreck/

minutes late, and Colfax at 5.04 a.m., 19 minutes late, and was travelling at more than 55 miles an hour when the engine and 10 cars left the track at about 5.10 a.m. The engine ploughed into an embankment and it and the leading eight cars were badly damaged. Eight passengers and Engineer Graham were killed and 194 passengers, Conductor Bohle, Brakeman Glen Peterson, a baggageman, five dining car employees, two train porters, a train service agent and another employee were injured. There had been 792 passengers on the train.

Fifteen ambulances went to the scene from the DeWitt General Hospital and nurses and doctors set up a temporary hospital at the site in the undamaged dining car. The injured were later taken to the DeWitt hospital except for four seriously injured who were taken to Colfax Hospital. Wrecking trains were sent from Roseville, California, and Sparks, Nevada, to clear the line. [229]

Reporters obtained brief accounts of the accident from a couple of passengers. 'Ralph Haley, U.S. army lieutenant of Sacramento, said he noted three distinct jolts, the first one waking him up in time to see suitcases flying through the air and people tossed from their seats. "It was as if our car hit something going at full speed," he said.

'R.H. Stalpt, Benton Harbor, Mich., navy aviation cadet, said he, "felt something queer as the engine pulled around the curve. It was as if the oil tender swayed and the cars were leaving the tracks."

'His companion, R.H. Snell, of Ainsworth, Neb., also an aviation cadet, said, "When the coach stopped whirling, it ended up crosswise of the roadbed. There were a few screams, but not many, because it was all over so quick. Our first idea was to get out and help the injured."' [230]

The Interstate Commerce Commission (ICC) found that the engine had traversed a 9-degree curve to the right and was on a 10-degree curve to the left when it became derailed. Trains between Truckee and Loomis, to the east and west respectively of the derailment point, were limited to 35 miles an hour on tangents and 30 miles an hour on curves.

The train crew said that the train had been running smoothly up until the accident and none of them knew of anything being wrong. They could not speak as to the speed.

Fireman Hurley stated that he had been working on the injector for several minutes before the derailment and paying no attention to how fast the train was running. He had noticed Engineer Graham making one brake reduction after leaving Colfax. His first warning of the derailment had been when the engine had lurched to the right and become derailed, his engineer moving the brake handle to the emergency position as the engine began to overturn.

The ICC noted that there was no evidence of any problems with the track, of dragging equipment or of an obstruction on the line, nor of any problem with the engine brakes.

The mechanical engineer stated that the overturning speed of the locomotive was 57¼ miles an hour. It was evident that the train must have been travelling at at least this speed when it left the track as it had overturned on the outside of the curve without marking the

[229] *Oakland Tribune*, Oakland, California, United States, 8th November 1944
[230] *Reno Evening Gazette*, Reno, Nevada, United States, 9th November 1944

rails and slid on its right side until it stopped.

The ICC concluded that the accident had been due to Engineer Graham failing to control the speed of his train in accordance with the speed restriction on the curve, for reasons which could not be determined. [231]

A passenger train ran into a rockfall and was derailed between Mikawa-Makihara and Mikawa Kawai, Japan, on Saturday 17th February 1945. Twenty people died. [232]

Very brief reports in a number of American newspapers mentioned the derailment of a passenger train near Andimeshk in southern Iran on Monday 25th June 1945. More than 50 people were said to have died and many others among the 260 passengers on the train were injured. The cause of the derailment was not immediately known. [233] [234]

A Ferrocarriles Consolidados de Cuba Havana to Santiago de Cuba passenger train was derailed on a curve less than a kilometre from the station at Victoria de las Tunas (Las Tunas from 1976), Cuba, on Thursday 12th July 1945. The train reportedly jumped the track on a switch and there was a suggestion that the track had been tampered with. Excessive speed was another possibility while one report stated that the engineer had 'faced a technical problem,' without explaining what that meant. The locomotive, two baggage cars and two passengers cars overturned and were crushed. At least 19 and possibly 25 people were killed and perhaps 75 injured. [235] [236] [237] [238]

On the evening of Sunday 30th September 1945 the London Midland and Scottish Railway issued a statement regarding an accident which had occurred that morning: -

"The L.M.S. Railway regret to announce that the 8.20 p.m. passenger train from Perth to Euston was derailed at Bourne End, near Hemel Hempstead, at 9.5 this morning. A number of passenger were injured, a few fatally. Details will be issued later. Two lines were reopened at 5 p.m. this evening, and the delays in main line trains to and from Euston will not be serious. A special inquiry is to be held to determine the cause of the accident." [239]

The accident was in fact much more serious than might be concluded from the first official statement. In terms of passenger fatalities it was in fact the more serious accident to have occurred in Great Britain since the overwhelming catastrophe at Quintinshill 30 years earlier. Forty-one passengers lost their lives, together with the driver and fireman,

---

[231] Interstate Commerce Commission report
[232] Author's personal notes. Source mislaid.
[233] *Chester Times*, Chester, Pennsylvania, United States, 25th June 1945
[234] *The Muscatine Journal and News-Tribune*, Muscatine, Iowa, United States, 25th June 1945
[235] www.juventudrebelde.cu – Historia Sobre Rielos
[236] *The Montana Standard*, Butte, Montana, United States, 13th July 1945
[237] *El Paso Herald-Post*, El Paso, Texas, United States, 14th July 1945
[238] *The Hammond Times*, Hammond, Indiana, United States, 13th July 1945
[239] *The Times*, London, England, 1st October 1945

and 124 passengers were injured, 64 seriously.

The up express consisted of 4-6-0 Royal Scot class steam engine No. 6157, *The Royal Artillerymen*, and 15 coaches, all electrically lit and all with timber or timber-framed bodies on steel underframes except the second, which was all steel. The train was in charge of Guard W. Horwood with Driver Sidney Swaby and Fireman W. Jones and it was carrying 398 passengers. It had left Crewe at 5.22 a.m., an hour and 22 minutes late. It stopped at Rugby at 7.33 a.m., passed Tring at 8.57 and Berkhamstead at 9.02, travelling on the Up Fast line. From Bourne End signal box, however, 1¾ miles south of Berkhamstead, the Fast lines were out of use due to Sunday engineering work being carried out in Watford Tunnel, and all Up trains were being diverted there, onto the Up Slow line, via a mains crossover through which speed was restricted to 20 miles an hour. The train reached Bourne End at between 9.03 and 9.04 a.m., travelling much too fast to safely negotiate the crossover, and the engine and all but the last three vehicles of the train were derailed. By chance the line at Bourne End was carried on an embankment, and the engine fell about 9 ft down the bank to the soft soil of a mangold field, into which it was partially buried. Six of the seven leading coaches piled up behind it and were destroyed. [240]

Assistance was given at once after the disaster and there chanced to be a doctor on the train. Other doctors soon reached the scene, and it was noted that, 'Prominent among them were the elderly Dr. Burnett, of Hemel Hempstead, and a woman doctor from the same town, Dr Bellamy. The latter gave many hours' assistance to a sergeant of the R.A.F. who was trapped by his legs.

'The local villagers threw open their houses to give all possible assistance, for it was only on the previous evening that they had celebrated V Day and many of the refreshments left over were brought out for the workers.' (VJ Day was 15th August, but victory parties were apparently still being held in various places at the end of September.)

The first casualty was apparently admitted to the West Hertfordshire Hospital already at 9.30 a.m., but some were stuck in the wreckage much longer. 'Numbers of people were trapped for periods of up to nine hours. One of the last to be got out was an old man about 70 years of age, whose greatest worry as he was on the stretcher was the whereabouts of his false teeth.' [241] A special service of buses took uninjured passengers to Watford.

A reporter described the rescue work in detail. 'One coach was flung across the engine and its tender, another reared up on end against them, and a third was partly telescoped by the force of the impact. The huge mass of wreckage which thus confronted the rescue workers presented a difficult problem. Before the three heavy cranes and the five and seven-ton "wreckers" brought from a neighbouring airfield could be put to work pulling apart and removing the shattered coaches from the summit of the pile, each portion had to be carefully searched, and at every stage endeavours were made to penetrate the solid mass of debris beneath, in the hope that some succour might be brought to any passengers who might still survive. The main weight of the coaches being in the bogie wheels, these had to be cut clear of entanglement by oxy-acetylene burners and ripped away by cranes before the coaches could be lifted bodily.

[240] Ministry of Transport report

[241] *The Times*, London, England, 1st October 1945

'By this morning the work of dismantling the pile was nearly complete, and some sort of grim order had been imposed on the scene. Two of the coaches, their roofs like crumpled tarpaulin, had been moved clear. To one side was heaped the splintered wood, the torn cushions, and the smaller fragments, many bearing distressing evidence of violence death. In another section, bogie wheels in their frames rested one on top of the other. Mangolds were scattered everywhere, mingled with the rubble, and crunching under the feet of the workers. Round the single coach that remained lying on its side on the heap of earth and debris were sombre collections of crushed luggage and personal belongings, which were being steadily augmented as the workers delving beneath the coach passed up now a forage cap, now a baby's bottle, unbroken and still full of milk, and now a paper package of sandwiches.

'Railway workers and engineers, rescue squads, first-aid parties, policemen, men of the N.F.S., and men of the U.S.A.A.F., as well as service men and civilians, worked steadily throughout the day, moving to orders with an efficiency which gave no hint of the fact that many of them had been working for more than 24 hours without pause. Private D. Fossey, of the R.A.O.C., who had been on leave in the locality, was early on the scene on Sunday, and was still working this evening.

'There are still some passengers not accounted for, including three of the party of dining-car attendants on the royal train who went to Balmoral on Friday. One of them, who was killed, was Mr. E.A. Hill, who had been an attendant on the royal train for over 30 years. Earlier in the year the King had decorated him with the Silver Medal of the Royal Victorian Order. Mr. W.H. Tyler, a personal attendant in the royal coach, and Mr D. Williams, an assistant chef, are in hospital. The Queen sent a message of sympathy from Balmoral to Mr. W.P. Keith, L.M.S. dining-car controller, on the death of the attendants.' [242]

The first witness whose evidence was given in the Ministry of Transport inquiry into the accident was Signalman R. Dimbleby at Berkhamstead, who stated that the Perth express had passed him travelling at normal speed, like the preceding Glasgow express. It had been a clear morning, with no sign of fog. He saw Driver Swaby and Fireman Jones, who appeared normal. Instead of the "Out of Section," signal he had received the "Obstruction Danger," signal at 9.05 a.m.

Signalman H.G. Nash at Bourne End testified that the Glasgow express had passed him at 8.54 a.m. Like the Perth express it had arrived from Berkhamstead on the Up Fast line. It had negotiated the crossover at the authorised 20 miles an hour and proceeded, clearing the next cabin at Boxmoor at 8.57. He had then accepted the Perth express from Berkhamstead, had received the "Train Approaching" signal from that box at 8.57 and had offered the train to Boxmoor and had it accepted. He had then cleared his signals, including No. 6, the Up Slow to Fast distant signal. This was a colour light which when cleared showed a double yellow aspect rather than a green to indicate to the driver that he must reduce speed for the crossover. As the express came into sight he could see that it was coming faster than usual for a train which was being crossed over, though not as fast as a through train on the Fast line. He did not see the train crew, did not hear a whistle nor notice the brakes being applied. He saw the engine rock badly as it went through the points and then go down the embankment.

[242] Ibid, 2nd October 1945

As was often the case the key witnesses, Driver Swaby and Fireman Jones, were of course not able to give evidence. Instead, the inspecting officer, Lt-Col Sir Alan Mount, had to rely on indirect evidence, especially regarding the conditions of visibility on the morning of the accident.

Driver G.F. Frost of the preceding Glasgow train, referred to "an exceptionally fine morning with good visibility, and there was no sign of fog at this point." He said that he had approached Bourne End at 60 miles an hour, and "the sun was just above the Distant signal, and to the left of it, and it did not stop me seeing the signal."

Fireman Herward had also seen it and remarked, "Two Yellows."

Driver Frost said that had he been unable to see the signal he would have treated it as a danger signal. He was aware of the diversion from the Fortnightly Notice. Frost added, "I rely on these double yellow signals at this particular place, and the double yellow signal takes the place of double semaphores that used to be there. I rather like these colour light signals, and think they are a big improvement on the semaphores, particularly at night and in foggy weather. I have no difficulty in finding them whether they are on the ground or higher up. This distant signal has been put further out, and it is an advantage as regards stopping, as we have more distance to stop in; in fact, the signal is in a better position today than the old semaphores used to be."

Fireman Herward confirmed his driver's evidence, except that he thought that they had been going a little faster approaching Bourne End.

On the other hand Driver G. Marr, who had had charge of an Up fish train from Aberdeen which had crossed from Up Fast to Up Slow at Bourne End 20 minutes before the Glasgow train and half an hour before the Perth one, had found it a problem to see the Bourne End distant signal due to the bright sun and a slight haze, and had commented on this to Driver L. F. Butcher at Watford before they were aware of the accident. Driver Marr told the inquiry, "There was a ground fog and the sun shining sort of blocked out the signal; the sun was more in front of me, and it would be behind the signal from my cab window. I could not really tell you whether the sun was right or left, but it was just in front of my face. The fireman saw the signal first, and he held his two fingers up, indicating two yellows. He saw them before me, and I was just underneath before I saw the signal, but I was looking out all the while... I just got a glimpse of one yellow, but my fireman said there were two yellows... It was a sort of haze and the sun over the top of it made it worse. We were not very far off when the fireman told me of the position of the signal. I had no difficulty in seeing the Outer and Inner homes."

Driver Marr's fireman, H. Weaver, stated, "Approaching Bourne End, I did not see the Distant signal before I was nearly on it, it was a bit hazy, and the sun was shining right behind it... It started getting hazy about Tring, and I could not see very far there, but it was not exactly foggy, it was a ground haze. I looked as I went round the bend approaching Bourne End and I could not see the distant signal. I first saw it as I got on top of it... I had no difficulty in seeing the Outer Home Signal."

Driver Butcher, who piloted Driver Marr from Watford into London Broad Street, said, "I had the same trouble at Willesden No. 5 the same morning. When I came to Willesden No. 5 it was a red one, and the sun was in a similar position in the rear of these signals. I had three attempts to see them when the change to yellow came."

Driver Butcher added, "It is common knowledge among Watford drivers that the sun interferes with the sighting of these colour lights at certain times of day. I have come up with the Northampton coaches, and when I have sighted the Distant signal at Castlethorpe, which is a colour light signal, I have had to brake to make certain. I prefer the semaphore type of signal during day-time."

The report on the accident continued with the statements of several passengers who had been on the express and then of Guard Horwood, who related, "The first indication I had of anything wrong was a feeling as if the rear part of the train was surging into the front, after which we very quickly came to a stand. Prior to the surging referred to, I did not observe that the brakes had been applied. I knew that we were to be diverted from the Up Fast to the Up Slow at Bourne End, but it did not occur to me to watch for the reduction of speed, as I had full confidence in the driver, and did not think for a moment that it would be necessary for me to take action."

Guard Horwood said that he had joined the train at Crewe at 5.30 a.m. He had spoken to Driver Swaby, asking his name, and informing him of the number of vehicles and the tonnage. Although he (Guard Horwood) was aware of the diversion at Bourne End he had not mentioned it to the driver as the train had been late and there was little time for conversation.

Guard Horwood was reminded of Rule 148(a), which read:

When passenger trains are approaching important junctions, terminal stations, and stations at which they are booked to stop, the Guards must carefully watch the running of the trains and take any action that may be necessary. They must also keep a good lookout when leaving stations, and, as far as practicable, on other parts of the journey.

Guard Horwood contended that even had he seen the Bourne End distant at double yellow he would have waited until he could see the outer home before he applied the brakes and by then it would have been too late. Concluding his evidence he repeated, "...where a train is put over a crossing in the Working Book, you naturally take it for granted the driver is acquainted with that as well as you are."

Crewe Running Shed Foreman J.F. Bolton spoke of Driver Swaby as "a very good conscientious driver and very willing." It was he who had asked Swaby to take the express from Crewe to London. He said that the driver had been quite well and in his normal state of health.

Crewe Assistant Foreman J.F. Ellis produced the Fortnightly Notices relating to the diversion, which he said Driver Swaby had signed for. The notice in question read in part:

Sunday Sept. 30 Between Boxmoor and Watford, 23 and 18¾ m. p.

Up and Down Fast 7 a.m. to 5 p.m.

Unloading new sleepers and loading rails and repairs to bridgework in Watford Tunnel

BLOCKED (except for ballast trains).

All trains to travel over the Slow lines

Between Bourne End and Watford No.1

Locomotive Platform Regulator F Parker said that he had spoken to Driver Swaby at

Crewe on the morning of the accident. He related, "I climbed on to the engine and Swaby at that time was sitting on his driver's seat, which is on the left hand side of the footplate in the running direction, and Jones, his fireman, was sitting on his seat. Immediately I arrived on the footplate I noticed Swaby was reading something... The firebox door was open at the time, and this illuminated the footplate, and I had no difficulty in discerning that he was reading a copy of the W 2. Fortnightly Notice. Swaby actually held the notice out towards me, and in this position, especially with the light of the fire from the firebox, I had no difficulty in seeing what he was studying and it was the Rugby-London page of the same, Rugby-London being in heavy type... During subsequent conversation, Swaby expressed satisfaction at having been given this particular job, and remarked that if Camden Loco was short of men for the Sunday evening trains he would be prepared to book off and cover a job back to Crewe. Fireman Jones said that he was also contented to do whatever his mate did and to make a big day of it. Noting that Swaby was studying his Fortnightly Notice I did not enter into any general conversation with him, although I remained on the footplate a matter of 20 minutes, having no other duties to attend to at that time... It has been a matter of comment at Crewe Motive Power Shed that Driver Swaby has always been very methodical in the way he looks after his fortnightly notices, &c., and that signing on or off duty he could usually be seen with them in his hand... He was the very best man at Crewe for reading the Notices, and I have heard drivers go to him and ask him if there are any permanent way slacks. He was so well known as one who studied his Notices."

As to Swaby's condition that morning Parker said that he was, "In his usual state of good health, and was in good spirits, just as I should expect to find him... He was a temperate man, very temperate. He would take a bottle of tea and sandwiches on the footplate with him... He was very active that morning. I do not think he was tired, not in the least. He was not a man whose mind was likely to wander, and he was not subject to fainting."

Parker said that the engine coal had been fine and the engine in good condition, though he mentioned possibly causes of the accident such as the brake valve jamming. Regarding the signals he thought that the colour lights were good but there should be a diverting arm or a route indicator rather than two yellows and pointed out that Rule 35 (b) ii was conditional. This part of the rule read:

The indication (YY) means - pass next signal at restricted speed, and if applicable to a junction may denote that the points are set for a diverging route.

Driver Swaby's adult son was questioned regarding his father's general health and said that the driver's three periods of absence in the previous three years had been due to colds and, "There was nothing at all which would have prevented his taking action." He was not diabetic, affected by any similar condition or suffering from anxiety. His father would have rested for about 7½ hours before reporting for duty at 1.13 a.m.

Beginning his conclusions as to the cause of the disaster, Lt-Col Mount stated that the acceptance of the Perth express had been in order, there could be no doubt that signal No. 6 had been showing a double yellow and that no responsibility rested upon Signalman Nash.

The inspecting officer noted the similarity of the disaster to an accident at Leighton Buzzard in March 1931, which had likewise occurred when a Sunday express was diverted from a Fast to a Slow line due to engineering work and had failed to reduce speed for the crossover. Here, again, the engine crew had been killed so there was no explanation for their mistake.

Driver Swaby's mistake was, the inspecting officer said, baffling. He was 'a most reliable and conscientious man with an exemplary character and was supported by an equally reliable and experienced fireman.' There had been a minute and a half in which he could have taken action, but it was doubtful if he had braked at all, and he had certainly not whistled. There was no indication of any problem on the engine in the minutes before the derailment.

No good purpose, Lt-Col Mount went on, would be served by speculating as to why Driver Swaby had made his mistake, but certain features could be noted:

(1) Driver Swaby was working his fourth consecutive Sunday turn and had worked 26 days without a break, though the actual hours he had worked had not been excessive.

(2) The bright low sun might have made it harder to see the signals, though it was hardly conceivable that a driver of Swaby's calibre could have mistaken a double yellow for a green and even less that he would have continued at full speed having missed the signal.

(3) The use of a double-yellow colourlight signal in a semaphore area did not, for whatever reason, have the desired cautionary effect.

Guard Horwood had been drinking tea in the rear of a sleeping car at the moment of the derailment but even if he had been looking out from his van it was doubtful if he could have appreciated the situation in time to prevent the accident and the inspecting officer did not consider that he should necessarily be blamed for not having kept a lookout. It will be noted that the inspecting officer did not say that, like Signalman Nash, Guard Horwood was entirely free from blame, but he merely conceded that the guard had been in a difficult position. Other inspecting officers in other reports have been much more critical of guards in similar situations.

The inspecting officer then embarked on a lengthy discussion regarding the use of a double yellow signal in a semaphore area to indication a need to reduce speed for a diversion. He noted that 52 such signals had been provided by the L.M.S. replacing splitting semaphore distant signals and was clearly not in favour of them as he regarded them as confusing. He wrote, 'I attach great importance to uniformity of meaning, and for this reason I do not favour the employment of YY as an advance indication of a junction ahead, leading to G at the junction itself, and so departing from the well understood conventional sequence of aspects. In my opinion, a preferable course in 2-aspect territory, at a junction set for a diverging path, over which there is a severe speed restriction, is to maintain the Distant Signal (or the Inner Distant if a YY Outer colour light Distant is provided) at Caution, or Y, and to delay clearance of the home signal or signals, until the approaching train reaches a point at such distance from them as may be selected, depending on all the relevant circumstances.'

Turning to remarks and recommendations, the inspecting officer remarked that given the rarity of serious accidents of the Bourne End type there could be no drastic criticism of operating methods or signalling. Nevertheless the human element could not be eliminated. As for the responsibilities of guards, he referred to Rule 2 (i), which read:

Employees must see that the safety of the public is their chief care under all circumstances.

In view of the Bourne End disaster, and the likelihood of an early return to pre-war train speeds, he considered that Rule 148 (a) should be strengthened by the addition of the words:

Particularly when they know that the train is being diverted from its usual path.

Apart from the guard the accident illustrated the need for some form of Automatic Train Control, whereby an audible warning was provided to a driver when he passed a distant signal at caution and the brakes were applied unless the driver cancelled the warning. It was possible that such a system would have prevented both the Bourne End disaster and that at Castlecary in December 1937 (dealt with in the Rear Collisions chapter) though any such apparatus which could be overruled by the driver was not infallible, as was shown by the accident at Norton Fitzwarren in November 1940 (see above). Statistics suggested that between 12 and 13 per cent of pre-war accidents into which inquiries had been held would have been prevented by automatic train control. Automatic train control should therefore be introduced on all main lines, of a type suitable for use in multiple aspect colourlight areas and of a standardised type. [243]

The first fast train to be run between Berlin and Hamburg, Germany, since the end of the war, was derailed on Friday 12th October 1945 at an unstated location. An initial report stated that not less than 80 people were killed and injured. The brief report on the accident in a Swiss newspaper speaks of the derailment having occurred between Dortmund and Hamburg, which is not between Berlin and Hamburg. Possibly confusion has arisen with the rear collision at Nauen on Wednesday 10th October (see Rear Collisions chapter)/ [244]

A train was derailed and overturned on the Sinyu line of the Kobe – Dentetsu Railway, in Japan, on Sunday 18th November 1945. Forty-five people died and 131 were injured. [245]

A train was derailed on the Inbaragi line of the Inbaragi – Kotsu line, in Japan, on Monday 26th November 1945. Ten people died and 142 were injured. [246]

In the third disaster in Japan in quick succession a derailment on the Kouya line of the Nankai Railways Company on Thursday 6th December 1945 killed 27 people and injured 224. [247]

---

243 Ministry of Transport report
244 *Journal de Geneve*, Geneva, Switzerland, 13th October 1945
245 Webpage: A Brief Survey of Railway and Aircraft Accidents in Japan and Investigations Thereof by Seiji Abe and Takemune Sato
246 Ibid.
247 Ibid.

CHAPTER 4

REAR COLLISIONS

The risks associated with one train running into the back of another had been low in the formative years of the railways when trains were few and train speeds not very high, but as the years progressed train speeds and traffic intensity often grew faster than advances in brake power, rolling stock construction and signalling and as a result by the early 20th century serious rear collisions had become common. The rear collisions covered by this volume include amongst others the overwhelming catastrophes at Lagny, France, in 1933 and Genthin, Germany, in 1939. This chapter is therefore a long one, even though a great number of less serious accidents have not been included.

Early on Friday 7th August 1931 a northbound slow train stopped, 'due to a mechanical defect,' near Bialystock, Poland. While it was stationary due to a misunderstanding between two signalmen a Warsaw – Vilna express was admitted into the occupied section and ran into the rear of the stationary train. Two carriages of the slow train were wrecked, and the engine of the express was overturned. Five people died and 17 were injured. The line between Warsaw and Bialystock was blocked by the accident and international expresses to and from Moscow were diverted via alternative routes. [248]

A goods train stopped outside of the station at Henterberg-Göss in Styria, Austria was run into from behind, 'due to an error of judgement,' by a Roma-Merano-Villach-Bruck express travelling at high speed at 3.50 a.m. on Sunday 16th August 1931. The four rearmost waggons of the goods train were thrown into a dry riverbed 60 ft below and a first-class carriage and a Pullman car were telescoped into one another. Twelve of the Austrian and Hungarian passengers died and 11 were injured. Among those who lost their lives was a Hungarian Protestant clergyman whose wife had to have both of her legs amputated and after the accident was reportedly lying in hospital at Leoben in a precarious state. Many of the passengers were children going on their summer holidays,

---

[248] *The Times*, London, England, 8th August 1931

among whom there was a wild panic when the accident happened. [249]

A passenger train stopped at an unstated location 100 miles south-east of Harbin, Manchukuo, on the Chinese Eastern Railway, 'because of a derailment.' The report does not make it clear if it was the passenger train itself which was derailed or whether another train in front of it had left the track. At any rate while it was stopped it was run into from behind by a goods train. Forty people died and over 100 were injured. [250] The accident happened on or about Monday 23rd May 1932 near Yaplonya. [251]

At 7.30 a.m. on Tuesday 10th January 1933 an express train from Timişoara ran into the rear of an omnibus train about a mile outside of Bucharest Nord station, Romania. The express was travelling at full speed at the time, so the collision was a very violent one. The colliding engine was derailed, and five carriages of the omnibus train were wrecked. A rescue operation was immediately organised and eight dead and thirty injured were extricated from the wreckage. [252] All traffic in the Cariova direction was interrupted. [253]

An inquiry established that the collision was apparently due to a signalman's mistake. The signalman and the conductors of the two trains were all arrested.

Mr Maniu, the transport minister, went to the scene of the accident. [254]

A passenger train which had been stopped by a derailment ahead was run into by a freight train between Chengchiatun and Ssupingkai, Manchukuo, at about midnight on Thursday 16th March 1933. Fifty passengers were said to have died with over 70 injured. [255]

The *Atlantic Express*, a crack Erie Railroad train between Chicago and New York, was involved in a bad accident at Binghamton, New York, United States, on the evening of Tuesday 5th September 1933, when it was run into from behind by a milk train. The engine of the milk train crushed the parlor car at the rear of the express, but the main damage was caused by the telescoping of the third coach from the rear by the penultimate coach. Fourteen passengers were killed and 31 passengers and a porter in the parlor car were injured. The accident happened at about 7.23 p.m., about two miles east of Binghamton passenger station. [256]

The track was cleared five hours after the collision and a special train took the surviving passengers on to the Jersey City terminus.

Newspaper reports of the disaster carried stories of misfortune and confusion.

---

[249] Ibid, 17th August 1931
[250] *The Berkshire Evening Eagle*, Pittsfield, Massachusetts, 24th May 1932
[251] *Journal de Geneve*, Geneva, Switzerland, 25th May 1932
[252] *Excelsior*, Paris, France, 11th January 1933
[253] *The Times*, London, England, 11th January 1933
[254] *Le Journal*, Paris, France, 11th January 1933
[255] *The Times*, London, England, 18th March 1933
[256] Interstate Commerce Commission report

'Mrs Thomas S. Keane, of Susquehanna, was one of those killed. She almost missed the train at Binghamton. It was getting under way when she raced to the platform and was assisted on by another passenger who was standing at the open door of the wooden coach.'

'A one-year-old baby was taken from the wreckage apparently dead and was listed as such on first casualty lists. Medical examination, however, revealed that the baby was not only alive but not seriously injured. By a similar error, the official death list for a while carried the name of Mrs Catherine Harrison of Binghamton. She was not hurt and continued on to New York in a special train the railroad made up to carry passengers on to their destinations.' [257]

Four trains were involved in the accident. The first was eastbound freight train No. 90, consisting of engine No. 3370, 92 cars and a caboose, in charge of Conductor Norton with Engineer Griffin and an unnamed fireman.

Switch engine No. 78, in charge of Switch Foreman Gaffney, followed train No. 90 with 30 cars to be added to it.

The *Atlantic Express,* eastbound train No. 8, consisted of engine No. 2916, three express cars, a mail car, a combination baggage car and coach, two coaches and a parlor car, in charge of Conductor Burlingame and Engineer G.O. Grantier and Fireman R.A. Washburn.

As usual the milk train was running as a section of passenger train No. 2, on the day of the accident as third No. 2. It consisted of engine No. 2929, six milk cars, an express car, 11 milk cars and a coach, in charge of Conductor D.H. Shea and Engineer M.H. King and Fireman Arey.

Before arriving at Binghamton the crew of train No. 90 were aware that they were to pull into a siding east of BD tower, beyond the station. When their train passed Binghamton at 6.59 p.m. they were given a message informing them that switch engine No. 78 would add 30 cars to their train and take 10 cars from it.

General Yardmaster Reap told the Interstate Commerce Commission (ICC) that train No. 90 had passed Binghamton at 6.55 p.m. (as opposed to the 6.59 recorded on the train sheet) and that he had therefore considered that there was sufficient time to carrying out the shunting operations before the arrival of train No. 8, due to depart Binghamton at 7.15 p.m.

Switch Foreman Gaffney testified that his engineer had whistled out a flagman (Switchman Noonan) who had flagged train No. 8. He did not remember train No. 8 ever having been delayed previously by the shunt movement.

Switchman Noonan said that he had flagged train No. 8 and told the engineer that he was from the switch engine which was adding car to train No. 90 and to be careful. He had then walked around the rear of train No. 8 and saw its flagman going back. He estimated that only a minute had elapsed between train No. 8 coming to a stand and the collision.

Engineer Grantier of train No. 8 said that Yardmaster McAndrews had advised him that work was being done at the east end of BD siding. The home signal at Liberty Street

[257] *South Haven Daily Tribune*, South Haven, Michigan, United States, 6th September 1933

tower (7,980 feet west of the point of collision) had cleared before he had passed it and the home signal at BD tower (5,962 feet west of the point of collision) had been displaying *Approach*, indicating that automatic signal No. 212.2 (893 feet east of the point of collision) had been at *Stop*. He had forestalled the train stop apparatus, reduced speed, seen Flagman Noonan, reduced speed further and picked the flagman up. Advised of the shunting work ahead he had stopped his train about 500 feet short of automatic signal No. 212.2 and whistled out his own flagman. His train had not been stopped over a minute when it was run into.

Fireman Washburn more or less agreed with his engineer's statement, as did Baggagemaster Williams of train No. 8, who added that he had seen the flagman signalling to the milk train with red and white flags. The baggagemaster and Head Brakeman R.D. Donovan thought that the flagman had gone back as far as was possible in the limited time available.

Conductor Burlingame spoke of seeing the flagman alight and of having had time to collect one more ticket before his train was struck about 45 seconds later. After the collision, the flagman had told him that he had gone back as far as he could.

Flagman A.C. Morris stated that he had flagged train third No. 2 which had stopped behind his own train at Binghamton station. After leaving there he had been sat writing out reports until his engineer had whistled him out. Then he had picked up his red and white lanterns, but had not stopped to take any fusees, being in a hurry to get back. Very shortly he had seen the headlight of train third No. 2, which was working steam, and had begun to run back, waving his lanterns. His signals had not been acknowledged but fire had been flying from the wheels of the engine of the milk train as it had passed him. Asked whether it would not have been better to have thrown off lighted fusees before his train had stopped and reminded of the rule regarding doing so if a train was likely to be overtaken by another train, he replied that he had not thought it necessary because the milk train was running under interlocking rather than automatic signals and had received a restrictive signal meaning that it must proceed prepared to stop short of any obstruction. He added that he had also not expected his own train to come to a dead stop.

Conductor McLaughlin, deadheading on train No. 8, put the time between the train he was on and the collision at 45 seconds, considered that Flagman Morris had gone back as far as he could and that the engineer of the milk train should have been able to see the flagman from 900 feet away.

Engineer King of train third No. 2, testified to pulling up behind train No. 8 at Binghamton, setting off 8 or 10 minutes after that train had departed, and being checked but not stopped at Liberty Street tower. At BD tower the home signal had been at *Stop* but the calling on signal had then been cleared. He said that he had cancelled the automatic train stop device and had gradually increased speed to about 20 miles an hour. When he had seen the cars of train No. 90 in the siding, and then automatic signal No. 212.2 at *Stop*, he had assumed that train No. 8 was in advance of that signal. Only when he got closer had he seen the rear of train No. 8 and its flagman. He had braked and opened the sanders and had reduced the speed of his train, he thought, to 5 or 6 miles an hour. The degree of damage caused he attributed to the run-in of the cars of his train. Never before did he remember train No. 8 being stopped at that point. At the coroner's inquest and at a separate inquiry he had maintained that he had only to stop according to his range of vision, that he had been able to see 80 or 100 car lengths, and that even had he passed BD tower at 65 miles an hour he would still have had his train under control.

Fireman Arey of train third No. 2 put the speed of his train at 15 miles an hour at Liberty Street tower, at 20 miles an hour at BD tower, that it had then increased to perhaps 25 miles an hour and had still been 15 or 18 miles an hour on impact. He had not seen the flagman. The rear markers of train No. 8 had been burning brightly. His engineer had appeared normal.

Conductor Shea put the speed of train third No. 2 at 10 to 12 miles an hour at Liberty Street, 15 or 18 miles an hour at BD tower and not over 25 miles an hour when the brakes had been applied. He agreed that Engineer King, whom he had known for 25 or 30 years, had appeared normal.

Brakeman Angell of the milk train put its speeds slightly higher than his conductor; at 25 or 30 miles an hour when the brakes had been applied.

The estimates of speed of Flagman Friend agreed with those of Conductor Shea.

Operator Packer at Binghamton recorded trains Nos. 8 and third No. 2 as passing at 7.16 p.m. and 7. 18 p.m.

Signalman Wixson at Liberty Street stated that trains Nos. 8 and third No. 2 had both passed him travelling at about 15 miles an hour. He did not record times. The rear markers of train No. 8 had been showing a good red indication.

Signalman Mace at BD tower said that he had given train third No. 2 a calling-on signal as the line ahead was blocked. This was the usual practice to avoid blocking the Lackawanna track at Liberty Street. He recorded the two trains as having passed at 7.18 p.m. and 7.21 p.m.

Road Foreman of Engines McCarthy stated that he had known Engineer King for 42 years and the engineer neither smoked nor used alcohol. When the engineer, who had previously worked freight trains, had applied to work the fast milk train he had been examined as to his knowledge of the Westinghouse brake and general suitability for the job and passed. After the accident McCarthy had asked Engineer King how it had happened, and King had replied that had got too bold with his speed and had been unable to stop when he had seen the flagman and the markers of the train ahead.

Trainmaster Murphy considered that Flagman Morris should have thrown off a lighted fusee before his train stopped.

As to the running of freight train No. 90 and the movement of switch engine No. 78 close in front of train No. 8, which had delayed it, General Yardmaster Reap, supported by Superintendent J. W. Graves, stated that Rule No. 86, relating to inferior trains keeping clear of superior ones, was not in effect between Hornell and Susquehanna, as train movements were governed by Rule 251, that "on portions of the road so specified in the time table, trains will run with the current of traffic by block signals whose indications will supersede time table superiority."

The ICC concluded that the accident had been caused primarily by the failure of Engineer King properly to obey a signal indication requiring him to proceed at 'restricted speed', defined as:

prepared to stop short of train, obstruction or anything that may require the speed of a train to be reduced.

He knew the meaning of the signal indication but had allowed himself to be misled by what he had assumed to be the case, that signal No. 22.2 was at *Stop* because train No. 8 was beyond it. The evidence also indicated that he had not been keeping a proper lookout or he could have seen the rear markers of train No. 8 in time to stop.

Train No. 8 had also clearly been running under circumstances in which it could be overtaken by a following train, within the meaning of the last portion of Rule No. 99 of the Standard Code, which read in part:

When a train is moving under circumstances in which it may be overtaken by another train, the flagman must take such action as may be necessary as to insure full protection. By night, or by day when the view is obscured, lighted fusees must be thrown off at proper intervals. When day signals cannot be plainly seen, owing to weather or other conditions, night signals must also be used. Conductors and enginemen are responsible for the protection of their trains.

In the circumstances Flagman Morris should have thrown off a fusee instead of continuing with his clerical duties.

Engineer Grantier had shown poor judgement in not having whistled out his own flagman when he had picked up Flagman Noonan and his not having done so was open to criticism.

As Engineer King had not been disabled, Fireman Avey had not been obliged to take over control of the engine as per Rule No. 875. However he had 20 years' experience on engines, and it was worthy of comment that it had not occurred to him to intervene when he saw that his engineer was going too fast.

The practice of using main lines for switching operations in the time of important passenger trains, while not strictly against the rules, was open to criticism and there had not appeared to be any necessity for them in front of train No. 8. The ICC commented that, 'operating officers of railroads are assuming a grave responsibility for which no excuse can be found, when they authorize movements of this kind.'

Signalman Mace was not criticised for using the calling-on signal, the ICC commenting that that was what the signal had been installed for, and 'it is necessary in the operation of a railroad to keep interlocking plants free of traffic in order that they may be free to perform their proper function, and the present case constitutes no exception to the usual practice throughout the country.'

Having operated the forestalling device of the automatic train stop it was incumbent upon Engineer King to control his train so as to be able to stop short of any preceding train or obstruction.

The ICC remarked that, 'The record of this case does not do credit to the Erie Railroad, for there have been too many employees who failed in one way or another to perform their duties to the fullest possible extent. Such a situation is a reflection on those responsible for the proper functioning of the operating department and justifies the recommendation that greater efforts should be made in the way of instruction and supervision.' [258]

---

[258] Interstate Commerce Commission report

A local train carrying a large number of schoolchildren was run into by a following train just outside the main station at Poznan, Poland, on Friday 15th December 1933. Two carriages of the front train, which had been stopped at a signal, were thrown down an embankment and a third was left suspended in the air. The driver of the colliding train was unable to see the train in front until he was very close to it because if a dense fog. Eleven people died and 80 were injured, four it was feared fatally and 20 more seriously. The rescue work was complicated not only by the fog but by the bitterly cold weather. A signalman who had allegedly admitted the second train into the occupied section was arrested. [259] [260]

The second worst railway accident in French history up to the present time (November 2021) occurred on the Chemins de fer de l'Est at Pomponne, near Lagny, France, on the evening of Saturday 23rd December 1933. To cope with the expected extra Christmas traffic the Est company had, like other railway companies in France and elsewhere, put on extra trains, but adverse weather conditions with fog and frost combined with a lack of staff at the Ourcq depot where the trains were prepared had resulted in inevitable delays. Thus it was that on the evening of Saturday 23rd trains were leaving the Paris Est terminus extremely late. Train No. 55, a Paris – Nancy express, due to leave at 17:49, did not get away until 19:22. It was followed at 19:31 by train No. 25 bis, a Paris – Strasbourg express which should have left at 18:00.

As far as Vaires there was quadruple track, and this section was equipped with modern colour light automatic block signals. Beyond Vaires the line was only double track, and the signalling was of the older type with semaphores and board signals. In front of train No. 55 was an omnibus train which terminated at Pomponne, where it had to be shunted into a siding. The signals were thus at danger against train No. 55, which was brought to a stand. The shunting of the omnibus train into the siding having been completed the signal was cleared and train No. 55 was just restarting when it was run into from behind by train No. 25 bis, which was still running at high speed despite the foggy conditions.

Train No. 25 bis consisted of modern steel coaches, hauled by a 4-8-2 *Mountain* class locomotive, in charge of Driver Lucien Daubigny, with Fireman Henri Charpentier. Train No. 55 on the other hand was equipped with old wooden coaches with metal panelling. The results of the collision were disastrous. The heavy *Mountain* engine ploughed its way through almost the whole length of the Nancy train, 'not merely telescoping the wooden coaches, but reducing them literally to fragments, sweeping their fragile superstructure, with their human freight, before it like snow before a snowplough, hurtling their flimsy roofs aloft and their partitions and panels to right and left, bending their metal chassis double, twisting them into corkscrew shapes, bursting them asunder and scattering their heavy wheels like cotton reels along the line. When at last it came to rest, the express engine had completely destroyed four coaches crowded with people. The front of the express engine when the breakdown train dragged the first wreckage away presented a strange appearance. Buffer beam, frame-plate, smokebox door and funnel were planed and levelled into the likeness of a gigantic ploughshare. The lofty boiler was plastered

[259] *Syracuse Herald*, Syracuse, New York, United States, 15th December 1933
[260] *The Times*, London, England, 16th December 1933

with debris, human and material. Long after the collision a child's white beret still hung on the handrail alongside the boiler. Of the wrecked coaches only one chassis, entirely shorn of wheels and coachwork and bent like a bow, was recognisable as a single unit. To be sure how many coaches there had been it was necessary to count the wheels and bogie frames among the wreckage.

'Both trains were crowded with passengers, mostly holidaymakers, on their way home for Christmas. The demolished coaches of the slow train must have contained about 300 people, including many children and soldiers and sailors on Christmas leave. Of these 300 none escaped injury. Three-quarters were killed. The steel express train, on the other hand, was hardly damaged, so little resistance had the wooden train offered to its impact. Of the mail van next to the engine one window was broken and the front panel was pierced in one corner. In the next coach two cases of mineral water left on the end platform stood intact after the collision amid the indescribable wreckage of the other train. In the express train no one was seriously injured, and only a few passengers complained of shock and bruises. The driver and fireman escaped unhurt.

'The disaster occurred in a thick mist accompanied by a severe frost. These conditions impeded the work of rescue and added to the misery of the injured. The first rescuers – firemen and villagers from Pomponne and the staff and uninjured passengers of the two trains – groped among the wreckage by the light of a few lanterns unable to see more than a yard or so in front of them in the swirling fog, guided only by the desperate cries of people pinned in the wreck. With numbed hands and dazed minds the rescuers dragged out dead and injured as best they could and laid them in a field close to the line, on iron-hard ground, slippery with hoar frost. Clouds of steam from the express engine, blown back along the wreck by a bitter east wind, condensed in sheets of ice on the wreckage, and mingling with the fog, added to the blindness of the rescuers.

'After a time doctors and ambulances began to arrive from Meaux, Lagny, and Paris. Bonfires were lit on the permanent way – there was plenty of fuel at hand – and their flames threw fitful light on a terrifying scene. For a distance of 70 yards the track was torn up and the ground heaped and littered with twisted metal, splintered wood, bodies, and parts of bodies, burst and scattered luggage, papers, Christmas provisions, and torn clothing. The passengers and staff of a third train (which was travelling on the up line and narrowly missed running into the wreck) joined the earlier rescuers, and the Paris hospitals rushed ambulances and surgeons to the scene; but fog and ice made road traffic as difficult, if not as hazardous, as moving by rail. For many hours before help arrived terribly injured people, trapped and half frozen in the wreckage, lay in helpless suffering.

'The hero of the disaster at this stage was Maurice Ancel, a young sailor belonging to the submarine flotilla at Cherbourg, who was travelling home on leave. All through the night he worked devotedly to rescue the injured and to relieve the suffering of those who could not be extricated, crawling to them until tangled heaps of metal, feeling his way in blood and darkness. At dawn he collapsed and was taken to hospital. Many of the rescuers were by then worn out and hardly able to speak or move.

'As soon as possible after the disaster the company sent a relief train to bring the injured back to Paris. The first survivors arrived at the Gare de l'Est in coaches detached from the express train. Some, however, were found to have died of their injuries; many were half naked in the biting wind, their clothes having been torn off in the collision. Among them were children who had lost their parents; a colonel of artillery had brought away a boy of five, whom he took home with him. At the first news of the disaster the stationmaster at the terminus, fearing an unmanageable rush to the station, had issued a reassuring statement. When the real news became known the Gare de l'Est was besieged by a great crowd of people, some clamouring for news of their friends and relatives, others moved only by curiosity; but there had been time to organize police cordons and inquiry offices. It was many hours, however, before even approximately lists of the dead and injured became available.

'The company decided on Sunday to bring all the bodies to the Gare de l'Est where they were placed in coffins and laid in an improvised mortuary. Over 150 coffins lay there last night, and some 40 deaths were reported from hospitals in Paris, Meaux and Lagny. All the bodies recovered from the wreck have now been identified, but it is feared that some will never be found. Among the dead are M. Schleiter, the Mayor of Verdun, and M. Henri Rollin, the Mayor of Saint-Dizier, both of whom are Deputies; and M. Paul Morel, the Mayor of Vesoul, a Deputy and former Under-Secretary of State, M. Poittevin, a Deputy for the Marin Department, was badly injured.

'The casualty lists issued so far contain only one British name, that of Miss Veronica Manning, with an address at Metz, who was injured. M. Lebrun, the President of the Republic, accompanied by Ministers and officials, visited the mortuary at the Gare de l'Est yesterday. M. Chautemps, the Prime Minister, and M. Paganon, the Minister of Public Works, hurried to the station when they heard of the disaster and spent Saturday night there with the railway officials. Expressions of sympathy have been received from many countries. Lord Tyrrell, the British Ambassador, was among the first on Sunday to offer the condolences of his Government.' [261]

No less than 230 people lost their lives in the accident and another 300 were injured.

Driver Daubigny was immediately assumed to have been at fault. M. Gaston Albucher, the public prosecutor, had him and his fireman arrested, and the driver was reportedly made to stand beside his locomotive for several hours between two policemen while the bodies of the dead were carried past him one by one. [262] He insisted that all the signals that he had received up to the point of collision had been clear, showing white lights. In the four kilometres preceding the collision point the train had passed three semaphore stop signals, Nos. 13, 15 and 17, each of which at night showed one red and one green for danger and a single white light for clear. Each semaphore was preceded by an *avertissement* or distant signal which showed two greens for caution and one white for clear. Under the new Verlant signalling code adopted in 1930 green was substituted for white as an all-clear signal and yellow for green as a caution signal but the railway companies were given up to five years to make the change and the Est company had yet to do so. The driver of train No. 55 had observed the double green of the *avertissement* for signal No. 17 and had stopped at that signal until its red and green had changed to a white. Signal No. 15 must therefore have been showing red and green against train No. 25 bis, but Driver Daubigny had not stopped there.

In addition to the fixed signals there were two additional warning systems. One was the so-called crocodile, an electrical contact between the rails similar to the Great Western Railway's automatic warning system, though there were differences. One was

[261] Ibid, 27th December 1933

[262] Ibid, 28th December

that the crocodile had no mechanical component, it was purely an electrical system. Another was that it was not failsafe. It did not apply the brakes if its warning whistle of a signal at caution was not acknowledged. Its main purpose was as a test of the vigilance of the drivers, as the aspects of the signals and the acknowledgement of the crocodile would be marked on the Flaman recorder. The crocodile was also liable to fail if its horizontal contact was insulated by hoar frost, such as was present on the night of the accident, in which case it would give a false all-clear signal. The driver of the Nancy train stated that the crocodile had been working normally but it could presumably have failed between the two trains.

The stop signals were also equipped with automatic detonator placers. Driver Daubigny and his fireman denied hearing either the siren of the crocodile or the explosion of any detonators. Fireman Charpentier had been on the tender trimming coal when the crucial signals were passed so could not have seen them anyway, but he could have heard the siren and the detonators. It also transpired that two passengers who had been riding in the last carriage of the Strasbourg train stated that they did hear the detonators explode beneath their coach and a crossing keeper stationed at a level crossing close to signal No. 15 agreed, implying that the detonator placer, frozen by the frost, had only worked belatedly when shaken by the passage of the train. [263]

The summary arrest of Driver Daubigny and Fireman Charpentier, both of whom were said to have been overwhelmed by the magnitude of the accident, was met with protests, including a notice of interpellation in Parliament, and after a conference between the *juge d'instruction* and the public prosecutor they were released on Boxing Day evening, though both still faced manslaughter charges. Both had excellent records.

By 28th December two inquiries were in progress regarding the catastrophe and Driver Daubigny was again questioned for three hours. He alleged that he had several times met with 'wrong side' signal failures on the Paris Strasbourg line which he had reported to his superiors (a wrong side signal failure would be when, for example, a distant signal was clear when its corresponding stop signal was at danger). Fireman Charpentier enlarged upon this. "They tell us the block system is infallible. It is so little infallible that there is no counting its failures. About 18 months ago Daubigny and I were driving a train to Strasbourg. At Noisy-le-Sec we found the green disc open though the next signal was closed. That is wrong. It happened in daylight so we could stop in time, but if it had been dark or foggy, and if a train had been stopped in that section, disaster would have followed. The next day at Rainey we found two discs open and the semaphore after them closed. There have been many cases. We have never had an acknowledgement or reply. But then look up our old records; they will see that the block system is not as perfect as they think. Perhaps the frost had something to do with it. Daubigny and I had noted that the electrical contact of the crocodile was often bad in frosty weather." [264]

On the morning of 29th December the Minister of Public Works, M. Paganon, had a long conference with the chairmen and directors of the principal railway companies and technical experts to examine a series of measures aimed at increasing the safety of the railways. Later the same day he received a deputation from the representatives of the railwaymen's union, who made various practical suggestions for the safe working of

263 Ibid, 27th December

264 Ibid, 29th December 1933

trains in fog and adverse weather conditions such as the wider introduction of the Rodolausse apparatus. (This was a system of automatic train control which like the Great Western system not only warned the driver if he passed an adverse signal but applied the brakes if the warning was not acted upon. It had already been tried on some local lines between Paris, Limours and Secaux and apparently performed satisfactorily but for whatever reason was never widely adopted.) The union also called for the construction of new steel rolling stock to take the place of the remaining wooden coaches.

On the same day, the Director of the State Railway, M. Dautry, stated that the Est was one of the best equipped railway systems in the country, and added that there was in France a tendency to avoid new expenditure as long as possible, and this prevented the effective modernisation of railway equipment. In 1933 the State Railways, in spite of all their efforts, had only 75 miles out of their 5,700 miles of track equipped with automatic block signals. The manual block signals of 30 years ago were still used on 1,200 miles of track, and over 4,000 miles still had no block system and depended upon precarious telegraphic communications. Similarly, they still had to use 1,000 wooden coaches owing to the difficulty of getting money for new rolling stock. Controversially, he suggested that out of 26,000 miles of railway in the country 10,000 miles of secondary lines should be closed, the rest being fully re-equipped with the latest signalling devices.[265]

On 3rd January 1934 M. Reboul, the Meaux *juge d'instruction*, issued charges of culpable negligence against six more Est officials, in addition to the driver and fireman. They were the chief engineer, M. Merlin; the assistant engineer, M. Martilot; Workshop Foreman Montignaul; Technical Inspector Mougeot; Assistant Foreman Piétremont; and Inspecting Mechanic Caron, these being the officials and employees responsible for the dispatch of trains and the supervision of the signalling and safety equipment.[266]

In the end it was Driver Daubigny alone who stood trial at Meaux on 19th December 1934, thus almost a year after the catastrophe. Asked by the judge if he had encountered any difficulties before the accident Daubigny replied that apart from the slight fog he had encountered none and he had been driving at his usual speed.

M. Godin, Est engineer, testified that there could be no doubt that Daubigny had passed three signals set against him. The automatic signals had been removed and tested after the disaster and the device which reset the signal to danger after the passage of a train had shown no defect after 40,000 operations. He added that travelling at 65 miles an hour Driver Daubugny would have had between a half second and a second to observe each signal and that travelling at such a speed in fog was "playing with fire," especially as the signals were not set at uniform heights.

In Daubigny's defence two engine drivers and other witnesses, including a signalman, mentioned problems they that had found with the signalling system.[267]

Prosecutor Albucher did not exactly go for a conviction with any great fervour, declaring to the judges that if they believed Daubigny's version of the events leading up to the disaster or had the slightest doubts they must acquit him.

It was another month before the Criminal Court announced its verdict. On 24th

[265] Ibid, 30th December 1933
[266] Ibid, 4th January 1934
[267] Ibid, 21st December 1934

January 1935 it stated that there was a slight doubt as to whether the automatic signals had operated properly after passage of train No. 55 and in view of this doubt, however slight, the case must be decided in Daubigny's favour. He was acquitted. [268]

A collision occurred five miles from Moscow, USSR, on Sunday 4th March 1934. One train had stopped due to defective brakes when a second crashed into it, telescoping the rear two coaches of the front train. Nineteen passengers died and about 50 were injured. [269]

Less than three weeks later Driver Fedunin and Fireman Chikoff of the colliding train were sentenced to death by the Moscow Supreme Court, it having been determined that the driver had deliberately disconnected the Westinghouse brakes to save energy and thus win a reward for economical driving. Driver Fedunin was found to have been a class enemy, as presumably was his fireman. State Prosecutor Akuloff declared that all such cases would be "demonstratively" tried because they were due to criminal negligence. Meanwhile a court at Makeevka had sentenced nine railwaymen to imprisonment with hard labour for terms ranging from five to ten years for having wrecked another train by negligence. [270]

A mistake by a signalman in a busy box led to an accident at Winwick Junction, England, on the London Midland and Scottish Railway, on Friday 28th September 1934. At about 9.10 p.m. the 5.10 p.m. Down express from London Euston to Blackpool, travelling at about 50 miles an hour under clear signals, ran into the rear of the 8.55 p.m. Down local train from Warrington to Wigan via Earlestown. Five passengers and Guard James Turner in the local train and three passengers in the express were killed outright and two more passengers died later in hospital. Fifty-five passengers and three railwaymen were injured.

The local train consisted of 2-4-2 tank engine No. 6632 and three carriages in charge of Guard Turner with Driver Hope and Fireman Hayes. The express consisted of *Prince of Wales* class 4-6-0 engine No. 25648 and nine carriages, in charge of Guard John Jones with Driver Roger Steele and Fireman Partridge. The vehicles of both trains were electrically lit and equipped with vacuum brakes.

From Warrington northwards to Winwick Junction there were four lines, which from west to east were the Down Slow, Down Fast, Up Fast and Up Slow. At Winwick the double-track main line to Golborne Junction, Wigan and Carlisle diverted to the right from the double-track branch to Vulcan Bank and Earlestown. Trains taking the branch to Earlestown could regain the main line to the north via Parkside and Lowton Junctions to Golborne Junction. The local train was routed via Earlestown and the express via the direct line to Golborne Junction.

In the minutes leading up to the collision Signalman William Bloor at Winwick had had no less than eight trains to deal with: -

(a) A Crewe to St Helens down goods via Earlestown, which he held at the Down Slow home signal from 8.53 p.m. for:

---

[268] Wikipedia – Accident ferroviaire de Lagny-Pomponne
[269] *The Times*, London, England, 5th March 1934
[270] Ibid, 2nd March 1934

(b) A Llandudno to Manchester express, also via Earlestown, which he had had accepted by Signalman Perkins at Vulcan Bank box and which passed on the Down Fast to Down Branch about 8.56.

(c) An Up local passenger train from St Helens via Earlestown to Warrington which passed at about 8.56 on the Up Branch to Up Fast.

(d) An Up cattle empties train from the Earlestown branch which Signalman Bloor, after speaking on the telephone to Warrington, was going to send after the St Helens local train on the Up Fast line until Golborne Junction asked for:

(e) A Fleetwood to Crewe Up fish train on the Up Main, which was more important than the cattle empties. Signalman Bloor thus kept his Up Branch home signal at danger against the cattle empties, waiting till it was at a stand before he reset the points for the fish train. He also had to telephone Warrington to advise them of the altered sequence of trains and hold the cattle empties at the Up Branch home for the Euston – Blackpool express. At about this time (8.57 p.m.) Signalman Wheeler at Winwick Quay, the next box on the Warrington side, asked on the Down Fast for:

(f) The Down Warrington – Earlestown – Wigan local train, which Signalman Bloor accepted, received the 'Train Entering Section' signal for, and then forgot about, either because he was on the telephone to Warrington or on the telephone to Earlestown about:

(g) An Up goods train off the branch known as the "sugar" train, and:

(h) The 9 p.m. Up motor train from St Helens via Earlestown to Warrington, both of which trains had to follow the cattle empties off the branch.

The local train came to a stand at the Winwick Junction Down Fast home signal at 9.06 p.m. and Driver Hope sent Fireman Hayes to the box, having whistled twice for the signal. Meanwhile Signalman Wheeler at Winwick Quay, have been advised of the approach of the Blackpool express, gave the 'Call Attention' signal to Winwick Junction, intending to follow it up with 'Shunt train for following train to pass.' Hearing this, Signalman Bloor, forgetting about the local train and thinking that he had failed to clear out the Llandudno express, unpegged his block instrument, gave the 'Train out of Section' signal to his colleague and accepted the Blackpool train. He then offered that train to Golborne Junction, got it accepted and cleared his signals for it.

Assisting Signalman Bloor was Booking Boy E. Derbyshire, who told the Board of Trade inquiry into the accident that after booking the acceptance of the Down local passenger train from Winwick Quay he had been on the telephone regarding alterations to the running of excursion trains, then he had gone to the end of the box to add the alterations to the weekly traffic notices. At the moment the signalman had been at the north end of the box, watching the approach of the cattle empties and fish trains. Signalman Bloor had then remarked, "Goodness, I have not given 2-1 yet," had given the 'Out of Section' to Winwick Quay and accepted the Blackpool express. Booking Boy Derbyshire said that he had then assumed that the local train must have passed while he was on the telephone and had forthwith added the missing times in the train register book, guessing what they must have been.

Upon seeing the home signal clear for the Down Main rather than the Down Branch

Driver Hope said that he had whistled twice. The signal not being altered he had then begun to move slowly towards the box and was moving at walking pace when his train was run into.

The rear coach of the local train was demolished as a result of the collision and its middle coach was badly damaged, while the sudden stoppage caused the two leading coaches of the express to telescope one another. Signalman Bloor took immediate action after the collision to block all lines and call for assistance. Doctors and nurses from the nearby Winwick Mental Hospital arrived in a very short time and ambulance trains were sent from both Warrington and Wigan.

The Ministry of Transport inspecting officer, Colonel A.C. Trench, stated that responsibility for the accident must rest with Signalman Bloor, as the latter had frankly admitted. His failure to remember the local train could only be attributed to a mental lapse and his sending of the 'Train out of Section' siganal without verification was a most serious breach of the basic principles of Block Working. He had been much pressed at the time of the accident though he did not suggest that the work of the box was unduly heavy. He said that the evenings were the busiest time of the day. In his favour he had admitted his responsibility and had kept his head after the accident.

It was unfortunate that Booking Boy Derbyshire had not reminded Signalman Bloor that his entries for the local train were incomplete or asked him what time it had passed. It was, in the inspecting officer's words, 'desirable for the Company to ensure that all booking lads should realise that if, owing to any interruption, they have to make entries other than from personal knowledge, they should make quite certain of their substantial accuracy by inquiry from the signalman.'

Turning to additional factors which might have mitigated the effects of the accident, Colonel Trench stated that he did not consider that Driver Steele, running as he was under clear signals, could have seen the tail lamp of the local train soon enough to have reduced speed significantly, and he did not appear to have any responsibility for the accident. Nor, given that the train was running in clear weather on open track and clear signals, could Fireman Partridge, who had his own duties to perform, be criticised for not keeping a look out from his side of the engine.

From the entries in the train register books there were about seven minutes between the times that the local and express trains passed Winwick Quay box and the collision. Of these three minutes would have been consumed in the slow running of the local train to the Winwick Junction home signal. After obtaining a lamp Fireman Hayes would have taken over three of the remaining four minutes to walk the 180 yards from the home signal to the box in the dark at two miles an hour. There did not, therefore, appear to have been any unacceptable delay in the carrying out of Rule 55 (regarding a member of the train crew going to a signal box to remind the signalman of the presence of a train stopped at his signals).

It seemed likely that Driver Hope had, as he maintained, whistled when he was stopped at the home signal but that the preoccupied Signalman Bloor had not noticed it.

The provision of a track circuit in the rear of the Down Fast home signal, locking the block instrument, would have prevented the accident. The Company had considered providing one but had decided that other places had more priority in view of the good view from the box and the infrequency of trains being stopped at the signal in question. It was, in any case, intended to include track circuiting in the proposed renewal of the signal box.

Other than a reminder to Companies to see that signal box lads made accurate entries in their train register the inspecting officer had no recommendations to make.

The question of the use of all-steel rolling stock had, Colonel Trench continued, been raised in connection with the accident. For some years the standard practice of all the main line British companies had been to construct all new passenger stock with heavy steel underframes with wooden or steel and wooden bodies. Their general view, recently confirmed after the Lagny disaster in France, was that this was the most satisfactory solution for British conditions as regards public safety. Instead work should be concentrated on the conversion to electrical lighting and the construction of frames, buffers, couplings and vestibules to resist telescoping. The rear seven coaches of the express, which had been eqipped with shock-absorbing buffers, had been little damaged. [271]

At the inquest into the accident Signalman Bloor said that he had been a signalman for 33 years and at Winwick Junction for 21½ years. He had never had an accident before. There were about 310 trains daily passing his box, of which 173 were passenger trains.

Describing the events leading up to the collision Signal Bluer {the newspaper spelling of signalman's surname differs from that in Ministry of Transport report} related, 'There were 10 trains in movement going on about that time. The 7.35 from Crewe to St Helens was signalled at 8.48 p.m. He had another train at 8.55 or 8.56. He accepted the Crewe train. While that train was passing, the 7.55 from Llandudno to Manchester was signalled and he was offered a motor train from Earlestown. He had the fact on his mind that the 5.20 from Euston was due about 9.8. By going to the end of his box he could see about three-quarters of a mile down the line. He was most anxious to get a cattle train on to the side lines. "But somehow," he said, "I seemed glued on the Wigan side."

'He continued: - "Then the signal bell rang, and I knew the train from Euston was coming. I went to the instrument board and gave the call clear." Passing his hand across his forehead Bluer said, "I completely forgot about the 8.55 from Warrington. I saw the fireman coming up the line. The only train standing in Winwick Junction was that from St Helens, the slow down. I turned away for a moment and then I turned again, and I could see headlights of the Warrington train. I knew something was wrong and it struck me that the fireman was coming to the box to see what was the matter."

'The CORONER. - What was it you omitted to do?

'Bluer. - I omitted to signal the 8.55 from Warrington and take off the signal.'

The coroner's summing up was favourable to Signalman Bloor and it took the jury a mere three minutes to return a verdict of "Death by misadventure."' [272]

An international express from Vienna ran into the rear of a Gdynia – Cracow express train which was stopped by signals outside of the station at Krzeszowice, Poland, near Cracow, on the morning of Tuesday 2nd October 1934. One report alleged that the express had been wrongly admitted into the occupied section, another appeared to suggest that its

---

[271] Board of Trade report

[272] *The Times*, London, England, 3rd October 1934

driver had overrun signals in fog or that the signals themselves were not working properly. The rear carriage of the Gdynia train was not badly damaged, but the two next vehicles were completely wrecked. The locomotive of the colliding train was also considerably damaged, but the driver and fireman escaped. Fourteen passengers were killed and 35 injured. The Krzeszowice stationmaster and the driver of the express were arrested. [273] [274] [275]

Seventeen people were killed outright, six more suffered fatal burns, 56 were seriously injured and 23 slightly hurt when a Leningrad – Moscow express ran into the back of a Leningrad – Tiflis express near Torbino station, USSR, early on the bitterly cold morning of Sunday 6th January 1935. The Tiflis train had slowed down to pass through the station when the Moscow train ran into it at full speed. Several coaches of each train were telescoped and caught fire. The line was blocked for 12 hours.

The chief Soviet prosecutor had gone to Leningrad to take charge of the case and the driver of the Moscow train, his two firemen, the guards of the train, the Torbino stationmaster and two pointsmen were all arrested.

An inquiry was in progress to determine the cause of the collision, which was not immediately clear, though it was theorised that the intense cold might have frozen the points and/or the signals. [276]

A second report stated that a preliminary inquiry had shown that the intense cold had indeed been a factor in the accident because it had led to a broken rail. This had been detected and the Tiflis train stopped. Before it had got away again it had been run into by the Moscow train which had overrun signals, smashing a locomotive and 11 carriages. [277]

At the end of January, the various railwaymen involved in the disaster were sentenced to terms of imprisonment ranging from three to ten years. [278]

The promotion of a probably over-ambitious signalman to a busy main line box which was beyond his capabilities was the principal cause of a collision at Welwyn Garden City, England, on the London and North Eastern Railway, late on the evening of Saturday 15th June 1935. Three trains were involved in the accident; they left London Kings Cross terminus for the north in quick succession. The first was train No. 825, the 10.45 p.m. express direct to Newcastle via Northallerton and Durham. This was followed at 10.53 p.m. by train No. 825A, for Newcastle via Northallerton, West Hartlepool and Sunderland, and then at 10.58 p.m. by train No. 826, a passenger, parcels, and mail train to Leeds. The first train passed through Welwyn safely and continued its journey northwards, the second was unnecessarily checked by signals at Welwyn and was only travelling at about 15 miles an hour when it was overtaken by the third (Leeds) train, travelling under clear signals at about 65 miles an hour. The rear coach of train No. 825A

---

[273] Ibid,

[274] *Le Matin*, Paris, France, 3rd October 1934

[275] *L'Indépendant des Basses Pyrénées*, Pau, France, 4th October 1934

[276] *The Times*, London, England, 9th 1935

[277] Ibid, 10th January 1935

[278] Ibid, 29th January 1935

bore the brunt of the collision and it was stated that no one in it could have survived; the next coach was not that badly damaged. In the parcels train the damage was less but the third coach was partly telescoped by the following three brake vans and the seventh coach was partly telescoped by the following brake van. Thirteen passengers and Guard J. McIntosh were killed, 29 passengers were seriously injured and taken to hospital and 52 others were slightly injured but continued their journeys.

An emergency call was immediately placed by a taxi driver at Welwyn who witnessed the accident. The Welwyn stationmaster was on the scene in five minutes, an ambulance arrived in ten minutes and doctors and nurses also soon arrived. A special train with staff from the Royal Free Hospital, London, arrived at 1.14 a.m. Two small fires broke out in the wreckage of five gas-lit coaches in train No. 826, but neither was attributed to gas; one was due to splinters being ignited by hot cinders and the other to oil waste from a broken axle box. The Welwyn and Hatfield fire brigades attended but the fires had already been extinguished, one being stamped out by the stationmaster and the other put out with an extinguisher.

The line was blocked until 4.30 p.m. on Sunday afternoon but as trains could be diverted via the Hertford Loop there was little disruption to services. [279]

Passengers, railwaymen, and eye-witnesses told of their experiences in the accident. 'Mr V. Castle, of Welwyn Garden City, who was crossing the bridge when the accident happened, said: - "I climbed over the parapet and ran down the line. The scene that met my eyes was a terrible one. Injured people were trapped in the wreckage and were moaning piteously. In some of the compartments the lights were still burning. One man was flung out of the coach in which he had been travelling. I ran to him to give assistance, but he was dead. In another part of the wreckage was a little girl only two or three years old. Her face was injured, and we thought she was dead. We took her to the waiting room on the platform and were about to leave when I noticed her eyelids flickering. I called a doctor, and she was taken away in an ambulance. We also found a girl with her legs trapped in the wreckage. For some time we could not lift the obstructions, but at last this was done, and the girl was carried away on a seat wrenched from one of the carriages."

'Mr W.E. Godlington, manager of the Cherry Tree restaurant, said: - "We heard a crash above the sound of the dance music which was being played. I went along the line and found the bodies of a guard and a girl. My staff ran down to give what help they could, and we brought back some of the injured to the restaurant. It reminded me of War-time in France."

'Mr G.S. Lindgren, prospective Labour candidate for the Hitchin Division, who is in the railway service and who took part in the rescue work, said: - "We turned the waiting-rooms into temporary dressing stations for the injured. Where we could we rushed injured persons off to the hospital, and others we treated on the spot. The heroism of the doctors was amazing. One doctor from Hertford, who was rushed to the scene, worked throughout the night beside the wreckage attending to the injured and calmly giving directions to the rescuers how they should extricate others. It was terrible work getting some of the people out, and masses of woodwork had to be sawn through before they could be freed."

[279] Ministry of Transport report

'Miss Violet Miles, a nurse, of Sunderland, had a remarkable escape from death, and she afterwards worked tending the injured. "A few minutes before the crash," she said, "I left my place, as other people in the carriage wanted to sleep. To make room, I moved further along the train. This action saved my life, for the coach in which I had been travelling was wrecked with heavy loss of life." Miss Miles, although badly shaken and suffering from shock, immediately went to the assistance of others, and worked continuously throughout the night. During yesterday morning she collapsed and herself required attention. Those who were working with her described Miss Miles as "a real heroine" and "absolutely tireless." Miss Miles lost all her money and luggage in the disaster.' [280]

Train No. 825A consisted of 4-4-2 (*Atlantic*) engine No. 4441, a brake van, a brake composite carriage, a third-class carriage, a brake composite carriage, and seven third-class carriages, and was in charge of Guard J. McIntosh with Driver R.E. Morris and Fireman G. Glenn. Train No. 826 consisted of *K3* class 2-6-0 engine No. 4009, a brake van, a third-class carriage, a first- and third-class composite carriage, three brake vans, a brake composite carriage and four brake vans, in charge of Guards W.H. Wisbey, G. P. Mountstephen and George Gale (riding in the first, fifth and tenth vans respectively) with Driver Charles Barnes and Fireman F.L. Jones.

There were six lines through Welwyn, four main lines plus the single-line Hertford and Luton branches. The station immediately south of Welwyn was Hatfield. Hatfield No. 1 box, at the south (Kings Cross) end of the station, signalled all the running lines but at the north end of the station there were two boxes, No. 2 box dealt with the Up line and the Hertford branch and No. 3 box with the Down line and the Luton branch. Along the line though, at Welwyn, there was only one box of 65 levers (20 spare) working all six lines. It had eight block instruments and no less than eleven block bells (because the Up and Down Fast and Slow lines were worked by different Hatfield boxes and so each line needed a separate block bell while the Down Goods and Up Slow lines to and from Welwyn North box also had separate bells.)

As already stated, train No. 825 passed through Welwyn safely (at 11.20 p.m.). It is relevant to the accident only in that it added to the confusion of the Welwyn signalman.

Driver Morris of train No. 825A told the Ministry of Transport inquiry that he had had a clear run until he had found the Welwyn Down distant at caution. He had reduced speed, he said, until he saw the home signal clear and had increased his speed to about 25 miles an hour and was within 100 yards of the Welwyn Down advanced starter when his train was run into. His account was not very accurate here, Lieut. Colonel A.H.L. Mount the inspecting officer, decided, as the collision must have occurred about 580 yards short of the advanced starting signal, as stated by Fireman Glenn. The consensus of other witnesses, who were railwaymen, was also that the speed of the train was rather lower at the time of the collision.

Driver Barnes of train No. 826 stated that he had been checked by signals at Wood Green, five miles from Kings Cross, but not afterwards, and had passed through Hatfield at 60 to 65 miles an hour. He said that the Welwyn Down distant had been clear, but his evidence as to where he saw the home signal at danger, applied the brakes and saw the tail lamp of the train in front was a little indefinite. He thought he had begun to brake 600

[280] *The Times*, London, England, 17th June 1935

yards short of the home signal. Fireman Jones was of the opinion that they had been much closer to the home signal – which he had also seen was at 'danger' – and had shouted, "On." Guards Wisbey, Mountstephen and Gale stated that they had not noticed any brake application before the collision.

Unusually for an accident report by this date the evidence of several witness was reproduced in full, beginning with that of Signalman A. Crowe in Hatfield No. 3 box regarding the signalling of the three trains. He stated:

"At 11.11 p.m. I was offered the 'Is Line Clear' signal by Hatfield No. 1 for the first part (No. 825) of the 10.45 p.m., express ex Kings Cross and I accepted it at 'Line Clear.' At the same time I offered the train to Welwyn Garden City, and this was also accepted at 'Line Clear.' I received the 'Train Entering Section' signal at 11.17 p.m. The train passed my box at about the same time and I sent the 'Train Entering Section' signal forward at 11.17 p.m. and received 'Out of Section' from Welwyn Garden City at 11.20 p.m. At 11.17 p.m. I was offered the 'Is Line Clear' signal by Hatfield No. 1 for the second part (No. 825A) of the 10.45 p.m. express ex Kings Cross and I accepted it at 'Line Clear' at the same time. I offered the train forward to Welwyn Garden City at 11.20 p.m. and this signal was accepted. I received the 'Train Entering Section' signal at 11.21 p.m. The train passed me at the same time, and I sent the 'Train Entering Section' signal to Welwyn Garden City and the 'Train out of Section' signal was sent to Hatfield No. 1 at the same time. I received the 'Train out of Section' signal for this train from Welwyn Garden City at 11.23 p.m. At 11.21 p.m. I was offered the 'Is Line Clear' signal from Hatfield No. 1 for train (No. 826) the 10.50 p.m. express passenger, parcels, and mail train and accepted it. I offered the train to Welwyn Garden City at 11.23 p.m., and this was accepted. I received the 'Train Entering Section' signal at 11.25 p.m. The train passed me immediately afterwards and I sent the 'Train Entering Section' signal to Welwyn Garden City and the 'Train out of Section' signal to Hatfield No. 1 at 11.25 p.m. At 11.29 p.m. I received the 'Obstruction Danger' signal (six bells) from Welwyn Garden City.

"From the time I took duty at 10 p.m. until 11.23 p.m. I had no telephonic conversation with Signalman Howes of Welwyn Garden City. At 11.23 p.m., however, when I received the 'Train out of Section' signal for the second part of the 10.45 p.m. express ex Kings Cross and I also received 'Line Clear' for the 10.50 p.m. express passenger, parcels, and mail train, I rang up Signalman Howes of Welwyn Garden City and said to him, 'Is that out?' (Meaning 'Out of Section' for the second part of the 10.45 p.m. express ex Kings). Howes said 'Yes,' and I said 'Right.' At 11.29 p.m., as previously stated, I received the 'Obstruction Danger' signal from Welwyn Garden City. I called up Signalman Howes and asked him what was the matter and he said, '826 has run into the rear of 825.' I said, 'Why, you gave me out for him at 11.23,' and he made no answer; I distinctly heard the receiver put up on the telephone hook."

Signalman Crowe stated that he was sure of the time (11.23 p.m.) he had stated for the second section of the Newcastle train (No. 825A), adding that as he only had to deal with Down trains he was never so pressed as to have to delay his booking. He said that he had found Signalman Howes satisfactory to work with and that they rarely spoke on the telephone. He referred to Howes as "a quiet man who gets on with his work."

Signalman Crowe continued:

"When I received 'Out of Section' at 11.23 p.m. it occurred to me that it was rather smart, being only about two minutes. The train had not been checked in any and was travelling at normal speed, viz., 60 to 65 m.p.h. Although two minutes is nothing unusual, e.g. train No. 823 (which preceded No. 825) was only two minutes between the 'Entering Section' and 'Out of Section' signals, No. 825A was a much heavier train, and, as the preceding train (No. 825) was only a short distance in front, I thought probably that it (No. 825A) would get a distant check at Welwyn Garden City and for that reason I was more or less surprised that it was only two minutes. I therefore called up the signalman at Welwyn Garden City and said, 'Is that out Fred?' And he replied, 'Yes,' and I had then no hesitation and pulled off my signals for No. 826. I had already had the train accepted before I initiated this telephone call.

"It would have been contrary to the usual practice to initiate a telephone call in regard to No. 825A in between the receipt of the 'Out of Section' signal and the acceptance of the following train, No. 826, (which followed immediately). My initiation of this telephone call was a very rare occurrence. Perhaps it does not happen once a year. I was not at all suspicious and I had no hesitation whatever. What doubt existed at the time was entirely dispelled at once by Signalman Howes' reply. It did not occur to me to mention the number of the train specifically, as I went straight from the bell to the telephone, and there was no lapse of time, so my enquiry was obviously about the preceding train, and as there was no hesitation in Howes' reply, I did not hesitate to pull off.

"With regard to queries relating to the 'Out of Section' signal, I do not remember ever asking the man in advance to confirm an 'Out of Section' signal with regard to an express, though possibly I have done so for a goods or empties train, which may have been a long time in section owing to being short of steam, etc.

"I have no hesitation in saying that I received acceptance from Welwyn Garden City for No. 826, and I know from my experience I could not have made a mistake. I am quite satisfied that the train did not run through my signals but that he had a clear road and was not checked at all."

Signalman Crowe said that this had been the first time he had had to telephone Signalman Howes regarding block working. He agreed that a more precise phraseology would have been better. He was sure that Howes had referred to trains Nos 825 and 826.

Signalman H. Broughton in Hatfield No. 2 box spoke of accepting up passenger and goods trains Nos. 830 and 736 from Welwyn Garden City, of clearing the two trains back to there, of Welwyn Garden City then offering him another up goods train, No. 787, at 11.16 p.m. of his refusing it until 11.18 p.m. and of receiving the 'Train Entering Section' signal for it at 11.23 from Welwyn. This train passed Hatfield at 11.32. The times for the last train in the Hatfield No. 2 and Welwyn Garden City train register books agreed exactly.

Signalman Horace Ball at Welwyn North, the next box down the line, stated:

"At 11.18 p.m. I was offered the 'Is Line Clear' signal for the first part of the 10.45 p.m. express from Kings Cross (No. 825) and I accepted it at 'Line Clear.' I offered the train forward to Woolmer Green at the same time and it was accepted. I received 'Train Entering Section' for this train at 11.20 p.m., and it passed my box at 11.22 p.m. I sent Out of Section' signal to Welwyn Garden City at the same time, and 'Train Entering Section' to Woolmer Green. I received 'Out of Section' signal for this train from Woolmer Green at 11.23 p.m. At 11.22 p.m. I was offered the 'Is Line Clear' signal from

Welwyn Garden City for the second part of the 11.45 p.m. express from Kings Cross (No. 825A) and I accepted it at the same time. I offered the train forward to Woolmer Green at 11.23 p.m., and it was duly accepted. I received 'Train Entering Section' for the train from Welwyn Garden City at 11.27 p.m. At 11.29 p.m. I received the 'Obstruction Danger' signal, six bells, for the Up Main line. At 12 midnight I received the 'Obstruction Danger' signal on the Up Slow line."

Signalman Ball testified that he had received both the 'Out of Section' signal for No. 787 Up goods train and the 'Train Entering Section' for Down train (No. 825A) at 11.27 p.m., he could not remember which came first. Signalman Howes had seemed to be working normally up until the time of the accident. Immediately after Welwyn Garden City had sent him the 'Obstruction Danger' signal Howes had rung him and said, "The News has run into the second portion of No. 825…Oh, it's terrible!"

Signalman Frederick Howes at Welwyn Garden City had taken over there on 11th May, after five weeks learning the box. He had joined the railway in June 1908 as a Lad Learner, progressed to Telephone Lad, Porter, Signal Learner, and then Signalman (in July 1912), first at Navenby, thence Kirton, Ranskill and Doncaster. When he had applied for the vacancy at Welwyn Garden City he had been only fourth in seniority but two of the applicants above him withdrew their applications and the third was rejected because he was only a class 4 signalman and the jump to the class 1 box at Welwyn was considered too big. So Howes got the post.

The Ministry of Transport report went into Howes' training in great detail. Howes himself stated, "I had no difficulty in learning {Welwyn Garden City}, but the traffic in this district was quite strange to me after what I had been used to (at Doncaster A, which is a permissive box, except for express trains on the Up Main), and that is why I took five weeks to learn."

Mr V.M. Barrington-Ward, Western Section, Southern Area, Superintendent, stated that Howes' record was perfectly satisfactory and that quite good reports had been received from Doncaster about him.

Doncaster District Inspectors A. Shores and A. Chamberlain also spoke well of Howes. The former had watched Howes at work at Doncaster and stated, "During the time Howes was at A box I had no reason to think that he was in any way unfitted for the responsible position which he held." District Inspector Chamberlain agreed. He had examined Howes both for Doncaster A and for Ranskill. Chamberlain stated, "During the time Howes was under my notice he did not give cause for any complaint with regard to unsatisfactory signalling as far as I know; and he always struck me as being a very careful signalman and one to consider everything well before making a move."

Signalman F.C. Birch, who was one of the men who trained Howes at Welwyn Garden City, considered that Howes had understood his duties 'fairly well' and had been entirely fitted to take over. He knew the Block Regulations very well. Signalman Birch added that, "Howes was not very talkative; he would remain very quiet and probably go for an hour without speaking. I was in contact with him almost daily for about 10 weeks and did not know whether he was a member of the Union or not."

Signal Inspector Hook spoke of his examination of Signalman Howes on 11th May. He said that starting at 10.10 a.m. he had watched Howes work the box for an hour and a half, then taken him outside to show him the location of the points and signals, then examined him on Double Line Block Working, Single Line Block Working and Special Regulations. At 12.15 p.m., satisfied, he had passed Howes to work the box. He had been quite satisfied that Howes was competent, or he would not have passed him. He thought Howes steady and rather slow. Hook admitted that he had never refused to pass a signalman for a box, though sometimes it had taken two or three goes.

Signal Inspector Hook was clearly not all that impressed with Howes, stating that he did not appear to be the brightest and was not at all forthcoming, everything having to be dragged out of him. Personally he (Hook) would have preferred a more local man, but the appointments were not decided by him, but by the District Superintendent, on seniority. He also considered that men should be chosen on the grounds of merit rather than seniority. He then went on to say, somewhat illogically, that, "I had every confidence that Signalman Howes would carry on the job satisfactorily, and I felt that he was going to turn out to be a first class signalman."

Another witness, Mr A.G. Rickett, a former Signalling Superintendent, who had known Howes for some years, stated, "I did not think he was a very brilliant signalman, and I think the same today. I would not say he does not know his Rules and Regulations, but at the time I knew him at Kirton from 1922 to 1926 I would not have expected him to have taken over a Class 1 box, but since the new method of dealing with promotion, men have got to higher positions than one would expect."

The report then moved on to an event which happened a week after Howes had taken over, which it appears was indirectly a cause of the disaster. An Up passenger train was approaching, for which Howes had not got the road ahead. When he thought that the express was pulling up he had lowered the outer home signal to bring the train up to the inner home at danger, but he did so prematurely, and the driver then overran the inner home. No harm was done, and he and the driver agreed to hush the matter up. Unfortunately, Stationmaster Hodson had witnessed what had happened and spoke to Signalman Howes, who was then asked to send in a report (what would later, and possibly already was in 1935, be called a Form 1, or a 'Please Explain', asking the offender for a written explanation of his or her conduct). This Howes refused to do on the grounds that he did not consider it necessary as he had already explained matters to the stationmaster. This of course was the signalman jumping from the frying pan into the fire, as the well-known idiom goes, for a refusal to submit a report when asked to do so by a superior was a disciplinary matter, and he now had to go to Kings Cross on 4th June for an interview with the District Superintendent, Mr. F. Warriner. There he was told that the case would have to be referred to Superintendent Barrington Ward, and that he would hear further about the matter. On Thursday 13th Mr Warriner received from Superintendent Barrington Ward confirmation that Signalman Howes was to receive a severe reprimand (subject to a written appeal submitted within seven days) which would be entered upon his record and that were there any further instances of his disobeying block regulations and refusing to obey his stationmaster's instructions he would be liable to dismissal. This confirmation was forwarded to Stationmaster Hodson on the 15th and left in the box for Signalman Howes to see when he came on duty at 10 p.m.

Signalman Birch, whom Howes relieved on the fateful evening, said that they had been together for about seven minutes before Howes took over. Howes had read out his reprimand to him and he (Birch) had commented, "Don't worry about that. They can't hang you for it." Howes had not commented about the reprimand and had not appeared upset, but because of his quiet, reserved manner it was difficult to see when he was upset anyway.

Signalman Howes himself stated, "When I came on duty I received a caution notice in regard to a disciplinary irregularity. This was a surprise to me, but it did not affect me very much for more than a few minutes, although I may have thought about it several times I received it... I was hurt to think I had to receive the form. Though I had had the interview on 4th June, I did not know the form was coming and it came as a surprise to me. It was a surprise because I though the case was finished when I went to London, but the form conveyed to me exactly what was said by Mr Warriner, District Superintendent, at the interview."

As for the signalling of the three trains, Howes said, "I knew No. 825 in two portions as we had a circular to that effect and Signalman Birch reminded me of this. I was offered 825 at 11.11 p.m., and I received the 'Train Entering Section' signal at 11.17 p.m.; I then offered it forward and it was accepted. The train passed Welwyn Garden City at 11.20 p.m., when I gave the 'Train Entering Section' signal forward and the 'Train out of Section' signal to the rear, and I placed my signals to danger as the train passed them. I received 'Train out of Section' from Welwyn North at 11.22 p.m. I received the 'Attention' signal and four beats with regard to No. 825A at 11.20 p.m. and I accepted the train. I received 'Train Entering Section' signal for it at 11.22 p.m., at which time I offered it forward to Welwyn North and got it accepted, but I did not pull off immediately for the train as there was a telephone call from the station and I answered it at 11.22 p.m. To the best of my knowledge it was the Porter's voice; he asked me if I had asked Hatfield about a parcel that was left in the Hertford train, and I informed him that I had not been able to get the Inspector's office at Hatfield. This was the second telephone message I had had in regard to this parcel, the first telephone message having been some time previous, but I cannot remember the time, probably 10 minutes, this message also having been from the same man. I did not think this telephone conversation was taking me very long, and during the time I was replying I had another telephone call on the other circuit, namely Hatfield and Knebworth circuit, that was from the signalman at Hatfield No. 3, and he asked me if I had given 'Out.' I replied 'Yes' (I understood Signalman Crowe's reference to 'Out of Section' signal to be to 825 and that is why I replied 'Yes'). I thereupon immediately went and pulled my signals off for No. 825A. Perhaps a few seconds after I had pulled off he struck the track circuit in the rear of my home signal and the train was passing. When the train struck the track I placed the distant signal at danger, and when he had passed over the track and it had returned to 'Clear' I placed the home signal at danger, and as I was watching the train passing me I heard a crash just against the box. I did not know that the following train was approaching, and I did not actually see the collision. I was looking out at 825A passing the box to see his tail lamp as he passed me, and as near as I can say the speed was about 25 miles an hour. I am afraid that I am not a very good judge of speed. As the train was passing me I did not send 'Train Entering Section' signal because I was at the window looking for the tail lamp, nor did I transmit the 'Out of Section' signal. When the accident occurred I immediately sent the 'Obstruction Danger' signal in both directions.

"I had the conversation with Signalman Crowe of Hatfield No. 3 box. I have met Crowe, and this was some weeks ago. I got on all right with him and found him quite satisfactory. I do not know the name of the signalman at Welwyn North, but I know his Christian name is Horace, and I have been working quite satisfactorily with him.

"With regard to the Up road, about this time I do not think there were any trains. I remember the Up stock train on the Luton line, which passed me at 11.25 p.m., about the time of the accident. If my memory does not fail me, this train passed me just before the crash. There was also another Up train on the Up Main, No. 787, which passed me at 11.23 p.m., this would pass me two minutes before the Luton train, but I cannot remember it going by. I know from my register that it was refused, but I cannot remember it. I make it a point of making up my book immediately after making my signals, as soon as possible.

"The 'Train out of Section' for No. 825A is not recorded in my book because I did not transmit the signal. I transmitted the 'Out of Section' signal for 825 and my conversation with Signalman Crowe referred to that signal. The alteration made in my Block Book with regard to 'Entering Section' signal 11.22 was made when I entered the acceptance of the train at 11.23. I put in 23 in the first place because in the position where I was standing I took it for 3 on the clock, but when I got on the back against the clock I realised it was nearer 22 than 23.

"I remember No. 787 up passing me but I cannot say at what speed it was travelling. I cannot explain the fact that I recorded 11.23 p.m. as the 'Out of Section' signal for the Up train No. 827 and the fact that the Welwyn North signalman records this as 11.27 p.m. I feel sure I should book it after I had transmitted the signal. I do not think I would have transmitted the 'Out of Section' signal at 11.23 p.m. to Hatfield instead of to Welwyn North. I have never found myself doing such a thing. I made no record with regard to the Parcels train, and I do not remember that it was offered to me, nor have I any knowledge with regard to any signal for this train.

"The only way I can account for the accident is that I did not know anything about 826. I have never heard of a signalman giving 'Out of Section' on the wrong bell and I do not think I have ever done it myself. I went to the book and booked the signals as I got them and at this time I was able to as I was not over pressed with work."

Howes though that only a minute separated his acceptance of train No. 825A and his conversation with Signalman Crowe. He said he had not heard any whistling and the box windows were closed – his usual practice so as better to hear the bells and instruments. He continued:

"...I have no idea how 826 got into the section, as I know nothing whatever about it. I am quite sure I did not overlook 825A and I knew all about this train but nothing about 826.

"There are 11 block bells in my box, and I have at times not been quite certain which bell was ringing, and I have usually waited for the second bell. I did no booking at Doncaster and found it rather difficult after taking over at Welwyn Garden City. I have never found the telephone messages too much, as I let them wait, neither did they interfere with the signalling part of my job. I answered the telephone before the bell for 825A; I expect I happened to be nearer the telephone than the block instrument. I did not realise No. 825A was a long time in the section...

"When the collision occurred it was a complete surprise to me; I was looking out of the window at the time for the tail lamp of 825A, I had put the home signal back when the track was cleared."

"There are several telephone bells in my box, but I do not think they would interfere with my signalling. The work at Welwyn Garden City is heavier than I have been used to, but not a great deal heavier than at Doncaster. I did not feel that I wanted any assistance, but I found a little difficulty in the booking…

"I have no knowledge whatever of having taken No. 826. I do not suggest that Signalman Crowe is making a mistake, but if I did accept it, I have no recollection of doing so… The conclusion I have come to is that I knew nothing whatever of No. 826, neither do I think I belled on the wrong instrument; I think I should have noticed the difference in the tone of the two bells."

Leading Porter A. S. Jakes at Welwyn Garden City explained that the two telephone conversations with Signalman Howes had related to a parcel which had been left on the Hertford train, the first to ask Hatfield to look for it and the second to enquire if the signalman had heard anything from Hartfield.

Concluding his report, Lieut. Colonel Mount stated that the accident stated that the accident had arisen from the improper entry of train No. 826 into the Hatfield No. 3 – Welwyn Garden City block section when it was still occupied by train 825A as a result of a lapse of a most serious nature by Signalman Howes at Welwyn Garden City, though the exact details must remain uncertain.

The crucial time, the inspecting officer stated, was 11.23 p.m., when Signalman Crowe at Hatfield No. 3 received the 'Out of Section' signal for train No. 825A and had train No. 826 accepted by Welwyn Garden City. Although Signalman Howes denied clearing out train No. 825A and accepting train No. 826, thereby implying that Signalman Crowe had wrongly admitted train No. 826 into the section, the subsequent telephone enquiry by Crowe was admitted by Howes, and enquiry which would not have been made had Howes not cleared back train No. 825A and Crowe not been unhappy about how quickly the train had passed through the section. It was, however, to be regretted that Signalman Crowe, an experienced and entirely reliable signalman, had not phrased his question more precisely, and had so easily allowed himself to be reassured by Howes' reply.

The Company's officers were of the opinion that at 11.23 p.m. Howes had mistakenly transmitted the 'Out of Section' signal to Hatfield No. 3 on the Down main line block for up train No. 787 instead of sending it to Signalman Ball at Welwyn North, and then accepted train No. 826 from Signalman Crowe under the impression that it was another Up train. The alternative, which the inspecting officer thought more likely, was that at a moment of pressure with having to deal with a number of trains in quick succession, combined with worrying about his severe reprimand and having to deal with the telephone call from the porter regarding the parcel, Howes might have just forgotten all about train No. 825A and seeing that his block instrument still showed 'Train Entered Section' have assumed that he had omitted to give the 'Out of Section' signal for it to Hatfield No. 3.

Noting the similar remarks from Stationmaster Hodson, Signalman Birch and Inspector Hook regarding the character of Signalman Howes, the inspecting officer concluded that he was 'obviously a man of quite unusual temperament, of very reserved demeanour, and of marked hesitancy of expression,' and that, 'I doubt also whether he has the power and habit of rapid thought, so essential in such a responsible position on such a high speed main line with comparatively short sections on either side. In addition, there is the operation of two single lines, one of which (Luton) carries considerable traffic.'

Commenting on the work of Welwyn Garden City box the inspecting officer noted that on the day before the accident Signalman Birch had dealt with 21 trains in an hour and that the box had 45 levers, 11 block bells, ten block instruments, five telegraph instruments, the ordinary telephone for traffic control purposes and a telephone switchboard with five trembler bells. He did not think that the marks taken to rate the box in December 1934 fully reflected the amount of work there at busy times.

The question of the selection and fitness of Signalman Howes for the post must be considered, Lieut. Colonel Mount continued. His promotion had not been unusually rapid, but the time that he had taken to learn Welwyn Garden City box 'indicated perhaps that he was below average.' Inspector Hook, himself a former signalman, would have preferred another man for the job and should perhaps have obtained a second opinion from Mr Warriner before passing Howes.

There could be no doubt, the inspecting officer stated, that the enginemen of train No. 826 had been on the alert. In the rare case of a mistake by a signalman resulting in two trains being in the same block section the tail lamp of the first train provided a second line of defence, and he regretted the decision of the Company, taken back in 1924, to dispense with side lights on passenger trains, as a pattern of lights was more recognisable and conspicuous than a single one.

The inspecting officer recommended the provision of a control on the block instrument such that once 'Line Clear' had been given for a train on the block instrument a second 'Line Clear' could not be given to the box in the rear until the first train had occupied and cleared a track circuit in the rear of the home signal. (On the day of the accident train No. 825A had not reached the 200 yard berth track circuit in the rear of the Welwyn down home signal before it was wrongly cleared out and train No. 826 accepted.) This recommendation was adopted and became known as the Welwyn Control.

The block instruments might be coloured or otherwise treated to better differentiate them and the tone of the block bells accentuated, or indicator shuttles fitted. (The author would comment here that, although he is not tone deaf, he had enough difficulty working a box with four block bells, let alone 11.)

As in the Kings Langley accident (a rare triple collision – see Miscellaneous Accidents chapter) a misunderstanding in block working could have been cleared up with more precise phraseology on the telephone. The use of the telephone for block working should be definitely prohibited unless exact phraseology was used, and the messages recorded in the train register.

In view of the distraction caused by the porter's telephone calls to the Welwyn box either a telephone and booking boy should be provided or the telephone switchboard should be moved to the booking office.

As in the accident at Stoke Works Junction on the London Midland and Scottish Railway in 1924 the disciplinary notice which Signalman Howes received at the start of his shift had affected his efficiency. A confidential company brochure regarding disciplinary matters advised that disciplinary notices should only be given to signalmen in particular at the end of their shifts lest the stress of the notice affect their working, but it was pleaded that as this was merely a confirmation of a previous decision it was not incumbent upon the stationmaster to leave matters until the Sunday morning or Monday. In future, as a matter of principle, the inspecting officer recommended that all such disciplinary notices, even if only confirmatory, should be given to signalmen at the end of their shifts.

Although he had a good record it seemed clear that Howes was not really fit to shoulder the work and responsibility of the box at Welwyn Garden City. However satisfactory a man's reports it seemed essential that he should be assessed by a responsible officer such as the district superintendent or his assistant.

The introduction of well organised classes and/or correspondence courses for signalmen was recommended 'to have a mind-broadening effect and be beneficial in the technical education of signalmen.'

The outstanding feature in the behaviour of the rolling stock in the collision was the manner in which the Buckeye couplings of train No. 825A had prevented telescoping of that train while the intermediate vehicles of train No. 826, which were not fitted with such couplings, were wrecked. All ex-Great Northern Railway and East Coast Joint corridor stock had been fitted with Buckeye couplings since 1896 but a proportion of the stock of other constituent companies remained unfitted and still equipped with gas lighting. An acceleration of the refitting of such stock was recommended. [281]

At the inquest on the victims of the accident it took the jury only 35 minutes to return a verdict of "Accidental death owing to a temporary lapse of memory on the part of the Welwyn Garden City signalman." [282]

On about Sunday 30th June 1935 a railcar carrying members of a railway inspection committee ran into the rear of a freight train near Omsk, Siberia, USSR. When the motorman of the railcar realised that a collision was imminent he jumped from the railcar to save himself instead of applying the brakes, so that its speed was not reduced at all. Following the impact an explosion occurred, doubtless due to the rupture of the petrol tank, and six people died in the resulting fire. It was announced that 'the driver of the goods train and several other railway officials will be tried for negligence.' As the accident was blamed in news reports on the failure of the motorman of the railcar it seems strange that he was not named as one of the railwaymen standing trial. Possibly he did not survive his leap from a railcar travelling at speed. [283] [284]

A rear collision occurred at São Francisco Xavier station, Brazil, in the suburbs of Rio de Janeiro, on the Estrada de Ferro Central do Brasil on Wednesday 16th October 1935, during the evening rush hour. Interviewed by journalists as to the cause of the accident, Company Director Colonel Mendoça Lima stated that a commission of inquiry had been appointed and it was presumed that the driver of the colliding train had become distracted and had failed to notice that a signal was against him.

[281] Ministry of Transport report
[282] *The Times*, London, England, 6th July 1935
[283] *Gazette de Lausanne*, Lausanne, Switzerland, 1st July 1935
[284] *The Times*, London, England, 13th July 1935

A preliminary inquiry found that train S D 49, had stopped between Engenho Novo and Meyer due to 'lack of pressure,' presumably meaning a failure of the air brakes. The signals at Riachuelo were thus at danger and two following trains were halted in turn at signals behind the first one, the second of them, the 17:52 from Rio to Deodoro, at the platform at São Francisco Xavier, where the collision occurred.

The accident was said to have caused a painful impression in the city. An early report stated that there were eight corpses in the mortuary, some unrecognisable and which had not yet been identified, and there were also 30 injured. The investigation to determine the culpability for the disaster continued. Two of the bodies could be identified as factory worker Manuel Figuelredo Faria and Amelia de Souza, wife of dealer Nilo Moreira. Just prior to the publication of the report three more of the victims of the disaster had died in the Hospital do Prompto Socorro and the report concluded that at present the number of dead was unclear as the information from the Central do Brasil company and the police was contradictory. [285]

A further report to the press the following day stated that the commission investigating the accident had found that the trains and track had been in good condition, and had concluded that the accident was due entirely to the criminal negligence of the driver of train S M 37, the 17:58 from Rio to Paracamby, who had paid no attention to the stop signal at Derby Club, nor to the efforts of the signalman there to stop him, the latter having, like his colleague at São Francisco Xavier, done all that he could to avert the disaster. His warning enabled some of the passengers in the Deodoro train to jump out of it before the collision but as the train was seriously overcrowded only a proportion had time to do so. [286]

The note concluded that the government, in conjunction with the recently granted electrification contracts, had demanded the adoption of steel cars by the railway company, which would protect the lives of passengers in case of accidents, unlike wooden cars.

The number of dead was now put at nine. Two of them, a man, and a woman, were still unidentified. [287]

Fifty-one people died and 52 were injured in a collision at Karymskaya, USSR, near Chita, on Monday 22nd June 1936. Details of the disaster were only revealed a month later when the Soviet authorities announced that 13 railwaymen were on trial for causing it. The leading train had come to a stand on a steep gradient, but the following train was allowed to proceed. A stationmaster was sentenced to death for criminal negligence and eight other employees to between three and ten years' imprisonment. [288] [289]

Chemins de fer de Paris à Orléans et du Midi electric train No. 148 from Bayonne to Avignon via Toulouse and Sète left Lourdes, France, at its usual time of 13:45 on

---

[285] *O Estado de S. Paulo*, São Paulo, Brazil, 18th October 1935
[286] *The Railway Gazette*, London, England, 6th December 1935
[287] *O Estado de S. Paulo*, São Paulo, Brazil, 19th October 1935
[288] *The Railway Gazette*, London, England, Vol 65, pg. 161
[289] *The Daily Herald*, Biloxi, Mississippi, United States, 18th July 1936

Wednesday 23rd September 1936, but shortly afterwards encountered difficulties and by 14:00 it had still not reached the level crossing at Saux, only about three kilometres from Lourdes, when it was run into by omnibus train No. 339. The driver of the omnibus train had braked hard when he had seen the train ahead, but the speed of his train could not be much reduced in the distance available, and a violent collision was inevitable. The accident was witnessed by motorists waiting at the level crossing for the first train to pass. They raised the alarm, and the initial help was given by people in the neighbourhood, but it was a railway police agent who first apprised Lourdes of the collision at 14:25. A short time later ambulances and stretcher bearers arrived at the scene together with the parquet (state prosecutor). Thirteen people died immediately or soon after the accident and the final death toll as was 16, with about 30 injured.

The line was completely blocked but two teams of workmen, together with three powerful cranes, worked to clear it, and by 15:00 the following day trains were running again, with a speed restriction at the site of the accident.

The investigation into the collision began the following afternoon and continued into the evening. It was determined that Emile Sarcia, the driver of the omnibus train and the telephonist/telegraphist at Lourdes, Mme Ricaud, had caused the accident. [290]

Driver Sarcia was interrogated at length, then arrested and taken to Tarbes prison, it being charged that he had disregarded a very well-sighted closed disc situated 150 metres before the curve at Monge, in the vicinity of the accident.

In a statement recorded in the press Driver Sarcia said, "The train that I was driving was waiting in the station at Lourdes for express train No. 148 to clear the section. Then I received the order to depart and left four minutes late on my schedule. Some hundreds of metres short of the level crossing at Adé I found the disc open (clear). At the time my train was travelling at about 60 kilometres an hour. Suddenly as I came out of a curve I saw train No. 148 in front. It was only 50 metres away. It was impossible to stop. I did all I could to pull up, but it was hopeless."

Assistant Driver Claverie of the omnibus train corroborated the statement of his colleague.

The Flaman speed recorder indicated that the omnibus train had disregarded the Adé *disque avertisseur avancé (*equivalent to a distant signal in the UK*)*. He said, "Yes, the disc was closed, but I assumed that the express was still at Adé station, and that was why the disc had not been cleared. That was what I thought. This would mean that I still had 1,200 metres to go." [291]

Driver René Turbo Labar of train No. 148 declared, "I left Lourdes station at 13:45. After passing the sub-station the engine cut out. I tried to remedy the problem, but it kept cutting out again. I then sent my assistant (M. Laurenson) into the engine room to try and fix the problem. The speed of my train dropped markedly until we were only running at 5 or 6 kilometres an hour. I was going to stop to carry out repairs. I thought that I had time when suddenly the omnibus train appeared. Then came the collision, the catastrophe."

M. Delmonte, *juge d'instruction*, then went to the bedside of Mme Ricaud, an employee with 24 years' service with the company who was held in high esteem by her

---

[290] *L'Indépendant des Basses-Pyrénées*, Pau, France, 25th September 1936

[291] *La Dépêche*, Toulouse, France, 25th September 1936

colleagues and superiors, and she at once admitted her guilt. She had been very busy preparing for the departure of the omnibus train and thought that she saw the shutter of the instrument to Adé fall, indicating that the line was clear, whereas in fact it was merely a call from Ossun station regarding a passenger without a ticket. By the time that she had become aware of her terrible mistake it was too late. She was greatly affected and was in a state of prostration such that the magistrate did not order her arrest. M. Boïriot, the Company doctor, was greatly alarmed by her condition.

Mme Caussé, the Saux crossing keeper, asked by a reporter how the accident had happened, said, "I don't know anything, although I was standing in front of the crossing keeper's house. I suddenly heard a loud noise. I turned round and I cried, 'My God! An engine!' It was automatic and instantaneous. My husband, who was inside, came running, and seeing a car driver shouted to him, 'Quick! Quick! Get help! Go and tell Lourdes station.' He shouted the same to a second car driver. Stretcher-bearers, firemen and police soon arrived. Meanwhile railwaymen and uninjured passengers, rushed towards the smashed carriages from which the awful cries of the injured and moans of the dying were coming."

M. André Marestan, a painter from Valence d'Agen, who had been a passenger on the express, also spoke to a reporter, "We left Lourdes at 13:45 on the express. Between Lourdes and Adé, on the curve near the Saux level crossing, where our train was running very slowly, we were suddenly thrown about by a violent shock. It was omnibus train No. 339, which had left Lourdes ten minutes after our train, which overtook us. The collision was a very violent one and the van at the rear of our train was destroyed and pushed into first and second-class composite carriage ahead of it. {Other accounts indicated that a wooden third-class carriage had in fact been immediately in front of the van and was telescoped by it. The composite carriage probably came next.} We do not know yet if the brakesman was killed or if he was able to save himself.

"There was a panic in our carriage. The seats were crushed. The passengers were buried in a mass of wood, metal, and glass. Though I was thrown to the ground I escaped with some slight bruises. I at once went to help those less fortunate together with others who willingly assisted. Some of the injured were very badly hurt; about twenty were lifeless. Cars stopped at the level crossing took some of the injured to hospital in Lourdes. Ambulances arrived an hour and a half later and took the seriously injured.

"A relief train was made up and we were able to leave the accident site at 16:00 and reached Toulouse at 20:24." [292]

A test train was run about a fortnight after the accident in a re-enactment of the circumstances, and it was reported that this seemed to confirm the partial responsibility of Driver Sarcia for the collision.[293]

It has not been possible to trace a record of any court case against either Mme Ricaud or Driver Sarcia.

Tuesday 24th November 1936 saw a serious rear collision at the Granville Avenue station on the North Side Division of the Chicago Rapid Transit Company, usually

---

[292] *La Charente*, Angoulême, France, 26th September 1936
[293] *Le Petit Parisien*, Paris, France, 13th October 1936

known as the L (for elevated), in Chicago, Illinois, United States. At Granville Avenue there were four tracks, electrified with a third rail and running on an embankment raised above the city streets. The tracks were numbered from 1 to 4 from west to east, with tracks 1 and 2 being used by southbound trains and tracks 3 and 4 by northbound trains. A tower at Granville Avenue operated a crossover between tracks 4 and 3. A double-arm lower quadrant semaphore signal controlled movements over the crossover, the top arm, No. 20, applying to movements straight ahead along track 4 and the lower arm, No. 9, applying to movements over the facing crossover to track 3. There were no distant signals preceding the stop signals. Trains of the Chicago North Shore and Milwaukee Railroad also worked over this part of the line. At Granville Avenue northbound L trains were normally diverted via the crossover from track 4 to track 3, while North Shore expresses continued on track 4.

On the evening of the day of the accident northbound L express train R-5, in charge of Conductor Florczak with Motorman George Correll, passed Lawrence Avenue at about 6.10 p.m. and was brought to a stand at the Granville Avenue signals, which were at danger. Shortly after it had come to a stand it was run into by northbound North Shore express No. 725, in charge of Conductor Smedburg and Motorman Van Rensselaer Grooms. The colliding train was only travelling at about 10 miles an hour and the leading six cars of the L train were undamaged and the seventh car was derailed and slightly damaged. The wooden eighth and rear car, however, was demolished and telescoped. None of the cars of the North Shore train was derailed and only its front end was damaged. Eleven passengers were killed or fatally injured, and 57 passengers and Motorman Coombs were injured. [294] One of the passengers hurt lingered until 6th December before succumbing to his injuries. [295]

William Biesel, of Libertyville, Illinois, a passenger in the North Shore train, described the accident. "It was the most tragic and horrible sight I have ever witnessed. I was riding in the front coach of the North Shore with Champ Carry, vice president of the Pullman company. Just before we got to Granville the air whistle of our train let out an ear-piercing blast. Almost at the same moment the lights blinked, and the train began to sway drunkenly. I could feel the wheels grinding on the rails as the brakes were set. I sensed tragedy but could do nothing. I recall a horrible grinding crash, and something scraped the roof above us – I found out later it was the top of the coach we had struck. Splinters of the wooden coach flashed past our window. I thought I saw a body arch through the air off the elevated platform, but I wasn't sure until Mr Carry and I dashed out of our coach.

"It was like the scene of a big explosion. Wreckage was scattered all over. Men and women were scattered all over the tracks. Men and women were lying on the tracks and on the ground beneath us. Some of them were moaning and others were horribly quiet. Sparks were flying up from the third rail where pieces of steel fell across it. One woman was slumped across the third rail. A little wreath of smoke rose up from her dress. She was burning. Mr Carry and I lifted her off the rail, beat out the flames with our hands and carried her to the station platform. On our way to the platform we saw two men who had been thrown out of the el train clear across the adjoining tracks. We took them to the platform too.

---

[294] Interstate Commerce Commission report

[295] *The Hammond Times*, Hammond, Indiana, United States, 15th December 1936

"When I looked back I saw men and women pouring out of the apartment buildings adjoining the track. Some of them carried towels, bedsheets, and bandages. I looked down off the platform and saw four bodies crumpled in the wreckage. The people from the apartments were taking care of them. Four priests had climbed out of the North Shore behind us and were walking calmly into the smoking wreckage to do what they could for the injured and dying.

"We couldn't do much to pry the injured out of the wreckage, so we just had to stand by and listen to them shout and moan. In a few minutes, their cries were drowned out by the fire and ambulance sirens. Then we stood aside. [296]

Motorman Grooms was interviewed by police in Evanston hospital and said, "I was travelling about forty miles an hour, then I saw two red lights on the rear of the Evanston train. The lights were very dim. I put on the brakes and that's the last I remember." Police Captain Martin McCormick described the statement as "unsatisfactory." [297]

No less than seven different investigations into the accident were opened. They were conducted by the Interstate Commerce Commission, the Illinois Commerce Commission, the Cook County coroner, the Chicago police, the city council transportation committee and the L and North Shore railroad companies.

Reporting on the accident the next day newspapers highlighted the fact that a wooden coach was involved and a spokesman for the city council transportation committee asserted that, "antiquated and unsafe cars should be substituted by the new, modern type, now in use throughout the city." Howard Bernard J. Fallon, executive officer for the receivers of the Chicago Rapid Transit Company, pointed out that the replacement of 500 wooden cars would cost at least $15,000,000 and the company could not afford it.

James M. Slattery, chairman of the Illinois Commerce Commission, confessed that the commission was powerless to order the order the replacement of wooden with steel cars, because:

(a) There was no law upon which the commission could make such a ruling, and

(b) The L was under federal court jurisdiction because it was in receivership. [298]

As already mentioned, although the accident did not occur on a steam railroad it was fully investigated by the Interstate Commerce Commission (ICC). The first witness to give evidence at the ICC inquiry was Motorman Correll of the L train. He stated that he had observed signal No. 20 go to the stop position as his train had passed Bryn Mawr, about 4,000 feet from it, had reduced speed 2,000 feet from the signal and had stopped his train 50 feet from it. There had been no difficulty in stopping. His train had been run into almost at once.

Conductor Florczac said that his train had been diverted to track 4 at Lawrence Avenue. At Thorndale Avenue, about 1,600 feet from the Granville Avenue signals, speed had been reduced to 5 miles an hour, at which speed the train had continued to Granville Avenue. His train had been about 15 minutes late.

Rear Guard Mullen of the L train testified that prior to leaving the yard he had placed

---

[296] *Logansport Pharos Tribune*, Logansport, Indiana, United States, 25th November 1936

[297] *South Haven Daily Tribune*, South Haven, Michigan, United States, 25th November 1936

[298] *Logansport Pharos Tribune*, Logansport, Indiana, United States, 25th November 1936

the marker lamps in their proper positions at the rear of the train and had also lit a white lantern which was placed on a chain in the centre of the rear end. He had inspected the markers en route and they had been in good condition and burning brightly.

Motorman Grooms stated that he had tested the air brakes of his train when he had taken it over at Roosevelt Road. Due to heavy traffic it had been 15 minutes late at Wilson Avenue. He had seen the red signal at Granville Avenue when halfway between Lawrence Avenue and there he had begun applying the brakes, gradually building up to 20-pound reduction. He said that he had seen the L train when about 4,000 feet from Granville Avenue (it was actually about 2,600 feet). About 3,000 feet from Granville Avenue the wheels of his train had begun to slide. He had released the brakes, applied power long enough to cause the wheels to revolve, applied sand and then reapplied the brakes, beginning with a 5-pound reduction building up to a 20-pound one. The wheels had again begun to slide, and he had made an emergency brake application when about 500 feet from the train in front. Then, realising that he was not going to stop in time, he had applied sand, given a series of short whistle blasts, reversed his train and retreated into its interior just before the collision. He could not explain why his train had slid unless oil on the rails had made them greasy. The weather had been clear with no visibility problems, but he denied having seen the lights at the rear of the L train. He said he had been in good health, had had ample rest and had had nothing on his mind.

Conductor Smedburg stated that he had been in the smoking compartment at the front of the third car when the brakes had been applied in emergency. Looking out of the window he had seen a red light which he had assumed was the Granville Avenue home signal. Thinking that his train would pull up in time he had withdrawn from the window but had no sooner done so when there were blasts of the whistle. Looking out again he had seen the left marker of the L train and had ushered his passengers to the rear of the car, following them back. He thought that flat spots on the car wheels showed that the train had slid, as his motorman had stated.

Towerman Gadke at Granville Avenue stated that because train R-5 had been late he had permitted a local train to precede it on track 3. This local train had then been stopped by signals at Devon Avenue, two streets further north, and so he had had to hold train R-5 at his tower. He confirmed that it had barely come to a stand when it had been struck by the North Shore train.

Towerman Gallipo at Lawrence Avenue stated that he kept no record of trains during the evening rush hour from 5.10½ to 7.59 p.m. but to the best of his recollection the L train had passed him at 6.10½ and the North Shore train at 6.13. Both of the red markers of the L train had been burning.

The evidence of Engineer Serzow of the L shop department and of Superintendent of Shops and Equipment Daus was that the brakes of train No. 725 had been in good condition. The latter had found sand on the rails after the accident but no sign of oil or grease. Marks on the rails indicated that the North Shore train had slid for 100 feet from the rear of the L train, indicating that the train had been reversed over that distance. General Superintendent of Transportation Hardin of the L gave similar evidence.

General Manager Johnson of the L said that much of the North Division trackage was unprotected by signals, while the rules largely relieved the trainmen from protecting their trains by flagging. As many as 73 trains an hour were operated over one track during the rush hours, full responsibility being placed upon the motormen of following trains to drive so as to avoid collisions. During foggy or stormy weather motormen were relieved from keeping time and required to drive at all times so as to be able to stop within their range of vision. Working under this system only three passengers had been killed in train accidents in 19 years before the present collision, which was also the first to involve a North Shore train since they had begun running over the line in 1919.

A test conducted on 30th November using a 2-car L train stopped at Granville Avenue and an 8-car North Shore train under similar conditions to the day of the accident showed that an emergency brake application made about 2,600 feet south of the L train stopped the North Shore train travelling at 40 miles an hour in 743 feet, while a service brake application of a train travelling at 45 miles an hour made at the same point stopped it in 1,162 feet. It was noted, however, that the green indication of an automatic signal situated 1,000 feet north of Granville Avenue was visible from half to three quarters of a mile south of the L train, directly above the red marker of the latter, and could easily be mistaken for the signal at Granville Avenue displaying a clear route for track 4.

The ICC noted that Rule No. 54 of the company placed the responsibility for the safe running of trains entirely upon the motorman, while Rule No. 83 required the motorman to maintain such a distance and to run at such a speed as to be able to stop short of a preceding train should that train stop unexpectedly, and Rule No. 180 required a flagman only to go back to protect his train in the rear should it stop unexpectedly between stations. The statements of Motorman Grooms that he had begun braking 2,600 feet from the rear of the L train, that his wheels had locked, he had released the brakes, reapplied them, the wheels had locked a second time and he had finally applied the brakes in emergency, were not supported by the statements of Motorman Challender, deadheading on train No. 725, or by those of other employees on the train, all of whom only noticed the final emergency application. Examination of the track showed that the train had only begun sliding 240 feet from the point of collision, while the brakes of train No. 725 were in good condition and the marker lights at the rear of the L train, which Motorman Grooms had stated that he did not see, were burning brightly. No other conclusion could therefore be reached than that for some unknown reason Motorman Grooms had failed to take effective action to bring his train under control until it was too late.

The ICC commented adversely on the fact that notwithstanding the heavy passenger traffic on the line, with trains being operated at headways of less than a minute at certain times of the day, no means were provided to ensure the proper spacing of trains or to guard against errors of judgement or failures other than rules putting the responsibility for avoiding rear collisions entirely upon the motormen. The line was not worked on the block system and there were not even distant signals provided.

As for the wooden car at the rear of the L train, the demolition of which in the collision had resulted in the fatalities, the ICC remarked that the accident was a repetition of numerous others which had occurred on various railroads in the country where wooden cars were used. Because of these repeated accidents steel and steel underframe cars had largely replaced wooden cars on the railroads, the percentage of wooden cars in use having dropped from 69.6 in 1915 to 10.2 in 1935. Furthermore, wooden cars were not used on main-line trains and not combined with steel or steel-underframe ones in one train. On the L, on the other hand, 41.8 per cent of the cars were still of wooden construction and they were intermingled with steel and steel-underframe cars, as on train

R-5, where the first, second and fifth cars had had steel underframes, the sixth and seventh had been all-steel and the third, fourth and eight had been of wooden construction. The North Shore train, in contrast, had consisted of three all-steel cars. 'Extensive experience in railroad service has conclusively demonstrated the material increase in safety of employers and travellers in case of train accidents which is effected by the use of steel and steel-underframe cars instead of wooden cars, and also the great hazard to occupants of wooden cars when associated in trains with cars of steel and steel-underframe construction.' [299]

The resumed inquest on the victims of the accident was told that there might be a 12th fatality as a Miss Eleanor Hope had little chance of recovery. Motorman Grooms, who, it was reported, had only recently recovered from hysteria induced by the accident, wept as he told jurors that he had applied the brakes after seeing the train stopped ahead but his train had slipped on either wet leaves or oil on the rails, adding that, "There was no way to take the car out of the skid." The jury decided that he failed to exercise 'proper precautions,' but found a verdict of 'accidental death'.

The jury also found that the Rapid Transit and North Shore companies were at fault for 'failure to provide adequate safeguards for the public safety,' by the lack of block signals. It also recommended the replacement of wooden coaches by 'modern equipment', the more frequent examination of train officials and the closer supervision of express trains. [300]

The day before the conclusion of the inquest, two L trains had been involved in another rear collision on the South Side of Chicago, an Englewood express running into a Jackson Park train in a thick fog. Windows were broken but no one was hurt. [301]

Stormy weather, with strong winds, contributed to a collision on the Ferrocarril de Antioquia in Colombia, on Sunday 10th January 1937.

The accident happened at a location known as La Volcana, between 6 and 7 kilometres. The mixed train bound for Medellin consisted of engine No. 28, a first, a second- and a third-class carriage and five goods waggons, with Driver Oscar Jaramillo. It had left Caldas well filled with passengers, mostly holidaymakers and tourists, because it had the advantage that, like a tram, it stopped at any place where a passenger wished to board or alight, and also it ran quite slowly. According to reliable accounts from the site of the accident the train travelled quite normally and on time from Caldas to La Volcano, with numerous passengers getting on and off on the way. It had approached La Volcana at moderate speed and its driver observed that some trees and telegraph poles had fallen across the line. He was able to pull his train up short of the obstruction. A storm with very strong winds had uprooted the trees and telegraph poles and they had fallen down a slope onto the track. The conductor then immediately organised the brakesmen and other train crew to clear the line.

The mixed train had left Envigado, the previous station, at 16:00. Just before 18:00 that station permitted a goods train to follow it. This service consisted of engine No. 60

---

[299] Interstate Commerce Commission report
[300] *The Hammond Times*, Hammond, Indiana, United States, 16th December 1936
[301] *The Oelwin Daily Register*, Oelwin, Iowa, United States, 16th December 1936

and 12 goods waggons, with Driver Jesús Maria Valencia. The driver of the goods train was cautioned and told that the mixed train had not yet been reported as having reached Medellin, but that it was supposed that it must have done. Nothing was, of course, known of the La Volcano obstruction at Envigado. Driver Valencia duly set off cautiously towards Medellin. The line fell steeply at La Volcano. As the goods train reached there torrential rain was falling. When Driver Valencia saw the train stopped in front of him he gave the alarm whistle, made an emergency brake application, applied sand, reversed his engine, and did all that he could to stop. However, the engine wheels skidded and the heavy load, the inertia of the train and the steep gradient made a violent collision inevitable.

The passenger carriages of the mixed train were reportedly completely destroyed by the impact and the whole train was overturned towards the left side of the line. It was calculated that there had been about 90 passengers on the train. Some prominent local people were riding in the first-class carriage, some employees in the second-class and some peasants and people of the humblest means in the third-class. Initially it was reported that eight bodies had been recovered from the wreckage, all of them peasants or townspeople, and three of them women, and that it was certain that there were at least 13 dead, none of whom could immediately be identified. There were estimated to be over 50 injured. Some well-known Medellin people were feared to be still in the wreckage.

The rescue operation was difficult because of the wintry conditions and the incessant rain. The continual downpour meant the Medellin River, close to the line on the right-hand side, had flooded. The river was almost impossible to cross, with ditches and dangerous quagmires hindering movement while the left side of the line was blocked by the overturned carriages. The crews of both trains worked unceasingly after the accident and it was thanks to this that none of the seriously injured died without receiving treatment. The accident happened just after 18:00. Darkness was falling, worsened by the inclement weather. With no lighting available and no resources to hand the first moments after the accident were, according to witnesses, horrific. Everywhere there was wailing, desperate cries for help, clamorous pleas. Some called for relatives and missing companions and uninjured passengers prayed that they had been saved. The railway company became aware of the accident soon after it had happened and rescuers were sent to the scene, while the authorities sent firemen, police, ambulances, and doctors to attend to the injured and have them taken to hospital. Many ordinary people also made their way to the accident scene but they were unable to provide any assistance in the darkness. After a while, however, the town mayor, who was directing the rescue operation, was able to get a team of electricians to set up a lighting installation, by which the rescue work was made much easier.

News of the disaster came as a shock in Medellin where everyone was anxious to obtain news of the accident. The different radio stations broadcast information about it, confused at first, with more details later. The anguish of the townspeople increased because it was not known exactly who had died. Some people had gone from the city to the accident site in their cars but few of them had returned, increasing the anxiety of those left. [302]

The following day it was announced that there were 14 dead and 32 injured, of whom

---

[302] *El Tiempo*, Bogotá, Colombia, 11th January 1937

20 only had cuts and bruises.

One of those who died was Maria Dolores Gaviria, a third-class passenger travelling to Medellin who had been in an 'interesting state'. She suffered a severe shock, bruising and crush injuries in the collision and the trauma resulted in her giving birth prematurely. No one paid any attention to her, and she passed away a few moments later. Railway officials said that she had come from Belén, where she grew up, and was presumably heading for a maternity ward or one of the hospitals in Medellin. It seemed that she had alighted from her coach after the accident, realising that she was about to give birth, and had done so a few metres away. A priest found her baby, a boy, in a ditch. He was taken to Medellin and removed to an asylum for homeless children. The slightly injured passengers, of whom there were a good number, were checked by a doctor at the railway hospital and then sent home as their injuries were not considered serious. There were fortunately few passengers in the first-class carriage, most of them having alighted at previous stations. Most of the remaining complement of the train were urban workmen or peasants. The train was much used by street vendors who could buy food at lower prices than those charged in Medellin, which they could then sell at a small profit.

The drivers and conductors of the two trains were arrested and held incommunicado while the accident was being investigated. One of the points which was being investigated was the signalling of the two trains. Details of what had happened now became clearer. The mixed train had been allowed to leave Envigado, Medellin, having correctly reported that the line was clear. It was then stopped for a long time clearing the obstruction formed by the fallen trees and telegraph poles. Meanwhile the goods train, bound for Amagá with a heavy load of coal, had arrived at Envigado. The stationmistress there, Señorita Ana Clara Diez, had attempted several times to contact Medellin, but without success, clearly because the storm had brought the telegraph wires down, interrupting communications. Reflecting that it was almost two hours since the mixed train had departed the stationmistress considered that despite its slow progress it should easily have reached Medellin. According to this later account the goods train was not even cautioned.

Despite this apparent clearing up of the causes of the disaster it was pointed out that the investigation had to consider all possible theories as to how it had occurred.

Two circumstances contributed to the accident, a newspaper wrote. One was that at 17:40, the time that it happened, it was usually still daylight, but thanks to the heavy rain it was considerably darker than usual. Nevertheless, Driver Valencia had seen the mixed train when his own arrived at kilometre 6, sounded the alarm whistle and applied his brakes and taken other measures to stop as described earlier. The other, presumably, was that the rails had been 'like soap' and he had been unable to pull up his heavy train. He had, however, reduced the speed of his train somewhat, thus preventing an even worse accident. [303]

An important point which does not seem to have been mentioned in the newspaper reports was the failure of the crew of the mixed train to protect it in the rear, probably because they had been wholly preoccupied with trying to clear the line. That they should have done so must, surely, have been part of the company regulations.

---

[303] Ibid, 12th January 1937

The first fatal railway accident in the inner London area since 1924 occurred at Battersea Park, England, on the Central Section of the Southern Railway just after 8.02 a.m. on Friday 2nd April 1937. The 7.30 a.m. electric passenger train from London Bridge to Victoria via Tulse Hill and Clapham Junction was about to restart after being stopped at the Battersea Park Up Local home signal when it was run into from behind by the 7.31 a.m. electric passenger train from Coulsdon North to Victoria, travelling at 30 to 35 miles an hour. The leading coach of the Coulsdon train mounted and telescoped the rear coach of the London Bridge train, killing six passengers and Guard Maurice Brooker of the London Bridge train and fatally injuring three more passengers. Seven passengers were seriously injured, and ten further passengers were taken to hospital together with Guard Robert Adamson and Motorman Arthur Anthony of the Coulsdon train. Motorman James Spiers of the London Bridge train was able to signal to the motorman of the approaching 8 a.m. Victoria to Brighton train, which pulled up just short of the wreckage. The lines were blocked until 4.45 p.m. and electric trains were diverted to London Bridge until that time, though from 11 a.m. steam trains were diverted to and from Victoria via the non-electrified goods lines between Pouparts Junction and Stewarts Lane. Other trains were terminated at Clapham Junction and other stations. [304]

At the north end of Battersea Park station, at Battersea Park signal box, the Local lines were joined by the South London lines. In the Clapham Junction direction Pouparts Junction box was 1,474 yards from Battersea Park, while towards Victoria Battersea Pier A box was 534 yards further on. In common with the remainder of the Southern Railway system in the London area the Sykes Lock and Block system of signalling was used, whereby once a train had been accepted by the box in advance and the signals lowered for it the instruments were locked and it was impossible for the relevant instrument to be freed or the signals cleared for a second train until the first train had passed over a treadle at the box in advance. To cope with situations where a train which had been accepted might for some reason not proceed, or with a failure of the treadle to release the instrument, release keys were usually provided, this being the Achilles' heel of the system. The use of the release key normally required the co-operation of the signalmen at each end of the section. From Victoria out as far as Battersea Park box the line had been resignalled about 30 years earlier and the semaphore signals were motor operated and worked by slides, ordinary levers being retained to work the points. The cancelling apparatus in these boxes consisted of push buttons above the slides rather than keys. Because Pouparts Junction was the first box out with conventional Sykes Lock and Block, the Battersea Park signalman could release his instrument without the cooperation of his colleague at Pouparts. To deal with the situation where the push-button cancelling apparatus itself failed a special instruction had been issued in 1932, which read:

In the event of the special cancelling apparatus failing, the Signalman must break the seal of the instrument case and move the slides in order to restore the instrument and lever to the normal position, immediately calling the Lineman to put the instrument in order and re-seal the case. The failure must be reported in due course.

Following the electrification of the Brighton main line the number of trains dealt with at Battersea Park had increased from 712 to 838 daily, with from 54 to 61 between 8.30 a.m. and 9.30 a.m. and from 57 to 65 between 5.30 p.m. and 6.30 p.m. During the evening rush hour the signalman had to perform 972 operations, not including the use of

[304] *The Times*, London, England, 3rd April 1937

the train describers or telephones or lever movements. Around the time of the accident there had been ten trains in the area, as follows:

(a) 7.54 a.m. Down electric passenger train Victoria to Beckenham which passed on the Down Local at 7.57 a.m.

(b) A light engine, Stewarts Lane to Victoria, passed on the Up Main at 7.58 a.m.

(c) The 7.56 a.m. Down electric passenger Victoria to Epsom passed at 7.59 a.m.

(d) The 7.57 a.m. Up electric passenger London Bridge to Victoria via the South London line was held at home signal No. 64 from 7.58½ a.m. to 8.0½ a.m. and passed at 8.01½ a.m.

(e) The 7.30 a.m. London Bridge to Victoria via Tulse Hill train was held at home signal No. 67 from 7.59½ a.m. and was run into about 8.02 a.m.

(f) The 7.58 a.m. Down electric passenger Victoria to Epsom Downs was first stopped at the Down Local home signal, then through loss of power following the collision.

(g) The Coulsdon train passed Pouparts Junction at 8.01½ a.m. and collided with the London Bridge train.

(h) The 8 a.m. electric passenger train Victoria to Brighton which was stopped as described above.

(i) The 8 a.m. electric passenger train Victoria to West Croydon was stopped at Battersea Pier A box at 8.02 a.m.

(j) The 6.10 a.m. Up steam passenger train, Tonbridge to Victoria, was stopped at Pouparts Junction at 8.06 a.m.

Motorman C.R. Horsfield and Guard F.W. Kemp of train (d) gave evidence at the Ministry of Transport inquiry into the accident of their train being held at signal No. 64, which was not unusual.

Motorman Spiers of train (e) spoke of being stopped at signal No. 67 and waiting for it to clear, which it had done a few moments before his train was run into.

Motorman Anthony of train (g) testified that drizzle had interfered with visibility. The Pouparts distant signal had been at caution but the outer and inner homes (Nos. 8 and 18), and starter (No. 19) had been clear. He had then opened the controller. The Battersea Park distant below the Pouparts starter had been at caution and he had shut off power to coast to the home signal, No. 67. That signal he had seen to be clear. The London Bridge train had then come into his view but for a moment he had thought that it was an empty train on the Down Main line, probably one bound for Selhurst for repairs. "Almost simultaneously" he had realised that it was a train on his line and had made an emergency brake application.

Guard Adamson generally confirmed his motorman's account.

Signalman C.H. Hendrick, of Longhedge Junction box (close to the scene on the lower level line) said that he had witnessed the collision and had immediately telephoned Stewarts Lane box for assistance and ambulances, asked Clapham A box to get in contact with Waterloo and spoken to Relief Signalman T.G.P. Hillman at Pouparts Junction. He added, "Hillman was very surprised on the first occasion when I rang him up, and as far as I can remember he rang me up and spoke to me a second time, when I repeated my message."

Signalman T.W. Truelove at Battersea Pier A box stated that Relief Signalman Frederick Childs at Battersea Park, whom he had realised was not the regular signalman, had sent him a bell signal asking for a release on the Up Local line, so he had pressed the button to give the release. No bell signal was received to indicate that the release had been obtained, and he had assumed that it was no longer required. He was sure that there had been no confusion as to which line the release was required for.

Relief Signalman Thomas Hillman at Pouparts Junction stated that the 7.30 a.m. London Bridge to Victoria train (e) had been signalled as usual to Battersea Park and he had then accepted the 7.31 a.m. Coulsdon train (g) from Clapham Junction B box. At that moment he had also been dealing with trains (f) on the Down Local, train (h) on the Down Main and train (j) on the Up Main. He had been waiting to offer the Coulsdon train to Battersea Park when the block indicator had gone to 'Line Clear', when he had offered and had accepted the Coulsdon train, had cleared his signals for it, and offered the Tonbridge train (j) on the Up Main, but unusually it had not been accepted. A couple of minutes later the Tonbridge train had been accepted. At 8.07 a.m. the Longhedge signalman had telephoned him. He could not understand what was being said and had asked for the message to be repeated, when he had been told that there had been a collision on the South Western bridge. There had been time to throw his signals back to stop the Tonbridge train and he had advised Clapham and Victoria. He was certain that he had received the 2-1 'Out of Section' bell signal for the London Bridge train before he had offered the Coulsdon one. (He was closely questioned by the inspecting officer on this point.)

He said that he had never had a false clear release on the Up Local instrument. He continued, "The arm went up with the usual click and the 'free' tablet dropped down in the normal manner. I am quite sure I got the 'clear' for the up main at practically the same time as I got the 'clear' for the Coulsdon; the Coulsdon was a little before the Main line train. The Tonbridge had not left Clapham Junction when I pulled off my signals for it… I do not remember ever having forgotten to replace the starting signal, and if I did the home signal would be locked and could only be freed by using the key. I have never used the key except for a treadle failure or for the cancellation of a train… I last used the key on No. 18 for a treadle failure about three weeks previous to the accident."

Relief Signalman Childs had been working his first day at Battersea Park on 2nd April. A special (highest) grade relief signalman, he said that he had spent 17 days learning the box and had been confident regarding the use of the release buttons, though he had not needed to do so during training.

In his initial statement to the Company officers Childs said that the South London instrument had become locked at 7.58 or 7.58 because he had tried to clear signal No. 64 before resetting points No. 5. He had asked for a release from Battersea Pier A, which had been given, but it had not worked. He had then put his arm in the frame and manually released the lock. He had then pulled off for a Down Epsom Downs train and the Up London Bridge train. Thinking the matter over he could only think that he might have lifted the 'Train on' ticket with his elbow when he had put his hand in the case and pushed the plunger in with his body as he was replacing the case. He denied having been offered or accepting the Coulsdon train.

In a second statement he said that he had had no intention of giving the Tulse Hill train precedence over the South London train. He was very familiar with the cancelling buttons.

In a third statement he maintained that he was sure that he had put the switch-hook on when he had cleared the Up Local home signal (so as to maintain the block indicator in the raised position at Pouparts). He spoke of a man coming into his box with S.R. on his collar saying that there had been a collision between two trains, but he did not know what he meant. He knew that the London Bridge train was at his home signal but had no information as to what the second train was.

The explanation of the failure of Signalman Childs to get the release he desired from Battersea Pier A box became clear when Signalman F.W. Harvey, one of the regular signalmen, was asked to attempt the same thing. He pushed two buttons, one of which was different to that pressed by Childs, and duly obtained the release.

Regarding his actions after failing to obtain the release, Signalman Childs stated, "As I was unable to get the 'free,' I immediately went to the seals and cut them with a knife and opened the case; I lifted the top with my right hand, then pulled the front forward with my left hand, or vice versa. I lifted the top and pulled out the front to an angle, and then put my hand over the top to release the lock of 64 signal. I was unable to reach it. It was too high, and I was unable to get to the bottom of the shelf. I was unable to do that, so I pulled 63, the South London starter, and then lifted the front out altogether, and I was then able to release 64… I operated 63 as a safeguard… the next instrument to it was the one I wanted. I then pulled over 64 slide. It (the South London train) had been standing there a minute or two. I realised I was delaying the train… The train then arrived at the station. I put back 64, and, after the train had passed my Local starting treadle, I returned 63… After pulling 63 with my right hand, I pushed the spring lock on 64, and with my right hand I was then able to pull 64 slide. During that time I had the front of the case under my left arm. I immediately replaced the front and the top fell down by itself… There is an instruction to the effect that if there is a failure you are justified in releasing."

The inspecting officer, Lt Col Alan Mount, wrote that Childs' statements amounted to an admission of having operated inadvertently and without knowledge both the cancelling button, which would have released the plunger and lowered the indicator at Pouparts if the switch-hook had not been applied, and the plunger which had released the Pouparts starting signal, No. 19. The first the inspecting officer accepted but he thought the second, the release of the Pouparts starting signal, hardly credible, though he conceded that Childs being very short (he was only 5 ft 6 in tall) made the possibility more likely. Childs also admitted that he must have failed to put on the switch-hook.

Signalman Harvey said that Childs did not use the switch hook as regularly as he did. If the switch hook was not put on before the home signal lever was replaced the block indicator at Pouparts would fall. He (Harvey) always made a point of keeping the switch hook on until a train had passed the starting signal, then removing it and giving the 'Out of Section' signal in one movement. Harvey said that relief signalmen varied in their use of the switch-hook. He thought its use at Battersea Park essential to keep things safe. Harvey added that he did not consider it exceptional, in a busy box, for a signalman to offer another train on seeing the block indicator fall without getting the 'Out of Section' bell signal. "Some signalmen might not notice the bell. Of course, there are a lot of bells ringing."

Harvey said that he had never felt the need for a second man, but the removal of the booking boys four years earlier had made things more difficult. The booking boy could see the headcodes of the Up trains so that the signalman could describe them without looking. Sometimes five trains might be described and have to be memorised. A train register maintained by a booking boy would also give the signalman something to refer to. He had never needed to cut the seals.

Area Inspector G.W. Matcham stated that his examination of Signalman Childs to see if he was competent to work Battersea Park box had lasted 55 minutes. Childs had worked the box faultlessly. The inspector added that, "As our relief men are taken from the smartest signalmen… you expect him to be almost as good as a District Inspector, and the first thing he should do when he goes into a box is to read his instructions – otherwise he would be lost."

Inspector Matcham also put forward his theory as to how the collision had happened:

"I think myself that both signalmen are involved. I believe that Childs did drop his 'Train-On' tablet, but might have done it with his arm, or he might have done it when he was trying to get his release by pressing different buttons. Whichever it was he dropped the 'Train-On' tablet, that would lower the arm in Pouparts Junction box and would release the plunger in Battersea Park, if the switch-hook were not on, in which case the plunger could be used again. My impression is that Childs did not give the 'Out of Section' for the 7.30 a.m. London Bridge train, and that Hillman saw the arm down, the alteration to the 'Train-On' tablet at Battersea Park having put the arm down. With the continual ringing, Hillman might have got confused as to whether he had had the 'Out of Section' for the 7.30 a.m. train or not, and with the arm down he would have jumped to the conclusion that he had received the 'Out of Section' signal, in which case he would have offered the Coulsdon forward to Childs, and I think Childs was taken unawares and accepted it, and he must have plunged. I think it is a question whether he used the switch-hook. He is none too sure, and a relief man does not use the switch-hook in every case. He could have put over at any time. In an ordinary box, there is not much in the switch hook, but in this box it has a most important function. When watching Childs working before passing him, I noted he was using the switch-hook every time. When he accepted a train he put the switch-hook over. Some relief men put the switch-hook over as a reminder when they get the 'Entering Section' signal, but I am not referring to this particular box at Battersea Park, where the man usually puts it over as soon as he accepts, viz., as soon as he plunges. I watched Childs for 55 minutes, and he used the switch-hook on every occasion after plunging his acceptance, and if he had not I should have noticed."

From Childs' own statement though he did not use the switch-hook all the time when working normally. Doubtless he was careful to do so when he knew that he was being watched by the inspector.

Concluding his report, Lt-Col Mount stated that there were three possible ways in which the collision could have occurred:

(a) A failure of the Sykes instrument at Pouparts Junction such that it gave a false clear leading to the irregular admission of the Coulsdon train into the section to Battersea Park without it having been accepted.

(b) The failure of Signalman Hillman at Pouparts to replace starting signal No. 19 behind the London Bridge train and his irregular release with the key of his home signal, No. 18.

(c) An irregular release by Childs at Battersea Park of the Pouparts instrument for the Coulsdon train while the London Bridge train.

Tests carried out after the accident showed that all the apparatus was working correctly, eliminating possibility (a).

As regards (b) and (c) there was a conflict of evidence between Hillman and Childs. Hillman asserted that he had received the 'Out of Section' signal for the London Bridge train, his block indicator had been lowered and Childs had accepted the Coulsdon train. Childs on the other hand denied having given the 'Out of Section' signal for the London Bridge or having accepted the Coulsdon train. In view of Hillman's strong denial of using the release key and the fact that the working at Pouparts had been perfectly normal up until the accident the inspecting officer concluded that the fundamental actions which had led to the accident were the cancellation of the London Bridge train and the plunging to accept the Coulsdon train, both by Signalman Childs. It was possible that Signalman Hillman might have offered the Coulsdon train without having received the 'Out of Section' signal for the London Bridge train, which was incorrect, but could not be severely criticised in view of the pressure of traffic and the number of bell signals. A signalman working under pressure was likely to accept the lowering of the block indicator as justification for offering another train without paying too much attention to the bell signal. Hillman, who had worked both Pouparts and Battersea Park boxes, was also entitled to assume that Childs was using the switch-hook. The inspecting officer therefore excluded Signalman Hillman from even a small degree of responsibility.

It was obvious, Lt-Col Mount stated, that the onus for the accident rested upon Signalman Childs, a man with a clear record and who was apparently imperturbable, but who was lacking due sense of responsibility and had been overconfident. The events leading up to the accident had begun, it seemed, with what Signalman Childs had believed was a treadle failure five minutes after he had come on duty, which in fact had resulted from his replacing the starting signal slide before the home signal one. All had gone well then for two hours until he had been unable to clear signal No. 64 because he had failed to reset trailing points No. 5. Even when he realised what he had done and had reversed the points the instrument had then been locked and he needed a release from Battersea Pier A box. He had attempted to obtain a release in the proper manner but had been unable to do so because he had pressed the wrong button. He had then wrongly assumed that the releasing apparatus had failed. Instead of then flagging the South London train past the signal, which would have effected the release when the train had passed over the treadle but would have caused considerable delay at the height of the morning rush hour, he had cut the seals of the Sykes instrument to release the slide of No. 64 signal by direct manipulation. In doing so or in replacing the case of the instrument he must have made his fourth mistake of the morning in accidentally removing the 'Train-On' indication for the London Bridge train and lowering the block indicator at Pouparts (because he had not used the switch-hook).

In his remarks and recommendations the inspecting officer noted that had the bodies of the two coaches which had telescoped been stronger the results would have been different, but it was impossible to say if they would have been more or less serious, and he did not seem especially in favour of all-steel rolling stock.

The seals of the Sykes instruments had been replaced since the accident by padlocks so that they could only be opened by the lineman.

The co-operative feature requiring two signalmen to agree to a release should be introduced at Battersea Park as elsewhere.

The use of the switch-hook on acceptance or receipt of the 'Train Entering Section' signal should be made obligatory.

The latest form of manual block with track circuiting might be introduced in place of Sykes Lock and Block.

A mirror had been provided to help the signalman see the headcodes of up trains at Battersea Park. The re-introduction of a booking boy might be considered.

A track circuit in the rear of the Up Local home signal would have prevented the accident.

Given the heavy traffic the replacement of the manual signalling with colour light signalling and full track circuiting, which was under consideration by the Southern Railway, should be pursued as soon as possible. [305]

At the inquest on the victims of the accident the jury returned a verdict of 'Accidental Death', adding a rider that they were satisfied that the railway company had done everything possible to ensure the safety of the travelling public. [306]

An accident occurred in France on the evening of Saturday 10th July 1937 on the Chemins de fer de l'Etat. Train No. 583A, a relief train to Caen, left Le Mans station at 18:35. It was followed by omnibus train 583, from Le Mans to Mézidon, in charge of Conductor Raoul with Driver Bénard and Fireman Reverdy, which left Le Mans station at 18:39. This train was crowded, with many passengers bound for Granville for what promised to be a beautiful Sunday. It came to a stand at the *carré* at *poste* (signal box) No. 4, 900 metres from the station and near level crossing P.M. 108. Soon after the train had stopped Conductor Raoul in the rear van saw another train coming and apparently not likely to stop before striking his train. Anticipating a collision, he had jumped from his van and run down the line towards the oncoming train, waving his handkerchief. The third train was No. 1645, a Le Mans – Rennes express. Only at the last moment did Driver Stanislas Marchand of the Rennes train see the train ahead and jam on the brakes. The train was still travelling at 50 kilometres an hour when it struck the Mézidon train at 18:43. (The author believes that he has correctly identified the trains and the sequence of events, but the different newspaper accounts are very confused.)

The third-class coach at the rear of the front train was completely telescoped into the preceding one, a first and second-class composite. The noise of the collision attracted the attention of the locals, who were soon on the scene and began the first rescue work. They were soon joined by firemen, police, and a squad of soldiers. The injured and the bodies of the dead were all removed from wreckage in an hour and a half. Nine passengers died

---

[305] Ministry of Transport report

[306] *The Times*, London, England, 21st April 1937

in the accident and 70 were injured, 63 of whom were taken to hospital. [307]

M. Bricara, aged 17, a passenger who had escaped injury and who was very active in the rescue work, spoke to a reporter. "I was reading when the train stopped. I put my nose to the door, as they say, to find out why and saw the signal at danger. I was just going back to my book when I felt myself being lifted into the air. (I fell back between two seats.) I felt a great pain in my back but when I realised what had happened and that I was not badly hurt I hurried to help."

Train No. 1645 had followed train No. 583 from Le Mans station and had stopped just outside the station at the junction with the line to Angers at a double red signal. One of the reds had then been extinguished, permitting the driver to continue cautiously, prepared to stop short of any obstruction and at a speed not exceeding 10 kilometres an hour. Although it was still daylight at the time of the accident neither the driver nor the fireman appeared to have seen the obstruction until the last moment. The collision had occurred at the exit from a sharp curve, which would have made it more difficult from his position on the left side of the engine, to see the train in front, but it would have been clearly visible to the fireman. The Flaman track recorder was examined after the collision and indicated a speed of 50 kilometres an hour when the collision occurred. [308]

An investigation into the accident was begun already at 19:30. Juge d'instruction Tiébault interrogated the driver and fireman, who were overwhelmed by what had happened.

Driver Marchand stated afterwards, "I drive from the left-hand side of the engine. I couldn't see anything ahead of me at that point because of the curve in the line. I admit that I did not ask my fireman to keep a lookout. He was busy firing."

It was pointed out to him that driving on sight he should have been especially careful. He replied, "When you are driving a train, you have to consider its smooth running. That is to say, you must not waste time. When leaving Le Mans station I stopped at a stop signal, then I continued when instructed to do so by my conductor. I ran slowly so as to be able to pull up short of any obstructions. I believe I was going at 45 or 50 kilometres an hour. I did not look at the speedometer. I was still on the curve when I saw the rear of the train in front of me. I shut off steam and applied the brakes, but it was too late!"

Fireman Robert Filliatre said, "After stopping at the signal for half a minute my driver set off again. I don't know what the speed was. Then I was busy firing, and I did not realise that there was going to be an accident until we struck the other train." [309]

Driver Marchand admitted that perhaps he had failed to control his speed but pleaded that he had only been driving for two months and did not know the line very well. [310]

Conductor Jules Forget of train No. 853 declared that in his opinion his driver had been going too fast.

Driver Marchand and Fireman Filliatre were charged with causing death by negligence but were allowed to return to their depot.

---

[307] *L'Ouest Éclair*, Rennes, France, 12th July 1937
[308] *Le Petit Journal*, Paris, France, 11th July 1937
[309] *Excelsior*, Paris, France, 12th July 1937
[310] *Le Radical de Marseille*, Marseilles, France, 11th July 1937

At the end of February 1938 the driver and fireman stood trial. Both were found guilty, and Driver Marchand was sentenced to six months' imprisonment, while Fireman Filliatre was given a suspended sentence of one month's imprisonment. [311]

The Fuji Express was standing in Okayama station, Japan, on the San'yo main line between Osaka and Hiroshima, on Thursday 29th July 1937, when it was run into by a following train. Six people died and 64 were injured. The station entry signal had been prematurely cleared for the second train. [312] [313]

Two trains were involved in a rear collision on the Chemins de fer de Paris à Orléans et du Midi one and a half kilometres from Vars, France, at 22:30 on the evening of Monday 20th September 1937. Almost every evening for the previous two months, due to the quantity of traffic, regular train No. 26 from Bordeaux had been supplemented by a relief train, No. 1026, which left Bordeaux at 20:20 and was due in Paris Quai D'Orsay at 04:54.

On the day of the accident train No. 1026 was stopped at the site of the collision at a signal at danger. After some minutes the signal cleared, but the conductor had meanwhile made his way on foot to the nearest station to find why his train had been stopped at this unusual point. Before he had been able to rejoin his train and instruct his driver to proceed, his train was run into from behind by the regular train, travelling at about 45 kilometres an hour. The rear vehicle of the first train was a third-class brake composite of steel construction which was strong enough to withstand the force of the impact. Instead it was the penultimate coach, with wooden frames and steel panelling, which suffered in the collision, folding up like an accordion and with its rear two compartments crushed.

A rescue operation was quickly organised and the bodies of six victims were recovered from the wreckage, together with 28 injured, one of whom died on the way to Angouleme Hospital. Numerous local farmers and villagers, plus a detachment of the 107 Angouleme infantry regiment assisted in the rescue work. As there were insufficient ambulances taxis, private cars and vans were used to take some of the injured to hospital.

Three passengers who had been in the corridor of the wrecked coach realised their danger and were able to open a door and jump to safety, while a young girl lying in a hammock strung between two luggage racks also escaped injury, though her father was badly hurt and her mother and two brothers less badly so.

Driver Bellec, of the regular train, said afterwards that he and his fireman, named Aubry, had seen the relief train in front of them too late to pull up. Following an interrogation M. Perrin, the *juge d'instruction* (examining magistrate) charged Georges Condouret, the *facteur mixte* (porter signalman?) at Vars with causing death and injury through negligence, questioning having implied that he was principally responsible for the accident. [314]

The investigation had consisted of taking statements the drivers, firemen and

[311] *Excelsior*, Paris, France, 1st March 1938
[312] Webpage: Japanese Railway Safety and the Technology of the Day by Masao Saito
[313] *Twin Falls News*, Twin Falls, Idaho, United States, 29th July 1937
[314] *Journal de Ruffec*, Ruffec, France, 26th September 1937

conductors of the trains involved, of the *facteur mixte* at Vars (probably the only person on duty there at the time) and also Signalman Bugeaud at Puyrenaud. These statements were said to have been in part confused and contradictory, but the mechanism of the accident could be deduced. Train 1026 had been stopped outside of Vars station because freight train No. 9028 had been stopped in the station. At the same time as Head Conductor Grazillier had gone ahead to the station and Assistant Conductor Louis Lalanne had gone back with detonators to protect the rear of his train. The driver of the freight train had then whistled to indicate that he was ready to depart, which had been misinterpreted by Assistant Conductor Lalanne as indicating that *his* train was ready to leave, and he had returned to rejoin it, presumably picking up the detonators he had put down. Conductor Grazillier, becoming aware of the approach of the regular train, had then asked Conductor Nivet of the freight train (a younger man who could run faster?) to run back and try and stop it. The crew of the regular train had seen the tail lamp of the relief train and the hand lamp of Conductor Nivet at the same time. Exactly how the regular train had come to be admitted to the still occupied Puyrenaud – Vars section was not stated. Porter Signalman Condouret had presumably got confused and signalled the line clear when the freight train had left, forgetting the relief train standing outside of the station. [315]

Conductor Lalanne was later also charged with manslaughter. [316]

The two railwaymen stood trial at the end of June 1938. Driver Bellec stated at the trial that his train had left Angouleme 10 minutes late and had been stopped by signal at Puyrenaud 5 or 6 minutes. He had exploded no detonators and by the time that he had seen the train in front had only been able to reduce speed to 40 or 50 kilometres an hour.

Conductor Grazillier was reproached by the lawyers for his interpretation of the rules in connection with the accident, while Conductor Nivet was praised for his efforts to avert the collision.

Porter Signalman Condouret pleaded in extenuation that he had only been four months in his position and the defence lawyers blamed the Midi company for the accident for giving inexperienced employees such responsible jobs. The judgement of the court was reserved for a week. [317]

The sentence of the court was lenient enough when it was announced. Condouret was sentenced to a suspended fine of 100 francs and Lalalle to a suspended fine of 50 francs, while the Société nationale de chemins de fer français (as the Midi company had by now become) was found to be civily liable. Siding with the defence counsel the court criticised the employment of inexperienced staff in important duties and also the continued use of partly wooden carriages, which offered their passengers little protection in the case of an accident. [318]

At 02:00 on Tuesday 2nd November 1937 a special passenger train of 15 carriages (confusingly identified in newspaper reports both as SPA-1 and as RPA-1) left the

---

315 *Excelsior*, Paris, France, 22nd September 1937
316 *Le Radical de Marseille*, Marseilles, France, 27th October 1937
317 *La Charente*, Angoulème, France, 1st July 1938
318 Ibid, 9th July 1938

Maritima station of the Estrada de Ferro Central do Brasil in Rio de Janeiro, Brazil. It was carrying reportedly two thousand members of the Integralist Party (a right-wing nationalist party dating from 1932, also known as Greenshirts) who had attended a mass meeting in the capital and were returning to their homes in Rio and São Paulo states. [319] The journey started off badly for the train had not gone far when, presumably due to a signalman's error, it was diverted at a junction onto the wrong route, the mixed gauge track of the Auxiliary line, travelling as far as the São Francisco Xavier viaduct before its driver became aware that he had gone the wrong way. The train then had to be set back to the junction before it could set off again on the correct line, now considerably delayed. It was also noted that the train had no headlight.

Shortly after 03:00 the special left Deodora for Nova Iguaçu and it was soon travelling at 60 kilometres an hour. The same speed was maintained through Nilópolis station and on to the next station at Mesquita. Suddenly there was a violent shock as it collided with the rear of freight train KP-1. The engine of the special buried itself in the rear of the train in front, completely destroying waggons 674-VA, 287-VA, 221-VK and 326-VK. In the special, coaches 526-D, 531-D and 101-B were badly damaged. The passengers further back the train disembarked and went forward to help their colleagues in the front coaches. Initially it was reported that the bodies of three dead had been discovered and that there were a large number of injured. A Reuter report spoke of 7 dead and 70 injured.

Clerk Sinesilo at Mesquita telegraphed to his headquarters, "Train RPA-1 collided at the entrance to Mesquita station with part of train KP-1 which had been left on the line by its crew, who cannot be found. Two passenger carriages are overturned. There are dead and injured."

A second longer telegram was sent by Stationmaster Veiga de Figuelredo: "RPA-1 has collided with part of KP-1 which was left on the line by its crew. Both lines are blocked. There are dead and injured. Assistance required. Extranumerary Assistant Onofre Fonseca, Second Class Conductor Marcolina Garcia de Medeiros and Extranumerary Shunter Edgard José dos Santos have abandoned the station.

"I consider the driver of train KP-1 and Shunter Edgard José dos Santos responsible for the accident. The one for reporting the train as being complete, the other for shunting it without noticing that it was divided. The conductor of train KP-1 has disappeared. *Veiga de Figuelredo*."

Returning from the accident scene Colonel Mendoça Lima, the director of the railway company (shortly to be appointed Minister of Communications and Public Works), told the press what he understood had happened, though he said that the committee of inquiry would be able to give a better account in due course. He explained that train KP-1 had been shunted at Mesquita for the special to pass. The train had supposedly been put out of the way and was reported as being clear, but neither its driver nor those assisting him had realised, in the darkness, that it had become divided, and that its rear 11 vehicles had been left standing on the main line. It having been reported as in clear the clerk at Mesquita had given permission to his colleague at Nilópolis for the special to proceed, and the signals at Mesquita had been cleared. Travelling at the normal speed of 60 kilometres an hour the special had thus approached, and with no headlight to assist him its driver had been within a short distance of the obstruction before he had become aware

[319] *The Times*, London, England, 3rd November 1937

of a dark shape ahead. When he had seen the waggons, he had applied his brakes in emergency and put down sand, but the distance had then been too short, and a violent collision had ensued.

There were unsubstantiated rumours that the accident had been an act of sabotage. Brigadier General Newton Cavalcanti, a military commander and politician with Integralist sympathies, told a reporter in a telephone conversation that he was convinced that the collision was the result of a Communist attack, while Conductor Luiz Sinezio da Silva, who had been in charge of the special, said that he believed that the 11 waggons had been left on the main line with criminal intent because the air tap of the last waggon had been closed. The accident also came only two days after the fatal derailment at Barra Mansa (see below). [320]

'Scottish Train Disaster', '24 Bodies Recovered from Wreckage', 'Accident in Blinding Snowstorm', 'Rescue Work by Light of Fires.' Thus read the headlines in a London newspaper on 11th December 1937, announcing the disaster at Castlecary, Scotland, on the London and North Eastern Railway main line between Edinburgh and Glasgow, on the evening of Friday 10th December 1937. The 2 p.m. express from Dundee to Glasgow Queen Street had been run into from behind by the 4.03 p.m. non-stop express from Edinburgh Waverley to Glasgow Queen Street. Both trains had been crowded and 34 passengers were killed outright and 179 were injured or suffered shock. The driver of the Dundee train was also injured, and the guard of the Edinburgh train suffered severely from shock.

Several passengers described their experiences to reporters. Mr R.H. Kirk of Glasgow was in the second compartment of the Edinburgh express with five other passengers, one a child. He said, "There was a terrific crash, and I was thrown on top of the other passengers. I seemed to be jolted right up towards the roof of the compartment, and then, to my horror, I saw the side of the carriage cave in. Apparently, the carriage behind us became telescoped, and it was in this carriage, I think, that most of the victims were killed. I managed to scramble out from the wreckage, and after considerable difficulty the rescuers succeeded in reaching the child.

"The scene was terrible. I shall never forget it. The snow and the darkness added to the handicap of the rescuers. For over an hour railway officials and others, myself included, hacked our way through the debris. Men, women, and children were lying under the wrecked train moaning, while others apparently not so seriously injured were shouting frantically for assistance. They were trapped in their compartments, while some of their fellow-passengers had been thrown out on to the permanent way."

Mr John Venters, a commercial traveller, was returning home from Edinburgh to Glasgow. He related, "As we heard no screaming our first impression was that the engine had jumped the rails, but our coach was lying over at a steep angle, and we had considerable difficulty in opening the door. Snow was falling heavily, and in the darkness, it was difficult to see what had happened. We were amazed and horrified to find that the front two coaches were completely telescoped, and that portions of them were piled right up on top of the engine tender. I could see no sign of any other train, and my impression was that the engine, after leaving the track, had crashed into the

---

[320] *Correio de Manhã*, Rio de Janeiro, Brazil, 4th November 1937

embankment. The station lights were dimly visible about 50 or 60 yards away.

"The disaster occurred almost immediately opposite some brickworks, and the workmen swarmed over the embankment and took part in the rescue work. Some of the passengers who had travelled in the front coach told me they had to step over the body of the fireman as they clambered from their shattered coach piled on top of the tender."

Mr J. Calder, of Glasgow, was returning from Edinburgh with a friend. He said, "I was suddenly pinned beneath the seat. The whole portion seemed to collapse, and it was some time before I was extricated and assisted to the dressing station in Castlecary Station. The thing that impressed me most was the intense darkness." [321]

The Dundee express consisted of Class D 29 4-4-0 steam engine No. 4896 *Dandie Dimont*, seven eight-wheeled coaches and a six-wheeled fish van, in charge of Guard T. Inglis with Driver D. Macauley and Fireman Fleming. The Edinburgh express consisted of Class A3 4-6-2 steam engine No. 2744 *Grand Parade* and nine eight-wheeled coaches, in charge of Guard Cameron with Driver David Anderson and Fireman William Kinnear. The two trains had been running eight minutes apart; the Dundee train was two minutes late and the Edinburgh train on time.

The events which led to the disaster began with problems at Gartshore, 5¼ miles on the Glasgow side of Castlecary, where falling snow and the failure of permanent way staff to completely clear it from a set of facing points, had meant that the signalman there was unable to move a lock-bar and bolt said points. As a result, several Down trains were held up while the problem was fixed, including a goods train at Dullatur East, the next box on the Glasgow side of Castlecary. With the Dullatur East signalman not having cleared back the goods train to Castlecary, the signalman there had kept his signals at danger against the following Dundee train. Despite this the Dundee train had apparently run through all the Castlecary signals and into the occupied section, whereupon the Castlecary signalman had cleared it back to Greenhill Junction and accepted the Edinburgh train. In fact, however, the driver of the Dundee train had seen the red light exhibited by the Castlecary signalman and had stopped a short distance down the line. The Edinburgh train then likewise overran the Castlecary signals and was still travelling at about 60 miles an hour when it ran into the rear of the Dundee train. The fish van and rear two coaches of the Dundee train were destroyed, 'practically beyond recognition,' and the fifth coach was very badly damaged. The engine of the Edinburgh train was derailed and was overridden by the two leading coaches of that train, which were badly damaged. Thirteen of the fatalities were to passengers in this, the colliding train. The next three coaches of the Edinburgh train were damaged to a lesser degree but passengers in the rear of the train only experienced a slight shock. Some of the passengers from that part of the train left the station to catch a bus, unaware that there had been a serious accident.

As might be expected the disaster was the subject of a comprehensive Ministry of Transport inquiry, conducted by Lieut. Colonel A.H.L. Mount.

The inquiry found that the initial problem at Gartshore had resulted from the fact that the two lengthmen who had tried to clear the facing points leading to the down loop had had little experience of such work and had not appreciated that just a small amount of snow would prevent the point blades from closing sufficiently for the bolt to enter the

[321] *The Times*, London, England, 11th December 1937

stretcher bar. The matter was only fixed with the arrival of Signal Fitter R. Wilson at 4.30 p.m. Meanwhile the 3.41 p.m. passenger train, Polmont to Glasgow, was stopped at Croy from 4.09 p.m. to 4.37 p.m. Behind it the 2.50 p.m. goods train from Falkirk to Cadder was stopped at the Dullatur East Down home signal. It could not be shunted across onto the Up line until the 4 p.m. and 4.06 p.m. passenger trains from Glasgow had passed, which they did at 4.23 p.m. and 4.31 p.m. repectively.

Signalman Andrew Sneddon at Castlecary was aware of the hold-up at Dullatur when he accepted the Dundee train from Signalman J. Beattie at Greenhill Junction at 4.22 p.m. He received the 'Train Entering Section' for it at 4.28 p.m. As the train was approaching him, he said, he had realised that it was still steaming and not likely to stop at his home signal at danger. He had had time to show it a red light from the window at the east end of his box.

The Dundee train had been travelling at 50 to 55 miles an hour at the distant signal. Driver Macauley stated that the signal, "was standing clear for me. I have no doubt about that however." He failed to see the home signal but saw the red light from the box and had immediately applied the brakes in emergency. He said that he had stopped, "At the advance signal which was standing at danger." The tail lamp of the train would then have been about 294 yards from the box.

Fireman Fleming said that he had been having some difficulty with his fire, being short of steam and low on water, and he had seen no signals after Falkirk.

Driver Macauley had not whistled as he pased Castlecary box just before 4.30 p.m. and having seen no indication that the train was pulling up, Signalman Sneddon had sent the 'Train or Vehicles Running Away on Right Line' signal, 4-5-5, to Dullatur East, and had imagined, he said, that the Dundee train was likely to collide in a couple of minutes with the goods train at Dullatur. What he apparently did not do was look at the indicator of the track circuit which existed in the rear of the Down starting signal.

Even after he had stopped Driver Macauley had not whistled but he had sent Fireman Fleming back to the box. Meanwhile Signalman Sneddon had telephoned Probationary Clerk D MacDougall in the booking office, asking him to fetch Stationmaster William Scott, who was at that moment returning from the neighbouring brickworks. The stationmaster stated that he "could see the tail lamp {of the Dundee train} quite plainly."

So far Signalman Sneddon had not done anything directly to cause the accident but now, instead of telephoning Signalman D. Smith at Dullatur to find out if the anticipated collision with the goods train had happened, he instead rang Signalman Beattie at Greenhill Junction regarding the Edinburgh train. He told Beattie that the Dundee train had run past his home signal, that it had had a tail lamp, that his track circuit was clear and that he had sent the running away signal to Dullatur. He said that he had spoken to Beattie because he was "doubtful whether this entitled me to accept the 4.03 p.m. Edinburgh train, and asked his opinion on the matter."

As regards the track circuit indicator having been clear Sneddon said, "I am positive about this, because I certainly would not have told Beattie, when explaining the position to him, that the track was clear without first observing the position of the indicator."

Signalman Beattie stated, "On asking what was wrong, I was told the 2.58 Falkirk High and Cadder was ahead, standing at Dullatur East. I asked, 'Have you sent the Vehicles-Running-Away-On-Right-Line signal?'

"He said, 'I have done so.' I also asked if he had put down detonators. He said, 'No.'

"'You should,' I replied. He asked why, and I said, 'It is always a precaution.'

"This conversation took place about 4.31. About a minute or two after he rang me again and asked, 'Am I justified in clearing back the Dundee and taking on the 4.3?'

"I replied, 'There is nothing to hinder you provided you saw the tail light and the regulations applied.' I asked, 'Are you certain all your signals are at danger?'

"He said, 'Undoubtedly.'

"I then asked, 'Have you got your quarter mile clear ahead of your home signal. Are you certain of that?'

"He replied, 'Yes.'

"I then said, 'There is nothing to prevent your accepting the 4.3.' He then left the 'phone, cleared back the 2 o'clock ex Dundee, and accepted the 4.3 ex Edinburgh at 4.32. 'Train Entering Section' was sent by me at 4.36."

Fireman Fleming reached Castlecary box at 4.35 p.m., closely followed by Stationmaster Scott. Signalman Sneddon seems at this point to have realised for the first time that the Dundee train had stopped, but according to the fireman thought that it had done so at some point in the section, and the fireman had to correct him. According to the fireman Sneddon had then remarked, " I'll have to see about getting the 4 o'clock stopped," and he (the fireman) had replied, " It is time there were detonators on the track." Stationmaster Scott added that, "then the signalman told me, 'That's the Train Entering Section signal for the 4.03 passing Greenhill, and I ran for the detonators." Signalman Sneddon followed the fireman and stationmaster out of the box and they managed to put down three detonators in the short time available, but these were not laid very far back, and and only one was properly affixed, so that it exploded.

Despite the poor weather conditions the Edinburgh express was running to time. It was probably travelling at 65 miles an hour at the Castlecary Down distant signal. Driver Anderson testified that he had not been able to see the signal light, probably because it had been obscured by snow, nor the arm at any distance. "I was practically on top of it before I saw it." He was, however, positive that the signal arm had been clear. Like Driver Macauley he had failed to see the home signal but he saw the red light held by the signalman "held steady," and had remarked to his fireman, "This is something unusual." He had applied the brakes, heard one detonator explode, had closed the regulator and applied sand, but had no time to reverse the engine. Seeing the tail lamp of the Dundee train he had shouted to Fireman Kinnear to hold on.

Fireman Kinnear confirmed his driver's evidence. He said that he had seen the Greenhill Junction distant signal from 30 yards away. After that he had been busy breaking up lumps of coal and sweeping the footplate and had not seen any of the Castlecary signals, though he had heard the detonator go off.

Various other witnesses testified as to the conditions of visibility that evening; unfortunately the evidence of the various persons was not always in agreement. Ganger T. Bell said that it had been snowing from 1 p.m., though it had slackened off a lot by the time that they had left duty at 4.30 p.m. He had no reason to believe that the Castlecary Down distant was not working properly. Lengthman H. Harrison agreed. Lookout-Man Aitken thought that it had stopped snowing at 3 p.m. and that the distant signal could be seen from half a mile away. Lengthman Fearns agreed that the distant could seen for half a mile but thought that it had been snowing later. Lengthman J. McQuillan said that it had been snowing when they had left Greenhill but had stopped before they had reached Castlecary. He could not remember the distant signal. Lengthman A. McDonald stated that it had not been snowing at 3.40 p.m. at Greenhill, but was a little at 4 p.m.

Signal Fitter Wilson, who had travelled on the footplate of an engine from Greenhill to Gartshore, stated, however, that it had been snowing heavily between Croy and Gartshore, especially between 4 p,m. and 4.30 p.m., to the extent that he considered that the visibilty of signals would be affected.

Driver J. Cairns and Fireman J. Miller of the Polmont passenger train said that they had been able to see the Castlecary distant signal at 'Clear', and that at Croy it had been snowing hard. Driver Cairns stated that he had been unable to see more than one of the signal lights and that the arms had been covered with snow. Guard Stavert of the same train said that he had had difficulty in observing signals passing Greenhill Junction, due to snow blowing in his face.

Driver W. McDonald and Fireman C. Smith of the Cadder goods train said that it had been snowing all the way, especially after Greenhill Junction. The Castlecary signals had all been clear, but only their arms had been visible at very short range. Guard P. Penny of the Cadder train said that all the signal arms after Falkirk High had been covered with snow. At Castlecary, "there was heavy snow on the arms of the signals and also on the specs."

Driver Macauley stated, "The weather was very bad, and it was snowing when we left Thornton... I was seeing the signals, but I could not exactly say at what range. The signals were not bad till we got to the main Edinburgh – Glasgow line... It was daylight till we got about Falkirk. It was in the gloaming at Falkirk, and visibility was getting less and less as we went along... The snow was very bad at Falkirk and Greenhill. The flakes were heavy. The snow was striking it (cab window) and melting on it, and you could not see anything." He said that he had seen the Castlecary distant signal at clear and passing over the viaduct had been looking out for the home signal. "I got halfway over the bridge when a red light appeared on the left side of the line... I did not see the light of the signal box at that time... The signalman seemed to lift the lamp. The lamp moved. I thought that was for me. I just came inside and applied the brakes. My attention was attracted by the lamp, and I missed the home signal... I assumed it would have been off with the position of the distant signal, but after the red lamp of the signal box I knew there was something wrong... I came to a stand with the starting signal ahead of the tender."

Stationmaster Scott did not consider that it had been snowing very heavily except between 3 and 4 p.m. He did not think that visibility had been poor enough to necessitate the calling out of fogsignalmen. Ganger Bell had assured him that all the signals were working properly at about 4.20 p.m.

Signalman Sneddon confirmed that he had been able to the arm of his Down distant signal when he had replaced it to caution at 4.09 p.m. behind the Cadder goods train, also that he had been able to see its backlight when he had accepted the Dundee train at 4.22 p.m. and the Edinburgh train at 4.35 p.m. He said that the heaviest snowfall had been between 2.30 p.m. and 3.20 p.m. but that even then he had been able to see all his signals, and that, "if I could not have seen the position of the distant signal, I would have got the train stopped at the cabin in rear and advised them to take that signal as defective… I had a view of my back spec on my distant signal, and therefore it could not be hanging off." The signal had been working perfectly, plus he had been shaking them, because "it was one of the Rules that you must carry out."

Signalman Smith at Dullatur East said that there had been intermittent snow since he had come on at 2 p.m. but that he had been able to see his fog object (the Up home signal, 301 yards away).

Signalman Beattie at Greenhill Junction said that about 4.30 p.m. "there was practically no snow and the visibility from my box was quite good. Looking east I could see about half a mile, but in the westerly direction I could only see my signal lights as far as the marking point, which is the Up home signal." (250 yards away.) He explained that the wind was driving snow onto the signal spectacles while the backlights (looking east) were not dimmed.

Driver Anderson said that it had started snowing at Ratho. Beyond there, "I was seeing the arms of the signals in most places and sometimes the light… Sometimes you could see them a good distance away and next time you were practically up against them before you saw them."

Approaching Castlecary it had been snowing fairly heavily and it had been dark, with no moon. He had known where he was. The Castlecary Down distant signal he had seen "distinctly" and "properly" clear. He "was looking in the air at the time, and I would not say for the actual distance." The home signal he had missed seeing because, "I observed the red lamp and changed my glance to the red lamp on the ground." A detonator had exploded at the same moment.

Fireman Kinnear testified that, "snow was falling all the time from leaving Edinburgh, and visibility was very poor."

Guard Cameron said that he had last looked out at Polmont (4½ miles short of Castlecary) where it had been snowing and visibility "was bad, very bad." He had been sat in his van at the moment of the collision and had felt no previous brake application.

Mr C. S. Inglis, District Engineer of the Company at Carlisle, who had been a passenger in the Edinburgh train, riding in the restaurant car placed sixth in the train, said, however, that he had felt the brakes go on immediately before the collision. After the accident he had alighted from the train at once and it had been snowing heavily.

District Inspector E. Cunningham, who had reached Castlecary by car at 5.40 p.m., stated, "there was only a very slight fall of snow and a little wind coming from the north-eastern direction. I consider the general visibility was quite good and I could see the lights of the signal box 600 yds. away." He went to the box to question Seddon at about 5.53 p.m. He could see neither the Down starter nor the backlight of the Down distant signal. The district inspector went to the distant signal at 6.20 p.m. He said that from a point 300 east of it, "neither the arm nor the aspect was visible, and it was only when I approached to a point about 100 yards from the signal that I could see the light. On getting nearer I observed that the arm was fully in the 'on' position, although there was snow on it and also on the spectacle." Snow on the spectacle of the home signal meant that its light was only visible at 30 yards.

The first down train after the collision was a light engine in charge of Driver T.

Downs, and carrying District Inspector Henry Crorie, Shed Foreman T. H. Thompson, and Telegraph Linesmen L. Robertson and A. Dickson, which reached the Castlecary Down home at 5.40 p.m. The district inspector related that it had been snowing all the way from Polmont. "It blinded our view of the signals and prevented us seeing the lights until we were close on the various signals. The arms could be seen just as we were close on them by the light thrown out from the engine firebox; we saw the Castlecary distant signal, both arm and light, but only when we were close on it. The arm of this signal was definitely in the horizontal position." The shed foreman and the driver agreed. The latter said that "it was pretty heavy snow." He had returned from Castlecary at 6.10 p.m. when it had been still snowing but the arms and lights of the up signals had been little affected because the wind had been blowing from the east.

Signalman G. McLeod relieved Sneddon at about 8 p.m. He testified that he had not been paying special attention to the weather, but, "It was a soft snow. There were snowflakes. They were on my coat when I went into the box, and I shook them off."

Lt. Col. Mount visited Castlecary box on the evening of 16th December. He stated that at 4.08 p.m. (equivalent to 4.09 p.m. on the day of the accident) the arm of the Down distant signal had been just visible, at 4.12 p.m. he had had difficulty seeing it and at 4.15 p.m. he could not see its arm, though Signalman G. Preston said that he could still see it. At 4.24 Signalman Preston admitted that he could no longer see the arm, nor was the backlight visible. At 4.30 p.m. the backlight could just be discerned, at 4.45 the backlight was easily visible.

Regarding the visibility of the distant Signalman Sneddon asserted that when he had replaced it at 4.09 p.m., after the passing of the Cadder goods, despite the snow, "I could see my distant plainly go to danger." Despite having told District Inspector Cunningham that he had not been able to see the backlight of the distant signal at 4.32 p.m., Sneddon had repeated in cross-examination that he had been able to see the backlight when he had accepted the two expresses at 4.22 p.m. and 4.32 p.m., thus contradicting himself. He also admitted that the east window of his box had been obscured by snow, and that he had been looking out from the front window, which he had been keeping open. He was subjected to a grilling cross-examination by the inspecting officer regarding Rule 81 (h), relating to informing the box in the rear if the distant signal could not be replaced to Caution or its light was out, and succeeded in tying himself up in knots. Stationmaster Scott was likewise grilled.

In conclusion the inspecting officer stated that despite the evidence of Sneddon and Scott he very much doubted if the visibility at Castlecary in the open between, say, 3.30 p.m. and 5 p.m., had exceeded 400 or 500 yards (the Down distant was 777 yards from the box).

As regards the stopping of the Dundee train, it seemed that Signalman Sneddon had not been aware that the Cadder goods, which had passed him at 4.09 p.m. and would normally have been cleared back by Dullatur East at 4.14 p.m., had been held up there until he telephoned Signalman Smith at 4.22 p.m. Sneddon explained that he had not inquired earlier as to the delay in clearing out the goods train as he was "not in the habit of doing that for a train going to blocks in advance… The man the train was going to would be the man to ask if anyone was an unusually long time in the section, if he thought it had not appeared to him within the time allowed."

Signalman Sneddon had given the 'Shunt Train for Following Train to Pass', 1-5-5, signal to Dullatur East at 4.28 p.m. when he had received the 'Train Entering Section' signal for the Dundee train from Greenhill Junction, but Signalman Smith had not been able to shunt the Cadder goods until the passing of the two Glasgow expresses. He had in fact given the 'Train out of Section' signal to Castlecary at 4.37 p.m., just before the collision.

Signalman Sneddon was closely cross-examined as to the approach of the Dundee train. He maintained that having seen it run past home signal and box and its tail lamp disappear he had at once sent the 'Train or Vehicles Running Away on Right Line' signal to Dullatur East. He had not while doing so looked at the indicator of the track circuit in the rear of the Down starting signal, situated directly in front of him. "I can't explain why I didn't look, but I never looked... That is one of the things I will never be able to explain."

It was alleged by Signalman Sneddon that his colleague at Dullatur East had failed to answer his 'Call Attention' signal or to understand the 4-5-5 until he had telephoned him. Signalman Smith denied both allegations and said that Sneddon had rung him up to say that the Dundee express "had run through his signals." Fearing a collision with the Cadder goods Signalman Smith said that he had told Fireman Smith of the latter, in the box to carry out Rule 55, to "get back as quickly as possible and tell his driver to come inside." Asked why he had not telephoned Castlecary when the expected collision had not occurred Signalman Smith replied, "Oh, I did not think of doing that."

Signalman Sneddon was likewise cross-examined as to why he had not enquired of Dullatur East as to the anticipated collision. He answered that he had been excited at the time. In a later interview he said that he had thought that the Dullatur signalman had been out of his box stopping the Dundee train.

Telegraph Linesman Robertson had been in Castlecary box when Driver Macauley went there about 6 p.m. According to Robertson the driver "asked the signalman if he had a repeater on his distant signal, and the signalman said, 'No.' The driver said he had got a clear signal when he passed it, and the signalman had lost his temper. He said, 'Oh, don't come that stuff,' and the driver said, 'There's no use getting angry about it,' and he then went down the stairs."

Driver Anderson was questioned as to the wisdom of driving at full speed given the poor visibility. He could not recall when he had last had a Caution signal at Castlecary; it was more than six months previously. The day before the accident he had stopped to report a half off distant signal at Polmont Junction, the result of the arm sticking due to the frost.

If the distant signal had been showing a false clear, the inspecting officer remarked, it must have been due to its failure to return to Caution rather than the weight of snow on the arm. A very detailed discussion of the causes of signal failures followed in the report. It was stated that there had been a total of 101 failures in the Scottish area of the Company in 1937, of which 53 had occurred on the Edinburgh – Glasgow main line. These had been due to: -

1. Wire freezing in pulleys – 16
2. Arm spindle frozen – 13

3. Both wire freezing in pulleys and arm spindle frozen – 19

4. Wires off pulleys, or otherwise fouled – 5

Of these 48 had been upper quadrant signals, 35 of which had failed "On" and 13 "Off."

As regards the 5 lower quadrant signals, 4 had failed "On" and 1 "Off."

Twenty-six of the failures had occurred when traffic was light, or the box had been switched out. This included the lower quadrant wrong-side failure which had happened at Dullatur West when the box was out of switch.

Signalman Sneddon was then cross-examined at length about his acceptance of the Edinburgh express. He explained why he had rung Signalman Beattie at Greenhill Junction. "What I was not clear about in my mind was whether having sent the 'Vehicles Running Away on Right Line' signal to Dullatur East, I would be right in accepting the following train, even though I was satisfied I had the regulation clearance ahead of the home. I accepted the 4.03 p.m. ex Edinburgh at 4.32 p.m." He asserted that he had checked that all of his signals had been working properly and that his track circuit indicator had been clear. The only reason that he could give for accepting the 4.03 p.m. train was that it was a very important train, and that therefore it was imperative that he accept it, even though he was expecting a collision at the box ahead.

According to Signalman Sneddon, when Fireman Fleming of the Dundee express had arrived at his box, he had asked him who he was, then where the Dundee train had stopped and the fireman had replied, "Through the starting signal." The signalman said that he had rushed out to stop the Edinburgh train.

Fireman Fleming's version, however, was that when he had identified himself Seddon had said, "Thank God, you've stopped," and as the fireman had signed the train register book the signalman had told someone on the telephone that the Dundee had stopped in the section. "No," the fireman had corrected him, "we are standing at the starting signal." Signalman Sneddon denied having this conversation with the fireman.

Signalman Sneddon suggested to Stationmaster Scott, to Signalman Beattie at Greenhill and later to Telegraph Linesmen Robertson and Dickson and Signalmen Macleod and N. Marr (the third Castlecary signalman) that after initially having stopped beyond his starting signal the Dundee train must have set back, thus explaining why the track circuit indicator had first shown 'Clear' and later 'Occupied'.

As an alternative to the Dundee train having set back, which it clearly had not, Signalman Sneddon alleged that Telegraph Linesman's maintenance had been faulty and that the track circuit indicator must have failed. Investigation determined that there had been no reported wrong-side failures of the track circuit (that is, it showing 'Clear' when a train was on the track) since its installation in 1921 and only three right-side failures (showing 'Occupied' with no train present) from different identifiable causes. The indicator was minutely examined after the accident, and nothing was found to be wrong with it.

Concluding his inquiry, the inspecting officer stated that the main feature of the accident was Signalman Sneddon's acceptance of the Edinburgh express under full 'Line Clear' two minutes after he had apparently failed to stop the Dundee express, which was at risk of colliding with a goods train at the next box. 'It is difficult to believe that a responsible signalman could do such a thing, still less to plead justification because Block Telegraph Regulations do not prohibit such action in the circumstances which he alleged. Indeed, doubt naturally arises with regard to the integrity of the evidence relating not only to the operation of the goods train in the first place, but to the subsequent acceptance of the Dundee train, and to the circumstances in which Sneddon found it necessary to stop the latter.'

Prima facie there was also human failure on the part of Drivers Macauley and Anderson, but the case was complicated by two separate allegations of equipment failure, one of the Castlecary Down distant signal, and the other of the track circuit indicator.

As to the equipment failures, statistics showed that a wrong-side failure of the track circuit indicator was a much less likely possibility than a failure of the distant signal, especially during the prevailing weather conditions. The case of a track circuit failure, Lt. Col. Mount noted, rested entirely upon Sneddon's word, and had the collision taken place a short distance further west, beyond the starting signal, his allegation would have fallen to the ground, though he would have escaped criticism for disobeying the Block Regulations. He had stayed in the box for six hours after the collision but no mention of a failure of the track circuit had been made during that time, rather it had been suggested that the Dundee train had set back. It was only some days later, after it had been clearly shown that the Dundee train had not set back, that Seddon had come up with the idea of a track circuit indicator failure to justify his acceptance of the Edinburgh express.

The inspecting officer was also not happy with the evidence of Stationmaster Scott, for example his statement that he had not heard the conversation between Fireman Fleming and the signalman, despite having followed the fireman into the box. 'It seems {the inspecting officer wrote} therefore, that on this point {the attempt to warn the Edinburgh express with detonators} also little reliance can be placed on either Sneddon's or Scott's accounts, and I think it may reasonably be assumed that probably more time was taken up by discussion in the box than they represented, and that the placing of the detonators may actually have been an afterthought, instigated by Fleming's arrival, by his information as to the position of the Dundee express, and by his remark with regard to lack of action; also perhaps by observation of the track circuit showing *Occupied,* and finally also by the realisation that the Edinburgh express was actually passing the distant signal, and was about to repeat what had just previously happened. It is difficult, otherwise, to account for the feeling which appears to have existed in their minds that this train also would inevitably fail to obey signals.'

As regards the track circuit indicator Lt. Col. Mount concluded, 'of all the contradictory features in this case, I have the least indication in forming the opinion that neither the track circuit, nor its indicator, failed as he {Sneddon} alleged. Indeed, it seems very questionable whether he should be given the benefit of any doubt which may remain on the subject, and the only extenuating circumstance is the fact that, although 458 yards away, Driver Macauley unfortunately omitted to whistle when the Dundee train came to a stand.'

Signalman Sneddon had neither attempted, between 4.32 p.m. and 4.36 p.m., to go to the window to see if the Dundee train had pulled up, nor to telephone Dullatur East to find if there had been a collision. There could be no reasonable excuse for his failure in these respects, and he seemed mainly concerned with trying to shift some of the responsibility onto Signalman Beattie, who had not persuaded him not to accept the

Edinburgh train. Sneddon had a clear record, and his officers considered him conscientious and reliable, but it appeared that his outlook was limited by the Rules and Regulations, of which he had an acute mind. 'As a theorist rather than a practical signalman, evidently he could not be depended upon in an emergency in such a responsible position as a main line signal box.'

It was surprising that Signalman Beattie had acquiesced in the acceptance of the Edinburgh express, but the inspecting officer did not consider that the disaster would have occurred had Beattie been at Castlecary.

There was no suggestion that Signalman Smith at Dullatur East had forgotten the Cadder goods.

Fireman Fleming had lost no time in getting to the box and his account of what happened there was more credible than those of Seddon and Scott. Unfortunately, he had been just too late.

Stationmaster Scott, had he realised the position earlier, might have prevented the accident but could not be held to blame.

As regards the position of the Castlecary Down distant signal, the inspecting officer was very doubtful if, as he said, Signalman Sneddon could have seen its arm return to danger at 4.09 p.m. after the passing of the Cadder goods, or its backlight, when he had accepted the Dundee express at 4.22 p.m. It was unfortunate that neither of the drivers could be supported by their firemen. Both drivers had clearly been on the alert. Neither had seen the home signal but both had seen the red light from the box. Both drivers had only had a brief view of the signal, but each had stated that it had been clear rather than merely drooping. Undoubtedly the signal had been at caution an hour later when the engine had passed it, possibly the vibration of the heavier Edinburgh train had been sufficient to permit the counterweight to overcome the resistance due to frost.

Lt. Col. Mount stated that in the circumstances he was unable to say definitely whether the signal had been at caution, drooping, or showing a false clear, though on the balance he thought a wrong-side failure unlikely, as it was an easy-working signal, with a straight run of wire and the temperature had been little below freezing. There was also the exceptional circumstances of two experienced drivers running past the same signal in eight minutes, and Macauley and Anderson should be given the benefit of the doubt.

Driver Anderson, the inspecting officer considered, had taken unnecessary risk in approaching Castlecary at such a high speed in the conditions, in violation of the spirit of Rule 127 (iv) (regarding drivers reducing speed during fog or falling snow).

Driver Macauley had sent his fireman to the box promptly after he had pulled up but should have whistled as per the provisions of Rule 55 (a).

Turning to remarks and recommendations, the question of all-steel rolling stock again arose. The problem of fire was being reduced by the substitution of electricity for gas; the problem of telescoping remained. Older coaches had less resistance, as shown by the rear coaches of the Dundee express. On the other hand, there were financial limits to the provision of new rolling stock. All-steel coaches were also prohibitively heavy. The Company's policy of adhering to articulation, Buckeye couplings and Pullman vestibules had outstanding merits, and there could be no justification for pressing the main line companies from their present methods of carriage construction.

Had the brake compartment of the Edinburgh train been marshalled next to the engine there would have been fewer deaths. This had not been done to the inconvenience in operation. It was recommended that this be looked into.

Amendments to Rule 55 regarding whistling and to Rule 81 (h) regarding defective signals were suggested.

The completion of the provision of signal repeaters and of the interlinking of distant signals with block instruments so that they could not be placed at 'Line Clear' unless the distant arm was at Caution was recommended to be carried out as soon as possible.

Although it had no bearing on the present disaster block controls as recommended after the Welwyn disaster of 1935 (Welwyn Control) should be provided.

If the Castlecary Down distant had been at Caution the collision would have been prevented by some sort of automatic train control. This was of four types: -

(a) The Train Stop. The contact type of train stop as used on the London Passenger Transport Board was not suitable for high speed lines

(b) The Warning Control, with an audible indication combined with a brake application which could be cancelled, as pioneered by the Great Western Railway, and the similar Hudd apparatus used on the Southend line of the London Midland and Scottish Railway,

(c) The Cab Signal in which signal indications were repeated in the engine cab without speed control.

(d) Continuous Control, which required continuous track circuiting, which could include Train Stop and Warning Control, Speed Control and Cab Signals. This was a very expensive system.

Cab Signals were used in the United States, France, Belgium, and Italy. Germany used Speed Control and Automatic Braking without Cab Signals and the Swiss used a similar system.

Indirect forms of protection such as the interlinking of block instruments probably did more to increase safety for a given expenditure than any form of train control. However, the only positive form of protection against certain classes of accident on high-speed lines was some form of Automatic Train Control. [322]

Despite the conclusions of the Ministry of Transport inquiry it was Driver Anderson rather than Signalman Sneddon who found himself on trial on a charge of culpable homicide. After the evidence had been taken the judge, at the request of the lord advocate, directed the jury to return a verdict of 'Not Guilty'. [323]

No fare-paying train passengers were involved in an accident at West Burlington, Iowa, United States, on Saturday 26th February 1938 on the Chicago Burlington and Quincy Railroad, but five railwaymen being taken home from work lost their lives and 33 were injured.

West Burlington was situated on the double-track Galesburg – Ottumwa main line,

[322] Ministry of Transport report

[323] The Times, London, England, 31st March and 1st April 1938

which was worked by train order and timetable with automatic block signals. The area in the vicinity of the station was within yard limits, meaning that the track could be occupied regardless of second or lower class trains or extras, which must run through the yard at such a speed as to be able to stop short of any obstruction.

One of the trains involved in the accident was Extra No. 312, an eastbound shop train, which consisted of yard engine No. 312 and four wooden coaches, in charge of Engine Foreman Ben Gardner with Engineer Louis Mueller and Fireman Welch. The other train was eastbound Extra No. 5604, which consisted of engine No. 5604, 48 loaded and 27 empty cars and a caboose, in charge of Conductor Parsons with Engineer William Powell and Fireman Fox. The work train left West Burlington at 4.40 p.m. and proceeded eastwards along a yard track running parallel to the main lines. At a point 1,800 feet east of the station it was delayed several minutes awaiting the passage of a westbound extra. It was then taken out onto the westbound track, thence through a crossover onto the eastbound track and had travelled a short distance, attaining a speed of 18 to 23 miles an hour, when it was run into from behind by the freight train. The latter had left Ottumwa, 70¼ miles from West Burlington, at 2.30 p.m., had passed Danville, 9 miles from West Burlington, at 4.39, had received a clear indication at automatic signal S-2100, and was travelling at between 40 and 60 miles an hour when it passed the *Stop-and-Stay* signal protecting the mains crossover at *Stop* and collided with the shop train. The rear coach of the shop train was slightly telescoped by the engine of the freight train and the fourth coach telescoped the third coach about 10 feet. [324]

The men who died were William Cook, 37; Otto Langer, 59; and David Neder, 57 – all boilermakers; James Cowger, 69, mechanic; and Walter Whitford, 55, pipefitter. All were from Burlington. Cook and Whitford were among the men who had jumped from the shop train on a warning being given; they were struck by part of the wreckage. News of the accident was said to have spread like wildfire and several thousand people went to the scene, together with ambulances, firemen and police. In some cases uninjured shopmen went some distance to telephone relatives to say that they were safe, only to find that they could not reach anyone as the relatives had gone to the scene. Ironically the shop employees did not usually work on Saturdays, they had only done so on this occasion because the works had been closed on the Tuesday (which was George Washington's birthday).

Pipefitter Roy Robinson, who escaped with a sprained ankle but lost a neighbour, David Neder, was interviewed by a newspaper reporter. He said, "Everybody was tired after a day of hard work. We were looking forward to a restful weekend. I saw the switch foreman, Ben H. Gardiner, signal frantically, excitedly, for the engineer of our train to speed ahead, and at the same time he called for the men on the train to leap out. At about this time, some of the men in the rear (fourth) coach saw the oncoming freight train, as had the switchman, and all through the train we could hear the shouting of men. There was a rush of men towards windows and doors from the vestibules as they thought of leaping to safety…All my life I've thought of what I'd do in a train wreck, and I did what I'd always planned to do. In the first place, I always try to sit near the middle of the train, because of the thought I've given to wrecks, and was in the front of the third coach of the

---

[324] Interstate Commerce Commission report

train."[325]

The Interstate Commerce Commission (ICC) investigated the accident in depth. Engineer Mueller told the inquiry that he had received a clearance card form A giving him authority to proceed the 4 miles to Burlington and information about eastbound way freight train No. 86 and a westbound extra. The westbound extra having passed the foreman and two helpers operated the switches. He had asked his fireman if the latter could see any eastbound train approaching and the latter had said only extra No. 86 going into the siding. He had then begun to move through the connections when the foreman had suddenly indicated for him to hurry up and get away and the collision had occurred.

Fireman Welch said that he had not seen train No. 5604 until it was close to the station and that the smoke from train No. 86 must have hidden it.

Engine Foreman Gardner stated that his view westwards had been obscured by the cars as they went through the crossovers but then Helper Knopp had called out a warning and he had signalled to the engineer.

Helper Knopp said that as soon as he had seen train No. 5604 he had warned his foreman, signalled to Engineer Mueller and attempted to flag the freight train with his hands. He had never considered it necessary to provide flag protection, considering the *Stop-and-Stay* signal sufficient.

Helper Robinson said that when he had first seen train No. 5604 it had been 10 car lengths east of the station; smoke had obscured his view when the train was further away. He, too, had warned the engineman. He did not think that the mains crossover switch had been operated early enough to put signal S-2100 to caution in front of train No. 5604.

Engineer Powell of train No. 5604 said that signal S-2100 had been at caution and he had reduced speed to 20 miles an hour, but the signal had then cleared. He had been travelling at about 45 miles an hour through West Burlington. The *Stop-and-Stay* signal had gone from *Clear* to *Stop* when his train was 1,600 feet from it. He had applied the air brakes in emergency and opened the sanders and thought that he had got the speed down to 10 or 15 miles an hour when he had struck the shop train. He understood the requirements of Rule No. 93 (regarding yard working) but it had always been his practice to operate trains through West Burlington in accordance with signal indications and he believed that other engineers did the same. He had been travelling at the usual speed.

Fireman Fox more or less corroborated his engineer's statement.

Head Brakeman Netsell stated that smoke had obscured the view ahead for a time. When it had cleared he had seen the *Stop-and-Stay* signal at *Stop* and had shouted a warning to Engineer Powell, who had at once applied the brakes.

Conductor Parsons of train No. 5604 put the speed of his train at 45 to 50 miles an hour through West Burlington, which he thought was the usual speed there when the block signals were clear. He had never been criticised for maintaining that speed there. He understood the provisions of Rule No. 93.

Operator Stroup at West Burlington stated that the dispatcher had advised him that there was an extra out of Burlington at 4.28 p.m. and No. 86 had left Danville at 4.19; he (the dispatcher) did not think that train No. 5604 would interfere with the shop train. The

---

[325] *The Daily Hawk-Eye Gazette*, Burlington, Iowa, United States, 28th February 1938

operator said that he had cleared the shop train at 4.40. Shortly afterwards the dispatcher had advised him that train No. 5604 had left Danville. He (the operator) had then tried to flag the shop train, but failed, and he could not use the westbound *Stop* signal because the extra from Burlington was passing. It had not occurred to him to use the eastbound *Stop* signal.

Dispatcher Walter, also at West Burlington, stated that he had advised the operator that train No. 86 would probably go into the siding for train No. 5604. At 4.42 the operator had advised him that the shop train was waiting for the westbound extra. He had then informed the operator that train No. 5604 had left Danville and the operator had best hold the shop train for it. At about 4.48 the operator had told him that he could not stop the shop train.

Conductor Bradshaw and Engineer McMurray of train No. 86 and Conductor McCulloch and Engineer Dioner of Extra 5359 West also gave evidence.

The ICC asked several officials of their views regarding yard limits.

General Manager Aydelott stated, "It is possible to go through some yard limits at a pretty fair speed and still be safe provided the track is tangent, the visibility is good, and there are no trains in view that could possibly go out, but, on the other hand, trains or engines may enter the main track at any time without regard to an approaching train."

General Superintendent Grisinger said, "There isn't any rule itself that sets up speeds – only as the way is known or seen to be clear, and no one can attempt to set to speed limits whether you are going 4 or 40 miles an hour."

Train Rules Examiner Shults said, "You cannot go through a yard limit 45 or 50 miles an hour – this has been done before and always results in disaster."

The ICC noted that since the shop train had been carrying passengers Rule No. 93 imposed a definite requirement to protect it, while its crew were aware of the usual speeds maintained by freight trains through the yard. Proper protection of the shop train would probably have prevented the accident.

Rule No. 93 was designed to permit the movement of trains through yard limits at the maximum speed consistent with the range of vision, and the statements of various officials indicated that it was so interpreted by them; although one official expressed the view that a speed of 45 to 50 miles per hour was not permissible within yard limits, another expressed the view that while it was proper for trains to be operated at fair speed through yard limits, it was unnecessary for trains about to occupy the main track to give consideration to approaching trains.

Although the location of the crossover switches involved in this accident was plainly visible for a long distance, the curve at the eastern end of the yard limits prevented the engineer of an east-bound train from seeing to the extreme end of the yard limits; furthermore, within yard limits a clear indication of a block signal gave no assurance that the track governed by that signal would not be occupied immediately ahead of an approaching train. The evidence clearly showed that Extra No. 5604 was being operated at a relatively high rate of speed through the yard limits and that this was in accordance with common practice; the shop train had started across to the eastward track when the approaching freight train, in view of its speed, was less than stopping distance away.

According to the evidence, for a long period of time it had been the practice for other than first-class trains to be operated through the yard limits at West Burlington at speeds of between 45 and 50 miles per hour under clear block-signal indications, and this practice had continued apparently with the knowledge of, and without criticism by, operating officials of this railroad. A construction of the yard-limit rule which permitted high-speed main-line movements and at the same time did not impose compensating restrictions upon the use of main track by switch engines and other than first-class trains was bound sooner or later to result in accident. Had Extra 5604 been operated under control within the yard limits at West Burlington, this accident probably would have been averted.

In closing, the report stated, 'It is evident that there has been a lack of proper supervision of operating practices in this territory, and it is recommended, that operating officials immediately take such steps as may be necessary to provide for safe operation within yard limits.' [326]

A rear collision occurred in Mexico early on Thursday 13th April 1939, involving two passenger trains. Train No. 2, which had left Laredo, Texas, United States, the day before, had stopped near La Llave station, about 18 miles south of Queretaro on the Laredo – Mexico City main line of the Ferrocarril Nacional de México, due to a problem with its air brakes. Before it was able to restart it was run into by a Guadaljara – Mexico City passenger at 4.10 a.m. An official statement from Mexico City indicated that 32 people had died and 42 were injured. Most of the casualties were said to have been in the Guadaljara train. A relief train was sent to the scene from the Mexican capital with emergency eqipment and physicians.

This was the latest in a series of accidents on the national system and a newspaper reported, 'It was the latest of a series of wrecks on the worker-operated National Railways, the management of which recently charged there were efforts at sabotage to discredit the railway and the government.' [327]

A Pullman car at the rear of the Laredo train appears to have suffered less than some of the coaches of the colliding train, where a second-class wooden car was crushed by a heavy steel car. [328]

A major disaster occurred in India in the early hours of Monday 17th April 1939 at Majdia station on the Eastern Bengal Railway. Train No. 16 Down North Bengal express bound for Calcutta was standing in the station when it was run into from behind by train No. 8, the Down Dacca Mail. The accident happened at 3.23 a.m. Two luggage vans, a third-class bogie carriage and a brake van of the Bengal train were completely wrecked and the engine a two bogie carriages of the Dacca train were derailed. An early official report spoke of 31 dead and 27 injured. Those killed included the guard of the Bengal train and Mr Birendranath Majumdar, a member of the legislative assembly (M.L.A.). Another M.L.A., Mr Manoranan Banerjee, was among the injured. The M.L.A.s on the train were all travelling to Calcutta to attend the Bengal Assembly due to meet that day.

---

326 Interstate Commerce Commission report
327 *The Big Spring Daily Herald*, Big Spring, Texas, United States, 13th April 1939
328 *Standard Examiner*, Ogden City, Utah, United States, 14th April 1939

Two relief trains arrived at Calcutta at 11.10 a.m. and 2.25 p.m. The up line was reopened at 11.15 a.m., but trains were much delayed. The Darjeeling Mail, whose passengers included the Premier Mr A.K. Fazlul Huq, Mr Justice Costello, and Mr Justice Biswas, and which was due in Calcutta in the early morning, had still not arrived by 3 p.m.

A reporter wrote, 'Mr R. Rahaman, member of the Dacca University in the Bengal Assembly, and Dr Siddiqui, of Allahabad University, were asleep in a second-class compartment when the collision occurred. Hearing the cries of those injured they left the compartment and assisted in the extrication of those who had been entrapped in the wrecked bogies. Valuable aid, he said, was rendered by villagers who were soon on the scene with pickaxes and lanterns. In one bogie Mr Rahaman saw five dead. Mr Rahaman warmly praised the rescue work of the local railway officials.'

The Bengal Legislative Assembly was adjourned for the day as a mark of respect to the memory of Mr Birendranath Majumdar. [329]

The final number of casualties according to the official report on the disaster was 35 dead and 55 injured. [330]

In mid-November 1939, Driver Pearson and Guard Namey of the Dacca Mail were sentenced to terms of imprisonment for having caused the accident but released on bail pending an appeal. [331]

On 31st August 1939 the French government, anticipating the outbreak of war with Germany, began the evacuation of 30,000 children from the capital to the provinces. Early on 2nd September 1939 two of these evacuation trains were stopped by signals outside of the station at Les Aubrais, on the Paris – Orléans main line of the Société nationale de chemins de fer français, with train No. 10.585 stopped at signal No. 140 and train No. 10.531 stopped behind it at kilometre 116.200. At 04:25 a third train, *supplémentaire* (extra) No. 10.481, which had left Paris Austerlitz at 02:34, collided with the rear of train No. 10.531 and the force of the impact pushed this train in turn into the rear of train No. 10.585. The vans and rear carriages of the front two trains were crushed. Thirty-six passengers lost their lives and 77 were injured. The driver of the third train had reportedly failed to comply with his instructions to proceed cautiously. The accident received little attention in the press despite the heavy death toll, due to its preoccupation with the outbreak of the war. [332] [333]

France, together with Great Britain, had declared war on Germany on 3rd September 1940, following Germany's invasion of Poland on 1st September. On 7th September the French had begun the Saar Offensive and in connection with this attack refugees were being evacuated from the left bank of the Rhine. A train carrying these evacuees was involved in a collision at Brancourt, France, on the Toul – Neufchateau line, on Monday

---

329 *The Bombay Chronicle*, Bombay, India, 18th April 1939
330 Report by the Railway Board on Indian Railways for 1939-40
331 *The Railway Gazette*, London, England, 17th November 1939
332 *L'Echo D'Alger*, Algiers, Algeria, 3rd September 1939
333 Wikipedia – Liste des accidents ferroviaires en France au xx siècle

11th September 1939. It was stopped behind another train at a 'closed' signal when it was run into from behind by a third, empty military train at 22:40. Sixteen people died and six were injured. The collision was blamed on the failure of the conductor of the evacuee train to protect it in the rear and an error in the signalling. Again, the accident was ignored by the French press, preoccupied with the war.[334]

The firemen at San Francisco de Cordoba, Argentina, were alerted at about 1 a.m. on 1st October 1939 to a train collision in the vicinity of La Francia, and between there and El Tio. The extent of the accident was not stated but 20 men were sent. They made their way to the scene with their rescue equipment and medical supplies in a motor coach of the Ferrocarriles del Estrada. As time passed, more details of the tragedy became known. It was shortly established that two Ferrocarriles del Estrada trains had collided with great violence 10 kilometres on the El Tio side of La Francia. According to the initial information the collision involved a passenger train from Salta to Tucumán and a freight train.

The collision occurred at exactly kilometre 561 of the Córdoba – Retiro line of the former Ferrocarril Central Córdoba (which had been taken over by the Ferrocarriles del Estrada the previous year) at 12.50 a.m. (? – time illegible in newspaper) when passenger train No. 2, bound for Tucumán from Salta and in charge of Conductor Guevara, had collided at high speed with the rear of a freight train. The latter train, No. 810 (? – number unclear), had come to a stand between two stations. It was loaded with stones (ballast?) and was in charge of Conductor Eduardo Rayes. The noise of the impact awakened the locals who had of course retired to bed, and they rushed out to the street to find what had happened. Police and other officials were also soon on the scene and they quickly recognised the magnitude of the disaster. Large numbers of people from around the area went to the accident site. Cries of pain and for help could be heard everywhere, indicating the magnitude of the tragedy. The darkness and rain made it difficult to carry out the rescue work, but with the aid of torches the police and locals went through the wreckage, searching for victims, attending to them, and releasing them.

When ambulances and voluntary firemen arrived from San Francisco with rescue equipment the work of the rescuers was made somewhat easier. A sleeping car of the passenger train was one of the most badly damaged. A waggon of the freight train had fallen onto it, crushing the sleeping occupants within. Their cries attracted the attention of the firemen who worked to release them. This took a considerable time and was done amidst the moans of the injured as they were being removed from the wreckage and taken to the first-aid room at La Francia, where doctors were treating them. Nevertheless, the large number of victims meant that many were also taken to the houses of locals, which were converted into temporary hospitals. As time passed more helpers arrived who joined in the rescue of the victims. It chanced that a party of 19 medical students from Buenos Aires were passengers on the train, accompanied by four doctors. They all escaped injury and were able to lend professional help in the rescue.

The work of identifying the corpses was slow and difficult, as in the majority of cases the victims lacked identification papers. Also almost all the victims were strangers to the place and not locals who might have been recognised by one of the rescuers or locals who

---

[334] Ibid.

came to the place. With the victims having been taken to different hospitals and locations it was additionally difficult to estimate the number of dead and injured but it was initially estimated that there were 9 dead and about 60 injured. There had been 620 passengers on the train.

A newspaper reporter spoke of survivors were searching for their relatives and friends and a young boy with injuries to various parts of his body was babbling of the presence of his parents, but other passengers said that his parents had died in the accident.

After the accident the police questioned Driver Zelada of the passenger train, which it emerged was 50 minutes late and was running fast. His train had been coming around a bend when he saw the waggons of the freight train in front of him, and at the speed that his train was going it was impossible to stop. It was a frightening moment for Driver Zelada, the fireman of the passenger train and Conductor Guevera. Driver Zelada went for the brakes but given the proximity of the train in front this was of little use. He managed to close the oil tap (the passenger train was possibly a diesel railcar) and in despair he and his fireman hugged one another and waited for the inevitable collision. Meanwhile Conductor Reyes of the freight train, seeing the passenger train coming, began to signal to its driver, but his signals were not observed. He suffered a severe shock in the accident and required medical intervention.

Describing the aftermath of the collision the reporter wrote that all around there was twisted iron, with heaps of rubble, smouldering remains and despair on every face, plus the cries of the victims who sometimes had to wait a long time for help because there were so few doctors and because the darkness and rain did not allow them to quickly discover the bodies lying on the ground, between the destroyed coaches. Thus, for example, a group of locals looking for more victims eventually found a semi-clothed man in a wrecked coach who was almost lifeless and not complaining any more. In his sunken eyes he seemed to reflect the horrendous event; his legs were pierced by large splinters which, like daggers, had stabbed him, while near his body were the remains of one coach on top of another half-destroyed one, explaining the origin of his injuries. The fireman quickly removed him.

As to the cause of the collision, it came to light that the engine of the freight train had been giving trouble, and so its driver had decided to stop his train, with a view presumably of trying to fix the problem. It also appeared that he had advised Conductor Reyes of this so that the latter would protect the rear of his train. Conductor Reyes declared that he had done so, but the driver of the passenger train asserted that he had had no warning of the train in front. The newspaper report noted that this information, although lacking confirmation, showed that there had been a serious breaking of the rules by someone.

According to the railway company there were 8 dead and 49 injured; the police, however, stated that the number of dead has risen to 12, with 50 injured.

An official report was expected shortly. Meanwhile, the local railway authorities refused to give any more information on the accident to the press. [335]

The collision was later blamed on the gross negligence of the El Tio signalman, who had given permission for the passenger train to leave La Francia with the freight train still

---

[335] *El Litoral*, Santa Fe, Argentina, 1st October 1939

in the section, doubtless because he had assumed that the latter train had passed.[336] It was also, presumably, the responsibility of Conductor Reyes to at once go back and protect the rear of his train, and it is possible that his apparent nervous breakdown after the accident was due to a realisation on his part that he had failed to do so, and had thus caused the deaths and injuries, but this is, of course, only supposition.

On Sunday 8th October 1939 the driver of passenger train D 17 failed to observe two *Vorsignale* (distant signals) at caution, ran past a *Hauptsignal* (main signal) at danger and collided with the rear of passenger train P 411 at Berlin Gesundbrunnen, Germany. The last coach of the front train caught fire. Twenty-three people died and 31 were injured.[337 338]

Germany suffered the worst railway disaster in its history up to the present day on Friday 22nd December 1939 at Genthin on the Berlin – Magdeburg line. At 23:15 on the evening of the previous day, Thursday 21st, express train D 10 left Berlin Potsdamer station for Köln (Cologne). It was seriously overcrowded with the usual pre-Christmas traffic supplemented by soldiers going on leave from the front. It left on time, but with the very heavy traffic and the wartime blackout conditions it soon began to suffer delays. At Potsdam it was already five minutes late, at Brandenburg the delay had increased to 12 minutes and at Kade, because it was following a slower military train, M 176, it was held at the *Ausfahrsignal* (starting signal), and by the time it left there the delay had increased to 27 minutes.

Train D 180 from Berlin Potsdamer to Neunkirchen in the Saar started on time at 23:45, hauled by steam locomotive No. 01 158. It called at Potsdam and was then due to run non-stop to Madgeburg. It had fewer stops than train D 10. In addition Driver Wedekind in charge of engine No. 01 158 allegedly had a reputation for driving "hot", that is, approaching red signals without braking on the assumption that they would clear, in order to avoid unnecessary braking which would lose time, a risky strategy at any time and one especially so on a wet and foggy night with poor visibility. By Groß Wusterwitz, the block post before Kade station, the trains were only one block section apart.

Train D 10 was offered by Kade to Belicke at 00:34. At the latter post, 7 kilometres from Genthin, it was again checked by train M 176, which was still in the Belicke – Genthin block section. It then proceeded from Belicke at a speed of about 80 kilometres an hour rather than the permitted 105. Behind it train D 180 was offered to Belicke by Kade at 00:48.

At 00:45 the *fahrdienstleiter* at Genthin instructed the *weichenwärter* (pointsman) at Genthin Ost (Go) signal box to clear *Einfahrsignal* (home) A for train D 10 and he (the *fahrdienstleiter*) cleared *Ausfahrsignal* F.

As train D180 had rounded the left-hand curve beyond Kirchmöser station Fireman Nußman had called to his driver that the Groß Wusterwitz signals were at danger. Driver Wedekind had sounded the whistle but had not applied the brakes and as they had reached them the signals were cleared. When the train was still some distance from Kade,

---

336 *The Railway Gazette*, London, England, 24th November 1939
337 *Katastrophen der deutschen Bahnen Teil II* by H.J. Ritzau, pg. 96 (in German)
338 *Eisenbahnunfälle in Europa* by Erich Preuß pg. 120 (in German)

Driver Wedekind had remarked to his fireman that the Kade signals were also at danger, and he could not understand what train it was in front. He had closed the regulator a little but had not applied the brakes. Beyond the *Vorsignal* (distant) the *Einfahrsignal* was cleared, and the *Ausfahrvorsignal* and the *Ausfahrsignal* were then cleared as well. Beyond Kade the visibility became worse, though even at Brandenburg there had been some ground fog.

Fireman (Reserve Driver) Nußman testified later that at Potsdam the fog had been reasonably bad, such that one needed to strain one's eyes to see the signals. The fog came down from above. The light of the Vorsignal was discernible, those of the stop signals could be made out intermittently. He and his driver had noticed that steam from the engine chimney was clinging to the boiler of their locomotive worse than usual and he had remarked to his driver, "We'll probably get different weather. The wind is coming from the front."

Beyond Kade Fireman Nußman said that he had been occupied with his fire and his driver had not called out the signals, then suddenly had shouted several times, "Rudi, watch out!" The fireman said that he had interpreted this to mean that he must stop firing and at the same moment his driver had closed the regulator.

As train D 10 had neared the *Einfahrsignal* at Genthin, *Weichenwärter* (Pointsman) Jakob at Belicke had shouted over the telephone, "Stop train D 180, it has run through my signals." Hearing this, *Hilfsschrankerwärter* (Auxiliary Crossing Keeper) Adermann at *Bude* (Gatebox) 89, 850 metres on the Berlin side of Go signal box, had immediately seized his hand lamp, signal horn and some detonators and run towards train D 180. He saw train D 10 go past him at low speed, having been checked at Belicke by train M 176. Behind it train D 180 was approaching at a considerably higher speed, so that he had no time to put down any detonators in front of it. All that he could do was to wave his hand lamp at the engine, which Driver Wedekind failed to see. What he did see was the Genthin Einfahr and *Ausfahrsignals* at *Fahrt frie* (clear) because they had not yet been replaced behind train D 10.

Weichenwärter Seeger in Go signal box had most unfortunately, as it turned out, also heard the shout of his colleague at Belicke that train D 180 was overrunning signals. He at once picked up his hand lamp and turned it to red just as train D 10 was nearing his box. Fireman Sztuka, who had obviously been keeping a good lookout, saw the red light shining through the signal box window and naturally assumed that it was intended to warn them. He shouted, "Halt!" to his driver, who immediately applied the brakes in emergency. The train came to a stand at 00:51. Having thus stopped train D 10 unnecessarily Weichenwärter Seeger, who clearly had lost his head at this moment, also failed to replace his fixed signals behind it. Train D 180, travelling at 100 kilometres an hour, ran into the rear of the stationary train at 00:53.

The rear carriages of train D 10 disintegrated, after which those further ahead were telescoped into one another. The colliding engine stopped embedded in a coach and six coaches of the colliding train were derailed. According to the official figures of the Deutsche Reichsbahn 186 people died and 106 were injured. Other sources speak of 278 dead and 473 injured. That train D 10 had been badly overcrowded was shown by the fact that the bodies of 35 victims were found in the luggage van, where passengers were not meant to travel. Three passengers in the rear coach who saw train D 180 bearing down upon them jumped and survived.

The deafening noise of the collision woke the residents of Genthin, some of whom thought that it must be a blast at the local laundry detergent factory. It took a week to complete the rescue and clearance work. Because of the wartime blackout the station lay in complete darkness and permission to switch on floodlights was not always given. Car headlights were used, but these too were dimmed and so only provided a little light. The initial rescue work therefore had therefore to be undertaken in semi-darkness, and some badly injured victims were probably not discovered until they had died. With many men having been called up there was also a great shortness of medical staff and firefighters. Finally the temperature dropped during the night to -15°C, with the result that some of the injured probably died of exposure. As has happened elsewhere in other disasters, there was some looting. This being Nazi Germany, there was no mercy for looters; two employees of a local undertaker who were apprehended stealing from bodies were reportedly executed. [339]

In the investigation which followed the disaster Weichenwärter Seeger, who feared that he would be blamed for it, declared at first that, "As I was replacing signal A to danger behind train D 10 I was called on the telephone by Weichenwärter Jakob at Belicke, who told me that train D 180 had run past his signals and was 1,200 metres behind train D 10. I immediately took my hand lamp and gave a danger signal to the second train. I also went to the window with my signal horn and gave three short blasts with it five or six times. My stop signals were not observed, and the train continued at full speed until it went into train D 10…"

However the *Einfahrsignal* would only normally be replaced when the rear of the train had cleared a treadle placed about 100 metres in advance of it. To replace the signal would have taken, say, 3 seconds, to speak on the telephone to his colleague at Belicke would have taken 10 seconds and to show a red lamp to train D 180 and then sound the whistle another 10 seconds, a total therefore of at least 23 seconds. Travelling at an assumed 80 kilometres an hour train D 10 would have, in this interval, have covered 500 metres, putting it 400 metres beyond Go signal box. The collision, though, had occurred in the vicinity of the box, hence his story could not be correct.

Later Weichenwärter Seeger discovered that it was not he but Driver Wedekind who was to be blamed for the disaster, and on 26th December he made a further statement to the police: "After careful consideration, it has now become clear to me that I saw train D 10 even before the red signal was given. Behind it was train S 180. When I was on the telephone I reached for the hand lamp… I want to tell the truth now. When D 10 passed my box, I saw sparks flying from the last coach and realised that the train was braking. It became clear to me in an instant that the stop signals that I had intended to give to D 180 had instead been seen by D 10." He also admitted that in the heat of the moment it had not occurred to him to put back the *Einfahrsignal* against train D 180.

In his first interview in hospital on 12th January 1940 Driver Wedekind had stated, "As soon as I had exited the left-hand curve beyond Kade I saw the Vor and Hauptsignals at Belicke at Fahrt frie, that is at green. I saw the Belicke signals at clear soon after I had passed Kade. The Vor and Hauptsignals at Bude 89 were also at Fahrt frie."

The driver was asked if he could have been mistaken.

---

339 https://mx-schroeder.medium.com/risky-decisions-the-1939-genthin-train-collision-b7b804db37ef

"That is out of the question. When I found the signals at Wusterwitz and Kade initially against me, I knew that there must be something ahead… I realised that there must be another train in front of me, so then I paid special attention to the signals. I kept my eye on the Belicke signals until I passed them, and they remained green."

The next day Driver Wedekind made a further statement: "The signals first showed 'Halt;, then 'Fahrt frie.' As I came out of the curve beyond Kade I saw a red light. Whether it was one or two red lights I could not tell because of the hazy weather and the steam beating down from the engine. At times the red light was visible, at other times it disappeared because of the steam obscuring it. I took the red light to be the Belicke Vorsignal."

Reminded that the Vorsignal in the warning position showed a yellow and not a red one he replied, "We drivers do not distinguish in our speech between yellow and red lights. When I got closer I saw that there were two yellow lights inclined to the right. It was definitely the Belicke Vorsignal."

The driver had thus now admitted that the Vorsignal had been at caution and not clear. He was asked what he had done then.

"Nothing, because I also saw a green light."

He meant the Blocksignal (home signal) had shown *Fahrt frie*, something which the signalling and the time interval ruled out. It seemed likely that the fog had concealed the Blocksignal for part or all of the time that train D 180 had been approaching it, or that Driver Wedekind had let his attention wander at the crucial moment, and he had run past it at danger. From his position on the footplate the driver also failed to see the red light shown by Hilfsschrankerwärter Aderman, steam beating down perhaps also being a factor, and then he had seen the clear signals at Go box, which had confirmed to him that he had a clear road.

Driver Wedekind, Fireman Nußman and Weichenwärter Seeger stood trial in June 1940. The driver, whose hands were burnt and who was suffering a nervous breakdown, continued to maintain that the signals at Belicke had been clear for him. He was found guilty, presumably of manslaughter, and sentenced to three and a half years' imprisonment. His fireman and Weichenwärter Seeger were acquitted. Driver Wedekind had overrun signals previously in 1917, 1922 and 1931. [340]

Weichenwärter Seeger was perhaps lucky to escape all blame. Had he first not inadvertently stopped the first train and then failed to replace his signals behind the second the collision would have either been averted or at least much less serious. Driver Wedekind was in any case not wholly to blame. His engine, like the others of its class, had been fitted with the *Indusi* automatic train control system, which should have stopped the train as soon as it overran signals. A week before the disaster the equipment fitted to the engine had been found to be defective and had been removed for repair. Before the fault had been fixed the decision was made to return the locomotive to service because of the heavy Christmas traffic and the severe shortage of engines. For some reason this fact was apparently not brought up in the driver's defence at his trial, perhaps because of his

---

[340] *Eisenbahnunfälle in Europa* by Erich Preuß pgs. 55-64 (in German)

stubborn insistence that he had had clear signals. [341]

On the evening of the very same day as the Genthin catastrophe 101 people were to die in a collision in southern Germany between Markdorf and Kluftern (see Single-Line Collisions chapter).

Forty people were unofficially stated to have lost their lives in a collision at Torre Annunciata, Italy, on the evening of Saturday 30th December 1939. An early report in an Italian newspaper spoke of 14 dead and 40 injured while the Rome correspondent of an English newspaper, writing the next day, mentioned 29 fatalities and about 100 injured, adding that some of the injured were in a very serious condition so that the death toll might well rise.

What was described in the Italian newspaper as 'extraordinary train for travellers No. 4090', which was in fact a troop train, should have been shunted at Torre Annunciata station out of the way of the following express from Calabria. However, frozen points made the idea of shunting the train impracticable and it was decided to let it proceed instead to the next station, even though this would delaying the faster direct train No. 88. The weather was described as having been intensely cold, with snow. The signals had, presumably, been placed against the express but either the driver of direct train No. 88 overran them or else due to the weather they had not properly gone to danger, and the direct train ran into the rear of the troop train just as the latter was setting off. Railway officials opened an investigation into what had happened. [342] [343] [344]

On the evening of Saturday 6th January 1940 a train carrying soldiers on leave was run into from behind during a thick fog by a second leave train travelling a low speed, near Orly, France, on the Grand Ceinture line near Villeneuve-St-Georges, and about 16 kilometres from the Quai d'Orsay terminus. As a result of the initial collision a van was derailed foul of the opposite running line and was run into by a third train. Seven soldiers died and 18 were injured. [345] [346]

In what a newspaper reporter noted was the 11th serious railway accident in either Germany or one of the countries which it had occupied since 4th November 1939, a passenger train ran into a stationary freight train near Krakow, Poland, on Thursday 18th January 1940. Five people died, ten were seriously injured and a number less badly hurt. [347]

Germany launched a blitzkrieg attack on the Netherlands, Belgium, Luxemburg and

---

[341] https://mx-schroeder.medium.com/risky-decisions-the-1939-genthin-train-collision-b7b804db37ef

[342] *La Stampa*, Torino, Italy, 30th December 1939

[343] *The Times*, London, England, 1st January 1940

[344] *The Railway Gazette*, London, England, 5th January 1940

[345] *La Dépêche Corse*, Ajaccio, Corsica, 10th January 1940

[346] *Le Petit Journal*, Paris, France, 9th January 1940

[347] *La Charente*, Angoulême, France, 21st January 1940

France on 10th May 1940 and the Allies were soon in retreat. There was also a massive exodus of refugees westwards from Belgium and eastern France, and in the resulting chaos normal railway operations, at least on the Amiens – Rouen line, clearly broke down, with successive trains simply following one another on sight. It was in such circumstances that a collision occurred near Morgny station, France, on Saturday 18th May 1940. A train carrying Belgian refugees was run into by a train carrying French refugees which was travelling too fast for the conditions (52 kilometres an hour). The steel carriages of the first train withstood the impact of the collision well but the wooden carriages of the colliding train were crushed up against the locomotive by the sudden stoppage. Fifty-three passengers died and 128 were injured. The driver of the colliding train had been on duty for several consecutive days and on his engine for the previous 23 hours. In view of these exceptional circumstances he was freed by the Rouen Court of Appeal in a decision of 23rd December 1941, a judgement which was upheld by the Cour de Cassation (Supreme Court) in 1943, while the Société nationale de chemins de fer français were likewise found to be free from blame. [348]

Passenger train P 2250 ran into the rear of hospital train V 5632 at Rheinhausen Ost, Germany, on Tuesday 29th October 1940, killing six people and injuring three more. A signalman who had prematurely cleared back the first train and accepted the second was to blame for the accident. [349]

A major catastrophe in Germany on Wednesday 22nd January 1941 involved a train returning Flemish Belgian prisoners of war to their homeland from Stalag 1A in East Prussia. (Flemish Belgian prisoners of war were treated more favourably than their Walloon counterparts as in accordance with Nazi racial policies they were considered to be Germanic and hence superior to the Walloons.) [350] The train, W 94112, consisted of an engine and 20 French carriages. It was one of several such trains which had left Stablack at about noon on Monday 20th January. No timetables were arranged for these trains, which progressed as best they could when the line was free. Progress was slow but by 07:00 on Wednesday 22nd the train had reached Isenbüttel, on the main line between Berlin and the Rhineland. It was still dark, foggy, and snowing intermittently.

After some delay the signal had been cleared and the driver of train W 94122 had released his brakes and was just about to set off again when the train was run into from behind by through goods train Dg 6120. Travelling at considerable speed because it was not timetabled to stop at Isenbüttel, the freight train demolished the rear four coaches of the prisoner of war train.

After the accident it was the prisoners who carried out the initial rescue work. There were three Belgian doctors on the train, but they were hampered as they had no equipment with them. The German Red Cross and army (Wehrmacht) were, however, soon on the scene, later also firemen and the Hitler Youth. The prisoners were then stopped from doing any more of the rescue work as they were said to be too emotionally involved. At 14:00 a replacement train made up mostly of first-class German coaches

[348] Wikipedia – Liste des accidents ferroviaires en France au xx siècle
[349] *Eisenbahnunfälle in Europa* by Erich Preuß, pg. 121 (in German)
[350] Wikipedia – Belgian Prisoners of War in World War II

continued to Antwerp with about 800 prisoners who were either unhurt or only slightly injured.

The number who died in the accident is uncertain. One source states that there were 94 deaths, [351] another states that 97 prisoners died [352] while Wikipedia states that 71 identified prisoners died at the scene, another 20 dead were unidentified, 26 more prisoners died in hospital at Braunschweig, while four German guards and the driver of the freight train also lost their lives, thus a total of 122. [353]As, however, the source which states that only 97 prisoners died names them all, this figure would appear to be the more reliable.

The injured prisoners were apparently well treated and were repatriated to Antwerp when they had sufficiently recovered to make the journey.

The cause of the accident remains unclear. One source states that it was due to an error on the part of a *fahrdienstleiter* (signalman) [354] and two signalling staff were apparently arrested after the collision, though this may have been just a knee-jerk reaction. The other obvious possibility is that the driver of the freight train, in difficult conditions with darkness, fog and snow showers and probably not expecting to be stopped there, might have overrun the signals. An investigation was opened into the disaster, but its results were never published. News of the accident did appear in the newspapers but in a censored form. [355]

The London and North Eastern Railway main line from London Liverpool Street to Norwich had been equipped from Gidea Park, through Harold Wood and Brentwood, to Shenfield, with three- and four-aspect colour light signalling and continuous track circuits back in 1934, but on Monday 10th February 1941 this modern signalling failed to prevent a collision in broad daylight between two passenger trains. Between Liverpool Street and Harold Wood the gradients for Norwich-bound trains were generally rising, and from Harold Wood to Brentwood the line climbed for four miles, mostly at 1 in 100.

The trains involved in the accident were the 10 a.m. Down express from Liverpool Street to Norwich and the 10.04 a.m. Down stopping train from Liverpool Street to Southend. The Norwich train consisted of B17 class 4-6-0 engine No. 2828 and 13 bogie corridor coaches, in charge of Head Guard H.G. Adams, with Assistant Guard D.L. Appleby, Driver H.G. Hudson and Fireman R. Greenly. The train left Liverpool Street two minutes late and was steaming sluggishly from the start. Time was lost at Ilford station and after passing Gidea Park steam pressure fell from 200 lbs per sq. in. to 160. With water also low in the boiler Driver Hudson realised that he might not be able to get the top of the four-mile Brentwood bank, and he decided to stop at colourlight signal D.T. 17 to get up steam and advise the signalman of the situation by telephone. The express duly stopped at the signal at 10.38 a.m. and Driver Hudson spoke to Signalman E.J. Fenton at Brentwood. Told by Fenton that there was no large engine at Brentwood to assist the express, Driver Hudson decided that less time would be lost if he just remained

[351] *Katastrophen der deutschen Bahnen Teil II* by H.J. Ritzau, pg. 104 (in German)
[352] http://www.maisondusouvenir.be/isenbuttel_stalag1a.php
[353] Wikipedia – Eisenbahnunfall von Gifhorn (1941)
[354] *Katastrophen der deutschen Bahnen Teil II* by H.J. Ritzau, pg. 104 (in German)
[355] http://www.maisondusouvenir.be/isenbuttel_stalag1a.php

at the signal and got up steam, so told the signalman that he would call him again when he was ready to restart. There were two lines available between Gidea Park and Brentwood but by the time that Signalman Fenton had received the message it was too late for him to get his colleague at Gidea Park to divert the Southend train from the Down Through line upon which the express was standing to the Down Local line, as the Southend train had passed Gidea Park at 10.39.

Just as Driver Hudson finished on the telephone Guard Adams reached the engine and was told what was happening. The guard then returned to his van (the eighth vehicle of the express) and was just telling Appleby of the situation when they heard the Southend train coming. Guard Adams then proceeded towards the rear of his train and had reached the penultimate vehicle when he saw that the Southend train was on the Down Through line rather than the Local line. He ran towards the train, waving, and had time to put on detonator down 50 yards from the rear of his train. Adams timed the collision at 10.46 a.m.

Assistant Guard Appleby corroborated Guard Adam's evidence. He had followed his colleague back up the line and waved a red flag at the Southend train. Both men had narrow escapes from flying debris. [356]

The engine of the Southend train smashed through the reach coach of the express and the leading van and first coach of the Southend train were telescoped. Five passengers were killed outright, two were fatally injured and 17 others and the driver and fireman of the colliding train were more or less hurt. Fire broke out in the wreckage but with the use of extinguishers and a sailor climbing up onto the roof of a coach with a bucket of water it was put out before much damage was done. Signalman Fenton called the emergency services and police, Air Raid Precautions staff, ambulances, doctors, and nurses were soon on the scene, while soldiers, sailors and airmen among the passengers assisted in the rescue work. Most of the injured were taken to Warley Woods Hospital; a couple were taken to Brentwood District Hospital. [357]

The Southend train consisted of B12 class 4-6-0 engine No. 8556 and ten non-corridor bogie coaches, in charge of Guard P.C. Robinson with Driver F.C. Burkett and Fireman W.G. Porter. It left Liverpool Street one minute late and was thus three minutes behind the Norwich express. Guard Robinson told the Ministry of Transport inquiry that his driver had appeared to be in normal health. There had been repeated signal checks before Ilford, from which Guard Robinson had concluded that the Norwich train must be losing time. After Ilford, stops had been made at Romford, Gidea Park and Harold Wood. Signal D.T. 15 at the end of the platform at Harold Wood had been green, the guard told the inquiry. He had then written in his journal, shifted a mail bag and begun to get some newspapers ready to be put out at Brentwood. As he was doing so his train stopped abruptly, throwing him across the van.

Driver Burkett and Fireman Porter suffered severe burns and scalds in the accident, and it was three months before they could be interviewed by Major G.R.S. Wilson. Driver Burkett fully admitted his responsibility for the accident. He said that he could remember seeing signals D.T. 15 and D.T. 15B at green but was quite unable to account for his failing to see signal D.T. 16 at yellow or D.T. 16B at red, or to see the train ahead

---

[356] Ministry of Transport report

[357] *The Times*, London, England, 11th February 1941

until he was very close to it. He had driven trains between London and Southend for six years, was quite used to getting yellow signals on Brentwood bank; visibility had been good, and the sun had not been in his eyes. He had been fit and well rested, he had had no personal worries, and the engine had been steaming well.

Fireman Porter said that he had worked with Burkett for some months. His driver had appeared perfectly well. There had been no conversation after leaving Harold Wood. Leaving there he had a 'faint recollection' that his driver had been cleaning his pipe. He (the fireman) had put some coal on the fire, brought some coal forward in the tender, fired again and brought more coal forward. At the point there had been a rush of air as the Westinghouse brakes had come on in emergency. He had looked at his driver and the latter had appeared dumbfounded. The collision had followed. Fireman Porter added that twice previously he had noticed his driver become drowsy on the footplate, just for a few seconds. Examination of the driver by the company medical officer after he had recovered from his injuries failed, however, to find anything wrong with him.

Concluding his inquiry, the inspecting officer noted that there could be no question that the signals were not displaying their correct aspects and, as Driver Burkett himself had admitted, the general conditions had been entirely favourable. Major Wilson stated that he had no alternative but to conclude that after setting the controls of his locomotive for Brentwood bank Driver Burkett had allowed his attention to relax to such an extent that he had become drowsy, and perhaps had even fallen asleep. He must have remained in this state for two and a half minutes, presumably until he was aroused by the explosion of the detonator laid by Guard Adams. He had then applied the Westinghouse brake on the engine but had had no time to close the regulator or apply the vacuum brake on the train.

Such a grave lapse on the part of an experienced main-line driver was difficult to explain. He was in good health, had had adequate rest and had a good record.

No blame was attached to Fireman Porter. He had been fully occupied with firing and the signals had been on the driver's side. Nor could 'a man in his position' be criticised for having failed to mention his driver's occasional episodes of momentarily dozing off to his superiors.

Had Guard Adams managed to get back, say, 200 yards, the accident would probably have been averted but only eight minutes had been available, and the rules obliged the guard to go forward to the engine first. Although the rules did not specifically state that he should have done so it was, however, regrettable that Assistant Guard Appleby had not risen to the occasion and gone back earlier.

The inspecting officer had no recommendations. Automatic Train Control would have prevented the accident but the arguments for it were less strong where there was colourlight signalling. The accident was a reminder that when trains were stopped out of course they should be protected in the rear as quickly as possible. [358]

In what was presumably a rear collision on Sunday 16th March 1941, a light engine collided with special leave train SF 831 between Ars and Metz, Germany. Twenty-three people died and 38 were injured. A signalman had wrongly cleared back the special train

---

[358] Ministry of Transport report

before it had passed. [359]

Late on the evening of Thursday 24th July 1941 a light engine left La Pola de Gordon, Spain, heading south towards León. It was a very windy night and the engine had only covered three kilometres when it was stopped by a large tree which had been blown down across the line near a level crossing with the Nocedo road. The crew of the engine set to work to try and remove the obstruction.

Following the engine was train No. 524, the Gijón – Madrid express and mail train. This train obviously should have been held at La Pola until information had been received that the engine had arrived at the next station at La Robla, but it was somehow allowed to proceed and collided with the engine not long afterwards, probably at about 23:30. Both engines and several vehicles of the mail train were badly damaged. Eleven bodies were removed from the wreckage and about 20 people were injured, three of whom died in the next few hours. Those who lost their lives included the unnamed driver of the mail train and Fireman Alfredo Moran Uria of the light engine. A honeymooning couple and Conductor Santos Ruano of the mail train were very seriously injured, and Fireman Marcelino Vidal Garcia of the mail train was badly hurt.

After the initial confusion rescue work was begun with the aid of doctors from nearby towns and the local populace, while upon news of the accident being received there a rescue train was sent from León. [360] [361]

A mistake by a signalman led to an accident on the Crewe – Manchester main line of the London Midland and Scottish Railway early on Sunday 14th September 1941. The 12.35 a.m. Down express from Crewe to Leeds and York was just starting away from the home signal at Holmes Chapel, England, when it was run into by the 12.50 a.m. express from Crewe to Manchester. The rear coach of the York train was telescoped and demolished, and the two leading coaches of the Manchester train were seriously damaged. Six passengers were killed, three were fatally injured, 21 were detained in hospital and 24 were less badly hurt. The stationmaster was awakened by the crash and he, his family, and the sergeant of the local police force summoned medical assistance and treated the injured. Assistance was also provided by the Home Guard and the Air Raid Precautions services. The first doctor arrived at 1.40 a.m., the first ambulance at 2 a.m. and the last hospital case had been removed by 3.15 a.m. As was the case in other accidents the normal blackout regulations were disregarded, a searchlight being set up in an adjoining field to illuminate the scene to assist with the rescue operation. Some of the injured were taken to hospital in private cars and tradesmen's vehicles.

A newspaper report mentioned instances of heroism. 'Pilot Officer Evans was found shining a torch on his severed leg. He remarked, "My God, I am in a mess," and refused a cup of tea, saying, "There must be others more badly injured." He died later in hospital. A sailor who was trapped and suffering from arm and leg injuries urged that others

---

359 *Eisenbahnunfälle in Europa* by Erich Preuß, pg. 122 (in German)

360 *La Vanguardia*, Barcelona, Spain, 26th July 1941

361 Webpage: Federación Castellano Manchega de Amigos del Ferrocarril (in Spanish).

should be helped before himself.' [362]

The Leeds and York train consisted of 4-6-0 engine No. 5514 and tender, a van and six eight-wheeled bogie carriages, in charge of Guard Swann with Driver Walsh and Fireman William Stoddard.

The Manchester train consisted of 2-6-4 tank engine No. 2395 and nine eight-wheeled bogie coaches, in charge of an unnamed guard with Driver Edward Hall and Fireman George Jones.

The Leeds and York train had left Crewe at 1.05 a.m. and run under clear signals until Holmes Chapel, passing Sandbach at 1.11 a.m. Driver Walsh told the Ministry of Transport inquiry that he had been somewhat surprised to be checked by the distant signal there and he had found the home signal at danger. Fireman Stoddard had immediately set off to the box to carry out Rule No. 55. He had taken about 2½ minutes to reach the box and just as he got there the home and starting signals were pulled off. Nevertheless, he decided that having got that far he might just as well sign the train register book.

Signalman Perks had appeared surprised to see him and had asked, "Who are you?"

"I'm the 12.35," the fireman had replied.

Upon this, Signalman Perks had said, according to the fireman, "I thought you were the 12.50." Meanwhile, seeing the signals come off, Driver Walsh had moved forward slowly to pick up his fireman. Halfway along the station platform his train had been run into.

The Manchester train had left Crewe at 1.11 a.m. Like Driver Walsh, Driver Hall had clear signals, passing Sandbach at 1.19 a.m. The Holmes Chapel distant signal had also been off, but he had reduced speed from 50 to 30 miles an hour for a speed restriction. He did not see the tail lamp of the York train, which was on the fireman's (right) side of the footplate while Fireman Jones was temporarily dazzled by the glare from the firebox and thus also failed to see it.

Guard Swann was in the rear van of the York train with Goods Guard Reginald Gregory, who was travelling as a passenger to take up duty at Manchester. Guard Swann said he had seen the fireman going to the box, then, as his train began to move forward, he had heard a rumble and had seen the Manchester train coming. Shouting to his companion he had jumped from the van; Guard Gregory, who had been sat down, had not had time to do so and was one of those who lost their lives.

Between Sandbach and Holmes Chapel, and open only at night, was an intermediate block post at Bradwell. Signalman Fellows at Sandbach stated that he had obtained 'Line Clear' for the York train at 1.07 a.m. from Bradwell, sent the 'Train Entering Section' for it at 1.11 a.m. and received the 'Train out of Section' at 1.14 a.m. The Manchester train he had accepted at 1.12 a.m. and received the 'Train Entering Section' for at 1.13 a.m. As soon as Bradwell had cleared back the York train he had offered the Manchester train, had sent it on at 1.19 a.m. and had received the 'Train out of Section' at 1.22 a.m.

Signalman Miller at Bradwell had accepted the York train, according to his train register, at 1.07 a.m. (times adjusted to account for the Bradwell clock being two minutes

[362] *The Times*, London, England, 15th September 1941

faster than those at Sandbach and Holmes Chapel). He had had 'Train Entering Section' Sandbach at 1.11 a.m., obtained 'Line Clear' from Holmes Chapel at the same time, the train had passed at 1.14 a.m., when he said he had sent the 'Train Entering Section; to Holmes Chapel, and his train register showed 'Train out of Section' having been received from Holmes Chapel at 1.16 a.m. This last figure had, however, been substituted for another figure. His only entries for the Manchester train were 'Line Clear' given to Sandbach at 1.14 a.m. and 'Line Clear' given by Holmes Chapel at 1.16 a.m. Originally Signalman Miller had said that he had been "under the impression" that he had had the 'Out of Section' for the York train but he later stated that he had been certain that he had had signal, though he had not seen his block instrument go to 'Line Blocked'. He was confident that he had offered and had accepted the Manchester train, and shortly afterwards he had seen the block instrument at 'Line Clear' and his Down signal levers back in the frame, so he had pulled them off. The train had passed, and he had sent 'Train Entering Section' to Sandbach. He had had to send it a second time before it was acknowledged.

Signalman Perks stated that he had given 'Line Clear' at 1.09 a.m. and had received 'Train Entering Section' at 1.21 a.m., when he had asked for and obtained 'Line Clear' from the next box at Goostrey. Busy putting his kettle on for a cup of tea and then going down below to his cycle for his supper he had not noticed the long interval between the two bell signals. He had been downstairs when the 'Entering Section' signal had been given and it had been repeated before he had got back upstairs to answer it. He was sure that he had only had one 'Is Line Clear?' and no 'Entering Section' signal before 1.21 a.m. He had had no idea that there was a train waiting at his home signal until Fireman Stoddard had arrived at the box. It was not the practice to obtain 'Line Clear' from Goostrey until 'Train Entering Section' was received from Bradwell. Having got the road from Goostrey he had cleared his down signals. He could not remember his comment to the fireman about thinking he was the 12.50 or explain why he might have made it. After the collision he had sent the 'Obstruction Danger' to Bradwell and telephoned Signalman Miller and said, "What's this, there's two of them here?" but he could not recall the reply.

Colonel A.C. Trench, the inspecting officer, stated that clearly either Signalman Miller or Signalman Perks was responsible for the accident. Signalman Miller alleged that he had sent the 'Train Entering Section' to Holmes Chapel for the York train and had received the 'Out of Section' for it, though he could not remember his block instrument going to 'Train on Line' or back to 'Line Blocked'. He maintained that he had replaced his signals after the York train and had pulled them again for the Manchester train. He admitted that he had been much interested in a book be had been reading.

Signalman Perks on the other hand insisted that he had received no 'Entering Section Signal' until 1.21 a.m.

An Up train had passed Goostrey at 1.06 a.m., Holmes Chapel at 1.09 a.m., Bradwell at 1,11 a.m. and Sandbach at 1.13 a.m. The booked 'Line Clear' times for this train at Holmes Chapel and Bradwell differed by three minutes.

It was evident that Miller had been offered the Manchester train as soon as he had cleared back the York one to Sandbach. If, absorbed in his book, he had omitted to send the 'Train Entering Section' signal to Holmes Chapel for the first train before clearing it back to Sandbach, he was unlikely to remember to do so later, and when he received the 'Train Entering Section' from Sandbach for the Manchester train and had seen his block instrument at 'Line Clear' he might have cleared his signals for the Manchester train, if he had ever replaced them, sending 'Train Entering Section' to Holmes Chapel for this train at 1.21. If he had sent 'Train Entering Section' to Holmes Chapel for the Leeds and York train and it had not been answered he would probably have sent it again, as he had for the Manchester train.

Alternatively had Signalman Perks put his block to 'Train on Line' for the York train he must have failed to ask for it to Goostrey, forgotten about it and restored his block to 'Line Blocked' without the train passing for no reason, which seemed an unlikely combination of failures. It appeared, therefore, that Signalman Miller was responsible for the accident.

Colonel Trench stated that his opinion that Signalman Miller was responsible for the collision was confirmed by the alteration of the 'Out of Section' time in the Bradwell train register and by the discovery of serious discrepancies in the times of the bell signals for other trains compared with those recorded in the Holmes Chapel and Sandbach train registers, implying that Miller had made the entries some time later, relying on memory and guesswork. The locker on which Miller would have been sitting while reading his book was within reach of the block instruments and bells, and it seemed likely that he had been transmitting bell signals mechanically and without thought without getting up. Miller was 31 years of age, with 4 years' service, 2½ of them at Bradwell.

The remark allegedly made by Signalman Perks that he thought that the York train was the 12.50 was difficult to explain but it did not seem that the primary mistake was made by him. On the other hand given that the block sections on either side of Bradwell were only 1½ miles long he should have enquired as to why he had not had the 'Entering Section Signal' for the York train before 12 minutes had elapsed and certainly before going downstairs, even briefly. It seemed likely that Perks had been downstairs longer than he thought and that was why he had not heard the engine of the York train whistle.

The inspecting officer did not think it reasonable to have expected Driver Hall to have seen the tail lamp of the York train from his side of the engine when running under clear signals. It was unfortunate that Fireman Jones' vision had been affected by the light of the fire at that moment.

With regard to recommendations, the inspecting officer stated that he thought it undesirable that a man should be employed on permanent night duty of such a monotonous if responsible character, 'especially in the case of a young man with such comparatively short railway service that he cannot have become imbued with the spirit of responsibility and faithful performance of duties, often monotonous, which characterise the majority of signalmen, and others, with longer railway service.' He commended these considerations to the attention of the Company.

There were a number of modern safety devices which might have prevented the accident, ranging from simple interlocking between block instruments and signals to full Class C block controls with interlocking between blocks, signals, and track circuits. Such devices had been installed extensively before the war and were still being installed as labour and materials permitted, but priority must be given to four-track sections, junctions, and suchlike and there no special conditions at Holmes Chapel to justify it being given such priority. [363]

A major disaster occurred at Aboshi station, Japan, on the San'yo line, on Tuesday 16th September 1941. A passenger train for Kyoto stopped in the station was run into by No. 8 express from Tokyo, the driver of which had overrun signals. The express was travelling at 85 kilometres an hour. Sixty-five people were killed and 110 were injured. The driver of the express was arrested following an inquiry held on the spot. [364] [365]

Forgetfulness by a *fahrdienstleiter* (signalman) caused a collision at Rabsztyn, Germany, on the Kattowitz – Tunel line on Sunday 21st September 1941. After hospital train W 93 230 entered the station the *Einfahrsignal* was not replaced to danger and following passenger train P 1462 thereby came into collision with it. Seven people died and 42 were injured. [366] [367]

On Monday 17th November 1941 Agram – Wien express D 181 failed at Eichkogel, Austria, Germany, near Spielfeld-Straß station, on the Marburg – Graz line. The driver of following local freight train N 8415 was warned to proceed with caution, but failed to reduce speed sufficiently, while there was also a failure to protect the rear of the stalled train. Five people died in the collision that followed and 22 were injured. [368]

According to some accounts a rear collision at Leichholz, Germany (now Drzewce in Poland), could have been an even greater catastrophe than that at Genthin in December 1939. It happened at about 01:56 on Saturday 27th December 1941. Leichholz station was on the main line between Frankfurt (Oder) and Neu-Bentschen. Passenger train D 123 (or Dm W 123) from Berlin Charlottenburg to Warsaw via Frankfurt (Oder) and Posen was booked to depart from Reppen at 01:08 and run non-stop through Leichholz to Schwiebus where it was due at 01:52.

In front of the passenger train was freight train Dg 7053, consisting of an engine and a number of tank waggons, loaded variously with petrol, kerosene, and water gas (a mixture of carbon monoxide and hydrogen). This train should have been sidetracked at Leichholz out of the way of the passenger train but approaching that station it became divided and the rear tank waggons were left in the section. According to a statement issued by the Deutsche Reichsbahn it was snowing hard at the time of the accident, the signals were obscured by snow, and the passenger train overran them. There was a violent collision with the detached waggons, followed by an explosion and fire which involved six of the tank waggons and five coaches of the passenger train. An eyewitness described a loud explosion followed by a huge pillar of fire, so that it was almost as bright as daylight, even at Topper-Sorgee, five kilometres away. The official report stated

---

[363] Ministry of Transport report
[364] Webpage: Japanese Railway Safety and the Technology of the Day by Masao Saito
[365] *Railway Gazette* Vol 75. Pg. 611
[366] *Katastrophen der deutschen Bahnen Teil II* by H.J. Ritzau, pg. 104 (in German)
[367] *Eisenbahnunfälle in Europa* by Erich Preuß, pg. 122 (in German)
[368] *Katastrophen der deutschen Bahnen Teil II* by H.J. Ritzau, pg. 104 (in German)

that there were 41 deaths, including the driver and fireman of the passenger train, and 57 cases of injury.

An article published in 1990 in a model railway magazine implied that the accident was considerably more serious than stated by official sources and that no less than 284 people had died. This figure was allegedly quoted to the then station chief at Bentschen, Kurt Gollasch, by the accident officer at Schwiebus. The main van, a sleeping car and three seated passengers carriages of the crowded train were apparently burnt out, so a higher death toll is certainly possible, but a more recent writer has stated that it is not possible to verify more than 44 fatalities (the figure quoted in the Brandenburg State Archives).

The line was not fitted with the *Indusi* automatic train control equipment, nor probably was the locomotive. [369]

An express train from Bruxelles collided with the rear of an omnibus train on the morning of Tuesday 9th January 1942 at Saint-Denis-Bovesse station, Belgium. The collision was so violent that two carriages were destroyed, and three others were badly damaged. At the time of the report there were known to be seven dead and a number of injured. Traffic was maintained ‘by connecting trains’ – presumably trains working to and from the point of obstruction on either side. [370] The accident was reportedly caused by a signalling failure due to a severe frost.

A serious rear collision on the Chicago St Paul Minneapolis and Omaha Railway on the morning of Saturday 11th April 1942 near Savage, Minnesota, United States, was the sequel to a level crossing collision. The events which led to the accident began when eastbound first-class passenger train No. 210 collided with an automobile driven by Donald Albin Johnson, aged 18, a newspaper carrier, of Minneapolis, on the crossing a third of a mile from the station at about 6.55 a.m. The car was carried 50 yards down the line and Johnson was fatally injured and died later in hospital. [371] After striking the car the train set back a short distance and it was run into from behind by eastbound first-class passenger train No. 202, the *Nightingale,* at 7.12 a.m. As a result of this second accident three passengers, a news service employee, and Brakeman George Ogren were killed and 66 passengers, four mail clerks, four railwaymen, two dining car employees, a Pullman employee and an off duty railwayman were injured. Most of the casualties, including the dead brakeman, were in the colliding train, the leading baggage car of which was telescoped, and the following coach splintered.

Train No. 210 consisted of engine No. 513, a baggage car, a mail baggage car, a baggage car, two coaches, two Pullman sleeping cars and four baggage cars. It had left St James, 100 miles west of Savage, at 3.53 a.m., 23 minutes late, left Merriam, 15½ miles from Savage and the last open office, at 6.33 a.m., 17 minutes late, and was travelling at 40 to 45 miles an hour when it hit the car.

Train No. 202 consisted of engine No. 1654, a baggage car, two coaches, a Pullman tourist car, two Pullman sleeping cars and a café lounge car. It had left St James at 4.49

---

[369] www.blocksignal.de: Zum Eisenbahnunfall in Leichholz 1941

[370] *Le Réviel du Nord*, Lille, France, 10th January 1942

[371] *The Winona Republican Herald*, Winona, Minnesota, United States, 11th April 1942

a.m., 24 minutes late, passed Merriam at 6.54 a.m., 20 minutes late, and travelling at 25 to 40 miles an hour ran into the rear of train No. 210.

The engine crew of train No. 210 told the Interstate Commerce Commission (ICC) inquiry that the locomotive bell had been ringing and the whistle had been sounded for the level crossing with Route 16. The fireman saw Johnson's car approaching from the north at 20 to 30 miles an hour. It failed to stop and passed directly in front of the engine. The fireman told his engineer that they had hit a car, the train was stopped at 6.55 a.m., and the engineer immediately whistled for flag protection. A couple of minutes later the train was set back until its rear was about 200 feet west of the crossing.

Engineer Joseph Ginther of train No. 202 stated that his train had been travelling at about 70 miles an hour and he and his fireman had been keeping a good lookout, but the bright rising sun had made it difficult to observe objects ahead. He had first seen the flagman about 150 feet away; the latter was waving a red flag about 1,100 feet short of the point of collision and torpedoes had exploded 200 feet further on. Engineer Ginther said that he had immediately made an emergency brake application and reversed his engine, but the distance had been too short.

Fireman Julius Burbeski of train No. 202 said that he had seen the red fusee exhibited by the flagman at about the same time as his engineer.

The flagman alleged that he had put down torpedoes 2,400 feet from the rear of his train and shown stop signals to the advancing train with a red flag and a red fusee from a point 300 feet further on. The evidence of other witnesses indicated, however, that he had probably not gone that far.

The ICC noted that the flagman had had 17 minutes to protect the rear of his train. Tests showed that a person walking briskly could cover 4,660 feet in 15 minutes. Had the flagman covered that distance he could have provided ample flag protection.

Savage, the ICC continued, was a flag stop for train No. 210, which it was required to approach at a speed at which it could stop if necessary but was not a flag stop for train No. 202. When train No. 210 was approaching Savage, at 45 miles an hour, train No. 202 was already three minutes overdue at Barden, 3½ miles in the rear. Another factor was that the flagman had been riding in the fifth car from the rear because there was nowhere to station himself in the baggage cars, and he thus had more distance to cover before he even reached the back of his own train. Furthermore the engineer of train No. 210 should have waited until adequate flag protection had been provided before setting back. Finally, and in British eyes the most important factor, the line was not worked on the block system, only by telegraph and train order, a method of operation which would have been unacceptable on a British passenger railway 50 years earlier. Automatic block signals were provided between Mankato and Merriam, but not east of the latter point.

The ICC concluded that the accident had been caused by failure to provide adequate flag protection. [372]

What was referred to as the Seaboard Air Line Railway's proud streamliner, the *Silver Meteor* express from New York to Miami, Florida, was involved in a rear collision at

[372] Interstate Commerce Commission report

Kittrell, North Carolina, United States, early on Sunday 14th June 1942. It was run into by a freight train and the rear part of its rear coach, an observation car, was crushed in. Five passengers were killed outright, one died en route to hospital at Henderson and two more after arrival at the hospital. Eight passengers and the fireman of the colliding train were injured, the passengers only slightly so. The collision occurred at 12.50 a.m. and there was a slight fog. [373]

The *Silver Meteor,* or southbound first-class passenger train No. 43, as it was to the railway, consisted of diesel-electric engines Nos. Nos 3009, 3016 and 3100, a passenger baggage car, four Pullman sleeping cars, a dining car, two coaches, a lounge car, a tavern car, a dining car, four coaches and a coach-observation car.

The colliding train was southbound second-class freight train No. 89, which consisted of engine No. 408, 22 loaded cars, an empty car, and a caboose. None of the names of the crews of either of the trains seem to have been recorded either in the newspapers or in the official report on the accident.

The Kittrell operator told the Interstate Commerce Commission (ICC) inquiry into the collision that he had had a clearance card to deliver to the passenger train, which he had stopped by waving a white lantern. The engineer of the latter had made an emergency brake application when he saw the operator's hand signal, which meant that there was a delay of several minutes while he recharged the engine brake system.

The engineer stated that he had whistled out the flagman, then, after about 7 minutes the gauge indicated that the system was recharged, and he had whistled to recall the flagman. Two minutes later there had been a proceed signal from the rear of the train, but the brakes would not release, and the collision had shortly followed.

The flagman testified that it had taken him about a minute to pick up his equipment after the train had stopped and he had then proceeded northwards. About four minutes later, when he had covered 350 feet, he had been recalled. Placing a lighted ten-minute fusee he had returned to his train and signalled to the engineer to proceed. When his train had not restarted he had again gone back and about the spot where he had left the fusee had seen the freight train approaching. He had then lit a second fusee and hurrying northwards had got back about 700 feet when the freight train, travelling at 15 miles an hour, had passed him.

The engineer of second-class freight train No. 89 stated that his train had been travelling at about 45 miles an hour. There had been a slight fog restricting visibility. Automatic signal No. 1203 had been displaying *Approach*, he and the other people on the engine had called out the indication to one another, and he had made a 15-pound brake application. Speed had been reduced to 35 miles an hour, he had made another 10-pound brake application and speed had been reduced to 20 miles an hour. A third brake application of 5 pounds had followed, and he had then seen the flagman waving his fusee and automatic signal 1217 at red but was unable to pull up and ran into the rear of the passenger express 90 feet beyond the signal. He claimed that his brakes had not functioned properly but checks after the accident showed them to be working properly, while a test freight train which passed signal No. 1203 at the maximum authorised speed was able to pull up short of signal No. 1217.

---

[373] *The Burlington Daily Times*, Burlington, North Carolina, United States, 15th June 1942

The ICC noted that if the flagman had not been recalled until it was certain that train No. 43 could proceed he could have had time to get far enough back, while if the freight train had been properly controlled in accordance with the *Approach* indication of signal No. 1203, it could have pulled up in time. It noted that a rear collision had occurred on the Company's line at Moncure, North Carolina, less than a week earlier from the same causes, indicating there was a lack of proper supervision of the enforcement of the rules.

The ICC concluded that the accident had resulted from failure to provide adequate flag protection for the preceding train and failure properly to control the speed of the following train in accordance with signal indications. [374]

Five hours after the accident, at 6 a.m., the body of Raleigh Divisional Superintendent H.M. Terrell was found in the railroad yard there. It was clear that he had shot himself in the mouth. Apart from the Kittrell and Moncure accidents there had been other local accidents recently near Roanoke Rapids on 6th May, when an engineer had died, and near Petersburg on 28th May, when 20 persons had been injured. The spate of accidents had apparently been too much for him. Coroner Roy Banks said that Terrell had been talking about the accidents and had been overworked and unnerved. At Terrell's inquest a verdict of suicide was returned. [375]

Nine people reportedly died and 17 were injured on Wednesday 29th July 1942, when a freight train ran into the back of a passenger train near Llay Llay, Chile, on the line between Santiago and Valparaiso. [376] The electric passenger train had 'developed motor trouble', and stalled on a steep gradient in mountainous country. [377]

A serious collision at Stuttgart-Bad-Cannstatt station, Germany, on Friday 16th October 1942, resulted from a signalman's error. Passenger train P 1954 from Waiblingen was stopped at the *Einfahrsignal*, which the signalman had neglected to clear. He then forgot about the waiting train, cleared back his block instrument and accepted the following train, P 1666, also from Waiblingen. Twelve people lost their lives in the rear collision which ensued and no less than 268 were injured. [378]

On duty at Sliedrecht in the Netherlands on Friday 27th November 1942 were *Iste haltechef* (Stationmaster) van der Kreeft and *stationsambtenaar* (Clerk) A. Otten. The stationmaster was working a split shift from 07:45 to 11:45 and from 14:00 to 18:10 and the clerk was on duty from 16:00 to 22:15. The stationmaster was in charge of the signalling of the trains during his hours of duty but both men were fully competent to do the work and at times assisted one another. The Gemeenschappelijk Onderzoek (accident investigation report) called this a highly reprehensible practice which fortunately was not widespread. The report noted that this happened at the time when duties were being exchanged, and was therefore more explainable, but was still to be disapproved of. This

---

[374] Interstate Commerce Commission report
[375] *The Burlington Daily Times*, Burlington, North Carolina, 15th and 17th June 1942
[376] *The Billings Gazette*, Billings, Montana, United States, 30th July 1942
[377] *The Railway Gazette*, London, England, 9th October 1942
[378] *Eisenbahnunfälle in Europa* by Erich Preuß, pg. 124 (in German)

prohibited but ingrained habit led to a misunderstanding which caused the accident.

Sliedrecht station was on the Dordrecht – Geldermalsen line. West to Dordrecht the line was single, but from Sliedrecht east to Gorinchem it was double track. The line to Dordrecht was worked by telegraph block, with an intermediate block post at Het Wantij. At Sliedrecht there were two signal boxes, T in the station buildings and I at the west end of the station yard close to where the single line from Dordrecht became double track. Post I was a subsidiary post and when a train was to proceed west towards Dordrecht the signalman there had to obtain permission, a so-called *treinrichtingsvenster* (train direction window) from post T. This done, the Post I signalman operated the points at his end of the station. The signalman at Post T could only clear his home signal after the *treinrichtingsvenster* had been accepted. The (standard) operating instructions for Sliedrecht stated that the levers must be returned to normal position as soon as possible after every train movement. The intention was that after the arrival of each train the *treinrichtingsvenster* should be cancelled and permission obtained again for a fresh movement. Instead, the irregular practice had arisen of not cancelling the *treinrichtingsvenster* prior to the arrival of a westbound train from Giessendam (the next station to the east) at track II at Sliedrecht. This was because almost all the trains using this track continued towards Dordrecht. Only if a train was arriving on the parallel track III was the window cancelled.

Three trains were involved in the accident. They were train No. 3168 from Nijmegen to Dordrecht, train No. 3171 from Dordrecht to Nijmegen, and workman's train No. 10814 from Leerdam to Rotterdam Delftsche Poort. Train No. 3168 arrived at Sliedrecht at 18:03, four minutes late. Stationmaster van der Kreeft operated the block and signals for it while Clerk Otten collected the passengers' tickets. Train No. 3168 could not continue towards Dordrecht immediately as it must cross train No. 3171 at Sliedrecht and the latter train had only left Dordrecht at 18:06, 16 minutes late. Having finished with train No. 3168, van der Kreeft sat down at his desk behind the signalling equipment and Otten continued with his work at the counter. When the telegraph bell rang Otten went to answer it. He was expecting a telegram about train No. 3171 and moved the commutator of the instrument to 'Post Brug Wantij.' He got no writing on the tape; the train had already departed at 18:06 while he was still on the platform. Presumably he then set the commutator towards Giessendam; he could not be sure afterwards. Writing did then appear on the tape, but it was difficult to read. The telegraph instrument to Sliedrecht often did not work properly; it had been checked that morning.

The message on the tape was from Giessendam to say that train No. 10814 had left. Because he could not read the message Otten answered not with 'understood' but with 'telephone'. Giessendam replied with a question mark and Otten again sent 'telephone'. It was now 18:10. Otten then spoke on the telephone with Stationmaster Kuiper at Giessendam. According to Kuiper Otten asked, 'Chief, is it not 18:09?' and he (Kuiper) had replied, 'Yes, Otten, it's 18:09.' After the accident Otten could not recall the exact words he had used. His mind was wholly on train No. 3171. Because he had been on the platform when van der Kreeft had accepted train No. 10814 from Giessendam Otten knew nothing of this train. As he had not mentioned a train number, nor paid attention to which telephone he was speaking on, he believed that he was being advised by Het Wantij of the passing of train No. 3171.

Otten then had to send train No. 3171 on to Giessendam. It was usual as soon as a train was reported as passing Het Wantij for permission to be obtained from Giessendam for the train to proceed further then once it had been accepted the starting signal B1 at the east end of the station could be cleared from Post T. At the same time or a little later Post T would give the *aankomstvenster* (arrival window) to Post I so that the home signal on the Dordrecht side could be pulled for train No. 3171. Permission did not have to be obtained from Giessendam for train No. 3171 as when closing the block behind train No. 3168 Giessendam had accidentally released the block for train No. 3171. Stationmaster Kuiper told Otten over the telephone of this error.

Otten neglected to give the arrival window to Post I and the home signal at Post I was therefore not cleared for train No. 3171. He did not clear the starting signal B1 at the Giessendam end of the station either. Instead he pulled the adjoining lever and cleared the home signal A2/A3 from Giessendam. Presumably he just pulled the wrong lever. Because the *treinrichtingsvenster* from Post I had not been cancelled as it should have been the lever was free and the signal could be cleared.

Train No. 10814 was not booked to stop at Sliedrecht and was still travelling at 60 kilometres an hour when it ran into the rear of train No. 3168 in pitch darkness. A postal conductor and 11 passengers on train No. 10814 were killed outright. Many others were more or less injured, six of whom died on the way to or shortly after reaching hospital, so that the final number of dead was 18, with about 50 more injured. Four carriages were wrecked and the end walls of three others were driven in. All the tracks through the station were blocked by wreckage and derailed vehicles but by the evening of Sunday 29th the lines were all clear. [379]

Christmas was and is always a busy time on the railway, in Canada as elsewhere. On the evening of Sunday 27th December 1942 Canadian Pacific Railway local train No. 550 from Pembroke to Ottawa was crowded with happy men, women, and children. It consisted of a steam locomotive and several wooden coaches and was in charge of Conductor M.O. O'Connel, with Assistant Conductor J.H. Morris, Engineer Joseph Sauve and Fireman F.J. Dixon. The heavy traffic, combined with unfavourable weather conditions with rain, sleet and snow, and an engine which was not working very well, meant that the train was losing time. It left Arnprior at 7.55 p.m., passed the unstaffed station at Pakenham and arrived at Almonte station, Ontario, at 8.35 p.m., having taken 42 minutes from Arnprior instead of the timetabled 28. The station stop at Almonte was only meant to occupy three minutes but with the extra number of passengers it lasted twice as long.

Behind train No. 550 was a fast troop train, passenger extra No. 2802, conveying men from Petawawa Camp, in charge of Conductor John Howard with Engineer Lorne Richardson and Fireman S.C. Thompson. This train had arrived at Arnprior already at 7.59 p.m., only four minutes after the departure of train No. 550, although a time interval of 20 minutes was meant to be maintained between trains. The train took on coal and water and then started but was signalled to stop again by Arnprior operator C.M. McLean as the 20-minute interval had not expired. Finally it was allowed to proceed at 8.15, exactly 20 minutes after the departure of train No. 550. Unlike the local train, the troop train ran fast between Arnprior and Almonte, covering the distance in 26 minutes, and it

[379] *Spoorwegongevallen in Nederland* by R.T. Jongerius, pgs. 113-7

ran into the rear of the local train just as the latter was setting off at about 8.41 p.m. The results of the collision were disastrous. The rear three cars of the local train were shattered, and 36 passengers were killed or fatally injured, while 155 were hurt. All of the dead and the injured were on the local train.

A pair of passengers who had been on the local train but had escaped death or injury told their stories to a reporter. 'Two young stenographers in the Royal Canadian Mounted Police intelligence branch, their clothes torn, and their faces begrimed, walked out of Union station this morning, scarcely believing that they were alive and sound. They were Rita Burns of Renfrew and Rhoda Scoble of Vulcan, Alta' … 'The locomotive tore through their car, shattering it to pieces, and went on to wreck two other cars, but they escaped with scarcely a scratch.

'"There was a jolt, the lights went out, and there was a terrific noise, and we were flying through the air," said Miss Scoble. I was sure that I would die – it never entered my head that we would escape." But the girls landed skidding about 20 feet away from the track. "We skidded in the snow too dazed to know what was happening to us, and wreckage started falling on us. But it was in small pieces. None of the big pieces struck us and we were not hurt."

'With an idea there might be an explosion they ran from the scene, but soon returned to help soldiers remove wounded. Their clothes were in tatters, and they suffered from the cold until officials ordered them into a temporary dressing station where they were wrapped in blankets.

"I invited Rhoda to spend Christmas with me at my home in Renfrew," said Miss Burns. "We had a good holiday and were returning for work this morning. The train was crowded, and we went through car after car looking for seats where we could be together. Finally we found one in the very last car."

'Miss Scoble was suffering from a cold and reclined with her head in her friend's lap. "All at once there was no seat beneath us," said Miss Burns. "The floor, ceiling and walls just parted, and we were flying with pieces of wreckage all around us. Everything just seemed to go to pieces."

'After describing their feelings when they found that they had landed together and apparently unhurt, they told how they returned to the wreck. "No one appeared to have climbed out," Miss Burns said. "What we saw was a pile of debris, with bodies all around and an engine on top of one of the cars. We ran over to one car under which there were three people, one an old man who seemed to be dead. There was a woman with something wrong with her neck who wanted us to move her, and we were afraid to. Then we went to an airwoman who was lying in the middle of the road. She had been thrown there and it was a long way from the wreck. She apparently had both her legs and wrists broken and she wanted us to stay with her. We found her a bed-jacket and put it under her head and left her with another woman."

'The girls were then taken to a shelter which was soon so filled with the badly hurt that they again resumed their role of helpers, serving hot drinks and cigarettes to the injured. From a window of the shelter they watched soldiers from the troop train take charge of the rescue work. "They were positively wonderful. They were so well organised and so good at making splints and improvising stretchers," said Miss Scoble.

'The girls were ordered to go from the train to an Ottawa hospital, but they did not feel that necessary. They were given baths and fresh clothes by the wife of a newspaperman who lives close by, and they declared themselves fit for work.'

Describing the scene at Almonte immediately before and after the wreck the newspaper wrote, 'Police chief William Peacock, of Almonte, with 150 other citizens of Almonte saying goodbye to relatives and friends, was on the station platform as the engine crashed. "I didn't see the troop train coming," he said later. "There was a terrific crash as it hit the rear end of the other train." Except for the three cars telescoped in the wreck – they were left strewn across the main street after the accident – damage to the Pembroke-Ottawa train was slight. The forward coaches were able to proceed later to their destination, with a solemn load of travellers thankful for their escape. Inspector T.W. Cousans of the Ontario provincial police was in charge of the task of identifying the bodies taken to the town hall.

'Passengers unhurt in the wreck arrived in Ottawa today. Shortly after their coaches pulled into the capital, a hospital train with 100 patients bound for Ottawa hospitals left Almonte. Just before it left an unidentified Renfrew, Ont. woman died on the train and her body was taken off. Almonte was thrown into confusion by the crash, one of the worst in the history of the line. Citizens and soldiers from the troop train carried dead and injured in a steady stream into the city hall and the O'Brien theatre, turned into an emergency casualty clearing station.

'So crowded was the passenger train, carrying throngs of persons from Christmas holiday reunions and winter sports outings, and some passengers were standing in the aisles. Milton E. Smythe, of Pembroke, Ont., who was among the injured, told reporters he had been sitting in the rear coach of the passenger train. He was probably saved from death or more serious injury because he happened to move two cars forward just before the accident.

'The corridor of the town hall was crowded with injured. E.D. Balfour, sworn in as a special constable to help handle the crowds, said every truck and many in the surrounding district were commandeered for use as ambulances. Stores were opened to provide bed furnishings, used as stretchers and to keep the injured warm.

'Jack Foster, of Renfrew, Ont., and C.C. James, of Douglas, Ont., near the front of the local train when the crash occurred, said the impact was so great they thought the engine had blown up. Even at the front of the train, baggage showered down from the overhead racks onto the passengers. Ambulances from Ottawa made their to the wreck over snow-packed roads. Lights burned through the night in every building along Almonte's Mill and Bridge streets – the two main thoroughfares. Inside were injured from the train. Occasionally one of the doors would open and some person who had died would be carried out to temporary mortuaries, to await identification by relatives.

'The crash occurred when the local for Ottawa stopped at the Almonte station to take on passengers – estimated to have numbered some 200. The train were already crowded. The passengers were put into the last three coaches. Previously the train had stopped at Pembroke, Renfrew, Arnprior, and other points to pick up men, women and children returning to the capital from Christmas gatherings. It was reported that none of the passengers aboard the troop train was hurt. Scream after scream came from the wreckage of the three cars damaged by the crash. Some of the dead were killed instantly, while many died after being taken to hospitals and temporary hospitals. The railway line at the station was a shambles. Rails were pulled up like so much wire, ripped away from the ties and twisted as the heavy locomotive on the troop train ploughed into the passenger train. By 4 a.m. C.D.T., however, railway wrecking crews had cleared the line for traffic.

'For hours, the rescue operations proceeded. One injured man, trapped by tangled steel, calmly gave directions to workmen releasing him with acetylene torches. Not for several hours after the crash did rescue workers happen to find the body of a small child lying beside the station building. Whether she was thrown there by the impact or crawled there before she collapsed was not known.' [380]

The inquest on the victims of the accident took place in early January. One person who was not present at the inquest was Conductor Howard of the troop train. Two days before it began he had committed suicide by throwing himself into the Rideau River, having left a suicide note in which he had written that he was free from blame but feared that he would be made a scapegoat.

On Saturday 9th January, after a retirement of three hours, the five-man inquest jury returned its verdict. Conductor Howard was mistaken, at least as far as the inquest was concerned. The jury returned a verdict that none of the train crew or station staff was to blame for the disaster, but that the entire blame for it lay with the Canadian Pacific company. Their reasoning was that:

1. There had been no operator on duty at Pakenham who could have stopped the troop train because it was overtaking the local train and held it until 20-minute interval had been restored. The jury considered that when Dispatcher B.O. Heintz at Smith's Falls had found that the troop train was gaining on the local he should have made an effort to get the Pakenham operator on duty and to have advised Almonte of the closeness of the troop train.
2. There was no protective signal at a "most dangerous curve" on the approach to Almonte.
3. There had been a green signal displayed at Almonte station which had misled Engineer Richardson into thinking that the line was clear.

The jury made four recommendations:

1. There should be an operator "on steady duty" at Pakenham.
2. An automatic stop signal should be installed immediately west of Almonte.
3. A speed limit of 25 miles an hour should be introduced through Almonte and strictly enforced.
4. The block signal at Almonte station, which was only put to danger once trains had left the station, should instead be used to protect trains standing there.

The Coroner, Dr Smirle Lawson, noted that the order given to the troop train called for it to run at a faster schedule at certain points than the local train in front of it, while the crew of the local had no knowledge that the troop train was following them. The engineer of the troop train was ordered to run at the speed of the local. The coroner observed that the conditions under which the troop train engineer was operating, "seem to

---

[380] *Winnipeg Free Press*, Winnipeg, Manitoba, Canada, 28th December 1942

me liable to confuse any man." [381]

The report of the transport commissioners, issued on 22nd February, placed more blame on the crews of the two trains. It found that the crew of the troop train had failed to comply with the rules by exceeding the scheduled speed of the local train between Arnprior and Almonte, and that "the engineer of passenger extra No. 2808 did not have his train under control and prepared to stop as he approached Almonte station. It is also felt that the company's official who as riding this train erred inasmuch as he failed to take such necessary action as would ensure compliance with the rules."

The crew of the local train were also blamed for neglect for failing to provide protection by means of red or yellow fusees at the rear of their train when it became apparent that they were losing time, as required by the rule book.

The commission agreed with the jury that the west approach to Almonte station was dangerous as it lay on a curve and under certain weather conditions mist could arise from nearby falls. It stated, "The combination of these facts having been disclosed, it appears that the erection of a station protection signal west of Almonte would be an additional safeguard to a train standing in Almonte station," and that, "a direction to this effect will go to the Canadian Pacific Railway company immediately." [382]

The year 1943 started badly for the German railways with a major disaster on Monday 4th January. During a blizzard the driver of a special leave train for frontline troops (train SFR 2304) ran past a stop signal and into the rear of passenger train D 8 between Hannover Hbf and Wunstorf. Twenty-five people died and 169 were injured. [383]

A passenger train standing in Jaleswar station, India, on the Bengal – Nagpur Railway, at 02:30 on Tuesday 18th May 1943 was run into from behind by a goods train. The two rear bogies (coaches) of the passenger train (a third, luggage and brake and a first, second and inter) were telescoped. Fourteen people were killed and 32 or 34 injured, 12 seriously. The injured were removed to Kharagpur Main Hospital. [384]

The Jaleswar accident was followed less than three weeks later by a major catastrophe on the Great Indian Peninsula Railway. Early on Friday 4th June 1943, train No. 1, the Down Bombay – Calcutta Mail (also known as the Nagpur Mail) collided with a goods train nine miles from Akola, India, while travelling at full speed. An initial report stated that 50 people had been killed and 100 injured. [385]

By Monday 7th June it was known that in fact no less than 80 people had been killed more or less outright and 140 injured, of whom 66 had been taken to hospital and three had subsequently died. [386]

---

[381] Ibid, 11th January 1943
[382] *The Winnipeg Tribune*, Winnipeg, Montreal, Canada, 23rd February 1943
[383] *Eisenbahnunfälle in Europa* by Erich Preuß, pg. 124 (in German)
[384] *The Bombay Chronicle*, Bombay, India, 19th May 1943
[385] *The Times*, London, England, 5th June 1943
[386] *The Bombay Chronicle*, 7th June 1943

Mr Satyendranath Mazundar, Chairman of the Editorial Board of *The Daily Jugantor* newspaper, chanced to be a passenger on the mail train and related his experiences. He said that when the collision had occurred the engine, two third-class bogies, a dining car and part of the inter class compartment of the mail train had been smashed to pieces. He had been in the less damaged part of the latter vehicle of the train and hence had escaped. After the accident he had helped some European soldiers who had been travelling on the train and a few of the uninjured passengers with the rescue work; the soldiers had been instrumental in saving the engine driver. [387]

Following an inquiry into the catastrophe by the Senior Inspector of Railways Akola Assistant Stationmaster D.M. Dave and Signaller M.A. Jalil were arrested under Section 101 of the Indian Railways Act before being released on bail, so the collision was presumably a result of a signalling error rather than the train driver overrunning signals. [388]

A rear collision occurred between Korschen and Lötzen, Germany on Christmas Day, 25th December 1943. At the time of the accident the two places were in East Prussia; they are now in Poland and named Korsze and Gizycko respectively. A *fahrdienstleiter* (signalman) failed to replace the *Einfahrsignal* (home signal) after the passage of through freight train Dg 94 476 and fast passenger train E 32, which was following, ran into it. There were 15 deaths and 32 cases of injury. [389]

On Monday 10th January 1944 mail and passenger train No. 21 left Madrid Principe Pio station at 22:45 for Coruña and Vigo. It was followed at 23:00 by express train No. 1 for Irún. Early on the following day both trains were running a little late when the first train stopped in Arévalo, Spain, having had problems with its brakes binding. It was still there when the express reached the station, where it was not booked to stop. The signals at the entrance to the station were at danger but the express overran them and collided with the rear of the mail train. As chance would have it the last two coaches of the front train were old wooden third-class ones which were completely crushed in the collision and there were a high number of casualties. According to official sources 41 passengers lost their lives in the disaster and 78 were injured, though unverified reports in the foreign press put the number of dead at 96, with 100 injured.

The help provided by the local population after the accident led the Franco government to grant the town the title of Ciudad Muy Humanitaria (Very Humanitarian City) on 28th December 1945. [390]

An accident occurred at Novice, Texas, United States, in the Sweetwater District of the Gulf Colorado and Santa Fe Railway, in the early afternoon of Thursday 13th January 1944.

Southbound first-class passenger train Second No. 75 consisted of engine No. 3432, a box car, a baggage car, a baggage mail car, five coaches, a dining car, two Pullman

---

[387] Ibid, 9th June 1943

[388] Ibid, 10th June 1943

[389] *Eisenbahnunfälle in Europa* by Erich Preuß, pg. 125 (in German)

[390] Webpage: Federación Castellano Manchega de Amigos del Ferrocarril (in Spanish).

sleeping cars and a business car. It left Coleman, Texas, about 18½ miles north of Novice and the last open office, at 12.22 p.m., 4 hours and 39 minutes late, and stopped at Novice at 12.51 p.m. About a minute later it was run in from behind. It was snowing heavily at the time.

Following train Second No. 75 was Extra No. 1279 South, which consisted of engine No. 1279, three tourist sleeping cars and a baggage car, in charge of Conductor C.C. Schultz, with Engineer J.B. McCall and Fireman J.R. Sanderson. This train passed Brownwood, 48¼ miles north of Novice, at 11.35, passed Coleman at 12.32 and was still moving at 30 miles an hour when it struck train Second No. 75 at 12.52 p.m. The colliding train was a troop train which was carrying soldiers to the Military Training Replacement Center at Fort Barkeley, Texas. The front of the ninth car of the forward train telescoped the rear of the eight car for 25 feet and all the fatalities occurred in the latter coach. Initial reports put the number killed at 12 but it was later found to be only seven. No less than 105 people were injured though, including the conductor, engineer, fireman and two brakemen of the colliding train. [391]

A reporter described the tragic end of a journey for one family. 'The frail, blue-lipped little woman shivered in the sudden gust of wind that swept the snow-covered plains and huddled deeper down in the heavy Army blanket that draped her shoulders. "My boy already has been taken to the hospital. I'm waiting to go there now. I don't know how bad he's hurt."

'She didn't know that a few feet away husky Camp Barkeley soldiers were hacking with pickaxes at the caved in end of a railway car, trying to free the body of her 11-year-old boy. She didn't know that when his body was finally pulled out, it was minus an arm and leg.

'Marvin Riordan, 11, had wearied of sitting up in his chair in the long ride from Hattiesburg, Miss. He decided to help the porter, E.V. Moore, of Temple, sweep the car. Marvin had taken a great shine to the porter, and the feeling was mutual. And when the playful little lad's life was snuffed out in the body mangling crash of car into car, his mother, Mrs Ed C. Young, and a brother, Melvin Riordan, 17, sat a few seats back, reading and talking. The family was en route to Winslow, Ariz., where Mr Young is an employee of the Santa Fe Railway Co.

'The unknowing mother's poignant wire to Mr Young said simply, "Marvin has been hurt in a train wreck. We got to stay until he gets well."

'"We got to get to the hospital. We don't know how hurt he is," Mrs Young said, clinging to her elder son. The boy nodded affirmation. "We just don't know. We haven't seen him since it happened. They say he has already been taken to the hospital." Their faces bore that dull, expressionless look that goes with shock. The bone-chilling cold failed to penetrate. Repeatedly kindly soldiers passing by tightened the blankets enveloping the couple, tucked them closer.'

Others were luckier. 'A pair of youthful Texas sailors, cousins, fervently thanked the impulse that made them move from the rear of the "death-car" up about mid-way a scant few minutes before the accident.' J. L Bothwell, Santa Fe Traffic Manager, and his wife had been riding in the business car at the rear of the front train but had gone forward to

---

[391] Interstate Commerce Commission report

the diner before the collision, while 'Mr and Mrs Troy Lee Bean, Garvin, Okla., injured in the Santa Fe train wreck at Novice today lost their two-months old daughter in the confusion. The parents were brought to Hendrick Hospital here {Abilene}. They gave a description of their child's clothing and blanket to hospital attendants, who located the girl, uninjured in Overall Memorial Hospital in Coleman, 50 miles away.' [392]

The Interstate Commerce Commission (ICC), which investigated the accident, noted that according to the rules the flagman of train Second No. 75 was required to protect his train. As the rear vehicle of the train was a business car the flagman had positioned himself, as per the bulletin instructions, on the rear platform of the penultimate car. He stated that had dropped a lighted 10-minute fusee about 2,200 feet north of where the rear of his train had stopped. A little later he had seen the following train about 800 feet distant and had jumped off to give flagging signals but had fallen in the snow. The remains of a burnt fusee were later found where the flagman said that he had dropped it.

The ICC commented that on this line trains were operated by timetable and train orders, being spaced only by time intervals at open offices and by burning fusees dropped by flagmen. In this case the trains were 10 minutes apart at Coleman, but there was no means of maintaining that interval until the train reached Lawn, the next open office 12.9 miles south of Novice and 31.3 miles from Coleman. Despite this trains were authorised to run at 70 miles an hour between Coleman and Lawn. Also despite the requirement to maintain a 10-minute time interval the Santa Fe company only supplied its flagmen with 5-minute fusees. The flagman in the accident had himself obtained 10-minute fusees from another source. Unfortunately the fusee that he had dopped had been hidden by the deep snow, which tests showed was likely to happen, and had not been seen by the engineer of the colliding train. A trainmaster considered that the flagman had done his duty.

Had a manual block system been in use the accident would (or at least should) not have happened.

The ICC concluded that the accident had been caused by the failure of the Gulf Colorado and Santa Fe Company to provide adequate safeguards for the movement of trains. [393]

Ilford station, on the main line from London Liverpool Street to Norwich, England, had been the scene of a major disaster on New Year's Day 1915 (covered in a previous volume of this series). Just over 29 years later, on the evening of Sunday 16th January 1944, it suffered another bad accident. In conditions of darkness and dense fog the 2.38 p.m. from Yarmouth to London was run into from behind by the 2.40 p.m. express from Norwich to London at about 7.20 p.m. The engine of the Norwich train penetrated the first and third-class composite carriage at the rear of the Yarmouth train for half its length and the leading van and following first-class carriage of the Norwich train were telescoped. Nine passengers died, 28 were badly enough hurt to be detained in hospital and 10 others less seriously injured. Those who lost their lives included Lieutenant-Colonel F.F.A. Heilgers, Tory M.P. for Bury St Edmunds, and three members of the American forces.

---

[392] *The Abilene Reporter-News*, Abilene, Texas, United States, 14th January 1944

[393] Interstate Commerce Commission report

An American doctor and nurse who had been passengers on the Yarmouth train immediately began first-aid, Ilford Civil Defence personnel (the Home Guard) arrived at 7.36 p.m., an ambulance and a doctor arrived at 7.50 p.m. and the first patient left for the King George V hospital at 8.10 p.m., with the last of the casualties being removed at 9.20 p.m.

A passenger, Mrs Wightman, of Kingsbury, Middlesex, who had come from Ipswich, told reporters, "I don't know how we managed to get out. The American troops who were travelling with us immediately set to and did what they could. They were grand." [394]

The Yarmouth train consisted of B-17 class 4-6-0 engine No. 2828, nine eight-wheeled bogie coaches and a two-coach articulated set.

The Norwich train consisted of B-12 class engine No. 8564, nine eight-wheeled bogie coaches and a two-coach articulated set.

Some modernisation of the former Great Eastern Railway, now the London and North Eastern Railway, four track main line had taken place in the Ilford area since 1915, but the war had necessitated the postponement of a resignalling scheme involving continuous track circuiting, and Sykes Lock and Block was still in use between manual signal boxes. East of Ilford, though, Seven Kings East box had been abolished and Goodmayes West up home and Seven Kings (ex-East) outer and inner up home semaphore signals had been replaced by colour lights, the previous month. The signals to the east, the last of which was Goodmayes East up home, and to the west, beginning with Ilford Carriage Sidings up home with Ilford East outer distant beneath it, remained as semaphores. At the time of the accident the colourlight signals were not controlled by the occupation and clearing of track circuits but were lever operated from the respective boxes as if they were semaphores.

A passenger train from Clacton had preceded the Yarmouth train; both of these stopped at Shenfield, 13 miles east of Ilford, and then ran non-stop to Liverpool Street, whereas the Norwich train did not stop at Shenfield but ran non-stop to London from Ipswich, about 62 miles east of Ilford. The first two trains had been six minutes apart at Shenfield, and the Norwich train 14 minutes behind the Yarmouth one, but near Ilford the Norwich train was gaining on the other two. The fog began to get really thick, according to the trainmen, at Chadwell Heath, three miles east of Ilford. Fogsignalmen had been called out, except for the three colourlight signals, but did not arrive until after the collision.

Because of the fog the Clacton train was running very slowly, thus the Yarmouth train received a yellow at the Goodmayes West home and was stopped at the Seven Kings outer and inner homes for one and four minutes respectively, these signals being the three colourlights. It then proceeded slowly forward, having been accepted by Ilford East box, but not by Ilford West box at the west end of Ilford station platforms.

Driver Ernest Brown of the Yarmouth train told the Ministry of Transport inquiry that he had seen the three colourlight signals, the latter two at very short range, but had failed to see the Ilford Carriage Siding home or starter signals or the Ilford East home signal, nor had he heard the co-acting detonator signals which should have exploded at the latter signal as it was at danger. He had just caught sight of the red light of the Ilford West

[394] *The Times*, London, England, 17th and 18th January 1944

starter signal and stopped his train about 110 yards west of Ilford West box. Driver Brown went to the west box, where Signalman Thomas Cutts told him that he had passed the East box home at danger as well, which Driver Brown was surprised at. The Clacton train was shortly cleared back from the box ahead and Signalman Cutts told Driver Brown that he could proceed.

Almost as soon as Driver Brown had left his box to rejoin his engine Signalman Cutts said that he had received a telephone call from Signalman Sidney Barbrook at Seven Kings to say that the Norwich train had run past all his signals at danger. Station Inspector C.H. Davey had been in Ilford West box at the time, enquiring why the Yarmouth train had stopped, and Signalman Cutts said that he had at once asked the inspector to go back with detonators and stop the Norwich train. The inspector had gone back along the island platform but had not had time to reach the rear of the Yarmouth train when it was run into.

Because the Yarmouth train had not cleared the Sykes contact in advance of the Ilford West starter and with the short block sections between the Ilford boxes, the Ilford East signalman was not authorised to give the 'Train out of Section' to Ilford Carriage Sidings, and with the fog and the absence of fogsignalmen the Carriage Sidings signalman was likewise prohibited from accepting the Norwich train from Seven Kings. The Seven Kings signalman was, however, permitted to accept the Norwich train up to his (colourlight) outer home signal, which was at red, while the Goodmayes West home signal (colourlight) which was also the Seven Kings up distant signal, was showing yellow. The Ilford Carriage Sidings, East and West box semaphore signals were, of course, all at danger against the Norwich train.

The inspecting officer, Major G.R.S. Wilson, remarked that it was clear from the statements of the various signalmen that Signalman J.E. Sayward at Goodmayes West had correctly offered the Norwich train to Signalman Barbrook at Seven Kings when the latter had cleared back the Yarmouth train and that Sayward and Barbrook must therefore have replaced their signals behind the Yarmouth train. It was also clear that Barbrook had not offered the Norwich train to Signalman Walter Rolfe at Ilford Carriage Sidings. He therefore could not have cleared his inner home signal to yellow and there was no reason to disbelieve his statement that he had not cleared his outer home signal to yellow either.

Signalman Barbrook said that he had expected to hear the engine of the Norwich train blowing off at his outer home signal, instead he had heard it approaching his box so had shown it a red light with his hand lamp – he could just see the train through the thick fog about six yards away. He had then sent the 'Train Running Away on Right Line' to Ilford Carriage Sidings and telephoned Ilford West and then Ilford Carriage Sidings, hoping thus to give the Ilford West signalman more chance to get detonators on the line.

Signalman Walter Rolfe at Ilford Carriage Sidings said that the Norwich train had been almost passing him when he had received the emergency signal. There had been no time to put down detonators by hand and his box was situated too far away from the line for the driver to have been likely to see a hand signal through the fog, so he had shouted to the driver through a megaphone, then sent the 'Train Running Away' signal to Ilford East. He was sure that had there been an emergency detonator placer worked from the lever frame he could have got a detonator down beneath the engine.

Signalman Cecil Oakley at Ilford East testified that on receiving the 'Train Running away' signal from Ilford Carriage Sidings he had gone outside, shouted, and shown a red light to the Norwich train. This had been seen and acted on by the driver, but it had then been too late (see below).

Driver Thomas Latter had taken over the Norwich train at Ipswich, with Fireman F.W. Trout. Driver Latter worked regularly to Liverpool Street and was familiar with the new colourlights, which he had passed 11 times since their installation on 19th December. Fireman Trout, on the other hand, who was not Driver Latter's regular fireman, had not worked to London for five months and was unaware of the new signalling. Driver Latter had not thought to question his fireman regarding his experience of the route and hence was unaware that it was so limited.

Driver Latter said that they had run into a blanket of fog between Chadwell Heath and Goodmayes. He said he had known when they were passing Goodmayes West and Seven Kings but had seen none of the colourlight signals himself. He thought that Trout had said, "Righto," regarding the Seven Kings outer home and, "Green light," on passing the Seven Kings starter, adding later that at the latter signal Fireman Trout had said, "It is green – I should not have seen the signal if it had not been a colourlight." Subsequently neither he nor the fireman had seen any of the semaphore signals. His speed at Goodmayes had been, he thought, 20 to 30 miles an hour. After Seven Kings he had braked slightly and "kept rolling steadily", seeing no signals or signal boxes until he had heard a shout and seen a red light at Ilford East, when he had braked hard just before the collision. It was his idea that Ilford Carriage Sidings box was out of switch (in fact it was open continuously) and that if the Seven Kings inner home was green he had a clear road to the Ilford Carriage Sidings starter. Assuming that the fogsignalmen were out (which they were not) and hearing no detonator explode there, he had assumed that he had the road through Ilford station. Driver Latter said that it had not been his usual practice to rely on his fireman to observe signals and he would not have relied upon Fireman Trout had he known that the latter had not been over the road for five months. He had had a good night's rest prior to the accident and had had no personal worries.

The inspecting officer remarked that if Driver Latter's account of the accident had been vague then Fireman Trout's was incoherent and contradictory. The fireman stated that he had seen the Goodmayes West home (Seven Kings distant) and Seven Kings inner home but had missed seeing the Seven Kings outer home. At first he said that the two signals which he had seen had both been green and that he had called out, "Righto," to his driver. Later, he said that because of his fleeting view of the signals he had not been at all sure of their indications and that he had not called out their colours. He alleged that, "Righto," was a less definite response than, "Green," or, "Yellow." He was not aware that colourlight signals had been installed there. The colour vision of both driver and fireman was tested after the accident and found to be normal.

Signalman Cutts and Station Inspector Davey thought that the Yarmouth train had been stationary when it was struck. Driver Brown and his fireman thought that it had been moving off. Major Wilson doubted if there would have been time for it to be got under way.

Concluding his report, the inspecting officer stated that there could be no doubt that the fog had been particularly dense and was the primary contributor to the collision. According to the evidence it had been extending eastwards and the view of the colourlight signals might have deteriorated considerably in the few minutes between the passage of the Clacton and Norwich trains. It was most unfortunate that the fogsignalman had not arrived at the Ilford Carriage Sidings Up home until 10 minutes after the collision, which would otherwise probably have been prevented. There could be no reasonable doubt that the Seven Kings distant (Goodmayes West home) and Seven Kings outer and inner home signals had been yellow, red, and red and that all of the signalmen had acted properly and could have done little to prevent the collision.

Driver Latter apparently knew where he was at Goodmayes and Seven Kings and had he reduced speed and crossed the footplate on entering the fog to pick out the well-sighted colourlight signals his task would have been much easier. Instead he had left the job to his fireman, which would have been fine if his fireman could have been relied upon. It seemed extraordinary though that Driver Latter had not troubled to find out if his fireman knew the road. Nevertheless, he had been satisfied with his fireman's assurances and had continued past Ilford Carriage Sidings box at little reduced speed, assuming that the signals which he could not see must be clear as no detonators had exploded. Driver Latter had thus contravened Rule No. 127 (xxii) (that during a fog where signals were invisible drivers must assume that distant signals were at caution and stop signals at danger unless hand signals from a fogsignalman indicated otherwise). In addition, the driver's failure to run with caution showed, remarked Major Wilson, 'a very unimaginative state of mind,' and it was very difficult to find any feature which relieved him from serious responsibility, though some share of responsibility must be shared by Fireman Trout. Both men, the inspecting officer continued, had satisfactory records and the difficulty of operation under the conditions described should not be overlooked. The fog and darkness must have contributed to Driver Latter misjudging his speed over a considerable distance and he had not reacted quickly enough to the change of conditions.

Turning to recommendations, the inspecting officer stated that the fact that even colourlight signals could be overrun was further confirmation of the same conclusion by the 1927 Automatic Train Control Committee (and would be demonstrated again at St Johns, Lewisham, in 1957). The committee had also considered the question of co-acting detonator placers at stop signals, but in the case of trains closely following one another immediate replacement of the detonators was not always practicable if they were exploded by the first train. In the present case it was fortunate that the co-acting detonators had not exploded beneath the Yarmouth train, which otherwise would have stopped further back than it actually did, resulting in a more severe collision.

Detonator placers in a signal box were in a different category. Had there been one in Ilford Carriage Sidings box the collision would probably have been prevented. Such equipment was being added in boxes in conjunction with new or altered work and it was recommended that they be provided in the Ilford area when the postponed extensive resignalling work was undertaken in the near future.

Otherwise it seemed that nothing short of Automatic Train Control would have prevented the present accident, but the view of the 1927 committee, that Automatic Train Control was less necessary with colourlight signals, still, Major Wilson considered, seemed to apply.

The inspecting officer looked forward to the extended introduction of colourlight signalling on main and suburban lines after the war, which would not, like Automatic Train Control, require the installation of equipment on locomotives, but pointed out that this could not be regarded as a complete substitute for Automatic Train Control, especially as the majority of main lines were likely to retain semaphore signalling for some years. He looked forward to a fresh review of the policy of introducing colourlight signalling and/or Automatic Train Control once the war had ended. [395]

An inquest was held on six of the victims of the accident, the Coroner, Dr. P.B. Skeels, having stated that he had no jurisdiction over the three Americans killed. At its conclusion, the jury returned a verdict that they considered that the collision was due to the inefficient system of detonator working between Ilford and Shenfield. They were of the opinion that the impact would have been lessened if the driver of the Norwich express had been driving more slowly.

"You mean that the driver of the Norwich train was driving negligently, but that negligence did not amount to manslaughter?" Dr Skeels asked.

"We have no desire to use the word negligence. We consider it a case of accidental death," the foreman of the jury replied, and a verdict of 'accidental death' was returned. [396]

On Tuesday 18th January 1944 passenger train P 360 stopped in the station at Müncheberg, Germany. It then restarted but stopped 100 metres down the line to take water. Meanwhile express train D 52 had been offered from the preceding station at Obersdorf. Failing to realise that the first train had stopped again in the station area the *fahrdienstleiter* (signalman) accepted the express. He then compounded his error by misinterpreting the indication on the block instrument from signal box W2 at the west end of the station as applying to the express rather than the previous train and so cleared his signals. The pointsman in W2 box realised that a mistake had been made but not in time to stop the express, which was travelling at full speed when it rammed the still stationary front train. Five coaches of train P 360 and the colliding engine were destroyed. Fifty-six people lost their lives and 159 more were injured. [397]

A major catastrophe at Porta Westfalica, Germany, on Thursday 20th January 1944 involved two express trains bound for Berlin. Train Dm W 3 from Brussels was being followed by train DmW 103 from Aachen [398] with a timetabled 11-minute interval between the two. The Brussels train passed the *Einfahrsignal* at Porta Westfalica at clear at 18:38 but about this time someone on the train pulled the communication cord and the train, which was not booked to stop at the station, ground to a halt. It was never determined who had stopped the train, or why. The signal was restored to danger by the signalman who was then trying to determine what had happened to the train when he was advised of the approach of the Aachen train. Without thinking what he was doing he then cleared the *Einfahrsignal* again for the second train. At 18:51 the collision occurred.

The engine, luggage van and five carriages of the colliding train and the rear two carriages of the stationary train were reportedly thrown into the air by the force of the impact and caught fire. One of the carriages also buried itself in a lineside house. Seventy-nine people died and 64 others were injured, some seriously.

---

[395] Ministry of Transport report
[396] *The Times*, London, England, 22nd January and 4th February 1944
[397] https://buckower-kleinbahn.de/eisenbahnunfaelle/
[398] *Katastrophen der deutschen Bahnen Teil II* by H.J. Ritzau, pg. 111 (in German)

A railway spokesman stated afterwards that the signalman had been confused by the constant wailing of the air raid sirens due to repeated Allied bombing raids. [399]

A train which had left Weesp, the Netherlands, on Sunday 19th November 1944, for Germany, was bound for Bremen and its passengers included a number of Dutch men who had been rounded up in Rotterdam to undertake forced labour (*Arbeitseinsatz*) in Germany. Probably early on Monday 20th November it was involved in a rear collision at Bad Zwischenahn, Germany, on the Oldenburg – Leer line, whilst travelling at speed. Several carriages were destroyed, 29 people died and 67 were injured, 30 seriously. Twenty-five of those who lost their lives were Dutch prisoners. The injured were taken to hospitals at Oldenburg and elsewhere. [400]

On New Year's Eve 1944 the United States suffered a very bad accident on the Great Salt Lake, west of Bagley, Utah, on the main line of the Southern Pacific Company. Westbound first-class passenger train First No. 21 was run into by Second No. 21, a first-class mail express and baggage train, at 5.14 a.m. It was foggy at the time. The twelfth car of the forward train was telescoped by the thirteenth, the fifteenth car was telescoped by the sixteenth and the eighteenth and rear car of the forward train was telescoped by the engine of the colliding train. Forty-one passengers, a Pullman car employee, four dining car employees, two off duty railwaymen, the flagman of First No. 21 and the engineer of Second No. 21 lost their lives and 66 passengers, a Pullman employee, seven dining car employees, a train service agent, an off duty railwayman and five railwaymen on duty were injured. The casualties were a mixture of civilians and army and navy personnel. [401]

A newspaper reported that, 'The site of the wreck is on a high grade, several miles before the trestle across Great Salt Lake starts. For miles on both sides is boggy, gray muck, making it impossible to reach the disaster scene with motor vehicles. This severely handicapped rescue work.

'Passengers were milling around for hours after the crash, assisting in the removal of the injured and dead and viewing the grisly scene. Out of one large double window were draped nine dead soldiers, which some thoughtful person had covered with a coat. On the second track were three or four dead civilians, likewise, covered with coats. Out of one window hung the limp arm of a soldier, blood trickling down over a wrist watch and dropping off the ends of his fingers.

'The first rescue train, carrying railroad personnel, doctors and nurses from Ogden and representatives of the military, left Ogden about 7.55 a.m. Another one was made up and dispatched to the scene as quickly as possible.'

Mrs Clifford Moss, of Ogden, the wife of a railroad employee and a passenger on the forward train, described the crash as a "terrible screaming collision followed by screaming, sobbing, shrieks of the injured," and added, "I was in my berth in the last

---

399 http://www.hiergeblieben.de/pages/textanzeige.php?limit=20&order=datum&richtung=ASC&z=90&id=1093

400 Wikipedia – Eisenbahnunfall von Bad Zwischenahn

401 Interstate Commerce Commission report

Pullman car of the first section. I wasn't asleep. Suddenly I heard a terrible crash. The car twisted up into the air on its nose. Then for a while I didn't know anything. The air filled with steam and there was a screaming noise. When I came to, people were running about outside, some of them almost without any clothes on, in the snow and cold, screaming. Everything was in terrible confusion." [402]

The line on which the accident occurred was double track and equipped with two-position lower quadrant semaphore automatic signals. Situated 2,458, 8,156 and 12,655 feet of the point of collision were signals Nos. 7639, 7649 and 7659, which with train First No. 21 occupying the track circuits which it had when it was struck would have shown red, red, and green aspects indicating *Stop, Stop*, and *Proceed.*

Train First No. 21 consisted of engine No. 4425, a mail car, four baggage cars, two U.S. Army hospital cars, seven coaches, a dining car, a Pullman tourist car, and two Pullman sleeping cars, in charge of Conductor Victor Kovarik with Engineer W.B. Ware and Fireman Louis Campbell.

Train Second No. 2 consisted of engine No. 4631, 19 cars and a coach, in charge of Conductor J. W. Welch with Engineer James McDonald and Fireman M.E. Hardman.

A lighted fusee was found on the westbound track 1,000 feet west of the rear of train First No. 21 and a second fusee close to the rear of the train, together with the body of Flagman W.S. Duerdon of the front train.

Fireman Hardman told the Interstate Commerce Commission (ICC) inquiry that because of pockets of dense fog he had been unable to see signal No. 7649 and had told Engineer McDonald so. He understood his engineer to have replied that the signal was clear, though the engineer had not called out the signal's indication. Close to signal No. 7639 the fireman had seen that it was red and had notified his engineer. The latter had appeared to be normal but had taken no action to reduce speed. Because he had been regulating the firing and fuel oil supply valves Fireman Hardman had not noticed the fusees or red marker lights of the forward train until immediately before the collision.

Other members of the train crew spoke of a brake application about 12 seconds before the collision when the train had been moving at 65 miles an hour. From the evidence of witnesses as to the position of the engineer after the accident it seemed possible that he had become incapacitated after making the brake application.

Given the magnitude of the accident the ICC report on it was surprisingly brief. It noted that when the engineer had failed to react to the *Stop* aspect of signal No. 7649 safety had depended upon the fireman taking action, but in this case the fireman had not seen the signal and had not realised the danger until seeing signal No. 7639 at close range. Had a cab signal system been in use the fireman could have observed the indication of signal No. 7649 and taken action, while an automatic train stop, or train control system would have applied the train brakes when the engineer failed to do so.

The ICC concluded, 'It is found that this accident was caused by failure properly to control the speed of the following train in accordance with signal indications.' [403]

A post mortem conducted on Engineer McDonald found 'advanced coronary

---

[402] *The Salt Lake Tribune*, Salt Lake City, Utah, United States, 1st January 1945

[403] Interstate Commerce Commission report

arteriosclerosis with total occlusion of one branch of the left coronary artery. It also revealed fibrosis of the upper left ventricle. These defects and the fact that severely blistered parts of the body showed no cellular reaction meant McDonald had died prior to the moment he was immersed in steam from his wrecked engine' and 'the absence of red cells in the alveoli of the lungs is strong presumptive additional evidence that death did not result from external injuries of shock.' Acting on the pathologist's evidence an inquest jury returned a verdict that Engineer McDonald had died of 'sudden shock or heart failure' and had thereby caused the accidental death of the other victims.[404]

A major disaster occurred in Northern Ireland on Wednesday 10th January 1945 on the Bangor branch of the Belfast and County Down Railway. The 7.10 a.m. workman's train from Bangor to Belfast was stopped outside of Ballymacarrett Junction when it was run into by the 7.40 a.m. motor train from Holywood to Belfast. Although the colliding train was allegedly only travelling at about 10 miles an hour the rear coach of the workman's train was, in the words of a newspaper report, 'smashed to matchwood,'[405] and the next coach was penetrated for a distance of ten feet; the leading bogie of the motor train was badly twisted, and the front of the driver's compartment was pushed in, but otherwise the motor was little damaged. Thirteen passengers were killed instantaneously and nine more died later of their injuries. Twenty-four others, all but one in the Bangor train, were injured. Guard James Hewitt of the Bangor train and the conductor and porter on the rail motor were also hurt. Guard Hewitt was badly cut about the face but being a good railwayman had nevertheless run back along the line and laid detonators to stop a third train from Bangor.

The accident happened at about 7.50 a.m., when it was still dark as well as foggy, seriously hampering rescue operations. The wartime blackout was still theoretically in operation, but despite this bonfires were lit from the timber of the wrecked coaches and the drivers of motor buses on the parallel East Road shone the headlights of their vehicles on the scene. Uninjured passengers and shipyard workers on their way to Queen's Island helped in the rescue work, which was aided by the proximity of the road and the injured were promptly removed to hospital. Single-line working was opened up at 11 a.m. and normal working was resumed by 5 p.m.

The Bangor branch was continuously track circuited and provided with automatic signals, which between Ballymacarrett Junction and Tillysburn were upper quadrant semaphore stop signals with associated distant signals. At Ballymacarrett Junction itself there was a signal box operating inner and outer home signals and a distant signal on the line from Tillysburn and Bangor. The last automatic stop signal was at Sydenham station, which was maintained at danger when the track circuit extending between it and the Ballymacarrett Junction inner home signal was occupied. Telephones were not provided at the automatic signals and a note in the 1943 Appendix to the Working Timetable stated:

When a train has stood at an automatic stop signal at 'Danger' for two minutes, the Driver must give one long whistle and may then pass the signal at 'Danger,' proceeding with great care to the next automatic stop signal at which he must also stop for two

---

[404] *The Salt Lake Tribune*, Salt Lake City, Utah, United States, 17th January 1945
[405] *The Times*, London, England, 11th January 1945

minutes if it is against him and repeat this at each automatic stop signal found at 'Danger' and the train must be run at such speed that it can be stopped short of any obstruction and full particulars reported at next stopping station.

The appendix also indicated certain signals, of which the Ballymacarrett Junction outer home was one, at which it was not necessary to send a person to the box in accordance with Rule 55 if a train was stopped at the signal at danger, only to sound the engine whistle.

The Bangor train consisted of 4-6-4 tank engine No. 25 and 13 six-wheeled coaches, in charge of Guard Hewitt with Driver J. McGraw and an unnamed fireman. The Holywood train consisted of a motor coach, two six-wheeled coaches and 2-4-2 tank engine No. 5, which was running bunker first and propelling the coaches, in charge of Conductor C. Smith with Driver Isaac McQuillan and Fireman William Lamont. When the train was propelling, its driver sat in the control compartment of the leading motor coach at the far end of the train from the fireman on the engine. The driver could apply the train brakes using a brake valve in the motor coach, but they could be applied more quickly from the engine, with which the driver could communicate by whistling. Also significant to the accident was the fact that one of the two routes into and out of the motor coach for passengers was via the driver's compartment. Sliding doors, lockable from the driver's side, separated his compartment from the remainder of the coach and a notice above the doors read, 'Passengers are not permitted to remain in this compartment.'

Driver McGraw of the 7.10 a.m. Bangor train told the inquiry into the accident that he had departed one minute late and losing time en route had been four or five minutes late when he had stopped at Ballymacarrett Halt (that is, immediately in the rear of the Ballymacarrett Junction outer home signal). He had stopped there at 7.47 a.m. and the collision had occurred about four minutes later. He had frequently been stopped at the outer hone signal. His speed at the Ballymacarrett Junction distant signal, at Victoria Park Halt, had been about 10 or 12 miles an hour. The rails had been slippery, and it had been foggy, but he had been able to see 150 yards. It had been too early for the fogsignalmen to be likely to be out. He had heard no whistle.

Driver McQuillan of the Holywood train said that he had left Holywood on time. The Tillysburn automatic distant signal had been at caution, but its associated home signal had cleared as he was about to pull up at it. When he was ready to leave Sydenham station the automatic home signal there had been at danger. He had waited two minutes, then proceeded in accordance with the 'Stop and Proceed' rule. Approaching Victoria Park Halt he had seen the Ballymacarrett Junction distant signal at caution. Passengers had got on and off at the halt, then he had proceeded, still at caution. The weather had been clear at Sydenham, but it had become increasingly foggy as he had neared Victoria Park Halt. Fifty yards beyond the halt the fog had got very thick. He usually closed the regulator at this point, but as he was already driving at caution he had not considered it necessary to brake again. Shortly afterwards he had seen a red light thirty yards ahead and had jammed on the brakes but had run into the rear of the Bangor train. Driver McQuillan added that he had been particularly careful every morning with this train as he knew the Banger train was close in front, though he had never before found it stopped at the outer home signal. He was quite satisfied that he had been driving at caution, at 8 or 10 miles an hour, and could have pulled up within the length of his own train.

Fireman Lamont of the Holywood train said that the fog had thickened so much at

Victoria Park that he had not been able to see the length of his train (59 yards). He remembered steam being shut off after leaving the halt but was sure that the brakes had not been applied before the collision.

Porter James McMillan spoke of a slight jar just before the impact.

Mr R. Dundas Duncan, the inspecting officer who conducted the inquiry, stated that there was no doubt that the signals had all been working correctly and their lights burning. Driver McQuillan and Fireman Lamont were, the inspecting officer remarked, somewhat inconsistent as to where the fog had become thick enough to cause anxiety, while the fireman's estimate of speed was useless.

Conductor Smith put the speed of his train at collision at about 10 miles an hour.

Driver W. Mulhall of the 7 a.m. train from Bangor agreed that the fog had been thickest in the vicinity of Ballymacarrett Junction and said that he had remarked so to his fireman.

Signalman W.J. Miskelly at Ballymacarrett Junction said that it had been clear when he had come on duty but became very foggy at 7.30 a.m. He considered that fogsignalmen should have been out but as they reported automatically at 7.45 a.m. anyway there had been no point in calling them.

Questioned as to whether there had been passengers in the driver's compartment Driver McQuillan said that although he knew that they were prohibited from being there, he had not closed the sliding doors and passengers had ridden in the driver's compartment on previous occasions. He denied that there had been any passengers with him from Sydenham to Victoria Park Halt but agreed that someone might have been standing behind him from there.

Signalman Miskelly explained that the Bangor train had been stopped at his outer home signal because the signal had failed to clear. When it had failed to do so the first time, he had decided to instead give priority to a main-line train from Newcastle. When this train had cleared the junction he had again tried to clear the signal. When this had again failed he had set off down the line with a lamp to bring in the Bangor train, the crash occurring before he reached it. The same signal had given trouble previously, especially when it was wet or frosty.

Guard Hewitt said that after his train had been stopped three minutes he had gone back to protect his train, but the collision had happened almost at once. After the collision he had gone back to Victoria Park Halt with detonators, preceded by Fireman Lamont, and they had stopped the 7.35 a.m. train from Bangor.

Coming to the conclusions of his inquiry, Mr Dundas Duncan noted that the amount of damage done in the collision was not consistent with a speed of only 8 or 10 miles an hour for the colliding train, while it was doubtful if the brakes had been applied even 30 yards short of the front train. Tests carried out showed that a train starting from Victoria Park Halt could reach a speed of 27 miles an hour at the point of collision.

Driver McQuillan knew that the Bangor train was close in front and was not only running under the 'Stop and Proceed' rule but the Ballymacarrett Junction distant signal had been at caution and he was to stop at Ballymacarrett Halt. His replies to questions as to whether passengers had been in the driver's compartment had been evasive. It was clear that the sliding doors to the driver's compartment had been open and highly likely that one or more passengers in the crowded car were beside or close behind the driver, interfering with his concentration.

The inspecting officer stated that, 'I am of the opinion, therefore, that the direct cause of this accident was the failure of Driver McQuillan to proceed at such a speed as would enable him to draw up before any obstruction. Considering all circumstances I am forced to the conclusion that Driver McQuillan had completely failed to realise his responsibility as driver of a passenger train. I cannot accept his statement that he was unaware of the distance in which he could bring his train to a stand when running at the speed he stated.'

Discussing the question of automatic signals, the inspecting officer stated that if traffic were to be kept moving there was no alternative but to permit a driver to pass an automatic signal at danger and proceed at caution unless a telephone to the nearest signal box were provided at the signal. On the Belfast and County Down line, as in earlier installations in England, no such telephones were provided, and trains were allowed to pass automatic signals at danger under the 'Stop and Proceed' rule even though this amounted to an abandonment of the space interval between trains fundamental to the absolute block system. All automatic signals in England were now provided with telephones, whereby the driver, after confirming which train he belonged to, was advised whether or not the section ahead was occupied, and if not, authorised by the signalman to proceed. Such a measure had been considered in Northern Ireland but rejected because of the lighter traffic, lower speeds, shorter distances, and lesser incidence of fog. If use of the 'Stop and Proceed' rule continued it was essential that drivers proceed at a speed consistent with the visibility and able to stop short of any obstruction within their range of view. The precise rules might vary as to the time the driver must wait at a signal before proceeding and as to whether the speed was fixed at say, 5 miles an hour, or left to the driver's judgement.

As regards recommendations the inspecting officer suggested increasing the candle power of the colour light signals and the addition of telephones and illuminated P signs (worked by the signalman to authorise a train to proceed past a signal at danger if the line was clear) at all stop signals on the Bangor branch. In addition the remote-control gear of the rail motors should be modernised to preclude operation by the fireman except in emergency and a solid partition should be substituted for the sliding doors between the driver's compartment and the remainder of the motor car, with extra side doors provided instead for the use of passengers. The arrangement for fog signalling should be overhauled and matters relating to the reporting of defective signals and their repair tightened up.

A footnote to the report stated that the railway company had already instituted a control system under which the 'Stop and Proceed' rule would no longer be applied as a result of traffic delay. [406]

A coroner's jury returned a verdict that the accident had been caused by 'the failure of a faulty signal and fog which obscured the driver's vision.'

Driver McQuillan stood trial for manslaughter but was acquitted.

Express train D 8 came to a stand due to lack of steam in conditions of poor visibility

[406] Ministry of Commerce report

near Herzebrock, Germany, on the Rheda – Oelde line in Westphalia on Monday 15th January 1945. It is not clear if any attempt was made to protect it in the rear but at any rate it was run into from behind by a special passenger train. Twenty-eight people died, 32 were seriously and 61 slightly injured. [407]

A Ferrocarril Nacional de México special train carrying pilgrims from Mexico City to an annual religious festival to the Shrine of the Virgin Mary at San Juan de los Lagos in Jalisco state, which had left the Mexican capital at 5 p.m. on 31st January 1945, was run into from behind by a freight train at 12.47 a.m. on 1st February at Cazadero, Mexico. All nine coaches were thrown from the track and three of them caught fire. Initial reports spoke of 100 dead and 70 injured. A relief train was sent from Mexico City with doctors and medical supplies and some of the injured were taken to hospital at San Juan del Rio, 10 miles away, Cazadero being only a hamlet. [408] Other injured were taken on the rescue train to Queretaro. Most of the fatalities on the crowded train were in the wooden coaches which caught fire and many of the injured suffered burns. [409]

'Railroad workers, farmers and passengers pulled dead and wounded from the wreckage and laid them in three long rows beside the track. By torch and lantern light, doctors from neighbouring towns and later from Mexico City performed operations and dressed wounds.' About 100 of the injured were taken to San Juan del Rio but others were too badly hurt to be moved. As well as the rescue train (or more likely two trains, one from Queretaro and the other from Mexico City, but the report is unclear) ten road ambulances and 25 private cars went to the scene. The Mexican Red Cross and the Green Cross both attended the accident.

Railroad officials blamed a signalman for allowing the freight train to follow the passenger train too closely. [410]

A further report from Mexico City on 2nd February stated, 'Searchers continued to poke tonight among the bones, ashes, and splinters and charred debris of wooden railway coaches to determine the full toll of this country's worst train wreck {this is clearly incorrect, there had been several worse accidents already}. Latest official information lists 69 known dead, with a probable death list of around 100, and 169 hospital cases. Some unverified estimates reached as high as 200 dead.

'Many women and children were among the 1,500 festival-bound pilgrims aboard the nine-coach special train which was rammed from the rear by a fast freight at Cazadero early yesterday. Parents who missed children and children who could not find mothers and fathers were unable to give coherent information to officials. It may be days before there is a complete list of the injured.

'The freight engine spilled burning oil on the wreckage and two cars of sugar fed the flames.' [411]

Confusion over the number of lives lost in the accident continued. The following day

---

407 Wikipedia – Liste von Eisenbahnunfällen in Deutschland

408 *The Brownsville Herald*, Brownsville, Texas, United States, 1st February 1945

409 *Montana Standard*, Butte, Montana, United States, 3rd February 1945

410 *Galveston Daily News*, Galveston, Texas, United States, 2nd February 1945

411 Ibid, 3rd February 1945

(3rd February) it was reported from, Mexico City that, 'Death and injury of children under 6 is the factor that continues to confuse the casualty list of a pilgrim's train wrecked at Cazadero early Thursday. The Mexican National Railways has announced 61 dead and 130 injured. This calculation was based on the sale of 1,726 tickets and the safe arrival, unhurt, of 1,535 persons in Mexico. Children under 6 pay no fare and photographs show small children dead, and in hospitals. There has been no check of their number.

'Officials agree that 56 bodies were recovered but they differ widely on the number of persons burned to ashes in three of the wrecked wooden coaches. Civil and military authorities at the scene say the number of dead is 'much greater' than 56.

'Passengers say there were many small children and many babies. A usual condition in Mexico's third-class travel. Babies seldom are left at home by Mexican families.

'Because of the confusion, Gen. Lazaro Cardenas, secretary of national defence and former president, went to the scene to see if he could clarify the situation.' [412]

The final number of dead has been stated as 127. [413]

An accident occurred at about 03:00 on Thursday 22nd March 1945 at Jungshari station, India, about 33 miles from Karachi on the double-line section between there and Kotri. Up passenger train No. 191 from Karachi to Rohri on the North Western Railway was run into from behind by No. 571 Up goods train. As a result of the collision the rear two carriages of the passenger train were telescoped, and the leading seven waggons of the goods train were badly damaged. Sixteen people were killed outright, eight more were fatally injured and 43 less badly hurt.

Relief trains were sent to the scene from both Karachi and Kotri and a special train carrying the injured arrived at Karachi Cantonment station at 9.17 a.m.

The Chief Operating Superintendent of the North Western Railway, Mr W.T. Bascoe, left Lahore for the accident scene at once upon being informed of the accident and opened an inquiry ay Jungshari. [414]

The chaotic situation on German railways in the last days of the Second World War was the principal cause of a rear collision in Westphalia, near Verne block post, on Saturday 31st March 1945. Allied forces were close at hand, and railwaymen had been ordered to shift all movable material from the Ruhr to Nordhausen away from the advancing enemy. Bombing attacks had damaged or destroyed most of the signalling and communications, so trains were being run on sight. At about 23:00 on the last day of March there were two trains moving eastwards on the Hamm – Warburg line. The first train had probably been shot up by an aircraft in the Geseke forest and it came to a stand short of Verne, where the block post had been wrecked six days earlier when an American fighter aircraft had crashed into it. It was, of course, pitch dark. Two steam engines were pulling the train (or possibly the train consisted simply of a pair of locomotives) and there were also several passengers on the engines, but no one seems to

---

[412] Ibid, 4th February 1945

[413] *The Railway Gazette*, London, England, Vol 82 pg. 147

[414] Ibid, 27th April 1945

have thought of going back to protect the rear of the train with detonators and the train was very likely not provided with a tail lamp. Why it had stopped in the first place is not clear, possibly the engines had been damaged in the air attack.

Following the first train was a second, described as a Bauzug – a work train – which was crowded with refugees from the Ruhr area. Its driver probably remained unaware that the train in front had stopped until the last moment. The ensuing rear collision caused the front vehicles of the colliding train to telescope one another. At least seven people died, some of them soldiers and others were civilians; the driver and fireman of the second train were among the dead. Seven dead were buried by the Salzkotten Red Cross, but it is likely that there were other fatalities. Locals who tried to help were hampered by the lack of rescue equipment and the following day the cries of injured victims could be heard coming from the carriages of the second train.

American soldiers were already close at hand and soon occupied the area. They cleared the line with armoured recovery vehicles on 6th April, but an engine and some vehicles which had been too badly damaged to be moved remained at the site for at least three years. Local residents, suffering from the deprivations of the immediate post-war period, helped themselves to the coal and any other useful items from the wrecked trains. [415]

There was another rear collision in India on the evening of Monday 21st May 1945, this time on the East Indian Railway, on its Howrah – Burdwan chord line. A goods train was moving onto a loop line at Manirampur station but part of it was still on the main line when No. 83 Up parcels express ran into it, telescoping the brake van of the goods train. Three bogies (coaches) of the parcels express were derailed and fell down an embankment, though strangely the leading inter and third-class bogie remained on the rails. The accident happened at about 22:30. Twelve people were killed outright, 40 seriously injured and 20 slightly hurt. A relief train was sent to the site from Howrah with medical staff and divisional officers and those badly hurt were taken to Howrah. Three more relief trains reportedly followed and the Up Gomoh passenger train picked up the other passengers, including those slightly hurt. Two hundred and fifty of the passengers were given refreshments and tea at Burdwan. Two of the injured died in hospital, increasing the number of dead to 14. [416]

A train carrying about 1,200 German prisoners of war from an American camp at Bad Aibling was involved in a catastrophic rear collision on the Rosenheim – München line at Oberölkofen, between Aßling and Grafing, Germany, on Monday 16th July 1945. Their eventual destination was to be Hannover, where they were to be released and able to make their way home.

The ill-fated train of 30 wooden goods waggons set off in the late afternoon. At Rosenheim the steam engine pulling it was replaced by an E 75 class electric locomotive. The replacement engine had difficulty pulling the train even at low speed, especially as the line climbed gently but continuously from Rosenheim to Grafing. An uneven,

---

[415] Wikipedia – Eisenbahnunfall von Verne

[416] *The Bombay Chronicle*, Bombay, India, 23rd and 24th May 1945

fluctuating voltage may have been a contributing factor. At any rate at 21:15 a technical problem with the engine resulted in the train coming to a stand one and a half kilometres beyond Aßling on a left-hand curve. A cutting with vegetation further impaired the view of the standing train, and it was raining. Most of the soldiers were trying to get some sleep but their mood was good, the war was over, and they were on their way back to their loved ones. [417]

At this date only the most important signalling and telephone equipment had been brought back into use following its wartime destruction. After the Oberölkofen *fahrdienstleiter* (signalman) had waited some time for the prisoner of war train and it had not arrived he asked his colleague at Aßling where it was. The Aßling signalman apparently misunderstood the message as the Oberölkofen man reporting the prisoner of war train as having cleared the section and permitted a following train to proceed. This was an American transport train, loaded with tanks, and hauled by a class BR 94 electric locomotive.

In the collision the rear six waggons of the prisoners' train was telescoped into one another. Ninety-six soldiers reportedly died at the scene, together with four after they had been taken to Kirchseeon Hospital and two in Rosenheim Hospital. One American was also stated to have died. others were injured. The accident happened next to the house of a track worker, who was fatally injured when his home was struck by the wreckage. The total number who died was thus about 104; one source states 110 died. Over 100 others were injured.

The Aßling signalman was subsequently sentenced to a term of imprisonment, it having been judged that he was principally responsible for the accident. A more recent German writer doubts if the Aßling signalman was wholly to blame, and if he would have been so stupid or negligent to so misinterpret his colleague's question. This writer suggests that the signalman was being pressurised by the American commander of the tank train, impatient at the endless delays, who was perhaps shouting to the signalman while the latter was on the telephone. According to Wikipedia the signalman himself was an American transport officer. [418] It is also suggested that the probably drunk and noisy Americans in the tank train had distracted the signalman. There had also been a failure to protect the stalled train in the rear, and perhaps to send a message to the Aßling signalman that the train had failed in the section from the conveniently placed railwayman's house. [419]

The Great Northern Railway suffered a rear collision involving two passenger trains at Michigan, North Dakota, United States, on Thursday 9th August 1945. The trains involved were the first and section sections of the westbound *Empire Builder*, train No. 1. The first section had stopped at Michigan because of an overheated axle box when it was run into by the second section. The engine of the colliding train telescoped the rear car of the first train over almost its entire length and it was in this car that the fatalities in the accident were found. Thirty-three passengers and a Pullman porter lost their lives and 264

---

[417] https://www.merkur.de/lokales/ebersberg/letzten-toten-weltkriegs-2367140.html
[418] Eisenbahnunfall von Aßling
[419] *Die Katastrophenszene der Gegenwart* by Hans Joachim Ritzau and Jürgen Hörstel, pgs. 15-17 (in German)

passengers, a mail clerk, a Pullman porter, five porters, two dining car employees, one person carried under contract, one off-duty railwayman and the fireman, baggageman, front brakeman and flagman of the colliding train were injured. [420] The collision occurred at about 7.22 p.m. and rescuers worked through the night and into the following day. The last living victim to be taken from the wreckage was a woman who was finally released at 2 a.m. Unfortunately she died in an ambulance ten minutes later. After her release rescuers continued cutting through the wreckage with oxy-acetylene torches to extract the remaining bodies, their efforts made more difficult by steam escaping from the damaged boiler of the colliding engine. [421]

Carl Klemarud, of Devil's Lake, North Dakota, a passenger in the second section of the train, told of his experiences:

'The first indication that I had that there was going to be a wreck was when the train brakes seemed to grab. I was sitting beside an elderly lady holding a baby on her lap when the train suddenly jerked. "You'd better hang on to that baby, for something is going to happen," I said. I hardly got the words out of my mouth when the crash came. I was sitting with my back towards the engine and the back cushion of the seat protected me from injury. When I climbed out of the car to see what had happened I saw the observation setting on top of the engine. My first impression was that it looked like somebody had taken a big can opener and split the bottom of the car, then hoisted it on top the engine like a tent.

'The peculiar thing about this wreck was the lack of cries of the injured which I always thought went with train wrecks. There seemed to be a stunned silence on the part of the dazed people who had clambered out of the coaches to find out what was wrong. I looked up on the side of the dangling car and saw a woman's body hanging out of the wreckage. Her body seemed to be pinned down in the twisted steel from the waist down. She was still conscious and crying feebly for help. Her voice was the only one that came from the smashed car.' [422]

Train First No. 1 consisted of engine No. 2584, a passenger-dormitory car, a dining car, two Pullman tourist sleeping cars, a Pullman sleeping car, a dining car, four Pullman sleeping cars and a Pullman observation car, in charge of Conductor John Vaughn with Engineer Alfred Unwin and an unnamed fireman.

Train Second No. 1 consisted of engine No. 2588, a mail car, baggage car, a passenger-dormitory car, two coaches, a dining car, and five coaches, with Engineer H.C. Nelson.

Train First No. 1 had had the journal boxes of the engine tender inspected at Grand Forks, an hour and 20 minutes before the collision, where oil had been supplied and the journal packing dressed with a packing iron. Twelve miles east of Michigan, Conductor Vaughn had noticed the smell of burning oil and nine miles east of Michigan the train was stopped for 22 minutes while the journal was cooled and repacked, and a water line arranged to supply water to the journal. Another stop was made at Petersburg to adjust the hose, then at Michigan Engineer Unwin noticed that the journal was smoking considerably and stopped again. Here it was found that the valve of the water line to the

---

[420] Interstate Commerce Commission report

[421] *The Racine Journal-Times*, Racine, Wisconsin, United States, 10th August 1945

[422] *Moorhead Daily News*, Moorhead, Minnesota, United States, 10th August 1945

journal had become blocked with sediment.

Engineer Unwin told the Interstate Commerce Commission (ICC) inquiry that he had realised that his train was considerably delayed. He had waited until the rear of his train was past the train order signal at Michigan, being under the impression that that office was open (it was closed) before stopping as he thought that by doing so more favourable conditions for flagging would be provided. It was not his practice to whistle out the flagman in such circumstances. After the train stopped he had been attending to the journal, with his fireman on the engine and the conductor and front brakeman stood nearby.

Flagman Henry Bruso of train First No. 1 stated that in accordance with the regulations he had been riding at the rear of the penultimate car as the rear vehicle was an observation car. He noticed that speed was being reduced and was aware that in such circumstances lighted fusees should be dropped off at intervals if the train was likely to be overtaken. However he had not dropped any off, being under the impression that fusees dropped from a moving train would not stay alight. Just before the train had stopped he had alighted, and he had got back 500 feet and was giving danger signals with a fusee and red flag when the engine of train Second No. 1 passed him.

Engineer Nelson of train Second No. 1 stated he had been travelling at 57 miles an hour according to his speed recorder approaching Michigan, having received no train order restricting his speed. The first warning that he had had of the train stopped ahead was seeing simultaneously, when about 1,100 feet away, the fusee of the flagman and the red marker lights at the rear of the stopped train. He had immediately applied the brakes but had still been travelling at 45 miles an hour when the collision took place. He added that had there been a fusee burning east of the curve on which the accident had happened he could have stopped in time.

As in the Novice, Texas, accident a few months earlier (see above) the line was worked by timetable and train orders only, with following trains spaced only by a time interval system – a situation which would have been unthinkable in Great Britain for many years by this date. The interval had been increased from 10 minutes to 20 minutes for a passenger train following another passenger train. However, as the ICC pointed out, in the 200 miles between Grand Forks and Surrey there were only four continuously open offices, 27¾, 64, 88¾ and 146 miles west of Grand Forks, these being the only offices open at the time of the collision. The trains had left the last open office 31 minutes apart, but Michigan was still 10 miles short of the next open office at Lakota. The three stops made by First No. 1 had occupied about 25 minutes, nullifying any protection offered by the 20-minute rule. The operating rules of the Company included those for the operation of a manual absolute block system. Had such rules been in effect on the line in question the second section of the train would not have been permitted to enter the section occupied by the first section and the accident would not have occurred.

In conclusion the ICC found that the accident had been caused by failure to provide adequate protection for the preceding train and recommended that the Great Northern Railway establish an adequate block system and stated that a rule to show cause why this should not be done would be served upon the Company. [423]

At the inquest upon the victims at Lakota, North Dakota, the jury returned a verdict that, 'Thirty-four persons, whose bodies were removed from the observation car of

[423] Interstate Commerce Commission report

section 1 of train No. 1 of Great Northern railroad on the evening of Aug. 9, 1945, and on the succeeding day, came to their deaths through the joint and separate negligence of John S. Vaughn, Alfred Unwin, and Henry Bruso, being respectively the conductor, engineer and flagman of the crew of train No. 1 of the Great Northern railroad co., on the evening of Aug. 9, 1945.' [424] As the verdict was only one of negligence and not of gross or criminal negligence the crew presumably avoided any manslaughter charges.

Early on Monday 24th September 1945 an accident occurred on the Paris – Bordeaux main line between Port-de-Piles and Saint Maure, France. The trains involved were two sections of train No. 33, due to depart from Paris Austerlitz at 21:30 on Sunday 23rd and to arrive in Bordeaux at 08:15 on Monday morning. Due, no doubt, to pressure of traffic, the train was run in two portions, the first continuing to Hendaye and the second to Tarbes. The first portion was stopped for an unknown reason and the second portion ran into it, due, possibly, to the failure of a *disque de protection*. It was initially reported that there were six dead and a dozen injured. The Bordeaux line was blocked and delays to trains of up to eight hours were expected. [425]

A major disaster occurred in the suburbs of Berlin, Germany, on Wednesday 10th October 1945. The suburban line from the Berlin station of Spandau West had reopened following the end of the war, though with only one of its two tracks in use, the Russians having dismantled the other track in May for use elsewhere, and there were only three daily trains running. It was the last train from Spandau West which met with disaster on 10th October. This train, nicknamed *Lumpensammler* (rag picker) was very late that day and did not reach Brieselang station until about 22:00. Here a few passengers alighted and others made their way to the front coaches so as to be the first to get to the exit at Nauen, the next station.

Standing at the entrance to Nauen station at this moment was a long coal train but the Brieselang *fahrdienstleiter* (signalman), identified only as Reinhardt H., somehow let the passenger train into the occupied section and a violent rear collision ensued. The driver and fireman of the passenger train had time to jump and save themselves at the last moment, though they probably had little warning of the train in front as its tail lamp was likely unlit. The engine and leading two carriages of the train overturned and hot coals from the engine set the coaches alight. Sixteen people died from the impact or in the fire which followed.

The passengers who lost their lives included a married couple, Hans and Charlotte Dorn, who had gone into the city in the hope of tracing missing relatives. They left three young children as orphans. Luckier was 16-year-old Helmut Preuss, who was riding in the middle of the train. After the collision he alighted and stood with other passengers from the rear carriages, silent and helpless, unable to do anything to help those in the burning front coaches.

---

[424] *Reno Evening Gazette*, Reno, Nevada, United States, 15th August 1945

[425] *La Croix du Nord*, Paris, France, 26th September 1945

An investigation was at once launched into the disaster by the Soviet authorities, especially , perhaps, as two Russians were among the victims, but Signalman H., who was presumably the only person to blame, had fled. [426]

[426] http://www.14641-bredow.de/doku/zug.htm

CHAPTER 5

SIDELONG, HEAD-ON, DOUBLE AND FOULING COLLISIONS

Trains can come into collision with one another in a variety of ways. Rear collisions, that is ones involving one train running into the back of another, trains meeting one another head-on on a single track and trains colliding on rail/rail level crossings are detailed in other chapters. This chapter will deal with other types of collision such as those involving trains running into one another at junctions, meeting head-on on multiple tracks and colliding with vehicles left in sidings but which end up foul of running lines.

Five people died, 11 were seriously injured and over were 30 slightly injured in a collision on the morning of Monday 31st January 1931 at Gdynia, Poland, in the Polish corridor, involving an eastbound workman's train and the engine of a goods train. The driver of the goods train was apparently to blame as he had overrun a stop signal. [427] [428]

According to a report from Riga, Latvia, two passenger trains collided near Leningrad, USSR, on about Wednesday 19th August 1931, and an engine and 14 coaches were wrecked. Seven people lost their lives. [429]

At least five railwaymen were burnt to death and many others were feared to have lost their lives in the fire which followed a collision between two trains carrying petrol between Ploesti and Slobozia, Romania, on Monday 21st September 1931. [430]

A disastrous double collision occurred in the suburbs of Moscow on Saturday 2nd

---

[427] *The Times*, London, England, 20th January 1931
[428] *The Bee*, Danville, Virginia, United States, 19th January 1931
[429] *Gazette de Lausanne*, Lausanne, Switzerland, 20th August 1931
[430] *The Railway* Gazette, London, England, 25th September 1931

January 1932. A suburban train was stopped near Kosino, USSR, and while stationary was run into from behind by a following train travelling at 70 kilometres an hour. A train of empty waggons then ran into the wreckage. A fourth train was brought to a halt just short of the wreckage but even so, 65 people were killed outright and over 130 injured, three of whom later died. The first train had come to a halt after running down a drunken suicide. Preoccupied with trying to remove the body of the victim from beneath the engine, the crew of the train had neglected to protect it in the rear. All would still have been well had not the Kosino stationmaster somehow cleared the block for the following train while the first was still in the section, though it appeared that Driver Burok of the second train had also run past automatic signals at danger. Six coaches were telescoped in the first collision, after which many of the surviving passengers alighted onto the opposite running line to be struck by the freight train. [431]

The accident was blamed on gross violations of the railway regulations by the various railwaymen involved amounting to sabotage and carelessness, so that despite a considerable interval between the two collisions the empty train was not warned in time to stop it. Eleven railwaymen were brought before the Supreme Court charged with responsibility for the disaster. A reporter who was in the court at the men's trial described them as unkept, unshaven and looking glum. The outcome of the three-day trial does not seem to have been reported but in a similar trial a short time earlier at Khabarovsk of four railwaymen accused of causing an accident in Siberia, the accused were found guilty and sentenced to death. [432]

News of a major train disaster at Zelonaya, in the USSR, on Sunday 15th May 1932, only came to light towards the middle of June when it was announced that 16 railwaymen were on trial at Dneprovetrovsk (now Dnipro in the Ukraine) in connect with the accident, which was a head-on collision between two passenger trains. Scores of passengers were reportedly crushed to death in the accident, their bodies so badly mangled as to be unrecognisable. Among the accused were the Zelonaya stationmaster and signalman, who it was said had mixed up the signals while intoxicated. Heavy sentences were expected to be passed on those convicted. [433]

The Black Sea express returning to Moscow from the Crimea collided with a freight train at Lublino, USSR, seven miles from Moscow, on Sunday 16th October 1932. Thirty-six people died in the accident and a further 51 were injured. A fortnight later five railwaymen stood trial charged with having caused the disaster through negligence. At the close of the trial Stationmaster F.M. Bulkov at Lublino was sentenced to death by shooting and three others were given terms of imprisonment by the Supreme Court. [434] [435] [436]

---

[431] *The Evening Star*, Washington, D.C., United States, 20th January 1932
[432] *The Times*, London, England, 18th January 1932
[433] *The Daily Courier*, Connellsville, Pennsylvania, United States, 10th June 1932
[434] *The Evening Tribune*, Marysville, Ohio, United States, 29th October 1932
[435] *Ames Daily Tribune-Times*, Ames, Iowa, United States, 31st October 1932
[436] *The Times*, London, England, 31st October 1932

At about 17:40 on the evening of Sunday 16th October 1932, Stationmaster Renoux at Cérences, France, realised that he would have to shunt out of the way a goods train which was at his station to clear the main line for the passenger train from Lison and Dol which was due to arrive at 17:47. The shunt manoeuvre involved the goods train going forward about 400 metres along the single line before it could reverse into a siding. The forward movement had been completed, and the goods train had set back about 20 metres when the passenger train appeared around a slight curve, travelling at considerable speed, and the two locomotives met in a violent head-on collision. Three vehicles of the passenger train were telescoped, a van completely crushing the following third-class carriage.

A rescue operation was quickly organised by the Chemins de fer de l'Etat and a breakdown crane was sent to the scene. The work went on with the aid of acetylene lamps but it was not an easy task. Eight people died in the collision and 20 were injured.

An inquiry into the accident was immediately opened by the MM. Duhamel and Chauvel, the local *juge d'instruction* and his deputy. Stationmaster Renoux admitted that he had been late in carrying out the shunting operation but said that he had belived that he had time to do so before the arrival of the passenger train. Driver Tiphaigne, of the passenger train, maintained that he had approached the station with his train under control, and could have stopped it at the entrance to the station, as per the regulations. However, an examination of the speed recorder showed that the train had still been travelling at a speed of 82 kilometres an hour when the collision occurred, indicating that it could not have pulled up at the required point had the collision not occurred.

The stationmaster and driver were both charged with culpable homicide and released.

It was later stated that the stationmaster appeared to be free from responsibility as the shunting operation had been protected by a colourlight signal two kilometres from the station, which Driver Tiphaigne had passed without reducing speed. [437]

Two local trains collided in the USSR on the Moscow – Kazan Railway on Monday 13th February 1933. Details of the accident were not reported, but in early March it was announced that the Supreme Court had sentenced one railwayman to death, a second to 10 years' imprisonment and seven others to terms of imprisonment ranging between one and five years for responsibility for the accident. [438]

A very brief report from Riga, Latvia, stated that at least 15 people had died, and many had been injured in a collision between two trains midway between Leningrad and Moscow, USSR, on about Tuesday 1st August 1933. [439]

It was reported from Riga, Latvia, on 15th January 1934, that a very serious collision had occurred a few days earlier near Stalingrad, USSR. A passenger train bound for Stalingrad had been run into and telescoped by a freight train, and almost 300 people killed or seriously injured. It was noted that the Soviet government would have forbidden

---

[437] *L'Homme Libre*, Paris, France, 18th October 1932
[438] *The Times*, London, England, 7th March 1933
[439] Ibid, 3rd August 1933

its own press to make any reference to the accident. [440]

Yet another bad accident occurred in the USSR on Monday 12th March 1934 at Tavatui, a station in the Urals, when a local passenger train, reportedly travelling at high speed, passed a signal at danger and collided with a shunting goods train. Thirty-three people died and 68 were injured, 22 seriously. [441]

The trial of the railwaymen involved began a week later. It ended with the driver of the passenger train and the Tavatui stationmaster being sentenced to be shot, while five other officials, either members of the train crew or station officials, were given prison sentences. Quite why the stationmaster could be found at fault for the train driver overrunning signals was not explained. [442]

Due to a misunderstanding between a signalman and a shunter, passenger train P2302 and a shunting engine were involved in a sidelong collision at Pforzheim, Germany, on Thursday 17th May 1934. Seven lives were lost and 94 people were injured. [443]

Port Eglinton Junction, Scotland, on the London Midland and Scottish Railway, was the scene of a serious collision at about 5.38 p.m. on the evening of Thursday 6th September 1934, thus at the height of the evening rush hour. The 5.35 p.m. Up local passenger train from Glasgow St Enoch to Kilmarnock was crossing from the Up Main to the Up Fast Canal line when it came into head-on collision on a diamond crossing with the 5.12 p.m. Down local passenger train from Paisley to Glasgow St Enoch. Six passengers, the driver of the Kilmarnock train and both firemen were killed or fatally injured, 10 passengers and the driver of the Paisley train were seriously injured, 18 more passengers and two guards were taken to hospital and 27 passengers sustained trivial injuries. The cries of the people in the carriages were heard over a wide area, it was reported, and as the accident happened in the middle of a large city help was soon at hand. [444] At least nine ambulances and over 20 doctors were on the scene in a few minutes.

Port Eglinton Junction was the point not far from St Enoch station where the four lines from there divided into the main lines to Paisley Gilmour Street and the Canal lines to Paisley Canal (the double track to the General Terminus goods station also diverged here). There were crossovers between the different lines at Port Eglinton and an Up train coming from St Enoch on the Up Main line (as the Kilmarnock train was) had to cross the Down Canal lines to reach either of the Up Canal lines. Port Eglinton box was a bridge cabin carried on girders over the tracks. A busy box, at the time of the accident it was being worked by two signalmen plus a booking boy. Between St Enoch and Port Eglinton colourlight signalling had been installed with train describers. Semaphore signalling was in use outwards from Port Eglinton, with Sykes Lock and Block between

440 *Journal des Débats Politiques et Littéraires*, Paris, France, 18th January 1934
441 *The Railway Gazette*, London, England, 30th March 1934
442 *The Lincoln Star*, Lincoln, Nebraska, United States, 23rd March 1934
443 *Katastrophen der deutschen Bahnen Teil II* by H.J. Ritzau, pg. 84 (in German)
444 *The Times*, London, England, 7th, 8th and 10th September 1934

there and Shields Road box on the Canal lines and Tyer's two-position blocks, modified to provide controls similar to the Sykes Lock and Block instruments, on the Main lines to Pollock Junction.

The acceptance of trains at Port Eglinton was governed by local instructions as follows:

Regulation 4

When the Down Main or Down Canal line is clear to this Box and the "Train out of Section" signal has been given for the preceding train, the "Is Line Clear" signal may be acknowledged for the following train.

A Down freight train must not be accepted from Shields Box while a train is being crossed from the Up Main to the Up Canal Line.

Down trains must not be accepted from Pollock Junction when a crossing movement is being made or about to be made from the Down Canal Line or Shields Bank sidings and the Down Main Line. The Junction points and the block instrument to Pollock Junction are interlocked.

When a crossing movement is being made to or from the Canal and Main Lines via the junction immediately West of Cumberland Street Platforms the Up Main Line Rear Home Signal must not be cleared to allow a train to draw forward to the Inner Home Signal.

The Ministry of Transport inspecting officer who investigated the accident, Lieut. Colonel A.H.L. Mount, remarked that the regulations appeared to authorise the acceptance of a Down train under Regulation 4 only if the line was "*clear to this Box*" but he had been informed that since the installation of the Sykes Lock and Block system in about 1904, the term *box* had been used to cover the whole of the area in advance of the home signals. Full-line clear acceptance was therefore permitted on the Down Canal line for passenger trains and engines up to signals Nos 73/75. Freight trains had been excepted since one had overrun the signals in April 1928. The area was, in fact, worked as if it were a station yard. A speed restriction of 15 miles an hour applied on all lines between Clyde Bridge (near St Enoch) and Port Eglinton. About 36 trains passed through the junction between 4.30 p.m. and 5.30 p.m. and about 33 between 5.30 p.m. and 6.30 p.m. About 100 trains daily crossed between the Main and Canal lines or vice versa.

The Kilmarnock train consisted of 4-4-0 engine No. 639 and six coaches, in charge of Guard W. Kydd, with Driver Robert Balmer and Fireman William White. The Paisley train consisted of 4-4-0 engine No. 591 and six coaches, in charge of Guard W. Walker, with Driver David Kerr and Fireman Robert Goldie. The Kilmarnock train had left St Enoch on the Up Main line because there were carriages for another train on the Canal line at Clyde Junction and remained on that line because a special train from the Barrhead line was on the Canal line at Gorbals Junction. It thus had to cross to the Up Canal lines at Port Eglinton and was doing so to the Up Fast Canal line when it was struck by the Paisley train. In the collision both engines were badly damaged, the two leading coaches of the Kilmarnock train were wrecked, and its third coach partly telescoped while in the Paisley train the leading coach was wrecked and the second partly telescoped.

Guard Kydd told the Ministry of Transport inquiry that he had felt no brake application before the collision.

Guard Walker stated that he had not been looking out nearing Port Eglinton as he had been tying up letters handed to him a little earlier. He had felt the brakes come on in emergency, then the collision, which he described as, "just a swish and a bang, that was all."

Signalman H. King at Shields Road said that nothing unusual had transpired until he had received the Obstruction Danger signal from Port Eglinton at 5.39. He had then replaced his Up Canal line Fast signals and been informed of the collision. He had accepted the Paisley train from Bellahouston No. 1 at about 5.33 p.m. and had it accepted by Port Eglinton.

Driver Kerr stated that he had stopped at Shields Road station. Fireman Goldie had told him that the starter at the end of the platform was off. Leaving the station he, "turned the reverser up in expansion, and then I saw that the Shields starter was clear and Port Eglinton distant at caution. After that I gave the engine another notch up and, being tender-first, I had a good view over the tender, and I saw Port Eglinton Down home Canal (No. 75) standing clear. Then to get me over the incline into Cumberland Street I turned and opened the second regulator valve and gave the engine a little bit up in the reverse. By this time I would be an engine length from the signal, and when I looked up it was at danger and naturally, I put on the brakes and my first intention was to look over the side and then everything after that was blank."

Driver Kerr said that his speed had been, "rather over 20 m.p.h." when he saw the signal at danger. Because of the severe reverse curvature of the Canal line there was the possibility of a driver on the Down Canal line mistaking the Down Main home signals (Nos. 65/67) for his own, but Driver Kerr stated that he was certain that he had not done so. He said, "I have a good working knowledge of the signals at this place and know that with the curves the relative view of these varies. I have a difficulty in describing their conditions but know them perfectly well on the ground."

Signalmen R. Murphy and James Mckay were on duty in Port Eglinton box, together with Bookmarker J. Gracie. The two signalmen stated that the signals had been off for the Kilmarnock train to cross from the Up Main to the Up Canal Fast line and for a Down Ardrossan train on the Down Main when they had seen the Paisley train passing signals 73/75 at danger. Bookmarker Gracie said he had heard Signalman Murphy shout, "There's that fellow coming through that signal."

The signalmen emphatically denied that signal No. 75 had been pulled off for the Paisley train and then replaced. The Paisley train had been accepted at 5.32, the Ardrossan train at 5.34, and the Kilmarnock train had been shown as *Train Arrived* on the describer. At that moment there was uncertainty as to whether the Kilmarnock or the Paisley train would be given priority through the junction. Although the Paisley train had been accepted first it was still not in sight, though the smoke of its engine was visible, and it was clearly still stopped in Shields Road station. They had therefore decided to let the Kilmarnock train go first. Signalman McKay said that Kilmarnock train had approached at normal speed, and he had been just about to send it on to Shields Road when Signalman Murphy had shouted.

The inspecting officer remarked that there was what he called a definite contradiction in the evidence. The signalmen stated that they had waited to see which train to give priority to before lowering any signals. Driver Kerr on the other hand alleged that signal

No. 75 had been off when he had first seen it and that it must have been thrown back in his face. This would have involved, in the signal box, in the maximum interval of 53 seconds between Driver Kerr seeing signal No. 75 at clear and the moment of the collision, the cancellation of the back lock on signal No. 75 by key, the replacement of signals Nos. 75 and 64, the replacement of route lever No. 21, the setting of points Nos. 8,7, and 6 and the pulling of signals Nos. 42, 44 and 48. Signalman Murphy carried out this series of operations in 33 seconds but a subsequent test showed that it could be done in 29 seconds. There would therefore have been time for the signalmen to have replaced the signals and reset the road but not, however, without severely checking the Kilmarnock train. It was clear that no such check had occurred and there was therefore, in the inspecting officer's words, not the slightest doubt that no such resetting of the road had occurred, and that Driver Kerr could not have seen signal No. 75 at clear.

Driver Kerr, Lt. Col. Mount stated, must therefore be responsible for the accident, notwithstanding his sincere evidence that signal No. 75 had initially been clear for him. He had doubtless been over-confident with regards to his knowledge of the signals and had mistaken signal No. 65 for his own. Had he realised the situation as soon as he had suggested he should also have been able to either avoid or else minimise the collision.

Guard Walker also came in for severe criticism, 'for merely tying up letters when approaching an important junction, and particularly the one in question, in view of the curvature of approach,' and 'had he been on the alert and fully appreciative of the responsibilities of a passenger guard in such circumstances, I do not think he would have failed to have made himself aware, as soon as curvature permitted it, of the aspect of the home signals.' It was questionable, the inspecting officer continued, 'whether guards, particularly of suburban trains, should be permitted to allow anything to deflect them from the duty of observing signals when approaching junctions and terminals.'

The inspecting officer did not consider that the fact that Driver Kerr had been running tender first had any bearing upon the collision.

The fundamental weakness which had led to the collision, Lt. Col. Mount stated, was the method of operation which had persisted for many years whereby trains could be accepted on the Down Canal line under Regulation 4 up to signals Nos. 73/75 while trains were being crossed Up Main to Up Canal or Down Main to Down Canal in front of them. He did not criticise the signalmen for their actions but considered that the route for the Kilmarnock train to cross the path of the Paisley train should not have been set up until the Paisley train had come to a stand. Signalman Murphy had objected that this would cause delay but the special local instructions regarding acceptance had been cancelled since the accident and standard instructions for junction block working substituted. The arrangements at other junctions in Scotland would also be looked into, so as to obtain uniformity. The speed restriction, which it appeared drivers were not always obeying, might then be raised.

The report concluded with some observations regarding the older rolling stock of the trains involved (which included gas-lit coaches, though there had been no fire), pointing out that new rolling stock was being constructed but that it was most efficient to give precedence to main-line stock and that it was inevitable therefore that local trains must be composed of older coaches. It was also impracticable to expect any type of rolling stock to resist the tremendous forces generated in a head-on collision (estimated as at least 10,000 foot/tons).

Driver Kerr stood trial on a charge of culpable homicide at the end of January 1935. The jury returned a unanimous verdict of 'Not Guilty'. [445]

Christmas 1934 saw a major disaster in Canada. To cope with the extra holiday traffic on Tuesday 25th December 1934, the Canadian National Railway put on an extra train between London and Toronto, Ontario, eastbound passenger extra No. 5300. As this train was leaving Dundas station at about 9.10 p.m. Engineer Norman Devine became concerned about a hot crank pin on his locomotive and decided to run his train into a siding to examine the engine. He did not want to stop on the main line because train No. 16, the *Maple Leaf Express* from Detroit to Toronto, was following his train. Having brought his train to a stand in the rear of the connection to the siding Engineer Devine therefore gave four short whistles, the code for the front brakeman, Edward Lynch, to reset the switch for the siding. As it happened at this moment Brakeman Lynch was busy answering an enquiry from a passenger and he failed to hear the whistle. When he received no response to his whistle the engineer leant out of his cab and shouted that he needed a brakeman. His shout was heard by the Rear Brakeman, Charles Phelps, who at that moment was on the ground alongside the fourth car of the train, heard the shout and went forward and opened the switch. The train moved forward into the siding and Brakeman Phelps reset the switch for the main line, locked it and changed the marker lights at the rear of his train from red to green to indicate to the engineer of a following train that it was in the siding. While the extra had been on the main line it had been protected by a red signal at Dundas station, preceded by a yellow caution signal, about two miles further west, but as soon as it went into the siding and the switch was reset these signals went back to green (clear).

After the train had been stabled in the siding and Engineer Devine had had a look at his engine he decided that it was not in a fit condition to continue, and they would need a replacement engine. He then instructed Brakeman Lynch to go back to Dundas station with the message that a relief engine was required to take the train on to Toronto and the brakeman set off back to do so. As he was making his way along the track Lynch saw the headlight of train No. 16 approaching. Preoccupied with the passenger's query he had not noticed that his train had gone into the siding at all, and in the darkness, with several lines in the vicinity, he then became confused. Thinking that his train was still standing on the eastbound main line, and that the *Maple Leaf Express* was just about to run into its rear, he grabbed his flagging equipment from the rear coach and rushing to the switch, hurriedly unlocked it and reset it for the siding in the face of the oncoming express. In his panic he had thus set up the conditions for the very accident that he was trying to prevent.

Engineer Bertram Burrell had received clear signals as he approached Dundas but stated that he had reduced speed to six miles an hour for a speed restriction there and had only been running at 13 miles an hour when he had come to the fateful switch. His engine had suddenly lurched to the left, Fireman John Kennedy had shouted, "Soak her!" and he had applied the brakes in emergency. At almost the same moment the trains had collided at 9.21 p.m.

All the coaches of the extra train were of wooden construction, though the penultimate vehicle had steel reinforcing. The rear coach was smashed to pieces, the penultimate

[445] Ministry of Transport report

coach was partly demolished despite its extra strength and the coach in front of it was badly damaged. Fifteen passengers died and 30 were more or less injured.

One of the passengers in the extra, who was not badly hurt though he was taken to hospital, spoke to a reporter. Mr Rice himself was a Toronto advertising executive for the *Saskatoon Star Phoenix* and *Regina Leader Post*. He related, "I could hear the other train coming and I turned sideways in my seat to watch it go by. I was sitting about 11 seats from the rear of the second coach, and I saw the locomotive come plowing right through the end. It came as far as the sixth seat and stopped. I was thrown out of the window and landed on my side, but I was not badly hurt. I got up. The rear of the coach was all buckled up, and you couldn't hear a thing because of the roar of steam from the locomotive. It was terrific. The second car, the one I was in, was shot ahead and it telescoped into the third car from the back. It went into about the fourth seat of the third car. I began helping people out, and in that way we got six out of the window. Then we went up and got four men out of the smoking compartment."

Another passenger was Douglas Mackie, one of the staff of the *Sudbury Star* newspaper, who said, "I was seated in the fourth car from the rear of the London – Toronto train which was drawn up at the siding at Dundas. Just after receiving a telegram informing me that I was due back at work at Sudbury. I looked at my watch. It was exactly 9.22 p.m. A second or so later the entire train shuddered with such violence that all the people in the car were thrown forward in their seats. At the same moment the lights went out. Silence reigned for a few seconds and then the air was filled with shouts of confusion as the passengers realised that the train had been struck.

"One man who had entered the washroom was almost suffocated when the steam pipe burst. His shouts for help were heard by those outside. Attempts to open the door failed. Helped by other passengers I made my way through a small window and was able to free the trapped man. Every window in the fourth coach was smashed.

"Then I made my way out of the car to find the chair car a shambles. They literally had been telescoped together by the engine of the following express. I can't for the life of me see how anyone could have been in those coaches and lived. It was simply horrible. The screams of the injured were intermixed with the confused shouting of those from other coaches."

Brakeman Lynch quickly admitted his error. At the railway internal inquiry he said, "I went down the track waving a lantern. Thinking something was going to happen, I unlocked the switch and swung it. Number 46 took the siding and the next thing I heard was the crash."

Conductor M. Cherry of the extra told the inquiry that Lynch had volunteered to go back to Dundas station, and continued, "That was the last time I saw him until afterwards. Later, he said that he had the impression that our train was on the main line. He frankly admitted his mistake. He was asked his condition before leaving London, and he said that he had been all right. He had had a full sleep and there had been a full crew on his train."

Conductor E. Palmer of the *Maple Leaf Express* testified, "I got out of the train as quickly as possible. I met the flagman of the train ahead. I think his name is Lynch. He stated that, thinking his train was on the main track, he had thrown the switch to avoid a pitch-in." [446]

On 5th January 1935 a coroner's jury found Brakeman Lynch guilty of manslaughter. [447]

The brakeman came to trial already towards the end of January. When the jury was out considering the verdict he was described as having been under intense mental strain, which increased during their long retirement. Finally, after almost four hours, the jury returned with a verdict of 'Not Guilty'. Justice Makins then told the brakeman, "You have been fairly prosecuted on this charge and the jury has found you not guilty. It is a pleasure for me to discharge you. I hope what has happened is not going to spoil your happiness and usefulness in life." [448]

The day after the trial a Canadian National official announced that Lynch had been definitely negligent, and hence would not be reinstated. [449]

A somewhat complex accident occurred in the USSR on Tuesday 8th January 1935, on the North Caucasian Railway between Koyshug and Kayal, about 20 miles from Rostov-on-Don. A brief description of the accident stated that a train had entered a stretch of line in defiance of adverse signals, had become derailed and had blocked three following trains, also that there had been two separate collisions. The exact nature of the accident is thus not very clear; possibly it was a double or even a triple collision. Six people were reportedly killed and 23 injured, six seriously.

Ten railwaymen were arrested following the accident, and 'a demonstrative trial is being arranged.' [450] They were charged with criminal negligence, making them liable to execution by firing squad if found guilty [451]

At 4.16 p.m. on Thursday 31st January 1935, a train carrying 700 employees from the new Egyptian State Railway locomotive works at Abu Za'bel to Cairo collided with two light engines standing on a dead-end siding at the Abu Za'bel quarries on the Marg – Shebin El-Kanater line. Two coaches of the staff train were telescoped and Mr W.R. Ambler, the Works Manager at Abu Za'bel and five clerks were fatally injured, and several other employees were seriously hurt. [452]

The accident was blamed on the failure of the driver to stop his train at the points leading to the siding, which had not been re-set for the main line, because 'Line Clear' had not been received from the cabin in advance. He had thus infringed Rule No. 156 of the General Rule Book, which read:

No engine must enter upon, set back from, or cross any running line without the permission of the person in charge of the points and signals nor then until the proper

[446] *The Winnipeg Evening Tribune*, Winnipeg, Manitoba, Canada, 27th December 1934
[447] Ibid, 5th January 1935
[448] *The Lethbridge Herald*, Lethbridge, Alberta, Canada, 25th January 1935
[449] *The Winnipeg Evening Tribune*, Winnipeg, Manitoba, Canada, 26th January 1935
[450] *The Times*, London, England, 10th and 12th January 1935
[451] *The Lowell Sun*, Lowell, Massachusetts, United States, 10th January 1935
[452] *The Times*, London, England, 2nd February 1935

signals have been exhibited.[453]

Eighteen people died and nine were seriously injured in an early-morning head-on collision at Karian-Stroganovo, USSR, on the Ryazan – Ural Railway on Friday 8th February 1935. Both engines were destroyed, and carriages and waggons piled in a mass of wreckage. The station employees had reportedly negligently routed both trains onto the same line. A mail car and a passenger coach caught fire and burned.[454] Eight railwaymen were subsequently arrested and handed over for trial.[455]

An undetermined number of Manchukuan soldiers were killed on about Tuesday 16th April 1935 in a collision between their troop train and the Trans-Siberian International Express near Pogronichinaya, Manchukuo, close to the border with the USSR on the North Manchurian Railway.[456]

A collision occurred near Blagoveshchensk, in the Amur region of the USSR, on about 12th June 1935, involving a passenger train and a freight train. The bodies of five dead and 28 injured had been recovered from the wreckage, many of the injured being in a serious condition. Two locomotives and eight waggons or coaches were seriously damaged.[457]

Early on Friday 21st June 1935, express train S.D. 7 of the Estrada de Ferro Central do Brasil coming from Santa Cruz collided with suburban train S.S. 14 in the station at Deodoro, Brazil. The newspaper report on the accident, written in Brazilan Portuguese, is a little hard to decipher, but it would appear that during very foggy conditions the driver of the suburban train, Antonio da Silva Moura, unable to see the protecting signal in the fog, had set back his engine, No. 416, too far, so that the rear of his train was foul of the line on which the express was approaching. A collision had then ensued in which the three rear carriages of the suburban train were badly damaged.

After the accident soldiers had to protect the signalman, Joaquim Nunes da Silva, from angry passengers who blamed him for the accident, but the newspaper reporter pointed out that it was the fog, rather than the signalman, which was to blame, and that in fact anticipating a collision Signalman da Silva had rushed from his cabin frantically waving a red flag at the reversing train, but had been unable to prevent the accident. It was reported that after the accident, the passengers, who appear to have been a rough crowd, and who were fed up with the repeated delays of the previous few days, vandalised the stations at Don Pedro II, Engenho de Dentro and Deodora.

Driver Evarista Gabriel da Rosa was in charge of the engine of the express, No. 214.

The initial newspaper report spoke of three fatalities: a soldier, Vincente Ferreira de

---

453 *The Railway Gazette*, London, England, 1st March 1935

454 *Nebraska State Journal*, Lincoln, Nebraska, United States, 11th February 1935

455 *The Times*, London, England, 11th February 1935

456 *Oakland Tribune*, Oakland, California, United States, 17th April 1935

457 *Journal de Geneve*, Geneva, Switzerland, 14th June 1935

Silva Pacheco; João Domigos da Sousa, a workman; and José Bartolomeu de Sousa, a seaman on the cruiser *Rio Grande de Sul.* [458] In addition there were over 20 injured, some seriously, and a report in an American newspaper speaks of six dead. [459]

Two electric passenger trains collided near Noventana, Italy, not far from Padua, on Monday 9th September 1935. An express from Venice to Padua travelling at full speed collided head-on with a train coming in the opposite direction, and the leading coaches of the two trains were smashed to pieces. Three people were killed more or less outright and two died after being taken to hospital in Padua. About 50 others were less badly hurt. [460] [461]

Heavy snow was blamed for a collision near Asparuchovo, in southern Bulgaria, at about 20:00 on Tuesday 11th February 1936. Both engines and 35 vehicles were wrecked, many being telescoped. Thirteen or 18 people died, according to different reports, including four railwaymen, and others were injured. Incessant snow also hampered the rescue work and the clearing of the line. [462] [463] The accident was blamed on the driver of the colliding train being unable to see the signals due to the blizzard. [464]

A German passenger train from Berlin to Pila in East Prussia collided with a freight train at Lamberg, in the Polish Corridor, on Thursday 1st October 1936. Twenty passengers reportedly lost their lives and 150 were injured. [465]

Harrowing scenes in pitch darkness were reported after a collision at Myslowice, Poland, at 05:00 on the morning of Friday 15th January 1937. The accident occurred in the province of Upper Silesia, close to the German frontier. A crowded local train filled with workmen bound for the mines and foundries was running at 35 miles an hour when it collided, apparently head-on, with a stationary freight train. The leading three carriages of the passenger train, which were badly smashed, were old third-class German stock. Scalding steam and fire from the wrecked engines added to the horror of the situation. It was reported that the screams of the injured could be heard a mile away in the town. Eight passengers and the driver of the local train were killed, and 40 others were injured, 16 it was said very seriously.

A Warsaw correspondent commented, 'Hardly a week has passed this winter without an accident involving loss of life on the Polish State railways, and complaints have been frequent that much of the equipment is in poor condition, and that railroad employees are being overworked. It is stated that the signalman who is assumed to be responsible for today's disaster went on duty at 8 o'clock last night and was to have been relieved only at

---

[458] *Folha de São Paulo*, São Paulo, Brazil, 22nd June 1935
[459] *The Bakersfield Californian*, Bakersfield, California, United States, 21st June 1935
[460] *International Herald Tribune*, New York, United States, 11th September 1935
[461] *The Times*, London, England, 11th September 1935
[462] *Journal de Geneve*, Geneva, Switzerland, 13th February 1936
[463] *The Times*, London, England, 13th February 1936
[464] *La Dépêche*, Toulouse, France, 13th February 1936
[465] *The Railway Gazette*, London, England, 9th October 1936

8 this morning, but it is possible the accident may have been due to a mechanical defect caused by the extreme cold, the temperature having been at 15 degrees below zero centigrade.

'The former practice of using the profits of state railroads for improving working conditions was reversed a year ago, when the postal telegraph service and railroads were forced to turn over their profits to the Polish Treasury to help Finance Minister Eugene Kwiatowski in his determination to balance the national budget.' [466]

Despite the correspondent's comment a 12-hour shift would not be considered excessive in Great Britain in a reasonably quiet signal box, though eight-hour shifts were the norm by this date.

The signalman and several other railwaymen were arrested after the accident. [467]

Six people were killed and 13 were seriously injured in a head-on collision between two suburban trains at Dzerzhinsky, USSR, on the outskirts of Moscow, on Monday 12th April 1937. [468]

A crowded passenger train collided with a freight train near Ciucea, Romania, on Wednesday 22nd September 1937. It was reported that the bodies of ten victims had been removed from the wreckage and that there were 22 injured, twelve of whom appeared to be in a hopeless condition, so that the number of dead was likely to rise. The accident was blamed on the passenger train having left Ciucea station too early. [469]

Over 20 people were reportedly killed or injured, several of them Japanese soldiers, when a troop train collided head-on with a passenger train near Tsinghuayuan station, on the outskirts of Peiping (later Peking, now Beijing) 'on the Peiping – Suiyuan railway,' early on Monday 25th October 1937. [470] The Japanese troops were bound for unspecified points on the Peiping – Suiyuan line and the passenger train was heading for the capital from Tatung in Shansi Province. [471]

A collision occurred at the entrance to the Alameda station in Valencia, Spain, at 13:55 on Saturday 4th December 1937. The trains involved in the accident were the Tereul – Valencia mail train and a two-car electric multiple unit. The mail train consisted of a steam engine and 10 coaches and was running about 20 minutes late; it was due at Valencia at 13:30. One of the cars of the electric train was crushed by the impact with the mail train and the other was overturned. The engine and four leading vehicles of the mail train were derailed; the collision with the electric train had occurred at junction of the tender and the first coach. The engine and two leading coaches, after running about 20

---

[466] *The International Herald Tribune*, New York, United States, 16th January 1937
[467] *Excelsior*, Paris, France, 16th January 1937
[468] *Albert Lea Evening Tribune*, Albert Lea, Minnesota, United States, 13th April 1937
[469] *Journal de Geneve*, Geneva, Switzerland, 24th September 1937
[470] *Valley Morning Star*, Harlingen, Texas, United States, 27th October 1937
[471] *The Malaya Tribune*, Singapore, Malaya, 26th October 1937

metres fell into a ditch, overturned and were badly damaged.

Initial reports spoke of 20 or 30 dead and 80 injured, but it was later found that only seven had died and there were 21 injured, mostly seriously. [472 473]

A passenger train from Lérida to Tarragona was in what was probably a sidelong collision with a freight train at the junction of the Lérida – Reus – Tarragona and Zaragoza – Caspe – Barcelona lines on the outskirts of Reus, Spain, late on Saturday 5th or early on Sunday 6th March 1938. Nineteen people died and 168 were injured. [474]

April 1938 saw the celebration of the Kumbh Mela, a major Hindu pilgrimage and festival, at Hardwar, India, when trainloads of pilgrims poured into the city from all parts of the country for several days up to the 13th, the most auspicious day for bathing in the Ganges. The following day the mass of pilgrims began to make their way homewards again, and their departure was marked by an unfortunate accident on the morning of Friday 15th at the East Indian Railway station. At 9.35 a.m. a rake of empty coaches was being set back into No. 4 Mela platform when, presumably due to an error on the part of the driver, the brakes were not applied as soon as they ought to have been, and the shunted vehicles demolished the buffer stops, and crashed into an enclosure behind them which was thronged with passengers. Three passengers were killed on the spot and 31 were injured and removed to Hardwar and Dehra Dun civil hospitals. Three more passengers succumbed to their injuries in hospital and three others were seriously hurt. The railway authorities undertook to defray the medical and surgical expenses of the injured. [475]

Once all the injured had been cared for the arrangements for the clearance of the immense pilgrim traffic continued 'and clearance of traffic is brisk,' it was reported. [476]

Two hundred people were reportedly killed or injured in a head-on collision on the evening of Tuesday 17th May 1938 on the Canton – Hankow railway at Milokang, China. A southbound passenger train from Wuchang and a northbound freight train from Canton were involved. The cause of the collision was not stated 'but certain railway officials are reported to have been taken into custody.' [477]

A moment's thoughtlessness and confusion on the part of a railwayman led to a disastrous head-on collision between two express trains at Tortuga, California, United States, early on the morning of Tuesday 20th September 1938, the circumstances of which were similar to the 1934 Christmas Day disaster at Dundas, Canada (see above). The trains involved in the collision on the Southern Pacific Railroad were westbound train

---

[472] *La Vanguardia*, Barcelona, Spain, 5th December 1937
[473] Webpage: Federación Castellano Manchega de Amigos del Ferrocarril (in Spanish).
[474] *La Vanguardia*, Barcelona, Spain, 8th March 1938
[475] *The Railway Gazette*, London, England, 27th May 1938
[476] *The Bombay Chronicle*, Bombay, India, 16th April 1938
[477] *The Railway Gazette*, London, England, 27th May 1938

No. 5, *The Argonaut*, and eastbound train No. 44, *The Californian*.

Train No. 5 consisted of engine No. 4362, a combination mail and baggage car, a baggage car, a coach, two chair cars, a Pullman tourist sleeping car, a diner, a lounge car, five Pullman sleeping cars and a business car, all of steel construction, in charge of Conductor John Cantrell with Engineer Robert Richardson and Fireman H.R. Parsons.

Train No. 44 consisted of engine No. 4304, two baggage cars, four chair cars, a diner, three Pullman sleeping cars and a lounge car, of all-steel construction, in charge of Conductor Charles Rankin with Engineer S.E. Mason and Fireman P.E. Martin.

At Glamis, just over 19½ miles east of Tortuga, train No. 5 had received train order No. 405, which read:

No. 5 meet No. 44 at Tortuga and No. 4 at Iris. No. 44 take siding.

Train No. 5 left Glamis at 1.12 a.m., three minutes late, passed Amos, just over 6 miles east of Tortuga, at 1.30 a.m., on time, and travelling at between 30 and 40 miles an hour was diverted onto the siding at Tortuga into head-on collision with train No. 44 at 1.36 a.m.

Train No. 44 received train order No. 405 at Niland, 11 miles west of Tortuga, left there at 1.12 a.m., 4 minutes late, entered the west switch of the siding at Tortuga at 1.32 or 1.33 a.m., and was struck there by train No. 5.

Both engines were practically demolished in the collision. In train No. 5 the first car was badly damaged, the second car slid by it and was telescoped by the tender of the engine, the third car was badly damaged, the fourth demolished and the fifth badly damaged. In train No. 44 the leading car was slightly telescoped by the tender and the front of the third car was badly telescoped by the second.

Eight passengers, Engineer Robertson and Firemen Parsons and Martin were killed in the accident and 132 passengers, the head brakeman of train No. 5, three mail clerks and three Pullman employees were injured. All the passengers who lost their lives had been in the third car of train No. 44. [478]

One of the passengers on train No. 5, *The Argonaut*, was Mrs Rita Skousen, of San Bernardino, California, who had been returning from a visit to her husband, a contractor, in Van Horn, Texas, and escaped with cuts and bruises. She spoke to a reporter. "I was about half asleep when the crash happened. The day coach in which I was riding was the second passenger car on *the Argonaut*, and it was hurled into the air and reversed itself, landing heavily on the right side. The rear of the coach was in the direction the train had been travelling.

"A youth who was sat on the opposite side of the car to me was thrown onto me in the darkness. When he realized what had happened his first complaint was, 'Oh, I've lost my sweater.' I told him he'd better be thankful that's all he'd lost. Nearly every other person in my car was injured. I saw several bodies thrown through the windows. There was blood all over the ground." [479]

Engineer Mason of train No. 44 told the Interstate Commerce Commission (ICC) inquiry that Head Brakeman Eric Jacobson had opened the switch and derail to let them

---

[478] Interstate Commerce Commission report
[479] *Oakland Tribune*, Oakland, California, United States, 20th September 1938

into the siding at Tortuga and he had stopped his train 200 feet from the east switch, which had been properly set for the main line. Extinguishing the headlight, he had got down to oil and inspect the engine. Reboarding, he had seen train No. 5 approaching and from its headlight that the switch was open for the siding, and the other train had whistled. Shouting a warning to his fireman, he had jumped. After the accident Brakeman Jacobson, whom he considered very dependable and who had seemed perfectly normal, had told him that he had thrown the switch.

Head Brakeman Jacobson testified that he had fully understood that train No. 44 was to take the siding at Tortuga. After operating the switch to let the train into the siding he had ridden on ridden on its engine till it had stopped near the far end of the loop. Half a minute later he had set off towards the east switch. At some point between the engine of his train and the switch he had forgotten that the arriving train was to hold the main line and reaching the switch he had leisurely unlocked it and set it for the siding. As train No. 5 was close at hand, he had suddenly realised his mistake and had tried to reset the points but had had no time to do so. After the accident he had gone to help in the rescue work, and he could not remember resetting the switch for the main line, though it seemed that he must have done so. During his nine months as brakeman for train No. 44 it almost always met train No. 5 at Tortuga and train No. 5 almost invariably took the siding. He was sure that this was the cause of his mistake. He said that he was 63 years old, had been a brakeman for 18 years, had had adequate rest, had had no worries, very rarely drank alcohol, and did not use narcotics.

Conductor Rankin of train No. 44 agreed with his engineer that Brakeman Jacobson was safe and dependable and said that the brakeman had been in good spirits, alert and normal in every respect on the trip.

Conductor Cantrell of train No. 5 said that the speed of his train had been briefly reduced approaching Tortuga, presumably so that his engineer could identify train No. 44. The Tortuga home signal had been green. Close to the east switch his engineer had applied the brakes in emergency and the collision had occurred almost immediately. About 35 or 40 minutes after the accident he had examined the east switch, which had been undamaged and set for the main track.

Rear Brakeman Soutar of train No. 5 confirmed that the Tortuga distant and home signals had both shown green aspects.

Engineer of Car Construction Dailey confirmed that the lightweight streamlined steel cars of the trains (the fourth car of train No. 5 and the third, fourth and fifth cars of train No. 44) had suffered more damage in the collision, though he could not explain why.

The ICC noted that although train No. 5 was superior by direction it had taken the siding at Tortuga on 32 of 34 previous occasions since 1st May. The brakeman's assertion in this respect was thus confirmed. However, he had failed to comply with Rule 104 (C) which stated in part:

Trainmen and other employees must not unlock derails or main track switches to enter main track until the trains which are to be met or passed have cleared the switch and must not stand within twenty feet of the switch stand while a train is closely approaching or moving over the switch. Where practicable, they must take position on the opposite side of the track from switch stand when train is passing. When a switch is thrown, the employee using it must see that both points have moved to proper position. A switch must be fastened as soon as thrown either way, and, when locked, the chain pulled to insure the lock is securely fastened.

Had the switch been equipped with a light its red aspect might have alerted the engineer, fireman or conductor of train No. 44 in time for them to alert the brakeman to his mistake.

The ICC was concerned that all the fatalities had occurred in one of the lightweight streamlined cars which had been telescoped for 18 ft, that these cars had sandwiched between heavier steel cars, and that the Southern Pacific had carried out no tests itself on the strength or stability of the lightweight cars. Instead, it had relied upon the manufacturer's assurance that the cars met the requirements laid down for Post Office mail cars. Such tests had been made with a section of underframe rather than a complete car. Subsequently, another carrier had conducted tests with a complete car, had found a liability to telescoping and had strengthened the cars with reinforcing plates. No such plates were fitted to the Southern Pacific cars.

The ICC referred to a statement made in its annual report in mid-1936:

Equipment of this type {lightweight streamlined cars} *must be considered as still being in the experimental stage. As yet there is no case on record where a train of this type has been involved in a serious collision or derailment of such character as to furnish data as to its ability to withstand impact shocks and tearing and shearing stresses.*

That accident, the ICC pointed out, had now happened, and it raised a grave doubt as to whether lightweight cars should be permitted to run between or ahead of heavier all-steel ones until their ability to withstand shocks with safety had been proven by appropriate tests. It recommended that the use of such cars should be discontinued until such tests had been conducted. [480]

Brakeman Jacobson refused to testify at the inquests on the people who died in the accident on the grounds that he might incriminate himself (two inquests were held, at Brawley and Colton, California, because two passengers had died later from their injuries). He declared, however, "If I could take the place of any of those who were killed, I would be glad to do it, but that opportunity is now denied me." He was said to appear near collapse. The Brawley inquest on nine of the passengers returned a verdict of 'Accidental Death'. Deputy District Attorney R.I. French stated that his office would not file criminal charges against the brakeman unless requested to do so. [481]

On Monday 26th September 1938, a shunting locomotive collided with passenger train Kp 1026 at Borken, Germany, on the line between Münster and Empel-Rees. Sixteen people died and 14 others were injured. [482]

At least 11 people died, and nine others were asphyxiated when the overnight passenger train from Helsinki collided in fog with a freight train near Tainionkiski station in eastern Finland on the morning of 6th October 1938. Most of the casualties resulted

[480] Interstate Commerce Commission report
[481] *Arizona Republic*, Phoenix, Arizona, United States, 23rd September 1938
[482] *Katastrophen der deutschen Bahnen Teil II* by H.J. Ritzau, pg. 92 (in German)

from the escape of sulphur dioxide from a large tank waggon on the freight train. [483]

A hospital train carrying wounded Republican soldiers from the Extramadura front which was stopped in the yard of Ciudad Real station, Spain, was rammed by a shunting freight train, killing 12 people, on Saturday 15th October 1938. [484] Occurring, as it did, in the last months of the Spanish Civil War, the accident received very little notice in the press, though it was briefly mentioned in a handful of American and one Cuban newspaper.

Inattention on the part of the driver of a track machine led to its colliding with extra train 2631 between Kammer-Schörfling and Vöcklabruck, Austria, Germany, on Monday 2nd January 1939. Five people died and one person was injured. [485]

Two military trains reportedly collided between Bélmez and Villanueva del Rey, Spain, on Saturday 14th January 1939. Eleven people reportedly died and 24 were injured. [486]

A major disaster in Nicaragua occurred at León on Sunday 12th February 1939, the day that a new section of railway was opened, the first kilometre of a line which it was intended would run from La Sauce to Esteli (the line was never completed, the project being abandoned the following year). The inauguration of the new railway was attended by the President of the Republic, Anastasio Somoza Garcia, members of the cabinet and the diplomatic corps, other senior officials, and journalists. At the close of the inauguration ceremony the train carrying the President, the diplomatic corps, members of the government and the principal guests was the first to depart. This was followed by a second train carrying the other invited guests. The third and final train, carrying railwaymen and workers, left at 21:30. This last train was the one fated to be in the disaster.

The first news of the accident was received in the capital, Managua, by the *La Prensa* newspaper at 02:00 when its correspondent in León telegraphed the paper to say that there had been an accident of immense proportions involving a collision between the passenger train which had left La Sauce at 21:30 and an engine which was shunting. Carriages were overturned and workers were burnt by 'boiling oil' (possibly hot asphalt) from a tank. There were ten dead and 30 injured. All the victims had been removed from the train.

The railway management obtained from one of its employees the following version of the accident: Engine No. 11 had arrived at León station with a freight train of asphalt and oil. In order to place the waggons in the siding where they were to be stabled for the night it was necessary for the train to be turned on the triangle (or wye as it is termed in North America and Y Griega in Spanish) by Shunter Cervando Bonilia. To do this the train had to occupy the line from El Sauce and the stationmaster had warned crew of the freight

[483] *The Times*, London, England, 7th October 1938
[484] *Albuquerque Journal*, Albuquerque, New Mexico, United States, 17th October 1938
[485] *Katastrophen der deutschen Bahnen Teil II* by H.J. Ritzau, pg. 93 (in German)
[486] Webpage: Federación Castellano Manchega de Amigos del Ferrocarril (in Spanish).

train that a passenger train had left El Sauce and must be near at hand. “It’s all right, we have time,” the driver of the freight train had replied and the operation of shunting the train had begun. As the train had reached the triangle the passenger train, pulled by engine No. 15, had appeared, travelling at considerable speed, and a terrific collision had ensued.

A different version had it that the points at the triangle had failed to work properly. According to this version the freight train had been on the triangle clear of the line from El Sauce but the points being wrongly set the passenger train had been diverted onto the wrong line and into collision with engine No. 11.

The collision was so violent that the engines were wrecked and locked together, while the coaches of the passenger train were overturned. Such was the force of the impact that the rear coach ended up on top of the mountain of iron and wood in a vertical position. The happy and contented passenger coming from the fiesta at El Sauce suddenly experienced a tremendous shock. Some were thrown into the air, others were trapped between the coaches, crushed between wood and iron, killed, and injured, their bones crushed and burnt. The leading coach the passenger train, which had been full, suffered, however, very little in the accident and its passengers had miraculous escapes. Not so lucky were those in the second coach, which was buried beneath the following ones in the collision.

The León stationmaster had been stood waiting for the arrival of the latter when a boy had come running, tired, pale, and frightened, and had cried, “There’s been collision at the triangle, with people dead and injured!” The stationmaster and other railwaymen had immediately run to the scene and begun the rescue work.

Official sources stated that the dead had comprised seven passengers from Chinandega, two from Managua, and two from León, plus two women and three men who had not been identified. In addition Driver Aquillo Garcia and Conductor Sebastián Garcia of the passenger train had lost their lives. A report from the San Vincente hospital in León added the names of another ten passengers who had died. Four passengers and Fireman Juan Gregorio Palazio of the passenger train were said to be very seriously injured and many others were less badly hurt.

Señores Manuel Urbino and Nicolas Aguado were commissioned to open an inquiry into the accident. The investigation would have a number of points to consider. Was the real cause of the accident the conduct of the driver of train No. 15, who had approached León down a falling gradient at a speed of 35 kilometres an hour, a speed at which it would be less easy to pull up were there any problem with the points and which must have increased the force of the collision? Conductor Cordero of the freight train and the León stationmaster might also be considered responsible for allowing the driver of engine No. 11 to shunt when they knew that the passenger train was due. Were the brakes of the passenger train in order? Finally, had the driver of the passenger train been intoxicated? The regulations limited the speed of trains over the relevant points to 6 kilometres an hour.

Three days of mourning were declared in León.

Passenger Julio Lopez, who had not been named in the initial report as one of those who were worst hurt, nevertheless died of his injuries the day after the accident, raising the number of fatalities to 29. [487]

Córdoba – Málaga omnibus train No. 524 and Málaga – Séville express No. 527 collided near Campanillas station, Spain at 23:00 on Monday 22nd May 1939. [488] Seven people died and 40 were injured. Two of those who died were on their way home from the Nationalist concentration camp at Torremolinos. [489] The accident was reportedly due to a signalling error. [490] The line was reopened at 10:45 the next day.

Twenty-two people died in Romania on Wednesday 12th July 1939 when a freight train collided with a troop train consisting of wooden carriages at Clodul Marc, near Dez in Transylvania. [491]

Due to an error by the Powsin, Poland, stationmaster on Sunday 16th July 1939, two evening motor trains, travelling at 50 kilometres an hour, collided head-on with one another on the suburban line between Warsaw and Piasetzno and were telescoped. Three wooden carriages of one train and two of the other were destroyed and the other carriages were seriously damaged. The petrol tanks of the trains exploded and set them on fire. A scarcity of water at the scene of the collision made it difficult to extinguish the blaze, but after two hours' work, firemen, assisted by local people, were able to contain the fire. The injured were taken to a local hospital, the worst hurt then being transferred to Warsaw. It was initially reported that seven people had died reports and perhaps 200 had been injured. Many of the injured were in a serious condition and it was believed that the number of dead would rise.[492] [493]

The next morning an investigation was opened by the public prosecutor, following which the Powsin stationmaster and two of his assistants were arrested. [494]

A German troop train collided with one carrying refugees between Metzdorf and Wintersdorf, Germany, on Friday 8th September 1939, and 12 lives were lost. [495]

Several German books which contain comprehensive details of almost all other German accidents of any magnitude make no reference to this disaster. This is doubtless because it involved a troop train running on a line close to the front at a date when the Second World War was only a few days old, so that news of the accident was doubtless censored by the Nazi government. With the accident occurring close to the border with Luxembourg news of it could not be kept entirely secret, however, and it was briefly

---

[487] *La Prensa*, Managua, Nicaragua, 14th February 1939

[488] http://campanillasysuhistoria.blogspot.com/2014/07/el-ferrocarril-en-campanillas-la.html

[489] https://www.laopiniondemalaga.es/costa-sol-occidental/2015/04/19/pasado-oscuro-torremolinos-28542603.html

[490] *Dépêche corse*, Ajaccio, Corsica, 27th May 1939

[491] *Rhinelander Daily News*, Rhinelander, Wisconsin, United States, 12th July 1939

[492] *Journal de Geneve*, Geneva, Switzerland, 18th July 1939

[493] *The Times*, London, England, 18th July 1939

[494] *Le Peuple*, Paris, France, 18th July 1939

[495] *Le Petit Parisien*, Paris, France, 9th September 1939

reported in two French and a handful of American newspapers.

Electric passenger train No. 527, which had left Milano Centrale on time at 18:10 on the evening of Monday 30th October 1939 for Roma, had proceeded only as far as the junction of the lines to Bologna and Venezia, at Lambrate, Italy, a short distance from its starting point, when it came into collision with the direct train from Venezia, due in Milano Centrale at 18:15. The initial report on the accident spoke of eight dead and about 40 slightly injured. This, however, proved to be an underestimate, for the next day it was reported that the 16th body had been recovered from the wreckage. The dead included Motorman Mario Tavella of the electric train. Meanwhile there had been another collision on 31st October between passenger train No. 365 from Belluno to Treviso and a freight train, near Quero, in which two carriages had been shattered and 17 people injured, nine seriously.

A third report corrected the number of dead to 15. The investigation into the causes of the accident was by then under way and it was believed that the driver of the Venezia train had overrun signals. At the same time it was conceded that the junction, where many lines intersected and which was used by about 260 trains a day, was a dangerous one, and there was talk of rearranging and improving its layout to reduce conflicting movements.

On 3rd November it was stated that Driver Olivio Tomassoni and Fireman Alessandro Sartirana of the Venezia train had been placed under arrest, while its conductor, Carlo Bertoluja, had been placed under supervision in hospital. It was theorised that the driver had reduced speed to eight kilometres an hour after passing a yellow signal but had then mistaken a signal applying to another line for his own at the junction and had increased his speed again to 30 kilometres an hour, before he had caught sight of the electric train bearing down upon him. The driver and fireman repeatedly denied that they had passed any signals at danger.

The final report on the accident to appear in the newspaper stated that tests had shown that all the signalling at the junction had been working correctly, but that it would be three months before the final cause of the accident could be given. [496]

Two military trains, M 164, and M 219, collided at Karwitz, Germany, on the Dannenberg – Uelzen line on Monday 30th October 1939. One of the trains overran an *Einfahrsignal* (home signal). Nine persons died and 19 were hurt. [497] [498]

A collision occurred on the evening of Sunday 19th November at Spandau West, Germany, about 11 kilometres from the Berlin Lehrte terminus, between an express from Kiel, probably train D 66, and a Wustermark suburban train. Most of the sparse information on the accident came from the widow of one of the railwaymen who was killed in it, probably the driver of the suburban train, which may have been more crowded than usual because an event was being held at the radio station for a winter aid

---

[496] *La Stampa*, Torino, 31st October and 1st, 2nd, 3rd, 4th and 5th November 1939
[497] *Katastrophen der deutschen Bahnen Teil II* by H.J. Ritzau, pg. 96 (in German)
[498] *Eisenbahnunfälle in Europa* by Erich Preuß, pg. 97 (in German)

organisation. [499] The driver of the express had supposedly been drunk. According to *The Railway Gazette* nine people died in the accident. The line was equipped with the *Indusi* warning system. [500]

On receiving a proceed signal from the conductor the driver of parcels train Leig 5811 set off against an *Ausfahrsignal* (starting signal) at danger at Nieder Wöllstadt, Germany, on the line between Gießen and Frankfurt Main, and came into collision with passenger train P 772 on Sunday 26th November 1939. Eleven people died and 18 were injured. [501] [502]

Due to the premature operation of a set of facing points passenger train P 3567 was diverted onto the wrong line and came into a head-on collision with passenger train P 3862 on the Hagen – Dortmund line between Vorhalle and Wetter, Germany, on Tuesday 12th December 1939. Fifteen people died and 36 were injured. [503] [504]

On Friday 15th December 1939, military train M 133 was driven past a red signal into collision with another train at Vennebeck, Germany. A warning to the driver to stop came too late. Six people died and the same number were injured. [505]

A report from Tokyo stated that a crowded passenger train and a freight train had collided head-on at Matsui, Japan, near the border of the Tokyo and Saitama prefectures, on Wednesday 3rd January 1940. Ten people lost their lives and over 60 were injured. [506]

Early on the morning of Sunday 21st January 1940, a leave train collided with a mail train at a junction at Saint-Julien, near Troyes, France. Ten people died and 40 others were injured, none seriously. Those who died included two drivers and a fireman. A rescue operation was quickly launched and an inquiry into the accident was opened. [507] The postal train consisted of an engine and 15 vehicles. The troop train was carrying almost 1,000 men.

The subject of the accident and its causes was brought up in parliament in mid-March by M. Emile Brachard in an address to M. Anatole de Monzie, the Minister of Public Works. He noted that the mail train had come from Vitry-le-Francois via Brienne-le-Château to Troyes and that since leaving Vitry it had been running early and was ten or twelve minutes before time as it neared Troyes. He questioned why it had not been detained at some intermediate point until its booked departure or passing time. The leave

---

499 *Katastrophen der deutschen Bahnen Teil II* by H.J. Ritzau, pg. 104 (in German)
500 *The Railway Gazette*, London, England, 24th November 1939 and 5th January 1940
501 *Eisenbahnunfälle in Europa* by Erich Preuß, pg. 120 (in German)
502 *Katastrophen der deutschen Bahnen Teil II* by H.J. Ritzau, pg. 97 (in German)
503 *Eisenbahnunfälle in Europa* by Erich Preuß, pg. 120 (in German)
504 *Katastrophen der deutschen Bahnen Teil II* by H.J. Ritzau, pg. 97 (in German)
505 *Eisenbahnunfälle in Europa* by Erich Preuß, pg. 120 (in German)
506 *The Railway Gazette*, London, England, 12th January 1940
507 *Paris-Soir*, Paris, France, 23rd January 1940

train, on the other hand, had been running nine hours late. The signalman at the junction had therefore given the leave train priority and kept his signals against the mail train, the driver of which had overrun four signals at either caution or danger. Why, it was impossible to say because he had lost his life in the collision, but M. Branchard questioned whether the safety measures had been sufficient. It had been just after midnight. It had been bitterly cold with freezing snow. Furthermore, the driver had been a stranger to the Est region. He had belonged to the Montargis shed and had barely known the route, having only been over it once before. He had approached the junction at Saint-Julien, which lay on a curve, running before time, and had passed successive signals without reducing speed, including a *carré d'arrêt* (absolute stop signal). Possibly he had only seen the signals too late. The brakes seemed to have worked properly but snow covered rails had made them less effective.

It was doubtful, M. Brachard continued, if the accident would have happened if the engine had been in charge of a driver from the Est region, who would have been aware of the difficulties of the approach to the junction. Nor would an Est driver have been running early, such things only happened on the old Paris-Lyon-Méditeranée network. The essential fact to remember was that the postal train, like a number of those on secondary lines, had only had one conductor, placed at the rear of the train, so that he could not communicate with the driver. This conductor did not have a detailed timetable for the train and had been unaware of necessary modifications to its timetable imposed by the military. From his position at the rear of the train the conductor had been unable to see the signals on the curve approaching the junction in time to know if they were being passed at danger or had been routinely replaced after the engine had gone by them. A conductor in the leading van, with a proper timetable, would not have permitted the driver to run so much ahead of time but would have signalled to him to reduce speed, via a bell, nor would he have allowed the driver to run past four signals against him without warning him, and the collision would have been averted.

M. Branchard drew attention to article 19 of the railway regulations, which read:

During the journey conductors must pay attention to the running of their trains and the aspects of the signals all the time that they are not occupied with dealing with luggage or goods. They must be especially vigilant at the approach to stations and on double track at points where the line becomes single.

They must supervise the loading of waggons.

If, for any reason affecting the safety of the train, it is necessary to stop it, conductors must, without hesitation, act as follows: The conductor at the head of the train must, according to the circumstances, either operate the horn, bell, or whistle on the engine or else apply the brakes by opening the control valve.

In this case, M. Brachard pointed out, there had been no conductor at the head of the train. This was a very important point. Conductors at the front of trains were there to correct any mistakes by drivers. They had been done away with for reasons of economy. The conductor at the rear of the train had been retained to protect it in the rear should it stop out of course but a derailment, for example, could occur at the front of the train, obstructing the opposite running line. It was impossible for one conductor to at once protect both ends of the train. Two conductors had always been used on the Est network. They should be reintroduced everywhere, especially with the present intense daily wartime traffic.

M. Branchard questioned why the mail train had not been stopped at Thennelières station, the last before the junction. He also mentioned that drivers and conductors were often working 12, 14 and even 18 hours at a time. It was true that this was partly due to the war conditions, but the utilisation and distribution of staff could also be improved. Finally he pointed out that the detonators which should have exploded when the mail train passed the *carré d'arrêt* at danger had failed to work.

Concluding, M. Branchard stated, "I beg you, I use the term advisedly, I beg you, Minister of Public Works, in the name of the safety of the people travelling on the railway, civil and military, and of the railway workers, to draw from this painful accident the conclusions to which it points and to take immediately the necessary measures to prevent any further disasters, because this is the first duty imposed upon you and the Société national de chemins de fer Français." (Applause.) [508]

The Soviet Union had invaded Finland at the end of November 1939. For the first two months the Finns repelled the Russian attack but by the end of February 1940 the superior forces of the Soviet Union were beginning to tell. At the beginning of March 1940 Finnish children were being evacuated to neutral Sweden by train, and it was one of these workings which was involved in a major disaster at Iitala station, north of Tampere, Finland, at about midnight on Tuesday 5th March 1940. The evacuees train, which was bound for Tornio, collided head-on with a southbound express, the driver of which had overrun a stop signal during a snowstorm. A coach of the refugee train overturned and caught fire, burning to death 16 children, two mothers, and all three of those in charge of the evacuees. Several others were slightly hurt. One Swedish family in Stockholm, which had agreed to accommodate three children, had to be informed that all three had lost their lives in the accident. [509]

Five people died and 25 were injured when two trains crashed head-on near Santa Caterine, Mexico, south of Puebla, on Saturday 9th March 1940. 'Several American tourists aboard one of the trains escaped injury.' [510] [511]

On Monday 11th a conductor and a fireman died and four other railwaymen were injured in a collision in Oaxaca state, the third serious accident on the Ferrocarril Nacional de México in just over a fortnight [512] and the following day the directors of the worker's administration of the railway resigned en masse. [513]

Initially 14 people were reported to have died when two Estrada de Ferro Central do Brasil trains crowded with passengers travelling during the Easter holidays collided near Rio de Janeiro, Brazil, on Friday 22nd March 1940. Three of the victims were women.

---

[508] *Journal Official de la République Française*, 15th March 1940
[509] *The Times*, London, England, 8th March 1940
[510] *El Paso Herald-Post*, El Paso, Texas, United States, 11th March 1940
[511] *The Railway Gazette*, London, England, 22nd March 1940
[512] *Hanover Evening Sun*, Hanover, Pennsylvania, United States, 11th March 1940
[513] *Charleston Gazette*, Charleston, West Virginia, United States, 13th March 1940

The bodies of some of the dead were so mangled as to render identification impossible. The accident was blamed in early reports on the failure of one of the drivers to wait at a crossing place for the opposing train, implying that it was a single-line collision. [514]

Later reports gave the location of the accident more precisely as near Magé, stated that there were 16 dead and 100 injured, and that the accident had occurred at a junction 'due to a confusion over signals,' indicating that it was, in fact, probably a sidelong collision due to one train overrunning signals. [515]

The Bombay Baroda and Central India Railway Frontier Mail train from Bombay to Delhi collided with a shunting goods train in the station at Jekot, India, on the morning of Wednesday 15th May 1940. Eighteen people were killed outright and 17 injured, three of whom subsequently died. The engine of the goods train was still occupying the main line while the train was being set back into a loop line when the mail train reached the station around a curve. The driver of the express applied his brakes as soon as he saw the obstruction but could not avoid the collision. The goods engine was badly damaged and the leading three carriages of the seven making up the mail train were smashed. The first carriage was an intermediate class coach, the second a first and second-class composite and the third an air-conditioned coach with passengers for Peshawar.

The General Manager of the railway company and the Government Inspector of Railways arrived on the scene promptly after the accident. The line was cleared within 24 hours. [516]

It was decided that the Jekot stationmaster was mainly to blame for the accident by allowing the goods train to shunt with the signals cleared for the mail train, while the goods train driver was at fault to a lesser degree. Both were subsequently prosecuted and fined 300 and 100 roubles respectively by the Additional Sessions Judge, Godhra. [517]

A train bringing about 1,200 Belgian and Dutch refugees back from France was in collision with a freight train near Sambre, Belgium, on about Wednesday 4th September 1940. [518] Fifteen people died and about 50 were injured. [519]

Gale-force winds which affected Belgium during the morning of Thursday 14th November 1940 were blamed for a major disaster at Diegem station, near Bruxelles. A train which had left Tirlemont at 07:35 collided with a train standing in the station. The collision was so violent that several carriages of the passenger train coming into the station were completely destroyed. There were believed to have been 21 people killed, 30 seriously injured and 50 slightly injured, all Belgians. [520]

---

[514] *The Titusville Herald*, Titusville, Pennsylvania, United States, 22nd March 1940

[515] *Sarastota Herald-Tribune*, Sarastota, Florida, United States, 22nd March 1940

[516] *The Railway Gazette*, London, England, 24th May and 14th June 1940

[517] Annual Police Administration Report of the Province of Bombay, including the Railways, for the year 1940, pgs. 44-5.

[518] *L'Indépendent des Basses-Pyrénées*, Pau, France, 6th September 1940

[519] *The Railway Gazette*, London, England, 13th September 1940

[520] *La Charente*, Angoulême, France, 17th November 1940

Romania had joined the Axis Powers in late November 1940, following the Soviet annexation of Bessarabia and northern Bukovina, and a month later the Germans were pouring large numbers of troops into the country, with reports of 1,000 or 1,200 trains being requisitioned to transport troops there via Hungary. [521] It was very possibly two of these troop trains which were involved in a collision near Arad, Romania, on about Friday 27th December 1940, in which six people were killed. [522]

The failure of a signalman to check that the line was clear led to semi-fast passenger train E 174 colliding with a shunting movement at Kodersdorf, Germany, on the Cottbus – Görlitz line on Sunday 29th December 1940. The shunting locomotive was thrown onto the adjoining running line and struck by through freight train Dg 6973. Five people died and three were hurt in the double collision. [523] [524]

A misunderstanding between a *fahrdienstleiter* (signalman) and a *stellwerkwärter* (pointsman) plus serious errors in the issuing of train orders led to the dispatch of a train into an occupied section and a collision between special leave train for front line troops SF 119 and through freight train Dg 6004 between Rathenow and Nennhausen, Germany, on Monday 30th December 1940. Six people lost their lives and 11 were injured in the accident on the Berlin – Stendal line. [525] [526] [527]

Eight people died and 26 people were injured, 14 seriously, in a collision between an omnibus train from Chimay and a freight train at Villers-La-Tour, Belgium, near Charleroi, at about 20:25 on Friday 3rd January 1941. Several coaches were completely wrecked. [528] Due to a signalling error the passenger train had been diverted onto the line occupied by the freight train, which was a heavy one, loaded with 1,200 tonnes of iron for the construction of a bridge at Revin (or very likely reconstruction following wartime damage). One of those who died was the Chimay *chef-garde* (stationmaster?). The rescue operation had to be undertaken in the dark because of the blackout, and in freezing temperatures. [529]

Twenty soldiers were reportedly killed in a collision at Nerettya-Ujfalu, Hungary, south west of Debreczen, on 8th January 1941. A German munitions train collided with a second train and an explosion followed. A large part of the town was damaged, and it was thought that it would take weeks to repair the damage. Further details of the accident

[521] *The Sheboygan Press*, Sheboygan, Wisconsin, United States, 26th December 1940
[522] *The Railway Gazette*, London, England, 17th January 1941
[523] *Eisenbahnunfälle in Europa* by Erich Preuß, pg. 121 (in German)
[524] *Katastrophen der deutschen Bahnen Teil II* by H.J. Ritzau, pg. 103 (in German)
[525] *Eisenbahnunfälle in Europa* by Erich Preuß, pg. 121 (in German)
[526] *Katastrophen der deutschen Bahnen Teil II* by H.J. Ritzau, pg. 103 (in German)
[527] *The Railway Gazette*, London, England, 17th January 1941
[528] *La Dépêche*, Toulouse, France, 7th January 1941
[529] Wikipedia – Liste des accidents ferroviaires en Belgique

were supressed by Hungarian censorship. [530] [531]

Passenger train P 437 ran into express train D 137 at Dreileben-Drackenstedt, Germany, on the Magdeberg – Helmstedt line after overrunning the *Einfahrsignal* (home signal). Eight people died and 31 were injured. [532] [533]

Ten people reportedly died in a head-on collision between a Warsaw – Siedlce passenger train and a freight train in the neighbourhood of Warsaw, Poland, probably about the middle of February 1941. [534]

During a snowstorm the engine of local freight train N 8647 was moved without the permission of the pointsman at Gardelegen, Germany, on the Berlin – Hannover main line and came into sidelong collision with troop train W 94 111. Seven people died and 30 were injured in the accident on Thursday 27th March 1941. [535] [536]

The circumstances of a double collision at Rengershausen, Germany, on the Kassel – Guntershausen line, on Sunday 6th April 1941, are not entirely clear, at least to the author, from the brief descriptions of it in the German literature. It would appear that a signalman's error was to blame, and that he had failed to realise that an *Ausfahrsignal* (starting signal) had not been cleared for through freight train Dg 6079. He had then accepted special leave train for front line troops SF 137, which ran into the waiting freight train. To compound matters troop train W 99 091 subsequently ran into the wreckage fouling the opposite running line, so it was perhaps fortunate that there were only six deaths and 11 injured. [537] [538]

A misunderstanding between a signalman and the crew of local freight train N 8106 resulted in a sidelong collision between the shunting freight and troop train W 92 065 which was passing through the station at Gusow, Germany, on the line between Berlin and Frankfurt an der Oder on Saturday 3rd May 1941. A major disaster resulted, with 16 deaths and 45 injured. [539] [540]

The English Great Western Railway had a very good safety record throughout its long

---

[530] *The Lincoln Star*, Lincoln, Nebraska, United States, 8th January 1941
[531] *The Railway Gazette*, London, England, 17th January 1941
[532] *Katastrophen der deutschen Bahnen Teil II* by H.J. Ritzau, pg. 104 (in German)
[533] *Eisenbahnunfälle in Europa* by Erich Preuß, pg. 121 (in German)
[534] *The Railway Gazette*, London, England, 7th March 1941
[535] *Eisenbahnunfälle in Europa* by Erich Preuß, pg. 122 (in German)
[536] *Katastrophen der deutschen Bahnen Teil II* by H.J. Ritzau, pg. 104 (in German)
[537] *Eisenbahnunfälle in Europa* by Erich Preuß, pg. 122 (in German)
[538] *Katastrophen der deutschen Bahnen Teil II* by H.J. Ritzau, pg. 104 (in German)
[539] *Eisenbahnunfälle in Europa* by Erich Preuß, pg. 122 (in German)
[540] *Katastrophen der deutschen Bahnen Teil II* by H.J. Ritzau, pg. 104 (in German)

history, but this was marred by two bad accidents during the Second World War. The first, at Norton Fitzwarren in November 1940, has already been described in the 'Other Derailments' chapter, the second occurred eight months later, on Wednesday 2nd July 1941, at Dolphin Junction, just east of Slough, on the quadruple-track main line between London Paddington and Reading. The 6.20 p.m. Up express passenger train from Plymouth to London Paddington collided head-on with the 1.30 a.m. Down freight train from Old Oak Common to Severn Tunnel Junction. Five passengers, of whom three were Naval ratings, lost their lives; nine passengers, all Naval ratings, and the driver of the down freight were seriously hurt; and 15 passengers and four railwaymen were slightly hurt. The passengers who died were Chief Petty Officer George Antell, Seamen Harry Housley, and Frederick Ives; Mr John Bertram, of Earley, near Reading; and Mrs Bate, of Barnstaple, Devon, who had been on her way to meet her husband, an RAF chaplain. Soldiers and sailors who had been passengers on the train were the first people active in the rescue work and ARP and other defence workers from Slough were soon on the scene, as were Local Defence Volunteers (the Home Guard) who had been on duty nearby. The uninjured passengers on the train were forwarded by motor coach direct to London, later passengers were shuttled by bus between Slough and Langley (the next station on the London side of Dolphin Junction) with trains between there and London, until the line had been cleared. [541]

The Up express consisted of Castle class 4-6-0 engine No. 4091, *Dudley Castle*, and eight corridor coaches in charge of an unnamed guard with Driver J. Hoskins and Fireman F.C. Harbridge. The freight train consisted of 2-8-0 London Midland and Scottish Railway engine No. 8298 (on loan to the Great Western), 32 unbraked waggons and a guard's van, in charge of Guard J. Cudd with Driver S. Clay and Fireman F.C.W. Ratcliffe.

Dolphin Junction signal box controlled connections between the main and relief lines. The lines here were paired by type and from north to south were Up Relief, Down Relief, Up Main and Down Main. A train passing from the Up Main to the Up Relief therefore had to cross the Down Relief line.

On the morning of the accident the Plymouth express was running 32 minutes late on the Up Main line and was catching up a Government Stores (munitions) train. Signalman W. Welch at Dolphin Junction accepted the Government Stores train from Slough East box at 2.38 a.m., obtained 'Line Clear' for it from Langley box at 2.39 and cleared his Up Main signals for it. Slough East gave the 'Train Entering Section' signal for it at 2.46 a.m. and at about the same time he was advised by Slough that the express had left Reading at 2.32. Signalman Welch calculated that the express would reach him at 2.52 and rang Signalman W. Woodhouse at Slough East to ask how the stores train was running. Meanwhile he had accepted the Old Oak Common freight train from Signalman J. Adams at Langley on the Down Relief line and obtained 'Line Clear' for it from Slough East.

The stores train passed Dolphin Junction on the Up Main at 2.49, Signalman Welch cleared it back to Slough East and was immediately offered the Up express, which he accepted. He then became concerned lest the slow-moving stores train delay the express and according to his account attempted without success to find out the position regarding

541 *The Times*, London, England, 3rd July 1941

the Up Relief line on the London side of his box. The Langley signalman then cleared out the stores train and, according to Signalman Welch, accepted the express on the Up Main, but at this moment he (Welch) was advised by the West Drayton signalman that there was nothing on the Up Relief. Hearing this, he cancelled the express on the Up Main and instead obtained 'Line Clear' for it from Langley on the Up Relief line. He said that at this time (2.53 a.m.) the Up express had come to a stand at his Up Main home signals. Up to this point everything was in order, but the problem was that he had accepted the Old Oak Common freight train on the Down Relief line, which the express had to cross to reach the Up Relief, and furthermore he had already received the 'Train Entering Section' signal for it from Signalman Adams at Langley. According to the regulations Signalman Welch should either have held the express on the Up Main line until the Down freight train had passed his box or else waited until he was certain that the Down freight had come to a stand at his Down Relief home signal before letting the express across its bows. Signalman Welch maintained that he had looked out for the freight train and had judged from its headlights that it had come to a stand at his home signals before he had cleared his Up Main to Relief home signal for the express, which he said, "drew away sharply." It was only then that he realised that in fact the freight train was also still moving, and the trains collided on the Up Main to Up Relief crossover. He denied having cleared his Down Relief signals for the freight train and then replaced them, although he had had the train accepted by Slough East, and he could only suggest that the freight train had overrun his signals. Signalman Adams at Langley denied ever having been offered the express on the Up Main.

Driver Hoskins of the express told the Ministry of War Transport inquiry that he had been held 1½ or 2 minutes at the Dolphin Junction home signals, had restarted gently and had not been moving at more than 4 miles an hour when he had seen the headlights of the freight train. He had shut the regulator but had not had time to apply the brakes before the collision.

Driver Clay was badly hurt in the accident, and it was weeks before he could give evidence, but his account was straightforward. He stated that the Dolphin Junction Down Relief distant signal had been clear for him and he had been completely taken by surprise on finding the home signal at danger. Assuming that for some reason the signals had been 'thrown back in his face,' he had applied the steam brake but had been unable to stop his unfitted freight train in the distance available. He did not recall having seen anything of the Up express. He knew the line well and did not regard the absence of the automatic train control equipment (fitted to all Great Western engines but not to 'foreign' engines) as a handicap. Somewhere about Langley his fireman had had a problem with a leaky pipe on the fixed coal watering spray and having seen the Dolphin Junction Down Relief distant 'off' he had been checking that his fireman was dealing with the problem correctly, and so did not keep the distant in view, but had seen the home signal at danger in good time. He was able to correctly describe the relative positions of the Down Relief and Down Main signals. He was not sure if the Down Main distant signal had been on or off (it might have been off for another down freight train).

Fireman Ratcliffe said that he had been preoccupied with the leaking pipe and had seen nothing of the signals.

Guard Cudd did not see the Dolphin Junction Down Relief distant signal.

Major G.R.S. Wilson, the inspecting officer, by a series of calculations, deduced that the freight train must have been at least 1,006 yards from the point of collision when Signalman Welch had reversed the crossover, so that the signalman's statement that he had believed that the freight train had come to a stand was clearly false, as the freight train had been nowhere in sight when he had reversed the crossover. Signalman Welch had then cleared his signals for the express on the assumption, without justification, that the freight train would stop at his Down Relief home signal. In doing so he had acted in direct contravention of Block Regulation 4 (c), which stated:

After permission has been given for a train to approach in accordance with this Regulation (4) no obstruction of the line on which such train requires to run must be allowed within a distance of a quarter of a mile ahead of the home signal or other prescribed clearing point, neither must any points in the rear of the authorised clearing point be altered except at those places where special instructions are issued to the contrary, authorised by the Superintendent of the Line, are in force, until the train has been brought to a stand at the home signal, or has passed into the section in advance, or the "Cancelling" signal has been received.

Major Wilson stated that he could find no excuse for the signalman's failure. Although it was a busy box he had not been working under any great pressure at the time. There was no reason why he could not have diverted the stores train to the Up Relief, and having failed to do so, he had lost his head in his anxiety not to delay an important passenger train. His comparatively trivial mistake had then led to a complete disregard of safety. Welch had a previously clear record but, in the words of the inspecting officer, 'It appears that, in spite of his record hitherto, he is temperamentally unfitted for the responsible duties of a signalman, at any rate in a busy main line junction box.'

As for Driver Clay his responsibility depended on the position of the Dolphin Junction Down Relief distant signal. It was possible that in his brief glance he had mistaken the Down Main distant signal, still clear for the other freight train, for it. On the other hand, Signalman Welch's evidence had been shown to be unreliable in other important respects, and it was possible that despite his denial he had cleared the Down Relief signals, then replaced them. In any case the signalman was basically responsible for the accident and the inspecting officer considered it only fair to give Driver Clay the benefit of the doubt.

Major Wilson had little to say by way of recommendations, other than that the Company might improve the poor visibility of the small, centre-pivoted Down Relief home signals. [542]

A collision between a German troop train and a freight train near Warsaw, Poland, on about Friday 25th July 1941 reportedly caused 2,000 casualties. The Gestapo made numerous arrests after the collision, clearly believing that the accident had resulted from an act of sabotage. Without further information it is impossible to be certain but collisions as opposed to derailments due to sabotage are rare. [543]

Several American newspapers reported that a German troop train had collided with a

[542] Ministry of War Transport report
[543] *San Antonio Sunday Light*, San Antonio, Texas, United States, 27th July 1941

freight train at Cseka Trehova, Czechoslovakia, on or about Saturday 11th October 1941, killing 15 people. [544]

An S-Bahnzug suburban train ran past a red signal and collided with another train between Berlin Ostkreuz and Grünau on Wednesday 29th October 1941. Six people died and two were injured. [545]

At 20:05 on the evening of Sunday 2nd November 1941, Société nationale de chemins de fer français (S.N.C.F.) train No. 60, arriving at Paris Austerlitz, France, from Orléans and still moving at 30 kilometres an hour, came into sidelong collision with the rear of train No. 6, arriving from La Rochelle on an adjoining line. The force of the collision derailed five carriages of the Orléans train, two of which were overturned and one of these was cut in two. An initial report spoke of 19 fatalities and 48 injured. As the accident occurred at a Paris terminus a rescue operation was quickly organised and the president and director general of the S.N.C.F and other officials went to the scene. [546]

A further report on 6th November spoke of 20 dead, 100 seriously injured and admitted to hospital plus 73 slightly hurt but able to go home. Most of the victims were Parisians. M. Charles, the *juge d'instruction,* had opened an investigation into the accident. [547]

The final death toll was apparently 21.

Two trains were involved in a collision near Tokyo, Japan, probably in the vicinity of Kami-Nagatoto station, on the morning of Wednesday 19th November 1941. Twenty people died and a large number were injured. [548] [549]

Carelessness on the part of the railwaymen involved in a shunting operation resulted in the shunting engine coming into sidelong collision with passenger train P 2815 at Unterholfen-Solgen, Germany, on the Metz – Salzburg line, on Wednesday 3rd December 1941. Six people died and 14 were injured. [550] [551]

The negligence of a signalman was again the cause of a collision between passenger train P 1000 and an express freight train on the Kiel – Eckernförde line at Suchsdorf, Germany, on Wednesday 17th December 1941. Seven people died and 15 were injured. [552] [553]

---

[544] *Tucson Daily Citizen*, Tucson, Arizona, United States, 13th October 1941
[545] *Eisenbahnunfälle in Europa* by Erich Preuß, pg. 122 (in German)
[546] *L' Ouest Éclair*, Paris, France, 4th November 1941
[547] *L'Indépendent des Basses-Pyrénées*, Pau, France, 6th November 1941
[548] *Le Petit Journal*, Paris, France, 20th November 1941
[549] *The Paris News*, Paris, Texas, United States, 19th November 1941
[550] *Eisenbahnunfälle in Europa* by Erich Preuß, pg. 123 (in German)
[551] *Katastrophen der deutschen Bahnen Teil II* by H.J. Ritzau, pg. 104 (in German)
[552] *Eisenbahnunfälle in Europa* by Erich Preuß, pg. 123 (in German)

Because of the wrong position of a set of facing points, troop train W 96 031 was diverted onto a track occupied by a pair of locomotives. Twenty-seven people died in the resulting collision at Langhagen, Germany, on the Rostock – Neustrelitz line. Another 33 were injured in the disaster on Monday 29th December 1941. [554 555]

A collision at Eccles, England, on the London Midland and Scottish Railway, on Tuesday 30th December 1941, was caused partly by the difficulties of working trains in foggy weather and partly by confusion on the part of a signalman whose three fogsignalmen chanced to have different but similar surnames. The 6.30 a.m. passenger train from Rochdale to Pennington, travelling on the Down Slow line, collided with the 6.53 a.m. passenger train from Kenyon to Manchester Exchange, which had been crossing from the Up Slow to the Up Fast line. At the lines were paired, as at Dolphin Junction, by type rather than by direction, this meant crossing the Down Slow line. The engine of the Rochdale train ripped out the left-hand side of the leading carriage of the Kenyon train, demolished the next two carriages and ripped out the side of the fourth carriage. Twenty-three passengers, three of them railwaymen, lost their lives, 56 passengers and the fireman of the Rochdale train were admitted to hospital, while 27 passengers, and the driver of the Rochdale train were less badly hurt. The accident occurred at 8.18 a.m. Help was quickly on hand after the accident. Doctors, nurses, ambulance staff and the police were in attendance within 15 minutes and assistance was provided by railwaymen, uninjured passengers, local residents and Civil Defence staff (the Home Guard). Because of the difficulties of accessing the mass of wreckage the last victim could only be extricated at 1.30 p.m. [556]

The Rochdale train consisted of 2-6-4 tank engine No. 2406 and eight non-corridor coaches in charge of Guard R. Clucas with Driver P. Mountfield and Fireman R. Ashcroft. The Kenyon train consisted of 2-6-4 tank engine No. 207 and eight non-corridor coaches in charge of an unnamed guard with Driver George Ashcroft and an unnamed fireman.

The next signal box on the Manchester side of Eccles Station box was Cross Lane Junction, just under two miles to the east. In addition to the Down Slow inner home signal protecting the mains crossover there was an outer home signal, 440 yards in its rear, so that during clear weather a train on the Down Slow could be accepted under full line clear with a train crossing from the Up Slow to the Up Fast line in front of it. During foggy weather, as on the morning of the accident, such acceptance was only permissible if there were fogsignalmen station at both the Up Slow distant and the Up Slow outer home signals. The fogsignalmen involved were A. Patton at the Down Slow distant signal, W. Parrington at the Down Slow outer home and A.C.J. Pantling, who fogged the Up Fast outer home. The night before had also been foggy and Patton at the Down distant had only booked off at 11.30 p.m. and did not book on at 8.30 a.m., after the accident. Fogsignalmen Parrington and Pantling, on the other hand, had booked off at 10 p.m. and

553 *Katastrophen der deutschen Bahnen Teil II* by H.J. Ritzau, pg. 104 (in German)
554 *Eisenbahnunfälle in Europa* by Erich Preuß, pg. 123 (in German)
555 *Katastrophen der deutschen Bahnen Teil II* by H.J. Ritzau, pg. 104 (in German)
556 *The Times*, London, England, 30th December 1941

were back on duty at 7 a.m.

On the morning of the accident the Kenyon train was running 49 minutes late and due to congestion on the Up Slow line Signalman Charles Lowe at Eccles Station decided to cross it there to the Up Fast line. Signalman Thomson at Cross Lane Junction box duly accepted the Kenyon train on the Up Fast line at 8.09 a.m. and the signals at Eccles were cleared for it.

The Rochdale train was running 27 minutes late. Signalman Lowe accepted it on the Down Slow line at 8.15 a.m. from Cross Lane Junction under full line clear up to his Up Slow outer home signal at danger and immediately received the 'Train Entering Section' signal. An annunciator bell rang when it operated a berth track circuit in the rear of the outer home, but he did not realise it was overrunning signals until he saw it, when there was no time to do anything to avert the collision.

Signalman Lowe had been at Eccles Station for three months. This was the first time, however, that he had been on duty during a prolonged fog. He told the Ministry of War Transport inquiry into the accident that he had been aware when he took over the box at 6 a.m. that there were no fogsignalmen on duty and had resorted to double block working, not accepting a train from the box in the rear until the line was clear to the box in advance, as per Block Regulation 4(e) (i). At 7 a.m. he had assumed that all the fogsignalmen were at their posts and had reverted to normal working. He stated that Parrington had reported at his box just before 7 and gone to the Down outer home. A second man, whom he did not see, had passed the box about the same time, and had called out, "I am coming on my post, Charlie." Lowe thought this was Pantling, whose voice he knew, and told his booking boy, L Acton, to book Pantling on. A minute or so later he thought he heard his booking boy say that Patton had telephoned to say that he was going to his post at the Down Slow distant and had remarked, "I have all the fogmen on. We are all right now," and had resumed normal working.

What had actually happened was that Pantling had gone directly from his home to the Up Fast signal and had telephoned the signal box to report for duty from the Eccles station platform. Booking Boy Acton said that he had informed the signalman that it was Pantling, not Patton, who had reported. He did not remember anyone passing the box or his signalman remarking that he now had all the fogmen. Lowe was apparently expecting that Patton would report by telephone and Pantling at the box and had apparently overlooked that fact that Patton had not booked off the night before until 11.30 p.m.

Driver Mountfield said that visibility had been fair between Rochdale and Manchester but foggy beyond there. Leaving Manchester the water had been low in the engine tanks, and he had arranged with his fireman to take water at Eccles troughs if the Down Slow distant was "off." All the signals had been off between Manchester up to and including the Cross Lane starter. He had reminded his fireman to look out for the Eccles distant and when they reached it the latter had assured him that it was clear. Furthermore, he had not exploded any detonators laid by the fogsignalman. He had not seen the outer home signal but had seen, he said, a green hand signal from the fogsignalman there (Parrington), while no detonator had gone off. He had then instructed his fireman to lower the water scoop. The inner home he had also failed to see but seeing the signal box had realised that he was closer to the station, where he was booked to stop, than he had thought, and had braked fully so as not to overrun the platform. He had not seen the Kenyon train before he hit it, at, he thought, a speed of 20 miles an hour.

Fogsignalman Parrington directly contradicted Mountfield's testimony. He stated that he had shown a red hand signal to the train and the detonator which he had laid had exploded. Driver Mountfield, however, stuck to his story of getting a green hand signal and hearing no explosion.

Fireman Ashcroft said that although he had fired over the line for some years he could not claim a detailed knowledge of the signals. He agreed that his driver had told him to look out for the Down Slow distant signal and claimed that he had seen it showing a distinct green light when it was about 10 yards away and had told his driver that it was "off." He had not seen the outer home signal or the fogsignalman there or heard a detonator.

Guard Clucas said that because of the fog he had not seen any signals after leaving Manchester, nor had he heard any detonators go off.

Fogsignalman Parrington had long experience at his job. He said that his brazier had been burning well and that although it had been raw it had not been uncomfortably cold (he doubtless had plenty of warm clothes on). Because of the proximity of the Manchester Ship Canal the fog was always worse at his post than elsewhere, both thick and black (and so a smog rather than a simple fog). He was sure that his detonator had exploded but doubtful if the driver could have seen his hand signal through the fog as he (Parrington) had barely been able to see the train. Directly after a train had passed he had replaced a detonator on the rail and turned his lamp to red, as per the rules. When the train had passed without stopping he had tried to get out his whistle to warn the guard but had not had time.

Tail Lamp Man H. Bowden at Eccles box (the person who spotted tail lamps for the signalman in a fog) was sure that he had heard the sound of a detonator being exploded by the Rochdale train.

Major G.R.S. Wilson, the inspecting officer, concluding his report, stated that the interlocking and arm-proving controls of the signals made it beyond reasonable doubt that the Down Slow line inner and outer home signals must have been at danger and the Down distant at caution when the Rochdale train had passed them. Given the fog and the lack of a fogsignalman at the distant signal the Rochdale train should not have been permitted to approach Eccles Station while the Kenyon train was being crossed over and Signalman Lowe was therefore primarily responsible for the collision, as he himself had admitted.

There was no doubt, the inspecting officer stated, that Signalman Lowe had genuinely become confused between the similar names of Patton and Pantling, though his story of someone calling out as he had passed the box was probably something which he had persuaded himself had happened when it had not. In a busy box it would be unreasonable to expect the signalman to keep a track of all the fogsignalmen without a written record, hence the standing instructions for the fogsignalmen's booking on and off to be recorded in the train register. This had been conscientiously done by Booking Boy Acton but surprisingly Signalman Lowe had not referred to the book.

Booking Boy Acton, who was only 17, had, the inspecting officer said, been quite free from blame and had carried out his duties with care and attention.

As for Driver Mountfield, he had not suggested that he had had a false clear at the distant signal but had rather implied that he had been misled twice, first by his fireman and then by the fogsignalman. His excuse for not crossing the footplate himself to see the distant signal rather than relying on his fireman had been unsatisfactory.

Fogsignalman Parrington, described by Major Wilson as 'a straightforward, if slow-thinking, type of man,' told an account which was more likely to represent the true facts than that of Mountfield. It seemed likely that he had shown a red light to the driver and that a detonator had exploded, as heard by the tail lamp man.

The inspecting officer surmised that having been reassured by his fireman as to the distant signal Driver Mountfield had relaxed his attention, while the enclosed cab of his engine might have deadened the sound of the detonator. The driver had been, Major Wilson considered, 'taking an unjustifiable risk in so delegating his own responsibility under difficult conditions to his fireman, who appeared to be a man of no great capacity, and whose knowledge of the road was weak.'

Fireman Ashcroft's mistake, if he had seen the distant at all, was inexplicable, as there were no other signals in the vicinity, but the inspecting officer did not consider that any part of the driver's or fireman's statements could be relied upon, and that the fireman likely had not seen the distant signal at all.

Driver Mountfield, Major Wilson decided, must accept a considerable share of responsibility for the accident.

It was unfortunate that, due to a general shortage of staff, it was impossible to man all the fog posts continuously during an extended period of fog. All possible sources of extra men should be explored. Also unfortunate was the fact that the hours of the different fogsignalmen had not coincided, due to uncertainty on the part of Sub-ganger Unsworth, who had arranged the rotas, as to the instructions he had been given. The matter would no doubt receive attention.

Automatic Train Control would have prevented the disaster and had been installed experimentally in the London suburbs but its installation on the main lines, involving the wholesale fitting of locomotives and track equipment, 'would be a problem of considerable magnitude.' (Though the Great Western had managed it.)

A colourlight distant signal would have been unlikely to be missed. There was a plan to substitute colourlight signals when distant signals on important routes became due for renewal, but reasonable steps should be taken to accelerate the programme, especially in areas prone to fogs. [557]

A collision occurred at Banjaluka station, Bosnia, Yugoslavia, on Saturday 3rd January 1942. Due to the incorrect setting of some points a passenger train was diverted onto a line occupied by a freight train and a head-on collision ensued. Both locomotives and a number of vehicles were wrecked. Seven people lost their lives and 25 were injured. A relief train was sent to the scene from Zagreb. [558]

On Monday 5th January 1942 military train M 33 collided with a shunting movement which had occupied the main track at Ortelsburg, Germany, without the knowledge of the

---

[557] Ministry of War Transport report

[558] *Journal de Geneve*, Geneva, Switzerland, 4th January 1942

signalman. There were eight deaths and 22 injuries in the accident on the Allenstein – Lyk line. [559]

A major disaster occurred near Kausala, Finland, 100 kilometres north-east of the capital, Helsinki, on Saturday night, 17th January 1942. A steam-hauled express train from Helsinki, which had stopped at Kausala station to take water, on resuming its journey came into violent collision with a freight train. Nineteen people were killed, ten were seriously injured and 22 less badly hurt. [560]

The difficulties associated with wartime working were a major factor in the disaster at Cowlairs East Junction, Scotland, on the London and North Eastern Railway, on Friday 30th January 1942. Twelve passengers and two railwaymen were killed or fatally injured when the 4 p.m. down express from Edinburgh Waverley to Glasgow Queen Street collided with a light engine at 4.56 p.m. In addition 19 passengers were badly enough hurt to be taken to hospital and 44 passengers were slightly injured.

At Cowlairs East Junction the Edinburgh – Glasgow main line was joined by the Dumbarton line from Cowlairs North Junction, with the double line from Cowlairs West Junction to Cowlairs North Junction forming the third side of a triangle. Opposite the East Junction box on the Down (to Glasgow) side were various sidings and Eastfield engine shed.

At 4.42 p.m. on 30th January a troop train hauled by engine No. 4700 arrived at the East Junction on the Up Branch line from the North Junction and came to a stand at the Up Branch home signal. Engine No. 4700 was booked to be detached here, the train to be worked forward by another engine at 5.05 p.m. Driver McTear then sent Fireman Milligan to the East box to enquire as to the relief arrangements, but before the fireman had got there he saw an engine approaching on the Down Branch line, so he asked Driver MacMillan of this engine (No. 8543) if he was taking over the troop train. No, Driver MacMillan had replied, Driver Graham was following with another engine to do so. Upon being told this Fireman Milligan returned to his engine, uncoupled it, then headed for the box once more to carry out Rule No. 55 and ask for instructions. As he neared the box Signalman Clement saw him and shouted, “Are you uncoupled? Hurry up!” At the same time Driver McTear saw the home signal come off, moved forward, picked up his fireman and continued past the box and beyond mains crossover No. 23, over which the new engine for his train would have to come from the engine shed, and which crossover his engine would then take in the opposite direction en route to the shed. As he was moving forward Driver McTear saw the other engine, 4-4-0 No. 6378, with Driver Graham, on the Down Main line at the other end of the crossover.

Immediately before Driver McTear had moved forward, Signalman Clement had lowered his signals for the 4.46 p.m. Glasgow – Perth passenger train on the Up Main. At this time, Driver Graham on engine No. 6378 was waiting in the dead-end siding on the Down side, as was engine No. 6378, bound for Singer on the Dumbarton line. Driver Graham telephoned Signalman Clement to say he was waiting and did Fireman Davidson

[559] *Katastrophen der deutschen Bahnen Teil II* by H.J. Ritzau, pg. 104 (in German)
[560] *Le Petit Journal*, Paris, France, 19th January 1942

of the Singer engine did the same; the fireman was told that Driver Graham's engine was to go first. Signalman Clement subsequently cleared his disc signal to let engine No. 6378 out onto the Down Main line, where Driver Graham came to a stand on the Queen Street side of crossover No. 23, ready to be put across to the Up line. In hospital, before he died, Driver Graham said that after coming to a stand on the Down Main he had whistled. Signalman Clement denied having heard any whistle.

Just after engine No. 4700 came to a stand east of the crossover Fireman Milligan noticed that the Down Main home and starting signals had been cleared and realising that engine No. 6378 was standing on the Down Main line, alerted his driver. Driver McTear told the fireman to hurry to the signal box while he popped his whistle and began to move forward to warn the approaching Down train but he had barely begun moving when the collision occurred. Fireman Milligan likewise warned Driver Graham, who was just about to move ahead when he was struck.

Signalman Clement told the Ministry of War Transport inquiry into the accident that he remembered the telephone calls for the two engines and after accepting the Perth train from the South Junction he had cleared his disc to let engine No. 6378 out onto the Down Main. The Perth train had passed, and he had then let engine No. 4700 out from the Up Branch onto the Up Main. Bishopbriggs box had then offered him the 4 p.m. Edinburgh train, which he had at first refused as he was crossing engine No. 6378 over. A moment later he had seen that there was still an engine in the dead end, forgetting about the Singer engine and thinking that for some reason engine No. 6378 had been delayed he had replaced his disc signal, accepted the Edinburgh train from Bishopbriggs, had it accepted from the South box and cleared his signals for it. He had then turned to book his times in the train register, when he had heard someone shouting, had seen the engine on the Down Main and had thrown his signals to danger, but by then the Edinburgh train had been only a few yards away.

The inspecting officer, Colonel A.C. Trench, calculated that engine No. 6378 had been stopped on the Down Main line for 2½ minutes before the collision.

The Edinburgh train approached Cowlairs East Junction at 50 to 55 miles an hour, running two minutes late, with Driver Robertson beginning to reduce speed for a 10 miles an hour restriction beyond the West Junction. The Bishopbriggs and Cowlairs East signals had been clear. Fireman Broomfield had also been looking out for the signals. He said that his driver had suddenly shouted, "Jump!" applying the brakes in emergency at the same time. Fireman Broomfield said that he had then seen the engine close at hand and rather than jumping he had held on to the sand handle. He put the speed of the train on impact at 30 to 35 miles an hour. Driver Robertson was killed.

The Edinburgh train consisted of Director class 4-4-0 engine No. 6401 and nine bogie coaches in charge of Guard Bruce. Both engines were badly damaged in the collision and the leading halves of the first two carriages of the Edinburgh train were destroyed. Its next three carriages were slightly damaged, the remainder of the train escaped. After the accident assistance was quickly available. [561] There were doctors on both the Edinburgh and troop trains and passengers on the express and Polish soldiers on the troop train also lent a hand. Twenty-six ambulances attended the scene. Those who lost their lives

[561] Ministry of War Transport report

included two British soldiers, a sailor, and an airman and two Canadian soldiers. [562]

Colonel Trench noted that as was often the case the simultaneous failure of more than one railwayman had been needed to cause the collision. Signalman Clement had frankly admitted his failure, which he attributed, in the opinion of the inspecting officer with some justification, to the difficulties of working under pressure with the limited visibility from the box due to the box windows being almost completely obscured in accordance with the blackout regulations.

The other person at fault had been Fireman Paul of engine No. 6378. The fireman said that his understanding of Rule No. 55 was that he did not need to go to the signal box in clear weather until three minutes had elapsed. In fact this only applied to a running train (Rule No. 55(a)), whereas in the case of a shunting movement the requirement was for the person to go to the box immediately (Rule No. 55 (b)). It was a pity, Colonel Trench remarked, that Driver Graham had not prompted his fireman to go at once. Given the clear signals and the obstruction to vision (from an overbridge carrying the London Midland and Scottish line) the crew of the express could not have been expected to have seen the light engine any sooner. It was to be regretted that the prompt action of Driver McTear and Fireman Milligan had not been more successful.

The inspecting officer conceded that the blackout precautions were necessary at night during wartime but remarked that an improvement could be made in the daytime by the provision of curtains, blinds, or shutters. (At the time of the accident the box windows were completely obscured by blue paint, apart from four areas about 30 inches long and 9 inches deep at eye level in four windows). When the war began the need to rapidly apply blackout measures had made the use of paint or fixed paper the only practicable method but since then improvements had been made in a number of boxes. He recommended that at busy boxes not already dealt with, especially those dealing with shunting and crossing movements where there was no track circuiting, these improvements should be carried out as soon as possible. A track circuit at Cowlairs East would have prevented the accident but could not be specially justified there. In any case a general modernisation scheme for the area, with full track circuiting and colourlight signals, had been prepared but had been delayed by the war.

As regards Fireman Paul's misunderstanding of Rule No. 55, it could only be emphasised that railway staff must have an intimate knowledge of rules, which were inevitably extensive and elaborate to cope with all possible circumstances.

The express involved in the accident had been a fixed set. The leading coach, a brake third composite, had been damaged on 20th January and the replacement vehicle had been sent with the brake compartment at the wrong (rear) end. No action had been taken in the days between then and the day of the accident to have the coach turned so that the train did not continue running without a brake compartment or similar next to its engine, as per the Company instructions issued after the 1937 Castlecary disaster. The faulty formation of the train had increased the number of deaths in the accident. It was true that the leading compartments could have been locked out of use, but the train had already been full and to do so would have resulted in many passengers having to stand. It was to be hoped that the spirit of the Company instructions would be more closely followed in future. [563]

[562] *The Times*, London, England, 30th January and 2nd February 1942
[563] Ministry of War Transport report

There was a collision at the entrance to the station at Tyringe, Sweden, on Thursday 12th February 1942. Early that morning long-distance freight train No. 5663 had arrived at Tyringe from Hässelholm, and had been routed into track 1, where it was brought to a stand. Approaching from the opposite direction, from Västra Torup, had been troop train No. 10624 from Helsingborg to Hässelholm. [564] Due to a mistake by the *tågklareraren* (signalman) the troop train was wrongly also routed onto track 1, where it collided head-on with the stationary freight train. Five soldiers died in the accident, four were seriously injured and 11 others needed hospital treatment. There was major damage to the locomotives and rolling stock and the line was blocked until 16:18.

The driver of the troop train had received a clear signal at the *Försignal* (distant signal). When he reached the *Infahtssignal* (home signal) he saw that it was showing clear for the wrong line and applied the brakes in emergency, but too late to prevent the collision.

It was considered that several factors apart from the faulty setting of the points had contributed to the accident. It was dark and had been snowing at the time of the collision, the accident site was approached around a curve and through a cutting, and both the signals and the locomotive headlights had been shielded in accordance with the blackout which had come into force the previous day, despite Sweden being neutral. [565]

A collision between Neudietendorf and Ritchenhausen, Germany, on Tuesday 17th February 1942, was caused by the driver of class 97 engine No. 448 running past a red signal. Eight people lost their lives and five were injured. [566]

Only a fortnight after the Tyringe accident, on Friday 27th February 1942 another serious collision occurred in Sweden, this time at Södertälje Central. The *växlingsledaren* (signalman) began to cross the stock of train No. 181 from track 3 to track 5, forgetting that train No. 310, consisting of four coaches, had not yet arrived from Södertälje Syd. He also failed to observe the track blocking signal which was at danger to prevent movements from track 3. The annual report on Swedish railways for 1942 states that four people died and six were slightly hurt. [567] Another source states that a fire broke out in the wreckage and that five people died. [568]

According to a brief report from Switzerland a wrongly placed set of points led to a collision between two trains in the station at La Verdure, Algeria, on about Thursday 13th August 1942. Five people died and 15 were seriously injured. [569]

---

564 https://www.tornsvala.se/2015/11/18/tagolycka-i-tyringe-den-12-februari-1942/

565 Redegörelse för Verksamheten inom Säkerhetstjänsten vid Statens Järnvägar under År 1942 (in Swedish)

566 *Eisenbahnunfälle in Europa* by Erich Preuß, pg. 123 (in German)

567 Redegörelse för Verksamheten inom Säkerhetstjänsten vid Statens Järnvägar under År 1942 (in Swedish)

568 Wikipedia – Lista över järnvägsolyckor

569 *Gazette de Lausanne*, Lausanne, Switzerland, 14th August 1942

On the morning of Friday 11th September 1942 two railcars, both travelling at full speed, collided at Herstal, Belgium, near Liége. Telescoping occurred and 14 passengers, mostly workmen, were killed outright and 100 were reportedly seriously injured. The driver of the train coming from Liége was said to have overrun a stop signal in conditions of poor visibility. [570] [571]

The Baltimore and Ohio Railroad suffered a double collision on the morning of Thursday 24th September 1942. Train No. 18, the Cleveland night express, stopped out of course because of a problem with its air compressor, near Dickerson, Maryland, United States. Just as it was restarting it was run into by train No. 20, *The Ambassador*. As a result of the collision the rear Pullman car was thrown foul of the opposite running line, where it was struck by a Washington to Cumberland freight train. Fire from leaking diesel fuel then broke out and enveloped this car.

Eastbound passenger train No. 18 consisted of steam engine No. 5046, a baggage car, a baggage mail car, a passenger baggage car, two coaches, two Pullman sleeping cars, a dining car, and two Pullman sleeping cars, all of all-steel construction. It left Weverton, 17½ miles from Dickerson, at 7.06 a.m., a minute early, passed Point of Rocks about 7½ miles short of Dickerson at 7.18 a.m., two minutes early, and came to a stand near Dickerson at about 7.25 a.m. The compressor was fixed, and the flagman was recalled 6½ after minutes and 2½ later it was given the signal to proceed by the flagman. It had moved about 30 feet when it was run into at 7.35½ a.m.

Eastbound first-class passenger train No. 20 consisted of diesel electric engine No. 60, of the two-unit type, a baggage car, a Pullman sleeping car, a passenger baggage car, three coaches, a dining car and six Pullman sleeping cars, all of all-steel construction, with Engineer Raymond McClelland and Fireman Bowman. A terminal air brake test had been performed at Cumberland and the brakes had performed properly en route. It left Weverton at 7.15 a.m., a minute early, passed Point of Rocks at 7.28 a.m., on time, passed signal E 38-34, showing Approach, passed signal E 36-14, showing Stop and Proceed, and was still moving at 45 miles an hour according to the speed recorder on its locomotive when it struck train No. 18.

Freight train No. 4362 West consisted of steam engine No. 4362, 43 cars and a caboose. A terminal air brake test was made at Philadelphia, Pennsylvania, 165¾ miles from Dickerson, and its brakes had performed properly en route. It left QN tower, just over 33 miles from Dickerson and the last open office, at 6.23 a.m. It was travelling at 35 to 40 miles an hour and passing the two passenger trains when it collided with the two rear cars of train No. 18 and was derailed. The fuel tank of unit B of diesel engine No. 60 was punctured, the oil leaked and ignited, and the rear car of train No. 18 was damaged by flames, fumes, and heat from the burning oil.

Ten passengers, a Pullman employee, and Flagman J.M. Gilhart of train No. 18 were killed, two persons were described as 'missing' and 61 passengers, 11 dining car employees, three Pullman employees and the conductor of the freight train were injured.

---

[570] *Journal de Geneve*, Geneva, Switzerland, 12th September 1942

[571] Wikipedia: Anexo: Accidentes ferroviarios en España

All the fatalities were in the rear Pullman car of train No. 18. [572]

'Among the first of the survivors to reach Washington were Mr and Mrs Leo Zimmerman of Philadelphia. They arrived at 2 a.m., six hours late. The Zimmermans were asleep in the second from the last car of No. 18, the Capital-bound Cleveland night express, Mr Zimmerman said, when it was struck by the Ambassador, fast train from Detroit. The car behind the one in which the Zimmermans were riding immediately burst into flames and the fire began to spread rapidly through the Zimmermans' car. Almost simultaneously this coach was pushed into the path of a fast freight train and one end of the coach was battered into a mass of twisted steel. Miraculously, Mr Zimmerman said, none of the passengers in this car was seriously injured.

'Mr Zimmerman said a soldier dashed into the diner to get a table for use as a stretcher for one of the injured. A porter stopped the soldier as he was carrying the table out, Mr Zimmerman said, and demanded a receipt. The soldier knocked the porter down and carried the table out.

'Only partly dressed, Mr Zimmerman and an unidentified soldier dashed into the burning car behind them in an effort to rescue some of the trapped passengers. "We were able to drag out only two men," Mr Zimmerman said. "They were still alive when we pulled them out, but they were very seriously injured. One of them apparently had his back broken. Mr Zimmerman said a rescue party had been able to carry out three passengers from the other end of the train before the flames became too intense to allow any further rescue work. He said he was told by the conductor that there had been about 30 persons asleep in the car that burned. Mr Zimmerman and his wife were en route to Washington for a short vacation when the accident occurred. Mr Zimmerman is a heating foreman engaged in war work at the Wissahickon Tool Works, a subsidiary of Empire Ordnance, Inc.

'Mrs Zimmerman, still visibly affected by the tragedy, said she was surprised how comparatively calm the passengers in her car remained. "After the first shock of the collision," she said, "there was some milling around and people were trying to get dressed in a hurry. Then a trainman came into the car and told us all to leave just as we were because the car had started to burn, and they were afraid of an explosion. We all grabbed whatever we could get our hands on then and walked out quickly, but there wasn't any panic." The Zimmermans said there were one or two doctors on the train, and they did their best to help the many injured persons. "One man was suffering severely from pain," Mr Zimmerman recounted. "A doctor in our car managed to get his bag before he had to abandon the coach. He had no hypodermic needle to administer drugs, but he gave the man a large dose of morphine tablets. "There were also a couple of WAACs in our coach and they were very efficient in treating people with minor injuries."

'Two priests who were at the scene of the wreck described the tragedy. They were the Rev. Kelian Healy, of White Friar Hall, Brookland, and the Rev. James Bifield, pastor of St Mary's Catholic Church, Barnesville, Md. Father Healy was on the Capital Limited, from Chicago, which was stopped behind the wrecked trains. When the limited was halted, a call was made for all doctors and priests aboard, Father Healy said, and he reached the scene about 20 minutes after the wreck. "I gave the rites to one man who looked like a soldier," the priest said. "The doctor told me he probably would die from

---

[572] Interstate Commerce Commission report

shock. I then went under the locomotive, where I found a Brazilian who was pinned from the hips down. He pleaded to be taken out and talked for about an hour and a half. He died after two hours. There was also a blond-haired man pinned from the shoulders down. He was crushed, and dead when I saw him. Father Healy said rescue workers thought both the Brazilian's legs had been cut off. He said workers tried to get a jack for the engine but feared that the man would die immediately if the engine were lifted. He said he joined workers who had 10 small pumps in an effort to keep fire, which was spreading over the wreckage, from the injured man. He said it was not certain whether the men pinned under the train were passengers or workmen.

Father Bifield, who reached the scene of the accident about 15 minutes after Father Healy, said he administered last rites to two who were dying and aided as much as he could in the rescue work. "I first heard of the wreck when the operator called me at my home and asked that I come to the scene of the accident," Father Bifield said. "When I arrived at the scene the injured had been laid out on stretchers. I don't know how many there were, but I was told that some had already been taken to the Frederick, Md. Hospital. I administered last rites to two who were dying. They were conscious, but I could not tell who they were, or where they were from." Father Bifield remained at the scene of the wreck for three hours and said he did not believe as many were killed as was first thought.' [573]

'Mrs Nanny Lee Owens, who lives in a farmhouse near the scene, said she heard a crash, but did not realise immediately that there had been a wreck. Then she and members of her family noticed the glare of the fire cutting through the morning fog. As the Owens started toward the wreck, they met a man in a gray suit running across the meadow, a cut on his face.

'"My God! There's been a bad wreck! Get me to a telephone!" he told Mrs Owens. Her son, Arthur, drove the man to a nearby community to summon help. At the railroad tracks, she continued, she could hear victims screaming. Some passengers were walking up and down the tracks.

'J.E. McCurdy, of Pittsburgh, a mechanic, was in the rear of the Diesel locomotive when the crash came.

'"The train buckled, and windows broke," he said. "I picked myself up and crawled out a window to look for the others. I saw Bowman and McClelland (engineer and fireman) and they were all right. We tried to get others out, but the wrecked freight started to burn, and the fire spread to the baggage car. There wasn't anything we could do. We could see parts of bodies in the baggage car." [574]

'All the undamaged cars of the passenger trains were pulled away from the fire, and a railroad crane was rushed to the scene to help clear the tracks. Later in the day workmen cut at the wreckage and the twisted rails with acetylene torches to help in the gruesome work of finding the bodies. Investigations were begun immediately by the railroad and by the safety bureau of the interstate commerce commission. A statement from railroad headquarters said the "accident occurred on a double track railroad governed by automatic signals. The cause is not immediately available but will be developed by investigations. This is the first train accident involving injury to passengers on the

---

[573] *Washington Evening Star*, Washington, D.C., United States, 24th September 1942
[574] *The Frederick Post*, Frederick, Maryland, United States, 25th September 1942

Baltimore and Ohio Railroad for the past 23 years.'" [575]

The Interstate Commerce Commission (ICC) noted that the air compressor of the engine of train No. 18 had failed due to lack of lubrication. Train No. 20 had approached signal No. E 38-34 at a speed of 66 miles an hour according to the speed recorder. Visibility had been restricted by dense fog. Engineer McClelland had given different and somewhat conflicting statements but finally stated that he had failed to see the signal until his engine had been passing it, when he had seen the back of the semaphore arm of the signal from the side window, showing Approach. As the automatic signal rapidly returned to that position, he had assumed that that was what it had done and that it had previously been showing Clear. He had therefore not reduced speed until his engine exploded torpedoes and he had seen a red fusee 937 feet west of signal E 36-14. He had then made a service brake application, followed, when he had seen the signal at red from a short distance, by an emergency application. As noted above his train had still been travelling at 45 miles an hour when it ran into the rear of train No. 18.

Because they had duties in the engine compartments of diesel engines the firemen and train service employees of such trains were not required to observe signals unless they were specifically instructed to do so.

The speed recorder tapes of 100 eastbound trains between Point of Rocks and Washington in the 10 days preceding the accident were analysed. Thirty-five of the trains had exceeded the maximum authorised speed through the Point of Rocks interlocking, 55 of them had exceeded the maximum authorised speed between Point of Rocks and QN tower and all of them had exceeded the maximum authorised speed between QN tower and Washington Terminal. In 9 out of 10 of the days preceding the accident train No. 20 had received approach indications requiring speed to be reduced so as not to exceed 30 miles an hour but in no case had speed been reduced below 36 miles an hour. Train No. 20 was thus being operated on the day of the accident as on previous days.

Flag protection had been required when train No. 18 stopped. After the accident a freshly burnt fusee and the cover of a torpedo were found 1,159 feet from the rear of train No. 18. Two torpedo burns 1,700 back could not clearly be related to the accident. There had been 9 minutes available to provide protection. Tests showed that a person walking at average speed could cover 2,572 feet in 8½ minutes, and someone walking briskly could cover 2,986 feet. The conductor of train No. 18 said that given the fog the flagman should not have returned when recalled. Had the flagman not been recalled the train would have proceeded 2½ minutes and the accident would have been averted. Why Flagman Gilhart had not gone back far enough could of course not be determined because of his death.

There was, the ICC stated, a lack of common understanding between both employees and operating officials as to what constituted proper protection. Eight accidents on the Baltimore and Ohio in the previous 21 months had been investigated by the Commission which had been due to failure to comply with operating rules. Three had involved inadequate flag protection, five failure to obey signal indications, five lax enforcement of the rules and two were cases where the engineer of a diesel locomotive was the only person in a position to observe signals. All of these factors were present in the Dickerson accident. There was no evidence that the operating officials had taken any action to correct these matters and the Company must report to the Commission within 30 days of

---

[575] *Bluefield Daily Telegraph*, Bluefield, West Virginia, United States, 25th September 1942

the publication of this report what measures they had taken.

The ICC noted that engine No. 60 had been equipped with an automatic train stop device but there was no equivalent track equipment on the line. Had there been, the accident would undoubtedly have been prevented. [576]

On the afternoon or evening of the day of the accident Engineer McClelland was arrested on a charge of manslaughter. [577] He was less fortunate than some of his fellow engineers who had been in similar circumstances, for at Rockville, Maryland, in December, he was found guilty of the charge and was sentenced to two and a half years in the Maryland House of Correction at Jessup. [578]

Late on Monday 2nd or early on Tuesday 3rd November 1942 a passenger train collided with a goods train at Bor station, near Nigde, in central Turkey, on the Erzeroun – Iskenderoum line. Thirty passengers were reportedly killed, 20 seriously injured and another 50 slightly injured. The injured were transported to hospitals in Bor and Nigde. The driver of the goods train was arrested. [579]

Two trains collided head-on about six kilometres from Hanoi, Vietnam, French Indo-China, on Tuesday 17th November 1942. Twenty-five people died and 68 were injured. 'This is the third serious accident on the line since the beginning of the war in the Pacific. Sabotage is believed by the French in Indo-China to have caused them.' [580]

Passenger train No. 803 from Napoli collided with a freight train at San Nicola yard, near Taranto station, Italy, late on Sunday 22nd or early on Monday 23rd November 1942. A set of facing points had been left open and the passenger train was diverted onto the wrong line into collision with the stationary freight train. Six railwaymen on the colliding train were killed and five railwaymen and five passengers were injured, two seriously. [581] The injured were given first aid on the spot and then taken to the nearest hospital. The railwaymen who lost their lives were Driver Pietro Abruzzese, Fireman Egidio Lusso, Conductor Vito Carotenuto, Signalman(?) Francesco Di Trani and Labourers(?) Giuseppo Rubino and Nicola Gentile. Traffic was not interrupted. [582]

Troop train No. ML-3410 collided with freight train No. TM-1692 at Nocera Inferiore, Italy, on Friday 27th November 1942. There was severe telescoping of the coaches of the troop train and 28 soldiers lost their lives and an unknown number were injured. Information regarding the accident was censored by the authorities and no details as to how it had happened were given, other than a vague statement that it had been due

---

[576] Interstate Commerce Commission report
[577] *The Cedar Rapids Gazette*, Cedar Rapids, Iowa, United States, 25th September 1942
[578] *Centralia Sentinel*, Centralia, Illinois, United States, 4th December 1942
[579] *Journal de Geneve*, Geneva, Switzerland, 4th November 1942
[580] *Dixon Evening Telegraph*, Dixon, Illinois, United States, 18th November 1942
[581] *Journal de Geneve*, Geneva, Switzerland, 24th November 1942
[582] *La Stampa*, Torino, Italy, 24th November 1942

to the difficult wartime conditions. [583]

The precise details of a collision on the Hamburg – Uelzen line at Lüneburg, Germany, on Thursday 17th December 1942 are a little difficult to determine from the brief written material concerning it, but it appears that passenger train P 839 had stopped at the platform at Lüneburg station. It had then restarted and was rejoining the through line when it was struck by military train W 92 321, which a signalman had cleared his signals for, having failed to make use of a blocking device which would have prevented him making just that mistake. [584] Train W 92 321 was composed wholly or partly of tank waggons loaded with oil and an explosive fire followed the collision, enveloping both the passenger train and the station buildings. Ten people lost their lives and 19 were injured. [585]

A German radio broadcast stated that there had been a collision at an unspecified location in south Belgium on Sunday 27th December 1942, involving a speeding locomotive and a stationary train, in which nine persons had died and 23 were injured. [586]

Mistakes made by a signalman and a shunter led to a collision in the station at Tantow, Germany, on the Berlin – Stettin line, between an express and a leave train for frontline troops. Thirty-eight people died and 16 were injured in the accident on Sunday 31st January 1943. [587] (Another source gives the fate of the accident as 31st *December* 1943.) [588]

Due to a signalman's error special leave train for front line troops SF 76, en route from Western Russia via Warsaw, Lissa (Leszno, Poland), and Dresden to Karlsruhe collided with through freight train Dg 19 540 at Galkau, Germany* (before 1943 and after 1945 Galków Duży, Poland), on Sunday 28th February 1943. This accident, on the line between Litzmannstadt (Lódź, Poland) and Koluszki, was a major disaster, with 25 dead and 12 injured. [589] [590]

* The accident occurred in what before the Second World War had been part of Poland. In 1943 Galkau was situated very close to the boundary between that part of Poland annexed into Germany (Reichsgau Wartheland) and the occupied part of the country (Generalgouvernement). [591] It is not clear in which area it was situated so for the purpose of this book it has been treated as part of Germany.

---

583 https://www.risorgimentonocerino.it/attualita/5937-nocera-inferiore-75-anni-fa-la-strage-della-stazione-ferroviaria.html
584 *Eisenbahnunfälle in Europa* by Erich Preuß, pg. 124 (in German)
585 *Katastrophen der deutschen Bahnen Teil II* by H.J. Ritzau, pgs. 105-107 (in German)
586 *Lawrence Daily Journal-World*, Lawrence, Kansas, United States, 28th December 1942
587 Wikipedia – Liste von Eisenbahnunfällen in Deutschland
588 *Eisenbahnunfälle in Europa* by Erich Preuß, pg. 125 (in German)
589 Ibid, pg. 124 (in German)
590 *Katastrophen der deutschen Bahnen Teil II* by H.J. Ritzau, pg. 110 (in German)
591 Wikipedia – Polish Areas Annexed by Nazi Germany

Deutsche Reichsbahn locomotive No. 05 002 became famous in 1936 when it set a speed record for steam traction of 200.4 kilometres an hour on the Berlin – Hamburg line. Seven years later it had a less useful claim to fame when, on Tuesday 2nd March 1943, it was involved in a collision with a SFR working (a special leave train for frontline troops) at Ashausen, Germany, on the line between Hamburg and Uelzen, the accident resulting in six deaths and 25 injured. Engine No. 05 002 was meant to have been coupled onto the leave train, which was being hauled by sister engine No. 01 001, but in some manner train and engine collided. [592]

On about Wednesday 30th June 1943 an electric railcar(?) and a train were in collision with one another on the line between Orvieto and Chiusi in Italy. Seven people lost their lives and 24 were seriously injured. [593]

Because of a signalman's error express train D 24 and passenger train P 210 came into collision at Dortmund Hbf, Germany, on Wednesday 30th June 1943. Twenty-eight people died and 90 were injured. [594]

Seven passengers, reportedly all sailors, were killed outright [595] and 13 passengers and a railwayman were injured in a sidelong collision at Warden, Washington, United States, early on Wednesday 4th August 1943. Three of the injured passengers subsequently succumbed.

Warden was a siding on the single line of the Chicago Milwaukee St Paul and Pacific Railroad between Maiden and Othello. Both trains involved in the collision were westbound.

Mixed train Extra No. 849 West, a branch train from Moses Lake, consisted of engine No. 849, 13 freight cars, a baggage express car and a coach. Train No. 849 had stopped on the siding at Warden at 12.35 a.m. Seven minutes later it moved forward until it was foul of the main track at the west end of the siding. An attempt was made to set back but before this could be done it was run into by train No. 251 at about 12.52 a.m.

Passenger train Extra No. 251 West consisted of engine No. 251, a baggage car and 19 Pullman sleeping cars. The cars were of all-steel construction. It had passed Lind, 22½ miles from Warden and the last open office, at 12.12 a.m. It passed automatic signal 90-5 showing Proceed and was travelling at 48 miles an hour according to the speed recorder of its engine collided with train No. 849. Both engines and tenders were derailed and badly damaged. The first three cars of train No. 251 were derailed and the 11th and 12th cars of that train telescoped one another.

The conductor of train No. 849 told the Interstate Commerce Commission inquiry that the Warden operator had told him at 12.45 a.m. that train No. 251 would pass about 12.50

---

[592] *Katastrophen der deutschen Bahnen Teil II* by H.J. Ritzau, pg. 110 (in German)
[593] *Gazette de Lausanne*, Lausanne, Switzerland, 1st July 1943
[594] Ibid.
[595] *The Havre Daily News*, Havre, Montana, United States, 4th August 1943

a.m. The conductor said that he had expected his train to move forward to the west end of the siding and to stop clear of the main line and had so instructed the engine crew. The conductor and flagman had been riding in the rear car of the train. The engineer, fireman and front brakeman stated that they had misunderstood what the conductor had told them and had thought that they were to precede train No. 251. At the west switch the front brakeman had got off the engine to operate the switch and the engine had just moved out onto the main line when the fireman had seen the approaching train and shouted a warning to his engineer. The latter had tried to reverse to no avail. The flagman said that when he had realised what was happening, he had given warning signals to train No. 251, but it had been too close. [596]

Less than a month after the Warden accident, another with somewhat similar features occurred at Wayland, New York, United States, on the Delaware Lackawanna and Western Railroad.

Freight train Extra No. 1248 West consisted of engine No. 1248, 13 cars and a caboose, with Engineer Al Driscoll. It arrived at Wayland at about 4.30 p.m. on Monday 30th August 1943 and went into the yard there clear of the main line. It then began shunting operations and 53 minutes later the engine and six cars were foul of the main line when westbound passenger train No. 3 was seen to be approaching. As at Warden, an attempt was then made to set back into the siding clear of the main line, but the trains collided.

Westbound first-class passenger train No. 3, the *Lackawanna Limited,* consisted of engine No. 1151, a mail baggage car, two Pullman sleeping cars, four coaches, a dining car and three coaches, all of all-steel construction, with Engineer James LeRoy and Fireman Theodore Paladino. It had passed Cohocton, 10¼ miles east of Wayland, and the last open office, at 5.10 p.m., nine minutes late, passed automatic signal No. 3109, showing *Proceed*, and travelling at an estimated 50 miles an hour struck engine No. 1248 at about 5.23 p.m. As a result of the collision both engines and six cars of train No. 3 were derailed. The sixth car came to a stand opposite engine No. 1248. Escaping steam and hot water from the damaged engine entered the car and caused almost all the deaths and injuries in the accident. Twenty-six passengers and Supervisor of Locomotive Operations F.H. Meincke died and 107 passengers, three railwaymen, three Pullman employees and a dining car employee were injured. [597]

Mrs Henrietta Gibson, of Corning, New York, a passenger in the sixth car, spoke to a reporter. “Several soldiers in our coach went into action when hot, scalding steam shot through the car from the outside after the Lackawanna Limited jolted to a stop. They took off their shoes and began smashing the windows to break them. People were crying and shouting, and you could hardly hear yourself think above their cries and the noise of the steam.

“We had been riding along smoothly in the last coach when all of a sudden there was a terrible jolting. The whole train seemed to be jumping. There was a lot of noise, as though the tracks were being torn up. Then the steam came – and water. At one time the water was one to two feet deep in the aisles of the car. It came swirling about our feet.

[596] Interstate Commerce Commission report
[597] Ibid.

Some passengers were burned on the legs when they attempted to walk out of the car. Others climbed through windows, cutting themselves severely as they did so. The Negro porter was calm and assisted many to get out. As for myself, I just didn't know what to do. Finally I decided to climb over the seats and not take any chances with the hot water in the aisles. As it was, I burned one of my hands. I went out the rear door and one of the doctors said I was suffering from shock." [598]

The engineer of train No. 3 told the Interstate Commerce Commission (ICC) that his train had been travelling at 80 miles an hour as it had neared Wayland. The last automatic signal had been at *Proceed.* Approaching the station he had made an 8-pound reduction to test the brakes. His first warning of the obstruction had been when his train was 800 feet from the siding switch when he had received a restricting indication from the cab signal. He had immediately made an emergency brake application but had only been able to reduce speed to 50 miles an hour before the collision.

The conductor of the freight train said that he had been advised by the dispatcher that train No. 3 had passed Bath, 25¾ miles east of Wayland, about 10 minutes late. He had not informed the other members of his train crew as he had expected the passenger train to be more or less on time at Wayland. He had been about 1,000 feet east of the switch and had not been aware that the engine of his train had fouled the main track before the accident. The flagman and swing brakeman said that they had understood that train No. 3 was late. When the flagman had realised that the main track was fouled, he had attempted to warn the passenger train but had no time. The swing brakeman, on top of the fifth car, had signalled to Engineer Driscoll to stop, as had the front brakeman. The fireman, on the tender delivering coal to the conveyor, was unaware of anything wrong. Engineer Driscoll stated that he had been aware that train No. 3 was overdue but had assumed that other members of the train crew had received authorisation for the main track to be occupied. Not until he had seen stop signals from the front brakeman had he become aware of the danger, when he had attempted to set back.

The ICC noted that the operating rules of the carrier required that inferior trains clear the time of superior trains in the same direction by at least five minutes, while timetable instructions specified that train No. 3 and certain other trains must be cleared by at least 10 minutes. However, in practice dispatchers had been giving time on first-class trains either orally or by message instead of by train orders as per the rules, and this had been considered sufficient authority for inferior trains to foul the main track on the time of superior trains. The chief dispatcher and divisional superintendent admitted that they were aware of this practice. The general superintendent admitted that during the first eight months of 1943 in the Buffalo Division such messages had been issued by the dispatcher to inferior trains 347 times, not including oral advices. The crews of first-class trains had no knowledge of such messages issued to inferior trains. Had train orders been issued to both trains the accident would have been averted.

The ICC concluded, 'Similar accidents may be expected to occur as long as operating officials fail to conform to the practices prescribed by their rules. The unsafe conditions disclosed by this investigation should be corrected immediately.' [599]

---

[598] *The Troy Record*, Troy, New York, United States, 1st September 1943

[599] Interstate Commerce Commission report

Failure to ensure that train N 8702 had come to a stand clear of the running line during a thick fog resulted in its being struck in the side by train N 8713 at Falkenburg, Germany (now Zlocieniec, Poland), on the Ruthnoe – Neustettin line, on Wednesday 15th September 1943. Although from their descriptions both trains were local goods workings the collision resulted in 18 fatalities and 41 cases of injury. [600]

A serious accident happened at Denderleeuw, Belgium, on the evening of Monday 30th November 1943. The 17:50 ordinary passenger train from Bruxelles Sud to Zottingem was a little delayed. The station work at Denderleeuw having been completed, its 20-year-old conductor gave his driver the signal to depart, whereupon without further ado the train set off. However, a Bruxelles – Courtrai express was also running a little late and the signalman had given it priority over the Zottingem train. The signals were therefore against the latter and the platform line was set for the safety siding. The driver of the passenger train failed to notice this or that he had run past two signals at danger, and it was not until his engine was close to the end of the safety siding that he appears to have realised his mistake. It was then too late to save the situation and the engine, tender and leading carriage went through the buffers, plunged down the embankment beyond and ended up in a canal. The driver, fireman, conductor and three passengers were killed. A further tragedy happened immediately afterwards when a second fireman and a workman rushing to the scene of the accident were run down by the express, the fireman being killed, and the workman seriously injured. [601]

On Thursday 2nd December 1943 a major disaster occurred at Naumburg, Germany, when the driver of special leave train for frontline troops SFR 33 from Eisenach to Breslau overran a stop signal and collided with a similar train, SF 360. Thirty-five people died and 123 were injured. The *Indusi* automatic train control equipment on the colliding train was switched off. [602]

A mistake by the *Fahrdienstleiter* (signalman) at Neuhaus, Germany, on the Villach – Tarvis line, led to a major disaster on Saturday 18th December 1943. After the passage of through goods train Dg 90 021 he forgot to replace the *Einfahrsignal* to danger. As a result the following train, Dg 90 024, ran past it and into sidelong collision with train Dg 7533. Twenty-three people died and 22 were injured. [603]

Two freight trains collided in the station at Fiñana, Spain, on Monday 27th December 1943. Twenty-five waggons belonging to one or other train were wrecked and 10 people died and another was injured. [604]

---

[600] *Katastrophen der deutschen Bahnen Teil II* by H.J. Ritzau, pg. 110 (in German)
[601] *De Poperinghenaar*, Belgium, 4th December 1943
[602] *Katastrophen der deutschen Bahnen Teil II* by H.J. Ritzau, pg. 110 (in German)
[603] Ibid.
[604] Webpage: Federación Castellano Manchega de Amigos del Ferrocarril (in Spanish).

According to a broadcast from Free Yugoslav Radio, 100 Germans died in a collision between two troop trains near Skopje, Yugoslavia, on about Monday 14th February 1944. [605]

By early 1944 Italy was devastated by war and was divided between the south, occupied by the Americans and British, and the north, still occupied by the Germans. In the towns and cities of the south many people were starving. Desperate to make their way into the countryside in search of food, they boarded freight trains. It was thus that they became the victims of two railway catastrophes in March and April 1944. The first of these, in Armi Tunnel, near Balvano, on Friday 3rd March, will only be touched upon here as it was not really a train accident, inasmuch as the train itself was not involved in a collision, derailment or fire and suffered no damage. What happened was that an overloaded freight train, hauled by a steam locomotive with inferior coal, stalled in the middle of the long tunnel and a large number of people, possibly 600, died of carbon monoxide poisoning. Published reports of the accident often describe the passengers, very unfairly, as smugglers, rather than as people doing what they had to do to survive.

The second accident, almost unknown and about which it has been possible to obtain very little information, occurred on Saturday 8th April 1944, at Pontecagnano station, on the line between Salerno and Reggio Calabria. A freight train carrying a large number of unauthorised passengers, who, as at Balvano, were doubtless searching for food, was entering the station when it came into collision with another train which was shunting there. According to witnesses between 100 and 130 people died. [606]

On the evening of Thursday 28th September 1944 the Chicago and North Western Railway suffered a serious collision at Missouri Valley, Iowa, United States, between a passenger train and a freight train, which resulted in the deaths of nine passengers and injuries to 95 more.

The passenger train involved in the accident had started from Sioux City, Iowa, as eastbound first-class train No. 10. It consisted of engine No. 509, a baggage mail car, two baggage express cars, a passenger baggage car, three coaches, a café parlor car and a coach, all of all-steel construction, in charge of Conductor G.S. Finley with Engineer Burt Lowder and Fireman Walter Thompson. It arrived at switch No. 1 at Missouri Valley, at one point of a wye (triangular junction) at 7.23 p.m. From that point it rather confusingly became westbound first-class train No. 203 to Council Bluffs and proceeded as such along the west leg of the wye to a junction with the double-track line from Council Bluffs to Boone (designated switch 2). The wye track only made a junction with the eastbound line to Boone and to reach the westbound line to Council Bluffs the train had to negotiate a mains crossover. Train No. 204 initially stopped clear of the junction with the Boone line at 7.28 p.m. but two minutes later it moved forward and was negotiating the crossover at 10 miles an hour when it was struck by the freight train.

Eastbound second-class freight train No. 256 from Council Bluffs to Chicago, otherwise *The Calumet*, consisted of engine No. 3020, 28 cars and a caboose, with

[605] *Sarasota Herald-Tribune*, Sarasota, Florida, United States 18th February 1944
[606] http://www.macchinistisicuri.info/ms/inciferr/index.php#1944

Engineer Ray Bunyon. It had left Council Bluffs, the last open office, at 7 p.m. and was travelling at 15 miles an hour when it struck the fourth car of passenger train at 7.32 p.m. The third to eighth cars of the passenger train and the engine and first four cars of the freight train were derailed. The fifth, sixth and seventh cars of the passenger train were badly damaged. [607]

'Jack Larson, a University of Illinois student, a Pvt. Charles Bland, both of Sioux City, met at the railroad station there, renewed an old school friendship, boarded the train and found seats on the rear car of the train. "We were just crossing the switch at Missouri Valley," Larson related, "we were moving kind of slow when we felt the jerk. At first, I thought it was the train coming to a jerky stop. Then we heard the crunching sound as the cars went over." Bland said people were screaming. "First person I saw," he continued, "was a dead soldier. I picked up a civilian and carried him to the ambulance, but he died just as I got there."

'An earlier report that an unidentified year-old girl was among the dead in Council Bluffs was found to be an error. The youngster referred to was being treated for injuries, the extent of which have not been determined. Efforts are being made to learn her identity.' [608]

'Willard Dale, editor of the Missouri Valley Times, said he saw soldiers picking themselves up when he arrived at the scene shortly after the crash. "They cut the wires of fences nearby to permit ambulances and trucks to back up the embankment to carry away the dead and injured," he observed. "The injured were brought to homes in Missouri Valley. We have only two or three ambulances in town, so trucks had to be used."

'Army ambulances from Ft. Crook came here to assist in removing the dead an injured. The three soldiers who died in the crash were taken to Ft. Crook. The bodies of Mayer, White and the Steenhoven baby were removed to Council Bluffs and Spraiali and Mrs White remained here. The body of Mrs McLaren was taken to Moville. Ambulances also came from Omaha and Council Bluffs. All available doctors in Council Bluffs were called to help care for the injured.' [609]

The line was equipped with automatic train control and with visible and audible indicators on the engines. There were no wayside signals except at interlockings. When the main track was occupied, or a switch was opened, the visual indicators of a train within the restricted zone defined by the stopping distance would change from green to red over yellow and the brakes would be applied automatically if they were not applied by the engineer and the speed reduced to below 23 miles an hour.

Engineer Bunyon of the freight train told the Interstate Commerce Commission (ICC) inquiry that approaching the point of the collision he had been travelling at 50 miles an hour and the visual indicator had been green until a point about 2,800 from the switch. It had then changed to red over yellow when he had closed the throttle and made a 20-pound brake application. Trailing smoke had then obscured his view. When he had seen the passenger train crossing his path, he had applied the brakes in emergency.

Head Brakeman C.M. Dobyns stated that when train No. 203 had arrived at switch 2

---

[607] Interstate Commerce Commission report

[608] *The Daily Times Herald*, Carroll, Iowa, United States, 29th September 1944

[609] *Ottumwa Daily Courier*, Ottumwa, Iowa, United States, 29th September 1944

he had gone to a lineside telephone and reported the fact to the dispatcher. Soon afterwards he had observed the indicator at switch 2 in the vertical position. He had then operated switch 2, taking the train onto eastbound track towards Boone, crossover No. 1, taking the train onto the westbound track to Council Bluffs, and then signalled to Engineer Lowder to proceed. Before the train had started the head brakeman and Fireman Thompson saw the freight train, which they thought was a mile or more away. It was only when the passenger train was on the crossover that the front brakeman and the engine crew had realised that the freight train was in fact close at hand. The engineer then tried to get across more quickly but failed to clear the eastbound track in time.

The ICC noted that the control circuits were so arranged that as soon as switch 2 was opened the visual indicator on the engine should have gone to red over yellow if the engine was within 9,998 feet of the switch. As it had not done so until the train was 2,800 feet away it was clear that the switch had not been operated until that point. The indicator at switch 2 should at that moment have been horizontal, prohibiting the opening of the switch until flag protection had been provided, but the head brakeman insisted that it had been vertical.

The ICC considered that given that the maximum authorised speed through Missouri Valley was 60 miles an hour, additional protection at the switches was required. An electric switch lock at switch 2 would have prevented the switch being operated when a train was approaching and thus prevented the accident. [610]

A brief Reuters report stated that a military train and a freight train had collided in the station at Craiova, Romania, on Thursday 2nd November 1944. Numerous coaches were destroyed, 60 people were killed and over 100 injured, some very badly. [611]

A major catastrophe in Germany on Thursday 23rd November 1944 involved freight train G 6395 carrying 400 prisoners from Flossenbürg concentration camp, bound for a so-called satellite camp at Leitmeritz, where they were to be forced to work in a secret armaments factory. The prisoners were locked into freight waggons, the floors of which were covered with straw and the vehicles were equipped with small cast-iron ovens (*Kanonenöfen*) to provide some protection against the cold. The prisoners' train collided with a second train near Wiesau, Germany, and the waggons overturned and presumably caught fire. Fifty-two prisoners, four SS guards and a fireman lost their lives, and many others were seriously injured. It seems probable that none of the prisoners who survived the accident lived to see the end of the war. [612]

Two trains apparently collided near Rozières-sur-Mouzon, France, on the cross-country line between Nancy and Dijon, on Thursday 11th January 1945, killing 21 people and injuring nine. [613]

---

[610] Interstate Commerce Commission report
[611] *Journal de Geneve*, Geneva, Switzerland, 3rd November 1944
[612] Wikipedia – Liste von Eisenbahnunfällen in Deutschland
[613] *Railway Disasters of the World* by Peter Semmens, pg. 118

According to a Reuters report from Paris, a Dijon – Marseilles express collided with a stationary petrol waggon at a point about 12 kilometres from Lyon, France, on Friday 9th February 1945. Five passengers were killed and 50 injured. The express was 25 minutes late and was reportedly trying to make up time when the accident occurred. [614]

Due, it was thought, to a fault with the signals, an American leave train collided with a freight train near Champigneulles, France, just outside of Nancy, on Sunday 1st July 1945. The violence of the collision was such that two coaches of the leave train were wrecked. Twelve soldiers were killed and 23 injured according to initial reports. [615] [616]

A train carrying American soldiers who were being redeployed to the Pacific where the war with Japan was still in progress came into collision with a freight train near Weingarten, Germany, early on Friday 13th July 1945. Six soldiers were killed and 28 injured. The men were all members of the 65th Armoured Infantry Battalion of the 20th Armoured Division and had been scheduled for an early sailing from Le Havre, France.

Fahrdienstleiter (Signalman) August Lindenmeyer, a German, who had been on duty at Durlach, 8 kilometres away, and had wrongly given the freight train a clear track, was arrested and held for investigation. [617]

Seventeen passengers who were either travelling on the footboards or sitting at the doorsteps of No. 99 Up Gaya passenger train were killed when at 8.30 p.m. on the evening of Saturday 28th July 1945, it collided with a pair of light engines at the crossing of the main and loop lines near the junction signal cabin of the East Indian Railway, about three miles east of the station at Mughalsarai, India. Five other persons were injured, of whom four were admitted to the King Edward Hospital at Benares. None of the coaches of the train were derailed and the damage to them was slight. The injured were first treated at the scene before being forwarded to Benares on another train together with the uninjured passengers. [618]

Train No. 1037 from Lyon to Saint-Etienne had only covered five kilometres on Saturday 28th July 1945 when it collided with a munitions train at Saint-Fons, France, a populated area on the bank of the River Rhône. The collision was followed a few minutes later by a terrific explosion which could be heard ten kilometres away. The carriages of the train were thrown from the track and at once caught fire. The houses alongside the line were badly damaged by the blast and many of their inhabitants were injured; about 20 of them being admitted to the Edouard Herriot hospital in a serious condition. Several hundred residents of Saint-Fons were made homeless, and the gasworks was set alight.

Firemen, rescuers, Red Cross workers and others immediately went to the scene and

---

[614] *The Railway Gazette*, London, England, 23rd February 1945
[615] *L'Avenir Normand*, Rouen, France, 4th July 1945
[616] *The Ada Evening News*, Ada, Oklahoma, United States, 3rd July 1945
[617] *Kingsport News*, Kingsport, Tennessee, United States, 14th July 1945
[618] *The Bombay Chronicle*, Bombay, India, 30th July 1945

worked through the night to release and treat the victims despite further explosions. It was hoped that all of the train passengers who had not been trapped had had time to get away from the train between the collision and the subsequent explosion. The initial figures for casualties were seven dead and 30 injured.

The line was back in use by the morning. [619]

An investigation was opened into the accident, but it appeared likely that the driver of the express had run through signals and thus caused the collision. [620] This was very likely a rear collision.

Two trains collided on the Omuta line of the Nishi Nippon Railway, Japan, on Monday 20th August 1945. Forty people died and 83 were injured. [621]

An announcement from Radio Bratislava stated that two trains had collided early on the morning of Friday 31st August 1945 on the Diosek – Galanta line, in Slovakia, Czechoslovakia. Thirty-one people had died and five were injured. Those killed included nine Russian soldiers. [622]

The station at Kedange, France, near Thionville, was the location of a major disaster at 15:00 on Tuesday 4th September 1945. A long mixed train bound for the Saar entered the station, consisting of an engine, some goods waggons containing German civilians who were being repatriated, some carriages containing 180 French soldiers and 19 tank waggons loaded with oil for the French army. As a precaution against an accident the rear coach and the leading tank waggon were separated by four empty barrier waggons.

The signalling at the station had been largely destroyed in the war and had yet to be replaced, so there were no fixed signals for the train driver to be guided by. Due to a misunderstanding between the stationmaster and the signalman the facing points at the entrance to the station were left set for a siding and it was into this that the train was diverted and came into violent collision with eight waggons which were stabled there and were loaded with cement blocks. As a result of the shock of the impact with such a solid obstruction, six of the tank waggons exploded and caught fire and a layer of burning oil fifty centimetres high suddenly spread along the track. The four barrier waggons proved to be insufficient, and the flaming oil spread to the rear passenger coach. Forty French soldiers died and 33 were hurt; 26 were taken to hospital in Thionville and the other seven to Metz. None of the German civilians was hurt. A rescue operation was immediately organised.

Following an inquiry the Auxiliary Agent Hennequin, who was acting as signalman, and Facteur Mixte Muller, who was the acting stationmaster, were arrested. [623]

---

[619] *Ce Soir*, Paris, France, 29th July 1945
[620] *L'Aurore*, Paris, France, 29th July 1945
[621] http://www.ikjeld.com/japannews/00000147.php
[622] *Journal de Geneve*, Geneva, Switzerland, 1st September 1945
[623] *Combat*, Paris, France, 5th and 6th September 1945

In an accident at Sasago station, Japan, on the Chuo main line in Yamanashi Prefecture, on Thursday 6th September 1945, a passenger train overran a set of buffer stops and collided with a cliff face. Sixty people died and 90 were injured. An inexperienced train driver had failed to apply the brakes in time to stop the train safely. [624] Shortly after the accident a train carrying 80 former American prisoners of war en route from Chino to Yokohama arrived at the station. The railway authorities were concerned lest the ex-prisoner-of-war train be in any way delayed but the Americans themselves, upon learning of the disaster, at once left their train to give medical attention to the injured using whatever supplies they had, while others opened tins of foodstuffs which had been distributed to them and gave the food to the injured. Woollen blankets were also offered by the ex-prisoners and emergency bedsteads erected, while foodstuffs and blankets were thrown out of the windows of further passing ex-prisoner-of-war trains. [625]

On Tuesday 18th September 1945 the driver of passenger train P 1011 overran at night the disc which served as a temporary *Ausfahrsignal* (starting signal), the light of which was unlit due to a lack of oil, and collided with a Soviet transport train at Strausberg, Germany, a little east of Berlin. The driver, who was also unfamiliar with the line, was accused of causing the accident by the Soviet authorities, who maintained that as it was a moonlit night he should have been able to see the signal disc and that despite the minor consequences of the collision he was liable to punishment. What punishment he received was not recorded. [626]

A collision occurred at Köpenick, Germany, also in the Soviet-occupied zone, early on Wednesday 19th September 1945. The *fahrdienstleiter* (signalman) there had, at 03:37, accepted by telephone from the previous station at Erkner train 1021, but continued with some shunting operations which were in progress. When train 1021 arrived at 04:00 its driver failed to see the unlit *Einfahrsignal* (home signal), which was at danger, ran past it and collided with the shunting movement. He did realise the danger before the collision and applied the brakes, but serious consequences could not be averted. Five Russians were killed, 25 Russian soldiers were injured and material damage of the order of 11,700 roubles was caused.

The railwaymen judged to be at fault were subsequently brought before a Soviet court. All were found guilty of causing the accident through negligence and the driver was sentenced to ten years, the conductor of the train to five years and the signalman to two years' imprisonment. [627]

At 20:30 on Friday 21st September 1945, a Michelin railcar crowded with passengers which had left Grenoble at 18:30 collided with a workman's train coming from Valence on the pont de Vernaison near Valence, France. As a result of the collision, one of the

[624] Webpage: Japanese Railway Safety and the Technology of the Day by Masao Saito
[625] *The Nippon Times*, Tokyo, Japan, 14th September 1945
[626] *Deutsche Eisenbahn-Katastrophen* by Hans-Joachim Ritzau, Jürgen Hörstel and Thomas Wolski, pg. 207 (in German)
[627] *Katastrophen der deutschen Bahnen Teil II* by H. J. Ritzau, pgs. 281-2 (in German)

coaches of the workman's train reared up into the air, was pressed up against the engine and caught fire. Some of its occupants were able to get out of it in time. Others were less fortunate, and rescuers recovered the charred corpses of a number of victims from the wreckage. Blowtorches were needed to help free the dead and injured. Firemen and police from Valence and Romans attended the accident. With the aid of ambulances and the vehicles of a transport company the victims were taken to hospital as they were released. The cause of the accident was not immediately known. At least 20 and perhaps 30 were killed and about 100 injured. [628] [629]

Thirty and possibly 40 people died and about 30 were injured in a collision between a passenger train and a freight train (according to another report two passenger trains) near Erzurum, Turkey, on Monday 8th October 1945. [630] Five rescue trains were reportedly sent to the scene. [631] Only brief reports of the disaster appeared in the foreign press.

A little before 16:00 on Wednesday 7th November 1945, a freight train which had arrived at the station at Tronzano Vercellse, Italy, was shunted back into a dead-end siding to clear the main line for direct train No. 195 which had left Torino Porta Nuova at 15:00. The direct train passed through the station safely and the deputy stationmaster then directed the pointsman to let the freight train out of the siding. He did so, and the freight train left for Santhià.

The next train to arrive at Tronzano, at 18:13, was direct train No. 182 from Milano to Torino, which was not booked to stop there. As it passed through the station it was suddenly diverted into the siding which the freight train had left two hours earlier and before its driver could pull up, it had crashed into the buffer stops at considerable speed. The leading vehicle, a goods waggon which had been adapted to carry passengers, was telescoped against the engine and the other coaches were considerably damaged. The rescue operation was hampered by the darkness and by a lack of equipment. Six passengers lost their lives, and 21 passengers and Driver Dario Merlo were sufficiently badly hurt to be admitted to hospital. The deputy stationmaster disappeared after the accident.

The investigation into the accident found Deputy Stationmaster Giovanni Greggio and Pointsman Domenico Pezzano responsible for it. The pointsman had handed over the key of the points to the deputy stationmaster, having forgotten to reset them for the main line, while the stationmaster had just assumed that the points had been properly reset and had not bothered to check them in the almost two hours available. [632]

Ten passengers reportedly died and about 15 were injured when the Down fast passenger train which had left Lucknow the previous evening collided with a shunting goods train at Mala, India (near Philibit), at 5 a.m. on Sunday 25th November 1945. The bogie (coach) next to the engine of the Oudh and Tirhut Railway train was telescoped.

[628] *Combat*, Paris, France, 23rd September 1945
[629] *Ce Soir*, Paris, France, 24th September 1945
[630] *Monitor-Leader*, Mount Clemens, Michigan, United States, 9th October 1945
[631] *Paris-Presse*, Paris, France, 10th October 1945
[632] Wikipedia – Incidente ferroviario di Tronzano Vercellese

Major S.S. Gupta, Civil Surgeon, Philibit, attended to the injured and they were removed to the district hospital there. [633]

An express train and a freight train collided on the Berlin elevated railway (Stadtbahn) in Germany on Saturday 16th December 1945. Eight people died and eight others were injured. [634]

Matillas Station, in Guadalajara province, Spain, was the scene of an accident at 23:00 on Thursday 20th December 1945. Barcelona – Madrid fast train No. 801 was an extra working to cater for the Christmas holiday traffic and was making its inaugural run of the year. It had left Barcelona at 19:57 and consisted of an engine and eight coaches. A severe storm was raging as it reached Matillas and it was believed that it was because of this that its driver had failed to observe the signals, which had been at danger because a freight train was shunting there. The trains met engine to engine and the first four vehicles of the passenger train, a van, a salon car, a Pullman car, and a dining car were derailed. The wooden Pullman car was telescoped and completely wrecked by the dining car.

Nine passengers lost their lives in the accident and 21 people were injured. The injured included Conductor Juan Llobregat Martínez, Driver Fernando Gómez and Fireman Alturo Bañera of train No. 6822 (the mail train?) and Driver Raúl Guitarte and Fireman Delfin Beltrán of train No. 8822 (the freight train?). (The newspaper report is not at all clear as to which driver, fireman and conductor belong to which train or as to the train numbers.)

A relief train took the injured to Madrid and a breakdown train was sent from Atocha to clear the line. The blockage of the meant that Madrid – Barcelona mail train No. 1823, due to arrive in the Catalan capital at 20:00, was 12 hours late. [635]

---

[633] *The Bombay Chronicle*, Bombay, India, 26th November 1945

[634] *Journal de Geneve*, Geneva, Switzerland, 17th December 1945

[635] *La Vanguardia*, Barcelona, Spain, 22nd December 1945

CHAPTER 6

SINGLE LINE COLLISIONS

The operation of trains in opposite directions over a single line of rails carries the obvious danger of head-on collisions between them. By 1931 there were many hundreds of miles of single-track railways throughout the world, especially in places where the density of traffic did not justify the provision of a second line. Most of them were worked with the aid of the electric telegraph, with decisions as to the meetings between trains being either made locally or, as in United States, by a central dispatcher. In Great Britain, a few lines were worked telegraph supplemented by a staff and paper tickets, while an improved method of single-line control had been bought into in 1880 after the invention by Edward Tyer of the electric train tablet system and by the 1930s most were worked by this method or its variants (electric train staff and key token), though these systems were not used extensively outside of the British Empire. The invention of a failsafe track circuit in 1872 by the American William Robinson[636] had led to the introduction in the United States of automatic block signals, by which train movements on a single line could be safeguarded by signals warning of the approach of an opposing train, but the cost of installing such signals was high and as late as 1906 the Interstate Commerce Commission was stating that out of 194,727 miles of railroad there were still only 41,916 miles protected by them.[637] With the protection of trains running over a single line left solely to the exchange of telegraph or telephone messages, rather than by the more secure electric train tablet or its later equivalents, and not supplemented by some form of automatic signalling, collisions due to human error in the transmission or interpretation of telegraph messages were all too common.

Whilst it was not infallible, as was shown by the 1921 Abermule disaster, the electric tablet system and its variants, as used in Great Britain, was very safe, and it was almost ten years after Abermule before there was another single-line collision in the British Isles. Then on Friday 16th January 1931, one happened near Great Holland signal box, England, on the Clacton-on-Sea branch of the London and North Eastern Railway (LNER).

636 Wikipedia – Track Circuit

637 Wikipedia – Automatic Block Signalling

The trains involved in the 1931 accident were a light engine and the 6.08 a.m. newspaper train from Thorpe-le-Soken to Walton-on-the-Naze. The newspaper train consisted of B-12 class 4-6-0 engine No. 8578, running tender first and two brake vans, in charge of Guard E. Elliston with Driver Edward Sheldrake and Fireman Stanley Wright. The light engine was D-16 class 4-4-0 No. 8781, with Driver George Goodridge and Fireman L.A. Wilding. [638] The light engine was proceeding correctly along the single line from Clacton to Thorpe-le-Soken; the newspaper train should not have been going down the Clacton line at all, but down the other branch from the junction station at Thorpe-le-Soken to Walton-on-the-Naze. The two collided at Great Holland, where there was a crossing place which was out of use during the winter months, the signal box being switched out. Driver Sheldrake and Fireman Wright were killed and Driver Goodridge, Fireman Wilding and Guard Elliston were injured. Both engines were considerably damaged.

The line from Colchester was double track as far as Thorpe-le-Soken, beyond which it continued as two single-line branches to Clacton and Walton. Both branches were worked by Tyer's No. 6 tablet instruments. The tablets for the two lines were distinctive; apart from having different station names upon them that for the Clacton branch was made of fibre and had a round hole in its centre, while that for the Walton branch was made of brass and iron and had a square hole in its centre. The starting signals at Thorpe-le-Soken were locked with the tablet instruments so that they could not be cleared unless the relevant tablet had been withdrawn at that end of the section.

Thorpe-le-Soken Porter Signalman A.G. Appleton told the Ministry of Transport inquiry into the accident that he had accepted the newspaper train from Clacton at 5.53 a.m. and it had arrived at 6.01 a.m. (the train was timetabled to run from London first down the branch to Clacton, thence back to Thorpe-le-Soken and down the branch to Walton). Before the paper train arrived he had already obtained a tablet for the single-line section to Kirby Cross (the first station on the Walton branch) from the instrument. He saw the porter collect the Clacton – Thorpe-le-Soken tablet from the paper train and give its driver the Thorpe-le-Soken-Kirby Cross tablet, after which the engine of the paper train ran around its vehicles. Meanwhile, he had accepted the 12.10 a.m. down goods train from Spitalfields from Weeley box at 5.41 a.m. As he was watching the goods train approaching and had lowered his home signal for it he saw the paper train moving forward, he assumed just pulling up to the starting signal. With the home signal cleared for the goods train to come in, the interlocking prevented him setting the route towards Walton across the path of it and so the platform line on which the paper train stood still led to the Clacton branch. Appleton said that he had also received the 'Train Entering Section' signal for a light engine from Clacton at 6.03 a.m., having accepted it at 6.01 a.m. under the 'Warning Arrangement' ('Section Clear but Station or Junction Blocked') because the paper train was in the station. He had then realised, at 6.06 a.m., that the paper train was pulling out of the station and back down the Clacton branch whence it had just come. Running out of the box onto the platform, he had run after the departing paper train, shouting at it to stop. He had then returned to the box, telephoned his colleague at Clacton, and sent the 'Obstruction Danger' signal.

Porter W. Southgate testified to changing the tablets with the driver of the paper train and piloting it around its train. After the engine had been coupled up again its guard had

[638] *The Times*, London, England, 17th January 1933

given its driver a green light and it had started away. He had then heard the signalman shouting, "Where's he off to?" and had seen that the starting signal was at danger. He had showed a red light to the guard, without effect.

Stationmaster F.W. Taylor said that as soon as he had been advised of the accident he had gone to the signal box and had found that a tablet had been extracted for the section to Kirby Cross, while the Clacton instrument showed that a train was on its way from Clacton. He considered that the signalman should hand the tablet to the driver but that it was correct for the porter to do so if supervised by the signalman. The light of the back platform starting signal had never been lit in the three years he had been at the station. He did not consider that the driver had been given the tablet too soon. (He was supported in this view by Lieut. Colonel H.H. Mauldin, Superintendent of the LNER Eastern Section.)

Signalman E.W. Girling at Clacton stated that the paper train had been accepted by Thorpe-le-Soken at 5.42 a.m., it had left at 5.52 a.m. and been cleared back at 6.01 a.m. He had asked for the light engine at the same time, had cleared his signals for it, cautioned its driver and it had left at 6.06 a.m., though it was not booked to leave until 6.10. A telephone call from Thorpe-le-Soken at 6.09 a.m. had indicated that the paper train was coming back 'on the wrong road,' and he had received the 'Obstruction Danger' signal at 6.11.

Guard Elliston said that he had spoken to Driver Sheldrake before they had left Clacton and he had appeared normal, as had his fireman. He had seen the tablets exchanged, the engine had run around its train, had checked the brake gauge was reading correctly and had then given his driver 'Right Away'. It had never crossed his mind to look at the starting signal as his train had pulled out. He had then been busy sorting papers for Kirby Cross and so had not realised that they were on the wrong branch. When questioned as to what he considered to be his first duty, he had first answered that it was the sorting of his parcels before correcting himself to say that it was the safety of the train.

Driver Goodridge and Fireman Wilding, who was badly hurt, were unable to add much of importance.

Driver W.G. Ince and Fireman R. Newbery of the Spitalfields train spoke of seeing the paper train start away.

The inspecting officer, Lieut.-Colonel E.P. Anderson, stated that the immediate cause of the accident was clear. Driver Sheldrake had left Thorpe-le-Soken after obtaining the Kirby Cross tablet but against the starting signal at danger and with the road still set for Clacton branch. For some reason that could not be explained the driver and fireman had run two miles without realising that they were on the wrong branch, despite the physical features of the two lines being entirely different. Both men appeared to have been in good health and fit to work. Driver Sheldrake had worked the branch for many years.

No blame, the inspecting officer stated, attached to Driver Goodridge or Fireman Wilding. Other railwaymen were not to get off so lightly. Guard Elliston, Lieut.-Colonel Anderson said, had not been required according to the rules to check the starting signal before giving the 'Right Away', but he should have attempted to exchange signals with the fireman (Rule 171 (h)) and kept a good lookout (Rule 177 (a)) in which case he could hardly have failed to realise that they were on the wrong line. With reference to the guard's views of his most important duty, the inspecting officer commented, 'I am reluctantly forced to the conclusion that he is a man who has not cultivated the habit of regarding the safety of his train as of primary importance…'

Signalman Appleton, the inspecting officer continued, had broken the rules in allowing Porter Southgate to give the tablet to the driver. The practice of giving the driver the tablet before the train was ready to start was not actually forbidden by the rules but was undesirable. Appleton's failure to hold the Down goods train at the home signal until the paper train had departed for Walton was an error of judgement. When he saw the paper train departing Appleton had to some extent lost his head. His evidence that he had received the 'Train Entering Section' signal for the engine at 6.03 was in conflict with Signalman Girling's time of 6.06. Given that the train and engine had collided halfway between Thorpe-le-Soken and Clacton it was more likely that they had departed simultaneously at 6.06. Had Appleton immediately given Clacton the 'Obstruction Danger' signal on the bell rather than running out onto the platform in a futile attempt to stop the paper train it was possible that the Clacton signalman might have been able to stop the engine by throwing back his starting signal, which was 340 yards in advance of his box.

The Company would no doubt, the inspecting officer added, take steps to ensure the stricter observation of the rules at Thorpe-le-Soken, such as the non-lighting of the platform starting signal.

As regards recommendations, the Lieut.-Colonel Anderson suggested:

(a) The amendment of the rules to underline that the primary duty of the guard must be to ensure the safety of his train.

(b) A stricter enforcement of the rules regarding the unauthorised handling of tablets.

(c) A rule to the effect that a driver should not be given the tablet until he was ready to start.

The provision of any of the following mechanical devices at Thorpe-le Soken should have prevented the accident:

(a) A system of Automatic Train Control at the starting signal.

(b) Facing trap points with or without a sand drag at the starting signal.

(c) Locking between the tablet instruments and the points so that a tablet could not be withdrawn until the points were set for the correct line.

(d) Re-siting of starting signals Nos. 25 and 26 to the left of the line to which they applied.

(e) A detonator placer working in conjunction with the starting signal.

(f) An advanced starting signal on each branch with a detonator placer as in (e).

Of these the inspecting officer considered facing trap points plus a sand drag to be the preferable option. [639]

On Thursday 30th June 1932, a head-on collision occurred at Elk Creek Siding, Ohio, United States, on the Cincinnati and Lake Erie Railroad, an electric interurban line. Its single track was worked by timetable and train order without automatic block signals.

---

[639] Ministry of Transport report

Northbound passenger train No. 10 consisted on the day of the accident of a steel car in charge of Conductor-Motorman Joseph Brosey. At Lindenwald, the starting point of the train, 10¼ miles south of North Trenton, the latter had received a copy of train order No. 4, providing for a meet with southbound train Extra No. 34 at North Trenton. The train departed from Lindenwald at 6.05 a.m., stopped at North Trenton and took up a passenger, but failed to wait there for Extra No. 34, and travelling at 35 or 40 miles an hour collided with that train 1¼ miles further on.

Southbound freight train Extra No. 34 consisted of a motor car and a box car, both having wooden bodies and steel underframes, in charge of Conductor Harry Shank with Motorman Roy Bowman. At Cranes Run Siding, 13½ miles north of North Trenton, it had received a copy of train order No. 4, and left there at 6.12 a.m. It was travelling at 35 to 40 miles an hour when it collided with train No. 10. The collision killed Conductor-Motorman Brosey, Motorman Bowman, an operator, a car inspector, a freight house employee, a trackman, and a car cleaner, the latter five all railwaymen who were deadheading; and two passengers. In addition, Conductor Shank, a car employee who was deadheading and three passengers were injured. The freight train telescoped the passenger train and pushed it back 55 feet. Both trains were derailed.

Conductor Shank told the Interstate Commerce Commission (ICC) inquiry that when his train had arrived at Middletown, 4¼ miles from North Trenton, Motorman Bowman had remarked that it was going to be a good meet with No. 10. Their watches had both read 6.30 and 10 seconds, giving them 4 minutes and 50 seconds to reach North Trenton by 6.35, the time that train No. 10 was timetabled to leave there. He and the motorman had been standing up when, rounding a curve approaching Elk Creek Siding, he had seen train No. 10 about 1,000 feet distant. He had shouted a warning to the motorman and realising that a collision was unavoidable had then jumped. The motorman had sounded the whistle, he was not sure if he had braked. They met train No. 10 at or near North Trenton every morning, usually at North Trenton.

Dispatcher Schneider testified to issuing train order No. 4 to Conductor-Motorman Brosey at Lindenwald over the telephone, the order being made complete at 5.49 a.m. Conductor-Motorman Brosey had repeated it and spelt out 'North Trenton'. The first that he had known of the accident was a telephone message from Conductor Shank.

Night Foreman Hartman at Lindenwald spoke of watching Conductor-Motorman Brosey copy out the order and of hearing him repeat it to the dispatcher.

The ICC found Conductor-Motorman Brosey responsible for the collision by failing to obey a meet order. In view of his death it was impossible to explain his failure. The commission drew attention, however, to the fact that officials had not been making regular checks on whether trainmen were making carbon copies of orders received by telephone. This was a matter of importance which officials should keep under careful observation. It had not been possible to find a copy of Conductor-Motorman Brosey's order in the wreckage, though it might have been destroyed in the collision.[640] [641] [642]

[640] Interstate Commerce Commission report

[641] *Hamilton Evening Journal*, Hamilton, Ohio, United States, 30th June 1932

[642] *Hamilton Daily News*, Hamilton, Ohio, United States, 1st July 1932

Nine people died and 80 were injured when two trains collided head-on on the evening of Saturday 2nd July 1932 near Vlašim, Czechoslovakia, on the line between Benešov and Dolní Kralovice. The dead included the driver of one of the trains and the conductor of the other. The exact circumstances of the accident are not clear, but it involved a delay to one of the trains resulting in an alteration of the point at which they were to cross one another, followed by a failure of communications between the stations involved because a severe thunderstorm had brought down the telegraph wires. [643] [644] The storm also made it impossible to summon assistance by telegraph or telephone after the collision, and the surviving conductor had to make his way to the nearest station to obtain help despite his injuries. [645] A relief train with doctors and nurses was then sent to the accident scene. Several coaches were demolished and rolled down a 15-metre embankment. [646] In addition to those who had lost their lives 12 passengers were seriously injured and 20 were slightly hurt.

A mistake by a stationmaster led to a serious collision in Italy late in the evening of Thursday 20th January 1933. Direct train No. 813, which had left Napoli at 20:10 for Foggia, should have crossed accelerated train No. 1896, which had left Foggia at 19:15 for Napoli, at Solopaca, but the Solopaca stationmaster let the former train proceed before the arrival of the latter, and the two trains duly collided at Kilometre 126, between Solopaca and San Lorenzo Maggiore stations, at about 22:00, as one of the trains had just exited a tunnel. The closing speed of the trains at the moment that they collided was estimated at 85 kilometres an hour and the two locomotives, the luggage van of one train and the mail van of the other were wrecked. Eight people died, five railwaymen and three mail clerks, while 30 people were injured, four seriously. The names of the railwaymen who lost their lives are a little difficult to determine from the various newspaper accounts, but they appear to have been Driver Giovanni Paglionico, Fireman Ruggero di Bitonio and Conductor Pironti of train No. 813 and Driver Balestrieri and Fireman Scarano of train No. 1896. Chief Conductor Alfredo Favla and Conductor Cabulo Sansone, both of train No. 1896, were among the injured. Undersecretary of Transport S.E. Postiglioni and various senior railway officials went to the scene of the accident to open an investigation into its cause. The line was expected to be blocked for 24 hours. [647]

Sunday 18th February 1934 was the day of the annual carnival festival in the town of Piombino, Italy, which lies on the coast opposite the island of Elba, and the normal service of 12 trains each way over the single-track branch from Campiglia was increased to 46 trains in all. Train No. 959 from Livorno and Campiglia, which left Campiglia at 21:17, normally crossed train No. 974 from Piombino at the intermediate station of Populonia. On this day it should have waited at Populonia not just for that train but also for the following extra train, No. 974 bis, which had departed from Piombino at 21:13. Trains No. 959 and No. 974 were third-class railcars, train No. 974 bis consisted of a steam engine and four third-class carriages. Instead, as soon as the regular train No. 974

[643] Wikipedia – Liste von Eisenbahnunfällen in Tschechien
[644] *L'Homme Libre*, Paris, France, 3rd July 1932
[645] *The Times*, London, England, 5th July 1932
[646] *Le Journal*, Paris, France, 4th July 1932
[647] *La Stampa*, Torino, Italy, 21st, 22nd and 24th July 1933

had arrived at Populonia, train No. 959 from Campiglia set off again despite the vain efforts of the station officials to stop it by shouting and manipulating the signal on the Piombino side of the station. The trains met on a curve between Populonia and Portovecchio di Piombino stations and five kilometres from Populonia at about 21:45. The steam engine of the special train was more than a match for the railcar from Campiglia, which was reportedly travelling at 120 kilometres an hour. The railcar was pushed back tens of metres and caught fire due to the rupture of its fuel tank before the engine of the special train derailed. The magnitude of the accident was not immediately realised; when it was, ambulances and private cars took the injured to hospital in Piombino. Sixteen people died, all on train No. 959, and including its motorman, conductor and a third railwayman. Eleven others were more or less injured. A judicial investigation into the disaster blamed it on the dead driver and conductor. [648] [649] [650]

Two days after the disastrous single-line collision in Italy there was one in Spain involving the regular night train from Sevilla to Madrid and a special train carrying an estimated 900 football supporters, mostly from Sevilla, who had watched their team play Atlético Madrid in the Spanish capital. The trains collided near Villanueva de la Reina at 04:00 on Tuesday 20th February 1934.

The probable cause of the collision was quickly established. The football special had been reported to Villanueva de la Reina from the preceding station at Espeluy and the regular train from Sevilla had been notified from the next station at Andujar. The crossing of the two trains should have taken place at Villanueva de la Reina but the football special ran through the station there without stopping and met the Sevilla train 300 metres further on. The conductor of the Seviila train, Victoriano Bustillo, said that he had seen that his driver was running past the signals and had sounded the alarm. The speed of the special train had thus been somewhat reduced by the last-minute brake application and an even more serious accident was averted, especially as the ascending train from Sevilla had apparently been more or less brought to a stand, its driver having become aware that the football special was not stopping at Villanueva de la Reina. Despite this, 11 people died in the collision and more than 50 were injured. [651] [652] [653]

Spain suffered another serious single-line collision on Monday 21st May 1934. An ascending train from Barcelona should have crossed a descending train at San Baudilio de Llobregat station but left the station there before the opposing train had arrived. It had not proceeded very far before the two collided at about 20:10. The two engines were described as having ended up embedded in one another, while the wooden carriages behind them were badly broken up.

A municipal guard, walking along a road which ran alongside the line with a neighbour, witnessed the accident. His colleague pointed out that the two trains were

---

[648] *The Times*, London, England, 20th February 1934
[649] *Gazette de Lausanne*, Lausanne, Switzerland, 20th February 1934
[650] Wikipedia – Incidente ferroviario di Piombino
[651] *La Vanguardia*, Barcelona, Spain, 21st and 22nd February 1934
[652] Webpage: Federación Castellano Manchega de Amigos del Ferrocarril (in Spanish).
[653] *The Times*, London, England, 21st February 1934

coming towards one another on the single track and the collision followed almost immediately. He told a reporter that after the accident the panicking, uninjured, passengers had tried to find their way to the road that he and his neighbour were on, but a three-metre-wide canal separated the railway and the road. There was a small bridge across the waterway which the passengers found but in their panic some of them fell into the water trying to cross it. Meanwhile the uninjured members of two choirs who had been travelling on the train, *La Sirena* and *La Camelia do Vallbona*, were the first to render assistance to the injured.

In the city of Barcelona, about 20 kilometres away, the *Fiesta Mayor*, the principal festival of the year, was in progress, but as soon as news of the accident was received the festivities were suspended and cars and motor buses set off for the scene. The headlights of the buses were played on the scene to aid the rescue work in the darkness. Many of the injured were taken to a large mental hospital in Barcelona while the Catalan flag was hung at half-mast over the town hall.

Twelve people died in the accident and 18 were injured. Those who lost their lives included Conductor Juan Gómez Blay, Driver Anastasio Pardo Garcia and Fireman Luis Font of the descending train, which had been running fast downhill. Driver Jaime Boronat of the ascending train was seriously hurt. His fireman, Luis Almirall, had jumped when he had seen the other train and was less badly injured. [654] [655]

A railway official stated that the accident had been due to 'personal confusion', without further explaining the matter. The San Baudilio stationmaster denied having given the ascending train the 'Right Away' but a number of witnesses stated that a passenger had blown a horn very similar to that used by the stationmaster, thus misleading the driver. [656]

The trains had been more than usually crowded on account of the Feast of Pentecost.

Passenger train P 1978 and relief passenger train P 1973 collided head-on on Saturday 22nd December 1934 at Schleißweiler, on the single line between Murrhardt and Sulzbach, Germany, due to a signalman's error. The accident, on the Backnang – Crailsheim line near Stuttgart, cost nine lives and 58 others were injured. [657]

A passenger train from Peianchen to Heiho collided head-on with a southbound goods train near Chenching, in north-east Manchuria, on Tuesday 2nd April 1935, in what was probably a single-line collision. Seven people died and 16 were injured. [658]

The Cincinnati and Lake Erie Railroad suffered another bad accident on Saturday 10th August 1935. A southbound passenger train and a northbound work train collided near Siding No. 61, 4¼ miles north of Springfield, Ohio, United States. Six passengers and a

[654] *La Vanguardia*, Barcelona, Spain, 23rd May 1934
[655] Webpage: Federación Castellano Manchega de Amigos del Ferrocarril (in Spanish).
[656] *The Times*, London, England, 23rd May 1934
[657] *Katastrophen der deutschen Bahnen Teil II* by H.J. Ritzau, pg. 84 (in German)
[658] *The Times*, London, England, 4th April 1935

railwayman died and 20 passengers, two railwaymen and a person carried under contract were injured.

Southbound train First No. 401 consisted of steel motor car No. 128 driven by Motorman Estrep. (The Interstate Commerce Commission (ICC) report on the accident rather confusingly refers to the people driving the two trains as Operators, a term more commonly used in North America for the staff who received train orders from dispatchers and gave them to train crews, doubtless because separate operators were not employed on the line, but here the more obvious term Motorman will be used). This train left Lima, 65¾ miles north of Springfield, at 12.40 p.m. on the day concerned. After travelling just over 38 miles it met train No. 402, driven by Motorman Roy Zimmerman, and the motormen of the two trains swopped over. Motorman Zimmerman continued southwards at 2 p.m., 24 minutes late, and at Urbana, 14 miles further on, he received train order No. 30, which read:

Train No. X Car No. 204 No and Train No. 1/2/401 Car No. 128/121 will meet at 59. Train No. 1/401 Car No. 128 and train No. X Car No. 710 will meet at 61. Train No. 2/401 Car No. 121 and Train No. X Car No. 204 will meet at 59. Train No. 2/401 Car No. 121 and Train No. X Car No. 710 will meet at 61.

The train left Urbana at 2.26 p.m., 29 minutes late, passed Siding 61 and was travelling at 45 miles an hour when it collided with train Extra 710 at about 2.42 p.m.

Train Extra No. 710 was a work train consisting of line motor car No. 710, of wooden construction, driven my Motorman A.N. Allen. At Springfield Motorman Allen had received a copy of train order No. 31, which read:

Train No. 1/2/401 Car No. 128/121 and Train No. X Car No. 710 will meet at 61. Train No. X Car No. 710 No and Train No. X Car No.647 So will meet at 61. Train No. X Car No. 710 No and Train No. X Car No.204 So will meet at 61. Car 710 run extra Spfld to Belft.

Train Extra No. 710 left Springfield at 2.28 p.m. and was approaching Siding 61 at between 10 and 20 miles an hour when it collided with train First No. 401. Neither car was derailed but the front of each was telescoped about 12 feet. Motorman Zimmerman was killed, and Motorman Allen and a lineman were the railwaymen injured.

Motorman Estrup told the ICC inquiry that he had had no problems with car 128. He and Motorman Zimmerman had exchanged train orders and read them to one another. His colleague had appeared normal and to have understood the orders.

Train Porter Bull stated that he had changed cars at siding 38 along with the motormen. He agreed that Motorman Zimmerman had appeared normal. He did not know what orders were received at Urbana. Seeing the line car approaching, he had run to the rear of the car and so had saved himself.

Motorman Allen said that on the day of the accident he had first taken the work train south from Bellefontaine to Springfield, where there had been power trouble. This had been fixed and at Springfield he had received train order No. 31 by telephone. He had heard the motormen of four other trains repeat the order to the dispatcher correctly. The brakes of his car had worked correctly at two stops after leaving Springfield to remove tree limbs from telephone wires. On seeing the opposing train he had braked in emergency but had had no time to reverse his car.

Lineman Grubbs, who had been riding on the work train, said that he had shouted a warning to Motorman Allen and had jumped when the trains were 100 feet apart. After the accident he had noted that the brakes of train First No. 401 were applied.

Dispatcher Mason at Lima testified that he had issued train order No. 30 to both sections of train No. 401 (the superior train) before he had completed train order No. 31 to the motorman of train Extra No. 710 at Springfield, and he had had no difficulty in making any of the motormen understand their orders. Previously he had issued train order No. 22 to train First No. 401 at Lima, fixing a meet with train Extra No. 204 at the north end of Springfield; the meeting point had been changed to siding 59 on account of delay to train First No. 401 because of a storm. He considered that train order No. 30 superseded train order No. 22, though he had not used the words "instead of" as required by the rules. He did not consider that this omission might confuse the motorman. Given that Motorman Zimmerman had correctly met train Extra No. 204 at siding 59 as per train order No. 30 he could not understand why he had then failed to meet train Extra No. 710 at siding 61 as per the same train order. Dispatcher Mason said that he had not been examined on the rules for four or five years, but he was familiar with the proper manner of issuing orders and he admitted that his wording of the order had not been strictly in accordance with the rules.

Superintendent of Transportation Duncan stated that he had been in the dispatcher's office when train order No. 30 was issued but had not realised that it had been improperly issued until after the collision. A later examination of the dispatcher's record book showed that orders were generally issued according to the rules except that the words "instead of" were sometimes omitted and the numbers and wording were not always as per the rules. Identical orders were generally issued to all the trains concerned, but sometimes the order might be reissued in part under another number and with slightly altered wording, a practice which he considered safer than rewriting the entire order, although it was not provided for in the rules. The same practice had been in use for the 18 years he had been with the company. Motormen Estrep and Zimmerman had swopped trains at siding 38 so that each could return to their home stations the same day.

General Superintendent T. Curtiss stated that the orders issued to Motorman Zimmerman had been found in his possession by a State highway patrolman after the accident. The last general meeting of employees had been in 1931. Since then there had been a great reduction in staff and as all those retained had been with the company for many years he did not consider it necessary to go through the rules with all of them individually. He described Zimmerman as "an old and trusted employee." [659]

The ICC concluded that the direct cause of the accident was the failure of Motorman Zimmerman to obey a meet order, however there appeared to be 'a general lack of understanding by some officials and trainmen of the operating rules in connection with the movement of trains by train orders.' Thus, orders were sometimes issued under different numbers to the different trains involved, where an order was superseded the words "instead of" were not always used, and a train order of a lower number was sometimes used to supersede an order of a higher number provided that the lower number order was the last to be made complete. The ICC commented that, 'Train orders which were not in conformity with the requirements of the operating rules have been issued

[659] *Lima Sunday News*, Lima, Ohio, United States, 11th August 1935

frequently; a situation of this kind should be given immediate attention and corrected by the officials of this railroad.' [660]

A major disaster occurred in Italy on the morning of Saturday 3rd October 1936 on the Terni – Sulmona line. The accident took place between Contigliano and Rieti stations and involved a passenger train due at Terni at 10.21 and a postal train. Both trains consisted of Littorino diesel or petrol railcars, built from the beginning of the 1930s by the Fiat motor company, the postal railcar was one of only three of the ALDb 200 class. The trains met at Km 196 between the two stations at 09:45 in a collision which badly damaged both vehicles, which ended up wedged into each other. Initial reports spoke of seven dead [661] but it appears that at least 15 persons died, including two railwaymen, probably the motormen of the two railcars. All the other 69 people on the passenger railcar were injured. Prominent in the rescue work was Carabiniere Raffaele Betti, who later received a commendation for his efforts. Undersecretary of State for Railways Mario Jannelli went to the scene to investigate the accident.

The passengers involved in the accident included the entire team of the A.S. (Associazione Sportiva) l'Aquila Calcio football club of Abruzzo, who had been on their way to Verona to play in a Serie B game. Attilio Buratti, the team's coach, was one of those who died in the disaster, and all but three of the other members of the team were injured. Players Brindisi and Michetti were not on the train because they had been suspended, nor was the team's goalkeeper, named Stornelli, who had overslept and missed the train. Predictably, with their normal coach dead and with most of their team recovering from the accident the team, which had been doing well, was badly affected, and it was relegated to Serie C at the end of the season.

Contemporary reports placed the blame for the accident on the motorman of the postal railcar for leaving Contigliano before the passenger train had arrived there, but more recent sources suggest that a mistake made by the Rieti stationmaster was to blame. [662]

A major disaster occurred in Sicily, Italy, probably early on Sunday 13th March 1938. It involved a head-on collision between what was described as a very long freight train and a Littorina passenger railcar. The trains met on a curve between Bicocca and Passo Martino stations on the Messina – Syracuse line. The railcar was said to have been travelling at high speed and the collision was a very violent one. On news of the accident becoming known a rescue train was sent from the nearest large station at Catania and teams of railwaymen, assisted by *carabineri* (police) and soldiers worked through the night at the scene extricating the victims from the tangle of wreckage while it rained incessantly. By the morning 12 bodies had been recovered, including those of Conductor Costo Giovanni and Motorman Tadduni Sebastiano of the railcar and Driver Cauda Pietro of the freight train. Eight others were injured, five seriously, and Conductor Usignolo Clemente of the freight train succumbed to his injuries after reaching hospital. The passengers who lost their lives included Dr Leone Francesco, director of the Syracuse psychiatric hospital, Professor Tusisco Michele of the University of Bari, two Swiss

---

660 Interstate Commerce Commission report
661 *La Stampa*, Torino, Italy, 4th October 1936
662 Wikipedia – Incidente ferroviario di Contigliano

tourists and a Swedish tourist.

A mistake made at Bicocca station, where the passenger train had been given a clear signal, was believed to have caused the disaster and the railwaymen on duty there, Stationmaster De Biasio Antonino and Signalman Rosario, were placed under arrest. [663]

What was described at the time as one of the worst accidents in the history of the railways of Southern Africa occurred at 3.30 p.m. on Monday 4th April 1938. Train No. 66 Up, a passenger and mail train from Bulawayo to Johannesburg and Cape Town, came into collision with goods train No. 45 Down at a point 72 miles south of Bulawayo, between the stations of Plumtree and Tsessbe, Southern Rhodesia, and close to the border with the Bechuanaland Protectorate. At least 26 people died in the accident and 22 were injured. Eighteen of those who lost their lives were passengers, the others were both drivers and firemen, a locomotive inspector and the occupants of a guards van carrying the second crew who were carried for the 484-mile run between Bulawayo and Mafeking. Both engines were, as might be expected, badly damaged. In the passenger train the two leading guards vans were wrecked, two carriages were telescoped and wrecked, and a third carriage was damaged and derailed, while several waggons of the goods train were overturned and derailed. The remaining carriages, dining car and rear guards van of the passenger train escaped damage. The uninjured members of the train crew and passengers were thus able to give the initial first aid. Hampering the rescue work was the fact that the accident happened at a point some distance from the nearest road, added to which the roads themselves were very poor. A storm and the lack of adequate lighting added to the difficulties. [664] The trains also met in a cutting. The line was blocked for 48 hours before the cutting could be widened sufficiently to build a deviation line around the wreckage.

A departmental inquiry was immediately opened by the South African Railways administration, which worked the Bulawayo – Mafeking section of the Rhodesia Railways.

The cause of the accident was revealed at the inquest on the victims. As the goods train was running early the stationmasters at Plumtree and Tsessbe decided to alter its crossing with the passenger train, which usually took place at Tsessbe. The new crossing place, arranged by telegraph, was meant to be the siding at Ramaquabane, but the crossing orders issued by the Tsessbe stationmaster to the goods train named the crossing point as the siding at Vakaranga, a point nearer Plumtree. Both trains thus had rights to the single line between the two sidings, where they collided. With their views restricted by the cutting the two drivers could only have had a momentary glimpse of one another but the driver of the passenger train had applied his brakes. [665]

The man in charge at Tsessbe, referred to only as Stationmaster S, had had a stressful day on 4th April 1938 before the accident occurred, for he was due for a transfer to another post and was expected to explain matters at Tsessbe to his successor, Stationmaster Dirk Crafford, that very day and to depart for his new position at once. At about 2 p.m. Stationmaster van Niekerk at Plumtree had telephoned Tsessbe regarding the crossing of northbound train No. 45 with train No. 66. Stationmaster Crafford had

---

[663] *La Stampa*, Torino, Italy, 14th March 1938
[664] *The Times*, 5th and 6th April 1938
[665] *The Railway Gazette*, 20th May 1938

replied that he would ring back when train No. 45 arrived. It did so at 2.30 p.m. and with its engine blowing off noisily outside Stationmaster S wrote out the order, then realising that he had failed to add the information regarding other, cancelled trains, he had discarded it and rewritten it. By then Guard Harold Botha had come into the office. Hurriedly finishing the order the stationmaster had handed it to the guard, telling him, "Jong, jy moet gou maak." (Man, you must hurry up.) The guard had given his driver the order and the train had left.

About 3 p.m. Stationmaster S had handed over to Stationmaster Crafford. At 4.10 p.m. Stationmaster Crafford had been surprised that train No. 66 had not arrived and had conferred with Stationmaster S. The latter had then gone to the office and found the carbon copy of the order which he had given to the guard. The new stationmaster had at once got on the telephone to Plumtree and asked, "Please repeat the crossing place arranged for train 45 and the Mail?" Plumtree had answered that the Mail had an order to cross train No. 45 at Ramaquabane. The original order, retrieved from the waste paper bin, and the telegraph tape, had agreed with this. The carbon copy of the order given to the guard of the goods train had named the crossing place as Vakaranga. Due to Stationmaster S's carelessness and rushing to finish the job what in the United States would be called a lap order had been issued, giving both trains the right to the line between Ramaquabane and Vakaranga.

Train No. 66, the Mail, had left Plumtree at 3.12 p.m. with Driver C. Reynecke on the engine together with Fireman Esterhuysen and Locomotive Inspector Marshall. Its passengers had included four of the five daughters of Stationmaster Burger of Figtree, a man and a woman travelling south to get married, a nurse and Stationmaster Benjamin Weakley of Plumtree and his wife, also transferring to a new post. Guard J.J. van Wyk had been sat in his van when he had heard two short whistles, then a violent shock had thrown him off his feet. Getting up and seeing what had happened he had gone back to protect the rear of his train.

Standing in the corridor of the Mail, 11-year-old Frederick Welner had been looking out of the window and had seen the goods train coming. He had shouted to his mother that a collision was imminent and watched the two engines collide and plough into the side of the cutting. The carriage he was in was considerably damaged, but he and his mother escaped unhurt.

After the accident off-duty guard Harry Moolman went to the engine. The badly injured Driver Reynecke said to him, "God, Moolman, my brake was hard on," and asked for a telegram to be sent to his wife. The guard gave him some brandy and helped him from the cab. Driver Bernadus Coetzee of the goods train was crushed beneath his engine and killed, it was hoped instantly, for his body was burnt beyond recognition.

Ganger Jose Isaacs, working about a mile away, heard the sound of the collision. It was he who took Stationmaster Weakley on his trolley to Ramaquabane, from where the stationmaster confirmed the disaster at 4.35 p.m.

Driver Reynecke died shortly afterwards of his injuries, as did Fireman Esterhuysen. Locomotive Inspector Marshall was killed outright. Other railwaymen who lost their lives were Firemen Combrinck and van Rensburg and Guard van der Heever. Drivers du Toit and Rausch and Fireman van Wyk were injured. Two of the four Burger girls lost their lives, the other two were injured.

A nurse on the Mail was active in the initial rescue work and help was on its way as soon as Stationmaster S realised what he had done, a road ambulance being sent from Bulawayo, followed by a second, and then a relief train, which reached Plumtree about 7.30 p.m. A breakdown train came down later and reached the accident site about 1 a.m.

It was Thursday 7th before all the wreckage was cleared. Meanwhile, Stationmaster S had been arrested on a charge of culpable homicide with a bail of £100, only to be released unconditionally the next day.

Stationmaster S duly stood trial at the beginning of June. The charge was that on 4th April he "did wrongfully, unlawfully and negligently make and deliver to Harold Botha and Bernardus Rudolphus Coetzee, guard and driver of goods train 45 Down, a token authorising them to conduct the said train to Vakaranga, by reason of which the train proceeded beyond Ramaquebane, and a collision occurred with passenger train 66 Up, which was proceeding from Vakaranga towards Ramaquebane." (Incoming trains coming from Beira were 'Up' and those going towards Beira were 'Down'.)

Evidence was given at the trial regarding the events leading up to the collision, then Stationmaster S gave evidence in his defence. He said that he had learnt of his transfer to another station ten days earlier but only at the last minute had he learnt that he was to move on the day of the accident and not two days later as he had expected. He had then had to sort out a problem regarding some missing money in the accounts, due to an entry being made twice. He had arranged over the telephone with Stationmaster van Niekerk at Plumtree for the goods and mail trains to cross at Ramaquebane and had made out a red order (a crossing order) but having done so had realised that in his haste he had not filled in the 'not running' section (listing train numbers and the sidings they would have crossed the goods train at). He had thrown the order in the wastepaper basket and made it out again, this time filling out the 'not running' section but writing Vakaranga rather than Ramaquebane. He had said earlier that he had never been north of Tsessbe, and that Ramaquebane and Vakaranga were just names to him. He did not know why he had made the mistake. Having rewritten the order he had handed it to Guard Botha of the goods train and had told him to hurry up to avoid delaying the mail. He had then changed out of his uniform and prepared to set off for his new post. About an hour later when he had returned to the office, he found that the mail was overdue and on checking the orders the mistake had been discovered.

Major Roper for the prosecution subjected the stationmaster to a fierce cross-examination.

Only one witness was called for the defence. Dr W.A. Willemse, Professor of Psychology of Pretoria University, who attempted to prove that a reasonable man who was rushed and worried might make such a mistake as the stationmaster without being criminally careless, but his evidence was not very convincing.

At the close of the case it took the court only 13 minutes to find Stationmaster S guilty of negligence. He was fined £50, with the alternative of six months' imprisonment with hard labour. Having lost his job as a result of the accident he was permitted to pay the fine in £5 instalments. [666] [667]

---

[666] https://issuu.com/hennieheymans/docs/sas-sar_vol_6_no_4

[667] https://issuu.com/hennieheymans/docs/sas-sar_vol_6_no_5

A head-on collision in darkness and during pouring rain near Martorell, Spain, at about 8.30 p.m. on the evening of Sunday 25th September 1938, had as its background the ongoing Battle of the Ebro, in which the government forces were being defeated by the Fascists with their greater manpower and air superiority. One of the trains involved was what was nominally a freight train from Sant Vicenç de Calders to Barcelona, which was packed with soldiers and refugees fleeing the Fascist advance. The other train, which was also crowded, was a passenger train from Barcelona bound for Villafranca del Penedès. The trains collided just outside of Martorell on a curve. The rescue operations were hampered by the fact that due to the civil war Martorell station was without electricity and hence had no lighting, while no vehicles were available to transport the injured.

News of the accident was all but completely censured by the Republican government, presumably because of its feared effect on morale, but it was reported by the foreign press. A brief reference to the funeral of one of the victims was the only mention of the accident in a Barcelona newspaper. [668] According to a foreign correspondent in Barcelona over 40 dead had been extricated from the wreckage by the following day and the total was expected to reach 65. The following evening survivors and villagers were still searching for survivors, while at least 120 were being treated in Barcelona hospitals, while the more seriously injured were being treated at the scene of the accident or in a hastily equipped first-aid station.

A private inquiry into the cause of the accident was opened the following morning. Surviving train crews blamed the heavy rainstorm and electrical disturbances for interfering with the proper operation of points of signals. [669] [670] An alternative version was that due to an error by the Martorell station staff the freight train had been permitted to leave after the one from Barcelona had already been accepted. [671]

A brief report received in Paris from Valencia, Spain, stated that two trains had collided head-on between Kilometres 328 and 329 of the Madrid – Valencia line, between Bonete and Alpera, Spain on Friday 16th December 1938. Eleven coaches of the two trains, in which there were numerous passengers, were embedded within one another. At the time of the report the bodies of seven dead had been recovered, together with 18 seriously injured and 25 less badly hurt. The injured were taken to Albacete. [672]

Overnight passenger train N-2 (*Nocturno Minetro – The Night Miner*) and freight train C-65 collided at Kilometre 354 between João Ayres and Sitio stations, Brazil, at 01:55 on Monday 19th December 1938 on the Estrada de Ferro Central do Brasil.

The passenger train consisted of engine No. 372, a mail van, a luggage van, a second-class carriage, two first-class carriages, a buffet, a *dormition commum* (second-class sleeping car?), two *cabines* (first-class sleeping cars?) and a reserved carriage. The

---

[668] https://blocs.mesvilaweb.cat/elbarrinaire/laccident-de-tren-censurat/
[669] *The Times*, London, England, 27th September 1938
[670] Webpage: Federación Castellano Manchega de Amigos del Ferrocarril (in Spanish).
[671] *The Railway Gazette*, London, England, 30th September 1938
[672] *La Prensa*, Santa Cruz de Tenerife, 17th December 1938

freight train consisted of engine No. 805, three waggons, four open waggons, a coal waggon and four closed waggons. The empty waggons were much damaged in the accident.

Early reports stated that 30 people had died in the four leading vehicles of the train – 27 passengers, two mail clerks, and a baggagemaster. Both of the greasers, and one of the firemen also died. Twenty-four injured were removed to Barbacena Hospital and 16 to Santos Dumont. Two of the coaches were filled with Bahian emigrants (Bahia is a Brazilian state) and another with Boy Scouts.

The first relief train was sent from Santos Dumont, followed by a second from Lafayette, the latter being accompanied by railway officials, doctors, and police. A special train carrying various senior officers of the company left Rio de Janeiro for the scene in the morning.

As to the cause of the accident it was established that a third train, freight train CEC-8, had passed through João Ayres prior to train C-65, and its driver had failed to collect his line clear ticket there. The driver of train C-65 had neglected to check, as he should have, that the ticket permitted him to proceed, instead of waiting at João Ayres to cross the passenger train there. Other circumstances which might have contributed to the accident were being investigated.

The uninjured passengers from the *Nocturno Minetro*, and those from other ascending and descending local and express trains, were transferred between trains on either side of the obstruction pending the reopening of the line.

The railwaymen who died were Fireman Oscar Pinto dos Santos, Greaser Antonio Mendes and João Ferreira da Silva and Baggagemaster Benedicto Conceção Luz. There had been 96 first- and second-class sitting passengers, 44 sleeping passengers, two mail clerks, the driver, fireman, baggagemaster, two greasers and three other unidentified persons on the passenger train.

Porter Warneck Rodrigues at João Ayres station, who, a reporter noted, appeared to be in possession of his mental faculties, though shaken, declared, "As soon as I saw C-65 pass I ran to the telephone to warn the foreman who was situated 300 metres beyond the station with the idea that he might to signal train C-65 to stop. It was no use. He ran after the train waving his handlamp violently, but as he was doing so the oil lamp blew out. Disaster was then inevitable."

Stationmaster Sizenando Santiago at João Ayres explained, "There is no electric lighting at our station. It was dark when the driver of train C-65 went through, so he probably wasn't able to properly read his line clear ticket."

Questioned as to who was responsible for the accident, Waldemar Luz, the director of the Central do Brasil, confirmed that, as previously supposed, there were three employees principally to blame for the accident, namely the station official at João Ayres, the driver of train CEC-81 who had left without his line clear ticket, and the driver of train C-65 who had failed to notice that the line clear ticket which he had picked up did not apply to his train.

Driver Franklin Carlos da Silva of train N-2, who was seriously injured, spoke to reporters from his hospital bed. "You know, of course, that I had the proper line clear ticket to proceed from Sitio station. Well, having crossed train CEC-81 there, I had no idea that a second freight train was going to be coming from João Ayres immediately behind it. As I proceeded from Sitio towards João Ayres I saw, about 500 metres away, what appeared to be lights coming towards me. At first I thought that it must be a car on the highway but then the thought came to me that there was no road near the railway at that point, so it could not be car headlights. I whistled twice and when I realised that there was a train coming towards me I jammed on the brakes and at the same time shouted to the fireman and the greaser to jump, 'Jump because we are going to die!' But it was too late. There was a tremendous collision and I saw my mates crushed in the wreckage and buried by the coal. Both died at once without knowing what had happened. You know the rest."

A further report indicated that the number of known dead had increased to 42, with the possibility that there were still other victims buried in the wreckage.

Rumours that overcome by grief at what had happened, Warneck Rodrigues at João Ayres station had attempted to kill himself were denied, though it was stated that he had disappeared.

Driver José Rabello Alves of train C-65 was seriously injured and unconscious for some time. When he regained consciousness he was able to make a statement. "As we were entering a curve my fireman, José Moysés Alves, saw a powerful light coming towards us. He told me that train N-1 was coming and that the line clear ticket that we had was not meant for us. At the same moment he jumped from the speeding engine and thus escaped certain death thanks to his quick thinking."

Another official stated that the blame for the disaster had not yet been confirmed, though it was clear that train C-65 had passed João Ayres, where it should have stopped. This station was at the summit of that stretch of the railway. Only with the line being clear could the freight have continued its journey, and the line could not be clear because the passenger train was expected. All the theories revolved around the fact that train C-65 had not waited at João Ayres for the passenger train. It was stated that another freight train, CEC-81, with Driver Fernando de Freitas, had passed through João Ayres at 01:32 but its fireman had not collected the line clear ticket, which was hanging at the station as usual, having been placed in the hoop by Porter Werneck Rodrigues. Train C-65 had then arrived at João Ayres at 01:40 and should have waited there for train N-2. However, Driver Rabello Alves or his fireman, seeing the ticket left by train C-81, had taken it and continued their journey, colliding in due course with train N-2. If so, the blame lay principally with the driver of train C-81 and the official at João Ayres, as well as with the driver of train C-65, who had failed to check that the ticket applied to his train. He added that these were, however, only assumptions based on the facts as reported in the newspapers.

According to witnesses, as soon as Werneck Rodriguez had become aware of the disaster he had suffered a severe nervous breakdown and was running to and fro screaming.[673]

At Christmas 1938, Romania was affected by a severe blizzard. On the afternoon of Christmas Day reports began reaching the capital, Bucharest, of a serious train accident

[673] *O Estado de São Paulo*, São Paulo, Brazil, 21st December 1938

but snow and ice had brought down many telegraph wires and poles and it was not until 9 p.m. on Boxing Day that the first official communication of the accident was received. Most of the passengers on the two trains involved had been making local journeys, so that no suspicion was aroused of an accident in Bucharest by the non-arrival of expected passengers.

It transpired that two trains had collided on the single line between Frecăţei and Etulia, on the line between Galatz and Chişinău, late on Christmas Eve, 24th December 1938. The accident was a very serious one. By 27th December the number of known dead was 93, with another 147 injured. Among the 1,500 passengers said to have been on the two trains were a number of soldiers going on leave, and a general commanding a brigade, four other officers and 13 lower ranks were among those who lost their lives. The lack of telephone communications meant that it was a considerable number of hours before news of the disaster emerged, when two relief trains were sent to the scene. It was stated that the injured passengers had had to lie unattended and without shelter through the bitterly cold night.

A student who had been a passenger on one of the trains and who had been fortunate enough to escape gave his account of the disaster to a reporter:

'It was a cold, stormy night, and the fields were covered with snow. About half an hour after we left Galatz there was a terrible crash, and I felt myself under our carriage. Terrible screams sounded through the darkness. I managed to crawl out of the wreckage of our carriage, in which there had been about 30 passengers. Many of them were now dead and others, badly injured, were imploring help. Once out of the carriage I saw that almost all the cars were derailed, some 30 lying on their sides. The snowstorm raged; no signal lamps were burning, and the few uninjured passengers and the railwaymen could do little for those whose cries for help came from the wreck. Finally one of the conductors decided to go to the nearest railway station, Reni, 10 miles away. He had to go alone, as no volunteer could be found to accompany him, everyone feeling that the attempt would be suicide in the snow and darkness. After some hours, the first rescue train arrived from Galatz. By this time many of the injured had frozen to death. It was terrible to see the long line of dead lying in the snow, surrounded by hundreds of badly injured.'

The accident allegedly occurred because the assistant stationmaster at Galatz (who appears to have been acting as a dispatcher, as per American practice) had decided to alter the crossing of the two trains from Etulia to Frecăţei. Notice of this alteration was either not received, not understood, or forgotten by the Frecăţei stationmaster, who forwarded one of the trains to cross the other at Etulia as per normal. Meanwhile the other train had been despatched from Etulia towards Frecăţei in accordance with the altered crossing arrangements.

An investigation into the disaster was ordered and the Minister of Communications, the Minister of Health and the Autonomous Railway Administration ordered the dismissal of the Galatz assistant stationmaster, the Frecăţei and Etulia stationmasters and an inspector. [674] On learning of the disaster King Carol II personally ordered the arrest of any of the railwaymen who might have been responsible. [675]

---

674 *The Times*, London, England, 27th and 28th December 1938

675 *The Advertiser*, Adelaide, South Australia, 29th December 1938

Later reports spoke of the hospitals at Braila and Galatz being full of injured and of 150 dead and 300 injured. [676]

At about the same time four passengers died in a collision near Blaj, Transylvania, Romania. [677]

A breakdown in telegraphic communications between two stations on the East Indian Railway led to a head-on collision between Mohammadganj and Garhwa Road, India, at 12:40 on Wednesday 25th January 1939. An Up engine and rest van, which had been detached from a material train, collided with the detached engine of a Down goods train. Both locomotives were badly damaged and the rest van was derailed and caught fire. Seven railwaymen were killed outright and three were injured, one of whom died later in hospital. The line was blocked until the afternoon of the following day. Because of the lack of communication both trains were running on special permission to proceed without 'Line Clear'. It would have been thought that the regulations would stipulate that in such a situation trains must proceed with great caution but if so, sufficient caution was clearly not exercised. [678]

Early on Friday 24th February 1939, two passenger trains collided head-on near Vanegas, Mexico, not far from San Luis Potosi and on the main line between Laredo, Texas, and Mexico City. The two locomotives were telescoped and the two engineers, their firemen, a baggage assistant, and a labourer lost their lives, while about 40 people were injured, probably mostly Mexican passengers in the second-class coaches. [679]

Mr Joseph Mullen, of Richmond, Virginia, United States, Mexico manager for Warner Brothers, who finally reached Mexico City 12 hours late in a special train, said, "The trains were travelling at about 50 miles an hour. The two engines were destroyed, and some lighter second-class cars were crushed, but the Pullmans remained on the tracks. No Americans were injured. We were told the engineer of the northbound train failed to follow orders to take a siding to let the southbound train pass." [680]

Due to the censorship of such information which began in the Stalin era and continued until the Glasnost reforms of President Gorbachev, details of railway accidents in the USSR between about 1937 and 1985 are almost unobtainable, but in 1939 a Swiss newspaper reproduced a report from the Polish *Express Poranny* newspaper regarding a disaster in the Ukrainian Soviet Socialist Republic 12 kilometres from Vinnycja (Winnica in Polish) on about 19th April 1939. A passenger train and a freight train collided, probably in a head-on single-line collision. Two locomotives and 20 vehicles were destroyed, and 29 people were killed and 90 injured. Most of the victims were soldiers. [681]

---

676 *The Northern Herald*, Cairns, Queensland, Australia, 31st December 1938
677 *Idaho Times*, Twin Falls, Idaho, United States, 26th December 1938
678 *The Railway Gazette*, London, England, 24th February 1939
679 *San Antonio Express*, San Antonio, Texas, United States, 25th February 1939
680 *Oakland Tribune*, Oakland, California, United States, 25th February 1939
681 *Journal de Geneve*, Geneva, Switzerland, 21st April 1939

On Tuesday 31st October 1939, a military train collided with two assisting engines between Berent and Sonnenwalde, Germany. Due to an oversight the military train was not held until the engines had cleared the single-line section. Five people died and 41 were injured. The line on which the accident occurred had formerly been part of the Preußischen Staatsbahn (Prussian State Railway) Berent – Bütow – Rügenwalde secondary line. Following the creation of the Polish Corridor after the First World War the line between Lippusch and Bütow had been closed at the end of June 1920 as it was then in Polish territory and about 7 kilometres of track between Lippusch and Sonnenwalde had been removed. With the outbreak of the Second World War and the German occupation of Poland the line had been restored, for use by military and freight traffic. [682] [683]

Two passenger trains collided on the single-track line between Heydebreck and Bauerwitz, between Rosengrund and Langlieben stations, Germany, on Sunday 12th November 1939. Train P957 bound for Heydebreck was a regular service while train P950 was a service which only ran as required. The regular train was running slightly late. The accident was the fault of the *fahrdienstleiter* at Rosengrund, who it appears had forgotten about train P950, which should have crossed train P957 at Rosengrund, and allowed the latter train to proceed. The trains collided while they were travelling at full speed at 19:25, and their coaches were partially telescoped. Forty-three or 48 people died, according to different sources, and 77 were injured.

The official responsible was sentenced to four years' imprisonment as well as losing his job. The senior officials of the company appear to have regarded this as a harsh punishment, which it perhaps was, though at one time in the USSR and in Francoist Spain railwaymen who had made similar mistakes faced death sentences, which in the USSR at any rate were also carried out.

There was also a tragic aftermath to the accident. The driver of train P950 was not hurt in the accident but the trauma of it proved too much for him. Six months later, overcome by grief for an accident which had not been his fault, he committed suicide by stepping in front of a train at Bauerwitz. [684]

In terms of safety, Friday 22nd December 1939 was the worst day in German railway history. On that day there was the catastrophic rear collision at Genthin, the worst accident at any time in Germany, and in addition there was a disastrous single-line collision in the south of the country, between the stations of Markdorf and Kluftern. The trains involved in this second accident were regular through coal train Dg 7953, bound for Lindau, and a special passenger train, variously designated as Sp 21 154 and P Kar 21 154. The special train from Oberstdorf to Müllheim was what pre-war had been a regular passenger train, but now only ran as required, and infrequently. On the day of the accident it was carrying 700 people from Weil am Rhein and the surrounding area, who

---

[682] *Katastrophen der deutschen Bahnen Teil II* by H.J. Ritzau, pg. 96 (in German)
[683] Wikipedia – Bahnstrecke Lipusz – Korzybie
[684] *Von Siegeldorf nach Aitrang* by Hans Joachim Ritzau, pg. 71 (in German)

had been evacuated at the outbreak of the war for fear of possible French air raids, and were now returning home as there had been no fighting in their area. In its pre-war schedule it crossed the freight train at Markdorf. It consisted of a class 57 steam engine, 11 carriages and four luggage vans.

The course of events which led up to the accident was clearly documented in a subsequent court case:

At 22:06 the freight train was at Grasbeuren, west of Markdorf, and was offered to the latter station by Bermatingen-Ahausen and accepted by the *fahrdienstleiter* (signalman), who was busy typing, but he neglected to offer it to Kluftern. The special train at that moment was at Eriskirch, east of Kluftern.

At 22:07 the Markdorf *fahrdienstleiter* cleared the *Einfahrsignal* and *Forsignal* (home and distant signals) for the freight train and instructed the *wärter* (pointsman) in signal box 2 to clear the *Ausfahrsignal* (starting signal) for the same train, having completely forgotten the special train, which was booked to cross the freight train at his station. The running of the special had been advised to the station by telegram the previous evening but the message had not been written down.

At 22:08 the pointsman cleared the *Ausfahrsignal* and cycled off to lower the barrier of a level crossing, 500 metres away. The freight train had advanced to Mimmenhausen and the special train was approaching Friedrichshafen.

At 22:12 the special train was between Friedrichshafen and Windhag. It was offered to and accepted by Kluftern by Fischbach but Fahrdienstleiter Huber at Kluftern did not offer it forward to Markdorf. He did, however, clear his signals for it, relying on the fact that it was timetabled to cross the freight train at Markdorf.

At 22:14 Kluftern belatedly offered the special train to Markdorf, but no one there answered the telephone. The freight train was at Bermatingen and the special passenger at Manzell.

At 22:15 the Kluftern *fahrdienstleiter* rang the bell at Markdorf to contact them on the telephone but found that the line was engaged. Hearing the bell the Markdorf *fahrdienstleiter* suddenly remembered the special train and tried to speak to Kluftern on the telephone, but the line was engaged because Kluftern was at the same time trying to ring him. The freight train was halfway between Bermatingen and Markdorf, the passenger special was at Fischbach, about two kilometres short of Kluftern.

At 22:16 the Markdorf *fahrdienstleiter* looked for his hand lamp to warn the driver of the freight train but could not find it.

At 22:17 the freight train passed through Markdorf. The *fahrdienstleiter* vainly tried to wave for it to stop but his gestures went unnoticed. According to the regulations the train should not have left Markdorf, even if the signals were clear, until the conductor, who should have been in the rear brake van, had exchanged hand signals with the *fahrdienstleiter*, but because it was cold the conductor, with the knowledge of the driver, had been warming himself in the front van next to the engine. The Markdorf *fahrdienstleiter* then tried to contact the pointsman at the Kluftern end of the station but the latter was away at the level crossing. At the same moment the special train was passing through Kluftern.

At about 22:17¼ the Markdorf *fahrdienstleiter* managed to get hold of his colleague at

Kluftern and told him that the freight train was on its way towards Kluftern. The Kluftern *fahrdienstleiter* replied that the special had passed him on its way to Markdorf. The collision was now unavoidable.

At 22:19 the trains met at km 43.190, 300 metres from the railway bridge at Lipbach, in an explosive collision. As well as the fact that it was dark and somewhat foggy, the engine headlights were, because of wartime blackout restrictions, only showing a small amount of light through narrow slits, so although the trains were on a two and a half kilometre straight stretch of line the drivers did not become aware of one another until the train were very close, and the driver of the freight train possibly not at all. The brakes of the passenger special were applied at the last moment. The first two carriages of the passenger train were destroyed, and others were badly damaged, while in the coal train the leading van and 15 waggons were piled up in a heap. Locals who heard the collision thought that they were experiencing the first bombing raid of the war in their area.

Ninety-eight passengers lost their lives in the collision, together with both drivers and one of the firemen. Thirty-eight people, including the surviving fireman, were injured, some seriously. The fireman was very badly hurt, and spent months in hospital, often in great pain according to a nurse who cared for him.

Locals ran to the scene on becoming aware of what had happened, but rescue work was at first difficult in the darkness and without the necessary tools. Later cars were arranged in a semicircle with their headlights on and an air defence unit brought floodlights and cutting equipment. In the face of the disaster the normal blackout regulations were ignored.

Both of the signalmen were members of the S.A. (the Sturmabteilung, the paramilitary division of the Nazi Party). They stood trial in July 1940 at Constance district court for negligence in causing the disaster. They were found guilty but, perhaps because others, such as the crew of the coal train, were also clearly at fault, they were given fairly lenient sentences; Huber was sentenced to a year's imprisonment and his unnamed colleague at Markdorf, who was judged to have been more at fault, was sentenced to three years' imprisonment. In contrast a railway policeman who had been found guilty of stealing two parcels was gaoled for ten years. [685] [686] [687]

On Sunday 25th February 1940, a Mexico City – Juarez express collided with a freight train [688] near Las Adjuntas, Mexico, on the Ferrocarril Nacional de México. Attempts made by railwaymen to flag the passenger train and prevent the accident were apparently frustrated by the darkness. According to different accounts nine, 13 or 15 people died in the collision, and about 60 were injured. The dead included both of the engineers and one of the firemen. Most or all of the passenger victims had been riding in the second-class cars near the front of the train, the leading car being telescoped. Some of them were members of the train's military escort. Riding in the Pullman car were seven Americans, none of whom was hurt. Three of them were consuls Lee Blohm, stationed at Chihuahua; Julian Pinkerton, stationed at Torlean; and Lewis Boyle, stationed at Agua Prieta. The

[685] http://www.geschichtsverein.de/pdf/AHK4-100-Jahre-Eisenbahn.pdf
[686] *Katastrophen der deutschen Bahnen Teil II* by H.J. Ritzau, pg. 101 (in German)
[687] Wikipedia – Eisenbahnunfall bei Markdorf
[688] *The Daily Illini*, Urbana-Champaign, Illinois, United States, 27th February 1940

others were Pat Hurley, of the American Smelting Company and three men named Zacharias, of Bay City, Michigan. [689]

Mr Boyle, who with the other consuls had been returning from a conference in Mexico City, told a reporter, "There was only a slight shock, and an abrupt stop. We in the Pullman thought the engine had struck a cow. It was the two second-class cars, near the engine, in which the military escort was riding, that caught the impact of the crash. They burned after the smash. All survivors were removed before the flames spread." [690]

Preliminary investigations indicated a mistake by a dispatcher, who had issued a lap order, giving both the opposing trains the right to be on the same stretch of single track at the same time. [691]

Having suffered a major disaster at Iitala on 5th March 1940, Finland suffered another just a week later, on Tuesday 12th March 1940, when a troop train and a freight train collided head-on on the single line between Harviala and Turenki. Thirty-four passengers were killed outright, five were fatally injured and 61 were more or less badly hurt, making it the worst accident in Finnish history up to the present (December 2021).

Troop train No. 4136 involved consisted of an engine, a third-class carriage, possibly two other passenger carriages and 40 freight waggons (details of train's composition unclear). It had left Loimaa at 23:00 the previous evening with men of the 71st Freight Company bound for Vyborg. The first-class carriage and horse boxes had been at the rear of the train when it left Loimaa but at Toijala the train had to reverse direction and from there the third-class carriage ended up immediately behind the locomotive.

Freight train No. 53 from Riihimäki to Tampere consisted of 53 waggons and a conductor's van and had started its journey at 04:32.

The troop train did not leave Toijala until 03:45, almost an hour and a half late, and passed Hämeenlinna at 05:06. The next station to Hämeenlinna was Harviala, where on duty that night were Trainee Officer Paavo Raikas and Officer Otto Stålberg. There were only a few trains to deal with and the two men spent the time in conversation, then later Raikas rested on a couch next to the telephone, relying on the ringing of the latter to awaken him when necessary, a procedure with was apparently permitted. It was Raikas who answered the telephone call from Hämeenlinna and accepted the troop train from there, telling Stålberg what he had done. Then at 05:22 there was a telephone call from Turenki, the next station along the line, and Raikas accepted freight train No. 53 from there, at the same time informing Turenki that the freight train would cross the troop train at Harviala. Trainee Raikas then updated Stålberg, who probably had been out of the office when Raikas had accepted the freight train, as to what has happening. "Another troop train?"

Stålberg had replied, "Where is that one coming from? We have only just had one pass through."

At this it had occurred to Raikas that the troop train, which he had assumed was still on its way from Hämeenlinna to Harviala, must have passed through his station while he

---

[689] *El Paso Herald-Post*, El Paso, Texas, United States, 26th February 1940
[690] Ibid, 28th February 1940
[691] *The Corpus Christi Times*, Corpus Christi, Texas, United States, 27th February 1940

was half asleep. Two telephone calls to Turenki confirmed that they had not given permission for any troop train to approach them. Stålberg said that the troop train had arrived from Hämeenlinna at 05:19, when he had asked Raikas if it could proceed, and Raikas had mumbled something to the effect of, "Let her go." With the freight train on its way from Turenki it was obvious that a collision was now inevitable.

The trains collided at 05:28 close to the 95 kilometre post. Due to the war both trains were running with hardly any lights showing but the drivers of each said that they had noticed the other train at the last moment and both had applied the brakes. It was still dark, of course, and the weather conditions were not the best. The troop train had been travelling at 40 kilometres and hour and the freight train at a rather lower speed. Fire broke out following the collision and although the blaze was extinguished before it had spread too far, there were reports of some of the ammunition exploding. In addition to the human casualties several horses being carried by the troop train were killed and other animals had to be shot.

As usual in such situations there were complaints that it was a long time before help arrived at the scene. Eventually a rescue train arrived from Hämeenlinna. The injured were first taken to the small hospital at Turenki by horse and cart and then, when that was full, to Hämeenlinna. The uninjured survivors and horses were first accomodated locally and then returned to Loimma. Ironically, the day after the disaster the Winter War ended anyway, with the signing of the Moscow Peace Treaty, though Finland and Russia would be at war again before very long.

Those who experienced the accident thought at first that it was a Russian air attack and later there were persistent rumours that it had resulted from an act of sabotage.

Raikas duly appeared in court after the disaster charged with having neglected his duties and thereby caused it and its attendant deaths, injuries and damage. He made no attempt to deny his responsibility for it, admitting having accepted the troop train from Hämeenlinna and later the freight train from Turenki, and that he must also, in his half-asleep state, have let the troop train proceed towards Turenki before the freight train had arrived at Harviala. He was sentenced to two years, one month and 10 days' imprisonment, in addition to which he must pay 30,000 markka compensation to the victims, plus any further compensation which might be demanded. The verdict was final and could not be appealed. It is possible that Raikas had been on duty continuously for 36 hours at the time of the disaster.

No news of the disaster was published at the time due to wartime censorship. [692]

Germany had invaded Norway at the end of the first week of April 1940. British and French forces intervened to support the Norwegians, but the aims of the Allied troops were confused and poorly planned, while Norway itself had no standing army to counter the German forces. The fighting in southern Norway ceased on 3rd May and the Helgeland Battalion of infantry regiment 14 was ordered by General Fleischer, the supreme commander in Norway, to proceed northwards from Nord Trøndelag to Mosjøen. The battalion consisted not of regular soldiers but of farmers, teachers, shopkeepers, miners, fishermen, and so on who had been called up to defend their

[692] http://www.vakkamedia.fi/harviala.htm#Laitilan_Sanomat_12.3.1990

country. Early on the morning of Friday 3rd May 1940, in response to this order, the part-time soldiers set off from Tangen station, riding in open flat-waggons which were being propelled by a locomotive. Their train had only gone a short way when it was run into by an empty coaching stock train coming southwards from Valøy, at about 05:15. Seven soldiers were killed outright, one was fatally injured and about 40 others were hurt.

How the accident had happened was never explained. It was not clear why the empty train had been sent southwards from Valøy along the single line when the stationmaster there should have known that the troop train was on its way from Tangen. There was much speculation about sabotage, not helped by the distrust that the troops had in their battalion commander, Major Sundlo, whose brother, Colonel Konrad Sundlo, was a member of the Nasjonal Samling (the pro-Nazi party headed by Vidkun Quisling) and who had surrendered Narvik to the Germans on 9th April without a shot being fired.

The uninjured members of the battalion continued their journey northwards and took part in further fighting until the final surrender of Norwegian forces on 10th June. [693] [694]

A serious single-line collision occurred in the Netherlands on Monday 29th July 1940. Train No. 2581 from Haps was timetabled to cross train No. 2592 from Gennep at Beugen. Instead it ran through there without stopping and collided with the opposing train between Beugen and Oeffelt. Six passengers lost their lives, and 50 passengers and two railwaymen were injured. [695]

The United States experienced a disastrous collision involving a gasoline railcar and a freight train on Wednesday 31st July 1940. Forty-two passengers, the whole complement of the train, and one railwayman were killed, and five railwaymen were injured in the accident on the Pennsylvania Railroad near Cuyahoga Falls, Ohio.

The freight train involved in the accident was Extra No. 4454 North, symbol FC-4. It consisted of engines Nos. 4454 and 3353, 30 loaded and 43 empty cars and a caboose, in charge of Conductor Collier with Engineers Orlan Lodge and Kropf and Firemen B.E. Reynolds and Huntsberger. At Arlington, the last open office 3¾ miles south of Cuyahoga Falls, they received train order No. 99, Form 18, which read:

Eng. 4454 run extra Arlington to Hudson and meet No. 3380 Gas Eng. 4638 at Switch 1, Silver Lake.

Silver Lake was a siding a mile and a quarter long, the south end of which was just over a mile north of Cuyahoga Falls, Switch 1 was the north end of Silver Lake siding; its south end was designated Switch 3 and there was an intermediate connection between the siding and the main line, designated Switch 2. The order thus permitted the freight train to proceed as far as the north end of Silver Lake siding to meet the railcar. The freight train left Arlington at 5.49 p.m. and moving at 30 miles an hour was approaching Silver Lake when it collided with the railcar.

---

[693] https://www.ytringen.no/ytring/helgelandsbataljonen-ir-14-og-deres-innsats-under-2-verdenskrig-har-igjen-opplevd-a-ikke-fatt-noen-plass-i-jubileumsfeiring-hvorfor/reptem!TL8H5ypuQPktQteyM3BFJw/

[694] http://v1.steinkjerkunnskapsportal.no/egge/index.php?art_id=450

[695] *Spoorwegongevallen in Nederland* by R.T. Jongerius, pg. 173 (in Dutch)

Southbound first-class passenger train No. 3380 consisted of gasoline-electric railcar No. 4638, in charge of Conductor Henry Shafter with Motorman Thomas Murtough. On the southbound trip it was being operated in reverse from the leading vestibule. At Hudson, just under 8 miles north of Cuyahoga Falls, it received a copy of train order No. 99. It left Hudson at 5.49 p.m., 4 minutes late, passed Switch 1 at Silver Lake, where it should have taken the siding, and travelling at about 25 miles an hour collided with the freight train. As a result of the collision the railcar was pushed back 537 feet, its front end was telescoped about 12 feet, its fuel tank was punctured and the gasoline caught fire, enveloping it in fire. Conductor Shafter, Motorman Murtough and deadheading Track Labourer Tod Wonn jumped immediately before or at the moment of the collision; they were the only survivors from the railcar. [696]

A motorist using a level crossing at Cuyahoga Falls described his experiences of the disaster. 'I was returning from a business trip going towards Akron when I noticed the flasher signals going and stopped about 10 feet from the track. The doodlebug was coming right along at what I judge to be about 50 to 60 miles an hour. No whistles were blowing, nor do I recall a screech of brakes as if an attempt to stop might have been made.

'I didn't see the freight train at all. The doodlebug whizzed by and as it was a single track I naturally put my '35 sedan in gear and started to cross immediately. I don't know what made me do it, but I jammed my foot down on the gas and the car jumped forward a little and then stalled. It was right next to the track.

'Then in a split second I thought I was dreaming and in another world. There was a terrific crash about a hundred feet to my left and the next thing I knew the coach was coming right back past me. Just as it was in front of me a rain of flame flew all around me and the coach was white with fire. The flames covered an area {of} about 75 feet on each side of the track. My car was enveloped.

'A terrific explosion sounded simultaneously and then the double header freight with its cars – the first 10 or 12 of 'em on fire – ground to a stop about 200 feet on my right. The doodlebug appeared like a crushed matchbox, and all this seemed like ages, but I now realize it didn't take over three seconds.

'Then in front of me I saw a gasoline tank car – part of the freight train – and fire was all around it. I jumped out of my car and by this time a few people came screaming out of nearby stores and houses. There were people lying on the ground but suddenly I felt that terrible heat and remembered the tank car. I was afraid that it was going to explode so I just left the car where it was and went over the hill, so to speak. But that "hill" was a ravine ten feet deep and there were rocks at the bottom. The next thing I remember was a lady laying a cloth around a gash in my left leg which was bleeding badly. Someone took me to St Thomas Hospital where I was treated and X-rayed.

'I've spent 15 months in the World War in the battle of St Mihiel and the Argonne and today I saw the worst five minutes of damage and death I've ever seen.' [697]

Motorman Murtough told the Interstate Commerce Commission (ICC) inquiry that his daily assignment had been operating several round trips over the 13 miles between Akron

[696] Interstate Commerce Commission report
[697] *The Sandusky Star-Journal*, Sandusky, Ohio, United States, 1st August 1940

and Hudson. Conductor Shafter had given him the train order at Hudson and told him, "We meet at switch 1, Silver Lake." The order had been legible and correctly written, and he had understood it. About three and a half miles north of Silver Lake the thought had occurred to him that the meet with train Extra 4454 would delay his train so that it would not be able to make Akron in time to connect without delay with the Baltimore and Ohio Railroad train there. He could remember nothing more until he had found himself passing Switch 3 and whistling for the level crossings at Payne Avenue and Front Street. At the same moment he had seen the freight train coming around the corner. He had then remembered that he should have taken the siding at Switch 1. Applying the brakes in emergency he had jumped when the trains were close together. He could not explain why he had forgotten the meet, though he mentioned problems with exhaust fumes on previous trips and the day had been hot and humid.

Conductor Shafter stated that it was customary for him to take both copies of the train order from the operator and give one copy to the motorman. He understood order No. 99 and believed that his motorman had. He could not recall if he had shown the order to Baggageman Charles Bilderbeck (who was killed). He had then been collecting tickets and carrying out clerical duties when he had realised that his train was passing Switch 3. He thought that he had operated the conductor's valve, then realising that a collision was imminent he had jumped.

Track Labourer Wonn said that he had been in the baggage compartment of the railcar. He had known nothing of train order No. 99. He had heard the railcar whistle for the level crossings, then the brakes had applied in emergency and seeing the conductor jump from the train he had followed suit. [698]

According to a newspaper report Track Labourer Wonn had told a reporter that the conductor had rushed into the baggage room, crying, "We're going to crash." Wonn continued, "The conductor leaped, and I followed him. My clothing caught fire. I rolled in some bushes at the side of the track and put out the flames. My pal, Bruce Kell, who lives at the same place I do, was with me in the car and I did not see any more of him." [699]

Engineer Lodge testified that a satisfactory brake test had been carried out at Columbus, 134¼ miles south of Cuyahoga Falls. His train had stopped normally at Arlington, where train order No. 99 had been received. Approaching Silver Lake his fireman had called out that the railcar was south of Switch 3, and he had applied the brakes in emergency. The collision had then occurred.

Fireman Reynolds spoke of seeing the railcar and shouting a warning to his engineer.

Engineer Kropf stated that he had received a copy of train order 99 and understood it. Because of the curvature of the line he had had no notice of the railcar before the brakes had been applied in emergency.

Conductor Collier of train No. 4454 stated that the brakes of his train had been tested at five points en route where cars had been added or removed. He had received and understood train order No. 99.

Dispatcher Weisz stated that he had issued train order No. 99. In order to prevent delay he had made the freight train superior as the single railcar could enter and traverse

[698] Interstate Commerce Commission report
[699] *The Logan Daily News*, Logan, Ohio, United States, 1st August 1940

the siding easier than the long freight train.

Operator Rickey at Hudson said that he had delivered both copies of train order No. 99 to Conductor Schafter. He understood Rule 211, which directed that the operator must deliver copies of the order personally to each person it was addressed to, but he had considered the conductor capable of delivering the order properly. The block signal at Hudson only gave the train block authority to Silver Lake. He had reported the departure of the train to the Arlington operator, who had reported the departure of the freight train.

Operator Whitt at Arlington recalled giving train order No. 99 to the engineers and conductor of train No. 4454. Extra No. 4454 had block authority only to Silver Lake, which was an unattended block station under the control of the Hudson operator. He said that when it was necessary for a train to obtain authority at Silver Lake to enter another block section the block records must be checked by the Arlington and Hudson operators and if the block condition was favourable the Hudson operator gave block authority. When a train stopped at an unattended block station to obtain authority and it was necessary for that train to report that it was clear of the block section in the rear, the engineer was instructed to pull up so that the conductor could report the train in clear. He pointed out that should an emergency arise at an attended station the operator could take action; this was not possible at an unattended block station.

Discussing the accident the ICC noted that since Motorman Murtough had begun operating the short turn-around service in April he had suffered from headaches and loss of appetite and lost about 15 pounds in weight. It was apparent that he had not been in full possession of his mental facilities at the time of the accident, and it was possible that he had been suffering from carbon monoxide poisoning. Research showed that carbon monoxide poisoning was most severe during the first hour of exposure, in hot and humid weather and in obese people – the motorman had been on duty an hour and a half, the weather was hot and humid, and he was 5ft 8in tall and weighed 195 pounds, giving him a body mass index of 29.7, thus on the borderline of obesity. He was also sitting in a small compartment with only a small opening for ventilation.

The conductor was equally responsible for the safety of the train but had occupied himself wholly with collecting tickets and clerical duties until it was too late to prevent the collision.

The fact that at the time of the accident the railcar had been carrying about 150 gallons of highly inflammable gasoline resulted in an increased hazard to the passengers and had contributed to the disastrous consequences of the accident.

Had Silver Lake been an attended station with a block signal the conductor might have observed the signal in time to avert the accident, or the operator there might have been able to stop the train.

The ICC added that on 20th August the Pennsylvania Company had applied for approval for a system of Centralised Traffic Control and Automatic Block Signal system between Hudson and Arlington in lieu of the manual block system. The Commission had approved the application on 12th September.

The accident had resulted, the ICC concluded, from failure to obey a meet order and occupation of the block without authority. [700]

Two passenger trains collided head-on on a stretch of single track at Gubbero, not far from Göteborg Central station, Sweden, just before 12:00 on Tuesday 17th September 1940. The trains which featured in the accident were train No. 507, the 11:46 departure from Göteborg Central to Borås and train No. 1456, the 11:46 train from Kungsback to Göteborg, which met on the Gullbergsbro viaduct. According to the timetable train No. 1456 should have waited at the end of the double track at Gubbero for train No. 507, which on the day in question was running six minutes late. Instead train No. 1456 entered the single line before train No. 507 had arrived and the trains had collided at a combined speed of about 40 kilometres an hour. [701] Both drivers escaped almost unhurt, Driver John Ekelin of train No. 507 jumping for his life after having shut off power and jammed on the brakes. The passengers were not all so lucky. There was considerable telescoping between the leading van and the following coach of train No. 1456, and nine passengers died and 29 were injured. The line was blocked until 21:10 the next day.

The driver of train No. 1456 alleged that he had had a clear signal to proceed despite the non-arrival of the opposing train. This was not accepted and on 29th April 1941 he was found guilty by a local court of causing death and serious injury through negligence and sentenced to two months' imprisonment and ordered to pay Statens Järnvägar (the state railway company) SEK 219,085 plus their legal costs. However, on appeal the conviction was quashed thanks to the efforts of a defence lawyer named Dellborg who argued that the possibility of a 'wrong side' signal failure could not be excluded. This was despite the fact that Signalling Inspector Ture Hård categorically denied that any such 'wrong side' signal failure could have occurred. [702]

On Saturday 16th November 1940 train B Dg 6916, a run-as-required freight train, was timetabled to cross early morning passenger railcar Kp 441 at Schönfließ station. Despite this and the *Ausfahrsignal* at danger the driver of the freight train entered the single line and collided with the railcar between Schönfließ and Carzig, Germany. Nine people lost their lives and 21 were injured in the accident on the line between Werbig and Frankfurt Oder. [703]

Early on the morning of Tuesday 19th November 1940 passenger train No. 2453 from Trondheim to Oslo was due to cross train No. 2456 from Kopperå to Trondheim at Hommelvik, Norway. Train No. 2453 consisted of class 31b steam locomotive No. 419 and ten assorted vehicles. Train No. 2456 consisted of a diesel railmotor and trailer. Norway was under occupation by the Germans at this date, of course, and they had given orders that it was especially important that train No. 2453 not be delayed, as it carried many workmen from Trondheim who were employed building a new airfield at Værnes, needed for the coming invasion of the Soviet Union. Upon its arrival at Hommelvik

[700] Interstate Commerce Commission report
[701] Wikipedia – Gubberoolyckan
[702] https://www.postvagnen.com/forum/index.php?mode=thread&id=625399
[703] *Eisenbahnunfälle in Europa* by Erich Preuß, pg. 121 (in German)

therefore Telegraphist Øynstein Nordtømme, who was in charge of the station at that hour, had given *Overkonduktør* (Head Conductor) Jacob Breivik an order reading:

Clear for train 2453 upon the arrival of train 2456.

It being wartime, blackout conditions were in force. There were no exterior lights in the station area and all the windows of its buildings were blacked out, as well as the train windows. Such lights as the train could display were confined to a 7 x 1 cm horizontal slit. At the time concerned (about 07:43) it was also pitch dark. Conductor Breivik looked across to the opposite side of the line on track 3, imagined that he saw the railmotor standing there and gave his driver the signal to depart. Telegraphist Nordtømme thought at first that the train was just moving forward a little, then realised that it was departing. His first thought was to jump on board and apply the brakes, then he changed his mind and decided that it would be better to contact Pointsman Rolf Bye at the Oslo end of the station. The pointsman answered promptly but said that the train had just passed him. Nordtømme next tried telephoning Nygården but could get no answer. Finally he rang Muruvik, to be told again that the train had gone by. He could do no more to avoid the accident and telephoned Trondheim to tell them what he expected to happen at any moment.

The trains met at Nygården at 08:03. The big steam locomotive was more than a match for the lighter railmotor and crushed in the front of it as well as pushing it back 70 metres. The front of the railmotor's trailer was also crushed. Motorman Einar Frigård was killed outright while Driver Oskar Solberg was knocked unconscious. He said later that he had seen the railmotor about 250 metres away and had just had time to put down sand and apply the brakes.

In all, 22 people, including the motorman, were killed, or died later of their injuries and another 45 were injured. Three doctors came from Trondheim by motor car and the most seriously injured were taken directly to Trondheim Hospital by ambulance, others reaching Trondheim on a railmotor from there.

The 'conditional departure order' issued by the telegraphist was an irregular procedure, not provided for in the regulations, but an inquiry sent out by the head of the district to various stationmasters after the accident revealed that a large number of them, faced with the irregular and relatively heavy wartime traffic, admitted to having issued such orders.

An inquiry into the accident was opened the next day. While it was in progress Head Conductor Breivik. Driver (Passed Fireman) Solberg, Passed Cleaner Gunnar Wardehaug, Brakesman Roald Bye, Telegraphist Nordtømme and Pointsman (Numbertaker) Rolf Bye were all suspended. All were re-employed after some months, but Breivik was reduced to Foreman Shunter and Nordtømme was given a 'final warning'. A Norwegian writer considers that Head Conductor Breivik was reduced in rank because the Germans were looking for a scapegoat and that the real cause of the accident was the abnormal blackout conditions. Breivik was reinstated as a conductor after the war. [704]

Spain suffered a major disaster at about 04:30 on the morning of Tuesday 3rd

---

704 *På Sporet* (magazine of Norsk Jernbaneklubb), No. 35, pgs. 32-35 (in Norwegian)

December 1940, involving two overnight express passenger trains; train No. 802 from Barcelona, which had left there at 21:00 the previous evening and was due in Madrid at 11:15, and its opposite number, train No. 803, which had left Madrid at 19:00 the day before and was due in Barcelona at 09:00. The accident happened at the entrance to the small station at Velilla de Ebdo, Spain. Train No. 802, from Barcelona, was running half an hour late. It should have stopped at Velilla to cross train No. 803 but ran through the station and was still travelling at 90 kilometres an hour when it collided head-on with the one from Madrid, which was stopped at the entrance to the station. As a result of the collision a third-class coach was thrown down a two-metre embankment and a sleeping car and a van were telescoped into one another. The drivers and firemen of the two trains escaped death. An official at the station signalled to the driver of the train from Madrid train with his handlamp and because that train was only coming slowly into the station it was able to pull up and even try to set back when it was struck by the one from Barcelona.

Manuel Corredor Castellón was the driver of the Barcelona train and his fireman was Jaime Massaguet Vidal. The driver of the Madrid train was Amadeo Saiz Avances; the name of his fireman does not seem to have been recorded. José Valmaya, described as a 'chief of machinery' and perhaps a locomotive foreman, was also on the footplate of the train from Barcelona. The usual crossing place of the two trains was at Puebla de Híjar, almost 20 kilometres closer to Madrid, but this had been changed due to the lateness of train No. 802. A probable cause of the accident would seem to be that the driver of the Barcelona train, perhaps distracted by the presence of the third person in the cab, had forgotten about the altered crossing and was hurrying to get to Puebla de Híjar to meet the other train.

The rescue operation was hampered by the extreme cold (-10°C) and witnesses believed that some of the victims had died from the cold before they could be rescued. Fires were lit by the lineside to provide some heat and light. By the day after the accident it was known that there were 47 dead, many of them soldiers, with about 60 more injured. The final number of fatalities appears to have been 53.

An investigation into the disaster was quickly opened and the drivers and firemen of the two trains were interviewed. [705] [706]

On the afternoon of Friday 27th December 1940 an express train from Keeling collided head-on, possibly on the single line, with a goods train from Takao near the station at Kyujo, Formosa, a few miles north of Takao. The engine and leading three third-class carriages of the express were overturned and there were over 200 dead and injured. A signalling error was blamed for the accident. [707] [708]

At about 09:00 on the morning of Tuesday 23rd September 1941, direct train No. 307 and passenger train No. 1970 collided head-on in thick fog near Kiesen station, Switzerland. The line was normally double track, but one line was out of use due to

[705] *La Vanguardia*, Barcelona, Spain, 4th, 5th and 6th December 1940
[706] Webpage: Federación Castellano Manchega de Amigos del Ferrocarril (in Spanish)
[707] *The Malaya Tribune*, Singapore, 30th December 1940
[708] *The Railway Gazette*, London, England, 30th May 1941

engineering work and single-line working was in operation.

Omnibus train No. 1970 which had left Berne at 08:10 was probably a little late. It was stopped at a red signal when it was run in by the direct train coming from Thoune at 08:47. The impact was terrific, and the express engine crushed that of the omnibus train into an unrecognisable mass, while the leading van of the express telescoped the second-class carriage following it. According to a statement from the Chemins de fer fédéraux suisses (CFF) the direct train should have been held at Kiesen to cross the omnibus train there but owing to a misunderstanding the signal was cleared for it to proceed and it thence collided with the omnibus train outside of the station.

The Director General of the CFF stated that there were eleven dead, including Conductor Christian Schenk of the express and Driver Walther Gertsch of the omnibus train. [709] Four of the others who died were permanent way staff who by an unlucky chance had been working on the out of use line at the point of the collision. [710]

A passenger on the omnibus train related his story to a reporter: "I was quietly reading my newspaper and I was surprised when, after a long stop at the station, our train stopped again betweem Wichtrach and Kiesen. Through the dense fog I heard the sound of a shrill whistle. Then I felt a shock, followed by several more, increasingly violent ones which threw the passengers one on top of the other. Then the coach stopped with a squealing of wheels and breaking of wood. Well, a collision, I said to myself, but it hasn't been too bad yet, but barely had I taken a few steps outside of the carriage when I realised that there had been a disaster. My first concern and that of my fellow passengers was not to touch the high voltage cables, which had been torn off and were lying on the ground.

"After wandering in a wet meadow, in which we met pale, frightened passengers, some with bloody faces and hands, I stood in front of our locomotive, which had, so to speak, buried itself in a coach of the direct train with which we had collided.

"The spectacle was indescribable. The tangle of wood and metal was unimaginable. But what struck me was the extraordinary speed with which the workmen, the soldiers, the doctors, the ambulances, the other improvised ambulances, consisting of lorries with straw bales, appeared on the scene. And, shortly after the collision, the soldiers and workers worked to free the injured trapped in the wreckage, and, alas, the bodies of the dead. I stayed at the scene for an hour, and by the time the sun had pierced the thick fog the bodies of those who had died had already been laid on biers in view of those of use who had lived through those painful minutes." [711]

In addition to those who lost their lives there were about 20 injured, who were taken to hospitals at Munsingen, Berdiesbach and Thoune.

At the end of April 1943 three railwaymen stood trial accused of being responsible for the collision. The principal accused was a 26-year-old man who had been in charge of the expedition of trains at Kiesen. On the fatal day he had wanted to report to Wichtrach station the crossing that should have taken place at Kiesen between the Thoune-Berne direct train and the omnibus train from Berne. To do so, he had telephoned the Wichtrach stationmaster, but there had been a misunderstanding, from which he had wrongly

[709] *Journal de Geneve*, Geneva, Switzerland, 25th September 1941
[710] *Gazette du Lausanne*, Lausanne, Switzerland, 24th September 1941
[711] *Journal de Geneve*, Geneva, Switzerland, 25th September 1941

understood that the trains were instead to cross at Wichtrach. Because of this misunderstanding the Kiesen employee, after the telephone conversation, had permitted the express train to depart from his station for Wichtrach, and it had thus collided with the omnibus train stopped at the Kiesen home signal at danger.

The question of the arranging of the crossing by telephone was crucial as, according to the regulations, crossings should only have been arranged by telegraph, unless there was an equipment failure. [712]

On the morning of Sunday 28th December 1941, passenger train No. 2407 from Armentieres to Berguette, with Driver Huart and Fireman Hennion, left Laventie station, France, had covered about two kilometres and reached the level crossing at Grand Chemin (also called Roi de Trèfle) when at 07:30 it came into head-on collision on the single line with train No. 2402 from Berguette to Armentieres (Driver Nattier), which had left the next station at La Gorgue some minutes previously. The trains normally crossed at La Gorgue and it was supposed that there had been some mistake in the signalling, or in the orders given. The results of the collision were terrible; the locomotives were crushed into one another, and the following carriages formed a mass of wreckage. Coaches were telescoped. The horror was added to by the fact that it was impossible to at once help the victims as both stations were several kilometres distant. An alert was, however, immediately put out from the nearby level crossing and soon doctors, firemen, French and German ambulances, German medical staff and so on were arriving at the scene, together with both motor cars and a relief train. The rescue work began in the early morning darkness with the aid of electric lights, car headlights and torches. French and German rescuers worked together throughout the day. Various officials, including the Hazebrouck prefect and railway engineers, were also soon on the scene.

The rescue operation went on tirelessly despite the difficulties posed by the location of the accident, which had happened amidst fields and alongside the decantation basin of the Nouveau Monde distillery, at a place almost inaccessible to vehicles. It was only with the aid of firemen's ladders that the rescuers were able to reach the wreckage and extricate the victims.

French and German cars took the injured to hospitals at Béthune, Hazebrouck, Armentieres, La Gorgue, Estairs, Laventie and Lille. The bodies of the dead, who were mostly workers from Armentieres bound for jobs in Lestrem, were taken to a mortuary in La Gorgue. The rescuers and the injured had to endure severe cold. One victim was trapped and had to have a leg amputated to free him, others had suffered traumatic amputations of limbs and hands in the accident itself.

At 14:00 the number of dead was estimated as 45, with 50 injured; by 15:00 two more of the injured had succumbed. Driver Nattier was amongst the dead. Conductor Louis Jonkière of train No. 2407, Driver Huart and Fireman Hennion were among those injured. [713]

By 1st January 1942 the number of dead had risen to 51, with 42 injured. [714] The final

[712] Ibid, 29th April 1943

[713] *Le Réveil du Nord*, Paris, France, 29th December 1941

[714] Ibid, 1st January 1942

death toll was possibly 56.

British Railway Disasters of the World, taking as its source *The Railway Gazette*, refers to a second disaster near Hazebrouck on 30th December, in which 50 people died. It has not proved possible to find a reference to a second accident in any French newspaper and it seems certain that there was in fact only one accident, on 28th December, as detailed above.

Japanese forces had invaded the Dutch East Indies (now Indonesia) on 10th January 1942, and British troops were sent to fight them as part of the American-British-Dutch-Australian Command. In early February officers and men of the 77th (Welsh) Heavy Anti-Aircraft Regiment disembarked at Batavia (now Jakarta) in Java after having spent several weeks at sea. They then had to make a 500-mile journey across the island to Surabaya. A small proportion of the troops made the journey by road but the majority left Batavia on 5th February by a train of the Nederlandsch Indische Spoorweg Maatschappij.

The troop train left Batavia at 6.30 a.m. and continued all day and on through the night until at about 3 a.m. on Friday 6th February it came into head-on collision with a freight train carrying bombs, petrol and perhaps rice near Lawang, about 55 miles short of Surabaya. The trains apparently met on a bridge over a ravine but there is no record of any part of the train falling from the bridge, nor did the collision cause any of the bombs to explode. Nevertheless, there were numerous deaths and injuries. Brief reports of the accident in Australian newspapers spoke of 21 deaths (18 'Europeans' and three 'Indians') and 75 injured. [715] It seems likely that the final number of dead was 26, with about 100 injured.

By dawn a party of Dutch military and some nurses had arrived on the scene and an ambulance train was also sent, probably from Surabaya. [716]

A month later the Dutch forces in Java surrendered to the Japanese. The British were initially ordered to fight on but quickly surrendered as well, and the men who had survived the accident spent the rest of their war in prison camps in Java or Japan or working on the Siam-Burma 'Death Railway'.

The cause of the accident remained a mystery, though there were suggestions that it had been deliberately engineered by a local railwayman sympathetic to the Japanese. [717]

The Caribbean island of Jamaica suffered a major disaster on Saturday 28th February 1942, when a passenger railcar collided with a freight train between May Pen station and Sevens Siding on the Frankfield branch. 'Kalamazoo Trolley Coach No. 22', the diesel railcar, was en route from the terminus at Frankfield to Kingston, the capital, when it collided with the freight train at 6.40 a.m. The railcar, with a capacity of 45 passengers, was almost full, and both trains were moving at some speed as they met on a curve. The freight train, extra No. 18, was moving sugar cane traffic from May Pen to the Sevens sugar estate and was reversing, that is, the engine was propelling its waggons. Seven passengers were killed outright, as was Brakesman A. Christian, riding on a car at the

[715] *The West Australian*, Perth, Western Australia, 9th February 1942

[716] https://www.cofepow.org.uk/armed-forces-stories-list/the-1942-java-railway-tragedy

[717] https://greglewisinfo.com/tag/from-java-to-nagasaki/

rear of the freight. Motorman Harry Stephens driving the railcar saw the danger in time to jump at the last moment and escaped unhurt. Rather more than 30 injured passengers were taken to the Clarendon and Lionel Town hospitals, four of whom had died before the end of the day. There had been no further deaths by Monday morning, but several passengers were gravely ill. Most of the victims came from Clarendon. After the accident the freight train was shunted to Sevens, the railcar brought to May Pen and the line was reopened by noon Saturday.

The railway company attributed the accident to a mistake in the delivering or receiving of orders (trains were worked as in the United States with a dispatcher and orders sent by either telegraph or telephone). Mr J.C. Atkinson, Traffic Superintendent and Acting General Manager, announced that a joint departmental inquiry would be opened at Kingston and attended by the stationmaster at Suttons (a place on the Fairfield branch). [718]

By Tuesday 3rd March the number of dead had risen to 15. [719]

How the accident had happened was fully explained at the inquiry in mid-April. Motorman Stephens told the inquiry that he had left Frankfield at 5.45 a.m. with a written order to proceed to Suttons. After several stops to pick up passengers he had arrived at Suttons at 6.23. There, Stationmaster Cyril Patterson had given him carbon copies of two clearance cards. He had started from Suttons at 6.24, gone a short distance, returned for a woman passenger, then proceeded. At Sevens Estate he had slowed up in case there were any passengers there, but there were not. Rounding a right-hand curve at 25 miles an hour he had seen the rear of the freight train coming towards him, had braked, and jumped. Having recovered himself he had gone back to Sevens Estate, from where he had telephoned Suttons station and asked Stationmaster Patterson how it was that he (Stephens) had had a clearance to May Pen. The stationmaster replied that he had written out the card giving the railcar clearance to May Pen before he had spoken to the controller at Kingston. Kingston had instructed him only to clear the railcar to Sevens but being in a rush he had forgotten to either alter the clearance card or to verbally tell Motorman Stephens that he only had clearance to Sevens. After the railcar had left, he had discovered his mistake and contacted May Pen, but nothing could then be done.

Mr Atkinson gave similar evidence to the inquiry. He said that having interrogated first Motorman Stephens, who had been able to produce a clearance card giving him authority to proceed to May Pen, he had questioned Stationmaster Patterson. The latter had admitted that he had assumed that the railcar would be cleared to May Pen and had written out the clearance card accordingly. When he had been informed by Kingston that it was only to go to Sevens he had been busy selling tickets for the market train which followed the Kalamazoo and had forgotten in the rush to change the clearance card or to deliver the verbal message to stop at Sevens. After the railcar had left, realising his error, he had run out of his office and frantically waved to the departing train, but had not been seen.

Mr Atkinson added that he was of the opinion that Brakesman Christian must have failed to open the emergency valve or else the collision would have been averted or been very minor. He also admitted that the position as regards staff had been acute since the

---

[718] *The Daily Gleaner*, Kingston, Jamaica, 2nd March 1942
[719] Ibid, 3rd March 1942

gasoline restrictions and that during the cane season some stationmasters were working 20-hour days. [720]

Stationmaster Patterson gave evidence at the resumed inquiry and swore on oath that he had not spoken to either Motorman Stephens or Brakesman Joseph Lewis of the railcar that morning, the clearance card had simply been taken through the office window by someone he did not know. He had received a Morse message from dispatcher Howard Mills at Kingston to hold the railcar at Sevens but had been too busy to write it down. He described his hours of work as 'extraordinary', for example on Monday 23rd February he had been on duty from 6 a.m. until 1 a.m. on Tuesday 24th, then on again at 6 a.m. on Tuesday until 12 midnight. He had not, however, complained about the long hours.

Cultivator George White, who had been intending to catch the Kalamazoo at Suttons but had had too much luggage (tobacco) to get on, corroborated the stationmaster's account of the motorman taking the clearance card through the office window.

Schoolmaster Javin Brown, a passenger on the railcar, heard the driver (motorman) speak to someone but he could not be sure who. Passenger Alfred Phillips gave similar evidence. Both he and Brown were injured in the accident.

Train Controller Mills stated that the Line Clear system was foolproof (!). He considered that the freight should have reached Sevens Siding in ample time before the Kalamazoo got there but had sent the message to Suttons as an extra precaution. [721] He was questioned as to why he had not held the freight train at May Pen until the railcar had arrived there or held the railcar at Suttons until he had been advised that the freight was in clear at Sevens but persisted in stating that it should have been perfectly safe to permit both trains to approach Sevens. Had he heard from Suttons that an error had been made he was almost certain he could have got someone at Sevens to stop the railcar there, though no railwayman as such was stationed there. The freight had taken longer than he had anticipated due to the excessive slipping of its engine. He had not considered a meeting or crossing order necessary under the rules in view of the clear directions of the Line Clear order. [722]

It has not been possible to determine the final outcome of the inquiry.

Through freight train Dg 26 969 overran a stop signal and collided, presumably head-on, with semi-fast passenger train E 140 on a single-track temporary bridge between Kehl and Straßburg, Germany, on Monday 18th May 1942. Eleven people died and 79 were injured. [723 724]

The Interstate Commerce Commission does not seem to have concerned itself with an accident near Norristown, Pennsylvania, United States, on the Lehigh Valley Transit on the evening of Wednesday 8th July 1942. A crowded interurban passenger car collided head-on with a freight trolley three and a half miles north of Norristown. The heavy high-

---

[720] Ibid, 16th April 1942
[721] Ibid, 18th April 1942
[722] Ibid, 21st April 1942
[723] *Katastrophen der deutschen Bahnen Teil II* by H.J. Ritzau, pg. 105 (in German)
[724] *Eisenbahnunfälle in Europa* by Erich Preuß, pg. 123 (in German)

floored freight car, running downhill, smashed through almost a third of the passenger car, which was the first section of the *Liberty Bell Limited*, on its way from Norristown to Allentown and Emaus. Ten passengers and Motorman Harry Strunk of the passenger car were killed more or less outright and about 20 passengers and Conductor Harry Custer and Motorman Grover Meckley of the freight car were injured. [725] An 11th passenger, Edne Lysinger, a 23-year-old waitress, died of her injuries on Sunday 12th. [726] The passenger car was filled with factory and office workers returning home from jobs in Norristown.

A local newspaper carried an account of one of the rescuers who had attended the accident. 'Prosper War, wholesale meat dealer of Colwell lane and Mill road and prominent member of the Lions' club, was among the first to render aid at the Allentown Limited disaster.

'Driving east on DeKalb street on business, Mr War reached the site of the catastrophe shortly after it happened. With another motorist he was the first to enter the telescoped car. He remained there two hours, working feverishly throughout that time to extricate the victims wedged in tight against seats and among debris.

'"It was a terrible sight. Men and women were screaming with pain, and members of their family called to the scene were crying out from the roadway to get them out," Mr War said. Trained in first aid, the local business man was able to give expert assistance, in one instance aiding in placing a splint on a man's badly fractured leg.

'Patrolman Harry Shear, also on the scene shortly after the crash, rendered extensive aid to the dying and injured.'

As to the cause of the accident, Motorman Meckley of the freight car told officials afterwards, "I was running under orders. I had stopped at Lansdale, and I was instructed to proceed." Rounding a curve he had seen the passenger car coming towards him. "I threw the lever to put the car in reverse and about that time, boxes and crates threw me to the front. I don't remember getting out of the car but my conductor, Harry, came up and stood alongside me. I felt the other car raise up when we hit and when I looked I saw my car had sliced through to the fourth window of the other one."

'His voice was the dazed voice of a man exhausted by the horror of seeing people die. As he wiped the sweat and blood from his face he said, "I don't think it was my fault."

A reporter asked, "You were late on your run?"

"Yes, but I was still running under my orders," Meckley replied.

Despite it appearing probable early on that Motorman Meckley was not at fault, Coroner Winslow Rushong ordered that he be lodged in Norristown jail overnight. Conductor Custer, who had suffered a head injury, was taken to hospital. [727]

On 14th July it was alleged by Edward Hawes, Lehigh Valley Transit counsel, that Motorman Strunk had ignored a red light and semaphore at Brush Siding and had failed to wait there for the freight car, train C-14. Having failed to wait at Brush he had also failed to deliver an order to James Gerhart, the motorman of a southbound car following

---

725 *Conshohocken Recorder*, Conshohocken, Pennsylvania, United States, 10th July 1942
726 Ibid, 14th July 1942
727 Ibid, 10th July 1942

the freight car, directing Gerhart to exchange trolleys at Brush with the northbound one following Strunk's. These orders had been transmitted by telephone to Motorman Strunk at Rink Call Box, Norristown, by Dispatcher Harvey Weikel at 4.57 p.m. As proof that the order had indeed been sent the carbon copy of it was retrieved from the roll in the call box. Motorman Meckley of the freight train had been proceeding under orders to cross Strunk's car at Brush. Gerhart's trolley had been running seven or eight minutes late due to having had to make numerous stops between Quakertown and Lansdale and Strunk had been notified of this.

The order issued to Strunk instructing him to meet C-14 at Brush had been found in his clothing. Dispatcher Weikel stated that Strunk had subsequently called him from the Philadelphia and Western station in Norristown and had again been told, "Don't forget to wait for C-14 at Brush," though there was no written record of this message. It was also denied that Strunk could have misunderstood that he was to wait for a freight car as passenger cars were designated by numbers only. [728]

In mid-September, the Montgomery County Grand Jury indicted Motorman Meckley, Dispatcher Weikel and William McGorum, vice president and general manager of the Lehigh Valley Company, for voluntary manslaughter. [729] By what logic Motorman Meckley could be guilty of manslaughter when he was simply abiding by the orders he had been given it is hard to see, and the case against the dispatcher and the general manager was hardly much stronger. The case, nevertheless, came to trial in early November, but on Wednesday 4th November Judge George Corson dismissed the charges on the grounds that the evidence clearly showed that the dead Motorman Strunk was to blame for the collision. [730]

Early on the morning of Friday 2nd October 1942 goods train No. 643 from Renens to Bienne (Biel in German) collided head-on with passenger train No. 1508 which had left Bienne at 05:44 for Neuchâtel just on the Bienne side of the station at Daucher (Tüscharz in German), Switzerland, on the Chemins de fer fédéraux suisses (CFF). The accident happened at 05:50, on a part of the railway which ran alongside the Lac de Bienne, with vineyards between the railway line and the lake shore.

The collision was a very violent one as the goods train was travelling at 60 kilometres an hour and its Ce 6/8 class locomotive was more than a match for the light passenger train, which consisted of an Ae 3/6 class locomotive, three carriages and a mail and parcels van. The second carriage of the passenger train mounted and telescoped the leading one, in which most of the deaths and injuries occurred. Fortunately there were only 20 or 30 passengers on the train from Bienne, most of them workmen. Ten of them lost their lives and about the same number were injured. Driver Camille Roulin, in charge of the engine of the passenger train, was fatally hurt, while Driver Henri Vuille and Conductor Gottfried Bigler of the freight train were among those seriously injured. [731]

According to one source ten of the 50 waggons of the goods train were wrecked, two of them falling into the lake, which was only shallow at that point. The same newspaper

---

[728] Ibid, 14th July 1942
[729] Ibid, 18th September 1942
[730] Ibid, 6th November 1942
[731] *Journal de Geneve*, Geneva, Switzerland, 3rd October 1942

reporter spoke of luggage, mirrors, running boards and benches being scattered about, while thousands of Red Cross packages destined for prisoners of war, had by the time he arrived been carefully piled up, though some were damaged.

A rescue operation was quickly put in hand, with the injured being taken to hospital in Bienne. The work of clearing the line took considerably longer due to the difficulty in separating the locomotives, which were jammed into one another, and the retrieval of the waggons which had fallen into the lake. Initially it was hoped to reopen the line in two or three days; in the end it was not back in use until 05:00 on 7th October, five days after the accident. In the meantime passengers were conveyed by bus between Bienne and Daucher and goods trains were diverted over alternative routes.

By the end of October Driver Vuille had recovered sufficiently to be interrogated by the railway company. He admitted that he had fallen asleep before his train had passed Daucher, where he had run through the signals at danger, and had not awoken before the collision. The conductor had been riding in the 12th vehicle of the train and Brakeman Thommen was at its rear. It was their duty to observe the signals at the stations that the train passed through and to apply the brakes if necessary, but they had done nothing.

Due to the length of time which would elapse before the judicial inquiry would be completed, the driver had been released from preventive detention and the brakeman was reallocated to a job directly not connected with the running of trains. The conductor had had a leg amputated.

The three men had appeared fit in the days leading up to the accident and none of them had appeared to be suffering from overwork. Like every other railwayman they had been affected to some degree by the abnormal conditions – extra traffic, worries about the outcome of the war and so forth. The fact that the freight train had been running late had meant that its crossing place with the passenger train had had to be altered. The single manning of engines had been questioned. There was no intention of double manning every train, but the question might be investigated for the small number of freight trains which ran during the night-time. [732]

The case eventually came to court a year later. It was brought out in evidence that the *signal d'entrée* (home signal) on the Douanne side of Daucher station had been cleared for the goods train to run into the station, and the *Signum* automatic train control magnet associated with it had therefore not operated. That the home signal (No. 3) could be cleared prior to the arrival of a train which had to stop in the station to cross an opposing train, was apparently in accordance with the regulations. Likewise there was no automatic train control at the starting signal (No. 5), it being argued that too many safety devices would make drivers too reliant on them (!). It was also pointed out that such a device would not have prevented the accident in question which occurred not very far beyond the starting signal.

Driver Vuille, asleep, had run past the starting signal and back out onto the single line, oblivious of the gesticulations and shouts of the station staff. The conductor had been busy making out a report and the brakeman compiling a timetable for an acquaintance.

---

[732] Ibid, 28th October 1942

The passenger train had reduced its speed to 8 kilometres an hour. [733]

For the defence it was argued that the single manning of engines placed a great strain on drivers, also that the gentle rocking motion of the Ce 6/8 class engines, which led them to be nicknamed *berceuse* (lullaby) made them especially liable to send drivers to sleep. Against this the prosecution stressed that the driver had failed to observe Rule No. 6:

During the journey locomotive personnel are to carry out their duties standing. The sitting position during the journey is permitted only on vehicles specially built for the purpose.

At the end of the trial Driver Vuille was sentenced to 12 months' imprisonment suspended for four years, and half the court costs. Brakeman Thommen received 20 days' imprisonment suspended for two years and quarter of the court costs. Surprisingly, and perhaps in view of his permanent injury, Conductor Bigler was acquitted. The remaining quarter of the court costs were to be met by the State. [734]

What was probably a single-line collision occurred on Thursday 8th October 1942 at or near Neustadtl, Germany (now Stráž, Czech Republic) between Ronsperg (now Poběžovice, Czech Republic) and Haid (now Bor, Czech Republic). The line involved was in pre-war Czechoslovakia, at this date incorporated into the German Reich. Passenger railcar Tp 3205 collided with engine No. 19 916 because owing to a misunderstanding the driver of one of the trains, probably the railcar, had left a crossing place without permission. As a result, 11 people lost their lives and 35 were injured. [735] [736]

Less than 24 hours after the Almonte disaster, Canada suffered another bad accident, late on Monday 28th December 1942. No passengers were involved this time, but five railwaymen lost their lives when two Canadian National Railway freight trains collided two miles from Callander, Ontario. The two locomotives ended up locked together and seven freight cars were derailed and caught fire, but firefighters from North Bay were able to prevent the fire from spreading.

Those who lost their lives were Engineer Elmer Hayes, Fireman Duncan McKenzie, and Brakeman Fausto Caverson of the eastbound train and Engineer Nelson Sweet and Fireman Grant Montcrieff of the westbound train. [737]

The noise of the collision was heard, a newspaper noted, by the Dionne quintuplets, who had been sleigh riding near their home (the Dionne quintuplets, born May 1934, were the first quintuplets to survive infancy). [738]

As to the cause of the collision, a Canadian National official in Toronto stated that, "It

---

[733] *Railway Accidents of Great Britain and Europe* by Ascanio Schneider and Armin Masé, pgs. 42-46

[734] *Gazette de Lausanne*, Lausanne, Switzerland, 18th October 1943

[735] *Eisenbahnunfälle in Europa* by Erich Preuß pg. 124 (in German)

[736] *Katastrophen der deutschen Bahnen Teil II* by H.J. Ritzau, pg. 105 (in German)

[737] *The Brandon Daily Sun*, Brandon, Manitoba, Canada, 29th December 1942

[738] Wikipedia – Dionne Quintuplets

has not yet been determined whether an order to meet the eastbound freight at Astorville had been delivered to the crew of the westbound freight." [739] Astorville was the next station east of Callander. The eastbound freight was presumably proceeding from Callander to Astorville as per the order and the westbound freight, having for some reason failed to receive the order, had failed to wait at Astorville for the eastbound one.

A major catastrophe occurred on the Roma Nord suburban railway from Rome to Viterbro in Italy on Monday 15th November 1943. Italy had entered into an armistice with the Allies at the beginning of September and on 10th of that month, Rome was occupied by the Germans. Conditions were very hard for the inhabitants of the city with bombings, curfews, round-ups and a lack of electricity, gas, and food. The food shortages led many of the residents of the city to travel out into the surrounding countryside to buy food directly from the farmers.

The 06:05 train from Viterbo that day consisted of an engine and three carriages driven by Motorman Giuseppe Galadini. The train stopped at Bagnaia and Soriano and reached Catalano at 07:09, its passengers and crew relieved that they had not suffered any Allied air attacks en route. By now it was well filled, and the decision was made to strengthen it there with a fourth car to accommodate the passengers who were sure to join it further along the line. In addition, a second engine was attached to the rear of the train to cope with the severe gradients ahead, this engine was driven by Motorman Emilio Pedica. Continuing the train arrived at Civita Castellana at 07:16 and at Sant'Oreste at 07:48.

Meanwhile in the opposite direction, proceeding along the single line towards Viterbo from Rome was another passenger train, which had left Rome Plazzale Flaminio at 06:52, consisting of an engine driven by Motorman Carabellesi (or Carabellese) pulling three very crowded carriages. The two trains normally crossed at Sant'Oreste but with wartime disruptions to traffic their crossing was often shifted elsewhere. The Rome-bound train, its station work completed, instead of waiting at Sant'Oreste, set off once more towards the capital. Its opposite number stopped at the next station at Rignano Flaminio at about 07:55, did its work there and then it proceeded as well. Exactly why neither train waited for the other is not clear; crossings were arranged over the telephone and there presumably was a misunderstanding between the different stations, a mistake was made by one of the stationmasters, or perhaps by a conductor, or one of the motormen.

The train from Rome cannot have got very far because the collision occurred on a curve opposite the Rignano Flaminio cemetery. To a peasant woman working in a chicken coop nearby the noise of the collision sounded like a bomb explosion. When she saw what had happened she ran for a kilometre along the road into town to get help. By the time that this had arrived from Rignano, a German military column heading for Monte Cassino had stopped, and was already giving first-aid. A terrible scene was said to have presented itself, with the colliding engines reared up into the air, the carriages crushed into one another, screams coming from injured passengers trapped within them and blood dripping from carriage doors. It was raining incessantly, and a red river of blood and water was flowing along the roadside.

Some of the injured were taken by northbound lorries to the Civita Castellana

---

[739] *Medicine Hat Daily News*, Medicine Hat, Alberta, Canada, 29th December 1942

Hospital; southbound trucks took other injured to hospitals in Rome. At the Civita Castellana hospital, Professor Vincenzo Ferretti operated on the injured for days on end with barely any rest, despite being almost 70 years old, impressing the German doctors with his incredible stamina. The German military placed the dead in the churchyard of the Church of Santa Teodora. General Kesselring, the German Commander-in-Chief in Italy, chanced to be at Sant'Oreste and ordered all his men to assist. To begin with there were at least 52 known dead (one account says 84), and with the deaths of a large number of the injured the final number of fatalities rose to over 100. Motormen Galdini and Pedica on the Rome-bound train were among the dead but Motorman Carabellesi on the outbound service, somehow survived. Among others who died were a prisoner being taken to the capital and the guards to whom he had been handcuffed. Another victim was Riccardo Lo Russo, aged 42, the father of seven children. After his death his son Giuseppe, aged 14, had to work to provide for his mother and six younger brothers. The final number of dead is unclear.

As in other accidents there were some unsavoury incidents. The Germans noticed a young man who seemed very keen to help the rescuers with the bodies of victims. He was searched and found to possess a number of wallets looted from the dead. He was put on a lorry bound for Rome along with the injured – his fate is unknown. As bad, if not worse, were the insurance companies who paid derisory sums to poor people who could not afford to pay for the funerals of their relatives. [740] [741]

Czechoslovakia (strictly the Protectorate of Bohemia and Moravia, as the country was designated between March 1939 and May 1945) suffered a major disaster just before 06:00 on Monday 20th December 1943. This occurred on the line between Turnov and Praha, between the stations of Praha-Kbely and Praha-Satalice, and involved a head-on collision between two passenger trains, Nos. 1111 (or 1101) and 1160, which were bringing workers to the Čakovice, Letňany and Vysočina factories. Both trains were crowded, and the casualties were high; 23 people were killed outright and 110 were injured, of whom 19 died later in hospital. (Some sources state that there were only 35 dead.) A new station had opened on the day of the disaster at Praha-Kbely together with a crossing place (?) and the accident was attributed to a misunderstanding on the part of the Praha-Kbely signalman, Antonín Pospíšil.

The funeral of the victims was attended by ministers Richard Bienert and Emanuel Moravec. Only a Roman Catholic priest was allowed to speak over the coffins.

The construction of the new Praha-Kbely station had involved the demolition of a Roman Catholic chapel and superstitious locals blamed the accident on Divine retribution for this act of sacrilege. [742] [743]

Early on Thursday 14th September 1944 the southbound *Dixie Flyer*, a crack Chicago – Miami express of the Chicago and Eastern Illinois Railroad, was involved in a

[740] www.ferroviaromanord.altervista.org/una_terrible_disgrazia_dimenticata.htm
[741] www.ferroviaromaord.altervista.org/disastro_ferroviario_di_rignano
[742] Wikipedia – Železniční nehoda ve Kbelích
[743] https://historie.praha19.cz/index.php/zeleznice

disastrous head-on collision with a northbound mail train near Dewey, Indiana, United States, a little north of Terre Haute. Twenty-six passengers, two railwaymen and a Pullman employee lost their lives and 32 passengers, five railwaymen, four mail clerks and a Pullman employee were injured.

The single line on which the accident occurred was worked by timetable and train order and was also equipped with semaphore upper-quadrant automatic block signals whose night-time indications were green (*Proceed* – proceed), yellow (*Approach* – proceed at a speed not exceeding half the maximum speed and at not more than 30 miles an hour, prepared to stop at the next signal) and red (*Stop* – stop). Northbound signals 174-8 and 173-8 were 6,298 and 6 feet south of the site of the collision and southbound signals 171-9, 172-7 and 173-7 were 9,587 and 5,592 feet north of and 6 feet south of the scene of the accident respectively. Signal 174-8 was continuously lit, and the others were approach lit.

Train No. 90, the northbound first-class express mail train, consisted of engine No. 1011, three express refrigerator cars, an express car, five mail cars, two express cars, three express refrigerator cars and a passenger baggage car, in charge of Conductor Ross Sears, with Engineer Charles Rohlfer and Fireman Osborn. At Terre Haute, 3½ miles south of Dewey, the train was given a clearance card and three train orders:

Train order No.3 Form 31

1st and 2nd No. 95, engs. 1019 and 1012 have right over No. 90 eng. 1011 Clinton to Terre Haute

Train order No. 11 Form 19

1st No. 95 eng. 1019 meet No. 90 eng. 1011 at Dewey 1st No. 95 take siding 2nd No. 95 eng. 1012 gets this order and wait at Clinton until 2.40 AM.

Train order No. 13 Form 19

1st No. 95 eng. 1019 take siding and meet No. 90 eng. 1011 at Atherton instead of Dewey.

Clinton and Atherton were just under 11 and 6½ miles north of Dewey respectively. Train No. 90 passed Haley, 2¾ miles south of Dewey, at 2.14 a.m., one hour and one minute late, passed signal 174-8, showing *Proceed*, and stopped at signal 173-9, showing *Stop*, at about 2.18 a.m. About two minutes later it restarted and had proceeded about six feet when it was struck by train First No. 95.

Train First No. 95, the *Dixie Flyer*, consisted of engine No. 1019, a baggage car, a baggage-mail car, a Pullman tourist car, two Pullman sleeping cars, five coaches and four Pullman sleeping cars, in charge of Conductor H.E. Harrington with Engineer Frank Blair and Fireman Louis Rausch. At Cory, 50½ miles north of Dewey, this train was given a clearance card and three train orders, one of which was order No. 3, detailed above. At Clinton the train was given three more orders, one of which was a reworded order No. 13, which read:

Train order No. 13 Form 19

1st No. 95 eng. 1019 take siding and meet No. 90 eng. 1011 at Atherton

The dispatcher told the Interstate Commerce Commission (ICC) inquiry that in order to reduce the number of orders delivered to train First No. 95 at Clinton he had issued order No. 14 to the operator at Clinton, annulling order No. 11, and told the operator to omit the words "instead of Dewey" from order No. 13. Surprisingly the ICC does not seem to have found fault with the dispatcher for not issuing train order No. 13 in an identical form to each train, which was certainly a violation of the rules.

Train First No. 95 passed Clinton at 2.06 a.m., 37 minutes late, passed Atherton, where it should have met train No. 90, passed Otter Creek Junction, just under two miles north of Dewey, at 2.18 a.m., still 37 minutes late, passed signal 171-8, showing *Proceed*, passed signal 172-7, showing *Approach*, and travelling at an estimated 35 miles an hour collided with train No. 90 six feet north of signal 173-7, which had been showing *Stop*. There was a dense fog at the time.

The force of the impact pushed train No. 90 back 68 feet. Both locomotives were badly damaged, the first four cars of train No. 95 were badly damaged and the fifth and sixth cars less badly damaged. In addition the second and third cars of this train were telescoped and all the passengers who died were in the third car. In the main train, No. 90, the second, third, fourth and eighth cars were considerably damaged. [744]

Many of the passengers on the train were soldiers who had fought in Italy and who were on their way from Fort Sheridan, Illinois, to Florida. One of them, Lieut. E.B. Owens, of Chicago, spoke to a newspaper reporter after the accident. "I was asleep in the second car and all I remember is waking up in the aisle. Our car was derailed and tipped over on its side but the other cars on the *Dixie Flyer* remained upright. There were 48 soldiers in the first car, and I counted 15 bodies as they were removed from the wreckage. There were at least two or three in there. I helped with the injured until I became too ill to continue." [745]

Engineer Rohlfer told the ICC inquiry that shortly after passing signal 174-8 at *Proceed* he had seen signal 173-8, some distance ahead, showing *Approach,* and had made a service brake application. The signal had then changed to *Stop,* and he had stopped his train at it. He had then proceeded under control as per the regulations but had barely started moving when he had seen the headlight of train First No. 95 and stopped again.

Copies of the train orders issued to the operator at Clinton were delivered to a member of the engine crew and to Head Brakeman Charles Riggles, riding in the sixth car of train First No. 95. The latter read and understood the orders. He did not attempt to give them to Conductor Harrington or other members of the train crew but instead made his way towards the front of the train ready to open the siding points at Atherton with the least delay when the train stopped. In the darkness and fog he was unaware that the train had passed Atherton.

Conductor Harrington and the flagman had been riding in the rear car of train First No. 95. They were aware that orders had been received at Clinton but made no attempt to find out what they were.

The Clinton operator stated that he had seen the orders picked up by a member of the train crew. They could not be traced after the accident. The train order signal at Clinton had indicated that there were orders for the train so if they had not received them they should have stopped at once. Why they failed to wait at Atherton could not be determined as Fireman Rausch was killed instantly and Engineer Blair only lived a few hours. None

[744] Interstate Commerce Commission report.

[745] *The Wisconsin State Journal*, Madison, Wisconsin, United States, 14th September 1944

of the surviving members of train First No. 95 noted any slowing of their train until its brakes had been applied in emergency about 30 seconds before the collision.

The ICC noted that signal 172-7, which had been showing *Approach*, had an oil signal lamp converted to electric lighting which gave a much less intense display than other signals fitted with reflector backs. It was possible that the engineer had failed to see the less intense signal in the darkness and fog.

An automatic train stop system was in use on sections of the line but ended 11½ miles north of the site of the accident. The engines of the two trains were fitted with the equipment but there were no lineside devices at the location of the collision. Had there been the accident would probably have been averted.

The ICC also found deficiencies in the signalling installation, which did not provide that a train should receive at least one signal showing *Approach* before it arrived at a signal showing *Stop*. The track circuits were arranged so that signals 174-8 and 173-8 had displayed *Proceed* and *Stop* respectively to train No. 90. Because train No. 90 passed signal 174-8 before train First No. 95 entered the limits of its control circuit Engineer Rohlfer had received no warning of the opposing train until through the intermittent fog he had seen signal 173-8 showing first *Approach* and then *Stop*. Had he not been able to see signal 173-8 until he had reached it he would have passed it at considerable speed and the results of the collision would have been more serious. Furthermore, the provisions of Rule 509 modified, which permitted train No. 90 to enter the occupied block section immediately after stopping and proceed under control practically nullified the protection provided by the automatic block system where first-class trains were involved. When train No. 90 passed signal 173-8 under control with train First No. 95 having passed signal 172-7 at *Approach*, each train was authorised to pass through the block occupied by the opposing train.

The ICC concluded that, 'This investigation disclosed that the signals involved did not provide an adequate margin of safety in stopping distance for trains operated at maximum authorized speed. Pending such modifications of the signal system as may be required to provide adequate protection for trains being operated at presently authorized maximum speeds, the carrier should at once reduce the maximum authorized speeds to the limits for which the existing signal installation will provide adequate stopping distances.' [746]

At the inquest on the victims of the accident, held at Terre Haute at the end of September by Coroner D.M. Ferguson, a verdict of 'Accidental Death' was returned, the coroner commenting that the deceased engineer and fireman were the only persons who could have explained how the accident had happened. [747]

Poland suffered a major catastrophe on Friday 24th November 1944 at Barwald Średni, at a time when the country was, of course, under German occupation. The single-track Kalwaria Zebrzydowska – Bielsko-Biala line was busy with trains carrying supplies to the Russian front. On the day of the accident the *fahrdienstleiter* (head signalman) at Wadowice arranged with his colleague at Kalwaria Zebrzydowska Lanckorona station for an eastbound military train to run non-stop between the two stations. The *fahrdienstleiter*

[746] Interstate Commerce Commission report.
[747] *Manitowoc Herald Times*, Manitowoc, Wisconsin, United States, 27th December 1944

at Kalwaria Zebrzydowska Lanckorona, having agreed to the despatch of the train, then left his office without having recorded its acceptance in the train register book.

Meanwhile a northbound express from Zacopane to Krakow had arrived at Kalwaria Zebrzydowska Lanckorona on the Żywiec – Skawina line. Because resistance fighters had blown up a bridge between Kalwaria Zebrzydowska Lanckorona and Skawina it had to be diverted via Wadowice and Spytkowice. While the *fahrdienstleiter* was absent the Kalwaria Zebrzydowska stationmaster, who naturally saw no entry regarding the military train in the train register, authorised the diverted express to continue westwards towards Wadowice. When the *fahrdienstleiter* returned he discovered what had happened. A message was sent at once to Kleczy Górna to stop the military train, but it had already passed there. At the same time an official at Kalwaria Zebrzydowska attempted to stop the departing express with a red flag but no one on the train saw it. In a last attempt to prevent the accident a light engine was sent in pursuit of the express, in the hope that its whistling might warn it, but to no avail.

The trains collided between 15:15 and 15:20. Over 130 people died in the collision and between 100 and 200 were injured. [748]

A collision occurred on the single-track Holzkirchen – Rosenheim line between Bad Aibling and Kolbermoor, Germany, on Monday 28th May 1945. A military train and an empty train were involved and there were five deaths. [749]

The circumstances which led to a catastrophic collision between the stations of Baschi and Attigliano, Italy, on Wednesday 13th June 1945 were remarkably similar to those which led to the accident near Goch, Germany, exactly two months later (see below). On the evening of the disaster southbound freight train No. 5641 from Chiusi was stopped in Biaschi station, awaiting the arrival over the single line of northbound passenger train No. 5646. Train No. 5641 consisted of 19 or 20 vehicles, 15 of which were loaded with goods and drums of petrol, the others were carrying Italian deportees returning from Germany and four British ex-prisoners of war, so it was really a mixed train. The northbound train was carrying a large number of British and Italian troops going on leave.

The Baschi stationmaster motioned for the driver of the southbound train to move forward a little, probably to clear a set of points, at a moment when there was a considerable amount of shouting from the soldiers, and the driver misinterpreted the gesture of the stationmaster as leave to proceed and set off into the single-line section. His train had covered about two kilometres and was passing through a cutting when it met the northbound passenger train. Both drivers braked but a violent collision was unavoidable. Twelve waggons of the southbound train were derailed, the fuel drums were overturned and damaged and both trains were set on fire. A reporter wrote, "There was a series of frightening explosions and soon the accident site was transformed into a bedlam of fire."

---

748 Wikipedia – Eisenbahnunfall ved Barwald Średni

749 Wikipedia – Liste von Eisenbahnunfällen in Deutschland

Two relief trains and a hospital train were sent to the scene, which was also visited by the Minister of Transport, Francesco Cerabona. It was impossible to say exactly how many people lost their lives, but 70 bodies were recovered, most of them charred, and about 100 injured were, according to an Italian source, taken to hospital in nearby Orvieto, though a British newspaper states that an officer and 26 other ranks were brought to a Roma hospital; possibly they were transferred from Orvieto. It was believed that between 10 and 20 British soldiers were among the dead. It was not stated if the engine crews survived.

The accident happened at about 21:30. The Roma – Firenze line was blocked for two days, traffic being diverted. [750 751]

An ambulance train and a passenger train collided at Povrly station, Czechoslovakia, on the line between Praha and Děčín, on Friday 15th June 1945. Forty-six people died and 42 were injured. From the brief reports on the accident it was a head-on collision during temporary single line working. [752]

A disastrous head-on collision in Germany on Monday 13th August 1945 involved two Allied military trains. It happened close to Goch, a few kilometres from the town of Kleve, on a line which had only been brought back into use the week before. The signalling installations on the line having been wrecked in the war, the points were being worked by hand and trains were handsignalled with flags and lamps. A telephone link had been re-established between the stations and was used to control the operation of trains over the single track, with the drivers being given written orders.

The accident happened at 02:30 and involved a train from Münster to Brussels carrying British, Canadian, and Polish soldiers going on leave and a corresponding train from Brussels to Münster carrying Belgian soldiers. One of the trains was composed of steel carriages and only the front coach of this train was badly damaged. The other train had at least one wooden coach, which was telescoped. An Army captain who had been a passenger on the train from Brussels complained to the press that the trains had been "criminally overladen with troops." [753]

When the trains met there was, a British officer in the train from Münster said, "a terrific crash, and we were all thrown into confusion, with the injured crying out." Three doctors on that train did what they could without having equipment available and were assisted by two Polish soldiers, the only women on the train. Meanwhile much confusion reigned in the darkness with passengers and railwaymen speaking in English, German, Polish and French. Several of the passengers in the telescoped wooden coach were trapped. After the accident German Shunter Peter Verfürth at Goch ran to the scene with a colleague to help but they were greeted at gunpoint by surviving soldiers who suspected sabotage, and they were not allowed to assist. There was also an unconfirmed rumour that one of the engine crew had run away from the scene in a panic and had been shot after

[750] https://www.vernicifirewall.it/incidente-ferroviario-di-baschi/
[751] *The Times*, London, England, 16th June 1945
[752] Wikipedia – Liste von Eisenbahnunfällen in Tschechien
[753] *Kalgoorlie Miner*, Kalgoorlie, Western Australia, 15th August 1945

the accident.

News of the accident was not censored, reports of it appearing in newspapers in various countries, but the military authorities attempted, at least at first, to conceal the magnitude of the disaster. *The Times* in London in its first report assured its readers that, 'Most of the British soldiers, who were going on leave to Brussels, were able to continue their journey in another train,' while a statement later in the day from the British general hospital in Bedburg, Germany, where the injured had been taken, said, 'The death-roll is not as heavy as was originally expected, although the number of injured is greater. Four British dead were taken to this hospital and 45 injured.' [754]

However, the next day the same newspaper stated that 17 persons had died at the scene, nine more had died in Bedburg Hospital and there were another 41 injured there, many seriously hurt. It added that the signalling was being operated by Germans, though the train crews were Dutch. An inquiry into the accident was shortly to be opened. [755]

Recent research into the accident has determined that in fact 48 or possibly 49 people died in the accident. In addition to 26 British and Canadian soldiers who lost their lives it was admitted by the military authorities after a request from Belgium that 19 Belgian soldiers had also died. Also killed were both Dutch drivers, whose names the author has not been able to determine, and one of the Dutch firemen, Wilhelmus Hendricks, who succumbed to his injuries two days later. The other fireman, Albert Ewijk, suffered severe burns but survived.

The results of the inquiry were not made public, but it has been theorised that the mistake which had led to the accident had happened at Pfalzdorf, the next station to Goch, where the trains were due to cross. It seems likely that a railwayman at Pfalzdorf asked the driver of the train from Münster if he could pull forward a short distance to clear a set of points, but the request was misunderstood as the order to proceed to Goch and the train set off into the occupied section. [756] [757] If the request was given in German to the Dutch driver that might have contributed to the confusion.

A collision occurred on the main line of the Assam – Bengal Railway between Calcutta (now Kolkata) and Siliguri at 03:10 on the morning of Tuesday 6th November 1945. The trains involved were train No. 15, the Up North Bengal Express and train No. 2, the Down Darjeeling Mail, which met one another at the southern end of the Atrai bridge, near Atraighat station, India. The express train was just leaving the bridge and the mail train was just entering it. The leading saloon and the third-class brake composite of the Bengal train were wrecked and the second and third third-class carriages of the Darjeeling train partially telescoped one another. The two locomotives were locked together, that of the Bengal train partly mounting that of the Darjeeling mail. Five passengers died at the scene, two on the relief train taking them to hospital and two more in hospital at Calcutta; all had been occupants of the third-class brake composite of the Bengal train. About 40 others were injured, among them Guard C.O.B Sankey of the

754 *The Times*, London, England, 14th August 1945
755 Ibid, 15th August 1945
756 https://rp-online.de/nrw/staedte/goch/gedenken-an-gochs-groesstes-zugunglueck_aid-21070667
757 https://rp-online.de/nrw/staedte/goch/50-tote-vor-70-jahren-ereignete-sich-gochs-schwerstes-bahnunglueck_aid-20108773

Darjeeling Mail, Driver A.J. Clubb of the same train, Driver W.J. Sinott of the Bengal train and two unnamed firemen.

The single-line bridge was blocked by the accident and six boats were employed to tranship passengers between the two trains. With the labours of 300 men the track was cleared, and communication re-established after a break of nine hours.

An inquiry into the accident was opened by Mr K.L. Ganguly, the Government Inspector of Railways. [758] [759]

At 01:10 on the morning of Thursday 15th November 1945 the mail train from Madrid to Almeria, train No. 1802, collided head-on with freight (parcels?) train No. 6831 at kilometre 211.300, between the stations of Gérgal and Fuente Santa, Spain, on the Linares – Almeria main line. As a result of the violent head-on collision the two engines finished up embedded within one another and the leading two carriages of the mail train, a first- and third-class composite and a third-class, were wrecked, as were three waggons of the freight train. To make matters worse the pile of wreckage was high enough to come into contact with the overhead line equipment, which set the vehicles alight, and they were eventually burnt out. The undamaged vehicles of each train were drawn back to the stations from which they had come, and the uninjured passengers of the mail train were able to proceed to Almeria at 08:00.

Seventeen people reportedly died in the accident; eight passengers, two civil guards and seven railwaymen, while 39 people, comprising 30 passengers and nine railwaymen, were injured. One source states that there were 18 dead, possibly one of those injured succumbed later.

The accident was due to a mistake by the employee on duty that night at Gérgal, Juan Licas Medina. The latter had accepted the freight train from Fuente Santa and then, after carrying out some shunting operations, and forgetting what he had done, he had cleared the mail train, which was running an hour and 23 minutes late, to proceed. When he realised his terrible mistake he fled the station. [760] Nine days later he committed suicide. [761]

A head-on collision between an ordinary passenger train and a military train occurred on the Berliner Nordbahn near the station at Neddermin, Germany, on the evening of Tuesday 4th December 1945. Until the end of the war this had been a double-track line but following the surrender of Germany and the occupation of the eastern portion of the country by the Russians one track had been seized by them as part of their reparations. Traffic over the remaining line was managed by telephone.

On the evening of 4th December the signal box at Neubrandenburg was manned by *Stellwerksmeister* (Chief Signalman) Herman Storbeck, *Fahrdienstleiter* (Signalman) Willi Schmidt, and a switchboard operator or telephonist named Erika Hübner (who was not qualified for signalling duties). Passenger train P 213 from Neustrelitz, which consisted of class 38 steam locomotive No. 3663 and a number of carriages, was in the

[758] *The Bombay Chronicle*, Bombay, India, 8th November 1945
[759] *The Railway Gazette*, 7th December 1945
[760] *La Vanguardia*, Barcelona, Spain, 17th November 1945
[761] Webpage: Federación Castellano Manchega de Amigos del Ferrocarril (in Spanish)

station, waiting to proceed to Stralsund. In the opposite direction a military train, Lzg 7266, consisting of class 38 steam locomotive No. 1605 and a single coach, carrying a Russian commander and his staff, was approaching from Altentreptow. At about 19:00 *Fahrdienstleiter* Schmidt was busy. He was engaged in a long telephone conversation and there were a number of people in his office, with three Russian soldiers wanting to know when the commander's train would arrive, and hassle from the driver of a freight train to Pasewalk. Meanwhile, the departure time for train P. 213 had arrived. The telephonist reminded the Schmidt of this and asked him if it could go. What he said in reply is not known but the telephonist interpreted it as, 'Yes, let the train go.' She conveyed the message to Pointsman Ernst Rickert, and he cleared the signals for the passenger train to depart.

At about the same time the signalman at Neddermin asked his colleague at Neubrandenburg if the military train could proceed and Schmidt, unaware of the departure of the passenger train, accepted it. Only when he was advised by the telephonist that the passenger train had left did he realise what had happened. There was no means of contacting either train and at about 19:05 they collided on a curve not far from Neddermin, each travelling at about 55 kilometres an hour. As is often the case the number of victims is unclear. On the day of the accident it was reported that there were 31 dead, the next day the number of dead was reported as 55 but two days later it was stated that 38 had died, with 68 injured. Drivers Rudolf Stuhr and Rudolf Waack and Firemen Erich Kruse and Johannes Tunk were among the fatalities. Which driver and fireman was on which train was not stated.

It was some time before a rescue operation was begun but at about 20:30 a relief train was sent from Neubrandenburg with doctors and nurses, while soldiers from the Red Army also assisted.

The three signal box staff directly involved were arrested after the accident and because Russian soldiers had been injured they were subjected to a court martial at Greifswald. All three were found guilty of having caused the accident through negligence and *Fahrdienstleiter* Schmidt and Telephonist Hübner were sentenced to ten years hard labour', and Pointsman Rickert to seven years. *Fahrdienstleiter* Schmidt was initially interned at the NKVD special camp at Fünfeichen; later he was transferred to the NKVD camp at Sachenhausen (the former Nazi concentration camp). Between 1945 and 1950 12,000 prisoners died at Sachsenhausen Camp of malnutrition and disease, and among them were Pointsman Rickert in early February 1947 and *Fahrdienstleiter* Schmidt on 7th April of the same year (from malnutrition). Telephonist Hübner was more fortunate. Her brother, who chanced to be a lawyer, managed to secure an appeal against her sentence, and it was quashed on 1st March 1947.[762] [763]

---

[762] https://www.nordkurier.de/neubrandenburg/warum-ueber-ein-zugunglueck-bei-neubrandenburg-kaum-etwas-bekannt-ist-0938346602.html

[763] Wikipedia – Eisenbahnunfall von Neddemin

CHAPTER 7

RUNAWAY TRAINS AND VEHICLES

A variety of situations can lead to a train or part of a train running away out of control. If there are railway staff in attendance the train or engine can only run away out of control if there is a failure of the brakes or if the brakes are incapable of holding it on a steep gradient. In the case of unattended vehicles, the action of gravity alone is sufficient to precipitate a runaway. In contrast to the situation on a road a train or vehicles can sometimes run for many miles without mishap, an accident occurring only if and when the runaway meets with another train, comes to the end of the line, or gets to a curve too sharp for the runaway to negotiate at the speed at which it is travelling.

In parts of the United States the night of Thursday 16th July 1931 was stormy and wet and seven unemployed men who had found temporary work as harvest hands decided to retreat to four boxcars stabled in a siding of the Missouri Pacific Railroad at Horace, Kansas, United States, as somewhere they could shelter from the bad weather and get some sleep. This proved to be an unfortunate decision as the wind was powerful enough to blow the box cars from the siding out onto the main line. The vehicles ran along the main line for about a mile and a half before they collided with a local passenger train, one of the cars mounting the engine of the train.

Three of the men were killed outright and four seriously injured, one of whom died not long afterwards. None of the crew or passengers of the local train was hurt. One of the dead men was identified from papers found in his pockets as George Miller, of North Platte, Nebraska. His identity was confirmed by his poor mother, who was able to say that one of the others who had died was Virgil Graber, 25, also of North Platte. The other two men were not immediately identified. [764] [765]

As Driver Defosse was bringing a freight train from Gisors into Beauvais, France, on the evening of Friday 21st October 1932, and was passing over some points at

[764] *The Lincoln Star*, Lincoln, Nebraska, United States, 17th July 1931
[765] *The Cattaraugus Republican*, Salamanca, New York, United States, 22nd July 1931

Penthemont, a few hundred metres from Beauvais, he felt his train shaking and realised that the brakes were not working properly. Concerned that something was amiss he stopped the train and jumped down onto the track with the conductor. Both thought that it was probably just the brake pipe having separated but in fact two of the waggons of the train had become derailed on the points. The conductor began to set off back down the line to protect his train by putting down detonators and at that moment the engine and nine waggons at the front of the train began to roll backwards. The conductor and driver attempted to reboard the locomotive but failed to do so and the runaway gathered speed until it disappeared around a corner. Meanwhile the Paris express was almost due.

The two men rushed to a pointsman's cabin and frantically telephoned Beauvais station to tell them to divert the runaway onto another line. They were too late. Travelling at 60 kilometres an hour the runaway tore through Beauvais station 'like a whirlwind' before the staff there had chance to do anything. A disastrous collision with the Paris express seemed inevitable.

By a stroke of luck another freight train was being shunted two kilometres from Beauvais. The signalman there, seeing the runaway approaching, shouted a warning to the men supervising the shunting operation and they jumped clear. There was a tremendous crash. Three of the waggons which were being shunted were completely wrecked and the nine waggons of the runaway piled up one upon the other in a pile of wreckage. The lines were all blocked, but no one was hurt. The express and a train from Creil were stopped and a 50 tonne breakdown crane was brought in to clear the line. [766]

At about 10 p.m. on the evening of Tuesday 11th July 1933, No. 21 Up mixed train of the Mysore Railways became divided between Sivapur and Tarikere stations, India, and its rear portion ran back down an incline until it collided with No. 17 Up mail train which was standing at the platform of Tarikere station. The train had run into difficulties on the long incline with its engine being unable to manage the load in the wet weather. In the collision one bogie and two four-wheeled coaches of the mixed train were smashed and the engine of the mail train was badly damaged. By Thursday 13th the known number of dead in the accident had risen to 15 with the same number injured.

Upon receiving news of the accident, Civil and Railway medical officers at Tarikere rendered first-aid and the District Medical Officer at Shimoga motored to the spot. The injured were taken to Shimoga Hospital. Passengers with minor injuries were treated at Tarikere and relief trains were sent from both Shimoga and Mysore. Senior railway officials also went to the scene, together with the Senior Government Inspector of Railways. The line was reopened to traffic by 2 a.m. the following morning. [767]

The subsequent report on the accident by the government inspector determined that after the train had stalled at mile 7 of the Birur – Arsalu branch its brakes had not been applied and insufficient steps had been taken to prevent its rear portion from running away. The final death toll was 17. [768]

---

766 *Le Populaire*, Paris, 23rd October 1932
767 *The Bombay Chronicle*, Bombay, India, 14th July 1933
768 Report by the Railways Board on Indian Railways for 1933-34 Vol 1.

Mount Vesuvius, in Italy, is an active volcano which by the 1880s had also become a tourist attraction. From 1903 the long climb to its summit had been made much easier by the provision of a railway, or more precisely two separate railways, the 7.7 kilometre Vesuvius Railway which ran from Pugliano at the base of the mountain to within half a kilometre of the summit, and a separate funicular railway which climbed 391 metres in 806 metres to the crater itself. The electric Vesuvius Railway was itself divided into three sections, with a central short rack railway, worked on the Strub system, with a maximum gradient of 25%, and two adhesion worked sections with maximum gradients of 8%. [769]

The Vesuvius Railway suffered a serious accident on Wednesday 28th November 1934. A car carrying tourists, guides and railway employees was descending the mountain and had almost reached the beginning of the central rack section at Eremo when it began to get out of control. The brakes were applied as hard as possible, but the speed of the car increased until it finally left the track and crashed into one of the poles supporting the electric catenary. Four people were killed outright, three more died later in hospital and nine were injured. Those who lost their lives included a honeymooning French couple from Tunisia and an Italian guide. [770] [771]

A Victorian Railways empty electric train ran away from Ringwood to Croydon, Victoria, Australia, on the morning of Sunday 10th February 1935. The line fell between these points, partly at 1 in 80. At Croydon the runaway was diverted into a siding where it collided with a stationary train stabled there and Motorman Gordon Young of the runaway, a second motorman on the stationary train and a third person had narrow escapes from serious injury or worse. Two of the coaches of the stationary train were telescoped and fire then broke out. Firemen were called and were soon on the scene but made no attempt to tackle the fire until they were sure that the electrical current had been switched off. Three cars of the stationary train and four cars of the runaway were wrecked or destroyed by the collision and fire.

When he found that he could not stop the train Motorman Young jumped from his cab and rolled down the embankment. The motorman of the stationary train had been sat writing notes when the crash came. He was unhurt but found himself trapped and had to be released. Finally, a Mrs J. Camm had been picking blackberries along the railway boundary fence at the point where one of the coaches fell following the accident. Fortunately she had moved away from the spot just before.

Operating Porter William Hailes, the only railwayman on duty at Croydon, told a reporter, "I heard the train approaching and set the home signal for the platform clear, and the signal in the siding against the train. The train should have stopped at the platform, but I knew something was wrong when I heard the roar it was making. It passed the signal room in a flash. The road to the siding was open. I had no chance of operating the points to send the train on the main line. I saw the train curve into the siding, and there was a terrible crash."

Initially there were fears that a girl who often rode on the train from Ringwood might

---

[769] Webpage – Japanese Railway and Transport Review No. 15: Thomas Cook and Son's Vesuvius Railway

[770] *The Times*, London, England, 30th November 1934

[771] *San Antonio Express*, San Antonio, Texas, United States, 29th November 1934

have been trapped in the wreckage, but she had not travelled that day. [772]

The board of inquiry into the accident placed the main blame for the accident on Guard George Wilsmore and Motorman Young of the runaway train, and to a lesser degree on two operating porters at Croydon. It was stated that almost £30,000 damage had been caused in the accident.

The inquiry determined that the stationary train in the siding at Croydon had been stabled there at 1 a.m. on the day of the accident but that the facing points had not been re-set for the main line as they should have been by the evening operating porter, nor had their position been checked in the morning when the early turn porter had come on duty. The other train had been due to leave Ringwood at 8.02 a.m. for Croydon. Because a gang had been working on a footbridge, power had not become available until 7.57 or 7.58 a.m., meanwhile the motorman had made a visual examination of the train. When power had eventually become available the motorman and guard had got the pantographs up and the equipment working in each of the motor carriages before the train had left Ringwood eight minutes late at 8.10 a.m. Crucially, probably because they were already late in starting, they had omitted to carry out a brake continuity test to ensure that the brakes could be applied and released from either end of the train. The train had still been travelling at about 35 miles an hour when it reached Croydon but had the road not been set for the siding the train would have come to rest on a rising gradient in the Croydon – Mooroolbark section. After the accident it was found that the brake pipe had been closed behind the second car of the seven-car train and there was no indication that this was a result of the collision. Had the brake continuity test been carried out at Ringwood as per the regulations this defect would have been discovered. Disciplinary action was being considered by the Railway Commissioners, but the guard and motormen were still suffering from the effects of the accident at the close of the inquiry. [773]

A runaway accident similar to that which had occurred between Midford and Bath, England, in November 1929 (see previous volume of this series), occurred on Tuesday 30th July 1935 in New Zealand. A northbound goods train had left Taumarunui at 11.10 p.m. the previous day and was due to arrive at Poro-o-tarao station at 1.55 a.m. It consisted of an Aa class engine, 46 waggons and a van, in charge of Guard H. McQuillan with Driver F. Westneath with Fireman J. Reidy. South of Poro-o-tarao station the line passed through the 1,172 yard long Poro-o-tarao tunnel, the longest on the North Island.

As the train was passing through the tunnel it was Driver Westneath who first lost consciousness. In the darkness of the tunnel Fireman Reidy did not realise what had happened to his driver before he, too, lost consciousness. After the train had exited the tunnel and was back in the open air the fireman regained consciousness and attempted to apply the Westinghouse brake, but it was then too late to prevent the accident and he lost consciousness again. The train was diverted into a siding, ran the length of the siding, collided with two empty waggons stabled at the far end of it and then demolished the buffer stops. The engine then overturned down a small embankment and the 15 leading waggons piled up behind it.

Guard McQuillan and two drovers riding in the rear van did not at first realise what

---

[772] *Weekly Times*. Melbourne, Victoria, Australia, 16th February 1935

[773] *The Age*, Melbourne, Victoria, Australia, 23rd February 1935

had happened as the train became divided. When they did they went to assist the driver and fireman, together with Porter H. Chote, on duty at the station. In contrast to the 1929 Bath accident no lives were lost; the driver and fireman were taken to hospital suffering from smoke inhalation but were not otherwise hurt; the guard and drovers escaped with a severe shaking. Thirty cattle and about 150 sheep were either killed or had to be put down because of their injuries. [774] [775]

The accident was blamed on the poor steaming of the locomotive, exacerbated perhaps by the weather conditions and the absence of wind. At 324 tons the weight of the train was below the maximum permissible load over that portion of the line. No previous problems of the same nature had occurred at that location. [776]

A runaway occurred on the narrow-gauge Denver and Rio Grande Western Railroad near Mears Junction, Colorado, United States, on the evening of Tuesday 6th August 1935. Eastbound freight train Extra No. 492 consisted of engine No. 492, 29 loaded cars, an empty flat car, four dead engines and a caboose, in charge of Conductor Ray Hallock with Engineer L.R. Allen and Fireman Berne Mitchell. It had started from Gunnison at 9.45 a.m., left Mears Junction about 6.45 p.m. and was derailed three quarters of a mile further on, at a point where the line hugged the mountainside with Poncha Creek about 55 feet below. The first 24 cars of the train and the front of the next car left the track, the cars piling up around and behind the engine. Head Brakeman Edmond Boots and three trespassers were killed, and eight trespassers were injured. Two of the dead trespassers were identified as J.A. Johnson, of Farmington, New Mexico, and Joseph Richardson of Denver, Colorado. [777]

Conductor Hallock told the Interstate Commerce Commission (ICC) inquiry that the train had started from Gunnison with 53 loaded cars. With cars being dropped off and picked up by Marshall Junction the load had been reduced to 36 cars. One car had been dropped off at Pocono, another at Grays, a third at Keene, and seven at Mears Junction. The brakes had been tested before leaving Mears Junction and had worked properly. Leaving there the air pressure had been 80 lb but it had dropped to 75 lb after about half a mile. He had assumed that the engineer was setting the brakes, but the speed of his train had continued to increase, so he had used the emergency valve. Before this, he had heard his engineer whistle but had thought that this was a whistle for a flag. He put the speed at 25 miles an hour at the moment of derailment. (Freight trains were limited over this section to 12 miles an hour.)

Rear Brakeman Wilson stated that he and the deceased Head Brakeman Boots had detached the waggons at Mears Junction. He was sure that he had opened the air cock before they left there and that the air had been working through the train. The only reason that he could assign for the accident was that someone had closed an angle cock after the train had left Mears Junction.

Engineer Allen testified that the brakes had worked correctly throughout the journey to Mears Junction. Leaving Mears Junction he had pulled the train a short distance, then

---

774 *The Argus*, Melbourne, Victoria, Australia, 31st July 1935

775 *Tragedy on the Track* by Geoff Conly and Graham Stewart, pgs. 171-2

776 *Waikato Times*, Hamilton, New Zealand, 31st July 1935

777 *Greely Daily Tribune*, Greely, Colorado, United States, 7th August 1935

shut off steam and let it drift. Its speed had been normal until its rear had passed the summit a quarter mile beyond the station and begun the steep descent of 3.75% (or about 1 in 27). Feeling the acceleration he had applied the brakes when he had only obtained a succession of short blasts, indicating a very short train line. An emergency brake application had produced a similar result. Realising that he had lost control of the train he had then whistled for brakes and reversed his engine. He had jumped when the train speed had reached, he thought, 35 miles an hour. He could only suggest that after leaving Mears Junction an air pipe had become kinked or an angle cock near the engine closed.

Fireman Mitchell had little to add to his engineer's evidence.

The ICC stated that there was conclusive evidence that the air brakes had been working properly until after the train had left Mears Junction. There were suggestions that an angle cock had been closed after the train had started from there. Thirteen trespassers had been riding on the train, seven of whom had been questioned and all of whom had denied interfering with an angle cock.

The commission concluded that the derailment had been due to excessive speed due to the engineer losing control of his train on a steep gradient, and that the cause of the air brake failure could not definitely be determined. It noted, however, that the average distance from the top of the rail to the angle cock was 22½ inches and that the angle cocks made convenient steps while moving between cars. The appearance of the angle cocks showed that they had frequently been used for that purpose and there had been previous cases of angle cocks thus accidentally being closed. [778]

Denver and Rio Grande Superintendent of Transportation R.K. Bradford was less cautious than the ICC and stated that he was certain that the accident had been caused by a trespasser stepping on the angle cock behind the fifth car. [779]

On Saturday 1st February 1936, No. 509 Up goods train from Raxaul entered the station at Sagauli, India, on the Bengal and North Western Railway at too high a speed and being turned into the dead-end dock siding, was unable to stop in time and crashed into the buffers. Continuing its course into the station buildings it killed four passengers in the waiting hall and the assistant stationmaster on duty in the booking office. A telegraph clerk escaped with slight injuries. The engine sustained little damage and the vehicles of the train none at all. [780]

Another accident occurred in England on the Somerset and Dorset Joint Railway in the summer of 1936. As usual, on the morning of Wednesday 29th July 1936 0-6-0 tank engine No. 7620 arrived at Braysdown signal box, a mile on the Bath side of Radstock station, at 9.10 a.m. with a train of 12 empty waggons and a brake van. This was the daily trip working whereby empty coal waggons were deposited in the sidings for Braysdown and Writhlington Collieries and loaded waggons picked up. Some of the waggons were placed in Braysdown Colliery sidings, on the Up (to Bath) side, and the remainder were then put into Writhlington Colliery sidings on the Down (to Radstock) side of the double

---

[778] Interstate Commerce Commission report

[779] *The Helena Independent*, Helena, Montana, United States, 7th August 1935

[780] Report by the Railway Board on Indian Railways for 1935-36 Vol. 1

line. A light engine passed towards Bath at 8.41 a.m., after which eight waggons were propelled from the Writhlington sidings onto the Up line and left there, the engine returning to Writhlington sidings. At 9.50 a.m. Signalman Haines at Braysdown accepted the 8.10 a.m. Up freight train from Evercreech Junction to Bath from Radstock. This train consisted of Class 7 2-8-0 engine No. 13803, 37 waggons and a van, with Driver (Passed Fireman) Brewer and Fireman (Passed Cleaner) Hiroms on the engine.

Signalman Haines then asked Shunter Ash to get engine No. 7620 back from Writhlington sidings to push the empty waggons further up the line with the intention, according to the signalman, of belatedly getting the 440 yards clearing point beyond his Up Main home signal. No sooner had the engine returned to the Up line when the Evercreech freight train appeared. It passed the home signal at low speed and a collision with engine No. 7620 appeared inevitable. Realising this, Driver Brewer on the Evercreech engine told Fireman Hiroms to jump and followed suit himself, without having closed the regulator. A shout from Shunter Ash alerted the crew of engine No. 7620 to the oncoming freight train. Driver (Passed Fireman) Rawlings opened his regulator fully to minimise the effects of the collision, then, shouting to his fireman, named Parker, to, 'Look after the engine,' jumped off. The freight train was now going very slowly and Driver Rawlings was able to board its engine and bring it to a halt. Whether or not there was actually a collision between the two engines was not clear. Unfortunately, seeing his driver abandon the engine, Fireman Parker followed suit. The Evercreech freight train was stopped but engine No. 7620 now set off towards Bath, propelling the waggons before it.

Signalman Haines sent the 'Train Running Away' signal to the next station at Wellow, but Signalman Banfield there had no facing points to divert the runaway and could only forward the running away signal to his colleague, Signalman Larcombe, at Midford. Here the double line became single onwards to Bath. Having passed through Wellow at an estimated 50 miles an hour the runaway caused much destruction at Midford. All but one of the waggons were derailed here and the signal box was half demolished, Signalman Larcombe having a lucky escape. The engine continued its journey for another three miles, propelling the remains of one waggon through the single-track Combe Down and Devonshire Tunnels, until finally it ran out of steam and was in any case derailed by part of the waggon getting under its wheels, midway between Devonshire Tunnel and Bath Junction. In contrast to the 1929 accident on the same stretch of line, this time no one was hurt.

Lieut-Colonel E. Woodhouse investigated the accident on behalf of the Ministry of Transport. His report was very critical of Signalman Haines. He pointed out that the 'Warning Arrangment' (Regulation 5, authorising a signalman to accept a train from the box in the rear when the line was obstructed within a quarter mile of his home signal, with the train being cautioned at the box in the rear) was not authorised at Braysdown. Signalman Haines had, in any case, accepted the Evercreech train under full 'Line Clear', pleading that he had been under the impression that the rearmost waggon only had been in the rear of his starting signal. He maintained that Signalman Croucher, who had taught him the box, had told him that the starting signal was his clearing point, which the latter signalman denied. The distances of the various signals were in any case marked on the box diagram, which showed the starting signal to be only 399 yards from the home signal. Furthermore it seemed that the waggons had, in fact, despite the statements of the signalman and the shunter, been left well in the rear of starting signal, and had thus infringed the clearing point to a greater degree. The action of Signalman Haines in then letting engine No. 7620 out from the Writhlington sidings onto the Up line, via a crossover only 155 yards in advance of the home signal, was also illogical. What he should have done was to send the 'Blocking Back Inside Home Signal' to Radstock East. He excused his not doing so by saying that he had been telephoned by Control asking him to shunt the Evercreech freight for a following passenger train, and this had caused him to forget to block back. It seemed likely that his motive for getting the engine to push the waggons further forward was not, in truth, to increase his clearing point, but to move the waggons beyond a crossover which the Evercreech freight would have to use to get to the Down line out of the way of the passenger train.

Turning to Driver Brewer of the Evercreech train, the inspecting officer commented that his conduct was 'particularly to be deplored.' The driver said that he had failed to see the Braysdown Up distant signal at all and had only seen the home signal at danger at the last moment. He had then applied the brakes, reversed his engine and opened the regulator, but the reversing lever had flown back to the forward position and he had been unable to fully close the regulator. Lt-Col Woodhouse remarked that although the driver could not be criticised for supervising the work of his inexperienced fireman, he was seriously at fault for allowing this to interfere with his observation of the signals, especially of the home signal after he had realised that he had missed the distant. Furthermore, his abandonment of his engine was 'quite indefensible' and suggested that he had entirely lost his head. His story that the reversing screw had gone back to the forward position was, however, accepted, as experiment indicated that it had a tendency to do so.

In view of his youth, brief experience, and the fact that he had been told to do so by his driver, the inspecting officer did not criticise Fireman Hiroms for jumping.

A considerable share of responsibilty for the accident also fell, the inspecting officer continued, on Driver Rawlings and Fireman Parker. The former deserved severe criticism for leaving his engine before assuring himself that his fireman knew what he was doing, and it seemed likely that both driver and fireman had also lost their heads. Rawlings' action in boarding the freight engine, while commendable, appeared to have been an afterthought, and it would have been better if he had remained on his own engine.

The only recommendation which the inspecting officer made was that in cases where signalmen receiving a 'running away' signal had no facing or trap points to divert a runaway, they should forward to signal to the next box *immediately*, before attempting to stop the runaway with detonators and hand signals, as Signalman Banfield had done at Wellow, before he had been informed that the train was unattended. [781]

Disaster struck a military train of the Ferrocarril de Cundinamarca in Colombia on Saturday 10th October 1936. It was climbing a steep gradient in a mountainous area and about to enter a tunnel when its rear two carriages, containing about 90 soldiers, became detached and began to run back down the bank. Gathering speed, they continued down the incline until they reached a curve where they left the track, struck some rocks, and were smashed, a number of the soldiers being killed and others injured. The accident happened at kilometre 52 at a location known as Santa Elena, between Albán and La Tribuna stations.

---

[781] Ministry of Transport report

The troops involved in the accident had belonged to a company under the command of Captain Jesús Lloreda, and were returning to Bogotá, having completed the pacification of the Quebradanegra region. (Exactly why the troops had needed to be sent to the region it has not been possible to determine, but it was presumably a right-wing uprising of some sort against the recent reforms of the left-wing government of President Alfonso López Pumarejo.)[782] They had begun their return journey on the morning of the disaster from Tobia in a train consisting of engine No. 23 and five vehicles, two of which were the coaches carrying the troops, two were goods waggons and the fifth was a waggon for the animals. According to the initial account of the accident when the train arrived at Villeta it was stopped to allow a passenger train to overtake it. Regarding this Captain Lloreda had told a newspaper reporter, "There was a small incident involving the Villeta stationmaster and myself. I asked him to give us priority because the soldiers had barely had time to have breakfast before they had set off and it was not possible for them to eat on the train. The stationmaster, however, spoke unsympathetically about the troops and said that he cared little if they went hungry. I protested, but he remained firm in his decision because our train would have to assist the passenger train when it arrived at the Albán – La Tribuna section, because of the steep gradient. But I repeat, we were forced to bank the passenger train."

"How many men were there in your Company?"

"One hundred."

"Do you think it was an act of sabotage?"

"I cannot see how it could not have been given how it happened."

"What can you say about this?"

"I can only tell you that as our train was entering a tunnel the two carriages became detached and began to run back down the slope."

The gradient at this point was, according to a railway employee, five per cent (thus 1 in 20) and if so was, of course, extremely steep. The coaches ran back about a kilometre at increasing speed before they left the track. When he realised what was happening the captain had called for his men to throw themselves from the windows and platforms of the coach and Sergeant Jorge Barreto had done the same. The latter said that there had been 60 men riding in the rear coach and 27 in the other car.

A reporter who arrived at the scene gave a detailed description of what he saw. He said that by 19:00 he had managed to count the corpses of 22 dead. Ten bodies had been lined up along the side of the line. He spoke of pieces of heads and bodies, pools of blood and mud and mangled bodies among the broken wooden boards of the carriages. In the darkness it had been impossible to identify anyone. Among the dead and injured there were rifles, cartridge cases, equipment, hats, blankets and so on. Help had soon come from people nearby though in the darkness it was at first difficult to do very much.

Travelling along the adjoining highway in two cars had been Dr Carlos Arturo Diaz with family and friends, on their way to San Juan de Rioseco. One of the two cars of the doctor was used to take several of the injured soldiers to hospital, the doctor waiting in the pouring rain until the car returned and the party were able to continue their journey.

---

[782] Wikipedia – Alfonso López Pumarejo

Other injured were taken in buses of the Flota Santa Fe company, which operated a service between Villeta and Bogotá.

A relief train arrived at Bogotá La Sabana station with some of the injured at 21:00. A few of the only slightly injured soldiers were taken to the barracks and the more seriously injured were admitted to the Hospital de San José. At the doors of the hospital a crowd of people had gathered, inquiring about their relatives. Inside Padre Sabogal, the hospital chaplain, was going from cot to cot of the most seriously injured patients, of whom there were about 30, administering the Sacraments and absolution. Sisters of Charity (Catholic nuns) were also in attendance, while about ten doctors performed operations, amputations, and set broken limbs. By 23:30 they had operated on 11 patients. The bodies of the dead were taken to a first-floor room. In addition to those taken to Bogotá another 21 of the seriously injured remained at Facatativá, not far from the location of the disaster. Doctors feared that many of the injured would succumb. Only 12 soldiers escaped unhurt and initial reports stated that there were 42 dead.

Cundinamarca Government Secretary Abelardo Forcera Benovides announced that everything indicated that the accident had been a monstrous attack on the army personnel going to contribute to the pacification of Quebradanegra. Mayor Indalecio Forero of Albán agreed. Thus it was that in addition to the rescuers seven detectives were sent to the accident scene and they ordered the arrest of the train crews of the troop and passenger trains. The director general of the police was also sending a special investigator.

A railwayman told Traffic Chief Cuéllar that both the couplings and the side chains of the carriages of the train had been broken, but experts stated that it was impossible that both could have done so while the train was moving at low speed up the incline. Nor was it explained how the couplings and chains concerned, which belonged to a third-class carriage marshalled second in the train, had come to be in such a state. [783]

By the following day it was clear that, though it was bad enough, the accident had not been quite such a catastrophe as had been at first feared. The number of dead was corrected to 23, 22 of them soldiers and the other fatality supposedly being Brakeman Luis Velásquez (but see below). Details of the accident had become a little clearer. It appeared that in fact three, rather than two vehicles had become detached; third-class carriages Nos. 2 and 326 and waggon No. 13, the latter vehicle weighing 8,600 kilos empty and loaded at the time of the runaway with 18,200 kilos of goods. The vehicles had run from kilometre 51 to kilometre 52, where carriage No. 326 had left the rails and overturned and carriage No. 2 had become derailed in turn and crashed into it, followed by the goods waggon.

The cause of the disaster remained unclear. Fireman Sanigo Zuolaga of the troop train stated that there had been no problem with the train brakes up to the time that they had arrived at Albán. As noted above the fireman, Driver Carlos Suescún and the other crew of the troop train had all been arrested on suspicion that they were involved in the suspected sabotage, likewise the crew of the passenger train and a number of other railwaymen, including the unnamed stationmaster who had held back the troop train, who it was stated was a former army reservist. Captain Lloreda, on the other hand, who had been prominent in the rescue work, had apparently changed his mind as to how the

[783] *El Tiempe*, Bogotá, Colombia, 11th October 1936

accident had happened and now said that he did not believe that the runaway had been due to sabotage.

The second day's reports of the accident included stories of the bravery of the troops involved. Captain Lloreda reportedly spoke to a soldier named Faustina Novos, who was very badly hurt, both his legs being severed. The captain asked the soldier how he felt and the injured soldier told the captain that he felt fine. It was then suggested that the soldier let the captain take care of his rifle but the soldier refused, saying that he would keep his weapon until they were back in the barracks. The captain then helped move the soldier to a more comfortable position, the man expressed his gratitude to his commander and then passed away in the captain's arms. Likewise Lieutenant Otriz Reyes urged his men to jump from the train but did not get the chance himself to do so, becoming one of those who perished. [784]

A more detailed account of the journey of the troop train, obtained from a further (?) interview with Captain Lloreda and appearing in the newspaper three days after the accident, varied somewhat from the initial account given. According to this second account the troop train had left Villeta before the passenger train but at Bagazal station, a few kilometres further on, it had been held by the stationmaster for the passenger train from Puerto Llévano to overtake it. Unhappy at the delay and wishing to reach Bogotá by 15:00 the captain had told the driver to continue and that he (the captain) would take responsibility for the driver doing so. The train had then proceeded as far as a siding called La Victoria, at which point the passenger train had been close behind them, so they had gone into the loop there to allow the passenger train to pass. They had then continued to Albán, arriving at that station at about 11:00. Here they had found the passenger train stopped, taking coal and water. It was at Albán, not at Viletta as had been initially reported, that the confrontation with the stationmaster had taken place, and where a request to give the troop train priority had been refused. The passenger train had eventually departed and the troop train followed it. By kilometre 50 the troop train had caught up with the passenger train.

Continuing his story the captain said that, mostly out of boredom, he had at this point been looking out of a small window as the speed of the trains had fallen. They had then reached a short tunnel, about 80 metres in length, by which point the engines of the two trains had been working very hard and had also been slipping badly. The tunnel had been filled with smoke from the two locomotives and the passage of it had taken two and a half minutes, implying a speed of only about 2 kilometres an hour, while the atmospheric conditions for the passengers had been dreadful. The engines had been supplied with automatic sanders but they had not been working. Beyond the tunnel the train had then stopped for a couple of minutes because the passenger train in front had stopped, perhaps to get up steam. The train had then restarted and he had been talking to Lieutenant Ortiz, remarking that once they got to La Tribuna station they might be able to overtake the passenger train and make some progress. Looking out of the window again he had seen that the front of the train was going forward while they were going backwards at increasing speed. Realising the danger that they were in he had at once shouted to his men, "Get off, guys, the cars have come off!" Going out to the platform between the cars he had pushed off a soldier who had been hesitating to jump.

---

[784] Ibid, 12th October 1936

It had then occurred to him that they must be very close to the tunnel which they had just passed through and he in turn had hesitated to throw himself off, but Captain Luis Guillermo Solano, behind him, had shouted, "Jump, Captain, there's still time." He had then done so, about ten metres from the tunnel portal. Solano had followed suit, jumping out in the tunnel itself and suffering a dislocated arm. In all, he said, 16 soldiers had succeeded in jumping out of the train, including Sub-Lieutenant Meneses and Sergeant Barreto. All had survived, though several were injured.

The captain went on to speak of the devastation after the accident, of the fears of the soldiers that the first helpers who had arrived had come to steal their weapons, and of Sergeant Barreta stopping cars coming along the highway for assistance. Later the mayor of Facatativa had arrived with a squad of police who had been able to establish a cordon around the wrecked vehicles. He said that there had been complaints that the front of the train had gone on to La Tribuna and had not come back to help them but he thought that the fact that they had done so was understandable as the soldiers had blamed the train crew who had no doubt feared for their own safety had they gone back. Almost all the weapons and ammunition carried on the train had been recovered and the remainder was probably buried in the wreckage. He admitted that immediately after the accident he had feared that he and his men would be attacked and/or their weapons would be stolen but nothing of this sort had happened. [785]

The investigations into the disaster continued in the hands of a judge, Doctor Carlos Riez. Driver Suescún of the troop train was interviewed and pointed out that the coupling between the two vehicles which had given way had been very rusty and was damaged to the degree that it was in a dangerous condition. The broken coupling was taken away for expert examination. There were also allegations that the two coaches involved in the accident, which had been wooden ones, were 20 years old and in poor condition.

The same day the Ferrocarril de Cundinamarca company issued a statement:

(1) The line at the point of the accident had been in a satisfactory condition

(2) The engine of the troop train had only just entered service. It had been in good condition and in charge of an expert driver

(3) The vehicles of the train had had Westinghouse air brakes and double couplings

(4) The locomotive and vehicles had been inspected both before and during the journey

(5) The transportation department did not authorise any measure not in accordance with the regulations and there was no evidence that and subordinate personnel had violated any regulation

(6) As soon as it had learnt of the accident the company had sent all medical aid and personnel necessary to attend to the injured and remove the bodies of the dead

(7) The causes of the accident were being investigated and would be announced in a timely manner

It also emerged at this time that Brakeman Luis Velasquez had not in fact been killed in the runaway but had disappeared afterwards. It seemed that he had been injured in the

[785] Ibid, 13th October 1936

accident and going up to the highway had been able to beg a lift away from the scene. [786]

On 14th October Judge Riez ordered the release of the crew of the passenger train – Driver Manuel Castro, in charge of engine No. 29, Fireman Salvador Mariño, Conductor Carlos Posada, Assistant Cupertino Correa, and Brakesmen Luis F. Gonzáles, Domingo Marino, Jorge F. Martinez and Luis F. Martinez. Meanwhile Driver Carlos Suescún, Fireman Sanigo Zuolaga, Helper Julio Rivas, and Brakesmen Neftali Cásceres, José Castañada, Celestino Urquijo, Campo Y. Velásquez and his brother Luis Velásquez remained under interrogation pending expert investigation of the broken coupling. (There had been no conductor as such on the troop train.)

Another point which came to light was that the troop train had started its journey with engine No. 2, which for unknown reasons had been replaced by engine No. 23 at Utica station, though it seems unlikely that this fact had any bearing on the accident. [787]

It was implied that the results of the investigation of the disaster would be announced in the next day or so but frustratingly nothing more appears to have been mentioned, at least before the end of the October, while the newspapers for November and December 1936 are not available to consult. To the author, at least, the likelihood that the disaster was the result of an act of sabotage seems doubtful and the accidental breakage of a worn out coupling much more likely. That the brakes of the runaway vehicles had failed to hold them was perhaps due to their weight and the steepness of the incline.

A double-headed mixed train of 24 vehicles on the Canton – Hankow Railway stalled on the steep northern ascent to the watershed between Honan (now Hunan) and Kwangtung (now Guangdong) provinces, between the towns of Hengyang and Shaoguan, China, on Thursday 22nd October 1936. The decision was then made to divide the train, with each engine taking 12 vehicles. If two engines were unable to cope with 24 vehicles it is difficult to understand how one engine could be expected to do any better with half that number and indeed, the leading portion stalled again further along the line. This front portion was then again divided, the four rear vehicles, full of troops, being detached and left behind. The second portion, that is the rear engine and rear 12 vehicles, waited for some time for the line to be reported clear, presumably by telegraph or telephone, but a military officer then reportedly became impatient at the delay, and allegedly threatening the railway staff with a revolver, forced them to proceed. Meanwhile the four detached vehicles, which clearly had not been properly secured, began to run back down the incline, and continued to do so at increasing speed until they met the advancing rear portion. Sixty people, all but seven of them soldiers, lost their lives and another 30 were injured in this disaster reminiscent of the 1889 Armagh tragedy in Ireland. [788]

An error during shunting operations led to a group of waggons running away down an incline on the Riesengebirgsbahn line between Hirschberg and Schmiedeberg, Germany, on Friday 11th June 1937. Before they could be stopped or derailed, they had collided

[786] Ibid, 14th October 1936
[787] Ibid, 15th October 1936
[788] *The Helena Independent*, Helena, Montana, United States, 26th October 1936

with passenger railcar 1441, killing five people and injuring 17. [789]

At 22:57 on Friday 19th November 1937 special train No. 5760 collided with freight train No. 5759 which was stopped in Alanis de la Sierra station, Spain, by this time in the part of Spain conquered by this date by the rebel (Francoist) forces. The accident was reported in the Seville press and by some newspapers in the rebel area but only in a brief note which did not give the number of victims. Republican sources implied that it was a deliberately engineered event rather than an accident.

Train No. 5760 had come from Bilbao and consisted of an engine, a van, eight waggons, a coach carrying Republican prisoners of war, a coach carrying an armed guard, a waggon of equipment and seven waggons loaded with between 260 and 270 tonnes of war materiel, the exact nature of which was not stated. It had five brakesmen. Train No. 5759 consisted of engine No. 1129, 26 loaded waggons and a brake van. The accident resulted, apparently, from a failure of the brakes to hold the special train on the steep falling gradient from Guadalcanal to Alanis (about 1 in 60), because of which it had gained a high rate of speed, ran into Alanis station, and collided with the rear of train No. 5759. The engine of the freight train was not affected as it had been detached from the waggons and moved to the main track before the collision.

The Alanis court civil registry recorded 57 dead, although, the official report, whose results were transcribed to the minutes of the Madrid – Zaragoza – Alicante Railway administration council, stated that there were 72 dead and 179 injured. Most of the victims were Republican prisoners being taken to a labour camp at País Vasco who were crushed in the collision. Four of the dead were members of the train crew and five or six were members of the military escort. [790]

A serious runaway accident happened at Randonnai, France, on the single-track branch between Mortagne and L'Aigle, in 1938. On the day of the accident, Sunday 6th February, two passenger railcars were due to cross one another at Randonnai station. The railcar from Mortagne was within 200 metres of the station when it was struck head-on by its counterpart from L'Aigle. The latter had suffered a brake failure on the descent to the station, was unable to pull up and shot through the station at 50 or 60 miles an hour, past the stationmaster, who realised that a disaster was inevitable.

The railcar coming from L'Aigle was a light one with a central motor; that from Mortagne was a heavier vehicle with its engine at the rear. As a result, the Mortagne car crushed the L'Aigle one for about five metres, pulverising it, and it was in the latter train that the deaths and serious injuries occurred.

Three passengers, M. and Mme Mercier and M. André Lemoine were killed outright, and a Mme Brugeron and a Mme Saillard died of their injuries after being taken to hospital. Eleven passengers and Conductors Beavais of the L'Aigle train and Jean-Baptiste Marcel of the Mortagne train were badly enough hurt to be taken to hospital; eight other passengers went home after having treatment. Motorman René Renet, of the L'Aigle train, and Motorman Albert Vaillant, of the Mortagne train, escaped with slight

[789] *Katastrophen der deutschen Bahnen Teil II* by H.J. Ritzau, pg. 87 (in German)
[790] Webpage: Federación Castellano Manchega de Amigos del Ferrocarril (in Spanish)

injuries.[791]

A runaway accident occurred on the Great Indian Peninsula Railway at 22:30 on Saturday 19th March 1938. Shunting operations were in progress at Jhukehi station, on the line between Katni and Satna, when a waggon began to run away down the bank towards the next station at Amdara, India. The engine which had been doing the shunting went in pursuit of the runaway but before it could retrieve it No. 28 Up Allahabad – Bombay express had arrived, running at full speed, and collided with both the stray waggon and the pursuing engine at mile 685½. Fireman Reid of the express was killed instantly, while Driver Noronha was grievously injured and died during an operation to amputate both his legs. Seventeen passengers and two railwaymen were slightly hurt. Both engines, the runaway waggon and the leading vehicle of the express were derailed. Relief trains were despatched from Jubbulpore and Satna and the Bombay and Calcutta mails were diverted via Jhansi and Manikpur. It was expected that the line would be blocked for 18 hours.[792]

A major disaster occurred in Vietnam, French Indo China on Friday 13th May 1938 on the branch from Dà Lat to the main North – South line between Hanoi and Saigon at Thap Cham.[793] The branch had very severe gradients in places and two rack and pinion sections. It was on the steepest section of the branch, with a gradient of 120‰ (thus about 1 in 8), that the accident occurred. The train involved was the 20:00 overnight passenger service from Dà Lat to Saigon, hauled by 0-8-0T cog locomotive No. 708. Descending from Bellevue (or El Gio) towards Krongpha the train got out of control, ran away, and derailed at K'beu (or Co Bo). It then rolled down a mountainside, a fourth-class carriage was crushed and there were many casualties, though it was said that if the train had rolled a little further or had derailed to the left instead of to the right and fallen over a cliff, the results would have been considerably worse. After the accident 19 passengers were reportedly trapped in the wreckage of the train for a period before they could be released, by which time 13 of them were dead. One account spoke of 15 dead and about 20 injured, another of 17 dead and 23 supposedly dying out of 44 passengers. At least one of the engine crew was amongst those killed.

A newspaper reporter who was sent to investigate the accident was very badly treated by the railway officials, having a strip of film ripped from his camera, being told that taking photographs was strictly forbidden, and being ordered to clear off. He had just spoken to a driver and fireman who had blamed the accident on the dilapidated condition of the engines and rolling stock, plus the excessive hours that they were forced to work for very low wages. Other problems highlighted were the fact that the line was only of metre gauge, that it had at least 50 sharp curves and that it was laid with lightweight rails (25 kg/metre) and had iron sleepers, liable to excessive expansion and contraction and giving poor adhesion to the ballast. The accident had followed, within 10 months, other

---

[791] *Le Peuple*, Paris, France, 8th February 1938
[792] *The Bombay Chronicle*, Bombay, India, 21st March 1938
[793] Wikipedia – Da Lat – That Cham Railway

serious ones at Ar-Lac, Da-Bac and Diêu Tri. [794]

After a long climb to Beattock Summit, 50 miles north of Carlisle, the London Midland and Scottish Railway (ex-Caledonian) main line fell almost all the way to Glasgow, with a few short climbs interspersed. From Craigenhill signal box, 28 miles from Beattock, the line fell continuously for 16 miles to just beyond Uddingston. On Thursday 4th August 1938 the 10.25 a.m. special freight train from Carlisle to Newton consisted of Class 2P 4-4-0 engine No. 14357, 47 waggons, loaded mainly with rolled steel sections, and a brake van, in charge of Guard D. Cameron with Driver W. White and Fireman M. Parker. Two waggons were detached at Beattock Summit.

Tables were provided detailing the maximum number of waggons of various types that different classes of engines could haul over different parts of the system, but the calculation of waggon loads was not easy. At Carlisle Guard Cameron told Yard Inspector G. Pearson that the load was 'equal to 45.' The latter man thought that the engine was of the more powerful 3P class and told the guard that the train was correctly loaded and must run as it was. The guard did not object. Driver White, told of the train's composition, had joked to the guard about its weight, but told the subsequent Ministry of Transport inquiry into the accident that he had been confident that his engine could manage the load. He said that his brakes had been working well and he had had no problems with the descent to Strawfank Junction, Carstairs. He had never had occasion to stop to pin down brakes before beginning the longer and steeper descent from Craigenhill signal box and had not done so on the day of the accident, though he said he would have done had he thought it necessary. Continuing beyond Craigenhill, with the tender brake partially applied to check the train he had passed Law Junction and Garriongill Junction signal boxes under clear signals at 20 or 25 miles an hour. The distant signal for Pather box, just over a mile beyond Garriongill Junction, had been at caution, and he had applied the Westinghouse brake, but the speed of his train had not dropped as he had expected. Before reaching the Pather home signal he had also reversed his engine, applied sand, and whistled for the guard's brakes, but his train had not stopped. It had run past the Pather home signal at danger and half a mile further on it ran into the rear of the 4.05 p.m. passenger train from Wishaw South to Glasgow Central at Wishaw South station. This train was empty and only its driver was on the engine footplate. The force of the collision threw him from the footplate just as he had opened the regulator to lessen the force of the impact. The result was that the engine of this train broke away from its carriages and with its regulator open it steamed away down the gradient, reaching a speed of 50 or 55 miles an hour, until after 3½ miles it overtook the 4.05 p.m. Motherwell to Glasgow Central passenger train at Motherwell Colliery signal box at 4.08 p.m. [795] In this second collision Guard Alexander Gillespie of the Motherwell train was killed. Assistant Guard Gavin Lindsay, riding in the van with him, was injured and two passengers suffered from shock. [796] The passenger train ran over 300 yards with the engine jammed into the guard's van, while its engine became detached in turn and moved forward a short distance before it was stopped.

---

[794] Facebook page: The Horrible Rail Accident in Kabeu: An impartial investigation by Xuan Tieu – HISTORIC VIETNAM.

[795] Ministry of Transport report

[796] *The Times*, London, England, 5th August 1938

Guard Cameron told the inquiry that passing Craigenhill signal box he had applied the brakes in his van fairly hard; he had applied them fully upon seeing the Pather distant signal at caution.

Signalman W. McCulloch at Pather stated that he had seen the freight train approaching his box too fast to stop and he had sent the 'Vehicles Running Away on Right Line' signal to Wishaw South box at 4.02 p.m. He put the speed of the train at 15 or 20 miles an hour.

Guard A. Lochhead of the Wishaw passenger train said that he had been in his van, making up his journal, when the freight train had approached. With him had been Assistant Chief Waggon Inspector G.A. Clark. Alerted by the whistles of the freight train engine and shouts from the station staff they had both jumped from the van and he (the guard) had run towards the engine, signalling to his driver to move ahead.

Driver (Passed Fireman) Simpson had been alone on the engine because his fireman, W. Fraser, was refilling their water bottles from a tap on the platform. The fireman heard the whistling and was running towards the engine to warn his driver, while the latter, by now aware of the danger, had opened the regulator and was looking back to see what was wrong when the impact had thrown him from his locomotive and stunned him. Fireman Fraser was near the engine but could not board it with his driver in the way, then the engine was running away from him too fast to catch.

Signalman J. McGlinn in Wishaw South box said that on getting the 'Running Away' signal from his mate at Pather he had tried to attract the attention of Driver Simpson. He had seen the engine of the passenger train break away from its train but had failed to notice that there was no one on it. He had then offered the engine to Wishaw Iron Works box, had it accepted and had cleared his starting signal for it. He had not sent the 'Running Away' signal, having assumed that Driver Simpson was on the engine and was just keeping ahead of his train. He had reported the events to Control at 4.15 p.m.

Signalman P. McLaughlin at Wishaw Iron Works testified that he had offered the engine forward to the next box at Shieldmuir Junction and had it accepted. When the engine had passed him at 40 or 50 miles an hour he had seen that it was unattended. Instead of sending the 'Running Away' signal to Shieldmuir he had telephoned the signalman there; he said that he had feared sending the 'Running Away' signal immediately after the 'Train Entering Section' would have been confusing. He had then tried for several minutes to telephone Control and Leshmagow Junction, but the lines had been engaged. He added that had he received the 'Running Away' signal from Wishaw South he could have diverted the engine into an unoccupied goods loop.

Signalman J. Grierson at Shieldmuir Junction stated that when the runaway engine had been offered to him by Wishaw Iron Works box he had offered it on to Shields Colliery, it had been accepted and he had cleared his signals for it. The engine had been past his home signal, travelling he thought at 70 miles an hour, before he had learnt on the telephone that there was no one on it. It had then been too late to divert the engine; all that he could do was send it on to Shields Colliery and follow up the 'Entering Section' signal with the 'Running Away' signal. He had then rung Control at 4.06 p.m., this was the first intimation that they had received of the accident.

Like his colleague at Shieldmuir, Signalman T. Reilly at Shields Colliery said that he had forwarded the 'Is Line Clear?' signal for the engine as soon as he had received it, but there had been no time to lower his signals before the engine was past them. There had also been no time to reverse his facing points before the engine had reached them, all that he could do was to send the 4-5-5 'Running Away' signal to Flemington.

Unaware that there was no one on the engine, Signalman S. Grierson at Flemington had shown a red flag to the engine as it had gone by him. He thought that had he had a little more warning he could have diverted the runaway to a loop line.

Signalman H. Greehin at Dalzell Works, where the facing points were in advance of the box, almost succeeded in diverting or derailing the engine but it had reached the locking bar when he had tried to unlock the facing points. He said that he would not have hesitated to derail the engine had it been possible.

Signalman T. McCarron at Motherwell South said that the engine had been going by as he had been sending the 4-5-5 to the north box and there had been no time to divert it into the Down loop.

Signalman J. Robertson at Motherwell North similarly said that the engine had passed him as he was sending the 4-5-5 to Leshmagow Junction.

By the time that the Control office at Motherwell had heard of the runaway engine from Shieldmuir it was too late to stop the 4.05 p.m. passenger train from leaving the bay platform there.

There were three sets of facing points at Leshmagow Junction but Relief Signalman J. Russell, working the frame, said that there had been no time to divert the engine into the sidings while there were trains on both the main and branch lines. His colleague, Signalman T. Twaddle, agreed and explained that due to the interlocking several levers would need to be shifted before the route could be reset. Signalman Twaddle had forwarded the 4-5-5 to Motherwell Colliery as soon as he received it.

Signalman R. McWha at Motherwell Colliery stated that he had received the 'Train Entering Section signal', closely followed by the 'Running Away' signal. As the passenger train neared his box he had seen the runaway engine close behind it and the collision occurred in front of him. He put the speed of the passenger train at 20 to 25 miles an hour and that of the engine at 60 miles an hour.

Driver J. Marshall of the Motherwell passenger train said that he had had no warning of the runaway. When he felt the impact he had looked back to see what had happened. He then found that his engine had become separated from the train, and he kept going until he had been sure that the coaches of his train were at a stand.

The Ministry of Transport inspecting officer, Lieut-Colonel E. Woodhouse, stated that the primary cause of the series of accidents had been the inability of Driver White to stop his train at Panther box. Having had no difficulty in stopping at Strawfank Junction he had been over-confident of his ability to control the train. He must therefore be considered primarily responsible for the first collision.

Guard Cameron appeared to have done what he could to control the train, though it was unfortunate that he had underestimated the load; had Driver White been given the correct load he might have driven more cautiously. However, the inspecting officer did not regard the guard's mistake as too significant as the estimation of loads was difficult, and the maximum loads were given more with regard to the hauling capacity of the engines than their braking ability.

More important was the fact that engine No. 17594 of the Motherwell passenger train had not been eqipped with a combination valve connecting the steam and vacuum brakes. Had it been, the severance of the vaccum pipes when the engine broke away from its train would have applied the steam brake and probably stopped it despite the regulator being open.

It was difficult to understand, Lt. Col. Woodhouse continued, how Signalman McGlinn at Wishaw South could have failed to notice that there was no one on the engine. Had he sent the 4-5-5 signal at once there would probably have been time for one of the signalmen in the following boxes to have diverted or derailed the engine, especially if they had been informed there was no one on it. By its being signalled forward as a regular movement the signals had been lowered for it at the next two boxes, preventing the prompt reversal of the facing points. It was also 12 minutes before Signalman McGlinn had alerted Control. Signalman McGlinn must therefore share a considerable measure of responsibility for the second collision, as he had failed to take the opportunity of averting it 'through lack of observation and alertness.'

The boxes between Wishaw Iron Works and Leshmagow Junction were all very close together and none of the signalmen in them had had any time to do anything after being warned of the approach of the runaway engine.

Turning to recommendations the inspecting officer suggested a simplification of the waggon load tables and the fitting of combination valves to all engines working passenger trains.

Finally, in order to provide as much information as quickly as possible to the signalman in advance regarding a runaway train, Lt. Col. Woodhouse suggested the introduction of a new bell signal – 'Unattended Vehicles Running Away on Right Line'. As far as the author is aware this novel suggestion was never adopted. [797]

On Saturday 11th February 1939, a crowded workman's electric train of the Ferrocarriles de Cataluña from Tarrasa to Barcelona, Spain, suffered a brake failure as it was leaving Les Planes station, ran away, and collided with another train stopped in Sarriá station at about 7.30 a.m. One of the three cars of the colliding train was wrecked in the violent collision. The train which had been run into then started away along the line in turn towards Barcelona, and quickly gained speed on the steep falling gradients until at Plaza Molino, between San Gervasio and Gracia stations, it collided with another train coming in the opposite direction. [798] The number of victims of the accident is unclear. A Madrid paper stated that there were 36 dead, a Cuban newspaper published the next day elevated the figure to 53, adding that of 22 of the injured had succumbed, while a Spanish newspaper of 2004, whose source was not stated, spoke of 32 fatalities and 160 injured. [799] Information on the accident is scanty due to strict censorship imposed by the railway authorities. [800]

---

[797] Ministry of Transport report

[798] *La Vanguardia*, Barcelona, Spain, 12th February 1939

[799] Ibid, 20th April 2004

[800] Webpage: Federación Castellano Manchega de Amigos del Ferrocarril (in Spanish)

A complex series of events led to the death of two railwaymen near Tennant, Iowa, United States, late on Saturday 18th February 1939 on the Chicago Great Western Railway.

Westbound second-class freight train No. 81 consisted of engine No. 755, 56 cars and a caboose, in charge of Conductor Walsh with Engineer John Anderson and Fireman Ed Sharpe. [801] It left Carroll, 48½ miles east of Tennant, at 7.59 p.m., and at Harlan, the last open office, just over 8¼ miles east of Tennant, it received train order No. 63, Form 19, which read:

No. 34, eng. 759, wait at Magill until 10.15 p.m., Tennant 10.25 p.m. for No. 81, eng. 755.

The train left Harlan at 7.35 p.m. and entered the siding at Tennant at 10.10 p.m. Stopping beyond the west switch leading to the parallel house track, the engine was detached and went into the house track and pulled out a car of corn behind a car of ties (sleepers). This car was pushed back onto the remainder of the train standing on the siding and the engine went forward again to return the car of ties back into the house track. As it was setting back into the house track the car of corn rolled forward and sideswiped the engine, breaking a steam pipe and forcing Engineer Anderson and Fireman Sharpe from the cab. With its throttle open the now unmanned engine ran back through the house switch to collide with the rear end of its train at the east end of the house switch. This second collision shifted the reverse gear of the engine and uncoupled it from the car of ties. It ran forward again through the house track, collided with the car of corn which had rolled to the west end of the siding, and pushing that car entered the main line. After propelling the corn car for nearly five miles on mostly falling gradients the runaway engine collided with train No. 34 a mile and a quarter short of Magill.

Westbound passenger train No. 34 consisted of engine No. 759, two baggage cars, a postal car, a coach and a Pullman sleeping car, in charge of Conductor Hart with Engineer Henry Miller and Fireman Art Sellars. It had left Council Bluffs at 9.42 p.m. and Magill at 10.28 p.m., having received a copy of train order No. 63. The collision occurred at 10.30 p.m. Engineer Miller and Fireman Sellars were killed while Brakeman James Conners and Porter Walter Taylor of train No. 34 and Engineer Anderson of train No. 81 were injured.

Engineer Anderson described at the Interstate Commerce Commission (ICC) inquiry the shunting operations at Tennant. He said that he had suddenly become aware that his engine had been struck by the corn car, then had felt scalding water in his face. He had attempted to reach the brake valve but could not close the throttle before he had been forced to abandon the engine. Not having seen the brakeman set any brakes or take any other precautions to secure the corn car, he had hesitated at first to move the ties car to the house track, but as the car had not rolled freely when he had pushed it back, and cars spotted for loading usually already had their brakes applied he had assumed that they were.

Fireman Sharpe said that he had been examining the coal conveyor and had not felt the corn car strike the engine. He had then seen the engineer on the ground and the latter had shouted to him to stop the engine, which had just started moving. He had tried to reboard it, but failed. After the collision at the east end of the house track the engine had

[801] *The Evening Democrat*, Fort Madison, Iowa, United States, 20th February 1939

come back towards him at what he thought was 30 miles an hour. It had not occurred to him to derail the engine.

Head Brakeman Robert Brown stated that when pulling out the cars of corn and ties he had not noticed whether the brakes were set on either, but as the cars would not moved freely he had assumed, like Engineer Anderson, that they were. He had not seen, in the darkness, the corn car strike the engine. When the engine had begun to move backwards he had boarded it, but the steam and hot water had made it impossible to get to the cab and as the engine had begun to gain speed he had jumped off again. He did not think that he would have had time to operate the derail when the engine began to run forward again, even if he had thought of doing so. He admitted that he had not complied with the rule requiring the hand brakes on all cars left in sidings to be securely set and that they also be blocked if necessary.

Conductor Walsh said that when he had learnt that there was a runaway he had notified the dispatcher via the local telephone exchange. He considered the head brakeman to be sufficiently experienced and careful at his work.

Rear Brakeman Mehren said that when he had heard the runaway engine steaming he thought that it was train No. 34 coming and then, when he had not seen either the corn car or the engine he had realised that the corn car had run away and thought the engine was chasing it.

Conductor Hart of train No. 34 said that his engineer had been on the alert as he had answered a communication from Brakeman Conners after they had left Magill. He thought, however, that his engineer would have had no warning of the runaway, as the box car ahead of the engine would have obscured its headlight.

The ICC concluded that, 'This accident was caused by failure to secure a car on a grade, which resulted in the car side-swiping the cab of the engine, damaging air, steam and hot water connections which compelled the engine crew to leave engine without closing the throttle, and allowing the engine to move out of control.' [802]

The internal inquiry by the railroad recommended the dismissal of Head Brakeman Brown.[803]

Train No. 4386, a heavy mixed passenger and freight service composed of 51 waggons and a passenger carriage, which included coal waggons en route from Germany to Italy, had left Brig in Switzerland and was bound for Domodossola in Italy on the Simplon route, on Friday 11th August 1939. As it passed Varzo the stationmaster there realised that it was going too fast on the long descent from the Simplon Tunnel and sent a warning to Domodossola for them to keep the line clear for it. His warning unfortunately was of no use as the train reached a very high speed, supposedly 150 kilometres an hour, and arriving at Domodossola it was going too fast to take a curve and became derailed, the locomotive and 20 or 25 waggons piling up in a heap of scrap metal. Conductor Edmond Gaillard, Driver Rodolphe Grieb, Assistant Driver Erb, two Italian border guards and two passengers all lost their lives. It was not known at the time of the report if there had been anyone in the two passenger carriages attached to the train. It was stated that

---

[802] Interstate Commerce Commission report

[803] *Ottumwa Daily Courier*, Ottumwa, Iowa, United States, 24th February 1939

beyond the obvious fact that its brakes had failed to control the train on the incline, the cause of the accident would not be known until there had been a thorough inquiry. [804] A further report stated that 11 passengers and Brakeman Arnoldo Zibetta had been injured (initially he was said to have lost his life) but Brakeman Imhof was almost unhurt. [805]

The official inquiry into the runaway found that six waggons had been added to the train at Iselle. The tyres of the first three of these waggons were found to have been blued by the action of the air brakes. No blued tyres were found on the waggons of the remainder of the train, though the hand brakes of three were hard on. The non-functioning of the brakes on the greater proportion of the train was clearly due to the blockage of the brake pipe between the third and fourth of the waggons attached at Iselle. It could not be determined if a brake tap had been closed or if a coupling had blocked the brake pipe. Other causes, such as damage or deterioration of the engine brakes, could be excluded. The maximum speed of the train had not exceeded 90 kilometres an hour. [806]

A major disaster occurred on a railway under construction between Harca and Avrio near Sibiu in central Romania on Friday 1st December 1939. A train carrying both material and 59 workmen, and which was possibly overloaded, reportedly suffered a brake failure and got out of control on an incline, ran away, and was derailed. Twenty of the workmen reportedly died and another 16 were injured. An inquiry was opened. [807]

Towards mid-January 1940 Europe was affected by a bout of very severe winter weather. Most of the trains in Romania were running late, sometimes many hours late. Owing to frozen brakes a heavy goods train ran into a siding near Brasov, Romania, on about Thursday 11th January 1940. Several waggons were telescoped, and four guards were killed. [808]

A Ferrocarril Nacional de México train made up of passenger coaches and repair cars, on its way from Mexico City to Chilpancingo in Guerrero State suffered a reported brake failure and ran away down an incline until it reached a sharp curve. There it was derailed and fell down an embankment near Iguala, Mexico, also in Guerrero State, on Saturday 14th December 1940. Initial reports put the number of passengers killed at 19, with 39 injured. [809] With the discovery of another body in the wreckage and the death of five of the injured the number who lost their lives increased to 25. [810]

A mixed train which had left Santander, Spain, at 08:50 on Monday 20th January 1941, came into collision with the mail train from Madrid which was due in Santander at 11:00. There had been no mail from Madrid the previous three days. Beyond Reinosa

---

[804] *Journal de Geneve*, Geneva, Switzerland, 12th August 1939
[805] Ibid, 19th August 1939
[806] *Gazette de Lausanne*, Lausanne, Switzerland, 30th August 1939
[807] Ibid, 2nd December 1939
[808] *The Times*, London, England, 13th January 1940
[809] *Altoona Mirror*, Altoona, Pennsylvania, United States, 16th December 1940
[810] *The Amarillo Daily News*, Amarillo, Texas, United States, 19th December 1940

there was a long, steep descent towards Santander, and it was hereabouts that Driver Aveltno Tormos of the mail became aware that the brakes of his train were not working. The train passed at high speed through Santirude and Pesquera stations and its passengers also realised that something was wrong, especially when, at the latter point, the driver whistled for brakes. The passengers tried to help by ringing the alarm bell (communication cord?) but the speed of the train was little reduced. A parish priest, Don Lucas Alonso Blanco, of Beranga, then persuaded the mail passengers to move to the back of the train as it ran through Montabliz station as well. In addition to the danger of the mail train leaving the track on one of the curves, due to its excessive speed, at one of the stations that it had run through it should have crossed the mixed train, the line being single track. The trains met between Montabliz and the next station at Barcena, between Tunnels Nos. 16 and 17. Thanks to the timely action of the priest the front passenger carriage of the mail train was almost empty when the collision occurred, but Driver Tormos and two mail clerks, Antonio Villada and Francisco Nogueira were killed. In the mixed train Conductor Francisco Herranz and passenger Joaquín Alonso lost their lives. Fourteen people were injured, five seriously, including Fireman Germán Guinea of the mail train. Injuries to the engine crew of the mixed train are not mentioned; possibly they had jumped when they saw the oncoming mail train. [811]

Two freight trains collided in Spain on Tuesday 28th January 1941 on the single line between Jimena de la Frontera and San Pablo de Buceite stations. Like the previous entry this accident is included here because it was due to one of the trains running away. For reasons which were not reported a train coming from Jimena towards Bobadilla, which consisted of an engine and 30 waggons loaded with flour, suffered a brake failure and was unable to stop at Jimena, where it was due to cross an opposing train from Gaucin, and instead continued down a falling gradient at high speed. The Gaucin train had already left San Pablo station and the trains collided on a curve between the two places. Six railwaymen lost their lives in the collision and another 11 were injured. A relief train was sent to the scene from Algeciras. [812] [813]

A month later it was noted that, 'In consequence of the increase in the number of accidents on the Spanish railways, the Madrid Government has issued an order to the effect that official inquiries shall in future be carried out by midland tribunals, with special powers to punish cases of extreme carelessness with the death penalty.' [814] General Franco was probably taking his cue from Joseph Stalin in the USSR. It is not known if any railwaymen were executed for causing accidents in Spain.

At about 15:00 on the afternoon of Tuesday 29th July 1941 a serious runaway accident occurred on the south slope of the slope leading to the Gothard Tunnel in Switzerland. Erstfeld – Bellinzona freight train No. 8622 was a relatively light one of about 300 tonnes. It consisted of Ce 6/8 III electric engine No. 14304, 21 Swiss, Italian, and Czech waggons and two brake vans. The engine was provided with Westinghouse, regenerative

---

[811] *La Vanguardia*, Barcelona, Spain, 21st January 1941
[812] Ibid, 28th January 1941
[813] Webpage: Federación Castellano Manchega de Amigos del Ferrocarril (in Spanish)
[814] *The Railway Gazette*, London, England, 28th February 1941

and hand brakes. The driver first lost control of the train about Rodi-Fiesso and it ran for 17 kilometres before it became derailed on a 300 metre radius curve on the Piano-Tondo viaduct, near the entrance to the spiral Piano-Tondo Tunnel, between Lavorgo and Giornico. All the waggons of the train left the rails. Some of them blocked the entrance to the tunnel, while six others fell into a ravine. A short distance north of the tunnel portal were a watchman's house and soldiers' barracks. Both buildings were struck by the derailed waggons, set on fire by the red-hot wheels and reduced to ashes. Five soldiers who had been in the barracks, the train conductor and a brakeman were killed. The soldiers were guarding the railway, which was a key link between Germany and Italy.

The engine of the train was not derailed and continued down the track at a speed of up to 120 kilometres an hour, through Piano-Tondo and Travi Tunnels, until, after four kilometres, near the old Giornico station, it collided with the rear of freight train No. 886, which had preceded train No. 8622, and was stationary because the derailed waggons had damaged the overhead line equipment, cutting the current. Part of the front train was derailed in turn. The colliding engine was completely overturned and burnt out, but its driver escaped unhurt. The brakeman in the rear van of train No. 886 saw the danger, jumped, and escaped as well. There was considerable material damage, 13 waggons being destroyed at Piano-Torno Tunnel and nine more at the old Giornico station; the locomotive was salvageable. It was thought that the line would be blocked for 48 hours. The injured were taken to hospital in Faido. While the line was blocked passengers were taken by road between Lavorgo and Bodio.

The investigation into the accident by the Chemins de fer fédéraux suisses concluded that a technical problem had caused a failure of the brakes as the train had passed through Rodi-Fiesso. [815] [816]

The line from Paola to Cosenza in southern Italy had steep gradients and included a section worked on the Strub rack system. Passenger train No. 4731, which left Paola for Cosenza at 16:30 on Monday 16th November 1942, was on the rack section between San Lucido and Falconara Albanese stations when it suffered a brake failure and began to run back down the gradient at increasing speed. The runaway journey continued until a sharp curve was reached, where the carriages of the train left the track and smashed up against the wall of a cutting. The engine, detached from its train, continued its course along the line until its driver could pull it up. Eighteen people died and about 50 were injured. Senior railway officials went to the scene to investigate the cause of the runaway. [817]

On the afternoon of Saturday 12th December 1942 a troop train from Piacenza, in northern Italy, composed of an engine and about 30 carriages, arrived at Potenza, in the toe of the country, after a day's journey. The soldiers were mostly from the Lombardy and Emilia areas, and all but one were young men. Continuing its journey through the mountainous and wooded Basilicata region, the train was soon passing through Tito station. Between here and the next station at Picerno it began to pick up speed on the steep falling gradients. The brakes were applied but they failed to check the speed of the

[815] *Gazette de Lausanne*, Lausanne, Switzerland, 30th July 1941
[816] https://second.wiki/wiki/eisenbahnunfall_von_pianotondo
[817] *La Stampa*, Torino, Italy, 16th-17th November 1942

train, which quickly increased over a dozen kilometres. The brakes did not seem to work on some of the coaches, while on others they were hard on, and the sparks began to set fire to the wooden vehicles. One of the survivors of the accident, named Mario Mussii, described later the increasing horror of the recruits as their train went faster and faster, and they were trapped in their carriages in total helplessness, awaiting the inevitable crash. The train was going too fast to jump and there was nothing to be done but to await the impact, and to pray.

The train passed through Picerno too quickly for the staff there to divert it into a siding. It tore past the halt at Franciosa and somehow managed to negotiate a sharp curve at the Jura Tunnel. Then, near the crossing keeper's house at kilometre 138 in the Serra area, the train broke into two and its rear three carriages left the rails and crashed into some oak trees. The front portion of the train was brought to a halt just beyond Baragiano-Ruoti station, where the gradients eased off. It was 17:00.

The inhabitants of Baragiano in general did all that they could to help the survivors, though there was also reportedly some looting.

Twenty-nine soldiers died in the disaster and 150 were injured. The dead were interred in the cemetery at Potenza after a solemn funeral ceremony. Later in the war and afterwards their remains were re-interred at their places of origin.

The accident was not reported in the contemporary newspapers. [818]

Italy suffered another, considerably more serious runaway accident in Sicily on 21st March 1943. It has only been possible to find one account of the catastrophe, in Italian, on the internet, and the author, whose knowledge of Italian is rudimentary, has had to rely on Google Translate. Some of the basic details of what happened remain unclear, but this is the best that I can do.

The events leading to the accident began with the departure from Castrofilippo, near the south coast of the island, of train No. 8864, carrying 800 soldiers of the 476th Coastal Battalion for Termini Imerese, east of Palermo on the north coast. The Allied invasion of Sicily was only a few months away and the island was subjected to repeated air attacks. It was only 130 kilometres between the starting point and destination of the train but due to the Allied attacks its progress was extremely slow. Only about 10 kilometres out the train engine was machine gunned by an aircraft and a day was lost before the train could proceed. Approaching Serradifalco, 14 miles further on, another aircraft again knocked out the locomotive. The train was stopped another day and replaced by train No. 8860, or possibly this was the same train given another number.

At 04:00 on 21st March the train arrived at Caltanisseta Centrale station, 19 kilometres further on. What happened next is unclear. It would appear that permission was given for the train to proceed at 05:10 but for some reason it did not actually depart until 15:15. The two kilometres from Caltanisseta Centrale station to Caltanisseta Xirbi station was all downhill and partly at least in tunnel, and it appears that the train got out of control travelling between the two, entered Caltanisseta Xirbi station at high speed and collided with another train standing there. Explosions followed the collision (of ammunition?).

---

818 www.comune.cinisello-balsamo.mi.it : La Pietro Raccontano – L'Incidente Ferroviario di Baragiano by Gianluca Barnesci

The consequences were very serious, with 137 soldiers dead and another 360 injured. After the accident, an attempt was apparently made to disguise its magnitude of the accident by covering the corpses with sheets 'so as not to alarm the citizens,' but local women took to the streets en masse to watch the passage of the dead, crying out repeatedly, "Figli, figli miei! {Children, my children!}" according to local custom. The soldiers were again mostly from northern Italy, from Veneto and Lombardy. The local population were also generous with visits and gifts to the injured in hospital. As often happened during wartime there were suggestions of sabotage, that a railwayman had deliberately closed the brake valves of the carriages, but nothing to this effect was proven. [819]

Four people had lost their lives in a similar accident at Xirbi back in December 1920. [820]

Since 1875 Istanbul, Turkey, had been unusual in having an underground funicular railway (Tünel), connecting Galata (now Karaköy), the neighbourhood of the city where the port, banks and stock exchange were located, with Pera (now Beyoğlu) at the top of a hill, where were found the embassies, hotels, and commercial markets. In 1943 the 573-metre-long line was double track and had a maximum gradient of 15%. Two wooden two-car trains were provided, one car of each being reserved for male and female passengers (in separate compartments) and the other for animals, carts, and goods. [821]

The funicular suffered a serious accident at an unknown date in early July 1943 due to the snapping of a cable. A train ran away and crashed through the buffers at Galata and out into the street. The initial report from Reuters spoke of two people being killed and 22 injured. [822] A later from German Radio stated that there were 18 dead and 44 injured. Possibly the lower figures referred to passengers and the higher figures also included residents and passers-by in the street. [823]

There was a lucky escape from a more serious accident at Edle Krone, Germany, on Sunday 12th December 1943. A freight train was descending an incline on the Freiberg – Dresden line in Saxony when it got out of control. When its driver found that the air brakes were not working he reversed his engine and did all that he could to stop but the engine left the rails on a curve, followed by the waggons of the train, which piled up in a spectacular heap. The only vehicle to remain on the line was the rear van in which a number of railwaymen were riding. The engine pulling the train had been changed at a previous station and someone had forgotten to open the brake valve at the rear of the replacement engine; there must also have been a failure to carry out a brake continuity test before restarting, and so the problem had remained undetected. [824]

---

[819] www.ilfattonisseno.it: La tragedia del treno dimenticata : Il 21 Marzo 1943 a Xirbi

[820] https://milocca,wordpress.com: Quel treno impazzito per Xirbi

[821] Wikipedia – Tümel

[822] *The Railway Gazette*, 16th July 1943

[823] Ibid, 30th July 1943

[824] *Eisenbahnunfälle in Sachsen* by Claus Schlegel, pg. 174 (in German)

The accident at Torre del Bierzo on 3rd January 1944 which involved a runaway train is dealt with in the Tunnel Accidents chapter.

One of several bad accidents in Norway during the Second World War happened at Breifoss, between Geilo and Ål stations on the Bergensbanen, early on Monday 28th February 1944. Two trains were involved, No. 607, the overnight service from Oslo to Bergen, and goods train No. 1428.

Train No. 607 consisted of type 31b engine No. 403, with Driver Johan Ellingsen and Fireman Leif Boesen and a train described in Norwegian fashion as consisting of '30 axles.'

Train No. 1428 from Bergen consisted of type 39a engine No. 169, two goods vans containing drums of herrings, destined for Germany, 10 tank waggons of oil and petrol, two flat waggons and a conductor's van, in charge of Conductor Kåre Kartevold, with Driver Olav Fossgård and Fireman Olav Vestheim.

The brakes of the goods train had worked normally on the descent to Haugastøl and had not been needed thence in the 11 kilometres to Ustaoset. Beyond the latter place Driver Fossgård suddenly realised that the brakes were not working properly. He at once gave two short blasts on the engine whistle (apply brakes) followed by a series of short blasts (danger). The train was gaining speed and the majority of the waggons only had hand brakes. The conductor reacted immediately. The goods vans made it impossible for either of the engine crew to make their way back along the train to apply the hand brakes on the waggons, but the conductor attempted to make his way forward along the train, a very hazardous procedure in the pitch dark and bitter cold. He tried at first to do so without his hand lamp, so as to leave both his hands free, but this proved too hard, and he return to his van for the lamp. His beloved, Gunda Aslegard, was accompanying him in the van and knowing that an accident was likely he told her to lie flat on the floor and brace herself. His train was due to cross the passenger train (No. 607) at Geilo at 02:56; he hoped that perhaps the crossing had been moved to the next station at Ål.

The train tore through Geilo at about 85 kilometres an hour. The maximum authorised speed was 30 kilometres an hour and sure that the train must become derailed the driver and fireman jumped from their locomotive at the entrance to the station. They proved to be wrong; the train held the rails and continued towards Ål, watched by Operator *(togexpeditør)* Dahl-Michelsen from the door of his office. Dahl-Michelsen presumably did not realise that the engine crew had abandoned their steed and telephoned the ganger (*baneforemann*) at his house at Breifoss. The ganger promised to try and stop the train.

Meanwhile, Conductor Kartevold had worked his way along his train and managed to pin down the brakes of five tank waggons. The last he remembered was seeing the waving of a hand lamp by the ganger at Breifoss, then came the collision. The impact threw him off the waggon and against the side of a cutting; there the snow cushioned his fall. He did not entirely escape though, for several of the tank waggons were smashed by the impact, their contents were released and caught fire, probably from overheated brake shoes, and he suffered burns to his forehead and back.

The crew of the passenger engine had no warning when the collision occurred at 02:53, especially as due to wartime conditions the engines were running without headlights. Driver Ellingsen was killed outright. Fireman Boesen, somehow emerged unharmed, until he suffered burns attempting to rescue his driver. Behind the passenger engine the heavier and stronger leading mail van telescoped the following sitting carriage, killing 25 of its 30 passengers (24 Norwegians and a German woman). In the next carriage there were many injured, mostly not seriously. How many Germans were killed was not stated; possibly, it would seem, only the aforementioned woman, though some sources spoke of German soldiers also losing their lives. Driver Fossgård, Fireman Vestheim and Gunda Alesgard all escaped unhurt. It was common practice during the war for the Norwegian passengers to be made to ride in the leading carriage as a precaution against sabotage. The burning oil soon set the vehicles of the passenger train alight but its conductor, aided by uninjured passengers, managed to uncouple the undamaged part of the train, and move it back out of danger.

A relief train was soon sent from Ål with a doctor, and another doctor came by car from Geilo. As soon as the undamaged carriages could be removed to Ål a breakdown train also went to the scene, with others later in the day from Oslo and Bergen. A large workforce was soon clearing the line, including 25 Russian prisoners of war. Traffic resumed 67 hours after the collision.

Sabotage was suspected by the Germans, and the driver and fireman of the goods train were arrested and interrogated. They were able to convince the Germans that the runaway was not their fault. Even so they were kept captive, first for 70 days at Bergen and then for 10 months at Grini (a concentration camp at Bærum) as hostages against railway sabotage.

Conductor Kartevold received 500 kroner from the state railway, the Norges Statsbaner (NSB) for his efforts to prevent the accident and after the war was also awarded a medal by the King.

An inquiry by the NSB was unable to find any direct failure of the brakes and concluded that instead they must have been working sluggishly due to a combination of the extreme cold and the poor quality wartime lubricating oil. It was also noted that the steam brakes on the engine and the brakes on a couple of the waggons (probably two German waggons, the brakes of which had not been serviced for 20 months) were not fully effective. A brake test which had been carried out at Voss had shown no problems, but there the temperature was only -5°C whereas at Finse it was -29°C. According to a newspaper report there had also been previous runaways in the Geilo area. [825]

On Monday 26th June 1944 on the line between Yssingeaux and Voulte-sur-Loire, between the stations of Bessamorel and Rosières, France, where the line runs along the side of a deep ravine, a well-filled passenger train began to run away down a hill. The brakes refused to work, the driver reversed his engine, but the carriages of the train crashed into one another and the whole train left the track. The alarm was raised and meanwhile the less injured passengers tended their less fortunate fellow travellers as best they could. The mayor and sub-prefect of Yssingeaux set up a rescue operation. The bodies of a dozen victims were retrieved from the wreckage and taken to the nearby Saint

[825] *På Sporet* (magazine of Norsk Jernbaneklubb) No. 90 pgs. 44-45 and No. 158 pgs. 4 -13 (in Norwegian)

Pierre school, where a *chapelle ardente* (chapel of rest) was established. [826]

When a Ferrocarril de Antofagasta a Bolivia freight train No. 811 reached the station at Sevaruyo, Bolivia, on Saturday 7th October 1944 its driver discovered that the distribution valve rings of the air brakes of his engine had fractured. He warned his conductor that that latter should be ready to apply the hand brakes if necessary to control the train on the falling gradient. Notwithstanding this, the conductor chose to ride in an empty covered goods waggon with the brakeman. With no help from the conductor the driver was unable to control the speed of his train and it was travelling at 50 kilometres an hour when it reached Huari station, where train No. 812 was entering a siding. Engine No. 404 and four waggons were damaged in the resulting collision. No one seems to have been hurt. [827]

Monday 6th November 1944 was election day in the Caribbean island of Puerto Rico and in the evening celebrations were in progress to mark the victory the victory of Jesus T. Pinero of the Populare (Popular Democratic) party [828] with a fireworks display. Many people had gone to the capital to cast their votes. At 9 p.m. the last train of the day, No. 3, left San Juan, the capital, for Ponce, on the far side of the island, via the American Railroad. It consisted of engine No. 72, six carriages and two freight cars. The carriages were crowded with people who had travelled to the capital to vote and were now on their way home, while some soldiers were also passengers on the train. Progress must have been slow, for it was 2 a.m. on Tuesday 7th when the train reached the Jiménez station in Aguadilla, about 132 kilometres from San Juan. By then it was an hour late.

At Jiménez station the passenger train crossed train No. 4, a freight train bound for San Juan, and the engine crews of the two trains swopped trains. The passenger train set off again at 2.20 a.m., with Engineer José Antonio Román and Fireman Mario Valentin Ruiz. Engineer Román was a qualified freight engineer but had no experience of working passenger trains, nor probably had his fireman. From Aguadilla the line fell steeply down the Cuesta Vieja (Old Hill) and it would seem that soon after restarting Engineer Román lost control of his train. The passengers, realising that their train was going too fast, became frightened, and at least two mothers threw their young children from the train to save them. At a level crossing at the foot of the incline the train left the rails, fell into a ditch and the boiler of the engine exploded. Police Chief Guillermo Arroyo reported that the engine, an express car and three second-class cars were destroyed. Fourteen passengers, the engineer and fireman lost their lives and about 50 passengers were injured. Fireman Ruiz survived long enough to speak to a newspaper reporter, to whom he said, “It was impossible to slow down, especially when we started to go down the hill.”

The noise of the crash was heard by Engineer Luis Fernández Lourido, another engineer, who was asleep at home. Hurriedly dressing, he had made his way to the station and commandeering an engine had taken it to the scene. He was certain that the blame for the disaster lay with the workshops at Mayagüez in delegating an inexperienced engineer

---

[826] *La Loire Républicaine*, Saint Etienne, France, 28th June 1944

[827] *From the Pacific to La Paz* by Harold Blakemore, pg. 254

[828] *Clovis News-Journal*, Clovis, New Mexico, United States, 7th November 1944

to drive the passenger train. Engineer Lourido also later related finding a year-old child, alive, in some bushes. The body of the child's mother, from Sabana Grande, was found in the second car.

As soon as news of the disaster reached Aguadilla the bells of the mayor's office and the church were rung to call out the citizens and cars and a small bus went to the scene and transported the injured to hospital. Engineer Lourido was occupied for three days with the rescue work. Almost everyone in Aguadilla and nearby towns united to support the relatives of the accident, which was easily the worst ever to occur on the Caribbean island. [829] [830]

A catastrophic runaway accident appears to have occurred to a Wehrmacht (German army) train on Wednesday 27th December 1944 in that part of occupied Czechoslovakia which had become the First Slovak Republic. A train consisting of an engine and vehicles loaded with soldiers, ammunition, and military equipment, coming from Vrútok, apparently suffered a failure of its brakes between Kremnica and Hronská Dúbrava. At the station of Stará Kremnička the Wehrmacht train had to cross both a freight train, bound for Kremnica, and a following passenger train from Hronská Dúbrava. The freight train was already in the loop at Stará Kremnička, with the passenger train following it along the single line. Aware of the approach of the runaway train and that if it continued along the line towards Hronská Dúbrava it would collide head-on with the oncoming passenger train, the station official on duty diverted it into a dead-end siding. Still travelling at a reported 100 kilometres an hour the runaway rammed the reinforced concrete buffers at the end of the siding and was derailed. Fire broke out in the wreckage and the explosion of ammunition caused further destruction. It was alleged that only seven of the soldiers on the train survived, while about 370, as well as the Czech driver and his assistant, lost their lives. Locals described a hellish scene which lasted for several hours, with ammunition exploding and many pieces of wreckage falling in the village. [831]

A major catastrophe in France at about 10:35 on Wednesday 17th January 1945 involved a train carrying American troops. The German counterattack in the Ardennes (the Battle of the Bulge) had been repulsed, but at a considerable cost, and General Patton, in charge of the American Third Army, was in desperate need of reinforcements. Troops who had disembarked at Le Havre from the troopship *Henry Gibbons* boarded troop train No. 2980, which consisted of a British engine and 45 boxcars, bound for a staging camp at Fort Lucky Strike, where they were to remain for six weeks before continuing to the front. Twenty-four of the boxcars were of wooden with poor brakes, the other 21 were of steel construction with better brakes. The British locomotive was not provided with a Flaman speed recorder like its French counterparts.

According to an account written at an unknown date after the event by an American sergeant who was on the train the French engine crew had at first refused to take the train out because of the poor condition of the engine and rolling stock and in particular the brakes. In a repetition of what allegedly happened in France in 1917 to the train involved

---

[829] https://magicinpuertorico.com/2013/04/22/the-train-that-never-arrived/

[830] Facebook page – Mi querido Aguadilla: La tragedia del 7 de noviembre de 1944

[831] https://ozkamenica.wbl.sk/Havaria-nem-voj-transportu.html#.YbKCBr3P1hE (in Czech)

in the Modane catastrophe an American colonel was said to have ordered the engine crew at gunpoint to proceed. Whether or not this was literally true Major General Frank Ross, European Theatre of Operations, certainly demanded that the troop trains get away as quickly as possible, and any delays had to be explained in detail.

The train set off from La Havre at 02:00 and travelled very slowly, taking five hours to cover the 51 kilometres to Motteville. It carried two drivers and two firemen, the second crew riding in the leading car behind the tender, and at Motteville they changed over. After a break the train then proceeded the junction station at St Vaast Bosville, from where it was to continue to St Valéry-en-Caux, a branch terminus on the English Channel coast, close to the Fort Lucky Strike staging camp. At both Motteville and again at St Vaast one of the drivers reportedly expressed his concerns about the ability of the brakes of the train to hold it on the steep descent to St Valéry-en-Caux but he was reassured/overruled and told to proceed. The line fell continuously over the 10 kilometres of the branch and the train soon began to pick up speed, despite the best efforts of the engine crew. The soldiers, fed up with the long, slow crawl riding in spartan boxcars were pleased that the train was at last going a little faster, but the speed increased more and more, and the boxcars were rocking violently until it must have been obvious to most that the train was out of control. The relief crew certainly realised the danger, wrapped themselves in their bedrolls, and lay waiting for the inevitable crash, while the townspeople of St Valéry crossed themselves as they saw and heard the train hurtling to destruction. Finally, with a tremendous crash, the train, travelling at an estimated 100 kilometres an hour, collided with the buffer stops at St Valéry-en-Caux station with disastrous results.

The engine of the train, after demolishing the buffers crossed the station forecourt, rammed the empty waiting room, and ended up with its boiler poking out into the street beyond. The vehicles of the train piled up behind up to ten high, only ten of them remaining intact, presumably those at the rear of the train. Sergeant Lowell Sells related later, "I sat stunned for a moment, not knowing what had happened. A large chunk of iron had come through the front of our car, right between Stafford and me. It was either a coupling or a buffer from another car in front of us. Stafford and I had both been hit on our shoulders, but we both said we were OK. I saw French people running toward the train, most of them with wine bottles. I could not see down the right-hand side of the pile, but I could see to the left. We were high up on the top of six or seven cars. As I climbed down I saw bodies and much blood. But I kept on climbing down. I felt helpless!"

Doctors and nurses from the staging camp rushed to the scene and almost the entire population of the town turned out to dig in the wreckage of the station for victims. Bodies were still being recovered on the evening of the following day. Initially it was stated that 120 had died, with 200 injured, 80 seriously. However, an account written many years later stated that only 89 soldiers had died with 152 suffering permanent injuries, many losing arms and legs. Both engine crews survived and the station itself had been empty, so all the dead were soldiers.

News of the accident was not censored at the time, accounts of it appearing in several French newspapers, in the London *Times*, and in the *New York Times* and other American newspapers, but the American government subsequently denied that any such accident had occurred, apparently to cover up the mistakes of the higher officials involved. At the time of the accident the military police were said to have confiscated the cameras of anyone seen to be photographing the wreckage, whether they be American soldiers or French locals. [832] [833] [834] [835]

The U.S. Army did not investigate the accident, but it was inquired into by the Société nationale de chemins de fer français, who on 21st February reported that the brakes of the train had been inadequate, the absence of a speedometer had not helped the French driver, used to being able to refer to one, and the driver had also not been very familiar with the branch and its gradients. [836]

Passenger train P 3221 ran away as it was approaching the lower station at Eibenstock, Germany, on the Aue – Adorf – Chemnitz line in Saxony, on Monday 29th January 1945 and at 06:39 its engine collided violently with the east wing of a building occupied by a pointsman and his family. A good part of the building was demolished, and the engine and two leading vehicles of the train were wrecked. The driver and fireman were killed, as were three occupants of the first floor of the building. The people in the train were injured.

Because of the steep (50‰ or 1 in 20) descent to the station the locomotives used on the line were equipped with additional counter-pressure brakes. According to the local regulations trains destined for the lower station must first stop at the *Einfahrsignal* at the upper station. Only when the train had thus come to a stand were the points to be reset from a safety siding ending in a sand drag to the incline down to the lower station. These wise safety precautions had, however, been neglected for a long time. When the train arrived at the upper station the points were already set to take it down to the lower station. In addition no brake test was carried out before it began the descent, and it was only when the engine crew came to apply the brakes to pull up that they found that they were not working. [837]

A runaway accident which could have been considerably more serious than it was occurred in Belgium on Friday 16th February 1945. With the Second World War still in progress and Belgium not long liberated, the railways were still under the control of the United States Army Transportation Corps. A train of waggons loaded with petroleum, oil and lubricants got out of control during shunting operations. It began running down a steep incline at increasing speed despite the best efforts of its crew, Driver Technician Fourth Grade Benjamin Tate, and Fireman Private Clarence Chandler, to stop it by applying the engine brakes and the hand brakes on the waggons. Finally, when they realised that they could do no more, the two enginemen jumped and escaped. The runaway crashed into a diesel shunting engine in the yard at Châtelineau and the load caught fire, causing a substantial blaze. Two other railwaymen, Sergeant Gerald Daley

---

[832] *Combat*, Paris, France, 19th January 1945
[833] *France-soir*, Paris, France, 19th January 1945
[834] *The Times*, London, England, 19th January 1945
[835] https://www.tribdem.com/news/local_news/area-soldier-survived-world-war-ii-train-disaster/article_98c73c4b-9bb4-525f-864b-538aa82f1d2f.html
[836] https://www.historynet.com/russell-c-eustice-recalls-the-troop-train-2980-tragedy-at-st-valery-en-caux-during-world-war-ii.htm
[837] *Eisenbahnunfälle in Sachsen* by Claus Schlegel, pg. 176 (in German)

and Lieutenant Edward Gongaware, managed to save 20 waggons from the rear of the runaway by commandeering the engine of a passenger train and dragging them away from the blaze; the 14 next to the engine had just to be left to burn. [838] [839]

At about 20:00 on the evening of Sunday 24th June 1945 the attention of the inhabitants of Merknes, French Morocco, was drawn to the large volumes of smoke emanating from a valley in the direction of Ouarzirha. A little later they discovered that a leave train, which was also carrying oil, had become derailed near Ouarzirha station, 4 kilometres from Merknes.

A rescue operation was immediately begun by both the civil and military services. In the cinema doctors, nurses, railwaymen, and others were urgently alerted to the accident. The Fez Parquet went to the site of the disaster the next morning.

Witnesses related what had happened. A mixed train from Casablanca which was a little late was stopped near Merknes at about 19:27. Its driver saw that a child had climbed up a pylon, probably trying to reach a bird's nest, and had electrocuted himself on the overhead wires. The driver feared that the body of the child, hooked on the wires, would cause a short-circuit if the pantograph of his engine touched it. Stopping his train, he therefore got down from his engine and attempted to remove the corpse of the young victim from the wires with a large wooden pole. Meanwhile about a hundred soldiers, fed up with being cooped up in the crowded train, had climbed down from their carriages. The removal of the body could not be quickly done and after half an hour, while the driver was still endeavouring to do so, the train, which had been standing on a considerable gradient, began to run backwards. As the train began to move many of the soldiers who had alighted tried to climb back into their carriages, while others, who had remained in the carriages, jumped out. The runaway quickly gathered speed and was rumoured to have been moving at 150 kilometres an hour as it negotiated a series of curves until finally it reached a curve which was too sharp for the speed it had attained, left the tracks, and fell down an embankment. A number of waggons of oil, which formed part of the train, immediately caught fire and an enormous blaze ensued, with the passenger carriages in the middle of it. Only the engine and three waggons loaded with tomatoes escaped intact. The remaining vehicles of the train were all involved in the fire from the spilt oil and a horrific sight met the rescuers when they arrived at the scene, with red hot twisted and melted carriages containing the charred remains of the victims.

It was not immediately known how many had died but thought that the number was likely to be high. Those who had escaped were those who had alighted before the train began to run back or perhaps jumped off as soon as it began to do so. [840] Only 85 were thought to have left the train before it began to run away, out of, it was said, over perhaps 500 passengers, but it was not possible, early on, to obtain any official statement as to the number of victims. It was hard to know how many passengers had been on the train. In addition to the ordinary passengers who had come from Casablanca or had joined the train at Rabat, Port-Lyautey or Petitjean, there were, because of a transport crisis, numerous passengers who were riding without tickets in the crowded coaches, in the

[838] https://milwaukeeroadarchives.com/MilwaukeeRoadMagazine/1945August.pdf
[839] https://milwaukeeroadarchives.com/MilwaukeeRoadMagazine/1945December.pdf
[840] *La Vigie Marocaine*, Casablanca, French Morocco, 25th June 1945

corridors or on the platforms at the ends of the carriages. The civilians, who were believed to include women and children, were riding on what was really a military train carrying a detachment of Senegalese tirailleurs. It was also believed that families who had been repatriated to France following the end of the war had taken the train on their way to the embarkation ports in Algeria. On the other hand the number of survivors, stated as 85, might actually have been greater because of those who had fled the scene following the accident. Officially it was stated that 100 people had died; unofficially it was rumoured that the number was 400. Identification of the dead was difficult or impossible. Only a dozen injured were taken to hospital, one of whom died during the night. The injured were passengers who had jumped from the runaway when it was moving at speed.

While the line was blocked by the accident passengers and luggage were transhipped by road between Merknes and Aïn Kerma stations. [841]

Further details of the disaster were announced about ten days later. It was stated that the train had first stopped at km. 252.3 and had been stood there for 33 minutes before it began to run away. It had run for about nine kilometres, attaining a speed of 115 kilometres an hour, before leaving the track at km. 243. Army sources stated that 110 soldiers were 'missing.' The number of civilians who had died was harder to determine but it was believed that 50 had lost their lives. The electric locomotive, of the 6 A.E. class, had belonged to the Chemins de fer Algériens. The engine was wrecked, seven passenger carriages, 14 tank waggons – a sixth of the total stock of the Moroccan company – and six goods waggons were destroyed, and four goods waggons were badly damaged. Two hundred metres of track, the same length of catenary and 220 metres of telegraph wires were destroyed. 268,600 litres of fuel oil bound for Oujda had been lost; no diesel or petrol was involved. The line was reopened on Wednesday 27th June.

The inquiry into the disaster was apparently unable to establish if the brakes of the train had been properly applied before it ran away. [842] If the air brakes of the train had been applied but insufficient or no hand brakes had, it is possible that during the half hour that it was stationary sufficient air had leaked away for it to begin running away.

On Saturday 6th October 1945 freight train No. 11, loaded according to different accounts with tungsten ore, *Räumgut* (room goods) or factory equipment and pulled by a single engine got out of control on the falling gradient approaching Geising, Germany, entered the station at excessive speed and the whole train left the rails. The driver, fireman and a Soviet transport officer were killed, and the conductor and rear brakeman were injured.

The approach to Geising was around a left-hand curve through a cutting, and the line fell towards the station at 1 in 28. Freight trains were not meant to cover the 5.3 kilometres between Altenberg and Geising in less than 18 minutes, passenger trains in less than 10 minutes. The runaway train covered the 5 kilometres between Altenberg and the point of derailment, just short of the Geising *Einfahrsignal* (home signal), in only 5 minutes.

---

[841] *Le Petit Marocain*, Casablanca, French Morocco, 25th June 1945
[842] Ibid, 4th July 1945

A German writer has pointed out that in normal times a BR 64 class engine which had been pulling the train would not have been in charge of such a heavy train on such a gradient without the addition of a second engine. The brake power of the train had also been also defective; it had included French and Italian waggons, one of them a four-wheeler.

The conductor, who had been riding on the locomotive, described the increasing anxiety of the deceased driver as the journey progressed. The Geising signalman said that he had heard the emergency whistles of the train when it was still some distance away and understanding that it was running away he had advised his colleague at the next station at Lauenstein to keep the main track there clear so that the runaway could have a free passage. The rear brakeman of the train stated that he had applied the brake in his van to help control it. The question as to whether he had heard the brakes acting or had seen smoke coming from the wheels, was apparently not asked, nor whether the wheels of the waggons had been examined for the effect of braking after the accident.

The inadequate braking effect of the light locomotive on wet and slippery rails was put forward as one explanation for the estimated 100 kilometres an hour speed of the train when it finally left the rails. Other possible factors were the heavy rain washing any sand put down by the engine crew off the rails, a failure to carry out the required brake test before departure from the starting point or some waggon brakes which were not applied or defective. Finally, there was the possibility that the statements of the surviving conductor and brakeman and of the Altenberg signalman were misleading or even deliberately false in order to protect themselves from being found in some way responsible for the accident (and perhaps ending up in a detention camp, like railwaymen found to be to blame for other accidents in the Soviet zone in this period). [843]

---

[843] *Deutsche Eisenbahn Katastrophen* by Hans Joachim Ritzau, Jorgen Hörstel and Thomas Wolski, pgs. 208-210 (in German)

CHAPTER 8

BRIDGE FAILURES

Some of the most serious railway accidents have resulted from the failure of bridges. Bridges may fail for a variety of reasons. The inherent design of the structure may be faulty, the materials used in its construction may be poor and inferior, the materials may deteriorate over time, or its foundations may be washed by a stream or river which has become exceptionally swollen following abnormally heavy rainfall. With improvements in bridge engineering failures due to faulty design had greatly diminished by the start of the 20th century, but accidents, including major disasters, continued to occur from causes such as bridge foundations being washed away. Later in the 20th century a new danger, mainly though not entirely affecting road underbridges, would be so-called 'bridge bashing' by vehicles which were too high to pass beneath the structure.

Missing from the 1881-1900 volume of this series is an accident on about Monday 8th May 1893. A locomotive and four freight cars carrying labourers broke through a bridge near Abrens, Cuba. An initial report stated that the bodies of two men had been recovered, 20 or 30 were believed to be still under the wreckage and nine had escaped with slight injuries. [844] According to further reports 20 men had been killed [845] but what was probably the most accurate final account stated that there had been only seven deaths. [846]

At about 21:30 on Sunday 23rd August 1931, the engine and six waggons of a Down special goods train fell from Bridge No. 78, between Godhra and Tuwa, India, on the Godhra – Anand section of the Bombay Baroda and Central India Railway. The 20 ft bridge and about 100 ft of embankment at mile 41/13 had been washed away by floodwater. The driver, fireman and four other railwaymen – probably three platelayers and the guard, were killed. [847] It was hoped that the line could be repaired within a week,

---

[844] *Daily Globe*, St Paul, Minnesota, United States, 9th May 1893
[845] *Tribune-Republican*, Union, Missouri, United States, 12th May 1893
[846] *The Ohio Democrat*, New Philadelphia, Ohio, United States, 18th May 1893
[847] Report by the Railway Board on Indian Railways for 1931-32

meanwhile traffic was being diverted via Baroda. [848]

The engine and four cars of a Pacific and Great Eastern Railway mixed train went through Bridge No. 130, 10 miles north of Lillooet, British Columbia, Canada, at 10.30 p.m. on Monday 28th March 1932. Engineer Minor Bazley, Fireman J.R. Duncan and Conductor T.P. Cumming lost their lives and Brakeman C.M. Conley was badly hurt but expected to recover. None of the train passengers was hurt; they included the dead conductor's two daughters, Jean, 15, and Ellen, 13. Aware of the possibility of trouble the train had approached the bridge slowly, but floodwater had washed away the foundations of the bridge, which had given way under the weight of the train. [849] [850]

A serious train accident, which also involved a crack express passenger train, *The Crescent Limited*, occurred just outside of the United States capital in the early hours of Thursday 24th August 1933. The Atlantic seaboard had been struck by a hurricane, the most severe, it was said, for some years, which caused widespread death and destruction and damage estimated at millions of dollars. It also resulted in the failure of the Anacostia River Bridge, near Washington D.C., and the wrecking of the train.

Southbound passenger train No. 147 of the Pennsylvania Railroad, otherwise *The Crescent Limited*, was en route from New York to New Orleans and consisted of engine No. 3855, two mail cars, a combination car, two sleeping cars, a club car, three sleeping cars and an observation car, all of steel construction, in charge of Conductor Wilson, with Engineer A.H. Bryde and Fireman J.H. Faye. [851] Beyond Baltimore, Maryland, it had lost time due to the severe weather conditions and slow running in accordance with a train order instructing it to run cautiously because of possible obstructions due to the high winds, so that when it reached the Anacostia Bridge at 3.23 a.m. it was 45 minutes late. The engine became derailed on the bridge and fell down the embankment at its south end; the first seven cars were derailed and stopped beyond the bridge, the eighth car fell into the river, the ninth car initially remained suspended from the bridge and the tenth car on the track. Later the ninth car also fell into the river, pulling the tenth car onto its side. Engineer Bryde and Fireman Faye were killed and Assistant Road Foreman of Engines Ben Johnson, four passengers, nine mail clerks, two off-duty railwaymen and four Pullman employees were injured.

Assistant Foreman Johnson told the Interstate Commerce Commission (ICC) inquiry that the train had run slowly as far as Odenton, Maryland. At Bowie, just over 11 miles from bridge 2.95 at Anacostia he had measured the level of water in the tank of the engine and in consultation with Engineer Bryde they had decided that they had enough water to reach Washington. The speed of the train had increased to 70 miles an hour at Lanham, 5½ miles from the bridge, but then the left injector had begun to gurgle slightly, and he had told the engineer to let the train drift into Washington. The train had been moving at 40 to 50 miles an hour when the crash had happened without a warning.

Conductor Wilson said he had been in the fifth car when he had felt it quiver and it left the track. He put the speed at 30 to 35 miles an hour and had noticed no brake application before the crash. After the accident he had recrossed the bridge to reach a

---

848 *The Bombay Chronicle*, Bombay, India, 25th August 1931

849 *Medicine Hat News*, Medicine Hat, Alberta, Canada, 29th March 1932

850 *Winnipeg Free Press*, Winnipeg, Manitoba, Canada, 30th March 1932

851 *Charleston Daily Mail*, Charleston, West Virginia, United States, 24th August 1933

telephone and the central pier of the bridge had been in position but about an hour later it had collapsed, leaving the track suspended. The water in the stream had been close to the bridge girders and the current very strong.

Rear Brakeman Douglass and Baggagemaster Selfridge also put the speed of the train at 35 or 40 miles an hour.

Engineer Hare said that he had taken southbound Baltimore and Ohio train No. 9 over the bridge an hour and a quarter before the accident. He had seen nothing wrong with the bridge, though the water had been very high.

Engineer Urban and Fireman Huckins of northbound train No. 400 had crossed the bridge at low speed half an hour before the accident. They agreed that the bridge had appeared normal though the water had been very high.

Track Foreman Bushey, whose length included the bridge, stated that as he had gone off duty at 3.30 p.m. the afternoon before he had asked the track walker if there were any problems and had been told there were not. He had had no fears for the safety of the bridge.

Track Foreman Reigle, in charge of the next section northwards, said that he had been called out by the assistant track supervisor because of high water. He had continued south to the bridge between 10 and 11 p.m. The water at the bridge had then been 3 or 4 feet below the tops of the piers, flowing freely and there had been nothing to indicate an unsafe condition.

Assistant Track Supervisor Smucker testified that he had had reports of the track being under water and had investigated with Track Foreman Reigle but had found that the tracks were not submerged. After giving instructions to the track foreman, he had then proceeded to a washout on the Camp Meade branch. He had felt no apprehension as to the safety of the bridge.

Track Supervisor Pevler said that he had passed over the bridge on the engine of train No. 158 at 7.15 p.m. to investigate reports of high water. At the bridge the water had been 4 or 4½ feet below the tops of the piers and he had seen nothing to cause apprehension.

Dispatcher Eicker said that a slow order covering the area including the bridge had been issued at 7.38 p.m. the previous evening, but annulled at 11.11 p.m.

Bridge Inspector Erdman stated that he had last inspected the bridge on 1st August. He had observed no cracks in the masonry or joints and had assumed that the bridge foundations must be all right, though he had made no underwater inspection.

Master Carpenter Benton said that he had inspected the masonry and steel work of the bridge in June or July, and it had been in good condition. Minor defects, such as the bridge needing painting, had not affected its safety. The semi-annual and annual bridge inspections had all been above-water checks; it was not the practice to make underwater inspections. Underwater soundings were taken after storms of bridges where washouts were expected but no such inspections had been made of bridge 2.95 and there were no periodical soundings to ascertain if a stream was scouring or changing its bed from year to year. Dredging operations had been undertaken since 1928 to within 100 feet of the bridge. Prior to the accident he had not considered that they had endangered the bridge piers; since the accident he had decided that changes in the elevation of the stream bed due to dredging below the bridge might have washed out the gravel underneath the structure and undermined its piers.

District Engineer Fox stated that soundings made five days after the accident had shown excessive scouring of the central bridge pier to a depth of 12 feet below the original bed of the river. There were no signs of derailment on the approach to the bridge and he attributed the accident to the collapse of central bridge pier from undermining.

Engineer S.H. Brackin said that he had supervised the construction of the bridge in 1904. A mid-stream boring had shown 8 feet of mud and sand, below which were at least 5 feet of sand and gravel. The correct depth for the sub-foundation had been left to his judgement and it had been approved by Mr Farnham, the assistant engineer. An open coffer dam method had been used and the sub-foundation had been of non-reinforced concrete. Were he to construct the bridge today he would use the same method except for the employment of an interlocked steel sheet piling casing. In his judgement periodical underwater examination was unnecessary.

Chief Engineer Robert Farnham, the 1904 assistant engineer, related that at that time there had been discussions with United States Army engineers, who had told him that there would be high waters, usually due to ice jams causing the water to back up from the Potomac River. When the bridge had been built he had considered the gravel stratum sufficient foundation for the piers because the gravel had been very compact and cemented. He added that streams were susceptible to changes in their beds and soundings should be made after periods of high water. He thought that such soundings had been periodically made at bridge 2.95. He agreed that dredging operations below the bridge would be likely to lower the bed of the stream at the bridge.

Bridge and Buildings Engineer A.R. Wilson said that the bridge had been built to the standard of the time and that he did not think that a pile foundation would have been necessary as there had been 5 feet of gravel upon which to place the foundations, an amount which a bridge engineer would be pleased to find. He said that soundings taken after the accident had indicated 8½ feet of scour on the south side of the pier and 10 feet on the north side. He was satisfied that the undermining of the pier by high water had caused it to collapse. He thought that the dredging might have had some small effect in causing scouring, but he did not think that the normal velocity of water in the stream would be sufficient to cause damage. Prior to the accident it had not been considered necessary to conduct an underwater examination unless there had been clear signs of a defect above water. From knowledge gained as a result of the accident he now thought that periodical soundings should be made, as well as drainage areas, and deeper foundations with sheet steel piling should be used.

The ICC noted that observations made after the accident showed that the centre pier of the bridge had broken in two about midway and the downstream half of it had disappeared. The girder spans which had supported the northbound track on either side of the pier were in the water. The south pier had broken off on the upstream side and examination by a diver had shown undermining. The north pier had been still in place.

The Pennsylvania company had no record of any soundings made between 1904 and the time of the accident to determine the effect of flood waters on the stream bed or if any dangerous conditions had begun to develop before the storm. Soundings taken by United States engineers after a severe storm in September 1928 had shown filling at the north end of the bridge and a scour of almost 5½ feet at its south end. Soundings taken by them after the accident showed that there had been severe scouring since 1928. In 1928 the stream bed had been about 1½ feet above mean low water, after the accident it was 15 feet below mean low water.

Formerly the Anacostia River had been 100 feet wide at low water, 3 feet deep and flanked by mud flats and marshes. Subsequent improvements had seen it widened to 300 feet with a depth of 6 feet below the bridge, the erection of dikes along its banks and the filling in of the flats and marshes behind the dikes.

In the day before the accident the Washington Weather Bureau had recorded a rainfall of 6.39 inches, mostly between 10 a.m. and 8 p.m. The water depth had increased to 10 feet by 10 a.m. with a maximum of 11 feet 9 inches at 1 a.m.

The ICC concluded that the accident had been due to the scouring of the central pier of the bridge. It was critical of the fact that the railroad company had made no underwater inspections and taken no soundings at any time since the bridge was built in 1904, despite the findings of scouring by the United States engineers in 1928. With the lack of inspections it was impossible to say if there had been any damage to the central and south bridge piers before the storm.

Inspections and soundings to discover evidence of undermining should have been made immediately after the occurrence of flood conditions, before the waters had had a chance to fill in the scouring which might have been created; there had been previous heavy rains in the territory drained by this river which might well have resulted in scourings, and careful inspection and the taking of soundings immediately following such flood conditions might have developed a necessity for action which would have prevented the occurrence of this accident. The commission recommended that such inspections and soundings be made in the future, for without them it would be difficult if not impossible to detect the existence of scour or the tendency of a stream to shift its channel and create conditions which might lead to a dangerous situation.

The American Railway Engineering Association rules for the inspection of bridges, trestles and culverts provided in part as follows:

MASONRY AND COMPOSITE STRUCTURES

(A) Points to be Examined:

(f) A careful inspection shall be made at the water or ground line for disintegration due to the action of water or ice.

(g) The inspection of foundations shall include an examination for indications of scouring or undermining and when necessary, soundings shall be taken for this purpose.

(j) Soundings shall be taken to determine the relative elevation of the bed of stream and the bottom of the masonry.

From the data available it appeared that the water at flood had attained a height 11 or 12 feet above mean low water. In view of the fact that six days after the accident the bed of the stream at the bridge had been scoured to a maximum depth of 15 feet below mean low water, this indicated that at the time of the accident the depth of water passing under the bridge might have been as much as 26 or 27 feet. The evidence indicated that the water had been impounded above the bridge until it was considerably higher than the water below the bridge, and the head of water thus produced undoubtedly accounted for much of the scour which took place. That there was a rush of the water through the comparatively narrow bridge opening was further indicated by the fact that five days after the accident the water both 500 feet above the bridge and 500 feet below had a depth at mean low water of only four feet, and then increased gradually toward the bridge opening until it was 16 feet deep at that point.

Some question was raised at the investigation as to whether the dredging operations carried on below the bridge, together with the confining of the stream to a maximum width of 300 feet instead of allowing it to spread out over the adjoining marsh land, might have had some bearing on what occurred. There was no positive evidence on this point, and even so, it was the duty of the railroad company to keep informed concerning changes which might have any effect on the safety of its bridge, and to take such precautions as were necessary under the circumstances. [852]

The Southern Pacific Company suffered a major disaster on Tuesday 29th August 1933 with the failure beneath a train of Bridge No. 1623.27, about a mile and a quarter east of Hargis, New Mexico, United States.

The bridge in question was a three-span deck-plate girder structure, 210 ft long, which crossed the normally dry Blue Water Creek. The train was eastbound passenger train No. 4, the *Golden State Limited*, which consisted of engine No. 4388, a mail car, a baggage car, a chair car, a Pullman tourist sleeping car, a club car, five Pullman standard sleeper cars and an observation sleeping car, in charge of Conductor Arthur DePew, with Engineer C.J. Crofts and Fireman James Randall.

Santa Rosa, 53½ miles west of Hargis, had been passed at 2.37 a.m., and the train was travelling at about 20 miles an hour when the bridge settled beneath the engine and its eastern span then collapsed at about 4.20 a.m. The front of the engine struck the far embankment and the tender, and first five cars of the train fell into the gap and piled up on the engine, the bridge girder, and the embankment. The sixth car fell into the stream bed and the front of the seventh car was derailed. Engineer Crofts, Fireman Randall, Bridge Inspector Paul Cook and eight passengers were killed and four railwaymen, 39 passengers, a mail clerk and two Pullman porters were injured. [853]

One of the passengers injured was Harvey Miller Sr., a New Rochelle, New York merchant, who had been returning from Los Angeles with the body of his son, Harvey Jr., 31, who had died in Los Angeles following an operation. News of Mr Miller being injured was being kept from his wife, who was already prostrated by the death of their son.

Another passenger, Mrs W.B. Clark, escaped with a wrenched back but was trapped for over an hour beside the body of a dead woman. She said, “The blood dripped on me from other injured persons. It was awful. Without warning the bags, seats and persons in the car began to tumble about. There were groans and screams. I couldn’t move. I knew the woman pinned at my feet was dead.” [854]

Conductor DePew told the Interstate Commerce Commission (ICC) inquiry that at Vaughn, 96½ miles short of Hargis, he had received a slow order for points west of Hargis plus an order which read:

Within storm limits between Pastura and Tucumcari run carefully, engineers keeping sharp lookout for high water and soft track reducing speed if necessary to insure absolute safety.

Pastura was about 74½ miles west of Hargis. At Santa Rosa train order No. 759 was received. This read:

Very high water from mile post 1610 to Tucumcari bridge just east of mile post 1613 and near mile post 1614 running full track washing in cut west switch Palomas spillway and pond just east of Palomas is dangerous. Stop and examine before passing over it to make sure track safe.

Palomas was 6½ miles west of Hargis. After leaving Santa Rosa the train was stopped and the conductor examined the bridge together with the engineer as per the order. The train was again stopped short of Hargis because boulders had fallen near the track. The

---

[852] Interstate Commerce Commission report

[853] Ibid.

[854] *El Paso Herald-Post*, El Paso, Texas, United States, 30th August 1933

train having passed Hargis station, Conductor DePew said that there had been a service brake application, followed by an emergency brake application and then a crash. He had arranged with Conductor Vaughn, who had been deadheading, to apply the hand brakes of the cars which were not derailed, then made his way to Tucumcari via an adjoining highway bridge. From that structure he could see that the water in the stream was four or five feet deep, but driftwood indicated that it had been considerably deeper. He had never considered the bridge unsafe.

Conductor Vaughn and Head Brakeman John Harvey corroborated Conductor DePew's statement, except that the head brakeman put the speed of the train somewhat higher.

Rear Brakeman McMechen said that the train had been handled very carefully. They had encountered no rain between Santa Rosa and the bridge.

Engineer Jacobs of westbound train No. 3 stated that it had begun to rain at 11.45 p.m. He had left Tucumcari at 12.32 a.m., and it been raining very hard between there and Palomas. He had crossed Bridge No. 1623.27 at 40 miles an hour; the water in the stream then had been from 18 inches to 2 ft deep. The bridge had appeared to be safe. Further west he had been flagged because of high water and at Santa Rosa he had sent a message to the dispatcher to caution train No. 4 regarding the conditions.

Fireman King, who had been deadheading on train No. 4 and had previously passed over the bridge the morning before, Engineer Alexander, who had passed over the bridge on the 26th and the 28th, and Engineer Wack, who had crossed it on the 27th, had noticed no problems.

Deputy Sheriff White, who had lived at Blue Water Creek for 27 years, said that the rain on the night of the accident had been the heaviest he had ever known. He had feared for the safety of both the highway and railroad bridges and had discussed with his family whether he should go and flag trains.

Rancher Louis Metzger, who had lived in the area 22 years, recalled a similar storm in 1924, before the erection of the highway bridge. There had been two previous storms in August and earth had been washed away from the bridge piers. He considered that the situation would be dangerous were more earth washed away but had not reported the matter to the railroad or highway authorities as he had assumed that they would attend to their own business.

Section Foreman Roberts, based at Palomas, stated that on his own responsibility he had started east on a rail motor car with his men to examine the line because of the heavy rainfall. They had passed train No. 3 at Hargis, passed over the bridge going eastwards at 12.40 a.m., returned to it at about 1.20 a.m. and had then examined it using a flashlight and a lantern. It had been raining and there was about 5 ft of water in the stream. He had not been able to see any sign of the washing away of the earth around the piers and deciding the bridge was safe had returned to Palomas, where train No. 4 had passed him. He had heard of the accident at 6.30 a.m.

Roadmaster Wells stated that he had examined some small pile bridges where he had thought there might be trouble on the morning of August 28th but not Bridge No. 1623.27, where he had never experienced any problems.

Assistant Signal Supervisor Garrison, who had been driving the motor car with Roadmaster Wells, said that he had not observed any washing or caving of the bank between the highway and railroad bridges.

Supervisor DeBord stated that he had made his annual inspection of the bridge on 14th July. He had not considered the use of rip-rap (material placed to protect the bridge from scour and erosion) necessary nor been concerned about the presence of the nearby highway bridge. Had he been aware of the conditions described by witnesses just before the accident he would have taken immediate action.

Assistant Divisional Engineer Dawson considered that the location and type of the highway bridge had had some effect on the erosion of the east bank of the creek, however, the need for rip-rap, wing walls or other erosion protection there had, as far as he was aware, never been discussed.

Divisional Engineer Stansbury thought that there had been no flood of such severity since at least 1913. He thought the highway bridge a significant factor in the washing away of the railroad embankment either because of its location or because a head of water had built up behind it. He said that had railroad officials been aware of the washing away of the bank in previous August storms protective measures would have been taken.

Bridge Engineer Van de Greyn of the New Mexico State Highway Department stated that in many cases when highway bridges were being constructed adjacent to railroad ones there had been a discussion with the railroad authorities regarding possible effects, but no such conference had taken place in this case. From the high water mark the water had risen 6 ft higher on the night of the accident than on any previous occasion. As to whether the highway bridge might have exacerbated the washing away of the railway embankment he considered that it might have offered some protection, on the other hand any impounding effect from the decreased opening under the bridge would increase the current and trees, logs or other material might have lodged against the bridge to deflect the current.

General Bridge Inspector Bober of the Chicago Rock Island and Pacific Railroad believed that the bridge had collapsed beneath the engine, not before. He considered the highway bridge a significant factor as it had greatly narrowed the opening for the water, thus increasing the velocity of the current, and probably had also imparted an oblique action to the current, forcing it against the railway embankment.

The ICC concluded that heavy rainfall had undermined the earth fill supporting the east span of the bridge, causing it to collapse beneath the train. Three conditions had contributed to the failure: very heavy rain leading to an extraordinary amount of water in Blue Water Creek, the highway bridge increasing the velocity of the current and probably directing it against the railway embankment, and the lack of any protection against the erosion of the railway embankment.

The railroad bridge had been constructed in 1914 and the highway bridge 150 ft upstream in 1931. Had a comprehensive survey been made by competent railway engineers following the addition of the highway bridge the changed hydraulic conditions resulting from the construction of the latter should have been recognised. Furthermore, the railway company had not made sufficiently frequent or thorough inspections to identify the washing away of the east bank of the stream between the bridges caused by storms earlier in August, which had been noted by local people, and which had indicated the need for protection of the railway embankment.[855]

A Southern Pacific Company fast freight train of 40 cars from Santa Rosa to Oakland fell through the Buchli (spelt Buckley in the newspaper report) Trestle, near Napa Junction, California, United States, on Saturday 2nd September 1933. The engine and

[855] Interstate Commerce Commission report

leading car crossed the short (25 ft) trestle safely, but the next 11 cars fell through it, were telescoped, and derailed while the shock derailed the engine. Engineer Martin Fairfield and Fireman J.H. Wolfskin were unhurt but five hoboes, all of whom said that they were from Detroit, Michigan, were treated for cuts and bruises. The cars which fell were a mixture of tank and box cars. Napa Fire Department attended to put out the blaze and a wrecking (breakdown) train was sent from Oakland. The damage caused was estimated at between $50,000 and $75,000. This time it was not floodwater which had caused the wooden trestle to fail but an undetected fire in its underpinning, perhaps caused by a spark from an engine. [856]

The train passengers and crew aboard the 11.20 a.m. mixed train from Collie to Narrogin had a lucky escape on Thursday 8th March 1934. As the train was crossing a bridge over a creek a quarter mile west of Dumberning, Western Australia, the structure collapsed. The engine had barely crossed the bridge when the embankment on the west side gave way and the engine fell back into the creek, followed by about seven waggons at the head of the train. Driver J.G. Hall and Fireman A. Gilbert were fortunately able to jump clear and the coal trucks, single passenger coach and guard's van at the rear of the train remained on the track. Four large trucks were described as forming a pyramid in the creek, with a water tank on each side and another truck, badly crushed, beneath them. The few passengers were later forwarded to Narrogin in a motor car. Other lines in the area had also suffered washouts following a severe storm. [857]

On the morning of Sunday 29th February 1936 a repair train left Rodinga, Northern Territory, Australia, to make good damage caused by recent heavy rain and flooding. A bridge only a quarter mile north of Rodinga over Camel Creek had previously been examined by permanent way staff and declared safe. Presumably in the interval between its examination and the arrival of the train floodwaters had washed away its foundations, leaving the track suspended over the gap. The train consisted of an engine and three trucks. Unaware of the danger its driver entered the bridge, which was about 30 yards long, and the engine and two of the waggons fell to the muddy bed of the creek. The driver and the guard, who had been riding on the locomotive, were thrown clear and not much hurt. When they recovered themselves they found that the engine was lying on its side with no sign of the fireman. He was found buried in the mud beneath the engine. The engine had not settled in the bed of the creek, so they were able to pull the fireman free. He had suffered a broken leg and with the driver and guard was taken to Alice Springs on a quadricycle and admitted to the Inland Mission Hostel. [858]

Early on the morning of Tuesday 17th May 1937 an engine in charge of Engineer O.C. Mosley with Fireman D. Taylor shunted a car of gasoline from the station at Nashville, Arkansas, United States, to the bulk station of the Gulf Refining Company, a short distance away, a trip with involved crossing a trestle over Mine Creek. The outward journey passed without incident but as the engine was backing to the station the central pier of the trestle gave way and the engine plunged into the creek, killing the engineer and fireman. A brakeman named Martin was only slightly hurt and a fourth railwayman,

---

[856] *Oakland Tribune*, Oakland, California, United States, 3rd September 1933
[857] *The Western Argus*, Kalgoorlie, Western Australia, 13th March 1934
[858] *The Age*, Melbourne, Victoria, Australia, 2nd March 1936

Clint Burleson, who was probably the conductor, was unhurt. The accident happened on the Peach Orchard branch of the Missouri Pacific Railroad at 5.25 a.m. according to the engineer's stopped watch. The central pier of the trestle had apparently been undermined. The same morning a bridge inspector arrived at Nashville to inspect the trestle, to be told that it had collapsed. [859]

Severe flooding affected the eastern Panhandle of Texas in mid-June 1938 and led to the wrecking of a Fort Worth and Denver City Railway freight train. On Thursday 16th June 1938 train No. 92 was a northbound freight working consisting of 2-8-2 engine No. 409, 30 cars and a caboose, in charge of Conductor Joe Harper, with Engineer M.V. Griggs and Fireman C.E. Burton. It had left Childress, about 42 miles south of the scene of the accident, at 1.20 a.m., passed Wellington, 10 miles south of the scene of the accident, at about 3.20 a.m., and fell through Bridge N-262.39, 4¾ miles south of Samnorwood, Texas, United States, at 3.50 a.m.

Bridge N-262.39 was a 46-span open-deck pile trestle, 920 feet long, across the Salt Fork, a normally dry branch of the Red River. The engine and tender fell from the trestle to the bed of the creek and were almost entirely buried in quicksand. The front end of the leading car of the train stopped leaning over the edge of the broken bridge and the remainder of the train remained on the track. [860] The engineer and fireman were killed, and Flagman Dick Brown was seriously injured. Flagman Brown was carried five miles downstream before he was able to scramble ashore. He went to the farm of a Mr A.Y. Bell and a neighbour of Mr Bell took him to hospital. [861]

Head Brakeman H.L. McKenzie told the Interstate Commerce Commission (ICC) inquiry into the accident that the brakes of the train had been tested before leaving Childress. It had been raining slightly there and at Wellington. At Wellington they had met southbound train No. 91 and had been given a note to the effect that there were heavy rains between there and Meldavis, that the water was high at Salt Fork, and not to leave Shamrock (about 6 miles beyond Samnorwood) until they had seen the section foreman. He (McKenzie), the engineer, and the flagman had read the note. Nothing had been said about stopping to examine Salt Fork bridge, which they had begun crossing at 5 miles an hour when he had seen that a rail was out of line. He had shouted a warning to the engineer, who had applied the brakes. Climbing back from the tender he had reached the first car when the engine and tender had fallen from the bridge. He thought that the water in the creek had been between 15 and 18 ft deep and the current very swift. Never previously had he seen high water in the creek.

Flagman Brown added that he had spoken to Section Foreman Collins at Wellington and the latter had mentioned that there was high water in Salt Fork creek.

Conductor Harper said that he had been told at Childress by both Superintendent Mode and the dispatcher to instruct his engineer to run carefully, as there was a great deal of rain up the road. He had understood, however, that the problem area was around Magic City (25 miles beyond Samnorwood). At Wellington the engineer of train No. 91 had called out that there was a lot of water up the river. His train had passed onto the bridge very slowly and he had thought when it had stopped that it was to inspect the bridge.

The members of train No. 91 stated that they had crossed Bridge N-262.39 at about

---

[859] *Hope Star*, Hope, Arkansas, United States, 18th May 1937

[860] Interstate Commerce Commission report.

[861] *Wichita Daily Times*, Wichita, Texas, United States, 16th June 1938

2.40 a.m., when it had been in good condition. The water in the creek had been the highest they had ever seen it, 5 or 6 ft deep, and the current swift. There had been some drift material up against the bridge piling. Heavy rains had been encountered north of Salt Creek (beyond Shamrock) and section foremen had preceded their train from Bellco (33 miles beyond Samnorwood). The note handed to train No. 92 at Wellington had read:

Hard rain Meldavis to Bellco. Have someone go ahead of you. High water at Salt Fork.

Dispatcher Henderson, on duty at Childress until midnight, stated that no rain had been reported south of Magic City and no water at Salt Fork Creek. Heavy rain had been reported north of Magic City and he had held train No. 92 for some time pending notice of a washout further down the line. He had never known of any problem at the Salt Fork bridge and had been unaware that the watershed at Clarendon on the main line, where a section foreman had reported flooding at 10 p.m., drained into Salt Fork Creek.

Dispatcher Bush, who took over at midnight, stated that a section foreman at Samnorwood had reported heavy rain at 2.30 a.m. and that he was patrolling the track ahead of trains Nos. 91 and 92, while the conductor or train No. 91 had also informed him that he would mention the weather conditions to the conductor of train No. 92. Like his colleague, Dispatcher Bush was unaware that the watershed at Clarendon, where 6 inches of rain had fallen, drained into Salt Fork Creek.

Section Foreman Collins at Abington, between Childress and Wellington, stated that he talked to the dispatcher at 9 p.m. about rain at Miles City moving southwards. His section ended 13 miles south of Salt Fork bridge, but he had continued northwards (presumably on a track motor car) until he had met Section Foreman Mills 11 miles north of Bridge N-262.39. Returning southwards at 2 a.m. he had examined the bridge with a flashlight; the water had then been about 3 ft deep, and he had not anticipated any problem there. The bridge had formerly been in his section, and he recalled three previous cases of high water; the current was usually swift. He had returned to Abington ahead of train No. 91.

Superintendent Mode testified that because of rain reported at Magic City he had instructed the section foreman there to go north and meet train No. 91 and precede it southwards. At 11 p.m. a washout had been reported at Lelia Lake on the main line to Amarillo (which left the branch to Wellington, Miles City and Pampa at Childress). As he had been leaving for Lelia Lake he had seen Conductor Harper of train No. 92 and had told him to warn his engineer to run carefully because of rain beyond Magic City. The superintendent said that he had only heard of the accident at about 4 a.m. He was of the opinion that the swollen creek had first knocked out highway bridges upstream of the railroad bridges and that it was the debris from these which had then demolished the railroad bridge.

The ICC's own inspectors noted that the remaining portion of Bridge N-262.39 was in good condition. About 400ft of an all wooden highway bridge 7 miles upstream of the railroad bridge had been washed away on the night of the accident and debris from this bridge had been found downstream from the railroad bridge.

The ICC concluded that rain of cloudburst proportions had washed away the highway bridge upstream of the railroad bridge and that the debris from the highway bridge had subsequently been carried downstream and struck the piling of the railroad bridge, knocking it out of line and weakening it to the extent that it had collapsed beneath train No. 92. There was no indication of scouring of the creek bed. It made no recommendations. [862]

Three days after the Salt Fork accident a major disaster occurred near Saugus, Montana, United States on the Chicago Milwaukee St Paul and Pacific Railroad. The accident involved *The Olympian*, the crack passenger train of the company, and resulted in the death of 41 passengers, two persons carried under contract and four railwaymen and the injury of 56 passengers, one person carried under contract and 18 railwaymen, five of them off duty.

Bridge AA-438 spanned Custer Creek 4,032 east of Saugus station. The creek was dry for nine months of the year and never usually carried more than three or four feet of water. The bridge was a steel and reinforced concrete structure 180 ft long, with six intermediate piers, two 50 ft plate girder spans and five concrete trestle slab spans. The piers were protected by rip-rapping (material placed to protect the piers from scouring).

The Olympian, otherwise westbound passenger train No. 15, consisted of engine No. 220, a mail and express car, a baggage car, two coaches, three tourist sleeping cars, a dining car, two sleeping cars and an observation car, in charge of Conductor McGee with Engineer Frank Merrifield and Fireman H.M. McCoy. At Marmarth, 97¾ miles east of Saugus, it had been given Form 19 train order No. 51:

Heavy rains reported from Bluffport to Bonfield. Section men patrolling track.

Bluffport was 20 miles east of Saugus and Bonfield six miles west of there.

At Mildred, just over 34 miles east of Saugus, Form 19 train order No. 395 was received:

Do not exceed 20 miles per hour between switches Calypso account of soft track.

Calypso was just over 6½ miles east of Saugus.

The train left Marmarth at 10.19 p.m., 19 minutes late, on Saturday 18th June 1938, left Mildred at 11.47 p.m., 20 minutes late and the accident happened at 12.35 a.m. on Sunday 19th June. The engine and tender came to a stand just beyond the bridge, the first car ended up on top of the engine, the second car on top of the tender and the next five cars in the stream bed. Engineer Merrifield, Fireman McCoy, and Baggageman Charles James were three of the railwaymen killed. [863]

A newspaper reporter underlined the human side of the accident. 'At the hospital, and at the Miles City morgue, where relatives came to claim their dead, stories of tragedy and of heroism were told by the survivors.

'Ten year old Anne Lallas, whose mother was dead and whose father was missing, related how she tried desperately to keep her mother's head above the swirling water. When the mother, Mrs Thomas Lallas, fainted from shock, the small girl grabbed her hair and held her above the water for several minutes. Anne's brother, nine-year-old George, suffered a broken arm. Unable to comprehend the tragedy, he kept crying, "I want my mamma and daddy."

'"But you haven't any mamma or daddy any more," the boy's sister tried bravely to explain. The family was en route to Bellingham, Wash., from a visit to Greece.

'A negro porter on the train, Lew Williams, was credited by survivors with possibly saving several lives by calmly directing passengers out of his car.'

'Ellis Lund and Thomas Thoreson, of Dawson, Minn., together helped "15 to 20" persons from another car. "It was like a dream, I hardly know how I got out," Lund said

---

[862] Interstate Commerce Commission report

[863] Ibid.

later. He and Thoreson were in the smoking car, one end of which was crushed between two coaches in the creek bed.

"The first I knew it was as if we were going through air," Thoreson related. "Then there was a terrible bump. I don't know what happened. The water started pouring in. People in the other part of the car screamed and fought to get out," Thoreson said.' [864]

Conductor McGee told the Interstate Commerce Commission (ICC) inquiry that the brakes had been tested at Marmarth and had worked properly. He had been in the fifth car when the accident had occurred. There had been no previous brake application. He could not recall any previous slow order regarding Bridge AA-438.

Brakeman Reece put the speed of the train at 45 miles an hour at the time of the accident. He thought that there had been about 20 ft of water in the main channel of Custer Creek; he had never previously seen more than 2 or 3 ft there.

Flagman Preston stated that he had gone back to flag immediately after the accident. He had seen no marks to indicate a prior derailment of the train. At that time it had been dark and cloudy but had not been raining. He put the train speed at between 40 and 50 miles an hour. Returning 45 minutes later there had been much water in the creek and it had been roaring such that one had to shout to be heard. He had never seen more than 2 or 3 ft of water in the creek.

Fuel Supervisor Griffith, riding as a passenger, stated that initially the sixth and seventh cars had stopped on the bridge before first the sixth and then the seventh had fallen into the stream. The water had been coming down in waves. No brakes had been applied before the accident.

Engineer Davis said that he had driven the last train (No. 263) over the bridge at 10.15 p.m. The water then had been 3 or 4 ft deep. There had been heavy rain at Terry, 12 miles east of Saugus, but none west of there.

Dispatcher Hagerty of Miles City, on duty until midnight, testified to being advised at 7.50 p.m. by the section foreman at Calypso of heavy rain there damaging the track there and he had issued a slow order. At 9.45 he had been informed of rains at Terry, Calypso and Mildred so he had issued another order covering the line between Bluffport and Bonfield. He had not issued an order about Bridge AA-438 in the last ten years.

Dispatcher Grogan, who took over at Miles City at midnight, likewise could not recall any previous problems at the Custer Creek bridge.

Section Foreman Goplen said that it had rained at Calypso from 6.30 p.m. to 8.15 p.m. At the latter time he had patrolled the line westwards to beyond Saugus station, accompanied by Section Labourer Eastwood. Returning, they had met train No. 263 at Saugus at about 10.5 p.m. and had reached Bridge AA-438 at about 10.20, when he had inspected it. The water had been 6 or 7 ft below the girders, and he had not anticipated any trouble. When he had returned after the accident at 4.15 a.m. the water had been 2 ft below the girders. He had seen no indication of a derailment or of dragging equipment.

Section Foreman Shear, in charge of the bridge between 1928 and 1937, remembered the greatest previous depth of water in the creek as being 5 ft in 1935. It had not harmed the bridge.

Roadmaster Fuller spoke of passing over the bridge in the caboose of a freight train on 16th June. The stream had then been dry and the bridge and rip-rapping in good condition.

Divisional Engineer Ring stated that rip-rap had first been placed around the bridge piers in 1935. It consisted of rough stone and should have fully protected the bridge from

---

[864] *The Muscatine Journal*, Muscatine, Iowa, United States, 20th June 1938

scouring or washing. He attributed the undermining of piers 2 and 3 to some obstruction brought down by the current causing a scouring action. When the water had subsided the rip-rap had been badly scattered and covered in silt.

Divisional Engineer Johnson stated that a complete inspection of the bridge had been made in June 1937. The rip-rapping had been adequate and in good condition. He was authorised to spend any money necessary to correct unsafe conditions. He considered that the failure had been caused by the undermining of the north side of pier 2 and the east side of pier 3.

Assistant Chief Engineer Middleton stated that in his opinion the bridge had been adequate in design and structurally safe. The trouble had been caused by a cloudburst much in excess of any previously experienced or which might be anticipated, and the volume and velocity of the water had scoured the rip-rapping, causing piers 2 and 3 to settle. He believed that the bridge had still been intact when the train had reached it, though with piers 2 and 3 undermined. When the engine had reached pier 2 it had tipped to the right, spans 3 and 4 had collapsed as pier 3 had tipped, the engine had struck the north slab of span 5 between its boiler and cylinder and the lower part of the engine had struck and destroyed piers 4, 5 and 6. He estimated that the water had been 12 ft deep.

Claims Adjuster Sleavin said that he had explored the watershed of Custer Creek and had found that the cloudburst had included practically the whole of the north portion of the watershed.

Rancher Frank Hamlick, who lived 19 miles north-west of the site of the disaster, spoke of a cloudburst at 9 p.m. which had been the worst that he had experienced in 33 years.

Rancher Henry Haughian, who had lived in the area all his life, spoke of three separate storms that evening and of seeing the highest water in his memory.

A third rancher spoke of being marooned all night by a creek because a highway bridge had been washed away.

The ICC concluded that as the engineer had made no brake application before the accident he had been unaware of any problem with the bridge before it had given way beneath his locomotive. The bridge had conformed to the specifications of the American Railway Engineering Association and had been structurally adequate for the traffic it was required to support. The accident had been due to the undermining of its piers following a cloudburst. [865]

Flooding due to excessive rain led to the collapse of a bridge over the Torre River between Udine and Remanzacco on the Udine – Cividale line near San Gottardo, Italy. The accident happened a little before 18:00 on Tuesday 22nd November 1938 and involved passenger train No. 593 of the Società Veneta Company.

The river involved in the accident had overflowed many times in the past and November was normally a wet month. It had begun to rain on Sunday 20th and from then there had been continuous rain up until the time of the accident. On the afternoon of the accident Platelayer Luigi Buiani, within whose length the bridge lay, had crossed the bridge at about 16:00 and seeing that the water was running high he had gone home, collected a handlamp and returned to the bridge at about 17:00. There he had met his immediate superior, Ganger Enrico Bersani, and the two had re-examined the bridge together. According to the ganger the water had been high, but he had seen it higher

[865] Interstate Commerce Commission report.

previously. The ganger said that he had left the bridge at 17:20, telling his subordinate to stay there a little longer and if the water level rose higher to come and notify him.

Train No. 593 consisted of engine No. 272, three loaded goods waggons, two empty goods waggons, two third-class coaches, a first- and second-class composite coach and a van, in charge of Conductor Angelo Candusso with Driver Antonio Gaigher and Fireman Guido Degano. Riding in the rear van was Conductor Ernesto Raghellin. The train started from Cividale at 17:29. It stopped at Remanzacco at 17:41 and passengers got on and off. Starting off again it reached the bridge as Platelayer Buiani was nearing his home at the gatehouse at kilometre 5. He watched the train approach the bridge, then suddenly there was a loud crash, and the telegraph wires shook forcefully. Realising that something must be wrong he had got on his bicycle and made for the bridge once more.

The engine almost made it across the bridge when the structure gave way beneath the leading vehicles of the train and the engine was pulled back, falling with the waggons to the raging waters below. The two third-class carriages fell on top of the waggons, smashing them, while the composite carriage was left hanging vertically half on and half off the bridge. Only the rear van remained on the line. Conductor Reghellin, bruised and frightened, climbed out of this van and made his way to the adjoining road, where he managed to flag down one of the few motor cars and told its driver to go for help. He then met Platelayer Buiani and told him what had happened. Meanwhile Labourer Mansueto Bergozza, who had been riding in the rear van with Conductor Reghellin, independently went for assistance, while another railwayman, Libero Nadalutti, who had been cycling from Remanzacco in the pouring rain and who had witnessed the accident, rushed back to the gatehouse, from where a message was telephoned to Udine station. He also alerted the local priest, who rang the tocsin (alarm bell).

With all these different people raising the alarm a rescue operation was soon in hand, firemen, ambulances, and soldiers rushing to the scene, together with a train from Udine. Prominent in the initial rescue were two young men, Luigi Greatti, who lived close at hand, and Ferruccio Petrini, who arrived on the train from Udine. Their attention was drawn to shouts from Sacher Galdino, who had made his way to a little gravel island in the middle of the rushing waters, along with two badly injured passengers – a young woman teacher named Athenaide Vatta, and Dr. Eligio Donatis. Without further thought the two men partially stripped and jumped into the torrent. With great difficulty they reached the little island but then realised that the task of rescue was beyond them. At this juncture some soldiers of the Royal Piedmont Cavalry arrived and ten of them, with the aid of a rope, were able to rescue, in the darkness and pouring rain, the five people now on the island, though by this time Doctor Donatis and Signorina Vatta had both succumbed to their injuries. The soldiers themselves then had to get out of the water. All but three were able to do so easily but then someone stumbled or slipped, the rope broke and the last three were in trouble. Two of them were able to eventually make it ashore but the third, named Enrico Cocci, a 21-year-old corporal, failed to make it and drowned. His body was found the next day.

By 23:00 two passengers had been rescued unharmed, 17 passengers had been rescued with injuries and three bodies had been recovered. [866] The following day a newspaper report spoke of eleven dead, ten of them passengers and the other Conductor Candusso. Several of the bodies had been carried some kilometres downstream by the

---

[866] http://www.storiastoriepn.it/wp-content/uploads//2018/09/Disastro_treno221138.pdf

fierce current. [867]

The final number of dead was hard to establish with certainty as the number of passengers who had been on the train was uncertain. Thirteen passengers were known to have died and five others were believed to have been on the train and had not been heard of since. In addition Conductor Candusso, Driver Gaigher and Fireman Degano lost their lives, as had Corporal Cocchi during the rescue operation. Seventeen passengers and Conductor Reghellin were injured in the accident and two soldiers named Bonfiglioli and Piga, plus rescuers Greatti and Petrini were hurt trying to save others. The latter two men were taken to hospital suffering from exposure and perhaps near drowning but soon recovered. Greatti was later awarded a silver medal for valour; it is not clear if Petrini received any award. The courage of Corporal Cocchi was also much praised.

The work of clearing the wreckage of the accident occupied several days. At first passengers and mail were carried by road but from 1st December trains ran as far as the collapsed bridge from either side with passengers and mail being transferred between them via the adjoining road bridge. The next improvement was the laying of a temporary railway across the bed of the river downstream of the bridge, a measure which remained in use for over a year, with train services being interrupted every time the level of the river rose too high. Finally a replacement bridge was brought into use.

An investigation was launched into the accident. It concluded that the failure of the bridge had been due to the undermining by the flooded river's powerful current of the seventh pier of the structure, which, left without support, had sunk about 60 centimetres, and had tilted to the left, causing the arches on either side of it to fail beneath the weight of the train. The report concluded that the disaster had been an unforeseeable and unpreventable 'Act of God'. A modern Italian writer seems to suggest that the report was something of a whitewash and that the bridge should have been constructed strongly enough to withstand the powerful currents to which it was exposed. [868]

Early on the morning of Monday 16th January 1939 a ballast train set off from Frias, Argentina, for Quiros, where ballasting operations were to be performed. It was in charge of Conductor Juan Avila (described as a *vaporista* or steamer in a Spanish report but presumably the conductor) with Driver Gerardo Asiares, Fireman Bartolome Maldonado and 15 workmen. The train only got as far as a bridge over the Albigasta River. The latter was swollen following heavy rains and apparently had undermined the foundations of the bridge, which collapsed beneath the train at 5.15 a.m. The engine crew and the workmen all died. All the victims were from Frias. The river was about 16 ft deep at the time of the accident and parts of the bridge were carried almost a mile downstream by the water. Rescue work was hampered by the fact that the area around the bridge was almost entirely flooded. [869] [870] The steel bridge had dated from 1893 and consisted of six 20-metre-long spans. [871]

A *Reuters* news report towards the end of March 1941 stated that recently a train had

---

[867] *La Stampa*, Toriono, Italy, 24th November 1938

[868] http://www.storiastoriepn.it/wp-content/uploads//2018/09/Disastro_treno221138.pdf

[869] *The Racine Journal-Times*, Racine, Wisconsin, United States, 17th January 1939

[870] https://www.facebook.com/164553183898052/posts/amanecia-el-16-de-enero-de-1939-bajo-la-escalofriante-noticia-de-que-el-puente-d/450285205324847/

[871] *The Railway Gazette*, London, England, 10th March 1939

fallen from a bridge into a ravine near Oporiise, Bulgaria, with the loss of ten lives. It has not been possible to trace any further details of this accident. [872]

In a typical accident of its kind, a trestle over Spring Branch Creek, two miles south of Marion, Kansas, United States, was weakened by a flash flood early on Saturday 6th September 1941. When southbound Chicago Rock Island and Pacific Railroad freight train No. 93 reached the bridge the weight of the train proved too much for the weakened structure. The engine and leading two cars passed over safely but then the trestle buckled under the strain and 12 cars of 'valuable merchandise' (motor cars, grain and wine) fell into the flooded creek. The train crew was unhurt but the Herington, Kansas, to Fort Worth, Texas, main line was blocked all day. Usually there were only three or four feet of water in the creek. [873]

A severe typhoon with heavy rain cause two bridge accidents in Japan on Wednesday 1st October 1941. Between Kumamoto and Oita on Kyushu Island a crowded three-coach passenger train fell from a bridge weakened by the storm. Dispatches from Tokyo stated that 100 passengers had been rescued, 50 bodies had already been recovered and more than 50 others were still being sought. The passengers were mainly secondary school pupils. [874] In a similar accident near Kochi on Shikoku island a mixed train fell from a bridge. Two passengers and eight railwaymen lost their lives in this accident. [875]

The usual culprit, heavy rain, was responsible for a bridge accident on the Great Southern Railways between Killarney and Farranfore, Éire, on Monday 10th November 1941. Heavy rain the previous day had flooded the countryside and the Farranfore River had burst its banks after sweeping away road bridges upriver from the railway. At a small bridge over a paved cattle underpass debris brought down by the river had accumulated and caused a blockage, whereby the water had washed away the approach embankments, leaving the rails sagging. The 3.35 a.m. goods train from Tralee to Valencia fell victim to the trap, its engine and most of its waggons falling into the gap, killing the unsuspecting driver and fireman. While the line was blocked a bus and lorry service was provided from Tralee while the main goods traffic was diverted via Limerick. [876]

A Baltimore and Ohio Railroad freight train was involved in a bridge accident on the evening of Sunday 23rd August 1942.

Westbound third-class freight train No. 87 consisted at the time of the accident of engine No. 2814, 14 loaded and 11 empty cars and a caboose in charge of an unnamed conductor with Engineer William E. Paisley and Fireman R.R. Smith. It left Gassaway, just under 50 miles east of Dorfee, West Virginia, United States, at 4.15 p.m., 2 hours and 45 minutes late. By the time it left Clay, 8 miles from Dorfee, at 7.30 p.m., it was 4 hours and 14 minutes late. It passed Dorfee and the accident occurred at Bridge 41.1, less than 2¼ miles further on, at about 7.55 p.m.

Bridge 41.1 crossed Big Sycamore Creek, which emptied into the Elk River 200 ft

[872] Ibid, 21st March 1941
[873] *The Hutchinson News-Herald*, Hutchinson, Kansas, United States, 7th September 1941
[874] *Oakland Tribune*, Oakland, California, United States, 1st October 1941
[875] *Journal de Geneve*, Geneva, Switzerland, 3rd October 1941
[876] *The Railway Gazette*, London, England, 21st November 1941

north of the bridge. The structure was 157 ft long and consisted of a steel open-deck girder span at its east end and a series of timber piers. It had been reconstructed as recently as 1939 and had a maximum height of 34 feet. The creek drained a watershed of 32 square miles. Usually it carried little water but there had been heavy rain before the accident. [877]

As the train reached the bridge the engine and leading 11 cars fell from it to the bed of the creek below. Engineer Paisley, Fireman Smith, and Front Brakeman C.E. Thompson lost their lives. [878]

The surviving train crew, the conductor and flagman riding in the caboose, told the Interstate Commerce Commission (ICC) that the train had been riding smoothly until an emergency brake application immediately before the derailment.

The ICC determined that at the time the engine moved upon the bridge the pier supporting the steel girder span had leant to the west sufficiently to unseat the west end of the girder, and the girder had slipped downward and been suspended between the east face of the pier and the east abutment. Apparently the engine truck had struck the upper part of the pier and the pony bent, and then the general derailment had followed. The investigation disclosed that floodwater had undermined the pier a considerable distance below its footing. The bottom of the footing of this pier extended about 3 feet below the normal ground line. There was an unobstructed view of the bridge over some distance, but the engine crew had apparently not become aware of anything being wrong until the last moment.

The section foreman had passed over the bridge four and a half hours before the accident. The water in the creek had been high but he had seen nothing wrong with the bridge. Half an hour later train No. 2912 West had crossed the bridge; its crew had seen nothing wrong. A local resident shad seen the pier leaning just before the accident and had attempted to warn the section foreman; the accident had happened before he could do so.

The ICC concluded that the accident had been caused by flood water undermining the bridge pier and made no recommendations. [879]

Apparently unnoticed by the press was an accident on the Northern Pacific Railway on the evening of Saturday 20th June 1943 near Bass, Montana, United States. The train involved in the accident was freight Extra 1356 East, which consisted of 4-6-0 steam engine No. 1356, 28 loaded cars and a caboose. The train crew are not named in the Interstate Commerce Commission (ICC) report on the accident.

Having left Darby, about 39½ miles west of Bass, at 5.55 p.m., the train had passed Bass and was moving at 25 miles an hour when it was derailed on Bridge 23.2 over the Bitterroot River, just under 3 miles beyond the station. The engine, tender, and first five cars of the train fell from the bridge and the engineer, fireman and front brakeman were killed.

The bridge had been reconstructed in 1925. It was slightly less than 429 ft long, with a pile and timber trestle at each end and a central span of three 120 ft steel truss spans.

The Bitterroot River drained a watershed of 2,100 square miles. Its level rose due to floodwater every spring, but the high water levels had never previously damaged the bridge. However, the water content of the snow in the watershed area had been considerably higher in May 1943 than in at least the previous seven years.

---

877 Interstate Commerce Commission report.

878 *The Charleston Gazette*, Charleston, West Virginia, United States, 25th August 1942

879 Interstate Commerce Commission report

The conductor and flagman of the train told the ICC inquiry that the train had been moving smoothly; the flagman said that the engine had suddenly disappeared from view. There had been no brake application. About five hours earlier the same crew had passed over the bridge going westward, this having been the last previous movement over the structure. They had noticed nothing wrong with the bridge then.

The ICC noted that there was an unobstructed view of the bridge for a good distance before reaching it. As the crew of train Extra 1356 had made no brake application the bridge must have appeared normal to them until it had collapsed beneath their engine.

The river had been in flood for several days before the accident. After the accident it was found that pier 3, which supported the east end of span 1 (numbered from the east end) and the west end of the east approach to the bridge, had collapsed and was demolished. The east end of span 1 was on the riverbed. Floodwater had undermined the pier several feet below the maximum penetration of the piles.

An examination of pier 3 on 2nd June had shown that the river channel had shifted eastwards and the east bank of the pier had been scoured by the floodwater to within 2 or 3 feet of the west face of the pier. Upon discovering this the section foreman had immediately notified his superiors and two carloads of small stones had been dumped around the pier on 7th June. The day before the accident the section foreman had inspected the bridge, finding the water within a foot of the west face of the pier. He had not known the penetration of the piles but had considered the bridge to be safe.

The ICC noted that over 15 years the channel of the river in the vicinity of the bridge had gradually shifted eastwards and the east bank of the river had receded about 80 feet to within a few feet of the west face of pier 3. The maximum penetration of the piles of pier 3 was seven feet. When the piles had been driven through hard gravel this was the maximum penetration which could be obtained, though after the accident the piles had been driven down another 13 ft in the same place. The gradual encroachment of the channel towards the east end of the bridge and the rising of the water level to flood proportions had caused the water to scour the footing of the pier below the lower ends of the piling. [880]

At the very end of the Second World War in Europe, just a few hours before the German surrender on 7th May 1945, the Nazis made a fruitless attempt to halt the advance of the Russian forces by blowing up the railway bridge across the River Elbe at Niederwarthaer, Germany, using tailless aircraft bombs. Two demolition attempts were made but not all the charges exploded and although the bridge was badly damaged it was not rendered completely unusable.

Rather less than two months later, on Tuesday 31st July 1945, an accident did put the bridge completely out of use for rail traffic for nine months. Because of the weakened condition of the structure trains were only permitted to cross it at a severely restricted speed but the driver of an empty train from Berlin failed to slow down, the stress proved too much for the bridge, and the surviving portion of it collapsed under the train. The fireman and a Russian soldier escorting the train were killed. The Soviet authorities accused the railwaymen of sabotage, several Reichsbahn officials were arrested, and one of them, a 72-year-old on the point of retirement, died a short time later in a Soviet detention camp. The fate of the driver is not recorded. [881]

---

[880] Ibid.

[881] https://www.vorschau-rueckblick.de/2005/05/im-archiv-gestoebert-die-geschichte-der-niederwarthaer-elbbruecke-schluss/

CHAPTER 9

OPENING BRIDGE ACCIDENTS

Accidents involving opening bridges appear to have occurred in only a handful of countries, with the majority occurring in the United States. It seems that most countries managed to a large degree to avoid building such bridges, which from a railway point of view are a constant source of expense and worry. Shaw states that the early railroads on the eastern seaboard of the United States were forced by opponents of the new form of transport to build such bridges over even insignificant waterways where they were never really needed. [882]

A Louisville and Nashville Railroad steel drawbridge across the Mobile River, 2¼ miles south of Hurricane, Alabama, United States, was the scene of a bad accident on Wednesday 25th February 1931. The total length of the drawbridge (Bridge No. 193) was just over 1,050 feet and it had five spans, of which the draw span was 330 feet long. Mounted above the draw span was the bridge operator's tower. Three oil-lit bridge lights mounted upon the draw span showed red to the railroad and green to the navigation when the span was open. The drawbridge was also protected by colourlight home and distant signals. The home signal normally displayed a double red stop indication and the distant signal a yellow indication. The home signal could not display a clear indication unless the drawbridge was locked, and the line was clear two blocks ahead, while the drawbridge could not be opened unless the home signal was at danger. The bridge operator was able to replace the home signal to danger at any time, but a time release then meant that it took 3½ minutes for the drawbridge to be fully opened. At the time of the accident the drawbridge had been open seven or eight minutes. Neither trains nor boats had superiority; the practice was for the bridge operator to give precedence to whichever whistled first.

The train involved in the accident was northbound passenger train No. 98, *The Pan American*, en route from New Orleans, Louisiana, to Cincinnati, Ohio. It consisted of engine No. 217, a combination baggage and mail car, four Pullman sleeper cars and an observation car, all of steel construction, in charge of Conductor Durkovitch with

---

[882] *A History of Railroad Accidents, Safety Precautions and Operating Practices* by Robert B. Shaw, pg. 187

Engineer Arthur Ingham and Fireman Ezekial Crumpton. The train left Sibert, the last open office before the drawbridge, on time at 12.32 a.m., passed the distant signal, which should have been at caution, passed the home signal at danger and travelling at a speed estimated as 10 to 35 miles an hour, the engine, tender, and leading car fell into the river at about 12.55 a.m. Engineer Ingham, Fireman Compton, Baggagemaster Nick Vaughn and Porter L. Woods were drowned. None of the passengers was hurt. [883]

A newspaper found a hero in the accident in the shape of a passenger named J.G. Johns, who had been an occupant of the first Pullman car, which had come to a stand half overhanging the drawbridge. He said that he had been awakened by the violent bouncing of the car. "There was a roar and a hissing outside almost unbelievable. The lights went out and the car became as dark as midnight. I swung out of my berth and found my shoes and slipping them on moved through the narrow passageway to the outside door. As my eyes became accustomed to the darkness I could see below me the river and could hear the water as it lapped about the piers. I instantly sensed what had happened. The grinding and swaying meant only one thing – the cars ahead of us had plunged into the water. I recalled there were several women on the car, who no doubt were terror-stricken. I called to Mrs W.I. Grubb (wife of Federal Judge W.I. Grubb), whom I knew, and told her that we were in a wreck and that our position, while precarious, was safe at least for a few moments. I asked her to gather up her wraps and to notify the other women in the car." [884]

The accident was investigated by the Interstate Commerce Commission (ICC).

Conductor Dukovitch told the ICC inquiry that the air brakes had been tested and had worked properly at Sibert. The train had braked and whistled at Three Mile Bridge and at the bridges at Chickasaw Bogue and Bayou Sara and had whistled at Menemoosha and Ortiz. The brakes had been applied as usual near the Mobile River distant signal and the engineer had whistled for the home signal. There had been no further whistle to indicate that the home signal had been cleared; the brakes had then gone on in emergency. After the accident he had checked the angle cocks, which had all been open, and had also looked to see if any sand had been applied. None had. Engineer Ingham had appeared normal and in good spirits.

Bridge Operator W.M. Lewis stated that a tug boat going downstream had whistled at about 12.35 a.m. Realising that if he opened the drawbridge he would probably delay train No. 98 he had waited until the tug boat had whistled a second time, then looked to see if it was towing logs or barges. Had it been towing logs he would have given the train precedence, but it was towing barges. This meant that because of the swift current of the river, if the drawbridge was not opened for an approaching tug there was a danger of it getting out of control and striking and damaging the drawbridge piers. He had therefore opened the drawbridge. The drawbridge had been open seven or eight minutes before an annunciator bell rang to indicate that the train was a mile and a half away. He had heard the train whistle for the home signal, then had been watching the tug boat just nearing the bridge when he had realised that the train was not stopping. He put its speed at 30 to 35 miles an hour. As the train plunged from the bridge he saw the tug boat change course to avoid it and start back upriver. During this manoeuvre one of the towed barges struck the drawbridge pier. The engine headlight had been burning brightly. No whistle was given by the engineer to indicate that the home signal had been cleared. The signals had been working properly.

---

[883] Interstate Commerce Commission report.

[884] *Hattiesburg American*, Hattiesburg, Mississippi, United States, 26th February 1931

Flagman Presley more or less corroborated the statement of his conductor.

Captain McDonald of the tug boat said that the drawbridge had been opened and the bridge lights had been green for him before the train had whistled. He put the speed of the train as it reached the bridge at 30 miles an hour and could not say if it had been braking.

Bridge Operator Beasley at Bayou Sara said that train No. 98 had reduced speed to 10 miles an hour at his bridge and given the usual whistle signals. Engineer Ingham had shouted a greeting to him.

The ICC came to the obvious conclusion that the accident had been due to the failure of Engineer Ingham to control his train. Why could only be conjecture, as both he and the fireman had lost their lives. Both he and the fireman were men of long experience, thoroughly familiar with the line. The train had been operated perfectly normally up to the point of the accident and the signals and lights at the drawbridge had given an unmistakable indication that the draw was open. Given that the accident occurred an hour after midnight a likely explanation which suggests itself to the author is that the engineer suffered a microsleep at the critical moment and that either his fireman was also asleep, or he was preoccupied with his firing.

The ICC noted that Engineer Ingham had been previously reprimanded more than once for exceeding the 15 miles an hour restriction over drawbridges.

During the 30 days preceding the accident it was noted that the drawbridge had been opened 211 times, delaying 32 trains for a total of 5 hours and 13 minutes.

The ICC suggested that the railroad company introduce additional safety measures, such as smash boards, automatic train control or automatic cab signals. [885]

Buffalo Bluff, Florida, United States, on the Atlantic Coast Line Railroad, was the scene of an opening bridge accident on Sunday 11th October 1931. The St John's River was here bridged by what was described as a bent frame drawbridge with a 228 feet pivoting steel draw span, which carried a single railway track. The line over the bridge was worked by timetable, train orders and an automatic block system. Northbound movements over the drawbridge were protected by a home signal, automatic block signal 7048 situated 5,081 feet of the home signal and automatic signal 7058 located 5,923 feet further south again. In addition a marine lamp on the draw span showed red to an approaching train when the draw was open and there was a smash board 10 feet north of the home signal. An operating tower on the east side of the draw span worked the bridge. To open the bridge the operator had first to place the home signal at danger and extend the smash board across the line, after which he could unlock the draw. A time release prevented the bridge being unlocked for two minutes after a train had passed signal 7058. The speed of trains of the class involved in the accident was limited to 45 miles an hour over the whole line and further limited to 20 miles an hour over drawbridges.

Northbound freight train Extra No. 1657 consisted of engine No. 1657, 25 loaded cars, six empty cars and a caboose, in charge of Conductor Terrible with Engineer R.P. Torrence and Fireman Williams. It passed DeLand, the last open office, 46½ miles south of the drawbridge, at 1.54 a.m., passed signal 7058 at *Clear*, passed signal 7048 at *Approach*, passed the home signal at *Stop*, broke through the smash board, and plunged from the open draw at about 2.57 a.m. Its speed had been reduced from 40 to 45 miles an hour to 15 to 25 miles an hour. The engine and first four cars fell into the water, which was about 40 feet deep. The engineer lost his life.

[885] Interstate Commerce Commission report.

Fireman Williams told the Interstate Commerce Commission (ICC) inquiry that after surmounting Satsuma Hill his engineer had eased off and worked light throttle down the bank. Signal 7058 had been green. Moving at 35 or 40 miles an hour the train had then neared signal 7048 and Head Brakeman Thomas, who had also been riding on the engine, had called out, "Yellow." At this point he (the fireman) had moved behind the engineer, who had been peering out of the front window, and had added, "Yellow saying board." The engineer had not acknowledged either call, just looked towards the brakeman, and after passing the signal had applied the brakes lightly. This having little effect on the speed of the train on the falling gradient. Seeing the home signal at red both he and the brakeman had cried out, "Red board." The engineer had made a slightly stronger brake application, but it was clear that the train was not going to stop at the home signal. By now the fireman was standing on the steps of the engine, the brakeman having already jumped off, and the engineer, who was still sat down, had looked across to him and said, "You had better look out." Fireman Williams said that he had then jumped off at about the smash board; the train was still travelling at 15 or 20 miles an hour at this point. His engineer had made no attempt to jump. At no time had the engineer made an emergency brake application or reversed the engine, the sanders had probably been used. The fireman added that every signal had been called and 7048 was the first to which the engineer had not responded. The engineer had been awake and seemed to be normal.

Head Brakeman Thomas corroborated the fireman's account except that he thought that the speed had been slightly higher passing signal 7048. He agreed that the engineer had been awake and alert. He had become afraid and jumped two or three telegraph poles south of the home signal because the engineer had not braked early enough.

Conductor Terrible said that the brakes of the train had been tested before leaving Sanford and again after 18 cars had been detached at Orange City Junction. They had worked properly at DeLand. He had spoken to the engineer at Sanford and the latter had appeared normal. There had been no emergency brake application approaching the drawbridge and he had not realised that there had been an accident until he had gone forward after the train had stopped. He had not seen the signals before the accident, afterwards the home signal had been showing *Stop* (red over red) and the smash board was broken.

Flagman Stephenson stated that he had only noticed one brake application, 10 or 15 car lengths north of signal 7048.

Bridge Tender Sweat testified that if a train was approaching or due within five minutes when a boat whistled it was his practice to give the train precedence but that a boat must not be delayed more than 10 minutes under the government navigation rules. On the morning of the accident a boat had whistled at about 2.38 a.m. He had opened the bridge and the boat had passed through at 2.55 a.m. He had seen the train coming when it was about two miles away but had not realised that it was not going to stop until it had passed the home signal. He had then stopped the operation of the draw, which was about two-thirds closed when the engine fell from the bridge.

Roundhouse Foreman Benson said that he had found no indication that sand had been applied. In tests conducted after the accident the air brakes of the train were found to be working properly. The signalling was also tested, and no problems were found.

A post mortem on Engineer Torrence found that he had died from heart failure before his engine had gone into the river.

The ICC concluded that the accident had been caused by the train failing to obey signal indications. Although the engineer had appeared to be normal just before the accident the fact that he had died of heart failure before his engine left the drawbridge (as demonstrated by the fact that there had been no water in his lungs) provided the probable explanation for his failure to bring his train under control. The engineer had been 65 years old. [886]

Having recommended in their report on the 1931 Hurricane accident that smash boards should be installed to prevent such events, the Interstate Commerce Commission (ICC) must have been less than amused to be notified of the Buffalo Bluff accident, just described, and another at Jupiter, Florida, United States, on Tuesday 12th February 1935. In both, such smash boards had been installed but they had failed to save the day.

The drawbridge over the Jupiter River, a mile north of Jupiter station on the Florida East Coast Railway, was 585 feet long with nine spans. The 55 feet drawbridge span was of the bascule, single-leaf type, operated by a gasoline motor from a room under the next pier north of it. The bridge was protected by continuously lit colourlight signals, with automatic signal 2809 situated 9,695 feet north of the point of the accident, distant signal 2821 situated 3,135 feet to the north and the home signal (number not stated) 507 feet to the north. When the drawbridge was open the home signal was red, the smash board at the home signal was horizontal, distant signal 2821 was yellow and signal 2809 was green. When the drawbridge was open all the signals worked automatically. A sign "Draw bridge one mile" was located 5,235 feet north of the point of the accident. Boats had precedence over trains 'under the requirements of the War Department.' Before opening the drawbridge the bridge tender replaced the north and southbound signals to danger from an interlocking cabin. Provided no train was within the approach locking section he could then release the mechanical lock on the lift machinery, walk to the bridge, make nine revolutions of a large wheel and the gasoline motor would then operate the bridge. The entire operation took at least 5 minutes. If a train had passed signal 2809 a time release prevented the draw being unlocked for 2 minutes.

Involved in the accident was southbound passenger train Second No. 87, *The Florida Special*. This consisted of engine No. 427, a baggage car, four Pullman sleeping cars, a dining car, three Pullman sleeping cars, a Pullman recreation car, and a Pullman observation car, all of steel construction, in charge of Conductor J.E. Kerr with Engineer F.W. Turnipseed and Fireman William Barton. The train passed Hobe Sound, the last open telegraph office, 7½ miles north of the drawbridge, at 3.59 p.m., 13 minutes late. It passed signal 2821 showing yellow or *Approach*, the home signal showing red or *Stop*, broke through the smash board, and struck the lifted draw span at a speed estimated at between 10 and 25 miles an hour at about 4.08 p.m. The draw span was damaged and fell into the river with the engine on top of it. The engine tender and all but the rear of the first coach left the rails. Forty-seven passengers, nine Pullman employees, 12 dining car employees, a newsagent, the engineer, fireman, a baggageman and a flagman were injured. No one was killed.

Engineer Turnipseed told the ICC inquiry that automatic signal 2809 had been at *Approach*, then had cleared, indicating that he was closing up on the train ahead. He had then seen signal 2821 at *Approach*. His statements as to what happened then were confused. It appeared that he had applied the brakes but had been expecting to find the home signal at *Approach* as well, because he believed that he was running close behind the preceding train, also that he had thought that he had more distance than there was

---

[886] Ibid.

between the distant and home signals. It was only when he saw that the home signal had not cleared that he had applied the brakes in emergency, telling his fireman to jump when they went through the smash board. He put the speed of his train at 10 miles an hour when it struck the drawbridge.

Fireman Barton said that he had called the yellow distant signal to his engineer, who had raised his hand in acknowledgement and made a 5- and then a 10-pound brake reduction, reducing the speed from 65 to 55 miles an hour. He did not see the engineer release the brakes. His engineer had not called out the home signal but had applied the brakes in emergency as he had neared it. He (the fireman) had jumped off as they struck the smash board, the speed at that point being 25 miles an hour.

Conductor Kerr stated he had been making out reports when he had noticed a service and then an emergency brake application. He put the speed at 10 to 15 miles an hour when they struck the drawbridge.

Superintendent of Air Brakes Love was riding on the train. He testified that at Mars, 2½ miles north of Jupiter, the train had been running at high speed, about 65 miles an hour. The train had still been running at 60 miles an hour at the yellow distant signal (2821) and the brakes had not been applied until the rear of the train was passing the signal. A service brake application was made when the train was about 1,000 feet beyond the distant signal, reducing the speed to 55 miles an hour. There was then an emergency brake application, which had reduced the speed rapidly, he thought to 10 miles an hour at the accident point. He had found nothing wrong with the brakes after the accident.

Bridge Tender Luther Cook said that he had been sat on the porch of his house, 930 feet from the bridge, when First No. 87 had passed. He had then set off towards the bridge as he expected the return of a small fishing smack for which he had opened the bridge that morning. [887] Halfway there he had heard the boat whistle, so he had hurried to the bridge, replaced the signals, no train being on the approach circuits, unlocked the drawbridge, gone to the engine room and operated the wheel to lift the draw, raising it to an angle of 30 degrees, sufficient for the boat to pass through. At least 10 minutes must have elapsed since he had left his house. As he left the engine room he saw the locomotive run through the smash board and the engine crew jump off and he realised that it was not going to stop. On average, he said, he only needed to open the bridge once a day, and seldom needed to stop a train.

Signal Inspector Cargile confirmed that the signals and drawbridge equipment had been working properly.

The ICC did not accept Engineer Turnipseed's statement that automatic signal 2809 had initially shown *Approach* as it was clear that First No. 87 must have by then have been too far ahead, given that the bridge tender had had time to walk from his house and open the bridge after it had passed. It was also clear that the engineer had not applied his brakes as early as he had maintained, otherwise he would have been able to stop in time. The commission criticised the operating officers of the company for employing an engineer on a fast passenger train who was neither thoroughly acquainted with the road nor apparently properly qualified as to the action to be taken when passing adverse signals. The ICC was also not happy that the maximum speed of trains had been raised from 50 to 65 miles an hour without the spacing of the signals being altered, thus reducing the margin of safety. [888]

---

[887] *Syracuse Herald*, Syracuse, New York, United States, 13th February 1935
[888] Interstate Commerce Commission report

No other accident involving an opening bridge occurred until Friday 11th April 1941. On that day. a Minneapolis and St Louis Railroad light engine was returning eastwards, running tender first, after having assisted a freight train to Marsh. It was being driven by Engineer G. Ray, with Fireman H.A. Williams. When the locomotive reached Keithsburg, Illinois, United States, the vertical lift bridge there had been raised for a boat. The engineer failed, for some reason, to observe the protecting signal in time and the engine struck the lifting span, stopping extending over the gap with two wheels derailed. The impact broke two of the cables, 318 feet long and ¾ inch thick, which raised and lowered the span. In addition two guides of the draw span were broken. As a result the draw span fell onto the tender, 'catching it like a rat in a trap,' as a newspaper put it. No one was hurt, but so entangled were engine and bridge that it was feared that the structure would be out of use for a month. [889]

Numerous American newspapers carried a sensational photograph of a near escape from disaster on Sunday 21st September 1941. [890] The Chicago and Alton Railroad *Midnight Express* from Chicago to St Louis, Missouri, had not progressed very far when it ran onto the bridge over the Chicago River, which had been opened for a freight boat. The engineer just managed to stop the train in time, but not before the front bogie and leading driving wheels of his locomotive had run off the end of the rails. The four sleeping cars of the train were drawn back to Chicago Union station, the passengers resuming their journey after the line had been cleared. No one was hurt. [891]

The third opening bridge accident of 1941 occurred on the Baltimore and Ohio Chicago Terminal Railroad (BOCT) at Indiana Harbor, Indiana, United States, on Tuesday 2nd December, at the railroad's drawbridge across the Indiana Harbor Ship Canal. This was one of no less than five drawbridges over the canal, all worked from HC tower. The drawbridge was described as being of the double-track, single-leaf bascule type and was worked by electric power. The bridge was protected by fixed signals and derails (trap points). On the eastbound track, that on which the accident occurred, *Approach* (distant) signal E253-25 was 7,276 feet from the drawbridge, the home signal 140 was 558 feet from the bridge and derail 131 504 feet. Signal E253-25 was an automatic three-indication, two-arm, upper quadrant semaphore while signal 140 was a semi-automatic, four-aspect colour position light signal. Both were approach lit. The night aspects of the signals at the time of the accident were respectively yellow over red staggered *(Approach)* – Proceed prepared to stop at next signal, and two horizontal reds – *Stop*. Levers in the tower operated the signals and bridge and rail locks. A time release for the eastbound track prevented the bridge being unlocked for 3 minutes and 35 seconds after the signals had been replaced to danger.

Baltimore and Ohio Railroad eastbound first-class passenger train No. 8, *The Shenandoah Limited*, consisted of two-unit diesel electric engine No. 57, two baggage cars, a passenger baggage car, three coaches, a Pullman sleeping car, a Pullman club car and four Pullman sleeping cars, in charge of an unnamed conductor with Engineer Jack O'Connor and Fireman-Helper Rush Moody. It had left Chicago Grand Central station at

---

[889] *The Fairfield Daily Ledger*, Fairfield, Iowa, United States, 15th April 1941

[890] *The Brainerd Daily Despatch*, Brainerd, Minnesota, United States, 24th September 1941

[891] *Reno Evening Gazette*, Reno, Nevada, United States, 22nd September 1941

10.04 p.m., 4 minutes late, and a brake test was made shortly afterwards. Whiting, the last open office, 2½ miles from the drawbridge, was passed at 10.54 p.m., 7 minutes late. According to the speed recorder tape of the engine, signal E253-25 was passed at 66 miles an hour; signal 140, at danger, was passed at 30 miles an hour and the two diesel units and first five cars of the train were derailed on derail 131. Continuing on the ties (sleepers) the train had almost stopped when it reached the drawbridge, but the leading ("A") unit fell into the canal, into 22 feet of water, at about 10.58 p.m. Engineer O'Connor drowned. Fireman-Helper Moody was in unit "B" at the time. He jumped and was injured, as was a mail clerk. None of the passengers was hurt.

The Interstate Commerce Commission (ICC) investigated the accident. Its report stated that a ship proceeding southwards along the ship canal had whistled for the drawbridges at 10.32 p.m. The BOCT signals had been at danger at the time and the HC signalman had opened the derails, unlocked the bridge locks, and raised the drawbridge. He was then occupied for some minutes opening the four other drawbridges. When the last bridge was opened the ship started up and at about the same moment the annunciator sounded, indicating the approach of train No. 8. As the ship cleared the BOCT drawbridge the signalman saw the train overrun the home signal and the first diesel unit fell into the canal.

A special agent and an operator who witnessed the accident thought that the train would pull up short of the drawbridge. The fireman-helper did not notice any brake application prior to the derailment. The diesel engine maintainer and a baggageman noticed a slight brake application at the derail, than an emergency application. The brakeman noticed several surges and had opened the conductor's emergency valve. The conductor noticed the brakes being applied in emergency when the speed was 40 or 45 miles an hour.

A post mortem upon the engineer determined that he had suffered a cerebral haemorrhage (stroke) which probably had happened shortly before the accident, as the train had been operated normally up to signal E253-25, including making station stops. The ICC noted that the rules required the fireman-helper to observe signals as far as possible, but with the diesels running to full capacity he was required to spend about 85 per cent of his time supervising their running in the engine rooms.

The ICC report concluded that the accident had resulted from a failure to operate the train in accordance with signal indications. It made no recommendations. [892] Two factors not mentioned in the report, presumably because they were not considered important, were that it had been somewhat foggy at the time of the accident and that Engineer O'Connor was 70 years' old.

A newspaper report on the accident was headlined 'Heroic Railroader Dies to Save 150', and went on, 'As the investigation went forward the missing engineer {his body had not at that point been recovered from the canal} emerged as a hero who apparently gave his own life to save those of the approximately 150 passengers aboard the train. Witnesses said that showers of sparks flew from the wheels of the locomotive as it ploughed up the track for several hundred feet before toppling into the river, and rail officials discovered by examination of the second portion of the engine that all brakes were set, and the speed reversed. From these facts it appeared that O'Connor had stayed with his train to the last, doing all in his power to halt the speeding streamliner. There was no question, investigators said, that O'Connor by sticking at his post had prevented

[892] Interstate Commerce Commission report

what might have been a major disaster.' [893]

On Monday 29th March 1943 it was the turn of the Norfolk Southern Railway to suffer a drawbridge accident, near Fentress, Virginia, United States (Fentress station was 2 miles south of the drawbridge; the Interstate Commerce Commission (ICC) gives the location as near Fentress, the newspaper reports give it as near Great Bridge). [894] In this case the drawbridge concerned was located where the single-track railroad crossed the Abermarle and Chesapeake Canal at right-angles. It was described as being of the single-leaf rolling bascule type with a draw span of 94½ feet and was operated by a gasoline engine. Protection for the bridge for northbound train was provided by a warning sign, an approach signal, and a home signal. The warning sign, 3,024 feet from the point of the accident, consisted of a rectangle mounted on a post and bearing the words '3,000 FEET TO DRAWBRIDGE' in black letters on a white background. The approach signal was a single fixed semaphore arm, 1,412 feet from the drawbridge, and the home signal a two-position lower quadrant semaphore, 396 feet from the drawbridge; both were oil lit. The daytime aspects of both signals when the drawbridge was open were horizontal; at night-time the approach signal displayed was a yellow light and the home signal a red light, the indications being *Caution*, and *Stop* respectively. A single lever in the drawbridge tender's cabin worked both the bridge lock and the home signal.

On the day of the accident train No. 64, a northbound second-class freight train, consisted of engine No. 602, 18 loaded and 13 empty cars and a caboose, in charge of Conductor A.L. Bonney, with Engineer R.E. Elliott and Fireman Burt Rankin. It passed Elizabeth City, North Carolina, 33½ miles south of Fentress and the last open office, at 4.45 a.m., an hour and ten minutes late, passed the warning sign and approach signal at 39 miles an hour according to the speed recorder on the engine, passed the home signal at *Stop* and was still moving at 10 miles an hour when the engine fell into the canal from the open span. Only the engine and tender fell from the drawbridge, the engine ending up at an angle of 45 degrees. The leading end of the first car was also damaged. The accident happened at 6.20 a.m., in clear weather, as dawn was breaking. Engineer Elliott was fatally scalded and hurt and died in hospital, Fireman Rankin was scalded, and Head Brakeman C.L. Gaylor, who had been riding on the locomotive, was also hurt. The fireman had jumped, and the head brakeman had escaped from the engine after its fall.

The ICC report stated that the drawbridge had been opened for a boat at 3.01 a.m. and not closed again. When the train had been 1,000 feet from the drawbridge the fireman had seen at the same time that the home signal was at *Stop* and the drawbridge was raised. He alerted his engineer and when the latter had not reacted had spoken to him again. About 150 feet south of the home signal, or 500 feet from the drawbridge, the engineer had finally applied the brakes in emergency, just before the fireman jumped. The front brakeman told the inquiry that being inexperienced, he had been studying the timetable and had not seen the signal at *Stop* until about the moment that the fireman had shouted. When the engineer had not reacted he (the brakeman) had touched him on the shoulder, whereupon the engineer had applied the brakes in emergency. The conductor and rear brakeman had been unaware that anything was wrong until the emergency brake application. When he saw that the train was not stopping Bridge Tender C.B. Wilson had attempted to lower the draw span but there had been no time to lower it more than slightly.

[893] *The Circleville Herald*, Circleville, Ohio, United States, 3rd December 1941
[894] *The Evening Star*, Washington, D.C., United States, 29th March 1943

The fireman said that he had not been alarmed when the engineer had failed to whistle for the signal as per the regulations, or to brake at the approach to the bridge, as it was on a right-hand curve and the engineer therefore had a better view of the home signal; furthermore the engineer had not always called out signal indications as per the rules. The ICC pointed out that if the fireman had warned his engineer more forcefully the accident would have been avoided. This was doubtless true, but as well as the usual reluctance of a fireman to criticise his driver there was the additional factor here that the driver was white and the fireman black, and therefore considered very much a second-class citizen in the South at this time. The engineer was not in a condition to be questioned as to his failure before he succumbed to his injuries.

In addition to the failures of the train crew the ICC criticised the lack of definite rules and instructions for the safe operation of trains. An August 1941 bulletin stating that the proceed signal was being changed from white to green was contradicted by special instructions of October 1942 which mentioned a white board. A bulletin of February 1929 required trains to be under control at distant signals without defining what that meant. Operating officials stated that the train was required to be under control such as to stop short of the home signal but there were no written instructions to that effect. Investigation showed that different engineers handled their trains differently approaching drawbridges, depending upon their judgement.

The ICC was also unhappy that there was no normal position for the drawbridge and that it had been open for three hours preceding the accident. The bridge tender said that this had been the normal practice for the last 16 years, to save wear and tear on the bridge and power operating it. Nor was there any annunciator to warn the bridge tender of an approaching train. He relied upon the train whistling and if the span was open and no boats were approaching he then closed it. It was not possible for him to have time to close the span after hearing a train whistle for it without stopping the train. Once the bridge had been closed for a train it remained closed until a boat needed to pass. An average of 6.7 trains and 8.96 boats passed the drawbridge daily.

The railroad, the ICC found, had ten drawbridges in all. Of these five were normally closed to boats, one was closed between 7 a.m. and 11 p.m., three were normally open and that at Fentress had no normal position. The distance between the approach and home signals varied between 893 and 1,283 feet. One bridge had no approach signals, and one had no signals or signs at all. The time taken to close the bridges varied between 1½ and 20 minutes. The home signals at the bridges which had them could be replaced to *Stop* and the bridges opened at any time unless the train was actually on the bridge. None of the bridges had approach or time locking. A train could thus receive a proceed indication at the home signal and there still be time to open the span sufficiently to derail the train.

In conclusion the commission found that the accident had resulted from a failure to provide adequate protection at the drawbridge and to the failure of the engineer to obey a stop signal. It recommended that the Norfolk Southern proved adequate protection for movements over its drawbridges.

A note in a British railway magazine described a minor opening bridge accident in early December 1943. Wartime censorship meant that the exact location was not stated, only that it was 'a well-known docks in the Norfolk area,' served by a line which at one point crossed a channel by a swing bridge, 'pivoted in the centre and worked from there by bevel gear.' At one time there were several swing bridges in East Anglia and the author has not been able to determine the exact location. A small London and North

Eastern Railway 40-ton tank engine was approaching the bridge, pulling seven waggons, when the driver and fireman, 'to their astonishment,' saw that the bridge was open. The driver applied the brakes in emergency and when he and his fireman realised that they were not going to be able to stop they jumped off; the driver grazed his knee and was shaken, the fireman was apparently unhurt. The locomotive and one waggon fell from the bridge, blocking the channel to an inner dock. A heavy breakdown crane retrieved them after 25 hours' work. Little damage was caused to the engine. There was no explanation as to how the accident had happened. [895]

The Norfolk and Portsmouth Belt Line, a freight-only short line connecting various railroads in Norfolk, Virginia, United States and jointly owned by the Atlantic Coast Line, Chesapeake and Ohio, New York Philadelphia and Norfolk, Norfolk and Western, Norfolk Southern, Seaboard Air Line and Southern companies, suffered an opening bridge accident at Berkley, Virginia, United States on Wednesday 23rd August 1944.

At the point of the accident, half a mile north of Berkley, the Belt Line, which at this point was only single track, crossed the South Elizabeth River at an opening bridge known as River Bridge. This was a steel through truss swing bridge, with an opening span of 277 feet, on either side of which was a steel truss span of 250 feet. The bridge was operated by a bridge tender from a cabin on the swing span; movement being powered by an electric motor. The span was normally closed for river traffic and open for the railway but at the time of the accident it had been opened for a tugboat and stood at right-angles to the track. There were no signals or signs protecting the bridge. The line over the bridge, like the remainder of the Belt Line, was operated under yard limits, meaning that the engineers of trains were meant to operate them at a speed which would enable them to stop within one half of their range of vision. The special instructions for the bridge were that all trains must stop short of it and not proceed until signalled to do so by the bridge tender exhibiting a green flag by day or a green light at night. Trains were also restricted to a speed of 10 miles an hour over the bridge. An average of 35.75 trains and 38.43 ships used the bridge daily.

The train which came to grief was Extra No. 21 South, a southbound freight train which had been made up in Baugh's Siding, just north of the bridge. It consisted of engine No. 21, running tender first, a caboose and 16 cars, in charge of Conductor E.J. Hatchell, with Engineer R.A. Chapman and Fireman T.E. Sawyer. At about 12.35 a.m. it entered the single line from the siding, proceeded southward about 1,300 feet at 10 miles an hour and the engine, tender, caboose and three cars fell from the bridge at about 12.40 a.m. The engine and tender ended up almost upright on the riverbed and the caboose and cars piled up on top of them. The water was about 40 feet deep, and Engineer Chapman, Fireman Sawyer, and Conductor Hatchell were all drowned. Brakeman K.M. Blake and Flagman T.J. Morris escaped. The former, who had been in the caboose, was slightly hurt; the latter, riding on the sixth car, not at all. [896]

The two surviving railwaymen stated that the engine had sounded four short blasts on the whistle as it pulled out of the siding, followed by two short blasts. The first was an indication to the bridge tender that they were waiting to cross and the second an acknowledgement that all was clear for them to do so. They had been unaware that anything was wrong until the accident occurred.

---

[895] *Railways*, February 1944, pg. 29

[896] *The Raleigh Register*, Beckley, West Virginia, United States, 23rd August 1944

About three minutes before the accident the unnamed bridge tender had operated the controls to open the bridge for the tug boat. As the bridge began to swing he had noticed that his lighted green lantern was being exhibited towards the approaching train and had belatedly moved the lantern to a position inside his cabin where it was out of sight from outside. Clearly this green light had been interpreted by the engine crew as permission to proceed and in the darkness they had not been able to see that the span was open. With nothing else, such as a fixed signal, to guide them, they were misled and lured to disaster.

The Interstate Commerce Commission found that, as in the Fentress accident, the previous year, the accident was due to failure to provide adequate safeguards for the drawbridge and recommended that the Belt Line Company do something about it. [897]

A 150-ton Navy crane was used to lift the locomotive, tender, caboose, and cars from the river.

The last opening bridge accident before the end of the period covered by this book occurred on the Hudson and Manhattan Railroad, a subsidiary of the Pennsylvania Railroad, near Harrison, New Jersey, United States, on Monday 17th December 1945. In contrast to the previous accident the train here was an electrically powered subway passenger train and the drawbridge was fully protected by fixed signals.

The precise location of the accident was designated Dock and was a third of a mile west of Harrison station. The double-track electrified line here crossed the Passaic River by an opening bridge designated Bridge 8.50-B. The bridge was of the steel, through truss vertical lift type, 237 feet long and was controlled from Dock tower. At the time of the accident its span was being lowered for railroad traffic after having been raised for river traffic. Movements on the westbound track were protected by signals 66R and 100R, 2,371 and 350 feet respectively east of the point of the accident. Both were continuously lit position light signals. In addition the trains running over the line were provided with position light signals in their cabs repeating the aspects of the outdoor signals and audible warning alarms.

At the time of the accident signal 66R was displaying three horizontal white lights over three diagonal white lights to the right (*Slow Approach*) – proceed prepared to stop at next signal. The corresponding cab signal would have been three diagonal white lights to the right (*Approach*) – proceed prepared to stop at next signal. If proceeding at more than medium speed reduce at once to the speed.

Signal 100R was displaying three horizontal white lights (*Stop*) over a horizontal smash board. The corresponding cab signal would have been two diagonal white lights to the left (*Restricting*) – proceed at restricted speed.

Electro-pneumatic interlocking in Dock tower ensured that signal 100R must display *Stop* and the smash board be in position before the bridge could be unlocked. Likewise the bridge must be locked shut before signal 100R could be cleared. The audible warning signal sounded in the cab of a train whenever it passed a signal not showing proceed until the warning was acknowledged by the motorman.

Westbound train No. 065 on the day of the accident consisted of seven all-steel multiple-unit coaches, with coach 1934 leading. It was in charge of Conductors Robert Kelly and Chester Middleton, with Motorman Michael Mullin. It passed signal 66R indicating *Slow Approach* – proceed prepared to stop at next signal, called at Harrison station, left there at 6.25 p.m., seven minutes late, passed signal 100R indicating *Stop* and

---

[897] Interstate Commerce Commission report

collided with the raised span of the lift bridge at an estimated speed of 12 miles an hour at 6.27 p.m. Both ends of the first car and the front end of the second car were badly damaged. Motorman Mullin was crushed in his cab and killed outright while Conductor Middleton and a passenger, Alice Horowith, 39, of Newark, were fatally injured. [898] Seventy-nine other passengers and three railwaymen were more or less hurt.

The Dock tower operator had operated the controls to raise the lift bridge at about 6.23 p.m. At this moment signal 66R was therefore showing *Slow Approach*, signal 100R was showing *Stop*, the smash board was horizontal across the track and no train was shown in the block in the rear of signal 66R on the track diagram in the tower. At about 6.27 p.m. the operator moved the control lever to lower the span again, the ship having passed through. The span was descending but still about 11 feet above rail level when it was struck by the train.

Surviving members of the train crew said that the train had been proceeding normally and they had been unaware of anything wrong until the brakes had come on in emergency. The brakes of the train had been tested and had functioned normally before the accident, and the interlocking between the lift span and the protecting signals was also found to be working properly after the accident. In a test made with a similar train travelling at 17 miles an hour, the application of the brakes in emergency at the moment that it passed to smash board resulted in it stopping in 95 feet, about 255 feet short of the lift span.

It could not be determined, the Interstate Commerce Commission stated, why Motorman Mullin had failed to stop his train. A post-mortem was apparently performed on him, presumably to see if he had perhaps suffered a heart attack or similar, but it does not seem to have revealed anything. Possibly the fact that after having passed signal 66R the train had to stop at Harrison station was relevant. [899]

---

[898] *The Evening Star*, Washington, D.C., United States, 18th December 1945

[899] Interstate Commerce Commission report

CHAPTER 10

TRAIN ACCIDENTS INVOLVING BRIDGES

Bridges could feature in train accidents in one of two ways. Either the bridge itself could collapse for a number of reasons such as faulty design, faulty construction or because its foundations had been damaged, or a train could fall from a bridge, the structure itself remaining intact. Bridge failures have been dealt with in the previous chapter, this chapter will concern itself solely with accidents where the bridge itself was not at fault. It may seem perverse to list such accidents separately instead of in the chapters dealing with primary causes of the accidents, the procedure adopted elsewhere in this book, but the fortuitous presence of a bridge has such a potential for transforming what might have been only a minor accident into a major disaster that it has been decided to treat such accidents separately, in the same way as accidents occurring in tunnels have been given a chapter of their own.

The night express from Shimonoseki to Tokyo was passing through Kochi station, Japan, on the San'yo main line at 4 a.m. on Monday 12th January 1931, when its driver failed to reduce speed through a set of facing points and passing through them at 80 kilometres an hour the train became derailed. The points adjoined a bridge over the Mukunashi River and the locomotive, and five third-class carriages fell 20 feet into the shallow water. Seven people died and 80 were injured. There had been 438 passengers on the train. [900] [901]

On Monday 21st August 1933, a train carrying 500 Chinese officers was taking them back to the Kiangsi front where they were to resume operations against Communist bandits after having completed a special course in mountain climbing at Kuling, the headquarters of General Chiag-kai-Shek, commander of the Chinese army. Between Kiukiang and Nanchang, China, the train became derailed, and two carriages fell from a bridge into the Siao River. There were 200 men in the coaches which went in the water. The passengers from the remainder of the train made desperate attempts to rescue them,

---

[900] *The Times*, London, England, 13th January 1931

[901] Webpage – Japanese Railway Safety and the Technology of the Day by Masao Saito

but failed, and rescuers from Nanchang and Kiukiang were unable to raise the coaches in times to save any of the victims. [902]

On Tuesday 24th October 1933, train No. 354, the Chemins de fer de l'Etat express which had left Cherbourg at 09:55 and was due in Paris at 11:26, was in charge of Conductor Prioult with Driver Benjamin Martin and Fireman Jean-Marie Roguel. The journey went normally until the train was between Conches and La Bonneville stations, when it left the rails with disastrous results. The engine, leading van and four following carriages fell from the Saint Élier bridge over the River Rouloir and the two following vehicles were derailed. An early report spoke of 30 dead and 32 injured; the dead included the driver and fireman.

Conductor Prioult told a reporter, "I felt my carriage swaying for about five metres and said to my colleague, Raguel, who chanced to be riding with me – he was not on duty – 'We are off the road, the train is derailed!' Straight away there were cries of pain and of terror, and a terrific commotion. I immediately went forward to protect the obstruction while Raguel went to telephone La Bonneville. You know the rest."

The train had been travelling at over 100 kilometres an hour when it became derailed (112, to be precise, according to the *Flaman* track recorder). The leading three coaches of wooden construction were telescoped, and it was in these that most of the loss of life occurred. Railwaymen and police were soon on the scene and began the rescue operation. Damaged sleepers were reportedly used to break into the carriages and pieces of lineside fencing were used as makeshift stretchers. The 7th regiment of Chasseurs was sent from Evereux to help in the rescue operation, along with a detachment of divers from Rouen. Both lines were blocked, and the telegraph wires were brought down.

By 18:00 27 bodies had been recovered, 19 injured had been taken to hospital in Evereux and a special train took 11 of the less badly hurt, together presumably with the uninjured passengers, on to Paris. [903]

One of the passengers, his face bloody from the rescue of an injured woman, spoke to a reporter. "I was sat quietly smoking a cigarette in the compartment that I shared with several other people, when before I even realised that something was wrong there was a violent shock and a terrible crash which threw me from my seat. My fellow passengers had also been badly thrown about. A woman near me screamed; another, huddled in a corner, her baby in her arms, was trembling. One of the (luggage) nets of the compartment had broken and several windows were smashed. It was all over so quickly that the exact realisation of what was happening did not hit me straight away; but the pictures of chaos, of fear, are at the moment fresh in my memory, and are very intense. A deafening din was literally breaking through my eardrums. There were howls of grief, anguished appeals…

"Derailment or collision? I did not yet know. As soon as I had pulled myself together, after enquiring as to how my fellow passengers were – only one was slightly hurt in the face and others were a little cut elsewhere from shards of broken glass – I opened the door and jumped down onto the ballast, as had other bewildered passengers. It was then that I realised the magnitude of the disaster. I swear to you, sir, that it was a sight which I will never forget! The twisted rails, the broken carriages, and the sobs that arose from the

---

[902] *The Mobile Times*, Mobile, Alabama, 21st August 1933

[903] *The Times*, London, England, 25th October 1933

enormous mass of wreckage!"

The reporter added that the horror of the scene could be seen in the eyes of the passenger he was interviewing; he was pale, and his hands trembled. The passenger continued, "I began to run, like the others, towards the overturned carriages. Bodies were already being removed, covered in blood, together with moaning injured. The rescue work was beginning to be organised, despite the panic, a panic which could not be placated. Some of the carriages had fallen 10 metres into the river; now I understood the reason for the terrible noise that I had heard." [904]

M. Joseph Paganon, the Minister of Public Works, and M. Raoul Dautry, the managing director of the Chemins de fer de l'Etat, arrived at the scene soon after the accident, and an inquiry into the cause of the derailment was begun by railway officials. Any suspicion of sabotage was soon discounted. It emerged, however, that a gang of platelayers had been shifting ballast near the scene of the accident, with the possibility that they had inadvertently disturbed the track prior to the arrival of the express.

By the following day 27 bodies had been taken to an improvised morturary at Everaux station and ten of the injured had died in hospital, bringing the number of dead to 37. Another 84 passengers were injured, 24 of whom were still in hospital. One line had been reopened to traffic with a severe speed restriction.

M. Ferdinand Maison, a retired chief engineer, who headed the investigation into the accident, stated that the displacement of the track had begun at the point where the ballast had been removed from beneath the sleepers by the platelayers before packing them with fresh material. The trackmen, however, maintained that they had witnessed the engine of the express rolling abnormally from side to side for several hundred metres to the rear of the point where the work on the track was in progress, implying that it was the abnormal motion of the locomotive which had displaced the track. That the train had been rolling abonormally would of course tie in with the evidence of Conductor Prioult. M. Diret, traffic manager of the State Company, stated that the main line from Cherbourg to Paris was as well maintained as any in the country, while M. Frédéric Sirleau, the engineer in charge of the line, said that another train had passed the spot a little earlier without incident. M. Maison suggested that the heavy express engine travelling at speed might have displaced the track at a point where it had been disturbed. [905] There was no speed restriction in place at the point where the track was being worked on because none was thought necessary for the minor work being performed.

By 27th October it was known that the number of dead had risen to 39. It was reported that the investigation into the cause of the disaster was continuing but was made difficult because of the death of the driver and fireman. It was clear, however, that the cause must lie either in a problem with the track or with the engine.

The track had been entirely re-laid in 1930. It was one of the strongest in France. It consisted of 18 metre-length rails weighing 50 kilograms per metre. Each rail was supported by 29 bolted sleepers.

If the track were excluded as a cause, then the locomotive must be to blame. It was a *Mountain* class (4-8-2) engine, commonly used on the Etat and Est networks. A month previously technicians had noted that its axles, which were made up of steel parts under a pressure of 200 m/m diameter were slightly bent. The axles had been removed and

---

[904] *Le Petit Journal*, Paris, France, 25th October 1933
[905] *The Times*, London, England, 27th October 1933

replaced with steel and nickel axles 210 m/m. Since then the engine had run 1,500 kilometres. The driver and fireman had been questioned very recently and had said that the engine had given them every satisfaction. However, the statements of passengers indicated that the train had been rolling abnormally after Serquigny. This tended to suggest that the engine had for some time had a problem, the nature of which remained to be determined.

Pieces of broken buffers were found on the track in the rear of the accident site, which also suggested that the engine was running abnormally. Moreover the *Mountain* class was, the investigators noted, one which had given concern earlier. A locomotive of the same class, a *Mountain* 241, was pulling the train which derailed at 110 kilometres an hour at Villepatour, on the Est system. That accident had been difficult to explain; it had happened on a curve similar to that at the Saint Élier bridge and the passengers in that accident had noticed an abnormal rocking of the whole train in the seconds before the derailment.

That the class of engines was susceptible to derailment at certain speeds and with certain track conditions was not impossible, but the investigators, before pronouncing a cause which would involve the rebuilding of the whole class, had to be certain. The investigators knew that it was important that they reach a conclusion as soon as possible but said that it would still be some time before they could do so. [906]

The final, 150-page report on the accident appeared nine months later. It stated that the work on the track had played a role in the disaster, whereupon the Evereux magistrates arrested Area Chief Jeanne, District Chief Loisel and Section Chief Imhoff of the permanent way department on charges of reckless homicide for non-observance of the regulations. There was something of an outcry at the arrest of three railwaymen whose competence and integrity were beyond dispute and no further legal action was taken against them. It was clear in any case that it was the locomotive which was largely responsible for the disaster.

The final verdict on the accident, delivered by Deputy Lucien Midol, was that it had been due to 'the work being done on the track without the imposition of a speed restriction and the operation of an excessive heavy engine with insufficient lateral play at too high a speed over such track.' The charges against the three railwaymen were dropped and no action was taken against anyone else. The maximum speed of *Mountain* class engines was restricted to 90 kilometres an hour pending their redesign. [907]

The accident has obvious similarities to the Sevenoaks disaster of 1927 in England described in the previous volume of this series, and the Bihta disaster in India, described in an earlier chapter.

On Christmas Eve, Tuesday 24th December 1935, Berlin – Basel express D 44 was running in three sections, Vorzug zum D44, D 44 and Nachzug zum D 44, because of the heavy traffic. Nachzug D 44 passed through Großheringen, Germany, on time, without any problem but the following train, D 44, running 20 minutes later, met with disaster. Approaching the station, it ran past an *Einfahrsignal* and came into sidelong collision with train P 825 from Erfurt to Leipzig. The trains collided with great force on a bridge over the Saale River. Several coaches of train P 825 were telescoped into one another,

---

[906] *L'Indépendant* des Bassses Pyrénées, Pau, France, 27th October 1933
[907] Wikipedia – Accidente ferroviaire de Saint-Élier

and their upper parts were ripped off. One passenger was thrown from a coach onto the railing of the bridge, held onto it briefly, then fell into the river, but was able to save himself by swimming to the bank. Thirty-three other passengers died.

A passenger on the express related, "We were travelling at speed when suddenly there was a brake application followed almost at once by a terrific crash. Suitcases, boxes and so on fell from the luggage rack and the train stopped. Windows and doors were opened, and no one realised that there had been a collision. A call went out: 'Keep calm.' Everyone obeyed."

On 22nd June 1936 the waiting room at Großheringen station was transformed into a courtroom. On trial were Driver Kurt Dechant in charge of the pilot engine of train D 44 and Driver Willi Baude, who had been in charge of the train engine. Driver Dechant denied having failed to see the signal. "I applied the brakes at once and whistled a warning," he maintained. However, there were signs of braking and sand on the rails only in the last 130 metres in the rear of the signal.

Dechant continued, "I felt that my braking was less effective because there were two engines. I shouted to my fireman, 'I can't pull up the train.'"

Driver Baude maintained that due to smoke and steam from the pilot engine he had been unable to observe the stop signal at all. Nor had he heard any whistle for the brakes.

The fireman of the pilot engine testified that he had gone to help Driver Dechant after the accident (in jumping from his engine Dechant had broken his ankle) but his driver had told him, "Help the passengers first. I shall be going to prison. I'm not a coward and know how to bear my guilt."

Dechant denied having said this.

The case was continued the next day in Naumburg, where Assistant Signalman Böhm testified, "I saw him sitting on the track, and he shouted to me, 'Go away mate, I ran through the signal.'"

The third day of the case saw a representative of the train driver's union, representing Dechant, criticising the actions of the signalman. The latter, he said, could have held train D 44 back at Bad Kösen until train P 825 had gone. The prosecutor described that idea as absurd. There was also a conflict between the Driver Dechant and the prosecutor as to how fast train D 44 had been going.

The case ended on 26th June with Driver Dechant being sentenced to a year and three months' imprisonment and Driver Baude to seven months' imprisonment.

Train D 44 needed a pilot engine because the train engine only had a little water, and its pump was defective. At Weißenfels, where the train stopped, there was no water crane. There was a replacement engine there, a P 8, but neither its driver nor his fireman had any experience of driving express trains. The fireman was not able, because of his inexperience, to maintain boiler pressure and Driver Dechant, realising this, had picked up the shovel himself to assist with the firing. It was while he had been doing so that he had missed the *Forsignal* at Großherringen indicating that the *Einfahrsignal* was at danger. When he had seen the red light of the latter signal, he had been momentarily paralysed. It was calculated that had he applied the brakes six seconds earlier the collision would have been avoided. Driver Baude had relied on his colleague and failed to see either signal. The driver's responsibility, above all else, was of course to observe signals.[908]

The derailment of a Pittsburgh and West Virginia Railroad freight train early on Friday 18th September 1936 on Pigeon Creek Trestle, near Sudan, Pennsylvania, United States, was considerably less serious than it might easily have been. The engine and 12 cars of the 60-car Connellsville to Pittsburgh fast freight train left the track and four cars fell from the 160 ft trestle at 1.30 a.m. Two of the cars fell within a few feet of the home of Clyde H. Douglas, where he was sleeping with his family. Three other cars hung over the edge of the trestle. No one was hurt but the damage done amounted to an estimated $30,000. The cause of the derailment was not stated and was probably not immediately known.[909]

Another spectacular derailment which was probably without casualties happened in the early hours of Thursday 16th September 1937 on the Chesapeake and Ohio Railway at Rocky Point, Virginia, United States. Freight train No. 92 was running along the single-track James River line when it became partially derailed, possibly due to a broken rail. Thirty-five large boxcars and coal gondolas left the rails. Some of them ploughed up the track, brought down telegraph poles and blocked the adjacent State Highway 622, while 15 of them went over a 40-foot-high trestle, falling onto the mill of the Liberty Lime and Stone Company below. The accident happened at 4.20 a.m. None of the railwaymen on the train was hurt but the wrecking crew who set to work clearing up the mess were looking out for the bodies of any transients who might have been "riding the rods." Given the extent of the derailment it was not thought that the crews from Clifton Forge and Gladstone would be able to clear it up before the following day.[910]

In late 1937 the Estrada de Ferro Central do Brasil company was unfortunate enough to suffer two accidents in quick succession, the first a derailment at Barra Mansa, Brazil, and the second a collision at Mesquita (see rear collisions chapter).

The Barra Mansa accident occurred on Sunday 31st October 1937. At 07:00 on that Sunday morning an express passenger train left Rio Alfredo Maia station for São Paulo. It was better filled than usual because many people in the capital were taking advantage of the three-day All Souls Day holiday to go and visit relatives. It consisted of engine No. 385 and an unspecified number of coaches in charge of Conductor Alberto Ramos de Poiva, with Driver Benedicto Monteiro dos Santos and Fireman Carlos Luiz de Sousa. The train had not covered that much of its journey when Driver Monteiro dos Santos realised that there was a problem with one of its vehicles and a stop was made at Barro do Pirahy (now Barro do Pirai) station for the train to be examined. It was found that the spring bracket of one of the bogies of the mail was loose and had been dragging along the ground. The option of detaching the mail car was considered, but in the end it was decided instead to carry out a temporary repair by welding the loose part in place. The train was delayed about an hour while this was being done. Resuming its journey the express passed through Vargem Alegre, Pinheiro (now Pinheiral), Rademaker and Volta Redondo safely and the nine kilometres between the latter station and Barra

[908] *Eisenbahnunfälle in Europa* by Erich Preuß, pgs. 96-98 (in German)
[909] *The Charleroi Mail*, Charleroi, Pennsylvania, United States, 18th September 1936
[910] *The Daily News-Record*, Harrisonburg, Virginia, United States, 17th September 1937

Mansa were mostly done, when Driver Monteiro dos Santos began to become aware of strange noise from a point just behind his engine. The passengers in the front of the train also heard the noise and mindful that the train had already been stopped for some time because of a problem they became alarmed. At this moment the train was approaching a bridge over the Rio Barra Mansa at Km 152, a small structure 22 metres in length over what was little more than a stream, a few feet deep. About 100 metres beyond the railway bridge the Rio Barra Mansa flowed into the considerably larger Rio Parahyba (now Rio Paraiba do Sul). The train was already derailed and although the engine crossed the bridge one or possibly two of the following coaches fell from it into the river, while two other coaches were left suspended in the air. The mail and conductors' vans were destroyed and two second-class carriages were badly damaged.

Passengers Lazaro Bastos, a merchant from Minas Geraes; Gastão Camargo, merchant from Barra Mansa; and Gummercindo Paria, employed in commerce in Rio de Janeiro, were killed, together with a passenger who could not be immediately identified as he carried no documents. There was also apparently a fifth fatality, as it was stated that the body of a child aged about three had been found under one of the carriages. Three passengers were seriously injured, as were Conductor Ramos de Poiva, Assistant Conductor Romão Pereira, and mail clerks José Ramos Brandão, Manuel de Souza and Octaviano Malta. Ten other passengers were less badly hurt, and 25 others sought help from chemists for minor injuries. [911]

Local inhabitants were soon on the scene, including several doctors, and the most seriously injured were taken in motor cars to the Santa Casa hospital in Barra Massa. A newspaper reporter wrote of a badly injured woman lying, half covered by water, in the riverbed, surrounded by five children. She turned out to be Señora Eliza Hildenfelder, bound for Passa Quatro and the wife of Señor Max Hildenfelder of Campos. The children, Noemi, 14; Eli, 11; Elmar, 6; Edelweis, 2; and Siegelsida, 1, were their offspring. The reporter noted that the girl Elmar had had the presence of mind, despite the horror of the scene, to grab her little brothers and support them, as her mother, badly hurt, was unable to do so.

When news of the accident reached nearby Resende a batch of 40 soldiers was sent from there to assist the police with the rescue work, while the Stationmaster Lino Oliveira at Barra Mansa telegraphed his headquarters that he "supposed that the cause of the accident was a broken axle which had been found on the right-hand side of the line about 50 metres from the bridge, and that the remaining bogies of the train were little damaged."

Examination of the line indicated that the axle had broken 240 metres from the bridge and that the sleepers had been torn up over a distance of 150 metres. Nevertheless, under the direction of Engineers Septembrino do Carvalho and Moraes Lacerda the line was cleared at 21:00, nine hours after the derailment. The badly damaged coaches had been burnt at the scene in order to expedite the clearance operation. Invariably, there was considerable disruption to traffic, for example train RP 2 from São Paulo, due in Rio Alfredo Mais at 18:00, did not reach there until 01:15 the next morning, and other trains suffered similar delays.

Speaking to the press, Colonel Mendoça Lima, the director of the railway company,

---

[911] *O Estado de S. Paulo*, São Paulo, Brazil, 2nd November 1937

attributed the accident to the breaking of an axle, though he stated that this conclusion must be treated with some reserve pending the results of the official inquiry into the accident. He added that Dr Itagiba Escobar, in charge of that inquiry, was already at the scene of the accident collecting all the necessary information. Continuing, the colonel stated that the company had already authorised the replacement of all the wooden rolling stock on its interior lines with steel vehicles within one year. He emphasised that it was certain that the accident would have been less serious had all-steel carriages been involved, as had been demonstrated worldwide. [912]

Two accidents occurred in Spain on Saturday 4th December 1937. One, at Alameda, involved a runaway and is dealt with in the Runway Trains and Vehicles chapter. The other, which was rather more serious, occurred between the stations of Freginals and Planes de Montsiá, at Kilometre 70,700. A waggon which had clearly not been properly secured was blown down the line from one or another of the stations by the wind until it came to rest on a bridge over the Galera River at the site of the accident. There it was struck by the Valencia – Barcelona mail train at about 03:00. The train was derailed, and a van and two third-class carriages fell about seven metres from the bridge. Twelve passengers died and about 60 others were injured. [913] [914]

A bridge between two tunnels on the Japanese Government Railway was destroyed by an avalanche on Tuesday 5th March 1940. At about 8.40 a.m. the engine, tender and part of the six coaches of a mixed train from Yonezawa to Saka-Machi, emerging from one of the tunnels plunged from the bridge, some 65 metres of which had been carried away, to the bed of the Arawaka River, almost 30 metres below. Of the 130 passengers on the train only 23 who had been riding in the rear coach survived and eight of them were seriously injured. Four of the five coaches which did not follow the engine into the water caught fire and their occupants were burnt. The rescue operation was hampered by deep snow. The accident occurred at a location in Yamagati Prefecture variously described as being near Mount Yokone Tunnel, and near Tamagaguchi. [915] [916]

Storm-force winds in Spain caused a railway disaster at Zumaya at about 23:30 on Saturday 15th February 1941. An electric train from Bilbao, mail train No. 21 of the Ferrocarriles Vascongados, bound for San Sebastian, where it was due at 21:00, was already running about three hours late due to the severe weather. It reached Zumaya station safely, but on leaving there only covered 150 metres, to the San Miguel Bridge over the Urola River, when it stopped due to loss of current, doubtless due to the wind having damaged the catenary. An exceptionally powerful gust then struck the train and overturned it, three of its 11 cars (a postal van, a third-class and a first-class) then falling from the bridge to the river below. The other cars remained on the approach embankment.

By 14:00 on Sunday there were known to be 15 dead. An hour later divers began to search for further victims and by 19:00 the number of known dead had risen to 22. About

[912] *Correio de Manhã*, Rio de Janeiro, Brazil, 2nd November 1937
[913] *La Vanguardia*, Barcelona, Spain, 5th and 7th December 1937
[914] Webpage: Federación Castellano Manchega de Amigos del Ferrocarril (in Spanish)
[915] *The Railway Gazette*, London, England, 8th March 1940
[916] *The Berkshire Evening Eagle*, Pittsfield, Massachusetts, United States, 5th March 1940

130 others were more or less injured. [917] [918]

Norfolk Southern Railway northbound second-class freight train No. 62, which had left Charlotte on Wednesday 24th September 1941 at 11.45 a.m. for Raleigh, consisted of engine No. 601, 16 loaded and four empty cars and a caboose. It passed the last open office at Oakboro, North Carolina, United States, at 1.15 p.m., 2 hours, and 5 minutes late, and travelling at about 25 miles an hour was derailed as it approached a 40-foot-high wooden trestle four miles north of Oakboro, at about 1.25 p.m. Following the derailment the engine, tender, five loaded and three empty cars fell from the bridge into a ravine. Both cars and bridge were 'reduced to kindling.' Engineer E.G. (Dad) McCollum and front Brakeman Preston Haywood were trapped in the engine cab and killed, and a hobo named Robert White, who was riding in an empty waggon next to the tender also lost his life. Fireman Lewis Moore jumped as the engine fell and was injured, the other members of the train crew escaped unhurt. The Interstate Commerce Commission investigated the accident and attributed it to the buckling of the rails due to the hot weather. At the U.S. Weather Bureau at Monroe, about 23 miles from Oakboro, the maximum temperature recorded that day had been 93°F (34°C). [919] [920]

Two St Louis – San Francisco Railway passenger trains, one of them the *Will Rogers* express and the other a troop train, collided head-on on a trestle at Granby, Missouri, United States, on Saturday 7th March 1942. Four passengers – all soldiers, a fireman, a mail clerk and a Pullman employee lost their lives and 60 passengers, eight railwaymen, seven dining car employees, five Pullman employees and two mail clerks were unhurt.

Eastbound first-class passenger train No. 4 from Oklahoma City, Oklahoma to St Louis, Missouri, the *Will Rogers*, consisted of engine No. 1528, a baggage-mail car, a baggage car, a passenger-baggage car, a chair car, a café-lounge car and two Pullman sleeping cars, in charge of Conductor Lester Powers with Engineer Frank Reed with Fireman Herschel Napier. It had left Tulsa, Oklahoma, at 2.25 p.m., left Neosho, about 7½ miles from Granby, at 4.50 p.m., 10 minutes late, passed signal No, 304.8, showing *Proceed*, passed signal No. 302.4, showing *Stop*, and travelling at 30 to 60 miles an hour collided with the troop train at about 4.58 p.m. [921]

Westbound Extra 1516 West, the troop train, consisted of engine No. 1516, a coach, a baggage car and six Pullman sleeping cars in charge of Conductor Ed Champion with Engineer Luther Ary and Fireman B.W. McAllister. At Monett, 20 miles east of Granby, its crew had received train order No. 138, Form 19, reading:

Eng. 1516 run extra Monett to Neosho has right over No. 440, Eng. 1332 and No. 4414 East Pierce City to Neosho and wait at Berwick until 4.33 p.m., Ritchey 4.38, Granby 4.43.

Train No. 1516 left Pierce City, just over 15 miles from Granby, at 4.44 p.m., passed signal No. 299.9, showing *Proceed*, signal 302.1, showing according to the engineer *Approach*, and moving at 5 to 40 miles an hour collided with the *Will Rogers* 413 feet east of the west siding switch at Granby, and on bridge G-302.5, both trains being on the

---

917 *La Vanguardia*, Barcelona, Spain, 18th February 1941

918 Webpage: Federación Castellano Manchega de Amigos del Ferrocarril (in Spanish)

919 *The Wilson Daily Times*, Wilson, North Carolina, United States, 25th September 1941

920 Interstate Commerce Commission report

921 *Joplin News Herald*, Joplin, Missouri, 9th March 1942

main track.

As a result of the collision the two engines were derailed and practically demolished, the second car of the troop train telescoped the third and the first car of the *Will Rogers* was derailed and its third car fell from the bridge to the dry riverbed below. The front end of the third car also ended up in the river.

The Interstate Commerce Commission (ICC) pointed out that the rules of the railroad specified that inferior trains must keep out of the way of opposing superior trains, while extra trains, such as the troop train, must clear the times of regular trains, such as No. 4, by five minutes or provide flag protection. Train No. 4 was due to leave Granby at 4.49 p.m. and Berwick at 4.59 p.m. Train No. 1516, the troop train, had left Pierce City at 4.44 p.m. and reached Berwick at about 4.50 p.m., where it should have taken the siding for train No. 4. Instead it had passed Berwick, had passed Ritchey and at Granby had collided with train No. 4 at 4.58 p.m., 14 minutes after the time when it should have been in clear there.

Engineer Ary told the inquiry that at Monett he had compared watches with his conductor. There he had told his fireman that the times specified in the wait order had expired. At Pierce City he had looked at his watch and had misread the time as 3.42 rather than 4.42 and had told his fireman that they had time to reach Neosho, 7¾ miles beyond Granby, to meet train No. 4. He said that at Monett he had thought only of the wait order, and at Pierce City only of the meet with No. 4.

Fireman McAllister said that he had known that No. 4 was due to leave Neosho at 4.40 p.m. but had not checked his watch when his engineer had spoken of meeting it there. He had also failed to consider that leaving Monett at 4.20 p.m. gave them only 15 minutes to cover the 27¾ miles between there and Neosho.

Engineer Ary stated that approaching Granby all the signals had been clear until he saw signal No. 302.1 at yellow (*Approach*). He had applied the brakes and entering the tangent track had seen signal No. 302.5 at red when he had made a further brake application to stop at it. Fireman McAllister had then shouted that a section foreman was giving them stop signals but he (the engineer) had thought that the fireman was referrring to signal No. 302.5. He had then seen train No. 4 approaching but had still not made an emergency brake application, thinking that train No. 4 would stop at signal No. 302.4. At the last moment he and the fireman had jumped.

The conductor, flagman and porter of train No. 1516 all admitted having forgotten the schedule of train No. 4.

Section Foreman Felix Arnold said that he had tried to brake train No. 1516 when he saw that the two trains were on a collision course.

Engineer Reed testified that he had passed signal No. 304.8 at *Proceed.* Signal No. 302.4 could not be seen very quickly from his side of the cab but Fireman Napier had shouted that it was at *Stop,* when he had applied the brakes in emergency but had been unable to stop before running past the signal.

The ICC noted that according to the evidence it was not unusual for a train to receive a clear indication at one signal and a stop indication at the next, although double-approach signals were now being provided at some points along the line. With the crew of train No. 4 receiving a clear indication at signal No. 304.8 it had been impossible for them to stop at signal No. 302.4, while the engineer of train No. 1516 had stated that he had only received an *Approach* indication from signal No. 302.1, though Section Foreman

Arnold said that the signal had changed to red immediately before the train had passed it.

The ICC remarked that the signalling at Granby was in violation of Section 207 of their order of 13 April 1939. This read:

207. On track signaled for movements in both directions, signals shall be so arranged and controlled that proper restrictive indications will be provided to protect both following and opposing movements

The commission concluded that this accident was caused 'by an inferior train occupying the main track on the time of an opposing superior train, and by an inadequate automatic block-signal system.'[922]

Local coroner J.R. Reynolds announced a few days later that he would probably not be holding an inquest on the deaths of the victims of the accident as several of the witnesses were out of state and he had no authority to summon them. Also the railroad, the army board and the ICC were already carrying out investigations. A local newspaper thought that whatever difficulties might lie in the way of holding an inquest, the coroner was mistaken in not holding one, that there were many aspects of the aspects to be explored, and that the public was entitled to know how such a terrible mix-up had occurred. [923]

A combination of worn-out and missing equipment and inexperienced staff were cited as the reason for a catastrophic accident at Tsuchiura station, Japan, on the Joban line in Ibaraki Prefecture, on Friday 29th October 1943. Due to a signalling error waggons being shunted were wrongly routed onto the Up main line, where they were struck by a freight train. As a result of the collision some waggons were derailed foul of the down line where they were almost immediately run into by a Down passenger train. To compound matters the fourth carriage of the passenger train fell from a bridge into a river. Fifty-seven people died in the accident and another 77 were injured. [924]

A series of events which began with the tragic death of a fireman ended in an accident to a goods train on Tuesday 12th September 1944 in Australia. The train involved in the accident was carrying wheat from Perth to Fremantle. As it was stopped at Karrakatta station on the outskirts of Perth, Fireman George Whitmore thoughtlessly alighted from the right-hand side of his engine into the path of a Perth-bound train. Badly hurt, he was taken on the locomotive the short way to Karrakatta station, thence by St John's Ambulance to Perth Hospital where he was found to be dead on arrival.

Driver R.J. Stafford then decided to take his 119-ton *Garratt* engine and train of wheat trucks weighing 700 tons onward the relatively short distance to Fremantle (about 14 miles from Perth) single-handedly. All went well as far as Claremont, Western Australia, but here his train was signalled into the "back road," a siding which ended in buffer stops about 20 yards short of a bridge over Davies Road. Driver Stafford, probably because he was preoccupied with also firing his engine, failed to realise that he was no longer on the main line and made no attempt to stop. As a result his engine, which was running tender first, smashed through the buffers at the end of the siding, traversed the 20 yards of embankment beyond, crossed the 22-foot-wide road and embedded itself in a

[922] Interstate Commerce Commission report

[923] *Neosho Daily Democrat*, Neosho, Missouri, United States, 17th August 1942

[924] Webpage: Japanese Railway Safety and the Technology of the Day by Masao Saito

wall on the opposite side, where it remained, suspended 18ft above the roadway. Eight of the waggons were damaged, six badly and some telescoped. Quantities of wheat were spilt, and the track was damaged. Driver Stafford escaped with a severe shaking. [925]

On Monday 27th August 1945 a lightning strike during a thunderstorm interfered with signalling communications on the single-track Hachiko line in west Tokyo between Komiya and Haijima, Japan. An inexperienced stationmaster and other staff then permitted a train to enter the section from each end at about the same time and the two trains collided while they were crossing the Tamagawa Bridge and fell into the river below. The accident cost 105 lives and another 67 were injured. [926]

Disaster struck *El Panamericano*, a long-distance train from La Paz, Bolivia, to Buenos Aires, Argentina, early on Sunday 9th September 1945. The train involved had started from La Paz on Friday 7th September and entered Argentina at the border town of La Quiaca. It consisted of a powerful engine, four sleeping cars, a second-class coach, a first-class coach, a mail van, a dining car and two luggage vans and was in charge of Conductor Maximiliano Alderete with Driver Carlos Barg and Fireman Regino Santos Paz. The border crossings and other stops on the long route had caused, it would seem, a considerable delay to the schedule, and the Driver Barg was running at high speed in order to regain some time when the train reached the Humahuaca valley (Quebrada de Humahuaca), a very dangerous point on the line due to the steep gradients, curves, and flood-prevention measures. Flooding from the Grande River (Rio Grande) in previous years had prompted the authorities to carry out these works since the railway was repeatedly badly affected by flooding throughout the year, but according to a newspaper report these works had made the location more dangerous from a railway point of view; the paper did not explain why, possibly a realignment of the track was involved with more severe curvature.

El Panamericano ran through the night with no warning for its passengers of the disaster which was to happen kilometres ahead. Clearly its speed was excessive, because the time taken between the two stations preceding the accident point, which the train was booked to cover in 35 minutes, was covered in just 16. At exactly 02:00 the train arrived at kilometre 1,313-400, between Iturbe and Tres Cruces stations, at just the point where the railway ran alongside the Rio Grande on an embankment of considerable height. As it was the early hours of the morning the passengers were all resting, though some noticed perhaps the sparks coming from the engine and the fact that the speed of their train had not dropped at all. Still travelling at excessive speed, the train then reached a point where there was a very steep descent of 60 in 1000 (that is about 1 in 17) and Driver Barg must presumably have realised the danger at this point and applied his brakes to try to avoid a disastrous derailment. The speed, however, was just too great, and arriving at a curve the engine jumped the rails and overturned, followed by most of the train behind. The engine and carriages then tumbled down the embankment to form a mass of wreckage, while part of the train fell from the bridge by which the line crossed the Rio Grande.

Many of the sleeping passengers had a horrific awakening when they were thrown from their beds through the train windows. These were the lucky ones in that they

---

[925] *Kalgoorlie Miner*, Kalgoorlie, Western Australia, 13th September 1944

[926] Webpage – Japanese Railway Safety and the Technology of the Day by Masao Saito

escaped certain death. Others were trapped in the shapeless mass of wreckage, some perishing shortly afterwards. Scenes of terrible drama followed in the darkness of the night with the cries of relatives searching for their loved ones mingling with the groans of the injured.

The surviving members of the train crew immediately set to work to help the passengers buried in the wreckage. They were joined by the passengers, including some who were themselves injured, with cuts and bruises in various parts of their bodies, but who set to work to care for their more seriously hurt fellow travellers. There was no lighting of any sort to help rescuers reach the places where cries indicated that injured passengers could be found. The victims were all eventually freed from the wreckage, though some died before they could be released.

Several of the uninjured passengers were dispatched to the nearest town to bring news of the accident and get aid. Help arrived as quickly as might be expected in the circumstances. Some reports stated that the first notice of the accident received in Jujuy, the provincial capital, came from the pilot of an aircraft who had spotted the wreck at about 6 a.m. and radioed a report of it to Jujuy Aerodrome.[927] A relief train was sent without delay from Jujuy, with doctors and nurses. It immediately returned with 16 injured whose condition was regarded as extremely serious, and then was taken once more to the accident site. Early on there were known to be five dead and 23 injured, many in a grave condition. Driver Barg was among the dead.

The federal administrator was informed of the accident at 6.20 a.m. and proceeded to the scene with the police commissioner and other authorities, while journalists reached the scene at about 10 a.m. They found some of the wreckage on the slope of the embankment, the remainder was in the bed of the river. A sleeping car rested on top of the dining car while the rear two vans of the train remained on the rails, undamaged.

The work of clearing the wreckage went on through the following morning with the aid of local residents. Those injured in the accident included Conductor Alderete and Fireman Paz.

The initial figures given by newspaper reporters as to the number of casualties proved to be overoptimistic; an announcement by the Ferrocarriles del Estrada company later in the day stated that 22 passengers and Driver Barg had lost their lives and there were 35 injured.[928]

The next day it was stated that the number of known fatalities had risen to 35, including Fireman Paz, who had succumbed to his injuries, while the number of injured had increased to 47. The final number of dead was probably 36. Most of the passengers who had lost their lives had been in the second-class coach which had been smashed to pieces and buried beneath other vehicles.

It also emerged that in an ante-mortem statement to a newspaper reporter, Fireman Paz had related that he had warned Driver Barg that he was going too fast, but that his warning had been ignored, thus confirming, if confirmation were needed, that the cause of the accident was excessive speed.[929]

---

[927] Wikipedia – Iturbe Railway Accident of 1945
[928] *El Litoral,* Santa Fe, Argentina, 9th September 1945
[929] Ibid, 10th September 1945

CHAPTER 11

TUNNEL ACCIDENTS

In his classic work on British railway accidents, *Red for Danger*, L. T. C. Rolt remarks on the acute trepidation felt by Victorian travellers when they were passing through a tunnel, and their oft repeated pleas to the railway authorities that all tunnels should be artificially illuminated; pleas which, probably for practical reasons if nothing else, fell upon deaf ears. He goes on to point out that these fears were by no means unjustified, especially in the period before the introduction of the absolute block system, when trains were merely spaced by the time interval system. If a train should chance to break down in the open there was a good chance that the crew of a following train might be able to see it in time to pull up, even if there was insufficient time for the guard of the first train to go back and protect its rear. In the smoke- and steam-filled darkness of a tunnel there was no such hope. Furthermore, should a serious collision occur in a tunnel the confined space would be likely to result in the wreckage being hideously compressed, and should a fire break out in a long tunnel provided with ventilating shafts, the latter could act as chimneys. In addition, the rescue work would be hampered by the darkness and the difficulty of working in a confined space where the use of breakdown cranes was impossible, and the only resort was to drag the wreckage bodily out of the tunnel.

A rear collision in a tunnel involving two electric trains on the metropolitan railway of Napoli, Italy, on Friday 18th March 1932, killed six people and injured a further 28, 15 of them railwaymen. Two coaches were completely crushed. A rescue operation involving railwaymen, firemen and soldiers was quickly organised. [930] Two English seamen from the British Mediterranean Fleet, which was on a visit to Napoli, and who were passengers on one of the trains, distinguished themselves in the rescue work and were thanked for their efforts by the Prince of Piedmont. [931]

An inquiry into the accident placed the blame for it on the driver and conductor of the colliding train, which had left two minutes early without any signal. [932] The two

---

[930] *Le Peuple*, Paris, France, 20th March 1932
[931] *The Times*, London, England, 21st March 1932
[932] *Journal de Geneve*, Geneva, Switzerland, 21st March 1932

railwaymen, who were seriously injured in the accident and had been taken to hospital, were placed under arrest. The damage caused by the accident was estimated at several million lira.

The north exit from the Swiss station at Lucerne (Luzern) station was (and is) something of a bottleneck, with all trains to and from Berne, Basle, the Seetal, Zürich and Arth-Goldau having to pass through the double-track Schönheim and Gütsch Tunnels, 199 and 326 metres long respectively. The tracks through the tunnels were operated as a pair of single lines, trains for the first three routes using the left-hand tunnel (coming from Lucerne) and those for the Zürich and Arth-Goldau (Gotthard) routes the right-hand bore.

At about 15:00 on Tuesday 13th December 1932 a serious head-on collision occurred in the right-hand track in Gütsch Tunnel, Switzerland, involving express train No. 264 from Zürich and stopping train No. 554 for Bellinzona on the Gotthard line. Train No. 264 had left Zürich at 13:50 and consisted of engine No. 10645, a third-class carriage (locked empty and used as a so-called 'protective coach' in the hope of shielding the passengers in the following vehicles), a first-, second- and third-class composite carriage, a luggage van, a first-, second- and third-class composite carriage and a third-class carriage. Train No. 554 consisted of engine No. 12335, a parcels van, a mail van, a luggage van, two third-class carriages, a second-class carriage and three cattle waggons.

Three passengers, both train drivers (named Willi and Weber) and Conductor Stevanoni died in the collision and 19 passengers were more or less injured. [933] The leading vehicles of each train were reportedly completely shattered in the collision and their wreckage blocked the width of the tunnel. Fortunately, no train had been passing on the other line at the time of the collision. The initial rescue work was impeded by the darkness; later searchlights were set up to provide illumination. Three and a half hours after the accident, rescuers were still busy cutting into the coaches with oxy-acetylene torches to free the bodies of the dead and injured. The tunnel was blocked for 36 hours, and buses and even private cars were used to transport passengers between Lucerne and stations on the far side of the tunnel.

An investigation into the accident was immediately begun by the Lucerne public prosecutor, assisted by M. Etter and M. Labhart, the general and regional managers respectively of the Chemins de fer fédéraux suisses (CFF) (Swiss Federal Railway). This established that had it left Lucerne station on time stopping train No. 554 would have diverged onto the Gotthard line at Sentimatt, just north of Gütsch Tunnel, before the Zürich express had arrived at that point, but on the day of the accident it had departed from Lucerne five minutes late. Meanwhile, shortly before, a steam-hauled freight train had passed through the tunnel, filling it with steam and smoke. The signal for the line from Zürich at Sentimatt protecting the junction with the Gotthard line had been at danger against the express but its driver (possibly M. Willi, though it is not entirely clear which driver was in charge of which engine) overran it, ran through the junction points, and entered the tunnel to collide with the local train. A post-mortem failed to disclose any medical reason for the driver's mistake. Criticisms were voiced in the press concerning the single manning of electric locomotives, but the measure was defended by CFF

[933] *Gazette de Lausanne*, Lausanne, Switzerland, 14th December 1932

officials on the grounds of economy. They stated that the locomotives of all single-manned trains were provided with extra safety devices in the form of so-called dead man's handles, while the accident also hastened the installation of automatic train control through the tunnel. A proposal was made to build a new tunnel for Berne, Basle and Seetal trains but nothing came of the idea. [934]

Hampstead Tunnel, England, on the London and North Eastern Railway (ex-Great Central) main line between London Marylebone and Neasden, was the scene of a relatively minor accident involving three trains on Monday 16th January 1933. A down empty coaching stock train from Marylebone was run into from behind by following passenger train and an up goods train ran into the wreckage. The guard of the empty stock was badly hurt, and two passengers and the fireman of the passenger train were slightly hurt. The collision occurred at about 7.20 a.m.

From London Marylebone station the Down line climbed, mostly at 1 in 196 and 1 in 100, through Lord's, St John's Wood and Hampstead Tunnels. This section of line was continuously track circuited with Hall type three aspect searchlight signals. On the Down line one of these was 792 yards from Marylebone at the south entrance to St John's Wood Tunnel, and the next, signal D.23A, was at the foot of an air shaft in St John's Wood Tunnel, and 1 mile and 31 yards from Marylebone station.

The Down empty stock train, 7.5 a.m. from Marylebone to Neasden consisted of 4-4-2 engine and nine eight-wheeled coaches, in charge of Guard Charles Chambers, [935] with Driver A. Norris and Fireman J. Armstrong. The Down passenger train, the 7.10 a.m. departure from Marylebone to High Wycombe, consisted of 4-6-2 tank engine No. 5154, running bunker first, two articulated suburban coaches, an eight-wheeled first-class carriage and two articulated suburban coaches, in charge of Guard W.E. Tregidden, with Driver P.W. Hands and Fireman G.R. Green. The Up goods train, 1 a.m. from Annesley to Marylebone Goods, consisted of 0-6-0 engine No. 5298, two open waggons, a plate waggon, and a brake van, in charge of Guard H. Bayliss with Driver L.G.L. Sanders and Fireman G.W. Coleman.

Driver Norris gave evidence at the Ministry of Transport inquiry into the accident that he did not consider nine coaches an excessive load for his engine; it was a normal load for a Monday morning. Nevertheless, his engine had begun to slip as it entered St John's Wood Tunnel and it had continued to do so until it was almost at a stand, at which point it was run into by the passenger train. As his engine had been running tender first only the sand pipes on the tender could be used, and his fireman had to feed sand down them, one at a time.

Driver Hands of the passenger train stated that the two colourlight signals which he had passed before entering St John's Wood Tunnel had been yellow, which he understood to mean that he must stop at the next signal. Signal D.23A, however, he had not seen, the tunnel being full of smoke and steam, and he had not realised that he had gone past it until he had reached the bridge over the London Midland and Scottish Railway (LMSR), between St John's Wood and Hampstead Tunnels. He had then

---

[934] *Railway Accidents of Great Britain and Europe* by Ascanio Schneider and Armin Masé, pgs. 35 to 41

[935] *The Times*, London, England, 17th January 1933

reduced speed, but as there were spring catch points at the south end of the bridge he had not stopped, lest on restarting his train slip back slightly and become derailed on the catch points. The curvature of the track, combined with the smoke and steam in Hampstead Tunnel, had prevented him from seeing the tail lamp of the empty stock train.

Fireman Green confirmed his driver's evidence and put the speed of their train at 15 to 20 miles an hour at the time of the collision.. He said that he and Driver Hands had spoken of one another about having missed the signal as they had approached Hampstead Tunnel.

Guard Tregidden likewise said that he had not seen the signal D.23A because of the smoke and steam, though he had been looking out for it.

Driver Sanders of the goods train testified that he had been running under clear signals when had collided with an obstruction in the tunnel. He added that although the weather was clear outside the tunnel was so full of steam and smoke that he would not have been able to see a signal within it. Stopping his train at the LMSR bridge he had telephoned the signalman to tell him what had happened and to get help for the injured Guard Chambers. He was commended by the inspecting officer for his quick actions.

Lieut.-Colonel E.P. Anderson, the inspecting officer, accepted that the much reduced visibility in St John's Wood Tunnel would have made the observation of signal D.23A very difficult, if not impossible. It was, however, the inspecting officer said, much to be regretted that when he realised that he had missed seeing the signal Driver Hands had not quickly brought his train to a stand, which there had been ample room to do clear of the catch points. Some form of automatic train control which applied the brakes of the train if it passed the signal at danger would almost certainly have prevented the accident. The difficulty of seeing the signal had already been recognised by the railway company, who had added a second low-level lamp to it. They were considering the use of more powerful signal lamps. The inspecting officer suggested re-siting the signal just inside the tunnel, and the addition of automatic detonator placers or a mechanically worked gong. He also thought that a maximum load should be laid down for empty stock trains, with the loads being reduced for engines working tender first. [936]

Considerable mystery surrounds an accident on Wednesday 21st June 1933 near Dupnitza, Bulgaria. A report in the London *Times* speaks of a collision between a passenger train and a trolley car, implying a level crossing collision. [937] However, a couple of American newspapers speak of a collision between a passenger train and a wildcat (runaway) engine in a tunnel on the line between Dupnitza and Radomir. The reports agree that there were 15 deaths, with five injured. [938]

As they were mostly built as economically as possible, railways in the United States tended not to feature as many tunnels as they might perhaps otherwise have done, but sometimes these were unavoidable, and there were two on the New York Central Railroad (Toledo and Ohio Central) line between New Lexington and Rendville, Ohio.

---

[936] Ministry of Transport report

[937] *The Times*, London, England, 22nd June 1933

[938] *New Castle News*, New Castle, Pennsylvania, United States, 21st June 1933

The more northerly of these, the 644 feet New Lexington Tunnel, was the location of a head-on collision on Thursday 2nd November 1933.

The line through New Lexington Tunnel was single track, worked by timetable and train order plus a manual block system. Northbound trains were superior to southbound trains of the same class unless otherwise specified. A train register book was provided in the office at New Lexington. In the United States these books were provided at points on single lines where the conductors of arriving trains noted their arrivals with the train numbers. The conductors of opposing trains were expected to check the train registers to ensure that all superior opposing trains had arrived at that point before entering the single line.

One of the trains involved in the accident was northbound second-class freight train First No. 81, which consisted of engine No. 40, 39 loaded and 61 empty cars, engine No. 9737, a caboose and engine No. 7989, in charge of Conductor J. Brotton, with Engineer J.W. Fleming and Fireman Murray. The other train was engine No. 9741, which was returning southwards to Corning after assisting a northbound freight train to Thurston, with Engineer John Turner, Fireman J.W. Parsons, and Flagman John Hoke. [939]

The light engine arrived at New Lexington at 3.45 a.m. Here Form 19 train orders Nos. 16 and 64 were received. Order No. 16 included the instruction to meet train Second No. 81 at Corning, order No. 64 was a slow order (an instruction to run at reduced speed over a designated section of track). There was also a clearance card, stating that the combined train order and manual block signal was at stop for the two train orders, but crucially the line on the clearance card giving the block indication was blank. A footnote at the bottom of the clearance card stated:

When Clearance Card is used when the block is not clear, the line giving block indication will be left blank, and Permissive Card used in addition to Clearance Card.

The engine left New Lexington, however, without a permissive card, passing the train order and manual block signal at stop, and travelling at 30 to 35 miles an hour, collided with train First No. 81, travelling at 10 or 12 miles an hour, in New Lexington Tunnel. Both front engines and the four leading cars of the freight train were damaged, Fireman Parsons was killed and Engineers Turner and Fleming, Flagman Hoke, Fireman and Head Brakeman Schreiner of the freight train were injured.

Driver Turner told the Interstate Commerce Commission (ICC) inquiry that he and Flagman Hoke had gone into the office at New Lexington, he had picked up the orders, given them to the flagman to read, and checked the train register, noting that it was complete except for train No. 81. Back on the engine he had given the orders to the fireman, remarking that they were to meet train No. 81 at Corning. He did not recall the operator saying that there was no hurry as they could not go at that time. Frequently they left the office with a clearance card with the block space left blank as they were permitted to run to the south end of the crossing loop. He had overlooked train First No. 81 (or perhaps failed to note that it was running in two sections). He admitted that he should not have left New Lexington without a permissive card. During the previous month he had not been feeling well and had slept poorly.

Flagman Hoke said that he had gone into the office first, but the operator had not had

---

[939] *The Van Wert Daily Bulletin*, Van Wert, Ohio, United States, 2nd November 1933

the orders ready, and he had left the office again to assist the fireman with repairs to the engine. He denied that the engineer had shown the orders either to him or the fireman. According to him Engineer Parsons had merely said, "We are going to Corning for everything. We have a clear block."

Engineer Fleming and Fireman Murray could only be interviewed briefly in hospital. The engineer said that his first warning of the engine had been a shout from his fireman. Fireman Murray said that the first thing he had seen was a marker on the light engine.

Flagman Nelson of the freight train said that the only orders he had found on Engineer Turner when asked to look for them were ones from Thurston to Bremen and from Bremen to New Lexington.

Flagman Wheeler of the freight train stated that Flagman Hoke had told him that neither he nor the fireman had been shown the orders at New Lexington, Engineer Turner merely telling them that they had a clear block.

According to Operator Robinson at New Lexington Engineer Turner and Flagman Hoke had come into his office together. He had asked the engineer if he had anything on Second No. 81 and was told he had not. He had then told the engineer, "There is no special hurry, they are leaving Clay Bank." (Clay Bank was the next station towards Corning.) Although he did not mention it by name he had meant train First No. 81. He had then enquired of Clay Bank if First No. 81 was ready and had received train order No. 16 related to Second No. 81 and given it to the engineer. He had not noticed when the flagman left the office, but it had been after he remarked that there was no hurry. When the engine had moved down the loop he had not been concerned as engines frequently did so (the end of the loop was a mile from the office).

The ICC concluded that the accident had been caused by operation of Extra 9741 against a superior opposing train without authority, and by its failing to obey the stop indication of the manual block signal. There was a discrepancy between the statements of Engineer Turner and Flagman Hoke as to whether the engineer had given the orders to the flagman and fireman to read. In the case of the engineer he admitted overlooking train First No. 81 and had disobeyed a block signal. In the case of the flagman and fireman they should not have been so engrossed with the repairs to the engine as to neglect to ask the engineer for the orders, if the flagman's version was correct, or enquire as to entering the block against a superior train which was long overdue (train First No. 81 was 3 hours and 8 minutes late leaving Clay Lake).

The ICC noted that the third car of the freight train was a tank car loaded with gasoline. The regulations stated that for safety reasons such cars should not be marshalled closer than six cars from the engine 'where practicable.' There seemed to be no reason which it would not have been practicable to place the tank car further back the train in this instance. [940]

At about 05:30 on the morning of Tuesday 23rd June 1936, the descending express from Madrid to La Coruña, train No. 405 collided head-on with freight train No. 1416 from Ponferrada to León at the mouth of Tunnel No. 27 (Fragas Tunnel) between Ponferrada and San Miguel de Dueñas stations, Spain. A third-class carriage was

[940] Interstate Commerce Commission report

wrecked, and initially it was reported that 13 passengers had lost their lives, together with the driver of the express. As soon as news of the accident was received doctors left for the scene by road and relief trains were despatched from Monforte and León. There were believed to be about 30 injured.

A second report stated that the collision had in fact happened within Fragas Tunnel. The number of dead was raised to 17 and those who had lost their lives were said to have included Driver Manuel Gonzáles Herrera and Fireman Rogelio Rodriguez of the express and Fireman Ignacio Fernández Fernández of the freight train. Driver Lorenzo Bayón of the freight train was seriously injured. The two locomotives were embedded in one another and one of the vehicles of the express was wrecked.

An investigation into the accident was immediately begun and the judicial authorities ordered the arrest of the Ponferrada and San Miguel de Dueñas stationmasters. One theory as to how the accident had happened was that the express, which should have had priority over the freight train, had left San Miguel de Dueñas station first and that the staff there had not been advised that the freight train had left Ponferrada. Another account had it that the driver of the express had left San Miguel prematurely and against the signals.

A third report indicated that the number of dead had increased to at least 20 and included five more railwaymen; Conductor José Montaves Toledo of the express, Conductor Manuel Rodriguez Arias, of the freight train, and Porters José Carreira Gómez, Emilio Ares Díaz and Severino Martinez López. As to the cause of the collision it was now alleged that the freight train had only been allowed to leave Ponferrada after instructions had been sent to San Miguel to maintain the disc signal there against the express, but for some reason the express had proceeded anyway, and that with the death of the engine crew of the express the mystery of why it had done so would remain. [941]

The following day it was confirmed that the cause of the accident had been the express leaving San Miguel without orders and prematurely. The signalman at San Miguel was held to be partly to blame for the disaster as he had failed to take any action to stop the departing express, either by placing detonators on the line or by waving a flag, although he must have known that it was heading into disaster. He was suspended without pay. Possibly, faced with an unexpected crisis he had just lost his head and not known what to do.

The drivers of the two trains apparently became aware of one another just before the collision despite the gloom of the smoke-filled tunnel, for both had braked and the collision had occurred at relatively low speed. [942]

In the aftermath of the disaster the Madrid newspaper *El Sol* called attention to the antiquated method of working single-line railways in Spain with verbal 'Line Clears', whereby the lives of passengers were dependent on the memory and accuracy of a single employee, the newspaper ignoring or unaware of the fact that the levels of traffic on many of the lines involved were very low and could not justify the installation of a more sophisticated signalling system. [943]

---

[941] *La Vanguardia*, Barcelona, Spain, 24th June 1936
[942] Ibid, 25th June 1936
[943] *The Railway Gazette*, London, England, 31st July 1936

Forty-two passengers reportedly died and 88 were injured, including women and children refugees, on Monday 3rd January 1938 when an express from Canton to Wuchung, China, was derailed and a mile-long tunnel collapsed following a Japanese air raid. The train ran into debris which had fallen from the tunnel roof and its derailment caused a further roof fall. Only a day earlier a train had reportedly become derailed due to track subsidence between Canton and Shaokwan. Possibly the different reports relate to the same accident. [944]

The catastrophic explosion of munitions in a Madrid, Spain, underground station on Monday 10th January 1938 is covered in the Chemical Explosions chapter.

The morning rush-hour on the Northern Line of England's London Underground system was disrupted on Thursday 10th March 1938 by an accident between Waterloo and Charing Cross stations. Train No. 65 was stationary in the tunnel between the two stations when the following train, No. 4, ran into its rear. The results of the low-speed collision were not serious; twelve passengers were taken to St Thomas and Charing Cross Hospitals with minor injuries and shock, all were later sent home. The passengers on the two trains and on five following trains had to be detrained in the tunnel and make their way on foot to the nearest station. Many thousands of passengers were delayed, and emergency omnibuses and tramcars were brought into use to help the situation. [945]

The line was fully track circuited and provided with two-aspect automatic colourlight signals equipped with train stops. A driver finding a signal at danger which was not provided with a telephone was instructed to wait one minute, whistle, then proceed past the signal until the train stop operated. He should then reset the train stop and proceed with caution.

Stationmaster A.J. Finlayson told the Ministry of Transport inquiry into the accident that the events which led to the accident had begun at 7.56 a.m. with an intermittent failure of the advanced starting signal at Waterloo. He had allowed two trains to pass the signal in accordance with the aforementioned 'Stop and Proceed' regulations and had reported the problem to the Controller who in turn had contacted two linemen. Power Signal Lineman C.A. Hardy was the nearest man to Waterloo, and he arrived there at 8.11 a.m. Having not been in the area more than three weeks Hardy was not familiar with the local equipment but had examined a local circuit diagram on the way to Waterloo.

Lineman Hardy told the inquiry that when he had arrived at Waterloo he had asked Stationmaster Finlayson to hold the next train and he and the stationmaster had then gone into the tunnel to the relay location. He soon traced the problem to a faulty top contact on Track Relay 94A which was preventing signal 94B form clearing. He had then returned to the station for a loop of wire with spring terminal clips and using this bridged the faulty contacts, allowing the signal to clear. Back on the station platform again he then informed Stationmaster Finlayson and Divisional Inspector S.W. Smallbone that he had the problem in hand.

Lineman Hardy realised that his temporary 'fix' had shortened the block section (the

---

944 *Defiance Crescent News*, Defiance, Ohio, United States, 4th January 1938
945 *The Times*, London, England, 11th March 1938

number of track circuits which had to be cleared by a previous train before the preceding signal would go to 'clear') but he did not realise quite how drastically it had been shortened, from the normal 615 yards (track circuits 94, 94A, 94B, 94C and 94D, to only 33 yards (track 94 only). His plan was to obtain another loop of wire and parallel the terminals of the defective contact with a spare contact, which would have restored the normal clearance of the signal. Unfortunately, he was delayed by his conversation on the platform and then by his desire not to cause delay to the frequent rush-hour trains. Meanwhile, he had allowed several trains to pass the signal showing 'clear', holding each for about half a minute with a red hand lamp. He thought that this was sufficient, overlooking the fact that a previous train might be stopped in the tunnel by, for example, a train in front being delayed at the platform.

Driver Green, of train No. 9, the one which preceded train No. 65, said that the starting signal at Waterloo had been green, but there had been a red hand lamp on the platform. He had asked the lineman, who had then changed the hand lamp to green, if it was all right to proceed, and had been told it was. Approaching signal S 96A at Charing Cross, it had been red; he stopped there two minutes, then, after a failed attempt to speak to his guard, had proceeded at caution into Charing Cross station.

Driver Thomas of train No. 65 stated that the starting signal at Waterloo had been green as he came into that station and Inspector Smallbone had told him that the signal had been failing and he could proceed with caution. He had then enquired about the red hand lamp and the lineman had turned it round to face the tunnel wall. Proceeding at 10 or 15 miles an hour he had seen the repeater of signal S 96A at caution, then the red tail lamp of train No. 9. With a full brake application he had managed to pull up a few yards short of it. He had been on the telephone to Guard Coe of his train when he had felt a bump and the guard had said that another train had run into them.

Driver Keber of train No. 4 likewise spoke of the starting signal at Waterloo having been green and of Inspector Smallbone telling him that there were reports of problems with the signals and to proceed cautiously. When he had pointed to the red hand lamp the lineman had turned it to green and told him that it was all right. Proceeding at 10 or 15 miles an hour around a curve he had seen the tail lamp of train No. 65, braked and reduced the speed of his train to about eight miles an hour before the collision.

The inspecting officer, Colonel A.C. Trench, concluded that Lineman Hardy was primarily responsible for the accident, as Hardy had himself admitted. In his desire to minimise delay in the rush hour he had not been sufficiently cautious, especially as he was unfamiliar with the local signalling.

Having been called away to the telephone, Divisional Inspector Smallbone had had little time to appreciate the situation but it was to be regretted that Stationmaster Finlayson had not obtained more precise information from the lineman and realised the need to caution the driver of every train.

Given that Driver Keber had not been passing a signal at danger and proceeding under the 'Stop and Proceed' rule, but only been cautioned about a possible problem with signals ahead, Colonel Trench did not think that he could be found fault with for failing to stop short of train No. 65.

Guard Coe was to be commended for his promptitude in going back to examine his tail lamps, and in waving one at the driver of train No. 4 until the last moment, at some personal risk, before retreating into his train.

The only recommendation that the inspecting officer made was that the London Passenger Transport Board 'should remind their staff of the necessity for the most rigid observance of the procedure laid down in regulations for dealing with cases of defective signals. Occasional signal failures are inevitable, and every effort of human ingenuity is directed to ensure that such failures shall be on the safety side, i.e. that any failure shall maintain the signal at danger. There are clear and well-established rules for dealing with such cases, and for keeping traffic moving slowly under such conditions of cautious operation as are necessary for safety. Provided these rules are followed, a signal failure does not involve risk of accident.' [946]

The above report was completed towards the end of April. Less than a month later there would be a considerably more serious collision on the London Underground, again the result of a lineman's error.

What was, at the date that it happened, the worst accident ever to happen on the London Underground system, or more correctly at this date the London Passenger Transport Board, occurred at 9.55 a.m. on Tuesday 17th May 1938. Eastbound Circle line train No. 59 ran into the rear of Ealing – Barking train, No. 21, which was stopped at a signal in the tunnel midway between Charing Cross and Temple stations, England. The rear two cars of the crowded Barking train were telescoped, and six passengers were killed or fatally injured. In addition 43 passengers and two railwaymen were injured, while a third railwayman suffered electrical burns shortly after the accident.

The passengers who lost their lives, all men, were Mr Arthur Jephson-Norreys, Lord of the Manor of Mallow, Ireland; Mr Eric Gallanty, head of a firm of London timber agents; Major R.C. Farden, of Burgess Hill, Sussex; Mr David Sturdee and Mr P.A.B. Drummond, dental students at Guy's Hospital; and Mr G. Wallis, barrister's clerk, of Hassocks, Sussex. [947]

The lines were fully track circuited and provided with two-aspect colourlight signals showing red and green. Every signal was provided with a train stop activated if a train passed the signal at danger, while repeaters, showing yellow and green, were provided if the view of the stop signal was restricted. When the signal cabin at Charing Cross was closed the signals all worked automatically, clearing from red to green as soon as a train had cleared the relevant track circuit in advance of the next signal. Signals 817A, 817B, 817C and 817D preceded Charing Cross station, signal EH 9 stood at the east (Temple) end of that station, and it was followed by signals 823, 825A and 825B, the last-named at the west (Charing Cross) end of Temple station.

On the morning of the accident Motorman John Longley of eastbound train No. 53 saw the tail lamp of the preceding train not far in front of him as he left Charing Cross. He stopped his train for a minute or so until the previous train had moved on, then proceeded to Temple station, signals 823, 825 A and 825B clearing for him. Upon reaching Temple he had whistled to alert the station staff. Station Foreman Arthur Foskew went to his cab and the motorman said, "Get on the Charing Cross at once, tell

---

[946] Ministry of Transport report
[947] *The Times*, London, England, 18th May 1938

them that their starting signal has failed in the off position for me."

The station foreman had replied, "They wondered at Charing Cross why you had stopped." Motorman Longley had not mentioned that he had almost run into the preceding train.

Guard Blake of train No. 53 saw the preceding train not far ahead and had made his way to the motorman's cab in time to hear the conversation between the motorman and the station foreman. This would have been at about 9.40 a.m.

Motorman Butler of eastbound train No. 24 told the Ministry of Transport inquiry into the accident that he had been stopped at signal 817A approaching Charing Cross. This had cleared and when he had entered the station signal EH 9 had been red, but it had cleared before he stopped, which had surprised him, as signal 817A having been red meant that the preceding train could not be far ahead. Leaving Charing Cross signal 823 had been red and the previous train only just beyond it. Waiting until signal 823 cleared for him he had continued to Temple station, where like Motorman Longley he had whistled for the station staff. Station Foreman Foskew had come to him, and he had told the station manager that "there was something wrong with the Charing Cross signals, as the starter seemed to come off too quick." He had added, he said, "When I got round the bend I found the next train about six or eight car lengths in front of me." It was now 9.49 or 9.50 a.m.

Motorman Read of eastbound train No. 52 said that signal EH 9 had been green as he entered Charing Cross. Soon after his train left Charing Cross he had seen a red light ahead which at first he thought was signal 823 but then realised was the tail light of a train ahead. By making an emergency brake application and reversing his train he had just avoided colliding with the train in front, though only, he thought, because the previous train had begun to move forward. He then waited until signal 823 cleared for him, moved forward to Temple station, and told Station Manager Foskew that he had almost run into the train ahead. He was told that other motormen had said the same thing and the matter had been reported. The time was about 9.54 a.m.

Motorman Berry of eastbound train No. 34 said he had also found signal EH 9 green as he entered Charing Cross station. As he had left there he had seen Motorman Longley signalling to him from train No. 53, which was now returning from Mansion House, and immediately afterwards he had seen the tail lamp of train No. 52 a short distance ahead. He had stopped his train, then drawn forward to signal 823, and when that had cleared for him, continued to Temple station, where he too had spoken to Foskew, who replied that Charing Cross had the matter in hand.

Motorman Cecil Holbourn of train No. 21 said that he had been stopped at signal 817D entering Charing Cross station and signal EH 9 had been red as he came into the platform. Leaving the station he had seen the tail light of train No. 31 in front of him, nearer to him than the signal though moving away. He had stopped at once, then drawn up to signal 823. His train had been stood at the signal half a minute and he had released the brakes, when he was thrown to the floor by the collision. Relighting his hand lamp, he alighted from his cab and stopped westbound train No. 39, which had just been leaving Temple station. Going back along his own train and train No. 59, which had run into it, he found train No. 49 drawing up behind it. He had then found a doctor in train No. 59 and conducted him to the telescoped cars and injured passengers.

Guard Diprose of train No. 21 had been badly hurt in the collision and had to be interviewed at home. He said that his tail lamp had been burning properly. He recalled seeing train No. 59 approaching and wondering if it would pull up in time. Only four seconds had elapsed between his first seeing train No. 59 and the collision, so he had had no time to give any warning.

Motorman Alfred McClean of train No. 59 said that he had been looking out for signal 823 as his train left Charing Cross. He had no clear recollection of seeing the tail lamp of train No. 21, only of seeing a "hazy red object," when he had released the dead man's handle and applied the brakes. Then he had attempted to retreat from his cab but had not got as far as the passengers' compartment before the impact, when he had been trapped for some minutes, though not much hurt.

Guard Parks of train No. 59 said that signal EH 9 had been green as they entered Charing Cross. His first warning of the accident had been the shock of the collision.

Motorman Mead of train No. 49 testified that signal EH 9 had been red as he entered the station. Coming into the station he had heard a loud report from the adjoining substation, and he had seen the gap indicator beneath signal EH 9 light up. This device indicated that his train was approaching a "dead" section which must not be passed. By the time the station work had been completed and signal EH 9 had cleared, the gap indicator had gone out and did not come on again, but suspicious that something was not right he had proceeded slowly and so had been able to pull up short of train No. 59.

Motorman Longley stated that his train, No. 53, had reversed at Mansion House. As he had been returning westwards between Temple and Charing Cross he passed train No. 52 standing at signal 823, and shortly afterwards train No. 34 leaving Charing Cross, which at least reassured him that he had been correct, and that there was some problem with the signalling. He had signalled to Motorman Berry that there was a problem. At Charing Cross he had found Inspector Barnes on the platform and said to him, "Look, the signal is still doing it, failing in the off position," and had also told the inspector that he had almost collided with the train ahead on his eastbound journey. The inspector had replied that he would "see to it." Having spoken to the inspector, Motorman Longley said he had not thought it necessary to warn the motormen of subsequent trains.

Stationmaster Barrand and Inspector F.H. Barnes had been on the westbound platform of Charing Cross station when train No. 53 eastbound had left there at 9.36½ a.m. They noticed that it had pulled up shortly after starting and had wondered why. The train had then restarted, and the inspector had telephoned Temple and told Porter Charles Hopkins there to ask the motorman why he had stopped. The Porter had rung back a little later with the information that Motorman Longley had said that, "The signal was on."

"What signal, what does he mean?" Inspector Barnes had asked.

Porter Hopkins merely replied, "That's what the driver says."

The inspector had been puzzled but the traffic seemed to continue to run normally, and he had assumed that nothing serious could be wrong. He had told the stationmaster what was said and as Motorman Longley would be returning from Mansion House at 9.50½ a.m., determined to question him then. The inspector also informed Assistant Controller Peters at Earls Court. A little later Porter Hopkins had telephoned again with the information that the eastbound starting signal at Charing Cross was "working sluggish," a statement which the inspector could make nothing of.

At Temple station at this time Station Foreman Foskew was on the eastbound platform and Porter Hopkins was on the westbound platform, where the telephone was sited. Messages between them were by shouting across the lines, provided no train was in the way. The station foreman said that in this manner he had received Barnes' instruction to question Motorman Langley, but before he had had time to do so the motorman had exclaimed, "Get on to Charing Cross, tell them that their starting signal has failed in the off position." Nothing was said by the motorman about a near collision with the preceding train. After train No. 53 and a westbound train had gone Foskew had shouted across what Motorman Longley had said to Porter Hopkins. The porter, however, either misheard or misunderstood him, and repeated the message to Inspector Barnes as the signal having failed in the *on* position, thus completely altering the meaning of the message.

Ten minutes later eastbound train No. 24 arrived, and Motorman Butler told Foskew that there was something wrong with the Charing Cross starting signal, which was clearing too soon. Foskew denied that Butler had also spoken of seeing a train close in front of him. Foskew had then shouted across to Porter Hopkins and thought that he had said to tell Charing Cross that the Charing Cross signal was "failing at the off." The porter recalled the wording as, "To ring up Charing Cross and ask them why nothing has been done because trains coming in were still reporting the same defect." Hopkins admitted that for some reason he had altered the words that he had used to Inspector Barnes to "the drivers are reporting that the signal is working sluggish."

Station Manager Foskew said that the first he had heard of a near collision was from Motorman Read (train No. 52). He had again shouted to Porter Hopkins to tell Charing Cross this but again Hopkins had passed on the message simply as that motormen were still reporting that the signal was not working properly. Inspector Barnes had not taken this message, which in any case came too late to have prevented the accident.

The Ministry of Transport inspecting officer, Lieut.-Colonel E. Woodhouse, criticised Foskew for not having informed the traffic controller of the reported signal problems, although he was aware of the potential serious consequences of a signal failure. The station foreman admitted that he ought to have done so but pleaded that he had thought the staff at Charing Cross would have done so. As for his not crossing to the westbound platform to use the telephone himself, he said that there had been many passengers on the westbound platform, he had thought Porter Hopkins quite capable of transmitting the message, and had had no suspicion of it being misinterpreted, even after Motorman Butler repeated that there was a problem.

It was only when train No. 52 returned to Charing Cross from Mansion House at 9.52 a.m. that Inspector Barnes had understood from Motorman Longley that it was signal EH 9 giving a false clear indication and not, as he had thought, signal 823, and that Longley's train had nearly run into the train in front. Unable at first to believe what the motorman was telling him he had questioned Longley further, then, convinced, he had telephoned the Controller. Having done so he had hurried up the stairs to the booking hall and thence to the eastbound platform to warn eastbound trains. The inspecting officer pointed out that if the inspector had crossed to the other platform via the permanent way he might have been in time to warn Motorman McLean or Guard Parks of train No. 59 but Inspector Barnes thought that train No. 59 must have left the station while he was talking to Motorman Langley. Asked why he had not got Longley to stop any eastbound train with his handlamp, the inspector said that he had not realised the urgency of the situation at that point.

Assistant Controller Peters stated that he had been busy when Inspector Barnes had telephoned and that his assistant, Rushton, had first taken the call, but the inspector had insisted on talking to Peters. When he had understood the matter he told the inspector to go across and keep signal EH 9 "under special observation." He had also attempted to ring the Charing Cross stationmaster to get him to open the signal cabin there to bring the signal under manual control but could get no answer for some time.

The immediate cause of the collision was clearly a 'wrong-side signal failure'. Until a short time before the accident direct current had been used for the track circuits. This could lead to problems with stray currents from the traction system and so, ironically in the interests of safety, and in accordance with more modern practice, a change was made to alternating current. At Charing Cross this was done on the night of 7th to 8th May. The work could only be done in the short interval between 1 a.m. and 5 a.m. when no trains were running. Because of the short time available each night, some non-essential jobs had to be left to be done later, and it was during the tidying-up operation on the night before the accident that a wrong electrical connection had been made. Normally signal EH 9 would clear only after track circuits H, G and GG, the successive track circuits in advance of the signal, were all unoccupied; the effect of the faulty connection was that signal EH 9 cleared as soon as the preceding train had left track circuit H, the first in front of the signal.

Chief Signal Inspector Francis Baker had tested the signalling after the initial changeover. The night before the accident a gang of 24 men under Chief Lineman Arthur Beer had been working at and near Charing Cross, with Signal Installer Charles Eeles working alone in the signal cabin. Chief Lineman Beer testified that he had been happy to leave Signal Installer Eeles working along in the cabin, considered him thoroughly reliable and could not imagine him making a mistake in reconnecting the wires. At 5 a.m. he had accepted Eeles' assurance that everything was correct in the cabin and had not checked it, despite a rule that every alteration must be checked. Such a test would have been a simple matter. Nor had he noticed that signal EH 9 had cleared prematurely after the first few trains.

Signal Installer Eeles stated that as part of his work he had had to disconnect several wires from their terminals in order to disentangle them and shorten some. There had been plenty of light for him to work by and he was certain that he had only unscrewed one terminal at a time, in accordance with his usual practice. He considered that the wrong connection must have existed since the initial changeover, which he had not undertaken.

In the conclusion to his report the inspecting officer said that he had no doubt that the initial changeover had been properly checked by Chief Signal Inspector Baker and that the mistake had been made by Signal Installer Eeles a few hours before the accident, though if, as Eeles maintained, he had only removed more than one terminal screw at a time it was hard to see how it could have happened. Regardless of this, Eeles was not competent to test his work and Chief Lineman A.G. Beer undoubtedly should have done so. His failure to check Eeles' work, due to thoughtlessness and perhaps preoccupation with other details of the work, was a primary cause of the accident.

Motorman McLean's view of the tail lamp of train No. 21 had been restricted to about

153 feet, a distance which would have been covered in slightly over four seconds at 25 miles an hour. Had he immediately made an emergency brake application he might have stopped in time or collided just slightly with the train in front, but he had doubtless momentarily confused the tail lamp with signal 823, as Motorman Read had a little earlier. Motorman McLean could be relieved of responsibility.

Most regrettable, Lieut-Colonel Woodhouse continued, was the failure of the station staff to prevent the accident despite ample opportunity. A man of Station Foreman Foskew's seniority and experience should have realised the serious implication of Longley's report regarding a false clear signal. If it had been essential for the comparatively inexperienced Porter Hopkins to communicate with Charing Cross, Foskew should have made sure that the porter understood what he was to say, and Foskew's excuse that he had felt it inadvisable to leave the eastbound platform to telephone himself was extremely feeble. Even the reminder from Motorman Butler that a dangerous situation still existed, had not roused Foskew to the urgent action required, nor had the third report from Motorman Read.

The inspecting officer considered that all in all Station Foreman Foskew had been more seriously negligent than Chief Lineman A.G. Beer, though after the accident he had shown commendable initiative in organising and conducting the rescue operation.

Porter Hopkins might have misunderstood the first message from the station foreman, but he had no excuse for garbling the second message into the meaningless statement that the signal was "working sluggishly." A correct second message might have prevented the accident and he deserved some of the blame for carelessness.

It was unfortunate that Inspector Barnes had not insisted on speaking to Foskew when he received a meaningless message from Porter Hopkins. Worse was the fact that Inspector Barnes, when confronted by an unexpected situation by the report from Motorman Longley on his return from Mansion House, had not risen to the occasion in the way that a man in his position might have been expected to do, as by more prompt action he could have warned Motorman McLean and prevented the collision. The inspector must therefore bear some responsibility for the accident.

Motormen Longley, Butler, Read, Berry, Holbourn and Mead were all to be commended for their alertness.

The inspecting officer noted that the accident was an unwelcome reminder that the introduction of automatic apparatus to prevent mistakes by operators increased the responsibility on the maintenance staff. He considered that Rule 81(*n*) should be amplified to make it clear that motormen should be advised to run with extreme caution should a signal failure be suspected or reported. The rule read:

Except in the case of an automatic signal failing in the Danger position, the Signalman at the box in rear or the Station Master at the station on rear must be informed of the circumstances, and he must stop all trains proceeding towards the signal concerned and advise the Driver accordingly.

A short circuit accidentally set up after the accident had caused serious arcing. There had been suggestions that the traction current should have been immediately cut off after the accident but had this been done 11 trains would have been brought to a stand, without lights, between Charing Cross and Mansion House and possibly 19 between Victoria and Mansion House, with the likelihood of causing a panic amongst their passengers.

Lighting in the tunnels, switched on automatically when the traction current was cut, as on the Tube lines, would be a solution.

A more distinctive method of marking the rear of a train should be investigated.

The heavy construction of the leading car of train No. 59 probably prevented it being more damaged. Apart from the telescoping of the rear cars of train No. 21 most of the injuries to passengers had been due to broken glass. New rolling stock being introduced would be of all-steel construction, have improved couplings, headstocks capable of resisting telescoping, safety glass and twin electric tail lamps. [948]

The inquest on the six passengers ended with a verdict from the jury that: "We find that the accident was caused by the serious negligence of Beer and the contributory negligence of Foskew, but in neither case does it amount to criminal negligence. We find that the mistake of Eeles amounted only to an error of judgement. We wish highly to commend the presence of mind of Holbourn and express our deepest sympathy with the relatives." It took the jury only 25 minutes to reach their verdict. [949]

Exactly 50 years later there would be a disaster at Clapham Junction, England, likewise due to an error relating to wiring during resignalling work.

A severe storm was raging in the Peak District of England on the night of Friday 2nd February 1940 and presumably because the storm had brought the telegraph wires down, block working was suspended between the signal boxes at the ends of the 1-mile, 1,224-yard-long Dove Holes Tunnel on the London Midland and Scottish Railway. The torrential rain which accompanied the storm also caused part of the tunnel roof to collapse. An Up train ran into the debris, but a guard managed to make his way past the obstruction in time to warn the driver of a Down goods train and a worse accident was thus averted. [950]

As a result of a signalling error local freight trains N 9010 and N 9011 collided head-on in the single-line Alme Tunnel, Germany, between Paderborn and Brilon Wald on Wednesday 21st August 1940. Four people died and eight were injured. [951]

An 'accident' in Knight's Hill Tunnel, England, on the Southern Railway, on Friday 27th December 1940, was the direct result of a German bombing raid on London. By an unlucky chance a bomb penetrated the roof of the 331-yard-long tunnel and a local passenger train ran into the debris on the track, killing its driver. [952]

Early on Saturday 11th January 1941 an eastbound Canadian Pacific Railway freight train ran into a rockfall in a tunnel at China Bar, British Columbia, Canada. The engine

---

[948] Ministry of Transport report

[949] *The Times*, London, England, 28th May and 3rd June 1938

[950] *Railways in the Peak District: A History*, by Christopher Nicholson and Peter Barnes

[951] *Katastrophen der deutschen Bahnen Teil II* by H.J. Ritzau, pg. 103 (in German)

[952] *London Suburban Railways: Mitcham Junction Lines* by Vic Mitchell and Keith Smith, Plate 18

and four cars were derailed but no one was hurt. [953]

The Santa Barbara – Los Angeles main line of the Southern Pacific Company passed between the Simi and San Fernando valleys via the 7,366-feet-long single-track Tunnel No. 26 (also known as Hasson or Santa Susana Tunnel), California, United States. At 11.15 p.m. on Tuesday 18th November 1941 westbound freight train No. 4193 left Los Angeles, 29¾ miles east of the tunnel, and passed Chatsworth, the last open office, 4½ miles from the tunnel and the last open office, at 12.25 a.m. on Wednesday 19th November. It consisted of cab-forward *Mallet* class 4-8-8-2 steam locomotive No. 4193, 51 loaded and 45 empty cars, in charge of Conductor Harry Hartley with Engineer John Dunne, Fireman S.E. Snodgrass and Student Fireman Boyd Bonner.

About a mile short of the tunnel the engine slipped, and the train stalled. At the second attempt it was restarted and entered the tunnel, travelling at 10 or 12 miles an hour, only for the engine to slip and the train to stall again, when it was about halfway through the bore. The train then began to slip backwards and became divided at the front of the 75th car, whereupon the brakes were applied in emergency and the train stopped. An excessive supply of fuel oil to the firebox then filled the tunnel with smoke and gas, and the oil became ignited, flames enveloped the cab and the fire caused considerable damage. [954]

Student Fireman Bonner, Head Brakeman Virgil Rippey, Conductor Hartley, and Brakemen O.B. Castle and E.J. Byrne were able to escape from the tunnel, though the first four were injured. Engineer Dunne, Fireman Snodgrass, Brakeman G.E. Baker and two unidentified trespassers lost their lives.

Student Fireman Bonner spoke to a newspaper reporter from his hospital bed. "In the cab with me were the engineer, the fireman and the head brakeman. When the engine stalled I went out on one side of the cab with Rippey while the other two climbed out the other side. Smoke was coming out of the smokestack and the air was so heavy that we were hardly able to breathe. Just imagine yourself being lost in hell – that was the feeling I had. Rippey and I made our way to the end of the tunnel – and after we got one breath we went back in to find out what happened to the engineer and fireman. They could have gotten out if they could have cut loose the engine from the rest of the train. But I suppose they were overcome trying to disconnect the coupling by shutting off the lever that controls the flow of oil from the tank to the firebox."

Smoke and gas prevented rescuers from getting to the other railwaymen for several hours until wind machines from movie studios were brought into use. Ordinary gas masks proved insufficient to enable rescuers to cover the half mile to the engine. [955]

Brakeman Rippey told the Interstate Commerce Commission (ICC) inquiry that when the engine had begun to slip in the tunnel, Engineer Dunne had had difficulty in closing the throttle and the engine had slipped for some time. Steam had then filled the cab, preventing the brakeman from uncoupling the engine from the train.

Student Fireman Bonner added that the engineer had had trouble with the throttle throughout the journey.

---

[953] *The Lethbridge Herald*, Lethbridge, Alberta, Canada, 11th January 1941

[954] Interstate Commerce Commission report

[955] Tucson Daily Citizen, Tucson, Arizona, United States, 19th November 1941

The ICC noted that the engine was rated to pull 3,650 tons and the weight of the train was 3,550 tons. With the train travelling at 10 miles an hour as it entered the tunnel it should have been able to negotiate it without difficulty had it not stalled. The run-in of slack after the engine had stopped in the tunnel had caused a jerk which had broken the knuckle between the 74th and 75th cars and separated the train, causing the brakes to apply in emergency. The conductor had then closed the angle cock at the rear of the 74th car, replaced the broken knuckle and told the second brakeman to go forward to the engine to tell the engineer to take the front portion of the train forward to Hasson siding, at the west end of the tunnel. Shortly afterwards the conductor and third brakeman had almost been overcome by fumes and had been forced to evacuate the tunnel.

The tender had contained 5,799 gallons of oil when the train left Los Angeles. About 1,000 gallons had been used between there and the tunnel and 1,834 gallons remained after the accident, implying that about 2,900 gallons of oil were burnt while the train was stalled. Had the fuel-regulating valve or the emergency drop valve been closed after the train had stalled, excessive smoke and gas would not have been produced and the accident would not have happened. Also, had the train crew asked for the respirators kept at Los Angeles they would not have been overcome by the smoke and gas. A probable contributary factor was that the fireman had had very little experience, having only been hired on 29th August. [956]

The shifting of a load of steel plates on a freight waggon led to its overturning and the derailment of 13 further cars at the east switch at Hermosa, Wyoming, United States, on the evening of Monday 4th January 1943. As a result single-line working had to be opened up over the Union Pacific Railroad main line between Dale Creek and Hermosa Junction. A further accident followed early on Tuesday 5th January in the 827-foot Hermosa Tunnel. The westbound *Los Angeles Limited* passenger train was passing through the tunnel at 10 or 12 miles an hour when it was run into from behind by another train travelling at 18 to 20 miles an hour. Six Pullman cars, the club car and a dining car of the front train were derailed but none of its 136 passengers was hurt. They were transferred to cars sent from Cheyenne. [957]

Two 'accidents' in Frejus Tunnel, Italy, on 29th September 1943, are covered in the Sabotage chapter.

A major catastrophe occurred in Spain on Monday 3rd January 1944. Just before noon on that day mail train No. 421 from Madrid to La Coruña arrived at Astorga, almost two hours late. This train consisted of Class 240 4-8-0 locomotive No. 2423, Class 141 2-8-2 locomotive No. 4532, a luggage van, two mail vans, a first-class coach, a first-class coach with a bar, a first- and second-class composite coach, five third-class coaches and a van, and was carrying an estimated 900 passengers. Amongst others travelling on the train were an estimated 300 sailors returning from Christmas leave.

There had been problems with the brakes of engine No. 4532, which was why engine

---

[956] Interstate Commerce Commission report

[957] *The Lincoln Star*, Lincoln, Nebraska, United States, 5th January 1943

No. 2423 had been added to the train. Nine more minutes were lost as Astorga while the brakes were checked. The train then proceeded, but when it made its regular stop at La Granja de San Vicente, about halfway down a long descent from Brañuelas, engine No. 2423 was found to have an overheated axle (a 'hot-box') and it had to be detached. The driver of engine No. 4532 drew attention to the fact that the brakes of his engine were not working very well but the decision was made to continue the journey with just his locomotive. In the roughly 21½ kilometres from Brañuelas to Torre del Bierzo (the location of the accident) the line dropped from 1,054 metres above sea level to only 730 metres above, giving an average gradient of roughly 1 in 66. With the brakes of the single remaining engine not working properly it was not therefore very surprising that upon leaving La Granja de San Vicente the train began to gather speed. By the time that it reached the next station at Albares de la Granja, where it was booked to stop, it was completely out of control, and at 13:10 it ran through the station at high speed. The stationmaster there at once telephoned the next station, Torre del Bierzo, and told them that mail train No. 421 was a runaway. The Torre del Bierzo stationmaster, on receiving the news, gave the order for sleepers to be placed on the track as a sort of makeshift barrier, but there was no time even for this to be done, though whether in any case such a feeble barrier would have been effective seems doubtful. In less than five minutes the mail train tore through the station, its whistle blowing continuously, and entered tunnel No. 20 at the far end of the platforms.

As fate would have it the single-line tunnel was occupied by shunting engine No. 4421, behind which, that is on the Brañuelas side, were attached three waggons. Warned by the stationmaster, the driver of the shunting engine was attempting to move backwards onto another line in the station when the mail train collided with the rearmost waggon in the tunnel. The force of the collision pushed the shunting engine and leading waggon out of the far end of the short tunnel, while the rear two waggons remained inside it, as did the engine and leading five vehicles of the mail train, while the first- and second-class composite coach was partly within the bore. The coaches inside the tunnel began the burn. All were of wooden construction. The fire had started a few seconds after the collision, apparently due to the rupture of gas cylinders, the coaches still being gas lit at this late date. [958]

This initial collision and fire were not the end of the matter, for approaching the station from the opposite direction was coal train No. 7242, consisting of an engine, 27 waggons and a guards van. Normally this train would have crossed the mail train further along the line at Bembibre, but with the late running of the latter train the crossing had been altered to Torre del Bierzo. The initial collision having destroyed the wires controlling the signals, they remained clear for the coal train. The fireman of the coal train was putting more fuel onto the fire as the train emerged from Tunnel No. 22 and its driver caught sight of the driver of the shunting engine, who had run back along the line in an attempt to prevent a further collision. The brakes of the coal train were applied but engine No. 4421 had come to a stand only 200 metres from the exit of Tunnel No. 21 and a second collision was unavoidable. The shunting and coal engines were derailed, and the leading waggons of the coal train overturned, killing the shunting engine driver. The damaged whistle of engine No. 4421 kept sounding until the steam in its boiler had been exhausted.

---

[958] Webpage: Federación Castellano Manchega de Amigos del Ferrocarril (in Spanish)

After the accident, the van and four rear coaches of the mail train were made up into a working which returned to Astorga and León with the surviving uninjured passengers and some of those who had been hurt.

A number of causes had led to the accident. Three years of civil war, followed by five years of the Second World War, which had affected the country indirectly even though it had remained neutral, had left the Spanish railways in a terrible state. Maintenance had been neglected, while the driver of the mail train had been pressured to continue the journey, even though he knew that the brakes of his engine were defective. Many railwaymen had left-wing views and were militant union members. Since the end of the civil war and Franco's victory they had been subject to much repression, which was probably another factor in the decision to continue the journey, the driver being afraid of reprisals if he had refused to proceed, delaying the mail train even more. The safer option, of course, though it would have further delayed the mail train, would have been, once engine No. 2423 was found to be unable to continue, to have held the mail train at La Granja de San Vicente until a replacement engine could be added, perhaps the engine of the freight train. The fact that the collision had occurred in a tunnel and fire had broken out in the wreckage contributed to its severity, though an even greater disaster was prevented by the actions of the driver of the shunting engine.

The exact number of people who died in the accident remains unclear. According to Red Nacional de los Ferrocarriles Españoles (RENFE) there were 78 dead and 75 injured. At the subsequent trial of the railwaymen involved figures of 83 dead and 64 injured were given, which as regards the number of fatalities would agree with the RENFE figures if five of the injured had died in hospital. Apart from RENFE employees and Post Office workers most of those who died had been first- and second-class passengers, while most of the third-class passengers escaped or were only slightly hurt. Two members of the Betanzos football team, returning from a match in Palencia, also died, while a third was seriously injured. Some sources have quoted figures of 500 to 800 dead, others 150 to 200 fatalities, but more recent research has come up with a figure of exactly 100 deaths.

News of the disaster appeared in some newspapers the next day, in which it was reported that 26 bodies had been recovered and it was likely that others remained in the tunnel. In the following days further reports appeared, increasing the number of dead to 59, and editorials stated that "there is still much to achieve normality in the Spanish railway service, and much more to improve it to the point of modern technology." The coverage of the accident would probably have been more extensive but for the desire of the Franco regime to minimise the magnitude of the disaster. After the accident, the military had completely vetoed the entry of newspaper reporters to the accident area.

The line was cleared and traffic through the tunnel resumed on Thursday 6th January.

Four railwaymen, including the driver of the mail train, were prosecuted, and stood trial for the manslaughter of 83 people, but all were acquitted. [959] [960]

Fire broke out in a Mersey Railway train about midway through the under-river tunnel

[959] La Vanguardia, Barcelona, Spain, 4th and 9th January 1944

[960] Wikipedia – Accidente ferroviario de Torre del Bierzo de 1944

between Liverpool James Street and Birkenhead Hamilton Square stations, England, on Wednesday 5th July 1944. The blaze began in the motorman's cab at the rear of the train, which was promptly brought to a halt. The passengers, who numbered over 100, were ushered through emergency connecting doors to the front of the three-car train and subsequently onto another train. No one was hurt. The cause of the fire was not stated; possibly it was an electrical fault. [961]

An eastbound local train from Rupert, British Columbia, to Jasper, Alberta, was within six miles of its destination when it collided with a work train in a tunnel near Geikie, Alberta, Canada, on Saturday 3rd November 1945, on the Canadian National Railway main line. An engineer was seriously injured, and two firemen and a number of passengers were shaken up. As a result of the accident and the blockage of the line passenger trains from eastern Canada due to arrive in Vancouver early on Sunday did not reach there until late in the day. [962]

[961] The Times, London, England, 6th July 1944
[962] Winnipeg Free Press, Winnipeg, Manitoba, Canada, 5th November 1945

CHAPTER 12

LEVEL CROSSING COLLISIONS

Wherever roads and railways intersect they must do so either by means of a bridge or on the level. Bridges have the advantage of allowing road and rail traffic to operate entirely without interference from one another, but they are much more expensive to provide, especially in flat countryside, due to the need to provide extensive earthworks for approach embankments or cuttings as well as the cost of providing the span itself. Where underbridges are provided there is often also the need to build retaining walls and there can be problems with drainage.[963] The provision of level crossings, especially if, as in many parts of the world, they are unstaffed, is the much cheaper alternative, but where they exist, they introduce an additional source of danger to both road and rail users. Robert B. Shaw in his outstanding book on American railway accidents, *A History of Railroad Accidents, Safety Precautions and Operating Practices*, writes that he has excluded ordinary grade (i.e. level) crossing collisions from consideration on the grounds that they are not proper train accidents, inasmuch as trains usually suffer much less harm from such accidents than the road vehicle involved. This series of volumes, on the other hand, will attempt to include, amongst others, all such accidents which resulted in five or more fatalities, even if, as was often the case, none of those who died were train passengers or railwaymen. The First World War provided a great stimulus for the development of the internal combustion engine and the number of motor vehicles in use, and consequently of level crossing collisions involving them, greatly increased following the war.

Nine of ten people riding in a car were killed on New Year's Day, 1931, at Harvey, Illinois, United States, a suburb of Chicago, when the car was struck by the eastbound International Limited passenger train of the Grand Trunk Western Railroad. The victims were Mr and Mrs Olason, their children Calvin, 6; Muriel, 5 and Vilborg, 4; and Mr and Mrs Novak and their children Adam, 8, and Lorraine, 5. The one survivor was three-month-old Lorraine Olason, who was found in a snow-filled ditch, crying but unhurt. It was theorised that her mother had thrown her from the car immediately before it was struck.

[963]Backtrack Volume 24 pg. 88.

The Olasons had left their home in Elmhurst at 1 p.m. for a drive to Indiana, stopping for supper with relatives at Dixmoor, another Chicago suburb. The accident occurred at 7.30 p.m. The crossing had neither gates nor signals, and low buildings obstructed the view of motorists. It was supposed that Mr. Olason had been engaged in conversation and had failed to see the train, or else had assumed that the line he saw was only a siding. A train had struck a lorry on the same crossing only a fortnight previously, seriously injuring its driver. [964]

Five motorists were killed on Sunday evening, 18th January 1931, when their car was struck by a train on a level crossing near Beba, in Upper Egypt, by an express bound for Cairo. [965]

At about 14:00 on Saturday 14th February 1931, a bus owned by the Kuypers firm of Venlay, operating the regular bus service between Venlay and Venlo, was struck by a train at Crossing No. 66, near Venlo, The Netherlands, on the Venlo – Roermond line. The bus was carrying 25 passengers who had been to the market. It was carried some 200 metres by the train. The crossing keeper had forgotten to lower the barriers. Three of the bus passengers were killed and eight were seriously and others slightly injured. Rail traffic was delayed for more than an hour. [966] [967]

Five people lost their lives when an eastbound Chicago Burlington and Quincy Railroad train collided with an automobile near Colfer, Nebraska, United States, late on Monday 9th March 1931. There had been ten people in the car, all members of the Pursley family. Those who died were children Herman Pursley, 15; Golda, 15; Fern, 8, and an unnamed Pursley, girl aged 7; the other fatality was the children's grandmother, Nancy Osborn, 70. The children's parents, farmer Elmer and his wife Ella Pursley, both 45, and children Glen, 13; Pazel, 10 and Elmer Eugene, 4, were all injured. Elmer Pursley was driving the car. The train crew said that they could see the crossing for a mile but did not see the car. [968]

On Thursday 23rd April 1931, a freight train collided with a bus carrying male and female textile workers on a level crossing at Wicres, France, between Fourmies and Franelles on the line of the Société Générale des Chemins de fer Economiques. Nine bus passengers were killed, and a dozen were injured, five badly (*The Times* (London) correspondent in Boulogne, writing on the day of the accident, speaks of 14 dead, but this appears to be an exaggeration, as the French newspapers all speak of eight or nine dead.) [969] The injured were taken to hospitals at Lille, Béthune and Haubourdin. Two of those who were killed were Poles who could not immediately be identified. The noise of the

[964] *Centralia Evening Sentinel*, Centralia, Illinois, United States, 2nd January 1931
[965] *The Times*, London, England, 20th January 1931
[966] *Leeuwarden Courant*, Leeuwarden, the Netherlands, 14th February 1931
[967] *Nieuwe Hoornsche Courant*, Hoorn, the Netherlands, 17th February 1931
[968] *Joplin Globe*, Joplin, Missouri, United States, 10th March 1931
[969] *The Times*, London, England, 23rd April 1931

collision could be heard far and wide and workers in nearby fields rushed to the scene to help rescue the victims. The engine of the train had been thrown off the line and overturned and the right side of the bus was smashed to pieces. It cushions were scattered on the ground amidst a large pool of blood and the bodies of the victims.

An initial investigation determined that the train had left Marquillies station and was running at low speed towards the crossing. The weather was clear, and it seemed apparent that collision was the fault of the bus driver, named André Servais. Driver Saint-Pierre and Fireman Marcy on the engine, both of whom were seriously injured, and one of the workers on the bus, who had been fortunate enough to escape injury, were all agreed that the train could have been seen for a least 500 metres.

Fireman Marcy stated that when he had seen the bus, which was travelling at 60 kilometres an hour, he had shouted to his driver to stop. Unfortunately, it was then too late, and the train had struck the bus. The two trainmen were found buried beneath their engine. The fireman was gravely injured, and the driver had half of his right hand torn off. Both were taken to Lille Hospital.

At the close of the inquiry M. Renant, the *juge d'instruction*, arrested driver Servais on a charge of manslaughter through negligence. During the investigation, the bus driver said that he had thought that he had time to get over the crossing in front of the train and added that he had been over it twice a day for four years and had never had an accident before. [970]

M. Servais stood trial at the end of July, when Driver Saint-Pierre, Fireman Marcy, and Conductor Marcel Cardon of the train were among those who gave evidence. Numerous other witnesses gave evidence, including an expert who put the speed of the train as 10 kilometres an hour and that of the motor bus at 47 miles an hour at the moment they met. For the defence it was argued that the trains ran at irregular hours, and not usually at the times that Servais used the crossing. It was also argued that the railway were also at fault, that Servais was not the only one to blame and had an excellent record. [971]

Servais was found guilty and sentenced to four months' imprisonment and two 5-franc fines. [972]

Early on Friday 24th April 1931, five young men from Libertyville returning home from a fish fry were killed when their car was struck by a Minneapolis St Paul and Sault Ste Marie (Soo Line) Railroad freight train near Antioch, Illinois, United States. They were George Eaton, 24; Gus Gustafson, 22; Herrick Harrington, 21, a graduate of Marquette University; Edgar Lindroth, 21, a clerk at a Park Ridge bank, and Daniel Wigham, 24. The car was apparently driven into the side of the train, striking it with such force that five waggons were derailed. [973]

The wife of farmer Clyde Jackson, 38, of Riley Grove, Texas, was ill on Sunday 26th April 1931 and fearing that she was running a temperature he set off for his neighbour's

---

[970] *Paris-soir*, Paris, France, 25th April 1931
[971] *Le Grande écho du Nord de la France*, 28th July 1931
[972] Ibid, 23rd April 1932
[973] *Alton Evening Telegraph*, Alton, Illinois, United States, 24th April 1931

house for a thermometer. His sons Foffre, 12 and Billie, 5 asked to accompany him and two men who were visiting the family, farmer Allie Hutchinson, 26, and Preston Beggs, 18, son of widow Gertrude Beggs, also came along for the ride. The party were on their return journey when their automobile was struck at about 9.45 a.m. by the *Lone Star* passenger train of the St Louis Southwestern (Cotton Belt) Railway on Stridhem Crossing, Texas, United States and all five were killed outright, the car being wrecked and carried about 50 feet. There were no witnesses to the accident. [974]

Thursday 7th May 1931 saw a tragic collision between an Atchison Topeka and Santa Fe Railway freight train and a school bus on a level crossing at Merced, California, United States. At 3.15 p.m. that day the bus, driven by Floyd Creggar, aged 59, and carrying 57 pupils from the first six grades of the John C Fremont grammar school, attempted to cross the Santa Fe tracks at G Street, four blocks from the school. Though there were no gates at the crossing, witnesses stated that the weather had been clear, the view of the line unobstructed, the wig-wag signal at the crossing had been operating, a warning bell had been ringing and the freight train had whistled twice. Despite all this the bus was driven onto the crossing in front of the train. Accounts of the speed of the train varied from 4 to 6 miles an hour to 25 miles an hour.

A newspaper wrote, 'With a shriek of the engine whistle and a grinding of steel brakes as the engineer fought to bring his long, heavy, train to a halt the engine crashed into the bus and carried it, like a broken toy, for more than 50 feet. As the engine, slowing down, dumped its burden and the bus rolled onto its side in a ditch beside the tracks, the screams of children rent the air.

'Among the nearby motorists and passers-by who witnessed the horror were several parents who had come to meet their children and take them off the bus. Some dashed into the confusion of splintered wood and metal in attempts at rescue, while others, too sickened by what they saw to make a move, sat in their automobiles with their hands over their eyes… Some of the children, including a few of the injured, struggled free of the wreckage and ran, hysterical with pain and fright, from the scene. It was hours before all could be located, and, meantime, frantic parents, in some instances, gave up their children as dead as search and investigation brought no information. One of those who ran was Leland Balenie, 6. With blood running into his eyes from a scalp wound, hobbling from a broken ankle, the child staggered back to the school yard to notify his teachers. Bobbie Rou, also 6, walked four blocks to the railroad station, found his father, and collapsed in his arms. Later it was found he had suffered a fractured skull and a broken arm.'

Some of the children who had been on the bus spoke to reporters. 'Fontella Prescott, 9, daughter of Mr and Mrs Walter Prescott, of the Lakeside district, was sitting next to the Epsen twins, Eloise and Roy, when the collision occurred. The twins were killed. "I went up and down like a jumping-pack," the girl said. "My head hurt so much and the engine poked right into the seat next to mine where Eloise and Roy were together. Eloise was crying but when the engine came through the side of the bus she didn't cry any more. Her brother was leaning over her, and he was bleeding.

'"I screamed and fell on top of some of my classmates and then I woke up near the track. I saw Rosie and heard her scream. She held her leg and Jacqueline was rolling

---

[974] *The Whitewright Sun*, Whitewright, Texas, United States, 30th April 1931

about. Richard held his nose and ran around in circles." Fontella was unhurt except for severe shock. Her brother, Walter Prescott Jn., 7, who likewise escaped almost without injury, gave his version of the tragedy. "I was in the back of the bus with the boys," he said. "When we got to the tracks, the engine was there too. I thought it was going to pass in front of us and I leaned forward because I like to see engines. But I guess the driver wanted to go first. The train hit us, and all the fellows and the girls began to holler, and I guess I did too. Gee, there was sure some mess. I got hit on the head. But the others were so much worse. I wish it never happened."

It was stated the next day that there were six children known to be dead, two more children and the bus driver were believed to be dying, 19 children were injured but expected to live and the remainder were unhurt.

No less than five investigations were opened into the accident by the Merced police, the school district officials, the railroad company, the district attorney and coroner and the state railroad commission. It was soon apparent that the bus driver was at fault and District Attorney F.M. Ostrander announced that if Creggar recovered from his injuries he would be charged with manslaughter.

The bus driver made several different statements as to what had happened. At the crossing itself he supposedly told a police officer that as he had approached the crossing he had reduced speed and attempted to change into neutral gear but had instead put his vehicle into second gear. This account would tie in with the statements of some witnesses that they had seen the bus slow down before apparently lurching forward onto the line. Others, however, declared that the bus had run onto the crossing without slowing up at all. At the hospital Creggar reportedly said, before lapsing into unconsciousness, "I didn't see the train at all. I never saw it!" He had regained consciousness by the following day, when he made another statement. "I stopped at the crossing, looked up and down the track both ways, and saw no trains. The wig-wag was not moving. I put the bus into low gear and started up. Just as I was shifting into second gear the train hit me." [975]

At the conclusion on 11th May of the coroner's inquest into the accident, by which time a seventh child had died, the jury returned a verdict of 'accidental death'. [976]

Floyd Creggar, the bus driver, died of his injuries not long afterwards.

Six members of the Collins family lost their lives in a level crossing accident at near Alvord, Texas, United States, either late on Sunday 10th May or early on Monday 11th May 1931. They were Mr Charles Collins, his wife Nora; their daughter Charlene, aged 5; Mr Collins' mother, Mrs Fannie Collins; his sister, Florine, 20, and his brother-in-law Cecil Ferguson, 18. Engineer J.K. Gillespie of the Fort Worth and Denver City Railroad said that he had watched the automobile approach the crossing, stop short of it and then lunge forward. It was theorised that Mr Collins had let his foot slip off of the clutch and brake pedals, throwing it into gear and making it shoot forward. The road sloped down slightly to the crossing, which was unprotected. The family were returning from a visit to Mr Collins' sister, Mrs P.J. Evans, who lived in Fort Worth. [977]

---

[975] *Oakland Tribune*, Oakland, California, United States, 8th May 1931

[976] *Nevada State Journal*, Reno, Nevada, United States, 12th May 1931

[977] *Corsicana Semi-Weekly Light*, Corsicana, Texas, United States, 12th May 1931

Five occupants of a car died when it was struck by a train, presumably on a level crossing, near Jaffa, Palestine, early on the morning of Saturday 16th May 1931. [978]

Seven teenagers died in a level crossing accident in Calamus, Iowa, United States, on Sunday 17th May 1931. Norbert Mumford, 17, had just acquired a new car and was taking it for a drive, showing it off to his cousins, Martha Berner, 20; Madonna Berner, 18; Bernadette Berner, 16; Monica Berner, 15, and Arnold Berner, 11; plus a family friend, Mildred Byer, aged 15. As they reached the level crossing in Calumus High Street an eastbound freight train went by. The red wig-wag signal was still flashing, the warning bell was ringing, and the crossing watchman tried to stop them. Mumford, however, clearly assumed that these warnings applied to the freight train which had just gone, and overtaking a car occupied by Norman Guenther and his wife, which was waiting at the crossing, drove onto the tracks in front of Chicago and North Western Railway passenger train No. 13, the westbound *Columbine*, en route from Chicago, Illinois, to Denver, Colorado. The front wheels of the car had just cleared the rails when the train hit it, hurling the bodies of the victims into the air; they fell over 150 feet along the right of way. Many people on the street witnessed the accident. [979]

Eight of a party of ten Egyptians in a car bound for the market were killed when a train struck it on a level crossing in Cairo on Wednesday 20th May 1931. The other two occupants of the car were seriously injured. [980]

Mr Charles Miller, his wife, and their children Wilbur, 12; Dorothy, 10 and Elsie, 2, were killed when a Pennsylvania Railroad train on the Main Street in Berne, Indiana, United States, on Saturday 30th May 1931. (The line was actually that of the Grand Rapids and Indiana Railway, which had been bought by the Pennsylvania company in 1918.) The family lived a short way away from the scene of the accident and had been on their way to Cleveland, Ohio, to visit Mrs Miller's sister. [981]

A train collided with a school bus on a level crossing in suburban Bucharest, Romania, on Tuesday 30th June 1931. The petrol tank of the bus exploded following the collision and the bus was destroyed by fire. Eight children lost their lives, 22 were seriously injured and six slightly hurt. The children had been returning from a distribution of prizes. [982]

Another level crossing accident resulting in the death of several members of the same family occurred near Rhome, Texas, United States, on Saturday 1st August 1931. Mr.

[978] *The Times*, London, England, 18th May 1931
[979] *Telegraph Herald and Times Journal*, Dubuque, Iowa, United States, 18th May 1931
[980] *The Amarillo Daily News*, Amarillo, Texas, United States, 21st May 1931
[981] *Miami Daily News-Record*, Miami, Oklahoma, 31st May 1931
[982] *L'Homme Libre*, Paris, France, 1st July 1931

J.W. Keene, 49; his daughters Evanta, 17, Lewellyn, 15 and Marie, 10; and his 75-year-old father, John, all lost their lives when their car was struck by Fort Worth and Denver City Railroad passenger train No. 6 from Denver, Colorado, to Fort Worth, Texas, travelling at 50 miles an hour. When she heard of the accident Mrs Keene, who had been at her father-in-law's house at Rhome, reportedly ran three miles along the right of way to the scene. She was later taken to the Baptist Hospital in Fort Worth, suffering from severe shock. The victims had been visiting relatives who lived a quarter from the crossing and had just left the relative's house. Wreckage of the car was carried a quarter mile along the track and the engine cowcatcher was broken.

The train was in charge of Conductor McHam, with Engineer J.E. Collins and Fireman Oscar Marley. The latter man, who had been positioned on the left side of the engine cab, was the only witness to the accident. He said, "It was a light sedan and was travelling east, apparently to get onto the main Fort Worth – Decatur highway. When we were a short distance from the crossing the car stopped, and I thought it would wait until our train had passed. An instant later it lurched into our path." [983]

Another instance of a car driver accidentally letting his foot slip off of a pedal and putting his vehicle into gear, perhaps.

Fog was perhaps a factor in the third level crossing accident of 1931 involving a school bus. It happened at about 8.40 a.m. on Thursday 6th August 1931, at Toodyay, Western Australia, Australia. Described as a charabanc, the bus had been bringing children from the Coondle, Culham and Dewar's Pool districts into Toodyay when it was struck by a train on the Clinton Street level crossing. Five children were killed and 12 injured, one critically. The bus had been only 100 yards short of its destination.

The train was in charge of Guard G. Proud with Driver Lionel Herrod and Fireman Garden Bates. It consisted of a Garrett type locomotive and 25 trucks and was bound from Toodyay to Piawaning. Driver Herrod and Fireman Bates told the inquest into the accident that it had been foggy, but they had been able to see about 100 yards. They had been going slowly and had braked because a butcher's cart was crossing the nearby Duke Street crossing. Speed had been reduced to about five miles an hour at Duke Street; the brakes had then been released and the train was moving at about eight miles an hour when the radiator of the charabanc, which had been moving at about 15 miles an hour, had suddenly come into view only four or five yards in front. Driver Herrod had immediately applied his brakes fully and attempted to put down sand. Speed had dropped momentarily, then the engine had skidded, and the collision occurred. The line fell towards the crossing and the rails were slippery. [984]

A reporter described the aftermath of the accident. '"They were like little Anzacs," said an ex-digger who was present when the terrible work of extricating the dead, dying, and badly injured from the wreck began. And that is why in the midst of their gloom the people of the district also feel a sense of pride.

'Just before 9 o'clock on Thursday morning thirty-four happy boys and girls sang as the charabanc took them in the fog along Clinton-street, Toodyay, on the way to school.

---

[983] *Lockhart Post Register*, Lockhart, Texas, United States, 6th August 1931
[984] *Daily News*, Perth, Western Australia, 25th August 1931

They were getting ready for a concert to have taken place Saturday, and some of them had their fancy dresses with them. As they neared the open crossing, their song was "God Save the King." And as far as five of them were concerned, it was the last time they would ever raise their childish voices in song. A few seconds later a wheat train bound for Piawaning and the charabanc with its happy freight met in the fog and steam at the open crossing. There was a crash and a confusion of cries as they came in to collision, the bus was dragged along for nearly twenty yards, tearing down a fence guarding a cattle-pit in its course and when the first of the horror-stricken residents reached the scene they found a tangle of wreckage, mangled children on the line, in the debris and even under the train, and at least four little ones beyond recall. Within minutes people gathered from everywhere, and after the first shock of the disaster came realisation of the urgent need for help. Digging, tearing, lifting, hauling bands of willing workers lent eager hands to the task of extricating broken and mutilated bodies from the wreckage, frenzied relatives ran hither and thither seeking their loved ones, and children sobbed. Pain and suffering were in evidence on all sides, but the kiddies bore up wonderfully. Children who usually cry at a small cut, saw great jagged wounds in their bodies and yet found smiles of gratitude for their rescuers.

'"I'm all right mum," breathed one little fellow as he was carried away to hospital. "Me too," said his cobber, who had an even more painful injury, but wasn't to be outdone. A little Chinese girlie scrambled from the wreckage and limped down to the convent school. "I fall down," she told the sisters. And it wasn't until half an hour later that they learned that she had narrowly escaped death, and that she had injuries to her shoulder, which she had not even spoken of.

"Don't let them blame the driver, mister!" said another kiddie, who was lucky to get away with a few flesh wounds. He couldn't see in the fog, and he couldn't hear because we were singing that loud. He's a bonzer chap."

'Perhaps the most pithy summing up of the horrible suddenness of the collision came from a boy who was thrown beneath the train with minor hurts. "One minute I was singing with the mob," he said, "and the next I got chunked out. There was all steam under me, and I could see big wheels just missing me. It was all too quick to see it when it was going to happen." And that conveys the average impression of the actual crash, for it came so suddenly through the fog that nobody seemed to see or realise the onrushing horror.' [985]

Oliver Page, the driver of the school bus, gave evidence on the final day of the inquest. He said that there had been 34 children on his charabanc as they came into Toodyay. There had been eight children sat on the front seat with him, three to his right and five of the smallest to his left. The curtains on either side had been up but he had had a clear view through the micre windows. It had been clear early in the journey though it was foggy in Toodyay. As he had turned into Clinton Street the children stopped singing and he had sounded his horn lest there were children playing on the corner. His speed had been about 12 miles an hour. At the crossing he had looked to the left, to the right, straight ahead and to the left again but had seen nothing. It was hard to see even when clear of the brick wall. He knew that the crossing was dangerous, but his charabanc had had good brakes, and he had not expected to see a train at that time. He had used the

[985] *Mirror*, Perth, Western Australia, 8th August 1931

crossing twice daily over the previous four months. He had not needed to change down through the gears as he could reduce speed to six miles an hour in top gear. He had not heard the train whistle and the children had not been obstructing his view.

The Coroner, Mr F. M. Reed, ruled that the train driver was blameless. The speed of the train approaching the crossing had been reasonable, a warning had been sounded and the train brakes applied promptly. It was the duty of the charabanc driver to proceed with the utmost caution, and the fog and the building obstructing the view meant that he must be even more careful. The evidence indicated that if Page had looked more carefully he would have seen the engine and in view of the fact that special trains had been running over the crossing for the past three months he should have been expecting to meet a train at any time.

Continuing, the Coroner remarked, "It seems to me that the driver had driven over this crossing so many times and not seen a train that he had been lulled into a false sense of security. I can come to no other conclusion than that the accident was due to negligence, but I am of the opinion that such negligence does not amount to criminal negligence."

Mr Reed added that he intended recommending to the government that a regulation be framed under the Traffic Act compelling all vehicles to stop before going over a level crossing, as was done in the United States, though crossings with a clear view could be excluded. All school charabancs should be compelled to stop before negotiating a crossing. He also thought it wrong for more than one person to sit to the right of a driver in a charabanc where the space was less than two feet. [986]

On Monday 10th August 1931 a car carrying two families tried to get across the line in front of a Chicago Milwaukee St Paul and Pacific Railroad passenger train on Waukegan Road Crossing, Glenview, Illinois, United States, a northern suburb of Chicago, despite the fact that the warning lights were operating. Having reached the crossing the car then stalled. Bruno Woods, formerly Dizeroucki, his wife Helen and their six-moth-old child David, of Chicago; and James Rutowski, his wife Martha, and their children Henry, 10 and Dolores, 8, of Milwaukee, were all killed. The two families were related. The adults had had time to get out of the car and could have saved themselves had they not returned to try and save their children. [987] [988]

Engineer George Eckerson and Conductor Paul Gessne were arrested on a technical charge of manslaughter before being released on bail following a collision on The evening of Friday 14th August 1931 on Magnolia Crossing, Montvale, New Jersey, United States, though it was hardly likely that the accident had been their fault. The Erie Railroad train, possibly a gas electric railcar, had been travelling at about 30 miles an hour. The crossing was protected by warning lights and bells but had no gates. George Mehle, his wife Cecilia, their 14-month-old daughter, Nancy; Mrs May Paulson, Mrs Mehle's mother; Mrs Krinke, Mrs Mehle's sister, and Mrs Krinke's son Jay, 3, were all killed outright or fatally injured. The crossing equipment was found to be working

---

986 *Daily News*, Perth, Western Australia, 26th August 1931

987 *The Manitowoc Herald-News*, Manitowoc, Wisconsin, United States, 11th August 1931

988 *The Daily Illini*, Champaign-Urbana, Illinois, United States, 11th August 1931

normally but the noise made by the party, on their way to spend a weekend holiday with relatives, might have drowned out the sound of the train whistle. [989] [990] [991]

Five or six people were killed and two critically injured when the eastbound *Ocean Limited* passenger train of the Canadian National Railway collided with what was variously described as a motor truck and a bus on Coyle's or Dupuis Crossing, about 18 miles east of Moncton, New Brunswick, Canada, late on Sunday 6th September 1931. There had been 11 persons in the vehicle. The party were from Dupuis Corner and Cape Bald (now Cap Pelé) in Shedlac Parish and had been returning from a religious celebration in St Joseph's. [992] [993]

In yet another 1931 level crossing accident involving a school bus, northbound Seaboard Air Line Railroad passenger train No. 108 from Florida to New York struck a school bus at Wake Forest, North Carolina, United States at 8.10 a.m. on Tuesday 8th September. The bus was completely demolished, its wreckage being strewn along the track for several hundred feet, and John Caddell Jr., 16, and Robert Garner Jr., 17, were killed. The bus driver, Cameron Sharon, was himself only 17 and probably another pupil. He survived the immediate collision but was in hospital with a fractured skull, broken ribs, and internal injuries. Witnesses said that the crossing bell had been ringing and the warning lights had been flashing but Sharon had tried to beat the train over the crossing. The accident would doubtless have been a lot worse had it not been for the fact that the bus was at the beginning of its journey, on its way to have its petrol tank filled before picking up the remainder of its passengers. [994]

Farmer John Demmel, his wife and their children Frederick, 4; Doris, 3; Marjory, 2 and Hazel Irene, 3 months, were all killed when their car was struck by Chicago Rock Island and Pacific Railroad train on a crossing near Vinton, Iowa, United States, on Sunday 11th October 1931. Charles Berger, 20, a hired man who was riding with the family in the car, was believed to have been fatally injured. The party had been returning home from an afternoon visit to John Demmel's father, Henry, when their vehicle was hit by a meat special from Waterloo to Cedar Rapids, which was supposedly running at 60 miles an hour. The car was demolished, and bodies were strewn along the line for 600 feet. The little-used road descended to the crossing, and it was supposed that John Demmel had failed to observe the train. The crossing was said to be one of the most dangerous in the area. [995]

Five Mexicans were on their way from homes in Ellinor to a store when their car was

---

[989] *Sarasota Herald*, Sarasota, Florida, United States, 17th August 1931
[990] *The Lake County Times*, Hammond, Indiana, United States, 17th August 1931
[991] *The Indianapolis Times*, Indianapolis, Indiana, United States, 15th August 1931
[992] *The Brandon Daily Sun*, Brandon, Manitoba, Canada, 7th September 1931
[993] *The Altoona Mirror*, Altoona, Pennsylvania, United States, 7th September 1931
[994] *The Southwest Times*, Pulaski, Virginia, United States, 8th September 1931
[995] *Waterloo Daily Courier*, Waterloo, Iowa, United States, 12th October 1931

struck by Atchison Topeka and Santa Fe Railway train No. 2 on a level crossing near Saffordville, Kansas, United States, late on Friday 15th October 1931. Juan Manpano, his daughters Camelia, 7 and Amelia, 5; José Cordanado and his daughter Josephine, 13, all lost their lives. The trainmen said that their engine whistle had been blown and its bell had been ringing, while passengers said that the train had braked before the crash. [996]

Dr. Theodore Tarasoff, a Phoenix physician, his wife Rima, their 17-year-old son Alexander, A. Dague, a Y.M.C.A. employee and Charles Mullett, a businessman, all of Phoenix, were killed when their car was struck by a Southern Pacific Railroad passenger train on a level crossing at Coldwater, Arizona, United States, late on Wednesday 2nd December 1931. Engineer Thomas Davenport said that he did not see the car, though he had heard screams as his locomotive reached the crossing. There were no other witnesses to the accident. [997]

Five teenagers and young adults died in an accident at Cummings Crossing, Maine, United States on the evening of Sunday 6th December 1931. They were Malcolm Doe, 22, and Douglas Burgiss, 18, of Auburn; Thelma Conant and Thelma Magette, both 17 and both of Mechanic Falls, and Francis Rattigan, 25, of Lewiston. The car was struck by a Canadian National Railway train and carried 1,600. Four of its occupants were killed outright and Roe died on the way to hospital. The train was en route from Island Point, Vermont, to Portland, Maine, and was running eight minutes late when it struck the car. Engineer George McKelvie said afterwards that he had whistled, and the machine had in response first slowed down, then had suddenly surged forward across the track. [998]

On Wednesday 9th December 1931, a train collided with a bus on a level crossing on the Celje – Vransko line in the village of Levetz, in Slovenia, Yugoslavia. The bus was carried over 100 metres along the track and wrecked. The bodies of nine bus passengers were recovered from the wreckage and all the others were seriously injured. Three more died on the way to hospital and six others were unconscious. [999] There was a thick fog at the time of the accident. [1000]

Nine people, the occupants of a crowded car, lost their lives when it was struck by a Southern Railway passenger train, the *Birmingham Special*, at about 1.15 a.m. on Boxing Day, Saturday 26th December 1931. The accident happened at Dowd Road Crossing, near Charlotte, South Carolina. The occupants of the car had spent Christmas Day with relatives at York and were returning home to Paw Creek. Those who died were Mr T.H. Holton and his wife, their children Wilbur, 4, and Maxine, 2; Caughan Holton, Dorie Cox, John Love and Raymond Sharpe, all of Paw Creek, and Mrs Loma Cox, of York, a

---

[996] *The Abilene Morning News*, Abilene, Texas, United States, 16th October 1931
[997] *El Paso Herald-Post*, El Paso, Texas, United States, 3rd December 1931
[998] *The North Adams Transcript*, North Adams, Massachusetts, United States, 7th December 1931
[999] *La Croix*, Paris, France, 11th December 1931
[1000] *La Liberté*, Paris, France, 11th December 1931

niece of Dorie Cox. Eight of the victims were killed outright. Mrs Cox survived a couple of hours but passed away in hospital. The train was in charge of Conductor Milton Clapp with Engineer C.R. Nesbitt. The latter said that crossing had been clear as the train had approached. It was believed that the fact that the side windows of the touring car had been drawn had prevented its occupants from seeing the headlights of the approaching train. [1001] [1002]

Five of six men riding to work in a truck were killed when it was hit by a Reading Company train on the outskirts of Philadelphia, Pennsylvania, United States, early on the morning of Tuesday 16th February 1932. John Adams, Michael Biehm, Frank Johnson, and Joseph Stuski and were all killed outright, and Thomas Martin died several hours later from his injuries. The five were all foremen of the stevedoring firm of Murphy Cook & Company and were bound for the Ford automobile plant in Chester. The sixth man, Daniel Murphy Jr., had been taken with Thomas Martin to the Mercy Hospital in West Philadelphia; unlike his companions he was not seriously hurt.

The accident happened at 6.20 a.m. The crossing was guarded by a watchman between 7 a.m. and 11 p.m.; at other times warning lights indicated the approach of a train.

Daniel Murphy Jr. spoke to a newspaper reporter from hospital. "I was in the back of the truck with two other men trying to keep warm. Two men were riding on the front seat with Frank Johnson, the driver. It was so cold that I guess Johnson was not watching for the train, and before I knew anything, I heard a tremendous crash. When I came to I found myself clinging to the cow-catcher of the locomotive, and I hung on until it stopped. I saw some of the other fellows being dragged under the wheels of the engine and crushed to death."

The train consisted of an engine and 34 cars. It was bound for Philadelphia and was travelling at about 30 miles an hour. John Jordan, its engineer, said that he whistled when he had seen the truck, which had approached the crossing travelling at high speed. The truck driver had then looked up but had been unable to stop. Police said that the warning lights had still been blinking half an hour after the accident.

The truck was one of three bound for the Chester plant but the other two had taken a different route and their occupants remained unaware of the accident until the third truck failed to appear and they made inquiries.

Apart from the six men the truck involved in the collision was loaded with rigging and reels of steel cable, which were scattered along the line. Officials said that the victims would have been enmeshed in the gear with no chance to save themselves. [1003] [1004]

Mrs Mollie Krause, her children Evelyn, 10; Ruth, 7; Richard, 5; and Catherine, 1; and her sister Miss Teresa Ott, were all killed when the car in which they riding was struck by a Pennsylvania Railroad passenger train on Bowers Crossing, two miles east of Monroeville, Indiana, United States, on Sunday 24th April 1932. The automobile was

[1001] *The High Point Enterprise*, High Point, South Carolina, United States, 26th December 1931
[1002] *Gastonia Daily Gazette*, Gastonia, South Carolina, United States, 26th December 1931
[1003] *Chester Times*, Chester, Pennsylvania, United States, 16th February 1932
[1004] *The Titusville Herald*, Titusville, Pennsylvania, United States, 17th February 1932

carried 200 feet, its wheels were torn off, its engine ripped out and its body was thrown against a telegraph pole. The women had been returning to Fort Wayne, Indiana, after a visit to their parents at Payne, Ohio. [1005]

The westbound *Soo Dominion*, crack Minneapolis St Paul and Sault Ste Marie Railroad passenger train from Chicago, Illinois, to Vancouver, British Columbia, was in charge of Conductor Charles Hinckley with Engineer Meagher on Sunday 1st May 1932. Their train was making its maiden trip and the journey was uneventful until they were near Spencer, Wisconsin, United States, when tragedy struck. Shortly after 5 p.m. the train, travelling at 55 or 60 miles an hour, collided with a car carrying seven people, killing six of them.

The victims were Irvin Reimer, his wife, and their children Donald, 12, and Lois, 3, and Irvin Reimer's brother Joseph and his wife. All came from Randolph, Wisconsin, and they were returning from a visit to another brother, W.C. Reimer. A seventh occupant of the car, Joseph Reimer's daughter, Gwendolyn, 7, suffered a broken arm and collarbone. She was rushed to Marshfield hospital by John Orgish, of Spencer, a motorist who came along just after the accident. It was thought that she would recover from her injuries. [1006]

Six people were killed instantly and a second, a woman, died later of her injuries after an express train collided with a car on a level crossing near Cluy, in the Transylvania region of Romania, on Wednesday 4th May 1932. [1007]

Just before 16:00 on the afternoon of Sunday 15th May 1932, a car coming from nearby Reims arrived at a level crossing at Vitry-les-Reims, France. The crossing was protected by lifting barriers, which were raised, and the car driver started across the line, only for his vehicle to be struck by a leave train travelling from Mourmelon to Reims. The car was thrown over ten metres and its five occupants were all killed. They were M. and Mme Henri Petelat; M. and Mme Robert Petelat and the latter couple's 2-year-old daughter.

The crossing was worked by a crossing keeper, M. Georges Legrand, a man with 26 years' service, who was arrested following the accident. He stated that the alarm which normally sounded to announce the arrival of a train had not operated, and hence he had had no warning of the special train. [1008]

Northern Pacific Railway train No. 31 from Morris to Brainerd collided with a train on a level crossing a mile west of Sauk Center, Minnesota, United States, early on the afternoon of Tuesday 12th July 1932. Martin Schneider and his wife, Winzel Stengel and Norman Weiser were killed outright, and Mrs Henry Schneider was fatally injured and died a short time later. Henry Schneider Jr., a small child, was seriously injured. The

[1005] *The Tipton Daily Tribune*, Tipton, Indiana, United States, 25th April 1932
[1006] *Wisconsin Rapids Daily Tribune*, Wisconsin Rapids, Wisconsin, United States, 2nd May 1932
[1007] *The Times*, London, England, 5th May 1932
[1008] *Le Peuple*, Paris, France, 16th May 1932

party had been on their way from their homes at St Cloud to visit a federal prisoner at Fergus Falls. Martin Schneider reportedly drove onto the crossing directly in front of the train. [1009]

Street car services in Mexico City had been suspended for a month due to a strike in late June and July 1932. They had, however, resumed on 19th July, but it was thought that the crew of a Ferrocarril Nacional de México train were unaware of this when they reversed their train at some speed over the street car tracks in the suburb of Tacuba, Mexico. Early on the morning of Wednesday 20th July 1932, only 12 hours after the strike had ended, a street car crowded with employees of the company was struck by the train, five waggons passing over the wreckage before the train could be pulled up. Sixteen street car passengers were killed and 12 injured. [1010] [1011]

On Friday 22nd July 1932 train No. 4, the Down Frontier Mail, collided with a motor lorry on a level crossing at mile 601/2 between Nowshera and Pir Piai, India, on the North Western Railway. One of the gates of the crossing had been open, either because it had been forced open by the lorry or because the crossing keeper had neglected to close it. Three of the passengers on the lorry were killed instantaneously and six seriously injured, two of whom later died. [1012]

A car carrying five schoolchildren home from school was struck by a Cincinnati Northern Railroad electric car on a level crossing at Hudson, Michigan, United States, on the evening of Thursday 8th July 1932. Reita Marry, 15, the driver of the car, her brother Peter, 10, her sister Mary, 9, and Robert Heinbacker, 8, Reita's half-brother, were all killed outright. The other occupant of the car, Mary Jane Young, 9, died of her injuries the next day. It was alleged that the warning lights at the crossing had been out of order. [1013]

Mr Walter Douglas, 70; his wife Mary, 50; their son Prentice, and their daughters Mabel and Matilda were all killed when their car was struck by a Louisville and Nashville Railroad train three miles from LaGrange, Kentucky, United States, on Wednesday 31st August 1932. Mrs Patsy Douglas, wife of Prentice, was injured. [1014]

Milwaukee, Oregon, United States, a rather confusingly named suburb of Portland, was the scene of a bad level crossing accident on Sunday 11th September 1932. Mr Elwood Lyon, a local man, had driven his car along a road which ran parallel to the railroad for several miles when he reached the crossing and swung it across the line directly in front of the locomotive of a train. He, his wife, their children Barbara, 11;

[1009] *The Brainerd Daily Dispatch*, Brainerd, Minnesota, United States, 12th July 1932
[1010] *The Evening Star*, Washington, D.C., United States, 22nd July 1932
[1011] *The Sandusky Register*, Sandusky, Ohio, United States, 21st July 1932
[1012] Report by the Railway Board on Indian Railways for 1932-33 Volume 1
[1013] *The Sandusky Star-Journal*, Sandusky, Ohio, United States, 9th September 1932
[1014] *The Chronicle-Telegram*, Elyria, Ohio, United States, 31st August 1932

Flora Lee, 6; and Billy Jean, 6; plus two Japanese girls, Akiko Sasaki, and her cousin Kikuye Sasaki, both 8, were all killed. Another Lyon daughter, Marian Ruth, was seriously injured and died the next day, while a fifth daughter, Rose Lorraine Lyon, 3, was less badly hurt. [1015] [1016]

On the morning of Sunday 20th November 1932 a party of young fascists travelled in three motor coaches from Tarquinia to visit an exhibition of the Fascist Revolution in Rome. On their return journey in the evening, on a wet and stormy night, the third coach, carrying 52 passengers, met with disaster on an open level crossing at Furbara, Italy, when it was struck by a Rome – Milan express. Eleven bus passengers were killed and another 22 were injured.

As well as having no crossing keeper the crossing was perhaps more than usually dangerous in that road and railway ran almost parallel for some distance before the road crossed the line at an acute angle, though the usual warning signs were provided. The three coaches were running close together in a procession, so the first two must have had fairly narrow escapes. In the darkness and bad weather the driver of the third coach clearly failed to spot the train. His vehicle was carried over 200 metres before being flung onto the opposite running line. Because of the darkness and the isolated location at which the accident occurred it was some time before the dead and injured could be picked up. [1017]

A group of negro employees of the Lawley lumber mill requested a lift into nearby Jacksonville from Frank Hartley, the driver of a lumber track on the evening of Saturday 14th January 1933. It proved to be a fateful decision for them. The lorry had only travelled a short distance when it was struck by a freight train on a level crossing at Baldwin, Florida, United States. Six of the employees were killed outright, two were fatally injured and a ninth was critically hurt when the load of lumber fell on them following the collision. Frank Hartley, 42, was also killed. The crossing, with either the Seaboard Air Line or the Atlantic Coast line, was a blind one and the night was foggy. [1018]

A level crossing accident which could easily have been much worse occurred on the Ferrocarril Oeste de Buenos Aires (Buenos Aires Western Railway) on the night of Wednesday 1st March 1933. A mixed train consisting of an engine, eight carriages and six waggons was approaching the station at Jauregui, Argentina, when it collided with a motor lorry and trailer loaded with drums of naphtha. The collision derailed the locomotive of the train, the drums of naphtha exploded, and a fire broke out which engulfed the train, several coaches of which were destroyed. Thanks to speedy action on the part of the train crew the passengers, many of whom had been asleep in their berths, were all evacuated and most only suffered minor injuries. A relief train with medical facilities was also sent to the scene and the Chief Engineer, Traffic Manager and other company officials went to Jauregui on a special train. Among the vehicles burnt was a

---

[1015] *The Charleston Gazette*, Charleston, West Virginia, United States, 12th September 1932
[1016] *The Daily Inter Lake*, Kalispell, Montana, United States, 12th September 1932
[1017] *The Times*, London, England, 22nd November 1932
[1018] *Thomasville Times-Enterprise*, Thomasville, Georgia, United States, 16th January 1933

postal van but the mails it contained were saved. Both occupants of the lorry were severely injured, and one died in hospital the following day. [1019] How the accident occurred and if the crossing was provided with gates or barriers was not stated.

Three men and two boys lost their lives when a Grand Trunk Western Railroad train bound for Detroit struck their car on a level crossing near Andersonville, Michigan, United States, on Saturday 18th March 1933. They were Stanley Mislak, 39, his son Robert, 4; Peter Nowakowski, 44, his son Norbert, 8, and James Lewandowski, 29, all of Detroit. The car was carried 1,000 feet by the impact. [1020] [1021]

A family of seven were wiped out when their car was struck by a train on Miller Crossing, near Utica, Missouri, United States, just before 11 a.m. on Sunday morning, 2nd April 1933. The Grimes family had set off from their home on the Vern Crookshank farm and were heading northwards along a farm to market road towards Highway 36 when their car was struck by Chicago Burlington and Quincy Railroad fast passenger train No. 14. The crossing was located at the foot of an incline. The noise of the collision was heard in Utica, a mile away, and the remains of the automobile were carried 50 yards before rolling into a ditch. All seven were killed outright. The train pulled up as quickly as possible and sheets taken from berths in the Pullman cars were placed over the corpses by the train crew, the bodies being left beside the track for the attention of the coroner.

Those killed were Mr Glen Grimes, 39; his wife Fernella, 33, and their children Larkin, 9; Robert, 7; Glenn Jr., 6; Charles, 3, and Virginia, 3 months.

Engineer G.E. Groce and Fireman W.C. Brown gave evidence at the subsequent inquest. The engineer said that the accident had happened at 10.42. He had whistled at the whistling post and had seen the Grimes' car slowly approaching the crossing. When his train had been 15 feet from the crossing he had seen the car stopped on it. He immediately applied the brakes, but his train had been moving at 45 or 50 miles an hour and it was impossible to stop. Fireman Brown corroborated his engineer's testimony except that he could only see the car at the last moment from his side of the locomotive.

D.S. Sprouse and his brother E.A. Sprouse were eye-witnesses to the collision. The former man said that they had approached the crossing from the opposite direction. The red warning light had been on. He had stopped the car, remarked to his brother that a train was coming from the south, and had heard the train whistle. He had seen the Grimes' car approach from the south and assumed that it would pull up, but it continued onto the crossing and then stopped. A moment later the train struck it.

E.A. Sprouse's testimony was practically the same.

The six-man jury returned a verdict that the victims had lost their lives through being struck by the Burlington train and that no one was to blame. [1022]

[1019] *The Railway Gazette*, London, England, 21st April 1933

[1020] *The Sheboygan Press*, Sheboygan, Wisconsin, United States, 18th March 1933

[1021] *Logansport Pharos Tribune*, Logansport, Indiana, United States, 18th March 1933

[1022] *The Chillicothe Constitution-Tribune*, Chillicothe, Missouri, United States, 3rd April 1933

Five young men, all graduates of Jackson, Minnesota High School, lost their lives across the state line in Okoboji, Iowa, United States, late on the evening of Saturday 22nd April 1933. They had been on their way to a dance at nearby Arnolds Park when their car was struck by Chicago Milwaukee St Paul and Pacific Railroad gasoline railcar at 11.25 p.m. They were Frank Cirhan, 23; Glen Lester, 21; Sam Smith, 26; Harold Teigan, 25, and Rudolph Vacura, 20.

Engineer A.J. Deitrich said that he had been coasting down the grade to the crossing and could have pulled up very quickly, but the graduates' car had appeared to be slowing down. He had given the proper whistle signals for the crossing. As his railcar had neared the crossing the automobile suddenly speeded up. Engineer Deitrich theorised that either the car driver had misjudged the speed and distance of his train from its headlight, or he had been seeking a thrill by beating the train over the crossing.

The car was thrown into the air and landed 30 feet away, while the left front of the railcar was crumpled in. Four of the occupants were killed outright. Cirhan was still breathing when rescuers reached him but died on his way to Milford Hospital. Teigan had been driving the car. Local inhabitants said that the noise of the crash had been terrific, some even thinking that a nearby railroad bridge had collapsed. Baggageman Earl Hartshorn and Conductor Nick McGrath of the railcar were among the first to reach the scene after the accident. A case of broken bottles found among the wreckage led to rumours that the occupants of the car had been drinking (prohibition was still in force) but it was impossible to tell if the bottles had contained new beer, home brew or had been empty and no smell of alcohol was detected by the rescuers. [1023]

On Saturday 6th June 1933 No. 5 Up mail train collided with a bus on the level crossing situated 400 feet outside of the Up outer home signal of Bhadohi station, India, on the East Indian Railway. Fifteen bus passengers were killed outright and 16 were injured, two of whom later died. Although the seating capacity of the bus was only 20 it was carrying 36 passengers at the time of the accident. The official report on the accident blamed both the crossing keeper for not closing his gates in time and the bus driver for failing to exercise reasonable caution, a verdict which seems somewhat unfair on the bus driver. [1024]

Mr. Sleigh, the senior government inspector, made three general recommendations for level crossings:

(a) Semaphore stop signals, interlocked with the gates, should be provided, the cost to be met from the Road Fund.

(b) To ensure caution on the part of road users at level crossings, the penalty under Section 124 of the Railways Act IX should be increased and the offence made cognisable {that is, an offence in which a police officer has authority to make an arrest without a warrant and to begin an investigation with or without the permission of a court – usually more serious offences}.[1025]

(c) Level crossing caution boards should be standardised and the cost of providing

[1023] *Spirit Lake Beacon*, Spirit Lake, Iowa, United States, 27th April 1933

[1024] Report by the Railway Board on Indian Railways for 1933-34

[1025] Wikipedia – Cognisable Offences

> them met from the Road Fund. Motorists should be examined as to the meaning of the caution boards and of other road signs before being granted a driving licence. [1026]

A party of six people who were probably going home after enjoying a very late night out were involved in a collision with Illinois Central Railroad passenger train at 4.18 a.m. on Tuesday 13th June 1933. The accident happened on Thirteenth Street Crossing, Centralia, Illinois, United States.

Adjoining the level crossing was the tower controlling the junction of the Illinois Central and Chicago Burlington and Quincy tracks and Towerman Ben Goad was a witness to the collision. He said, "A car was coming from the west going east at a terrific rate of speed. The engine headlight was bright. It looked like the car reached the centre of the track when the train hit it, knocking the car off to the west side of the track and dragging it along. It was not light enough at the time to see what type of car it was."

Mrs Beulah Cavanaugh, 30; Genevieve Dallas, 25; Leno Gaetti, 23; Samuel Maynor, 33; and Mrs Blanch Rollinson, about 35, were killed outright. Harry George, 24, who was carried a short distance on the pilot of the engine, survived long enough to be taken to St Mary's Hospital before he too died of his injuries.

At the inquest Engineer Paul Rickards of train No. 4 said that he had been going at 50 or 55 miles an hour. He had not seen the car before he struck it, but his fireman had called out, "Stop it!" when he had applied his brakes in emergency. After the accident he had found Harry George on the pilot of his engine. Finding that he was still beathing he had called to a coloured man, 'riding on the blinds,' to fetch an ambulance.

Fireman Walter Hafell testified, "I saw the car before striking it. The engine was about 60 feet away and as soon as I saw the car I called the engineer to stop the train. The car had to pass over three tracks before reaching the northbound main on which No. 4 came in. The proper signal was given as the train whistled for the five crossings coming into Centralia. I could not tell if it was a man or a woman driving."

Conductor Clifford Bird of train No. 4 was not called to give evidence.

Mrs Rose Farthing, Mrs Rollinson's mother-in-law, stated that the party had called at her home at 1.20 a.m., hoping to get her daughter, Mrs Leola Summers, to accompany them, but her daughter had declined (and doubtless thereby unwittingly saved her life).

The coroner's jury took only five minutes to return a verdict of 'Accidental Death'. [1027]

William Dooley, 38, described as a 'Bass county farmer', his wife Mattie, 35, and their children Arrie, 16; W.L., 11; and Thelma, 9, were all killed when the antiquated car they were riding in was struck by southbound Texas and Pacific Railroad passenger train No. 7 on a 'blind' level crossing, where an embankment restricted visibility, at Wayne, Texas, United States, on Sunday 2nd July 1933. The wreckage of the car was carried about 50 yards along the track. The family were returning from a visit to relatives in Louisiana. A baby,

---

[1026] *The Railway Gazette*, London, England, 16th March 1934

[1027] *Centralia Sentinel*, Centralia, Illinois, United States, 13th and 14th June 1933 left with neighbours for the day, was the only survivor of the family. [1028] [1029]

On Saturday 8th July 1933 Vernon Bogan, 50; his wife Norma, 26; Albert Dickey, 18, his wife Carrie, 16, and Mrs Margaret Reed, 40, all of Dayton; and Mildred Bogan, 10, of Middletown, all lost their lives when their automobile was struck by a Cincinnati and Lake Erie Railroad interurban electric car on a level crossing at Franklin, Ohio, United States. [1030]

At about 16:00 on the afternoon of Sunday 20th August 1933 train No. 2262 left Coimbra for Lousa, Portugal. It had only travelled a short distance, had emerged from a curve onto a straight stretch of track and was gathering speed when a lorry entered a level crossing at Vendas da Feira, only 30 metres in front of it. The inevitable collision shortly followed, the lorry was carried some distance along the line and smashed to pieces.

There were 26 passengers on the lorry, which had been filled with people returning from a pilgrimage to the Roman Catholic Sanctuary of Senhor de Serra. When the occupants of the lorry caught sight of the train an anguished cry arose, which was almost immediately drowned out by the tremendous noise of the impact. The bodies of four passengers who had been killed outright were found after the accident and it was believed that another person had been cut to pieces by the train. Almost all the others were injured and nine of them were taken, in ambulances and private cars, to Coimbra Hospital. Some of the women who witnessed the gruesome sights after the accident reportedly fainted.

Following the accident vehement protests were made about the number of level crossings on the Coimbra – Lousa line which had neither gates nor crossing keepers, but a newspaper article several days later pointed out that the accident was not in any way the responsibility of the railway company, the Caminhos de Ferro Portugueses, adding that that level crossings were assessed by the company as to whether they warranted full protection, and that article 67 of the highway code stated that it was the responsibility of road users to ensure that no trains were approaching before crossing the line. The crossing in question was provided with a warning sign, 'ATENÇÃO A PASSAGEM DOS CAMBOIOS' ('WARNING – TRAIN CROSSING'). [1031]

Seaboard Air Line Railroad train No. 14, en route from Hamlet to Wilmington and in charge of Conductor Buck Williams, collided with a tobacco truck carrying 12 men and boys on Fairmont Road Crossing, North Carolina, United States, on Tuesday 22nd August 1933. The truck, owned by J.R. Davenport, was bound from Deep Run to Fairmont market auction, loaded with tobacco. Haywood Smith, the truck driver, who was one of those injured, said afterwards that a house at the crossing had blocked his view and he had failed to see the train, a two-car gasoline unit. Several negroes at the crossing saw the danger and waved at Smith to stop, but he was unable to do so. The lorry crashed into the baggage car of the train and was demolished. Three of the farmers on the truck were

[1028] *The Daily Herald*, Biloxi, Mississippi, United States, 3rd July 1933
[1029] *The Salt Lake Tribune*, Salt Lake City, Utah, United States, 3rd July 1933
[1030] *The Evening Independent*, Massillon, Ohio, United States, 10th July 1933
[1031] *Diario de Lisbòa*, Lisbòa, Portugal, 21st and 24th August 1933

killed outright, two died as they were being taken to hospital, three died after reaching hospital, and three others were critically injured. It was decided that no inquest into the accident was necessary. [1032]

The same day as the North Carolina accident, 22nd August 1933, five Canadian tourists were, in the words of a newspaper report, 'hurled into eternity,' when their automobile was struck by a New York Central Railroad double-headed Cleveland to Chicago flyer near Porter, Indiana, United States, on a level crossing with the Danes Relief highway detour road, at 2.30 p.m. The tourists, from Alberta Province, had been on a two month tour which had included a visit to their old home in Nova Scotia and a stop at the *A Century of Progress Exposition* in Chicago (also known as the *Chicago World's Fair*).

Those who lost their lives were Mrs A.F. Massom, her daughter Magdalene, 11, and her son Kenneth, 8; Miss Mary Anne M'Donald and Miss Winifred Christine M'Donald. The latter two were both teachers and probably sisters.

A newspaper report explained the somewhat complex layout at the crossing. 'The crossing where the tragedy was enacted is known as "death crossing." The railroad runs between two parts of U.S. Highway 20, which are to be connected by a viaduct running above the tracks. The project has been uncompleted for two years while highway officials have negotiated with the railroad authorities. At that point the four lane highway, one of the heaviest travelled in the country, is detoured for a mile where the New York Central has failed to complete its share of the highway overpass. The two detours, one to the south for eastbound traffic and the other to the north for westbound traffic, are intersected by railroad crossings, which are protected only by red blinker lights.'

An eastbound passenger train had passed just before the party reached the crossing and it was believed that Mrs Massom, who had been driving, and the others of the party, had been watching the eastbound train and failed to see the westbound one. The highway ran alongside the railroad before turning abruptly to cross it. The blinker lights were not easy to see and although the view of eastbound trains was good, that of westbound trains was limited to less than 300 yards. The train had been reportedly travelling at 75 miles an hour. [1033]

A special train carrying F.R. Girard, Philadelphia superintendent of the Pennsylvania Railroad, collided with a car at Brooms Landing, Pennsylvania, United States, on Friday 25th August 1933. Mr John Gilbert, 21, the driver of the car; his wife Henrietta, 17; their son John Jr., 4 months; Margaret Greenswalt, 17; and Henrietta Hill, 21, were all killed. The Gilberts were from nearby Columbia and the other two women were from Morrisdale. Members of the train crew rescued Doyle Hewitt, 21, of Morrisdale, whom it was feared was fatally injured. The car was dragged 60 feet, then thrown into the adjoining Susquehanna River. A seventh, indirect victim of the accident was William Meisenbach, 42, of Columbia, who suffered a fatal heart attack while hurrying to the

[1032] *The High Point Enterprise*, High Point, North Carolina, United States, 22nd August 1933
[1033] *The Vidette-Messenger*, Valparaiso, Indiana, United States, 23rd August 1933

scene.[1034] [1035]

On Sunday 27th August 1933, a Pau, France, registration inspector, his subordinates, and their partners were enjoying a pleasure trip in a chauffeur-driven car. Coming from the resort of Vieux-Boucau-les Bains they were heading for Saint-Vincent-de-Tyrosse for a late lunch. The chauffeur was driving at high speed towards Soustons, when their day out ended in disaster. Between Vieux-Boucau and Soustons trees and bushes doubtless prevented the chauffeur from seeing an approaching train bound for Léon at a level crossing, and the hood of his car entered the crossing just as the engine of the train also reached it. The train caught the automobile and carried it 50 metres before it could be stopped.

Doctors Getten, Naudiac and Nougaro of Soustons and Drs Lasalle (two brothers) of Vieux-Boucau rushed to the scene. They found that they could do nothing for five of the seven occupants of the car; M. and Mme. Bordenave, M. and Mme. Cazenave and M. Castets all having been killed outright. The chauffeur, Pierre Delplan, was still alive but died on the way to a Dax hospital. The seventh occupant of the car, Mme Castets, was thrown from the car and escaped with multiple bruises.

The local police set up an investigation into the accident and the Dax parquet (prosecutor) and M.Cattin, the principal engineer in charge of local railways, went to the scene.[1036]

At 15:30 on Saturday 2nd September 1933, a car driven by M. Pierre Belin, 40, a Prêmeaux wine merchant, and carrying his wife, their children Viviane, 9; Madeleine, 7; Lucette, 6; and Daniel, 4, plus Mlle Picavey, 17, daughter of a Dijon English professor and a friend of the family, and Mlle Zokanto, 20, the Italian governess of the Belin children, reached a level crossing near the station at Nuits-Saint-Georges, France. As the car reached the crossing it was struck by train No. 11, a Paris – Marseilles express which had just left Dijon, travelling at 80 kilometres an hour. The shock of the collision was described as *effroyable* (appalling) and the car was smashed to pieces and thrown against the parapet of a bridge across a little river, the Meusin. The victims were thrown from the car, except for the body of Mme Belin, which was trapped in the wreckage. All eight persons in the car lost their lives.

An investigation into the accident established that it had been due to a mistake made by the relief crossing keeper, Mme Villeret, who replaced the regular crossing keeper, Mme Germain Persil, every day between 15:00 and 18:00. The barriers at the crossing were normally kept closed across the road and were only opened for the passage of vehicles.

Mme Villeret, in a statement, said, "When the car blew its horn for the crossing the warning bell had not sounded. Before I opened the barriers I looked along the track. The semaphore was lowered. I didn't see the train. Suddenly there was a very loud noise which made me step back. I saw it… and I ran away…"

[1034] *The Clearfield Progress*, Clearfield, Pennsylvania, United States, 25th August 1933
[1035] *The Muscatine Journal and News-Tribune*, Muscatine, Iowa, United States, 25th August 1933
[1036] *La Dépêche*, Toulouse, France, 28th August 1933

The crossing had been the site of many accidents. [1037]

Three days after the accident it was reported that Mme Villeret, the crossing keeper, had confirmed her initial statement to the Beaune parquet, that she had not heard the warning bell indicating the approach of the express and that she had thus felt free to open the crossing barriers for the passage of the waiting car.

Meanwhile M. Connesson, the mayor of Nuits-Saint-Georges, sent a telegram to the Minister of Public Works in which he stated, 'In view of the terrible accident which has cost eight lives and annihilated a whole family, and on behalf of the distressed local population, I must protest strongly regarding the deficiencies in level crossing protection and the attitudes of the railway and the Ministry of Public Works, who have systematically refused, through the years, to construct a bridge to replace three deadly level crossings. [1038]

Italy suffered a bad level crossing accident on Monday 11th September 1933 on the line between Arezzo and Chiusi. A car going towards Arezzo and carrying in addition to its chauffeur nine passengers was struck by a fast train travelling at 90 kilometres an hour and smashed to pieces. Eight of its occupants were killed. All were tradespeople bound for Arezzo on business. [1039] A police investigation determined that the barrier at the crossing had been raised by an unknown person and then wedged open with a stone. [1040]

The mail train from Santander, which was due at Llanes at noon on Wednesday 11th October 1933, collided with a car on a level crossing at Unquera, Spain, tore it apart and killed its six occupants. The train stopped was immediately and the train crew, aided by locals, removed the bodies of the victims from the wreckage of the car. Some of them were at that point still alive but they died shortly afterwards. The victims comprised two women, a man and three children. They were later identified as a Gijon family, who had been on their way to Bilbao. They were Antonio Martinez Azcoitia, his wife and their three children: and their maid, Olvido Nieto. They had been on their way to visit a Señor Salaberry regarding a child affected by infantile paralysis (poliomyelitis).

The accident was not easily explained because although it was an open crossing there was a very good view of approaching trains. The train driver believed that due to the fact that the car's windows were closed because of the rain, its occupants had failed to hear the whistles of his engine. His engine had caught the middle of the car, destroying it entirely. [1041]

Thursday 14th December 1933 saw an accident involving a train and a school bus which claimed ten (possibly eleven) lives at Silver Pond Grove Crossing, Florida, United States, some three miles south of Crescent City, from which the accident takes its name. The bus was making the last stop of its journey picking up children for the elementary

---

[1037] *La Charente*, Angoulême, France, 4th September 1933
[1038] *Excelsior*, Paris, France, 5th September 1933
[1039] *La Dépêche*, Toulouse, France, 14th September 1933
[1040] *The Times*, London, England, 13th September 1933
[1041] *La Vanguardia*, Barcelona, Spain, 12th October 1933

school in Crescent City and thus was full. It had travelled down State Route 3, which ran parallel to the Atlantic Coast Line Railroad track, and then turned onto a dirt track road to reach the hamlet of Silver Pond Grove. Originally it had gone right into the settlement, which involved it crossing and recrossing the railroad. Thinking that this was too many crossings of the railroad, the residents had constructed a turning area on the State Road side of the crossing, but on this particular morning the bus driver, the 63-year-old Mr D.R. Niles, decided to cross the line anyway, apparently because he was worried that the thick fog which prevailed that morning put the children at risk should they cross the line on foot. It did not, it would seem, occur to him that instead he was putting the lives of all the children on the bus at risk by taking the vehicle over the crossing.

As the bus (which was basically a truck chassis with a home-made wooden body mounted onto it) swung off the main road towards the crossing just before 7.45 a.m. a northbound Atlantic Coast Line freight train, in charge of Conductor W.J. Morrison, with Engineer R.A. Howell and an unnamed fireman, was also approaching it. The engine whistle was being sounded and the bell on the front of the locomotive was ringing, but the bus, crowded with young children, was doubtless full of noise, while the fog itself would tend to muffle the sound of the whistle and bell. It is likely that Driver Niles neither heard nor saw the train as he drove onto the crossing.

The train struck the bus broadside on, smashing it and throwing the dead and injured children onto the lineside. Seven children were killed more or less outright, two died shortly afterwards and a tenth passed away in Palatka Hospital.

By chance Sheriff R.J. Hancock came along shortly after the accident and he immediately took charge of the rescue operation. Another of those early on the scene was Louis Leach, of the Crescent City Motor Company, who described it as "the most horrible scene imaginable." He added, "There were at least six bodies strewn along the track. A quick survey of the scene revealed an arm here, a head there, or a foot or other portion of a child's badly mangled body elsewhere."

The children who lost their lives were Evelyn, Frederick, Merle, and William Smith, aged 9, 14, 16, and 12 respectively, all children of widow Mrs Bennie Smith; Eddie and Hazel McGrady, aged 10 and 13, brother and sister; Bertha Gorton, 15; Hilda Hardy, 6; Troy Tipton, 13, and William Owens, 12, who died in Palatka Hospital. Almost all, if not all of the children who survived were injured, as was Driver Niles.

The coroner's jury at the inquest on the children reached the obvious verdict that Niles had been negligent in crossing the track in a thick fog with a bus full of schoolchildren when there had been no need for him to do so. He escaped being charged with manslaughter, though it was said that he was a broken man for the remainder of his life. [1042] [1043] [1044] [1045]

Most of the family of Mr Henry Weigel Sr. lost their lives in an accident early on Christmas Day, Monday 25th December 1933, when the car in which they were returning

[1042] *Paris Evening News*, Paris, Texas, United States, 14th December 1933

[1043] *The Key West Citizen*, Key West, Florida, United States, 14th December 1933

[1044] *Laredo Times*, Laredo, Texas, United States, 15th December 1933

[1045] https://disasteroushistory.blogspot.com/2016/03/crescent-city-florida-trainschool-bus.html to his home from midnight mass at St Mary's Church was struck by an eastbound New York Central Railroad flyer on the Division Street Crossing in Vermilion, Ohio, United States. Mrs Weigel Sr., their son Henry Weigel Jr., Mr George Weigel and his wife of Reading, Pennsylvania, and Ruth Welles of Cleveland were all killed or fatally injured. A sixth occupant of the car, Robert Weigel, 12, of Vermilion, suffered a broken leg and severe cuts and bruises.

The accident was witnessed by a large number of people who were also on their way home from midnight mass at one or other of the Vermilion churches. A westbound passenger train was flagged down and took George and Henry Weigel Jr. to Sandusky, but both died of their injuries later. The less badly hurt Robert was taken to hospital by road ambulance. A 20-month-old Weigel grandchild was at home with Henry Weigel Sr.

Coroner John Yochem investigated the accident and returned a verdict of 'Accidental Death' the same day. [1046]

Level crossing collisions involving school buses seem to have been surprisingly frequent in the United States in the early 1930s and less than a month after the Crescent City tragedy there was another accident at Bassfield, Mississippi. At 8.12 a.m. on Monday 8th January 1934 a westbound Mississippi Central Railroad passenger train in charge of Conductor W.F. Smith, which was only travelling at low speed, collided with a school bus which was also going at low speed, it having stopped before it entered the crossing. Even so three children were killed, five seriously injured and 20 slightly hurt. Those who lost their lives were Itasca Stapleton, 9, who was killed instantly; Alice Altman, 16, who died a few minutes afterwards; and Tessie Roberts, 12, who died as she was being prepared for an emergency blood transfusion.

The bus had been carrying 42 passengers, including pupils of Bassfield consolidated school, and was being driven by 19-year-old Ernest Williamson, whose mother was also a passenger on the bus. Neither Williamson nor his mother was hurt in the accident; his father, Sam Williamson, held the contract for conveying the children by the bus.

Engineer R. Evans brought his train to a stop as soon as possible after the collision. The bus was not overturned but the engine of the train tore a large hole in its floor, through which it was thought two of the children who died had fallen. The vehicle was otherwise badly damaged, its wooden body being splintered.

Very fortunate in the accident was a small child named Kitty Gardner, daughter of "Doc" Gardner. She fell under the locomotive during the accident, only to crawl out from beneath it afterwards with only slight injuries. [1047]

After the accident Williamson simply stated that he had failed to see the train.

Three couples from Akron lost their lives in a level crossing accident at Barberton, Ohio, United States, early on Sunday 20th May 1934. Mrs Charles Alexander and his wife, Mr John Cuzzelin and his wife, and Mr William Maske and his wife were returning from a party. An Erie Railroad freight train passed over the crossing, after which the car

---

[1046] *The Sandusky Register*, Sandusky, Ohio, United States, 26th December 1933
[1047] *Hattiesburg American*, Hattiesburg, Mississippi, United States, 8th January 1934

entered the crossing and was struck by the westbound limited passenger train. Five of the occupants of the car were killed outright. Mr Alexander lived a short time but died on the way to the Citizen's hospital in Barberton.

There were no external witnesses to the accident but when the Barberton police arrived at 12.40 a.m. Mr Alexander, though dying, was able to tell them that it was the Maskes' car, but that Mrs Alexander had been driving. He explained that none of the people in the car had been aware of the passenger train. The crossing lacked barriers but had flashing lights which were working after the collision. There was a clear view for half a mile in each direction from the crossing, though the street crossed the tracks at an angle.

Engineer James Robbins stopped his train in 500 feet. Wreckage of the sedan was scattered on one side of the line and its occupants were all thrown out of it. [1048]

Two coachloads of excursionists from the Sahagún Winery Group of Gijón arrived in León, Spain, on Sunday 10th June 1934 and from there proceeded to Sahagún, where they spent several hours at the winery of Don Paulo Laso, who had organised the trip. Señor Laso himself stayed in town on business, thus probably inadvertently saving his life. At 14:00 the party returned to León and at a late hour they set off back to Gijón along the Puerto de Pajares road. At an early hour on Monday morning, 11th June, the coaches arrived at a level crossing with the Norte railway, adjoining an iron bridge over the River Bermeja, and a mile from Pola de Gordón. When they reached the spot the chains were apparently not raised across the road and the first coach proceeded to cross the line, to be struck by an ascending Asturias express, travelling at high speed.

The coach was thrown to one side and against the railings and parapet of the river bridge, while its petrol tank exploded. The bodies of the passengers were mostly thrown out onto the road, though three corpses ended up on top of the locomotive, which stopped 300 metres down the line. Eighteen of the coach passengers were killed outright and two of the injured succumbed after being taken to hospital. Eleven others were injured, only one seriously.

The crossing keeper asserted that the chains had been in place to block the road but as they were undamaged after the accident it was apparent that they could not have been run through by the coach, and he was arrested. [1049]

It was the turn of South Africa to suffer a level crossing accident involving school bus on the morning of Thursday 14th June 1934. The crossing, which was the scene of the accident, near Bloemhof, in the Western Transvaal, was provided with gates, which were apparently worked by users of the crossing rather than by a crossing keeper. One of the children who were passengers in the private motor bus alighted and opened the crossing gates, whereupon the bus driver, who was also a teacher at the Bloemhof school, started across. As he did so a train emerged from behind a clump of trees, struck the bus before it could get clear and dragged it 30 yards along the line. Five schoolchildren and their teacher/driver were killed outright, and 24 children were injured, seven critically. Two of

---

[1048] *The Republican-Courier*, Findlay, Ohio, United States, 21st May 1934
[1049] *La Vanguardia*, Barcelona, Spain, 12th June 1934

the badly injured children had succumbed before the end of the day. [1050]

Five residents of North Grafton, Massachusetts, United States, met their deaths on Nursery Crossing, close to their homes and half a mile east of the Boston and Albany Railroad's North Grafton station, at about noon on Wednesday 27th June 1934. The automobile in which they had been riding was struck by a Chicago express in charge of Conductor James Morey and Engineer Walter Taylor.

The victims were a Mrs Flansburg, aged 30, her sister Catherine Dowd, aged 28, and Mrs Flansburg's children Marie, 4; Paul, 3; and Ann, 1.

The train had been travelling at its usual speed. Due to its location at the top of the hill visibility at the unguarded crossing was not good.

Dr Frank Clapp responded to an emergency call but found all five victims beyond his aid. The Rev. John Hardiman also went to the scene and administered the last rites of the Catholic Church to the victims. [1051]

A tragedy which was not really anyone's fault occurred near Salerno, Southern Italy, on Wednesday 4th July 1934. Five small boys were riding in a horse-drawn van which was stopped waiting for an express train at a level crossing. The driver of the train whistled, and the sound frightened the horse such that it bolted across the line in front of the locomotive. The van was shattered, and the boys were all killed. The horse possibly escaped; the report does not say. [1052]

Careless driving on the part of another motorist cost five lives on a level crossing on the Michigan Central Railroad at Vermontville, Michigan, United States, on the evening of Friday 10th August 1934. A car containing Dale McClintock, his wife, Floyd Cook, and his wife had halted at the crossing for a passenger train to pass when it was rammed from behind by a car containing George Swift and two companions. The impact pushed the McClintock automobile into the path of the train, which was running at 60 miles an hour, killing its four occupants. The obstruction caused by the car also derailed the locomotive of the train, which plunged down a 10 ft embankment. Engineer George Howland and Fireman Floyd Dennis escaped with cuts and bruises but one of four transients who had been riding on the rear of the tender was crushed beneath the engine and became the fifth fatality in the accident. The other three transients were thrown clear. The 17 passengers on the train suffered nothing worse than a shaking and were forwarded to Grand Rapids in cars.

Happening during the celebration of the annual Vermontville homecoming, the accident attracted a large number of people. Smith was interviewed by Sheriff Spencer Cribb and stated that he had not noticed the car in front of him. The sheriff and the local prosecutor were reported at the time to be considering what action, if any, they were to

---

[1050] *The Times*, London, England, 15th June 1934

[1051] *The Lowell Sun*, Lowell, Massachusetts, United States, 27th June 1934

[1052] *The Times*, London, England, 5th July 1934

take against Smith. [1053]

At the end of August it was stated that an inquest on the victims had been postponed indefinitely because Engineer Howland had been quarantined, suffering from scarlet fever. [1054]

At the end of September the Interstate Commerce Commission pronounced George Swift responsible for the accident and stated, 'under such circumstances it was nothing but gross carelessness for Swift to be paying so little attention to where he was going as to fail to see the automobile standing at the crossing in time to avoid striking it.' [1055]

A mistake by a harassed crossing keeper contributed to a bad level crossing collision on 19th August 1934 at Pont Ardèche Crossing, near Pont Saint Esprit, France. Crossing Keeper Delarque was in charge of the crossing on the day of the accident, relieving the regular crossing keepers, M. and Mme. Vignac. After the accident he told M. Eymeric, the *juge d'instruction* (examining magistrate), that he had opened the barrier because he had been harassed by the car driver, M. Laffont, who had demanded that he be let across at once, even though the warning bell announcing the approach of a train had already begun ringing. Seeing the open barrier, a motorcyclist had started to cross as well. By some means the car, coming from Montpelier, and the motorcycle, coming in the opposite direction from Saint-Julien-de-Peyrolas, had managed to slightly collide with one another on the crossing. The train had then suddenly appeared around a sharp curve and had struck both the car and the motorcycle. Following the collision the car had burst into flames and been consumed. Killed in the accident were M. and Mme. Fernand Charpail, of Bollène; M. and Mme. Giffon, of Grenoble; M. and Mme. Paul Laffont, M. Auguste Laffont and Mme. Gayte, née Vignal, all of Saint-Léger-Brissac. [1056] The Charpails had been riding the motorcycle, the others had been in the car.

The crossing keeper stated that there would have been plenty of time for the car and motorcycle to have got across safely had they not met one another on the crossing and then stopped. He was arrested and was to be taken to gaol in Nîmes the next day. [1057]

A private level crossing near Bigwood, Ontario, Canada, was the scene of a collision between an automobile and a Canadian Pacific Railway freight early on Thursday 23rd August 1934. The six occupants of the car had been returning to French River from a party at Bigwood. Five of the party were killed. They were John Francis, of Huntingdon, West Virginia, United States; Miss Julia Johnson, of Cicero, Illinois, United States; E. Roberts, of Quebec; Donald Schnell, of French River, Ontario, the driver of the car; and Miss Edith Weston, of Chicago, Illinois, United States. Thomas Schell, father of Donald, was seriously injured and was taken to Parry Sound. A rock cutting at the crossing limited the view of trains to 100 feet.

In charge of the train was Conductor E.J. Brien, with Engineer Tom Gravelle and

[1053] *The Ludington Daily News*, Ludington, Michigan, United States, 11th August 1934
[1054] *Clare County Cleaver*, Harrison, Michigan, United States, 31st August 1934
[1055] *The Ludington Daily News*, Ludington, Michigan, United States, 27th September 1934
[1056] *Excelsior*, Paris, France, 21st August 1934
[1057] *L'Œuvre*, Paris, France, 21st August 1934

Fireman Allan McNicoll. The engineer and fireman said that neither of them had seen the car before they hit it but in any case they would not have had time to stop.

There was one eye-witness to the accident, Gerald Portier, of Bigwood, who said that he had crossed the line just in front of the car and had turned to watch the train. "Just as I turned I saw the car lifted into the air about 10 feet. I ran over to the car and found four of the people dead and Donald breathing his last." [1058]

On Tuesday 11th September 1934, a lorry carrying 24 soldiers was struck by a passenger train on a level crossing at a location given as near Nich – possibly Nish (Niš) in Serbia, Yugoslavia. The lorry was overturned, and five soldiers were killed and 18 seriously injured. [1059]

Nine people died and ten others were seriously injured when a bus was struck by a train on a level crossing on the Bergamo – Seregmo line at Terno, near Bergamo, Italy, on the evening of Wednesday 26th September 1934. The crossing keeper, who had apparently been responsible for the accident, fled. [1060] Another report stated that when the bus had arrived at the crossing its driver had found the barriers raised, so he set off over the line, only for his vehicle to be struck by a freight train. [1061]

Vermont Street Crossing, Blue Island, Illinois, United States, was the scene of a level crossing accident on Sunday morning, 14th October 1934. A car driven by Arthur Patno and containing his wife Ethel, Fred Hochsadter, his wife Helen, and the Hochsadters' children Jean, 14, and Herbert, 13, was struck by a Baltimore and Ohio Chicago Terminal Railroad wrecking train, on its way to a derailment in the company's yards. The Buick sedan was cut to pieces and five of its occupants were killed more or less instantly. Arthur Patno was very badly hurt and succumbed to his injuries Monday afternoon.

The crossing was provided with barriers, worked from an adjoining tower. At the inquest on the victims Towerman Alfred Hall testified, "I was looking north, and I saw the train approaching on Union street. I did not hear a bell or a whistle. I tried to put down the east gate, but it stuck – the west gate worked all right. I turned on the flashers on the Grand Trunk crossing to warn the auto which I saw was approaching. I saw it wasn't going to stop. I tried to hurry out of my tower to grab a warning flag. I fell down part of the steps and was just picking myself up when the crash came."

Engineer Alex Mackiewicz, Fireman Earl Riley, Conductor Edward Morrissey and Brakemen Elmer Blagg and Thomas Martin of the wrecking train were among others who gave evidence at the inquest. They stated that the train had not been travelling at more than 20 miles an hour and the whistle had been sounded for the crossing. [1062]

The inquest concluded with a verdict of 'Accidental Death' with a recommendation

---

[1058] *The Winnipeg Evening Tribune*, Winnipeg, Manitoba, Canada, 23rd August 1934
[1059] *Journal de Geneve*, Geneva, Switzerland, 13th September 1934
[1060] *L'Œuvre*, Paris, France, 28th September 1934
[1061] *Gazette de Lausanne*, Lausanne, Switzerland, 28th December 1934
[1062] *Suburban Star*, Blue Island, Illinois, United States, 19th October 1934

that better protection should be provided at the crossing. [1063]

Nine bus passengers died and 18 were injured when a bus taking a party of voters to the polling station was struck by a goods train on a level crossing near Rio de Janeiro, Brazil, on Monday 15th October 1934. [1064]

On Friday 19th October 1934 a large car driven by M. Georges Faure, a Cap Ferret, France, wine merchant, and carrying seven other passengers, was passing over an unguarded level crossing at Serilhan, four kilometres from Saint-Médard-en-Jalles, when it was struck by a Bordeaux – Lacanau Michelin railcar. It was intended to drop off car passengers at Souges Camp and Temple de Porje before continuing to Cap Ferret. The road vehicle was completely demolished in the collision and carried a dozen metres down the line. Two of its occupants were killed outright, by midnight M. Faure, the driver, and a woman passenger had succumbed to their injuries, and two other occupants were unconscious, while the remaining two passengers in the car were seriously hurt.

A newspaper reporter spoke to a M. Norba, a retired captain whose house was 50 metres from the crossing. "I was coming from Saint Medard," M. Norba related, "and I could see that an accident was likely to happen. I tried to warn the driver by shouts and gestures, but, alas, it was too late."

M. Latournerie, the driver of the Michelin, said that he had been travelling at 70 kilometres an hour at the time, and it had taken 147 metres to stop his train. [1065]

M. Faure's son, also named Georges, passed away the following morning and his daughter, Georgette, was among those badly injured and probably became the sixth fatality in the accident. [1066]

A minor railway in northern France, owned by the Compagnie Générale de voies ferrées d'intérêt local, was the scene of a horrific accident on Sunday 28th October 1934. At 19:40 a small saloon car coming from St Omer, and driven by M. Jacques Dell, of Calais, approached what was said to be a dangerous level crossing near Pont d'Ardres. Also in the car were his wife, Mme. Dell, their six-year-old daughter, a friend, M. Bonne, and the latter's wife, Mme. Bonne. The car reached the crossing at the same time as a steam engine drawing a train of empty waggons. The train was making its daily run between Pont d'Ardres and Ardres. The car apparently ran into the side of the train and was carried 30 metres. What would probably have been a bad accident anyway was made much worse by the force of the impact rupturing the petrol tank of the car. The petrol immediately caught fire and the car was enveloped in flames. The train was immediately stopped, and Driver Auguste Hansse and Fireman Albert Lefebvre jumped down from their engine. Aided by a M. Roussel, a passer-by, they attempted to rescue the occupants of the car but for some reason they could not open the doors of the vehicle. Vainly they smashed the car windows in the hope of pulling the occupants out through them, but the

---

[1063] Ibid, 2nd November 1934

[1064] *The Times*, London, England, 16th October 1934

[1065] *La Liberté*, Paris, France, 21st October 1934

[1066] *L'Œuvre*, Paris, France, 21st October 1934

flames quickly engulfed the whole car. The heat became unbearable, and the would-be rescuers had to retreat. The five occupants of the car were burnt to death.

The St Omer parquet, M. Pucheval, the police and railway company officials went to the scene of the accident and on the Monday afternoon the events leading up to the accident were re-enacted with a similar train and the same train crew. [1067]

A bus crashed through a lowered crossing barrier between Langwedel and Minden, Germany, on Friday 14th December 1934. Sixteen passengers died and four were injured. [1068]

The Great Depression was raging in 1934 and it was a car carrying seven employees of the Illinois Emergency Relief Station, part of a government unemployment relief programme, which was involved in a level crossing accident at Harvey, Illinois, United States, in the suburbs of Chicago, on Thursday 27th December 1934. The large Buick sedan was driven by John Burk, 31, who picked up the other workers every morning to take them to the relief station at Chicago Heights. His passengers were Harvey Freidman, 25; Mrs Ruth Greenburg, 24; Mrs Hope Hines, 25; Miss Mary McCann, 24; Miss Marian Riggs, 23; and Wendell Tarkoff, 19.

The accident happened at about 8.50 a.m. as the car was proceeding southwards on Halstead Street. It reached the crossing with the Grand Trunk Western Railroad at the same time as that company's *International Limited* express, in charge of Conductor William Murfield, which was running about an hour late. There were no gates at the crossing, though there was a crossing watchman, Edward Lines, and it was protected by warning lights. Exactly why the car failed to stop remained unclear, though it seemed likely that Burk was driving too fast for the conditions – at 45 miles an hour according to Lyon – and that he was unable to pull up in time on the slippery snow-covered road. The train struck the car with a noise which could be heard for several blocks, carried the vehicle some way down the line and killed all seven occupants outright. In addition the leading truck (bogie) of the engine of the train was derailed. The sight of the mangled bodies of the victims scattered about was apparently too much for Engineer George Henry, who reportedly collapsed after the accident. [1069] [1070]

At the inquest on the victims, Engineer Henry and Fireman George Carr testified that they had whistled repeatedly approaching the crossing, but the car driver did not seem to have heard the warning. Crossing Keeper Lines, who also had charge of the crossing at 155 Street, said that the signals, lights and bells at the crossing had been operating when the train was still 1,800 feet away. Having reportedly earlier put the speed of the car at 45 miles an hour, he said at the inquest that he could not estimate its speed.

The inquest jury returned a verdict of 'Accidental Death'. [1071] [1072]

[1067] *Le Grand écho du Nord de la France*, Lille, France, 30th October 1934
[1068] *Katastrophen der deutschen Bahnen Teil II* by H.J. Ritzau, pg. 239 (in German)
[1069] *Freeport Journal Standard*, Freeport, Illinois, United States, 26th December 1934
[1070] *Alton Evening Telegraph*, Alton, Illinois, United States, 26th December 1934
[1071] *The Pointer*, Riverdale, Illinois, United States, 28th December 1934
[1072] *Suburban Star*, Blue Island, Illinois, United States, 28th December 1934

A special train of the Chicago Milwaukee St Paul and Pacific Railroad running from Watertown to Milwaukee, in charge of Conductor A.C. Bishop, with Engineer R.F. McLaughlin, collided with a car bringing six children home from the Lincoln grammar school on the Washington Street Crossing in Oconomowoc, Wisconsin, United States, on the afternoon of Wednesday 16th January 1935. The children all lived on the southern edge of the city and the driver of the car, Eugene Chaffee, alternated with Jacob Schlachter, father of one of the other children in the car, in driving the children home from school. Ironically, because of the onset of a sleet storm. Chaffee had left his job as a chemist's assistant early in order to drive the children home safely.

The car was driven onto the open level crossing directly in front of the train, which carried it a block and a half. Chaffee was killed outright, as were Denis Bliesner, 7; Pauline Durnford, 13, and her brother Ross, 7; and Ernest Schlachter, 8. Marilynn Bliesner, 10, sister of Denis, was reportedly near death in Summit Hospital. Kitty Jean Chaffee, 8, the daughter of the car driver, suffered a broken arm, cuts and bruises but was expected to recover.

That the engineer of the train had whistled for the crossing was confirmed by Police Chief Carl Marquardt and Deputy Sheriff A.L. Morris, who chanced to be nearby surveying traffic conditions, while motorist L.J. Zuelsdorf, whose car was stopped waiting for the train, saw Chaffee drive in front of it, apparently unaware of its approach. It was theorised that ice on the windscreen of the Chaffee car had prevented its driver seeing the train. [1073]

An accident analogous to that at Harvey, Illinois in December 1934 (see above) occurred in Canada on Thursday 7th February 1935. The victims were engaged on highway construction as part of an unemployment relief programme undertaken by the Ontario government and were returning to Stanley at the end of their day's labours when the accident happened at 6.20 p.m. as darkness was falling.

The accident was described in a statement issued by Canadian National Railway acting Divisional Superintendent A. P. North. "Train No. 210, Sioux Lookout – Port Arthur local, with two coaches and 25 cars, approached mileage 16½ on the Kashabowie subdivision at 6.20 p.m. The engineer, James Strothers, saw two trucks proceeding east on Arthur St. At the regulation distance from the crossing he properly sounded the whistle. He noticed the trucks continue along the highway and he whistled more vigorously. One truck crossed the tracks just in front of the engine. The second one ran into the buffer beam on the front of the locomotive. The engineer applied the emergency brakes just before the collision. The impact lifted the front of truck up against the guard rail and unloaded the passengers against the side of the train. That is where the damage was done. No damage was done to the engine. The truck was a heavy duty one, with a canvas top behind the cab.

"It is considered a safe crossing and no major mishaps have ever occurred there. The crossing is an east and west highway, crossing the railway at an angle of 21 degrees. It was improved in 1933, grades on approaches were raised, widened, and equipped with

---

[1073] *Stevens Point Daily Journal*, Stevens Point, Wisconsin, United States, 17th January 1935 crossing signs, and red reflectors on each side. The bush has been cut on the south-west corner giving a fair view westerly along the railway for about 2,000 feet from a point on the approach 100 feet from the railway.

"It is evident the men heard the train whistle because one or more jumped out and escaped injury."

Mr North added, "There is no excuse for the accident. The view at the crossing is excellent and it is properly protected with signs. The locomotive whistle was sounded, and a bell was ringing."

The accident happened at Kakabeka Falls, Ontario. There had been 22 men in the truck, seated on planks and boxes. Nine of the relief workers were killed in the accident and 19-year-old Roger Rose, who was not a relief worker but a member of the Fort William Macoons junior hockey team who had accepted a lift on the way to a match, was fatally injured and died in hospital the following morning. Several other relief workers were injured. Conductor E. Dodd was in charge of the train. [1074]

William Lavelle, one of the survivors, spoke to a reporter. He said that the men had been happily laughing, planning what they were going to do that evening, when they had heard the shriek of an engine whistle and there was a terrible crash.

"I went flying through the air and had trouble keeping my stomach where it belonged, I seemed to be turned inside out. I found myself butted into a snow bank, pawing wildly. The truck was demolished, its engine buried in deep snow. I stumbled to the tracks, and in the uncertain light saw a number of grotesque forms scattered as far as 150 feet from the right-of-way. I did not recognise one body. A hand here – a head there – bits of clothing. We got straw from another truck and laid it on the canvas of our own. Those of us only shaken up gathered the dead and badly injured. Big 'Mike' McHeady lay painfully hurt but said, "Never mind me, get the others." A train crew arrived and soon other help. Our foreman, Ernie Perrie – he was dead; our second foreman, 'Mike' McHeady – he was injured. I was left in charge. I have seen a lot of life but nothing that would compare with the horror of that sight along the track. How we were not all killed, I can't figure out." [1075]

An accident happened near Ricardo Albuquerque station, Brazil, on the outskirts of Rio de Janeiro, on the morning of Wednesday 13th March 1935. The driver of a bus approaching the station from Cascadura failed to apply his brakes in time and the bus collided with a train bound for D. Pedro II station. The train was running at full speed and the bus was thrown a fair way down the track. The bus passengers were then in fear that their vehicle would be struck again by a train from Iguassu, but fortunately this did not happen.

After the accident the victims had to walk some distance before they could be rescued. The bus driver, Tiburcino Alves de Souza, could not be found after the accident and it was assumed at first that he had fled the scene.

Most of the bus passengers had been workers at the Tecidos Deodora factory, which

---

[1074] *Medicine Hat News*, Medicine Hat, Alberta, Canada, 8th February 1935

[1075] *The Winnipeg Evening Tribune*, Winnipeg, Manitoba, Canada, 8th February 1935 adjoined the station of the same name, the bus being timed to get them to work at the right hour. Seven of them were initially believed to have lost their lives in the accident and about 20 were injured.

The crossing watchman, in a report, stated that the accident was the fault of the bus driver, who had continued onto the crossing despite being signalled to stop. No gates, barriers or warning lights were provided at the crossing, protection being limited to a red flag or lamp displayed by the crossing keeper to road vehicles, despite the fact that the crossing was situated on a curve with abundant vegetation which limited the view of bus and train drivers of each other's vehicles.

It was later found that the bus driver, rather than having made his escape, had been seriously injured, and he died on the evening of the same day, by which time the number of dead had risen to nine. [1076]

The bus had a capacity of 20 passengers but was carrying 33. Buses were habitually overloaded as drivers received bonuses based on the number of passengers they had carried; this also encouraged bus drivers to speed. [1077]

A school bus accident late on Thursday 11th April 1935 was the worst of its type to occur in the United States or probably anywhere else up to the date. It happened on Baltimore Road Crossing, Rockville, Maryland, where two main tracks and a siding of the Baltimore and Ohio Railroad crossed the public highway. The bus involved was taking students of Williamsport High School on an after school trip to a science fair at the University of Maryland at College Park. There were 27 students on the bus, together with science teacher Louise Funk and bus driver Percy Line. The trip to the university itself took two and a half hours and by the time the bus was on its return journey it was 10.35 p.m. The accident occurred at about 11.28 p.m. The bus was struck by the train, which had been travelling at 58 miles an hour, about 14 feet from its front end, and thrown against an embankment. Fourteen students were killed outright or fatally injured and the other 13, the driver and teacher were injured. [1078] [1079]

Baltimore Road Crossing lacked gates but was provided with a standard crossbar marked 'Railroad Crossing'. A second sign read: 'NO WATCHMAN ON DUTY 10 P.M. TO 6 A.M.', though according to the law there should have been a watchman there until midnight. The watchman had of course left duty at 10 p.m., as he was permitted to by the railroad. At that hour he had placed a lighted red lantern in a box as a warning to motorists that the crossing was unattended. Warning bells, operated by track circuits, sounded on the approach of trains regardless of whether or not the crossing was manned.

The train involved in the accident was eastbound passenger train No. 12, consisting of engine No. 5550, two express cars, a postal car, an express car, two baggage cars, a combination passenger and baggage car, a coach, and two Pullman sleeping cars, in charge of Conductor Compton with Engineer James Shewbridge and Fireman William Bussey.

[1076] *O Estado de S. Paulo*, São Paulo, Brazil, 14th March 1935

[1077] *The Railway Gazette*, London, England, 17th May 1935

[1078] *The Morning Herald*, Hagerstown, Maryland, United States, 12th April 1935

[1079] *The Salisbury Times*, Salisbury, Maryland, United States, 12th April 1935

Engineer Shewbridge told the Interstate Commerce Commission (ICC) inquiry into the accident that his train had been drifting as it had approached Rockville; he had sounded the whistle at the whistle board and the bell on the front of the locomotive had been ringing. Close to the crossing an automobile had crossed from the right or south side, and he had whistled again lest another car was following but had seen no other vehicles. Just as they had reached the crossing his fireman had shouted a warning; he had applied the brakes and there had been a crash. His brakes had been working well.

Fireman Bussey said that when his engine was about 380 feet from the crossing he had seen the headlights of a motor vehicle which he later realised was the bus. At first it appeared to be stopping, then it seemed as if the driver had decided that he could not pull up in time and he had tried to cross in front of the train.

Driver Line testified that he had been running at 12 to 15 miles an hour as he had neared the crossing and had then reduced speed to 10 miles an hour. He had seen a red light at the crossing but had not seen any crossing signs, nor had he seen any train. Only as he had been entering the crossing had he heard a bell, which he thought had been the engine bell. Miss Funk had been sat on his left side and had pointed out dangerous points on the outward journey. She had not said anything as they had approached the crossing coming back, presumably because he had slowed down. He could not explain why he had failed to see the reflector warning sign at the crossing. He had not been over the road in question for four years and was not aware of a Board of Education requirement that buses must stop at all level crossings. The bus had been in good condition and he was familiar with driving it. There had been no noise in the bus, most of the students having been asleep. He had been working since 7 a.m. but had not felt tired, nor had he been ill.

Miss Funk said that the bus driver had reduced speed from 25 to 10 or 12 miles an hour as he neared the crossing, leading her to assume that he knew he was approaching a crossing, so she had not thought it necessary to caution him. The bus had not stopped at either that or another crossing on the outward journey.

Some of the students said that they had been asleep and had not known of anything wrong before the crash. Others said that there had been no noise in the bus and most of them said that they had heard neither the engine whistle, the engine bell, nor the warning bell at the crossing. One student spoke of hearing Driver Line say afterwards that he had not known of anything wrong until he had looked up the line and seeing an engine headlight had stepped on the gas.

Mr E.L. Stevens testified that he had been approaching the crossing from the opposite direction to the bus in a roadster with side curtains. As he had neared the crossing he had heard its warning bell ringing and had seen both the headlight of the express and the bus coming towards him, travelling at about 20 miles an hour. Having negotiated the crossing in his car he had stopped to see if the bus had got across in time. As he had opened his car door and looked back the train had hit the bus. He had not heard the engine's bell or its whistle. He said that he and other Rockville residents regarded the crossing protection as inadequate. Two other motorists said that they had heard the train whistle, as had two Rockville residents.

Mr H.P. Poe, the owner of the bus, stated that he regarded Line as a good driver. It was part of Line's job to transport students to and from school functions, and Line received no extra payment for doing so. Mr Poe said that he, too, was unaware that school buses were required to stop short of all level crossings. The ICC pointed out that the contract signed by Mr Poe with the school board by which he agreed to transport students contained a clause which read:

All buses must be brought to a full stop at least 25 feet from any steam or electric railway crossing. The driver must be sure that there is no danger from approaching trains or cars before attempting to drive the bus across the tracks.

A traffic check at the crossing showed that during the 16 hours that the watchman had been on duty 2,112 vehicles and 29 trains had passed over it. Between 10 p.m. and 6 a.m. the numbers were 149 vehicles and 8 trains. Since the beginning of 1920 there had been two accidents at the crossing, both of which had occurred at night when it was unmanned. It was noted that several loaded school buses had passed over the crossing during the traffic check and none had stopped; the only bus which had stopped had been empty.

Further tests showed that while an engine headlight could be seen for a good distance the engine could not be heard until it was 650 feet away, and its bell was not audible at all.

Subsequent to the accident, indistinct warning signs at the crossing were repainted and in response to a letter from the town clerk the crossing was manned throughout the 24 hours from the end of April.

Discussing the accident, the ICC stated that it was clear that the bell that Driver Line said he had heard had been the crossing bell and that he should have seen the reflector warning light and engine headlights which had been seen by several of the students. There had been no noise in the bus, he had not been blinded by the headlights of the car coming towards him, and there was 'no reason why he should not have known of the exact location of the crossing in ample time to enable him to take the precautions necessary to be taken by every motorist before passing over a railroad crossing at grade, and which were especially incumbent upon him when driving a loaded bus at night on a strange road and in rainy weather.'

The ICC was not impressed with Mr Poe's assertion that he knew nothing of the requirement that school buses must stop at level crossings, as seven other holders of school bus contracts all were aware of this requirement.

The view at the crossing, the commission continued, was not good, and the accident emphasised the limitations of crossing protection relying entirely on audible signals. Crossing bells might have sufficed previously, but were insufficient for modern protection, and much greater protection could be afforded by visual signals giving a positive indication of the approach of a train. There was, however, no substitute for the use of common sense and sound judgement on the part of road users.

Driver Line had maintained that the construction of the bus had not interfered with his view, but the ICC considered that it must have and that the design of the bus should be improved.

The ICC concluded that: 'The evidence developed as a result of the investigation of this accident clearly shows that there is no reasonable excuse to be offered for its occurrence. Even though the driver did not know anything about the requirement of the school board that he stop before passing over grade crossings, it was his duty to take the precautions which every driver should take before entering upon a crossing, and had he done so in this case, or had the protection required by law actually been provided, the accident would beyond reasonable doubt have been prevented. The bus should have been stopped and the driver should have made certain that it was safe to proceed; the railway company was required by law to maintain a watchman during that part of the day within this accident occurred, and responsible state or local authorities should have seen to it that this requirement of law was enforced.'

The ICC made five recommendations:

(a) That the responsible authorities should take steps to enforce existing laws, regulations, and ordinances regarding level crossings.

(b) That continuous protection should be provided at the Baltimore Road crossing by watchmen, flashing lights or other visual devices.

(c) That the responsible authorities should see that all holders of school bus contracts were aware of the need for buses to stop 25 feet short of crossings and that they should ensure that this was done.

(d) That the front ends of motor vehicles should be designed to give drivers an all-round view.

(e) That as far as possible the routes of buses should be arranged so as to avoid level crossings unprotected by watchmen or visual warning devices. [1080]

Nothing was apparently made by the ICC of the fact that Driver Lines had been working excessive hours and was probably very tired.

Lines, Engineer Shewbridge and Fireman Bussey were all arrested after the accident, though the two trainmen were soon released. Lines was charged with manslaughter, but a grand jury refused to indict him on the grounds that it was known to be a dangerous crossing and the weather might have contributed to the accident. [1081]

The accident was front-page news throughout the United States and President Franklin Roosevelt pledged between $100,000,000 and $200,000,000 for the replacement of grade crossings by bridges. [1082] The Baltimore Road Crossing was predictably one of the first to be eliminated, though many thousands of crossings remain in 2022.

'A night of happy gaiety and enjoyment spent at the "Indiana Days" celebration at Michigan City ended in tragedy early Sunday morning.' Thus did a newspaper begin its report on a level crossing accident at Chesterton, Indiana, United States, on Sunday 7th July 1935. A Chevrolet roadster carrying seven people, two from Chesterton and five from Valparaiso, was struck by a New York Central Railroad suburban train on a level crossing with State Road 49 in Chesterton. The automobile, travelling at 50 or 60 miles an hour, was hit by the rear of the train which was setting back on a wye (triangle) having taken water at the end of its trip from Chicago. The car was practically destroyed by the severity of the impact.

Glen Coach, 24, of Gary; John Holzer, 20, of Valparaiso, the owner and driver of the

---

[1080] Interstate Commerce Commission report

[1081] https://disasteroushistory.blogspot.com/2016/03/rockville-md-train-bus-crash-april-11th.html

[1082] *The Frederick Post*, Frederick, Maryland, United States, 13th April 1935 vehicle; Harry Lane, 18, of Valparaiso; Virginia Olsen, 18, of Valparaiso; and Charlotte Wiseman, 17, of Chesterton, were killed. Mary Flatz, 20, of Chesterton, and Richard Slagle, 23, of Valparaiso, were injured. Slagle was initially reported as having been killed. Also initially reported as a victim was Robert Bradshaw, whose mother was recovering from the shock of having been told he had died in the accident. When the party had been leaving the dance at the Oasis ballroom he could not be found, so they had eventually left without him.

Conductor L.E. Bickenstaff told the investigating coroner and police officials that he had been stationed at the crossing to warn road users of the backing train. He said that he had waved his lantern in a vain effort to stop the car, without success. [1083]

Less than 24 hours after the Chesterton accident five young men lost their lives in another level crossing accident a mile and a half east of Kennett, Missouri, United States. At 5.25 p.m. on Sunday 7th July 1935 they were riding in a new Chevrolet sedan when it was struck by St Louis – San Francisco Railway train No. 881, a gas-electric railcar nicknamed a 'bull moose', at the crossing with Highway No. 25.

The men in the automobile were Jas Attebury, 35, a taxi driver who was driving the vehicle; Thomas Bell, 35, a fireman; William Adkins, 33, farmer and labourer; John Farmer, 23, labourer; and Wendell Davis, 25, farmer and baseball player. All were residents of the Braggadocie area, and they were returning home from Campbell, Missouri, where the other four had gone to watch Davis play in a game, which he had won.

An inquest into the accident was quickly held and Coroner Rigdon examined Conductor Bert McAdams and Motorman W.H. Hargrove together with a man named E.C. Martin, who had witnessed the accident. The two trainmen said that their train had left Kennett at the normal time and had been travelling at 30 or 35 miles an hour as they neared the crossing. They saw the automobile approaching at a very high speed, estimated as 75 miles an hour, and by the time that they had realised that it was not going to stop it was too late to stop the train.

Martin testified that he had been at a filling station near the crossing and had heard the train whistle and seen the car coming at a very high speed. He had watched the two and saw the collision. The tracks at the crossing were somewhat depressed and trees somewhat restricted the view. He did not think that the car driver had been aware of the train prior to the collision.

The train had reached the crossing first, the car striking the rear portion of the front coach. The violent collision caused considerable damage to the train, making it undrivable, while the automobile was described as 'almost telescoped.' Four of the occupants of the car were killed outright and Adkins died while being removed to Kennett. None of the 15 train passengers was any more than shaken up. They had to wait at the crossing until an engine arrived from Hayti to take the railcar forward to its destination. Damage to the railcar was estimated at $2,500; the car was a write-off. The crossing had always been considered a dangerous one, partly due to the restricted visibility and partly, paradoxically, to the infrequent train service, which tended to lull

[1083] *The Vidette-Messenger*, Valparaiso, Indiana, United States, 8th July 1935 road users into a belief that they would not encounter a train. [1084]

The inquest jury returned the inevitable verdict of 'Accidental Death' while Kennett citizens agitated for the elimination of the crossing. [1085]

Saturday 27th July 1935 saw a disastrous level crossing accident involving a train and a bus at Coliseo, near Jovellanos, Cuba. A bus carrying about 32 people tried to cross the line in front of a freight train and failed. An early report stated that three men and a boy were killed instantly, and five men and a woman were trapped in the wreckage and burned to death. The bus driver initially at any rate survived the accident but was too badly hurt to explain how the accident had happened. [1086] The number of dead later rose to 15, with a further 17 injured, several of them in a grave condition. The bus had been en route from La Habana to Varadero to the weekend national regattas while the freight train was travelling from La Habana to Santa Clara. It carried the bus 400 yards before it could pull up. [1087]

Nine persons died in a collision between a light truck and a southbound Seaboard Air Line Railroad flyer en route from New York to Florida on a level crossing at Camden, South Carolina, United States, on the afternoon of Saturday 3rd August 1935. Five people were killed outright and the other four had died in hospital by the end of the following day. Most of those who lost their lives were from one family. They were farmer J. Van Baker and his wife, their sons J. Van Baker Jr., 18; Murray, 19 (the driver of the truck); Samuel, 21; and Steve, 35, and Steve Baker's wife (age not stated). Also killed were Gaylord Jordan, 13, and Hubert Jordan, 14.

The victims were all residents of farms in the Lockhart section of Kershaw County and had been on their way into Camden, probably on a shopping trip and/or to sell produce.

It appeared that Murray Baker had been driving too fast when he had reached the crossing and found the train going over it. He had been unable to stop and by swerving to the right, avoided a right-angled collision, but the truck hit the second or third car of the train and overturned. One passenger on the train was slightly hurt. Additional ambulances had to be called from Columbia to supplement the single local ambulance and pick up all the victims. [1088] [1089]

Seven members of a French-Canadian family died in a level crossing accident at Brosseau station (now Brossard), Quebec, Canada, on Saturday 10th August 1935. Hippolyte Talbot, of Waterville, Maine, United States, the owner of the car involved in the accident, had gone with his parents, Mrs Charles and Mrs Appoline Talbot, his brother Ovide and his cousin Peter Talbot, to collect two lay brothers of the Roman

---

[1084] *The Democrat-Argus*, Caruthersville, Missouri, United States, 9th July 1935
[1085] *The Sedalia Democrat*, Sedalia, Missouri, United States, 8th July 1935
[1086] *The Sunday Register*, Beckley, West Virginia, United States, 28th July 1935
[1087] *The Wisconsin State Journal*, Madison, Wisconsin, United States, 28th July 1935
[1088] *The Sunday Star*, Washington, D.C., United States, 4th August 1935
[1089] *The High Point Enterprise*, High Point, North Carolina, United States, 4th August 1935

Catholic Order of Les Freres de l'Education Chretienne: the Rev Brother Maximillian (Girard Talbot – Hippolyte's brother) and the Rev Brother Laurentien (Thomas Talbot – Peter's brother). They were returning to Waterville when their large car and a diesel-hauled Canadian National Railway train collided on a level crossing with the Montreal – St Johns (Saint-Jean-sur-Richelieu) highway. Both train and car were travelling at high speed and the impact was sufficient to derail the two-coach train. The crossing had warning and 'Stop' signs but neither gates not a wig-wag signal. [1090] [1091]

Sir Basil Blackett, a well-known British economist and civil servant, lost his life in a level crossing collision near the village of Allendorf, Upper Hesse, Germany, on Thursday 15th August 1935. Accompanied by a friend, Fräulein Eisenträger, he had been on his way to the University of Heidelberg, where he was to have given a lecture the following day on "Problems of English Economy." Finding the main road that he wished to take blocked by roadworks he had taken a detour, in the course of which he had driven along a narrow track through meadows and fields until he reached an open crossing with a branch line, guarded only by warning signs. He failed to either observe or hear the whistle of an approaching train, which struck his car and carried it 200 metres. Mr Blackett and Fräulein Eisenträger were taken to Marburg Hospital, where he died of his injuries about three hours later. His companion was seriously injured but there were hopes for her eventual recovery. [1092]

On the morning of Sunday 25th August 1935 a motor coach carrying 40 vintners from Grinzing, the wine-growing suburb of Wien (Vienna), Austria, who had been returning from a visit to the Dollfuss Memorial Church, had to negotiate a level crossing at Piesting station, on the Leobersdorf – Gutenstein secondary line. Only two trains a day used the line on Sundays but one of them, travelling at a moderate speed, collided with the coach, which was overturned and wrecked. [1093]

Seven bus passengers died and 18 were seriously injured. The *Times* correspondent in Vienna reported that among those who had lost their lives was 'Frau Hengl, the wife of the wine garden owner, known to almost every Englishman visiting Grinzing to acquaint himself with the institution of the *Heurigen*, or the testing of the new wine.' [1094]

Six relatives having a family reunion lost their lives in a level crossing accident on the afternoon of Saturday 14th September 1935 when their car was struck by a Reading Company train near Churchville, Pennsylvania, United States. They were Mr George Gaul, 70, a foreman of the New York Shipbuilding Company, his wife Ella, 69; Mr Otto Schoenfeld, 51, superintendent of the Fred Rumpf Sons' tapestry mill, and his wife Mabelle, 48 (the Gauls' son-in-law and daughter); Miss Marguerite Schoenfeld, 28, a nurse at Philadelphia hospital (a granddaughter of the Gauls); and Mrs Jennie Countryman, 41 (another daughter of the Gauls).

[1090] *The Winnipeg Evening Tribune*, Winnipeg, Manitoba, Canada, 12th August 1935
[1091] *Sarasota Herald*, Sarasota, Florida, United States, 11th August 1935
[1092] *The Times*, London, England, 16th and 17th August 1935
[1093] *Le Populaire*, Paris, France, 26th August 1935
[1094] *The Times*, London, England, 26th August 1935

The train was in charge of Conductor N. Taylor with Engineer Fred Rook and Fireman John Saunders. It had been travelling at about 40 miles an hour. The crossing was protected only by a crossbuck and advance warning signs. Prior to crossing the line the car driven by Mr Gaul had to cross a highway running alongside the railroad and it was theorised that he had been concentrating on the traffic on the main road and had been unaware of the train. [1095]

On Sunday 15th September 1935 passenger train P 2915 collided with a lorry on an unprotected level crossing between Kirchheim unter Teck and Oberlenningen, Germany. Five people died and 13 were injured. [1096]

Three adults and two children lost their lives when their car was struck by a Wabash Railroad train on a private level crossing three quarters of a mile west of Peru, Indiana, United States, on the afternoon of Wednesday 16th October 1935. The victims were Guy Haughtington, 24, the driver of the car; his sisters Elizabeth, 26, and Florence, 20; and Florence's son Marion, 4, and daughter Alice, 18 months.

The Haughtingtons lived on a farm close to the level crossing and had been on their way into Peru. To do so they had to cross the railroad to reach state road 24. The line ran on an embankment and the road climbed to the crossing. There was, however, a clear view along the track in both directions. Eastbound train No. 4 struck the car, it was said, with terrific force, and Florence Haughtington and her two children were killed outright. Doctors and three ambulances were sent to the scene and Guy and Elizabeth Haughtington were rushed to hospital, where both succumbed.

Coroner Newell interviewed Conductor C.R. McBride, Engineer Ed Irish and Fireman Fred Donat of the train. The engineer said that he had not seen the car from his side of the engine until immediately before the collision. Fireman Donat testified that he had seen the car a little earlier and had shouted and signalled to his engineer to apply the brakes, but by then the train, its engine whistling, had been practically on the crossing.

Mr and Mrs Thompson, of Brackenridge, Pennsylvania, who had been driving east on the state highway which paralleled the railroad, said that the train had just overtaken them, and they had been watching its progress when the car, a 1926 Packard, had suddenly appeared on the track and been hit by the train. Mr Thompson had been the first to reach the scene and helped in the rescue work until he was almost exhausted, and his clothes were soaked in blood. The father and brother of the adults, Ollie and John Haughtington also soon reached the scene and walked around in a daze as the victims were picked up and taken away. It was impossible to explain why Guy Haughtington had failed to notice the train. [1097]

Shortly after midnight on Friday 18th October 1935 a Cincinnati and Lake Erie Railroad freight train collided with a street car taking workers home on South Main Street

[1095] *The Bristol Courier*, Bristol, Pennsylvania, United States, 16th September 1935
[1096] *Katastrophen der deutschen Bahnen Teil II*, by H.J. Ritzau, pg. 239 (in German)
[1097] *Logansport Pharos Tribune*, Logansport, Indiana, United States, 17th October 1935

Crossing, Dayton, Ohio, United States, injuring five people and shaking up 13 others. [1098] This relatively minor accident was to be a prelude to a more serious one on another Dayton level crossing only a couple of hours later.

At about 2.33 a.m. the gates of the crossing at Third and Conover Street had been lowered for an eastbound Pennsylvania Railroad passenger train, *The Gotham Limited,* in charge of Conductor Bert Bonner, which was running about five minutes late. Despite the lowered gates and warning bells ringing an automobile travelling at high speed failed to stop and smashing through the barrier, was struck by the train.

Olive Hall, 27; Sallie Hunter, 25; John Mangan, 28; Daniel M'Fadden, 26; Mrs Eileen M'Fadden, 23, and Ira Rowel, 28, the owner and probable driver of the car, were all killed or fatally injured. All were from Dayton. [1099]

A collision with Chicago Burlington and Quincy Railroad passenger train No. 15 from St Louis, Missouri to Denver, Colorado, killed five people on a level crossing near the station at Breckenridge, Missouri, United States, on the evening of Tuesday 22nd October 1935. Those who lost their lives were George Dickey and his wife, their grandson Melvin Snow, and Charles Gray and his wife. All were from Breckenridge and had been on their way home from church. Mr Gray was the assistant station manager at Breckenridge.

Station Agent Chris Welch witnessed the accident and said afterwards, "I thought they were going to stop. The road runs along the track for a block before crossing the tracks. Gray was driving abreast of the train, and I don't see how he could help but see it. He slowed just before he came to the crossing and then speeded up right into the path of the train."

The small car was carried three blocks before the train, which had been travelling at some speed, could pull up. The train was delayed an hour and a half. Its engineer was George Bartlett. [1100]

The crossing was unprotected but there was a clear view of the tracks. [1101]

A Buenos Aires Western Railway excursion train to Buenos Aires collided with a micro-omnibus on a level crossing near Carlos Maria Naon, Argentina, on the branch line from Bragado to Timote, at an unstated date in October 1935. The driver of the bus and its five passengers all lost their lives, the vehicle being destroyed. 'An investigation showed the accident to have been due to the recklessness of the driver of the bus, who was an old offendor in this respect. As is usual in country districts where there is little road traffic there were no barriers at the crossing.' [1102]

A lorry carrying 11 passengers as well as its driver was in collision with a train on an unprotected level crossing some 11 miles from Coonabarabran, New South Wales,

---

[1098] *The Evening Independent*, Massillon, Ohio, United States, 18th October 1935
[1099] *The Times Recorder*, Zanesville, Ohio, United States, 19th October 1935
[1100] *The Sedalia Democrat*, Sedalia, Missouri, United States, 23rd October 1935
[1101] *The Messenger*, Meadville, Missouri, United States, 24th October 1935
[1102] *The Railway Gazette*, London, England, 13th December 1935

Australia, on Saturday 9th November 1935. The lorry driver, contractor James Birchell, 40; labourer Edward Forest Sr., 60, Birchell's father-in-law; Michael Harper, 30, and George Mueller, 23, both labourers and both nephews of the driver; Colin Chapman, 7, and his baby brother, aged 5 weeks, were all killed. The other six passengers were injured. They included Mrs Presnell, the grandmother of the two dead children, who had been taking them to see their mother in hospital at Coonabarabran. According to another of the surviving passengers, Edward Forest Jr., the people on the lorry had all been laughing and chatting and no one had seen the train until it was about ten yards from them. The lorry was wrecked. With no ambulance available the injured were taken to Coonabarabran and on another lorry, on which some bedding had been placed. [1103] [1104]

A level crossing between Firchau and Schneidemühl, Germany, was protected by lifting barriers but on Sunday 8th December 1935 the barriers were left open, and a train collided with a bus. Seven people died and 14 were injured. [1105]

The end of 1935 saw another level crossing accident involving a school bus in the United States. The location was Holbrook, New York, not far from Ronkonkoma. Just after 8 a.m. on Friday 20th December 1935 a westbound Long Island Railroad train from Greenport to New York collided with a bus carrying children to Sayville High School. By chance there were only five pupils on the bus when it reached the crossing, one of whom, Eleanor Nagel, had only boarded the bus 40 feet from the crossing. Two of the children, Robert Seekamp, 17, and Alice Bedell, 16 were killed outright, while Alice's sister, Edna, was fatally hurt and died four hours later. Eleanor Nagel was critically injured but eventually recovered while the fifth child, August Ring, escaped with bruises and shock and was well enough to be taken home. The bus driver, August Blaske, 33, was also injured. [1106]

The crossing was unprotected except by a 'Stop, Look, Listen' sign. Engineer L.G. Griffin stated that he had whistled for the crossing, though two witnesses denied hearing any whistle. [1107]

Driver Blaske was arrested on a charge of second degree manslaughter, but the case apparently never came to trial.

At a Public Service Commission investigation into the accident Supreme Court Justice John Vunk criticised the lack of protection at the crossing, declaring that, "If the Long Island Railroad cannot be operated in a safe manner it should not be operated at all." In reply, counsel William Colton for the railroad pleaded that the company had sustained a deficit of $1,407,800 during the previous year. [1108]

[1103] *Mudgee Guardian and North-Western Representative*, Mudgee, New South Wales, Australia, 11th November 1935

[1104] *Albury Banner and Wodonga Express*, Albury, New South Wales, Australia, 15th November 1935

[1105] *Katastrophen der deutschen Bahnen Teil II*, by H.J. Ritzau, pg. 239 (in German)

[1106] *New Castle News*, New Castle, Pennsylvania, United States, 20th December 1935

[1107] https://patch.com/new-york/sachem/tragic-train-crash-remembered

[1108] *The County Review*, Riverhead, New York, United States, 5th March 1936

Schoolteacher S.P. White, of Flora, Illinois, his wife Irene, their children Phillip, 12; Mary Ellen, 9; and Jean, 6; and White's uncle Hanson Montgomery, lost their lives when the car they were in was struck by a Baltimore and Ohio Railroad passenger train No. 1, *The National Limited*, near the station at Salem, Illinois, United States, at 2.18 p.m. on Boxing Day, Thursday 26th December 1935. The impact threw the car 35 feet against the side of station building and the baggage room was set on fire. The Salem fire brigade extinguished the blaze before much damage was done.

Station Agent C.M. Erwin witnessed the accident. He said, "It looked to me as if the car was trying to beat the train to the crossing." [1109] At the inquest he testified that the warning signal at the crossing had been operating. The car had been travelling at 30 or 35 miles an hour with its driver apparently unaware of the train. The inquest jury returned a verdict of 'Accidental Death'. [1110]

It was theorised that a level crossing accident at Grabill, Indiana, United States, late on Friday 24th January 1936, might have been due to a car being deliberately stopped on the crossing. The circumstances of the accident were especially sad. Marion McBride, 43, was a Works Progress Administration (W.P.A.) worker. (The W.P.A. was an unemployment relief programme set up the previous year by President Roosevelt to help the millions of unemployed Americans during the Great Depression.) [1111] Mr McBride, his heavily pregnant wife Susan, 40, and their nine children, lived in a chicken coop on a farm in great poverty and had apparently suffered much in the intensely cold weather of the days preceding the accident. The family had travelled to Fort Wayne on the Wednesday to attend the funeral of Mrs McBride's brother, William Connor, and their return journey had been delayed by a snowstorm.

The small 10-year-old car stopped at a store 200 feet from the tracks at Grabill and Mr McBride went in and bought some candy (sweets) for the children. He then set off again and the car reached the crossing at the same time as a Wabash Railroad train.

Grabill Station Agent George King saw the accident from the station platform a few yards away. He said, "McBride halted the car fifty feet back from the tracks with a clear view of the right-of-way. Why he did so I do not know, but he started his car forward despite the fact that the train was already bearing down on the crossing and reached the tracks simultaneously with the locomotive." [1112]

Engineer W.F. Rowley said, "The McBride automobile appeared on the tracks just as I blew the whistle. My vision was obstructed, and I did not see it before then. It was impossible to stop."

Fireman Vernon Myers probably had a better view from his side of the engine. He related, "The train was travelling sixty miles an hour when it struck the McBride machine. I was stoking coal when I looked out and saw the automobile on the crossing. We applied the brakes, but it was too late. Wreckage of the car and human bodies were

---

[1109] *Alton Evening Telegraph*, Alton, Illinois, United States, 27th December 1935

[1110] *Centralia Evening Sentinel*, Centralia, Illinois, United States, 28th December 1935

[1111] Wikipedia – Works Progress Administration

[1112] *The Hammond Times*, Hammond, Indiana, United States, 25th January 1936 scattered 1,000 feet along the right-of-way."

Another railwayman at Grabill, Glen Roth, suggested an alternative explanation for the accident. "The windows and windscreen of the McBride car were frosted, and this might have obstructed the driver's vision. The car seemed to stall on the tracks after making a quick climb up the steep grade leading to the crossing." [1113]

Mr and Mrs McBride and their children Marion Jr., 12; Mary, 10; Arthur, 8; Virginia, 6; Robert, 3; and twins Richard and Ramona, 17 months, died at the scene or on the way to hospital. The accident also resulted in Mrs McBride giving birth, but the child was either stillborn or died immediately afterwards, according to different accounts. Clara McBride, 14, died in hospital ten hours after the accident and Phyllis McBride, 5, the last survivor, succumbed the next day. [1114]

Coroner Walter Kruse would not comment on the suicide theory, which could, of course, not be substantiated.

An indirect casualty of the accident was Conductor Frank Bray, who died of heart disease two days after the collision. Physicians declared that the shock of the event had aggravated his condition. [1115]

In early February Grabill citizens were urged to sign a petition calling for warning lights at the wholly unprotected crossing. [1116]

A minor school bus/train accident happened just before midnight on Friday 31st January 1936 at Kenton, Ohio, United States. The bus was carrying 22 Dola High School pupils home from a basketball game at Ridgeway. Bus driver Irwin Rodabaugh said that he had mistaken a siding to the Ohio Power Company for the main line and the rear of his vehicle was struck by a slow-moving New York Central Railroad freight train. One pupil suffered cuts. [1117]

Five residents of Savannah, Georgia, lost their lives in a level crossing accident at Port Wentworth, Georgia, United States, on Saturday 1st February 1936. The collision was clearly the result of the reckless driving of Bert Varnedoe, one of those killed. His car collided in the early morning with the ninth car of a Savannah and Atlantic Railway freight train, which was backing over the crossing, protected by a flagman. The car driver had been braking hard when his vehicle struck the train. The hood of the automobile went under the waggon and its windscreen smashed into it. It was dragged a short distance by the train. The others who died were Fleming Thacker, Robert Stewart, his wife Elizabeth, and Mrs Ruth Farris. Two other people in the car, Aaron Fine and Miss Margaret McElvain, were injured. The party had been en route to an amusement centre in nearby South Carolina. [1118]

[1113] *The Vidette Messenger*, Valparaiso, Indiana, United States, 25th January 1936
[1114] *The San Antonio Light*, San Antonio, Texas, United States, 26th January 1936
[1115] *Somerset Daily Herald*, Somerset, Pennsylvania, United States, 27th January 1936
[1116] *The Hammond Times*, Hammond, Indiana, United States, 5th February 1936
[1117] *The Republican-Courier*, Findlay, Ohio, United States, 3rd February 1936, United States
[1118] *Thomasville Times-Enterprise*, Thomasville, Georgia, 1st February 1936, United States

Eleven people, two of them children, died, and 16 were injured when a goods train collided with a bus near Aguascalientes, Mexico, on Thursday 6th February 1936. [1119]

A train collided with a bus on an open level crossing between Celle and Gliesmarode, Germany, on Saturday 8th February 1936. Six people died and 15 were injured. [1120]

Japan suffered a serious level crossing accident just after 9 p.m. on Monday 10th February 1936 at Okamoto on the Osaka – Kobe – Hankyu electric railway. The crossing was of class B, which meant that it was manned by a crossing keeper between the hours of 5 a.m. and 9 p.m. During the night-time it was unmanned but protection was provided by an automatic alarm bell and a flashing light signal. The accident occurred just after 9 p.m. and so the crossing keeper had gone off duty. The accident involved a train and a fire engine and resulted in the death of 16 firemen and serious injury to six others. Also badly hurt was the motorman of the train. The train had been travelling at 100 kilometres an hour and the fire engine was dragged 200 metres. One man, however, had a very fortunate escape, as he was thrown clear and escaped unhurt.

The inquiry into the accident found that the audible and light signals had been functioning correctly and visibility was good but there was a presumption that the firemen had been supposing that the crossing keeper was still on duty, and not seeing any railway official had assumed that it was safe to cross. As an additional protection it was decided that in future B class crossings would be fitted with "No crossing keeper on duty" signs.

According to the normal procedures, once a fatal accident had occurred all traffic over the section of railway on which the accident happened must be suspended until a preliminary investigation had been completed by the police. In this instance the Up line had not been affected and the railway officials initially attempted to maintain a train service with single-line working over the Up line. A very serious exception was taken to this course of action by the local government authority, and the train running had to cease.

The day after the disaster a gift of 1,300 Yen (roughly £75) was made to each of the families of the dead firemen. This was not regarded in any manner as a legal admission of responsibility by the railway company but simply as a gesture by a wealthy concern in a situation such as this to provide relief to the less well-off victims of accidents, to tide them over during the period that inquiries were taking place and while awaiting possible compensation. [1121]

Two months after the accident a decision was announced to eliminate 2,000 level crossings throughout the country. [1122]

[1119] *The Times*, London, England, 7th February 1936

[1120] *Katastrophen der deutschen Bahnen Teil II*, by H.J. Ritzau, pg. 239 (in German)

[1121] *The Railway Gazette*, London, England, 3rd April 1936

[1122] *Sunday Tribune*, Singapore, 17th May 1936

At 23:40 on Thursday 19th March 1936, a lorry carrying 43 labourers broke down near Cairo, Egypt, on a level crossing at a location in Giza Province at a location variously given as Nikla and 'between Berkash and Zat-el-Kom.' A ballast train then ran into the lorry, killing 12 of the men and seriously injuring 17 more. The violence of the collision was demonstrated in that three of the victims were stated to be 'missing', the bodies being so mangled that they could not be identified. [1123] [1124] The crossing was provided with a keeper between sunrise and an hour after sunset but was left unprotected at night, though there were warning boards marked 'Beware of Trains' and according to a Ministerial Decree all road vehicles were compelled to come to a stand before entering the crossing and their drivers were to ensure that it was safe to do so. The lorry driver, who survived the accident, was arrested. [1125]

Achiel Verhecke, his wife (Christian name not stated), their daughters Georgette, 13; and Mairyette, 10; and a friend, Maurice Van Den Hende, 22, who was driving, were all killed when their car collided with a Wabash Railroad engine and caboose on a level crossing two miles from Delhi, Ontario, Canada, on Friday 20th March 1936. Although the car was badly damaged in the collision its motor remained running after the impact. [1126]

At 4.15 p.m. on Wednesday 25th March 1936 five residents of Lansing, Michigan, United States, were just starting a return trip from a church meeting in a temple house at nearby Haslett. Meanwhile, also approaching Haslett was the westbound Inter City Limited Montreal – Chicago express of the Grand Trunk Western Railroad, in charge of Conductor Orson McAuany with Engineer Louis Strickland and Fireman A.W. Silver. As the express reached the crossing at Haslett the automobile crossed directly in front of it but did not quite make it before the locomotive caught the rear of the vehicle. The car was thrown against a signal post on the far side of the tracks and its occupants were killed. They were Mr Mike Travick, 52, the probable driver of the car, his 48-year-old wife, Mrs Helen Puagash, 42; Mrs Betty Schodisky, 40; and Mrs Helen Trus, 44.

Two additional victims of the accident were 15-year-old schoolgirl bystanders Doris Mock and Helen Goff, who were struck by flying wreckage. Miss Goff was only slightly hurt but her friend had to undergo the amputation of her left leg, while her right leg was fractured. Her condition in hospital the next morning was described as 'fair.'

The police theorised that bright sunlight had prevented the car driver from seeing the flashing warning light at the crossing. [1127] [1128]

Byron, Georgia, United States, was the scene of a train/bus collision on Saturday 28th March 1936 which cost 11 lives. The bus, owned by Southern Stages Inc. was northbound from Americus to Macon, while Central of Georgia Railroad train No. 93, the

---

[1123] *The Times*, London, England, 21st March 1936

[1124] *The Helena Independent*, Helena, Montana, United States, 20th March 1936

[1125] *The Railway Gazette*, London, England, 10th April 1936

[1126] *The Brandon Daily Sun*, Brandon, Manitoba, Canada, 24th March 1936

[1127] *Marshall Evening Chronicle*, Marshall, Michigan, United States, 26th March 1936

[1128] *The Escanaba Daily Press*, Escanaba, Michigan, United States, 26th March 1936 westbound *Southland* express, was en route from Florida via Macon to Chicago, Illinois.

The bus was a White-Bender streamline Inter-City coach with a seating capacity of 21 persons, which was being operated by Driver O.H. Looney. The train consisted of engine No. 437, two express cars, a combination baggage car and coach, and a coach, in charge of Conductor J.H. Howell with Engineer C.H. Hall and Fireman Woods.

Arriving at Byron about three quarters of an hour late the bus stopped on Main Street on the south side of the railroad, then crossed the line via the Railroad Street level crossing to its regular stopping point by a filling station, discharged some passengers, reversed, and was re-crossing Railroad Street at low speed when it was struck by the train. The bus was demolished and carried 480 feet down the line, while the whole train was derailed. All the casualties, however, were on the bus. Ten passengers and Driver Looney were killed outright or fatally injured and the only other two passengers on the bus were seriously hurt.

The accident was investigated by the Interstate Commerce Commission (ICC). Engineer Hall told the commission that he had been whistling as his train reached Byron, using first the steam whistle and then the air whistle. Two passenger car lengths from the crossing he had seen the bus moving slowly onto it and had immediately shut off steam and applied the brakes in emergency. His engine headlight had been burning brightly, its bell had been ringing and the brakes had been working well. The brakes had reduced the speed of his train from 40 to 35 miles an hour when it hit the bus. He was aware that according to the timetable the speed of trains was restricted to 8 miles an hour. He had never obeyed this restriction since he had started the run in October 1924.

Fireman Woods testified that he had turned on the bell ringer before reaching the first of the three Byron level crossings. He said that he had not been able to see the bus from his side of the locomotive. He agreed that the 8 miles an hour restriction through Byron was never observed and added that had he been driving he would not have observed it either.

Conductor Howell said that he had been looking for the train order board when the train's brakes had been applied in emergency. He thought that he had seen a bulletin raising the speed restriction through Byron from 8 to 20 miles an hour. He did not recall any train crew being criticised for going faster than that.

Operator Lamar at Byron stated that he had gone outside to watch the express pass when he had heard its engine whistle. As he had emerged he had seen the bus moving towards the crossing and had begun to wave to its driver to stop. There had been no reaction from the bus driver, and he had then gone right in front of the bus, waving a white handkerchief, until the bus had been within two or three yards of the crossing and clearly not stopping, when he had had to leap out of the way to save himself.

Section Foreman Davis said that he had witnessed the collision from a store on Main Street. He thought at first that the bus would stop, then it appeared to speed up slightly and the collision had occurred. As a town councillor he had been aware of the 20 miles an hour restriction through Byron, but not of the earlier 8 miles an hour ordinance.

Superintendent Smith stated that local ordinances regarding speed restrictions through towns, which were shown in the working timetable, were not expected to be obeyed by train crews unless complaints had been made. The 8 miles an hour restriction shown in the timetable would be corrected to 20 miles an hour in the next issue.

Many other witnesses gave evidence, most saying that the train had whistled, the engine headlight had been burning brightly and the engine bell had been ringing. Likewise they agreed that the bus had failed to stop before reaching the crossing and a number had seen the operator attempting to flag it and the warning lights at the crossing flashing. All said that the crossing was well lit and maintained and that the train could have been seen 1,000 feet away.

Driver Looney survived for a short time after the accident but was not in a condition to make any statement. The doctor who examined him noted that there was evidence that he had been drinking. He was stated by President Wilkinson of Southern Stages Inc. to have driven carefully on the many occasions that the president had ridden with him, but the ICC noted that Looney had had several previous jobs as a bus driver in the preceding years and had been involved in several accidents, having been suspended for 15 days for one accident and sacked for another mishap with a different company.

It appeared that before December 1935 buses had stopped at the filling station to the north of the railroad, then used Railroad Crossing once to continue their journeys. At that time, a new highway had been opened running alongside the railroad and to the south of it for many miles. From then the buses had crossed and recrossed the railroad to get to and from their old stopping point at the filling station. Since the accident it had been arranged for them to stop at a point on the new highway, thus eliminating the need to cross the railroad at all.

Discussing the accident the ICC stated that tests carried out with a similar bus showed that the bus driver had an uninterrupted view of the track for over 1,000 feet, plus he was required by State Law to stop at all railroad crossings. It was noted that the engineer of the train should also have been able to see the bus from a much greater distance than he had said.

Predictably, the commissioners were not very happy with the superintendent of the railroad permitting trainmen to ignore speed restrictions set down in the timetable, commenting that, 'Time-table speed restrictions are supposed to be observed, and to allow open violations of any of them is to invite violations of other speed restrictions and of rules and regulations in general. It should not be left to the employee to find out for himself what rules are to be observed and what rules are to be violated with impunity, and any rules which are not to be observed at all times and under all conditions should be eliminated.'

The ICC went on to note the similarity of the Byron disaster with that at Rockville less than a year earlier (see above) and commented that 'it is unfortunate that this story of failure has to be repeated again and again.'

In conclusion the commission found that the accident had been caused by the failure of the bus driver to stop before negotiating the crossing as required by both the state law and the bus company rules. Contributory causes were the failure of the local authority to enforce an ordinance limited train speeds through Byron to 20 miles an hour and the failure of the railroad company to require its employees to observe the ordinance. [1129] [1130] [1131]

[1129] Interstate Commerce Commission report

[1130] *The Brownsville Herald*, Brownsville, Texas, United States, 29th March 1936

[1131] *Thomasville Times-Enterprise*, Thomasville, Georgia, United States, 30th March 1936

Willow Street Crossing, Bloomington, California, United States, near Colton, was the scene of a level crossing accident on Monday 6th April 1936 when a car containing five Mexicans was struck by an eastbound Southern Pacific Railroad passenger train, *The Apache*. Driven into the path of the speeding train the car was hurled 50 feet and destroyed, its occupants being all killed. Two of the men were identified as Conception Quiroz, of Anaheim, the driver of the vehicle, and Sol Fernandez, of Los Angeles. The others could not immediately be identified. [1132] [1133]

At an unstated date in about April 1936 a bus stalled on a level crossing at Hsingchi, China, and was struck by a Tientsin – Pukow express. Eighteen people, presumably bus passengers, were reported killed and 23 seriously injured. [1134]

Another level crossing accident due to a car which was being driven too fast being unable to stop when it came to a train going across the road, happened at West Austintown Crossing, Ohio, United States, on Saturday 16th May 1936. Guy Patrous, 25, had acquired a new machine from a dealer earlier the same day. He had attended the wedding of Augustine Patrous and Marjorie Carrozzino at St Dominick's Church in nearby Youngstown and afterwards had gone on a joyride with six other teenage and young adult men. Witnesses J.B. Cox and Henry Henning saw the automobile approach the crossing at high speed and noted that the warning lights and bell had been operating. Henning, who lived near the crossing, said, "I saw the car coming down the road a-whizzing. The car started to put on its brakes. The tires screeched. It almost turned over before it hit." The impact derailed one or more waggons of the Erie Railroad train which was being shunted on a siding. Some of the victims were thrown from the automobile, the bodies of others were wedged in the wreckage of car and waggons. John Carrozzino, 12; and Dominick, Patrick, and Peter Pachell, aged 19, 24, and 29 respectively, were killed outright. Sam Carrozzino, 18, Guy Patrous died in Youngstown City hospital several hours after the crash and Carl Pachell, 24, lingered a little longer before becoming the seventh fatality. [1135] [1136] All had been relatives of the bride. [1137]

Four women and two children, on their way to decorate graves for Memorial Day (31st May), lost their lives in a level crossing collision at Calumet City, Illinois, United States, on Thursday 28th May 1936. The car was close to Holy Cross Cemetery, its destination, when it was driven into the path of Pennsylvania Railroad local train No. 43, from Logansport to Chicago, killing Miss Mary Trembczynski, 24, the driver of the car; Lilian Cichy, 24; Patsy Fleming, 2; Mrs Eleanor Kunka, 29; her son Daniel, 3; and Regina Luczak, 24. It was left to Mary Trembczynski's father, Joseph, owner of a Calumet City tavern, to identify the bodies. He said that his daughter had been driving for seven years

---

1132 *Morning Herald*, Hagerstown, Maryland, United States, 7th April 1936

1133 *The Helena Daily Independent*, Helena, Montana, United States, 8th April 1936

1134 *The Railway Gazette*, London, England, 5th June 1936

1135 *The Charleston Gazette*, Charleston, West Virginia, United States, 17th May 1936

1136 *The Portsmouth Times*, Portsmouth, Ohio, United States, 17th May 1936

1137 *The Salem News*, Salem, Ohio, United States, 18th May 1936 and had never been in an accident before. The train was in charge of Conductor George Fahnestock with Engineer Frank Patterson. The crossing was protected by flashing lights and there was a good view along the lines. [1138] 'Plants which the women had planned to place on graves were scattered with the bodies along the rails. The automobile was destroyed.' [1139]

At 01:00 on Monday 22nd June 1936 a coach carrying 28 passengers was struck by a train on the Boucheville level crossing, 500 metres from the station at Bernay, France, on the Paris – Cherbourg line of the Chemins de fer de l'Ouest. Seven of those on the coach were killed and 16 seriously injured. The passengers on the coach were mostly members of the fire brigade and their families from Mesnil-sur-Estrée, who had been on a visit to Cabourg, on the Normandy coast. They were travelling in two coaches. When these vehicles reached Boucheville the barriers were lowered for Caen – Paris train No. 2376. Crossing Keeper Diesnis, who had only been doing the job a month, thought that there was time to let one vehicle across before the train arrived, not realising that there were in fact two coaches. The first coach made it across the line, the second followed and was hit by the train travelling at 80 kilometres an hour. Those who lost their lives included M. Dézillon, the coach driver.

The police opened an inquiry, and the crossing keeper was arrested on a charge of manslaughter through negligence. [1140]

M. Diesnis made a statement after the accident. "Alerted by a horn, I went out to the road and saw that there was a coach that wanted to cross. Was there time to let him over before the train got there? I thought that there was.

"We are warned of the approach of a train by a treadle over two kilometres from the crossing which is operated by a passing train. An indicator at the crossing which displays a white light when there is no train is approaching displays a red light on the depression of the treadle, plus a warning buzzer sounds.

"When I left the gatehouse the indicator was still showing a white light and I thought that there was only one vehicle. When I found that there were two coaches I knew that there would be an accident. I tried to signal to the driver of the second coach to stop but it was too late."

The express was running two minutes early. It also appeared that the warning indicator had failed to work; this was a matter which would be investigated. [1141]

At 11 a.m. on Saturday 27th June 1936 the driver of a sedan travelling west on Route 20, the Cherry Valley Turnpike, apparently attempted to beat a Delaware Lackawanna and Western Railroad train over a level crossing at Sangerfield, New York, United States. He failed to make it and the six occupants of the car lost their lives. [1142] They were later identified as Mr Oliver Andreotta, his wife Elena and their son, Bruno, 14, and Mr John

---

[1138] *Cedar Rapids Gazette*, Cedar Rapids, Iowa, United States, 28th May 1936
[1139] *Logansport Pharos Tribune*, Logansport, Indiana, United States, 28th May 1936
[1140] *L'Homme Libre*, Paris, France, 22nd June 1936
[1141] *L'Indépendant des Basses Pyrénées*, Pau, France, 24th June 1936
[1142] *Syracuse Herald*, Syracuse, New York, United States, 27th June 1936

Bettio, his wife Rosino and their daughter Lillian, 18 months, all from New York city. [1143]

Two Pennsylvania families were wiped out when the large sedan in which they were riding was struck by an Ann Arbor Railroad train on a level crossing at Dundee, Michigan, United States, on Saturday 18th July 1936. Mr Harry Schuster, his wife, and their daughters Vina, 19; Anna, 12; and Sandra Jean, 6 months, had just started home after a visit to Mr Carl Schuster at Dundee. Accompanying them had been Mr Sam Pisano, his wife, and their daughters Betty, 9, and Mary, 6. Mrs Schuster and Mrs Pisano were sisters.

The Toledo to Ann Arbor train had been running at high speed. The train dragged the automobile 75 feet to a high trestle across the Raisin River. The bodies of the victims were thrown 30 feet into the shallow river, while the wreckage of the car caught fire and in turn set fire to the trestle, though the blaze was soon extinguished.

Ray Wilson, who worked at a nearby filling station, was the first on the scene. He said, "I heard a terrific crash and rushed out. I saw the bodies in the river, and the flaming wreckage scattered along the trestle and far down the right of way. Parts of the car dropped off each side of the bridge and the bodies were hurled in all directions. All of them fell in the river in about a foot and a half of water."

Also early on the scene was Carl Schuster, who collapsed when he discovered that it was the bodies of his relatives which were being recovered from the river.

The crossing was protected by warning lights, which were found to be operating properly after the collision. The highway (M-50) climbed sharply on the approach to the crossing and buildings partially obstructed the view at the crossing. Crew members stated that the train had been travelling at 45 miles an hour when it hit the car. [1144]

Hamilton, Ohio, United States, was the scene of a level crossing collision with multiple fatalities on Sunday 26th July 1936. A car containing ten members of a family left the Butler County Sportsmen's Picnic at the Family Gardens on Federle Road at 5.30 p.m. Less than a minute later the vehicle was struck by a northbound Pennsylvania Railroad passenger train running between Cincinnati, Ohio, and Grand Rapids, Michigan. The car was carried a quarter mile and then thrown into a ditch. Mrs Edna Seward, 54; Norman Schweitzer, 27, the driver of the car; his wife Eva, 28, daughter of Mrs Seward; the Schweitzers' children Charlotte, 3, and Edward, 1; Mrs Vogan, 33, daughter of Mrs Seward; Mrs Vogan's son Charles, 3; Patricia Jennewein, 7, granddaughter of Mrs Seward; and Harland Seward, 3, grandson of Mrs Seward, were all killed or fatally injured. Seriously injured was Mrs Driscilla Jennewein, daughter of Mrs Seward and mother of Patricia.

At least two people witnessed the accident. Karl Bley said, "I was sitting on my front porch when the accident occurred. The car was driven onto the tracks and was struck in the center and carried along the tracks by the cow catcher of the engine."

Another eyewitness was Lieut. W.O. Clark, who had been driving along the Dixie

[1143] *The Kingston Daily Freeman*, City of Kingston, New York, United States, 29th June 1936
[1144] *Ironwood Daily Globe*, Ironwood, Michigan, United States, 18th July 1936

highway and had helped loaded the bodies into the ambulance. "I was an aviator in the World War, but that crash was one of the worst tragedies I have ever witnessed. One of the little girls, whose body I found about 200 ft from the crossing, had a doll in her arms. I couldn't determine if the train whistle was sounded as the radio in my automobile was turned on at the time."

The train was in charge of Conductor J.L. Nelson, with Engineer C.B. Garling and Fireman F.A. Mosel. Conductor Nelson put the speed of the train at the moment of impact at 35 miles an hour and said that the engine whistle had been sounded. Coroner Cook disagreed as to the train's speed, stating that the distance that the car had been carried suggested a speed of 60 miles an hour, and added that the matter would be further investigated. The coroner also described the crossing as dangerous, as it lay on a curve, and stressed the importance of installing a flasher signal, implying that it did not even have this degree of protection at the time of the collision. [1145]

Oconomowoc, Wisconsin, United States, which had been the scene of a bad level crossing accident in January 1935 (see above) suffered another bad one on Sunday 26th July 1936. A car being driven by Rev. Waldemar Dobratz, 28, and containing his brother Everett, 22, his sisters Norma, 18, and Verna, 16; his fiancée Helma Gartzke, and Ralph Rubel, 17, was struck by two Chicago Milwaukee St Paul and Pacific Railroad trains. The party had been hurrying to a band concert in which Waldemar Dobratz and Ralph Rubel were to play cornets. Travelling at about 35 miles an hour the automobile had crossed the paths of the two slow-moving trains, one of which had been pulling into the depot and the other just leaving it. The collision happened on a different level crossing to the 1935 one, this time the one with Concord Street. [1146] Newspaper reports do not indicate what protection, if any, there was at the crossing.

'It was a merry party of six people who started off from Blairmore on Sunday afternoon at about 2.30 to attend a wedding feast at Hillcrest,' a newspaper reporter wrote. 'Before nightfall all six were dead.' The party were riding in a car being driven, it was believed, by Mike Panck, 45, and in addition to Mr Panck comprised Joseph Kubik Jr., 42; Miss Anna Kubik, 21, his niece; Miss Katherine Kubik, 27, his sister-in-law; Miss Angelina Hovan, 19; and Miss Lillian Marcial, 23. They were on their way to celebrate the wedding of Mr and Mrs Lusaski the previous day. For some unknown reason Mr Panck took an old, little used road rather than the recently opened highway and the car he was driving was struck by eastbound Canadian Pacific Railway train No. 12 from Nelson to Lethbridge. The crossing, at Bellevue, Alberta, Canada, was a blind one, and visibility was further impaired on the day of the accident, Sunday 9th August 1936, by smoke from fires which had broken out in the nearby Crow's Nest forest reserve.

Superintendent W.H. Ruthven issued a statement regarding the accident. This indicated that the train had consisted of seven cars and had been in charge of Conductor Sam Jones, with Engineer W.J. Smith. Engineer Smith had sounded two long and two short whistles at the whistle board, made a light brake application and started the bell ringing. At a point about 100 feet from the crossing he had whistled again and at that

---

[1145] *Hamilton Daily News Journal*, Hamilton, Ohio, United States, 27th July 1936
[1146] *Manitowoc Herald-Times*, Manitowoc, Wisconsin, United States, 27th July 1936

point his fireman had told him that a car was approaching the crossing from the north. Realising then that the car was not stopping his fireman had shouted for him to apply the brake in emergency. The collision had followed almost at once. [1147]

"It was the most gruesome sight that I have ever seen." Thus did Emile Ferron, Liberal Member of Parliament for Berthier-Maskinonge, describe the scene at Louisville, Quebec, Canada, in the aftermath of a level crossing collision in which a 60-car Canadian Pacific Railway fast freight train collided with a farm truck packed with people. Both the M.P. and the occupants of the lorry had been returning, on Saturday 15th August 1936, from an election meeting at St Justin (the Quebec general election was held on 17th August 1936).

Mr Ferron continued, "We {Mr Ferron and L.J. Thisdel, M.L.A.} were returning from in my car and stopped as the freight train went by. We saw the truck, which was loaded to capacity. It was going about 40 or 45 miles an hour and as the freight train rushed on, it crashed into the middle of the truck. The truck turned over and burned into flames as bodies were hurled as much as 150 feet from the tracks.

"It was a horrible sight. Shrieks of the injured and dying mingled with the relatives, who quickly gathered on the scene, hampering our efforts at rescuing the victims. I saw one body without a head and three other bodies split in two. There were dead bodies, arms and legs piled up along the tracks. It was ghastly. Relatives of the victims are searching all along the line for their own. You see many of the bodies are so badly dismembered they are not recognisable. The people from here rushed 25 miles away to Three Rivers, where the injured were removed to St Joseph and Norman and Cross hospitals."

Textile mill worker, Fernand Tousignant, 17, a passenger on the truck, survived the accident. He said, "I was in the back of the truck. I heard voices in the front saying: 'Don't pass, don't pass,' but still I did not notice the train. At first I paid no attention to the shouts, but when I heard them repeated I figured for myself something must be wrong. So I pushed my way to the side in a hurry and jumped over."

Seventeen men and boys were killed more or less outright in the collision and five others soon died of their injuries. Another twelve were hurt to a greater or lesser degree. Four others had, like Fernand Tousignant, jumped from the truck and then witnessed the collision from the ground. The lorry was carried 50 feet, deposited in a ditch, and caught fire. Edmond Houle, 45, owner and driver of the truck, was one of those killed. [1148]

The inquest on the accident was barely reported in the press but one newspaper mentioned that the coroner's jury had recommended that level crossings should be abolished 'wherever possible' to prevent a repetition of the tragedy. [1149]

Farmer John Fields of Adamsville, near Zanesville, and his sons Raymond, 18; Cletus, 13; Clarence, 11; and Russell, 5, were all killed when a Pennsylvania Railroad passenger train struck their car on a level crossing two and a half miles east of Coshocton, Ohio,

---

[1147] *The Lethbridge Herald*, Lethbridge, Alberta, Canada, 10th August 1936

[1148] *The Winnipeg Evening Tribune*, Winnipeg, Manitoba, Canada, 15th August 1936

[1149] *The Drumheller Mail*, Drumheller, Alberta, Canada, 27th August 1936

United States, early on Friday 4th September 1936. The family were returning home from a melon patch owned by Mr Fields when their old sedan was struck by the train, possibly after it had skidded onto the crossing. The vehicle was carried about 75 feet and crushed, the bodies of the victims being badly mangled. [1150]

Labour Day Canada, celebrated on Monday 7th September in 1936, was marred by a level crossing accident which cost six lives, including three generations of one family. A Canadian Pacific Railway passenger train from Megantic to Sherbrooke collided with the car occupied by the Jacques family and brothers Leo and Henry Truhan at McLeod's Crossing, near Scotstown, Quebec, Canada. The two brothers, Thomas Jacques, 61, his 51-year-old wife, their son Alfred, 20, and a two-year-old grandson were all killed or fatally injured. A 12-month-old granddaughter was seriously injured but it was thought she would recover. [1151]

On Tuesday 8th September 1936, a train on the Ferrocarril Interoceánio de México collided with a bus on a level crossing at San Martin Texmelucan de Lambastida, in Puebla state, Mexico. The vehicle was described as having been pulverised by the train and it was also set on fire. Seven lives were lost, and five people were seriously injured. [1152]

A car travelling along a lane which had been blocked off by construction workers building an overpass was hit by a Michigan Central Railroad train near Inkster, Michigan, United States, on Sunday 13th September 1936. It was assumed that Mr John Talanshar had taken that route to avoid heavy Sunday traffic on the diversion which had been set up. Mr Talanshar, 39, of Ecorse, Michigan; his wife Irene, 39; their children Dorothy, 12, and Joseph, 5; Mr Talanshar's sister, Mrs Mary Jantoshik, 35; and Mr Talanshar's mother Mrs Anna Talanshar were all killed. The family had been en route to the nearby Eloise hospital where Mr Jantoshik's husband was a patient. [1153]

At 18:00 on Monday 21st September 1936 a bus en route from Martignes to Marseille was struck by a regional train shortly after starting its journey at the Talan level crossing at Châteauneuf-les-Martignes, France. The shock of the collision was very violent, the bus was 'reduced to crumbs' as a French newspaper put it, and 10 bus passengers were killed and about 26 injured, several very badly. A previous accident had occurred only a short time earlier at the same level crossing, which was protected only by a warning bell.

The bus was a large new vehicle with a capacity of 40 persons. It was on its regular run from Pont-Saint-Louis-du-Rhône to Marseille when it was struck by the train, which had left the Pas-des-Lanciers a few moments earlier. Visibility had been poor on the evening of the accident, with dust and fog. The bus had been already halfway across the line when its rear was struck by the train. The train driver did what he could to stop when

[1150] *Steubenville Herald-Star*, Steubenville, Ohio, United States, 4th September 1936
[1151] *The Lethbridge Herald*, Lethbridge, Alberta, Canada, 8th September 1936
[1152] *Le Grand écho du Nord de la France*, Lille, France, 9th September 1936
[1153] *The Bradford Era*, Bradford, Pennsylvania, United States, 14th September 1936

he saw the bus but it was too late. The front of the engine was also badly damaged. [1154]

On the evening of Thursday 1st October 1936 an Amsterdam – Groningen passenger train collided with a lorry carrying cylinders of compressed gas on a level crossing near Zwolle, the Netherlands. Following the collision the engine and the leading carriages of the train were enveloped in flames. The driver succeeded in stopping his train. Along with his fireman he then jumped into the ditch bordering the track, both with their clothes on fire. There was a moment of panic amongst the passengers, but all escaped injury. Both lines were blocked. The lorry driver was killed. [1155]

Six people lost their lives in an accident at Dayton, Ohio, United States, at 2.28 a.m. on Sunday 18th October 1936. The victims were: Ora May Hall, 27; Sally Hunter, 24; John Mangan, 24; Dan McFadden, 26; his wife Eileen, 24; and Ira Towles, 28. The car in which they were riding was struck by the *Gotham Limited* express of the Pennsylvania Railroad, en route from St Louis, Missouri, to New York, and carried 300 feet. The accident was entirely the fault of Towles, the driver of the automobile, who had ignored all the signals to stop and driven his car through the lowered crossing barriers into the path of the train. [1156]

The *Taurus Express* from Istanbul collided with a motor bus running between Brousse and Eskishehr near Eskishehr, Turkey, on Tuesday 27th October 1936. Sixteen or 17 bus passengers lost their lives and about 18 were injured. [1157] [1158]

A party of six children returning home from an outing in the country, and the adult driving their car, were killed on Friday 30th October 1936 when the car was struck by a Southern Railway fast freight train at Rockmart, Georgia, United States. The adult was Mrs Myrtle Ruth Maudlin, 27, a cotton mill employee, and the party had been on a visit to Mrs Maudlin's mother, Mrs Jesse Maxwell. The children were Mrs Maudlin's son Howard, 10; four children of Mr and Mrs W. Hill, the local Baptist minister: Sarah, 15; Eva, 12; Virdie Lee, 10; and Emma Jean, 7; and Frances Wilson, 12, daughter of Albertus Wilson, of Rockmart.

Mrs Maudlin's car had stalled on the track at the crossing, which was situated at the top of a small hill. The train struck the car in its centre, carried it 50 feet and threw it into a ditch. Witnesses said that the train crew had had no chance of avoiding the accident. [1159] [1160]

[1154] *L'Indépendant des Basses Pyrénées*, Pau, France, 23rd September 1936
[1155] *Le Grande écho du Nord de la France*, Lille, France, 3rd October 1936
[1156] *The Evening Gazette*, Xenia, Ohio, United States, 19th October 1936
[1157] *The Times*, London, England, 28th October 1936
[1158] *Le Petit Journal*, Paris, France, 28th October 1936
[1159] *The Daily Mail*, Hagerstown, Maryland, United States, 31st October 1936
[1160] *The Anniston Star*, Anniston, Alabama, United States, 31st October 1936

Five people who were returning from a family trip to Wichita, Kansas, and Kansas City, Missouri, were killed when their car, a new Chrysler sedan, was struck by a St Louis – San Francisco Railway passenger train at a location two miles east of Keighley, Kansas, United States on Tuesday 10th November 1936. The victims were Mrs Wilson, 27, of Wichita, and her son Elwood, 4; an aunt of Mrs Wilson, Mrs Hall, of Amarillo, Texas; Mrs Hall's daughter Muriel, 18; and a cousin of Mrs Wilson, Margaret Ruth Abderson, 39, of Amarillo.

The automobile had almost cleared the crossing when the train struck its rear fender and spun it around, throwing its occupants out. The impact broke the petrol tank and the occupants of the car were sprayed with burning fuel. Four were killed outright and Muriel Hall died in hospital two hours later. Building materials and box cars stabled on a siding at the crossing interfered with the view to some degree. Ironically at the time of the accident work was in progress at the crossing to build an underpass, presumably to replace it. Melvin Butts, one of the workmen engaged at the site, saw the likelihood of an accident and attempted to flag down the car, but failed to stop it. [1161] [1162]

The catalogue of bad level crossing accidents in the United States continued with one early on Sunday 15th November 1936 at Wichert, Illinois. Six young men were returning home from a party in what was described as an ancient Ford sedan. The headlights of the automobile were apparently very poor, for at the level crossing it was driven into the side of a Chicago and Eastern Illinois Railroad train which was already partly over the crossing. As in the Keighley accident the impact destroyed the gasoline tank of the car, spraying burning petrol over its occupants.

A resident of Wichert named Austin Cramer heard the noise of the collision and was the first person on the scene. He alerted his neighbours and found two of the victims, Robert Lyon and Cecil Royce, lying on the ground beside the car. He pulled them away from the automobile and extinguished their burning clothing. At first he thought that they had been the only people in the car, which was completely enveloped in flames, but later he discovered that there had been four others in the vehicle as well. Intense heat made it impossible for Cramer and his neighbours to reach the car and no cries came from it, suggesting that the other four victims had been killed instantly or at any rate rendered unconscious.

Arnold Anderson, 22; Clark Steele, 25, the owner and driver of the car; Edward Lyon, 22 (brother of Robert?) and Melvin Royce, 19 (brother of Cecil?) died at the scene and Robert Lyon succumbed in St Mary's Hospital, Kankakee, at 8 a.m. that morning. Cecil Royce was seriously injured but it was hoped that he would pull through. [1163]

At the inquest on the victims of the accident the jury ruled that the protection provided at the crossing was inadequate and recommended that the railroad company install electric warning signals. [1164]

1161 *The Hutchinson News*, News, Hutchinson, Kansas, United States, 11th November 1936

1162 *Amarillo Daily News*, Amarillo, Texas, United States, 11th November 1936

1163 *The St Anne Record*, St Anne, Illinois, United States, 19th November 1936

1164 Ibid, 26th November 1936

Southbound Southern Railway passenger train No. 43 collided with a car on a level crossing at Basic City, Mississippi, United States, just before noon on Monday 30th November 1936, killing its five occupants. Conductor C.P. Stroble theorised that smoke from a nearby excelsior mill, lying close to the ground in the wet weather, had interfered with the vision of the car driver. (Excelsior is a stuffing material made from curved wood shavings.) The fact that both road and railway were curved at the crossing might also have contributed.

Engineer Henry Schlager said that he had been pulling the whistle cord when the collision occurred and had struck the car almost before he saw it. Running at between 40 and 50 miles an hour, the train pushed the car 50 feet down the line. After the train had pulled up its crew went back to the crossing to find that all the victims of the accident were beyond help.

Those who died were Mr Adan Fairchild, his wife and unnamed 10-year-old daughter, and Ed and John Weens, all of Quitman, Mississippi. [1165]

Sweden suffered a level crossing accident involving a school bus on Saturday 5th December 1936. The location was Joxtorp, a place between Rönneshytta and Mariedamm on the Mjölby – Hallsberg main line. At about 12.30 that day the bus from Gålsjö was on its way to Pellasbygget and the farms west of Multen, a journey which took it over the unguarded crossing at Joxtorp. As it was negotiating the crossing it was struck by a northbound express train. In addition to driver Gunnar Karlsson from Multna, there were five schoolchildren aged between 8 and 13 on the bus at the time: sisters Anna-Lisa and Alva Carlsson from Stenbocksfors, Maria Gustavsson from Kullen, Elvy Lundin from Pellasbygget and Karl-Erik Karlsson from Multna. The driver and all the children but Karl-Erik were killed. He was taken to Örebro Hospital with serious injuries and eventually recovered. [1166]

What to all intents and purposes was another school bus crossing accident occurred exactly a week after the Swedish tragedy, on Saturday 12th December 1936, at Zuni, New Mexico, United States. A light truck belonging to the Indian day school at Cañoncito was carrying seven Navajo Indian boys from the school who were to play in a basketball game against pupils from another day school on the reservation. The truck was struck by a westbound Atchison Topeka and Santa Fe Railway fast mail train and carried half a mile down the track before the train could pull up.

Fireman Tully Pugh said that the driver of the truck apparently only saw the train at the last moment, and then in his panic had stalled the vehicle on the crossing. The truck driver and four children were killed outright and two of the three injured died after admission to Fort Wingate Hospital. [1167]

It was left to teacher Mrs Alice Marshall to identify the bodies of the victims in what must have been a harrowing experience for her. "That's Tom Wood, 12," she reportedly

[1165] *Hattiesburg American*, Hattiesburg, Mississippi, United States, 30th November and 1st December 1936
[1166] Webpage – https://mittmotala.nu/MittMotala/diverse/tågolyckor.html
[1167] *Evening News-Journal*, Clovis, New Mexico, United States, 12th December 1936

said. "He was so anxious to make the trip that he asked my permission to visit his grandfather and slipped away to join the boys on the truck." She went on to identify the other five children as John Apachito, 10; John Chavez, 10; Ray Chavez, 14; Wilson Platero, 15; and Megalino Secatero, 12. The driver of the truck was Luciano Platero, 30. [1168]

Saturday 19th December 1936 saw the death of five men who were on a hunting trip on a level crossing at Honey Bend, Illinois, United States, not far from Lichfield. Those who died were Stanley Funk, 21; George Gerling, 23; Frank Kutchi, 25; Walter Semptoliski, 31, the owner and presumed driver of the automobile; and Joseph Tobako, 21. Three hunting dogs were also killed. The car driver had stopped at the Wabash Railroad crossing for a southbound freight train, then started off again to be struck by northbound passenger train. [1169]

The year 1937 began with an accident on 1st January at Bloomer, Wisconsin, United States. Farmer Otto Schneer, 44, who lived seven miles outside of Bloomer, was taking home Mr Walter Misfeldt, 27, and his wife, aged 25; and Mr Rollin Smith, 35, and his wife, aged 32. The car in which they were riding was struck by a Chicago St Paul Minneapolis and Omaha Railway train in the middle of Bloomer and all five lost their lives. Pieces of the badly mangled car were scattered along the line for several hundred feet. Originating in Duluth, Minnesota, the train had been due to stop in Bloomer, though the station was on the far side of town. The road crossed the line at an angle of 45 degrees. [1170] [1171]

A mistake by a crossing keeper led to an accident near Burzaco station, Argentina, on Monday 11th January 1937. When a car containing seven people arrived at the level crossing with the Ferrocarril del Sud (Buenos Aires Great Southern Railway) the crossing barriers were lowered for a train bound for the capital and the car stopped. As soon as the train passed the crossing keeper thoughtlessly raised the barriers, the car set off across the line and was struck broadside by a second train coming from the opposite direction and travelling at 60 kilometres an hour. Three of the people in the car were killed instantly, two died shortly afterwards and the other two were gravely injured. [1172]

No. 35 Up passenger train on His Exalted Highness the Nizam's State Railway collided with a lorry on a "C" class unprotected level crossing between Shadnagar and Timmapur stations, India, on Thursday 14th January 1937. Sixteen passengers on the lorry were killed and eight injured. [1173]

[1168] *The Helena Independent*, Helena, Montana, United States, 14th December 1936
[1169] *The Billings Gazette*, Billings, Montana, United States, 20th December 1936
[1170] *The La Crosse Tribune*, La Crosse, Wisconsin, United States, 2nd January 1937
[1171] *Daily Capital News*, Jefferson City, Missouri, United States, 2nd January 1937
[1172] *The Railway Gazette*, London, England, 26th February 1937
[1173] Report by the Railway Board on Indian Railways for 1936-37

After a break of just over two months from major level crossing accidents in the United States, five people died when their car was struck by a gas-electric railcar on a branch line of the Chicago and Alton Railroad somewhere near Stanford, Illinois, on the evening of Friday 5th March 1937. The victims were Mr C.W. Razar, 42; Mrs Rose Willard, 55; her daughter, Mrs Lillian Hartman, 25; Mrs Hartman's daughter, Lillian, aged three months; and seven-week-old George Veach Jr. George's mother was the only survivor of the accident; she was said to have been critically injured. The shock of the impact with the car derailed the train; its 12 or 13 passengers escaped with a shaking. [1174] The wreckage of the car caught fire. Mr Razar, the driver of the car, was described as 'the operator of collection agencies at Peoria, Springfield, Pekin and Decatur.' Motorman William Diesel said that the automobile had been driven into the side of his railcar despite there being an uninterrupted view at the crossing for at least three quarters of a mile. [1175]

Sunday 28th March 1937 saw an accident at Charlotte, North Carolina, United States, which resulted in five fatalities. Those who lost their lives were cotton mill worker Robert Clarkson, 42; his son, Bryce, 17; Mrs Gertrude Walker, 30; her son Joe, 13; and L. D. Hawley, 28, all of Charlotte. The younger Clarkson was apparently driving the automobile and drove directly in front of the Southern Railway train. There were no witnesses to the accident. [1176]

A westbound Canadian Pacific Railway freight train collided with an automobile on the evening of Monday 26th April 1937, on the level crossing adjoining the station at Copper Cliff, Ontario, Canada. The car somersaulted, damaged a switch stand, was carried 290 feet, and burst in flames. Its six occupants, all from nearby Sudbury, were killed outright. They were Mr W.J. Stesco, his wife, their children Patrick, two, and Robert, seven months; Mr Stesco's brother Alex and Alex's fiancée, Miss Gladys Laverne McDonald, all from Sudbury. Alex had been driving the car, which had been travelling towards Copper Cliff on the Sault Ste Marie – Sudbury highway. Conductor J.A. Rainville was in charge of the train, with Engineer E. McAuliffe. [1177]

Four children were among the victims in a level crossing accident at Milton, Vermont, United States, on Saturday 15th May 1937. They were riding in a light sedan car which was doing service as a school bus and were apparently going to school on a Saturday to make up for time lost, perhaps because they had missed lessons due to sickness. Killed in the accident were Earl Fuller, 14; Rene Larochelle, 11; her brother Howard, 13; and Earl Murray, 10, all of Milton. John Vasseur, 37, the driver of the 'bus', was also killed.

A couple of factors were probably involved in the accident. One was that the express from Washington D.C. and New York to Montreal, Quebec, was running 45 minutes late, so Vasseur was probably not expecting to encounter the train at the time he reached the

[1174] *The Boone News-Republican*, Boone, Iowa, United States, 6th March 1937
[1175] *Bluefield Daily Telegraph*, Bluefield, West Virginia, United States, 6th March 1937
[1176] *The High Point Enterprise*, High Point, North Carolina, United States, 29th March 1937
[1177] *The Winnipeg Evening Tribune*, Winnipeg, Manitoba, Canada, 27th April 1937

crossing as it had usually passed some time earlier. A light rain was also falling at the time of the accident which might also, with the dark sky, have obscured his visibility. Distraction from one or more of the children was another possibility.

Engineer John Newton of the Central Vermont Railroad train said that he had whistled as usual approaching the crossing. The train had been stopped as quickly as possible, and its crew had hurried back to the crossing. They found that one of the victims, Earl Murray, was still alive. He had been bundled into an ambulance which set off for the hospital at Burlington, a dozen miles away, but he died shortly afterwards without regaining consciousness. [1178]

A paragraph in a newspaper in mid-May noted that part of the United States federal unemployment relief programme involved the elimination of 2,882 level crossings. By 31st December 1936, 955 projects had been completed, at a cost of $46,429,767. Work was in progress on 1,179 crossings at an estimated cost of $107,624,888 while 748 more projects, costing £18,606,144, had been approved. Nevertheless, even when all these projects had been completed no less than 234,231 crossings would remain in the country; 4,930 people had lost their lives in level crossing accidents in 1936, 105 more than in 1935. [1179]

A Braşov – Bucharest express collided with a crowded bus at an unstated location in Romania on Friday 28th May 1937. Six people were killed outright, seven were critically injured and 15 were slightly hurt. [1180]

On or about Thursday 17th June 1937, an express train collided with a lorry carrying Arab workmen near Baghdad, Iraq, killing 19 of them. [1181]

Four people were killed outright and a fifth was fatally injured when their car was struck by a Minneapolis St Paul and Sault Ste Marie Railroad train on the evening of Sunday 4th July 1937 at Paynesville, Minnesota, United States. Those who lost their lives were George Harris, 76, of Paynesville, a retired farmer; Frank Pitz, 40, a Works Progress Administration truck driver, his wife, and their son Wilfred, 7, from Litchfield; and Dean Eisenberg, 2, a grandson of Mr Harris. The other two occupants of the car, Mr Ernie Harris, brother of George, and his wife, were badly hurt. The seven had just started for home from a visit to relatives on a farm. The accident happened on a crossing with a private road leading to the main highway. [1182] [1183]

Very brief reports appeared in the American press of a level crossing collision involving an express train and a motor bus near Taira, Japan, in Fukushima prefecture, on

---

[1178] *Fitchburg Sentinel*, Fitchburg, Massachusetts, United States, 15th May 1937

[1179] *Gazette and Bulletin*, Williamsport, Pennsylvania, United States, 17th May 1937

[1180] *Journal de Geneve*, Geneva, Switzerland, 29th May 1937

[1181] Ibid, 17th June 1937

[1182] *The La Crosse Tribune and Leader-Press*, La Crosse, Wisconsin, United States, 5th July 1937

[1183] *The Redwood Gazette*, Redwood Falls, Minnesota, United States, 8th July 1937

Monday 12th July 1937. Ten people died and another 11 were seriously injured. [1184]

A major disaster at Los Prados, four kilometres from the town of Cieza, Spain, at about 01:00 on Thursday 15th July 1937, was almost totally ignored by the media which was preoccupied with the ongoing civil war. Cieza, in Murcia Province, was still well within the area controlled by the government forces at this date and had been little affected by the fighting, apart from odd attacks from rebel aircraft. The train involved in the accident was an overnight mail train from Cartagena to Madrid while the road vehicle was a military lorry, one of several of a convoy which was also bound for Madrid.

Los Prados was the point where the Cartagena – Madrid highway crossed the railway. The crossing was operated by a gatekeeper, or more accurately by a family of three gatekeepers, each working an eight-hour shift. The husband, Juan Antonio Téllez Salinas, worked from 10 p.m. until 6 a.m. and so was on duty at the time of the accident.

The remainder of the convoy passed over the crossing before the arrival of the mail train but the lorry which was to feature in the accident had been delayed in Cieza, where a stop had been made to try and find somewhere where a bottle of rum could be purchased at this late hour. Its driver had then been going fast to catch up the other vehicles. A short time before reaching the level crossing where the accident occurred the lorry, loaded with bombs and other munitions, had already almost been involved in a bad accident when its driver had failed in the darkness to realise that there was a bend in the road and had come very close to driving into a house. When this lorry belatedly reached the crossing the gatekeeper was in the process of stretching his chains across the road, prior to the arrival of the mail train. He had in fact already closed off the far or west side of the crossing and was in the process of putting up the other train when the lorry arrived. Seeing the red handlamp of the crossing keeper the lorry driver had braked but did not manage to stop until his vehicle was on the tracks. An argument then ensued between the lorry driver and the crossing keeper, the lorry driver insisting on being let across and the crossing keeper refusing to open the crossing because of the proximity of the train. The lorry driver apparently became so irate as to threaten to shoot the gatekeeper. With the noise of the lorry's engine and the heated argument they failed to notice the oncoming mail train.

The train struck the lorry, and the force of the impact was sufficient to cause the munitions that it was carrying to explode, making the accident much more serious than it might otherwise have been. Buildings near the crossing, including a school, were demolished, and the engine of the mail train was derailed and wrecked, as were the leading vehicles of the train. Parts of bodies and the remains of the lorry were scattered over a wide area.

Those living near the scene imagined at first that it was another rebel air attack, but when what had actually happened became clear help was soon on the scene and ambulances, doctors and nurses rushed to the spot. At 6 a.m. it was announced that five people had died and 101 had been injured. These figures were never updated but more recent research has shown that at least 14 people died and over 200 were injured. Those who lost their lives included the crossing keeper, Juan Antonio Téllez Salinas; Pascual Téllez Martinez and Juan Téllez Martinez – two of the crossing keeper's family; Félix

---

[1184] *The Daily Ardmorite*, Ardmore, Oklahoma, United States, 12th July 1937

Cabanillas Fernández, identified as the fireman of the train; Benigno Ramiro Ortega, identified as a train porter; and presumably also the train driver. The other fatalities must have included the unnamed lorry driver, any passengers on the lorry, and some of the passengers on the train. The wife and two daughters of the crossing keeper were seriously injured but survived. [1185]

A family who were probably returning home after a day out were killed when their car was struck by a Pennsylvania Railroad train at Dola, Ohio, United States, at about 8 p.m. on Sunday 25th July 1937. They were Cloyce Getz, 30; his wife Anna, 24; and their daughters Loretta Fay, 6; Marjorie Ann, 4; and Claudetta, 2. Travelling south along Anthony Street the Getz car reportedly was driven onto the tracks directly in front of a fast eastbound passenger train. The occupants of the automobile were thrown out of it in by the impact of the collision and the vehicle was hurled against a switch and burst into flames. Four of the victims were killed more or less outright, and Mrs. Getz died in an ambulance before it could leave the scene.

Loren Benjamin, of Dola, witnessed the accident. He said that he had heard the sound of what appeared to be screeching brakes immediately before the collision. It was raining slightly. The train engineer stated that the car had moved onto the tracks and then stopped. Had it not done so it would have had ample time to clear the crossing in front of the train. Coroner K.W. Preston recorded a verdict of 'Accidental Death'. [1186]

Twelve people died and 22 more were injured when a train crashed into a bus on a level crossing near Tampico, Mexico, on Monday 26th July 1937. Those who lost their lives comprised seven men, four women and a child. [1187]

An unprotected crossing in Canada was the scene of an accident at about 7.40 p.m. on Tuesday 27th July 1937. Five residents of Sudbury, Ontario, lost their lives when the car in which they were riding was struck by Canadian Pacific Railway train No. 28 on Long Lake Road Crossing, on the Sudbury – Sault Ste Marie branch. Mr. Aksil Lampila and his wife Lempi died in the ambulance taking them to hospital and their four-year-old daughter died a few minutes after admission. Their son Oliver, five, and Mr. John Tuuri were killed instantly. A search was made after the accident for Mr. Tuuri's wife, whom it was believed had also been in the car, but she could not be found. The two men were employees of Frood Mines.

Although the crossing, to the west of Sudbury, lacked gates or signals, there was a clear view along the track for a quarter mile. [1188]

1185 https://www.regmurcia.com/docs/murgetana/N119/N119_013.pdf : EL TERRIBLE ACCIDENTE FERROVIARIO OCURRIDO EN CIEZA EL 15 DE JULIO DE 1937 LUIS LISÓN HERNÁNDEZ

1186 *The Lima News*, Lima, Ohio, United States, 26th July 1937

1187 *The Daily Times-News*, Burlington, North Carolina, United States, 26th July 1937

1188 *The Lethbridge Herald*, Lethbridge, Alberta, Canada, 28th July 1937

Five people died in a level crossing accident near Casselton, North Dakota, United States, on the evening of Saturday 18th September 1937. They were Mr George Wadeson and his wife, farmers, of Alice, North Dakota; Mr Thomas Brocklin and his wife, who worked on the Wadeson farm; and Joe Warra of Alice, an employee of the West Fargo Armour & Co branch. Mr Brocklin and the two women had come from Alice to pick up Mr Wadeson from the North Dakota Agricultural at Fargo and Mr Warra from West Fargo and they were on their way home, with Mr Brocklin driving. The car was travelling along U.S. Highway 10 when it was struck by a Great Northern Railway freight train. [1189]

The accident was witnessed by Max Taubert, the owner of a nearby filling station, who said that the car driver had apparently begun to brake when he was about 80 feet from the crossing, and the vehicle had then swerved, doubtless when its driver realised that he was not going to be able to pull up in time and had slid into the train. Conductor R.L. Kellogg, riding in the caboose, saw the wreckage of the car as his train cleared the crossing, applied the brakes to stop the train and then back to the filling station to get help, but the occupants of the car were beyond aid.

The federal bureau of public roads and the state highway department had earlier planned to erect a bridge at the crossing, but the plan had been abandoned after the Cass County Taxpayers Association and Casselton area residents had baulked at the cost. [1190]

Towards the end of September 1937 Mr Ellory Brown, 36, and his wife Bertha, also 36, had moved with their family from Lincoln County, West Virginia, to the mining community of Whittaker, in the same state. A week or so later, on Sunday 26th September they had returned to their former home, probably to transfer more possessions and some produce, riding in a truck owned by Mr Willard James, aged about 60, a schoolteacher who either was a relative or friend of the family or possibly also operated a removal service. As the truck was on its way once more to the Browns' new home it was struck by a Chesapeake and Ohio Railway passenger train from Huntington and Charleston to Washington, D.C., on Hansford Crossing, near Montgomery, West Virginia, United States. The truck was demolished in the collision, its chassis being carried 120 feet and hurled into Paint Creek, while its body was carried about 420 feet. Mr James, Mr and Mrs Brown and their children Donnie, 17; Ruby, 12; Myrtle, 9; and William, 4, were all killed. The one survivor was 2-year-old Emogene. Witnesses stated that a moment before the impact they had seen Mr Brown rise up and throw the toddler from the truck. A negro picked up the little girl and handed her over to rescuers. She was taken to Montgomery Hospital, having suffered a broken arm in her fall, but it was thought likely that she would recover.

Charles Porter, a garage owner, said that the truck had just left his garage after being serviced. Another garage owner, D. Recco, said that he had heard the noise of the crash, and added, "I ran outside. Bodies were everywhere. It made me sick."

The road rose gently for 150 yards from the Porter garage to the crossing. There were no gates or warning devices but there was a clear view along the line in both directions for 200 yards. Witnesses said that the train engineer had whistled for the crossing and had then

---

[1189] *The Bismark Tribune*, Bismark, North Dakota, United States, 20th September 1937

[1190] *Moorehead Daily News*, Moorehead, Minnesota, United States, 20th September 1937 applied his brakes. The truck driver had apparently neither heard nor seen the train. [1191]

What began as a fairly minor road traffic collision involving two cars was transformed into a bad accident killing seven people on the evening of Friday 8th October 1937 when both cars were struck by a New York Chicago and St Louis Railroad passenger train, the vehicles having ended up on the track at the level crossing at Gary, Indiana, United States. One car contained Mrs Schoon, 36; Mrs Irene Cass, 40, the driver of the car; Mrs Cass's son Robert, 3; and her daughter Judith, 18 months. All but Judith were killed – she sustained a possible fractured skull. The other car was carrying seven men who were Public Works Administration (P.W.A.) employees. (The P.W.A. was another part of President Roosevelt's New Deal – a construction agency which provided work for the unemployed during the Great Depression). In this car were Sam Chappa, a Mexican; and six negroes – Roy Prince, Percy Spann, A.C. Weatherly, Frank Coleman, James Rucker, and Oscar Shealey. The first four men lost their lives but Coleman, Rucker and Shealey just had time to get out of their car and escaped. All were from Gary.

The three surviving P.W.A. men said afterwards that they had been returning home from their work project at Schererville, 15 miles from Gary, and had been travelling east on Ridge Road, a through highway. They had crossed the railroad when their vehicle was rammed by the Cass car, which had come from Grant Street, and had failed to give way.

By a strange quirk of fate the train was carrying 600 Civilian Conservation Corps (C.C.C.) workers from Camden, New Jersey, to California. The C.C.C. was another New Deal programme for unemployment relief. Engineer T.J. Leeter said that it had been impossible to stop his 12-car train in the few seconds which had remained, when he had seen the obstruction on the crossing. [1192] [1193]

On Sunday 10th October 1937 a Mrs Kebert, 37, was driving six of her seven children home after they had attended Sunday school classes in Neodesha, Kansas, United States. They had not travelled far when their automobile was struck by a Missouri Pacific Railroad passenger train. Mrs Kebert and the children – Elouise, 16; Orville, 11; Ethel, 9; Mary Jane, 8; and twins Richard and Robert, 3 – were all killed. Mr Charles Kebert and son Ralph, 15, had remained at home.

The crew of a freight train parked in a siding saw the Kebert sedan and whistled to try and alert its driver. They doubted if Mrs Kebert had seen the train at all. The passenger train stopped half a mile down the line and Engineer J.M. Demastus set back to the crossing. The first people to reach the victims found one of the twins still breathing but he died very shortly afterwards. [1194] [1195]

North Western Railway Down passenger train No. 44 collided with a lorry on a level crossing between Bahawalpur and Samasatta, India, on Thursday 14th October 1937.

[1191] *The Raleigh Register*, Beckley, West Virginia, United States, 27th September 1937
[1192] *Centralia Evening Sentinel*, Centralia, Illinois, United States, 9th November 1937
[1193] *Logansport Pharos Tribune*, Logansport, Indiana, United States, 9th November 1937
[1194] *Lawrence Daily Journal-World*, Lawrence, Kansas, United States, 11th October 1937
[1195] *The Helena Independent*, Helena, Montana, United States, 11th October 1937

Fourteen passengers on the lorry were killed and 11 others injured. The official report found that the lorry driver had been responsible for the accident and that he had been 'exceptionally careless.' [1196]

Friday 22nd October 1937 saw another level crossing collision in the United States involving a school bus, this time at Mason City, Iowa. The bus was not taking children to or from school but was on a field trip from Renwick, Iowa, to the Mason City Brick and Tile Works, one of a number of industrial and business establishments the pupils were to visit. Such trips were made by Renwick High School every autumn in the hope of giving the older students some experience of work, as most of them would be looking for jobs when they left school, rather than going on to university. On this day there were 26 students on the bus, together with teachers Lauren Morton and Dorothy Moss and the bus driver, Rex Simpson.

The crossing where the accident occurred was on a private road of the brick company, running east-west and connecting Plant 3 of the company to the north-west of the railway with Plant 2 to the south-east. The road was constructed of dirt and pieces of broken bricks. Protection at the crossing was limited to a pain of crossbucks and the view of a northbound train obtained by the driver of a westbound road vehicle was very limited by the presence of a pile of bricks adjoining the crossing.

Crossing the road was the single-track line of the Chicago Great Western Railway but the train involved was owned by the Chicago Rock Island and Pacific Railroad, which had trackage rights (running powers) over the line.

The bus was 22 ft long and of wooden construction with a thin metal covering. It had six double seats on either side in addition to the driver's seat.

The train involved was the northbound *Kansas City Rocket*, train No. 508, which consisted of diesel engine No. 605, a combination baggage-dinette car, a coach, and an observation lounge car, in charge of Conductor Charles O'Neill, with Engineer George Simpson and Fireman Ed Clements. It left Short Line Junction, 119¼ miles from Mason City, at 2.06 p.m. entered Chicago Great Western tracks at Clear Lake Junction, a mile, and a half from Mason City, at 3.55 p.m., 5 minutes late, and was travelling at between 20 and 30 miles an hour when it hit the bus, which was moving westwards over the crossing at about 5 miles an hour. The bus was demolished, the train was only slightly damaged. Seven students, both teachers and the bus driver died in the accident and the other 19 students were injured.

Engineer Simpson told the Interstate Commerce Commission (ICC) inquiry into the accident that the air brakes of his train had been tested before leaving Des Moines. Speed had been reduced to 10 miles an hour for the Mason City and Clear Lake Electric Railway crossing at Clear Lake Junction, then he had increased speed and sounded a long blast for the crossing with the Chicago Milwaukee St Paul and Pacific Railroad crossing (beyond the brick and tile works), over which speed was restricted to 25 miles an hour. He had then whistled for the private crossing. The first that he had known of the accident was seeing pieces of what appeared to be a lorry flying up against his windscreen, when he had applied his brakes in emergency. The automatic bell on his locomotive had been

[1196] Report by the Railway Board on Indian Railways for 1937-38

ringing and his train had been moving at 20 or 25 miles an hour. A few flat wheels were the only damage to his train, which was not derailed.

Fireman Clements said that he had started the automatic bell before Clear Water Junction. He knew that there was a 15 miles an hour speed restriction through Mason City, but not where it was meant to begin.

Conductor O'Neill put the speed of his train at 30 miles an hour at the accident point. He did not think that the 15 miles an hour restriction began until the public level crossings.

Flagman Voitel put the speed at 20 to 25 miles an hour at the accident site.

Leverman Williams, in the Chicago and North Western tower 750 feet north of the crossing, witnessed the accident. He put the speed of the train at 25 miles an hour and of the bus at 5 miles an hour. He said the bus had not stopped short of the crossing. On an average day he had seen about eight cars and four or five lorries using the crossing; the cars would be those of brickworks employees and the lorries those of farmers taking clay products from the different plants to their farms. He had never before seen a bus use the crossing.

Two of the students on the bus agreed that it had not stopped before entering the crossing. [1197]

The statements of other witnesses to the accident appeared in the newspapers. Student Kenneth Knudson escaped serious injury. He said, "I saw the Rocket train the moment before it hit us. It seemed to just jump right out upon us from between two piles of brick. There was an awful rending noise that seemed to tear and smash right into my head. Then the first thing I knew, I was crawling out of the wreckage, all blood-smeared and climbing over bodies. For a moment, inside that bus, nobody seemed to cry out. I guess we were all too stunned by the sight of the huge, red engine right on top of us. Some of those sitting beside me seemed to be tossed out as we were carried along. It seemed we were carried a block or more, but I hardly knew it, so fearful was the grinding noise and the screams. I was sitting in the middle of the bus. It appeared to me the train struck us toward the front end, and that's why I believe I am alive now – but I can't be sure." [1198]

Vera Mott, a Mason City coal dealer, was sat in his car 100 feet behind the bus. He related, "I heard the Rocket whistle and then I saw the streamliner coming from the south. It didn't seem to be going very fast. Vision was perfectly clear, and I was utterly amazed when I saw the bus continue toward the crossing, without slowing up. The Rocket hit the bus squarely in the center and the big vehicle, of wooden construction, popped open as if it were a watermelon dropped on the ground. Bodies hurled out of the bus; children screamed – it was the most horrible sight I've ever seen."

J.H. Leary, a 'Minneapolis real estate man,' was a passenger on the train. "We were pulling into the city and going slowly, when the train almost stopped. Almost simultaneously I saw pieces of a truck flying through the air as I sat in my seat. I went to the front of the Rocket and saw the chassis of the truck on the pilot. I counted seven bodies. Boys and girls were strewn all along the right-of-way. Some were moaning and groaning. One body was right across the rails; another was lying near the rail with the

---

[1197] Interstate Commerce Commission report

[1198] *The Evening Tribune*, Albert Lea, Minnesota, United States, 23rd October 1937 head crushed in. Another was some distance away with a leg cut off." [1199]

Mr C.F. Garland, Ceramic Engineer of the Mason City Brick and Tile Company, told the inquiry that he had been advised that a party of schoolchildren from Renwick were to visit the works at about 1 p.m. The bus had eventually arrived at 3 p.m. and had gone to Plant 3, but that had been closed for the day, so he had conducted the bus to Plant 2, driving his own car in front of the bus and stopping at both the Chicago Great Western crossing and at the Chicago and North Western crossing, then back over the North Western crossing to Plant 2. While he had been showing the school party around Plant 2 the southbound *Kansas City Rocket* had passed. He had drawn the attention of the party to the train and remarked that the northbound Rocket would pass in a few minutes. At the end of the visit he had suggested that the bus take an alternative route out of the plant, but the bus driver had replied that he was not sure that he could find his way out that way and preferred to return the way he had come. He (Mr Garland) had then warned the bus driver twice to watch out for the northbound Rocket.

Mr M.K. Sperry, Renwick Superintendent of Schools, said that he had discussed the trip with Driver Simpson and cautioned him to stop before negotiating level crossings. Driver Simpson had driven school buses for a year and was considered to be a very careful driver. He had not been to the Mason City brickworks before.

The ICC commented that the crossing was an exceedingly dangerous one because of the pile of bricks which bordered the road to the south and extended to within 15 feet of the track. For the bus driver to have been able to see an approaching train he would have had to stop his vehicle no more than seven feet from the rails, which was too close for safety. It appeared that in any case the bus driver had not stopped at all. Why he had not done so could not be determined as he had lost his life in the accident. The evidence showed that the train had whistled several times approaching the crossing. The speed of the train was in excess of that dictated by the timetable within the corporate limits of Mason City and it appeared that the train crew had not had a clear understanding as to what these limits were.

The ICC concluded simply that the accident had been caused by a school bus being driven upon a private road crossing directly in front of an approaching train.

The crossing was closed, at least temporarily, after the accident. [1200]

The jury at the inquest found unanimously that the accident had been caused "wholly by the failure of the bus driver to heed railroad crossing signs, and to stop before crossing the tracks." [1201]

The first weekend of November 1937 was a bad one for road traffic deaths in the United States, with no less than 181 fatalities. Included in this total were 11 people who lost their lives in two level crossing accidents on Sunday 7th November, one in Georgia and the other in Missouri.

The Georgia accident happened at Adairsville. The Lear family had been visiting friends in Adairsville and were just starting their trip home to Rome when the sedan in

---

[1199] *The Telegraph-Herald*, Dubuque, Iowa, United States, 24th October 1937

[1200] Interstate Commerce Commission report

[1201] *The Evening Democrat*, Fort Madison, Iowa, United States, 25th October 1937 which they were riding was struck by a Nashville Chattanooga and St Louis Railway passenger train on what was referred to as a 'blind' crossing, that is, one with poor visibility of the tracks. Mrs Bessie Lear, 50; her daughter Elsie, 18; her son Walter, 25; Walter's 20-year-old wife; and Walter's children Patricia, 10 and Sandra, 8, were all killed. Walter's son Joe, 6, survived the actual collision but was said to be in a critical condition in Calhoun Hospital. [1202] [1203]

The circumstances of the Knox City, Missouri, accident on Sunday 7th November 1937 were detailed in the statement of an eyewitness, a Mrs Miner, who had been standing in the doorway of her cottage, 50 feet from the crossing. She said that she had seen the northbound automobile approaching the crossing travelling at high speed, and added, "The driver apparently decided he couldn't beat the train and tried to stop. The car stopped directly on the tracks. I saw there was going to be a collision, so I slammed the door to hide the horrible sight."

Conductor J.P. Johnson and Engineer Joe Mulligan of the Quincy Omaha and Kansas City Railroad train said that they had done their best to avoid the accident, but it had been impossible to stop their train in the distance available.

The victims were all teenagers or pre-teenagers. Three of them were siblings: John, Marjorie, and James White; 15, 12, and 10 years old, respectively. The others were Mary Jane Turpin, 19; and Leroy McEvoy, 15. [1204]

Five teenagers and a young man, 'on a pleasure drive,' lost their lives in a level crossing collision at Vineland, Ontario, Canada, on the evening of Wednesday 17th November 1937. In contrast to the Knox City accident the car was only moving slowly when it was hit by a Canadian National Railway Niagara Falls – Sarnia through freight train. The accident could only be ascribed to the carelessness of the car driver, who clearly had failed to see the train, which had been travelling at 40 to 45 miles an hour, despite there being good visibility at the crossing, which railroad officials descried as "exceptionally safe." Engineer Frank Pattison said that he had whistled and the bell on his locomotive had been ringing but the driver of the car had failed to react.

Three of the victims were members of the Welland Sea Cadets troop; the other three were presumably their girlfriends. According to one account they were Robert Colley, 18; James Fisher, 16; and Steve Sutch, 20, of Welland; and Madge Jordan, 16; Sophia Nichol, 15; and Ruth Wright, 16; all of Thorold. Other accounts give slightly different names and ages. [1205] [1206]

Six people, two of whom were a bride and bridegroom returning from their wedding, lost their lives when the horse and cart in which they were riding was struck by a train

[1202] *The Butler Herald*, Butler, Georgia, United States, 11th November 1937
[1203] *The Anniston Star*, Anniston, Alabama, United States, 8th November 1937
[1204] *The Brookfield Argus*, Brookfield, Missouri, United States, 8th November 1937
[1205] *The Winnipeg Evening Tribune*, Winnipeg, Manchuria, Canada, 18th November 1937
[1206] *Middletown Times Herald*, Middletown, New York, United States, 18th November 1937

near Lodz, Poland, on or about Monday 22nd November 1937. [1207]

Homeward bound to Elmore, Minnesota, five men died when their sedan car was struck by a southbound Chicago and North Western Railway gasoline railcar on a crossing a mile north of Burt, Iowa, United States, at 6.30 p.m. on the evening of Thursday 9th December 1937. The train was en route from Fox Lake, Minnesota, to Eagle Grove, Iowa. The men were dealers in furs and bundles of pelts were found in the wreckage of the car. The men were identified as Harry Blondon, 28; Herbert Curtis, 28; Elmer Meissner, 21; Ervin Richter, 20; and Louis Shamberg, 26.

Motorman S.E. Billings related, "When we approached the crossing I tooted the horn. The driver must have heard me. He slowed up and apparently saw he couldn't make it to a stop before reaching the crossing and stepped on the gas." The view at the crossing was unobstructed but the car windows were coated with frost, impairing visibility. Richter, whom it appeared had been the driver, had a red handkerchief in his hand with which he had obviously been wiping the windscreen.

The train was derailed by the impact with the automobile but no one on it was hurt. It was delayed several hours. [1208]

A Bucharest – Prague express collided with a lorry on a level crossing between Bod and Soldiora, near Braşov, Romania, on Tuesday 14th December 1937. Five people died and seven were seriously injured. [1209]

An accident which was reported very extensively in the American press, because its victims were the teenage children of prominent local families, occurred on the evening of Sunday 2nd January 1938 near Butler, Indiana, United States. The seven who died had been on their way from their homes in Hicksville and Edgerton, Ohio, to watch a movie in Butler and were in a hurry in order not to miss the start of the film at 8 p.m. It was 7.55 when the collision occurred. The crossing where the accident happened was protected by barriers operated by a watchman from an adjoining tower, and the barriers were down when the car occupied by the teens approached, the watchman, John Figg, having just lowered them for eastbound New York Central Railroad passenger train No. 146. Despite this, the car failed to stop. "They crashed right through the gates, into the train. It was terrible," Figg said afterwards.

The victims were identified as Lowell Blythe, Marvin Miller, Merle Sarver, Jacob Stauffer, and Richard Staup, all 15 and all from Edgerton, and Marjorie Roan, 14, and her niece Gwendolyn, 15, both from Hicksville. Four of the boys had been members of the Edgerton High School basketball team.

The car was broken to pieces and the bodies of the victims were strewn along the line over half a mile. The locomotive of the train was badly damaged, and another engine had

---

[1207] *The Times*, London, England, 23rd November 1937

[1208] *The Muscatine Journal and News-Tribune*, Muscatine, Iowa, United States, 10th December 1937

[1209] *Le Petit Troyen*, Troyes, France, 15th December 1937 to be sent from Toledo to take the train forward. [1210] [1211]

Coroner B.O. Shook said that the inquest would not be completed for several days but indicated that there would likely be a formal verdict of 'Accidental Death'. [1212]

A level crossing collision at Cunningham Spur, Arkansas, United States, on Sunday 6th February 1938, resulted in six fatalities, none of whom was in the road vehicle struck by the train. Instead those killed were three passengers, an express messenger, a porter, and the engineer of the train. The train involved was northbound Missouri Pacific Railroad passenger train No. 26, *The Texan*, en route from Texarkana, Texas, to Little Rock, Arkansas with 77 passengers. This train consisted of 4-6-2 steam locomotive No. 6610, two baggage cars, a coach (probably the Jim Crow coach, reserved for black passengers), a chair car, two coaches, seven sleeping cars and a dining lounge car, in charge of Conductor Harper with Engineer John Brannan and Fireman G.H. Johnson.

The car involved in the accident was a 1929 or 1931 model Pontiac sedan, owned by a Lee Hicks, but which had been taken without permission by his brother-in-law Homer Wilson, an Army private on furlough.

At 7.45 p.m. on the evening before the accident Homer Wilson had left Poyen, Arkansas, accompanied by his brother, Frank, and at 8 p.m. they had arrived at Malvern, Arkansas, where each had drunk a bottle of beer. Between 8.30 and 11 p.m. they had driven around several towns in the area with other friends and at 10.15 both Wilsons had drunk some whisky. There had also been trouble with the car's headlights. Just after 11 p.m. they had decided to go to Traskwood and as Homer Wilson did not know his way there the others in the car had to direct him. The car passed over the crossing at Cunningham Spur and had proceeded somewhat further. Wilson decided that he had gone the wrong way and tried to turn around. At this point the vehicle became stuck in the mud on the dirt road and it took two hours or more to free it. The car was then turned and headed back towards the crossings. When it got there the car became stuck on the crossing. Several attempts were made to free it, without success, and when they saw the train approaching the two brothers abandoned the vehicle and stood clear of the line. Homer Wilson said that he had turned off the car's engine before they had abandoned it and possibly, also, its headlights. They had not made any attempt to flag the train. Wilson said that he had been driving the car all night and had covered 50 or 60 miles. They had not run out of gasoline, and he had not been intoxicated. Although he did not have a driving licence he had been driving cars since the age of 14 and considered that he was a good driver. He denied having driven the car off the crossing and attributed its stalling to the insufficiency of the crossing and its rough condition. The accident happened at about 3.07 a.m.

Frank Wilson corroborated his brother's account.

Because it involved the deaths of train passengers and railwaymen and the derailment of a train, the accident was the subject of an Interstate Commerce Commission (ICC) investigation.

[1210] *The Hammond Times*, Hammond, Indiana, United States, 3rd January 1938
[1211] *Morning Herald*, Hagerstown, Maryland, United States, 3rd January 1938
[1212] *Defiance Crescent-News*, Defiance, Ohio, United States, 4th January 1938

Fireman Johnson stated that the engine had been in good condition and its brakes had been tested before leaving Texarkana. The train had been travelling at 70 miles an hour when it met with the accident. Both he and the engineer had been looking ahead. He had seen nothing of the car before his engineer had applied the brakes in emergency at the crossing. Almost immediately the engine had begun to sway, and it had then overturned.

Conductor Harper and Flagman Crowley agreed with the fireman's estimate of the train's speed and said that they had had no warning of a problem prior to the emergency brake application.

Section Foremen Cam and Heath both stated that there had been no sign of dragging equipment on the train and that the derailment of the train had begun at a switch beyond the crossing and that a wheel hub of the automobile had become jammed in the crotch of the frog (common crossing) of the switch (points), causing the derailment of the train. Divisional Engineer Jaeschke and Assistant Engineer May came to the same conclusion.

With regards to the traffic on the Civilian Conservation Corps forest road, the crossing was extremely quiet. During a 24-hour traffic check it was used by only two vehicles and three pedestrians. Average railroad traffic over the crossing consisted of just over 26 trains daily.

Discussing the accident the ICC noted that tyre marks indicated that the automobile had run off the crossing on its return journey. Its occupants had not attempted to flag the train. It was unclear if the headlights of the car had been burning when the train reached the crossing but in any case they would have been directed away from the train and it was unlikely that the engine crew could have seen them. Horace Wilson had not had a driver's licence, in contravention of State law. He had taken the car without its owner's specific consent, but the owner had been present when it was taken and had not objected. The driver had consumed some alcohol earlier but three people who spoke to him after the accident had not thought that he was intoxicated.

The commission concluded that the accident had been due to the train colliding with an automobile which had become stalled on a crossing, probably due to careless driving. [1213]

The passengers who died in the accident had been riding in the leading coach. They were scalded by steam from the overturned engine, which entered through a carriage window.

The Wilson brothers were not detained by the deputy prosecutor but were recognised to appear before the grand jury the following month. Coroner Sims found that the six victims had died accidentally. [1214]

The day after the Cunningham Spur accident, Monday 7th February 1938, five people died in a level crossing collision in Salt Lake City, Utah, United States. A car driven by Mr Lewis Criddle, 34, and containing his wife Nadine, 30; his brother Fred, 33; his sister-in-law, Mrs Geneve Criddle, 45; and Roy Johnson, 45, all of Garfield, Utah, was in collision with a Western Pacific Railroad passenger train at the crossing on Redwood Road.

[1213] Interstate Commerce Commission report

[1214] *Hope Star*, Hope, Arkansas, United States, 7th February 1938

Engineer Glenn Morton and Fireman Laron Sorensen spoke to a newspaper reporter afterwards. The engineer said, "I saw them coming when they were about 400 or 500 feet and they should have had plenty of time to stop, but they were making about 65 miles an hour and apparently tried to beat us to the crossing. We didn't have a chance to miss them, they just kept on coming. We blew the crossing whistle three times and had the bell ringing, and the lights were on, but it didn't make any difference – they just kept on coming. They were going about as fast as the car could go and I guess they wanted to beat us, or else they weren't looking. Sometimes we beat 'em, sometimes they beat us, and sometimes there is a tie. This time it was a tie.

"I was travelling about 35 or 40 miles an hour when the accident occurred. Of course that's just a guess, the same as I'm guessing at the speed of the car. I've been railroading for 40 years but that's the first time I've ever killed anyone. It don't make a fellow feel very good; I can tell you."

Fireman Sorensen added, "He {Engineer Morton) jammed on the emergency brakes and yelled, 'Did they make it?' I felt the train pulling up and leaned towards the window just as the glass burst out in my face, and I ducked behind the boilerhead. It all happened so quickly I didn't realise we had hit a car. Apparently they had tried to beat us across the track and failed to make it. They seemed to be right across the track when we struck, as the whole wreck was thrown over on the north side of the track.

"These emergency brakes stop things in a terrific rush – stand you right up – and we had probably gone 200 or 300 feet before we came to a stop. We rushed back and saw what had happened... It appeared that all five were dead, and after about 15 minutes we came on in. There was no particular damage to the train, except the pilot was torn off. I've never been in anything like it before." [1215]

On Tuesday 29th March 1938, what was almost certainly a major disaster occurred at Meliana, Spain, not far from Valencia, in an area which was still under government (Republican) control. An electric passenger train of the Compañia de Tranvias y Ferrocarriles de Valencia collided with a petrol tanker on a level crossing with the old Barcelona road. The load of the tanker was splashed over the train carriages and immediately caught fire. Local residents and the drivers and passengers of other road vehicles were able in the first moments to rescue a few passengers, several with slight burns, and seven with severe burns who were taken to hospital. Quickly, however, the fire became too intense to prevent any further rescue work, despite the prompt arrival of firemen as well as police.

The exact number who died in the accident remains unknown. The Republican press stated that two of those who had been taken to hospital had died of their injuries but was silent regarding any other deaths. Rebel (Francoist) sources, on the other hand, maintained that 23 people had lost their lives and Reuters news agency, probably relying on the Francoist sources, gave the same figure. The fact that a mass funeral was held two days after the accident suggests that the higher figure is probably closer to the truth. The mechanism of the collision does not seem to have been reported. [1216] [1217] [1218]

[1215] *The Salt Lake Tribune*, Salt Lake City, Utah, United States, 8th February 1938

[1216] Webpage: Federación Castellano Manchega de Amigos del Ferrocarril (in Spanish)

What different accounts described as a goods train and a train carrying railway employees was in collision on the evening of Thursday 14th April 1938, near Suez, Egypt, with a lorry bringing home workers who had been engaged in the construction of military roads. Nine or 18 workers were killed and 19 or 15 injured. [1219] [1220]

A coach carrying pilgrims from the department of Maine et Loire who were returning from the shrine at Lisieux was involved in a bad accident on Petit-Saint-Jean Crossing, PN 168, France, at 02:18 on Monday 25th April 1938.

The vehicle, owned by the Haye company, was being driven by Eugène Cousin, 35. It was carrying 17 passengers and was the second of two coaches returning from Lisieux. The leading coach made it across the line safely but that driven by Cousin was struck by Paris – Bordeaux express, train No. 781 *bis*.

The Société nationale de chemins de fer français express was being run in duplicate because of the heavy post-Easter traffic. It was in charge of Head Conductor Mellet, with Conductor Moreau, Driver Geay René and Fireman Rene Peltier and was travelling at 88 to 90 kilometres an hour when it hit the coach. As a result of the impact the coach was thrown up against a pair of telegraph poles, while the express continued 600 metres down the line to Crossing PN 169 before it could pull up.

The crossing was operated by a gatekeeper, Constantin Ernest, 52, who had only been doing the job for eight days, in place of the regular gatekeeper, M. Marchesseau. M. Ernest had watched the regular express, train No. 781, pass, but had failed to observe the yellow lights at its rear which indicated that a second section was following. Train No. 781 *bis* was seven minutes behind train No. 781. M. Ernest was reported to have been absolutely appalled by what had happened.

After the accident Conductor Moreau went back down the line to protect the obstruction, continuing to Noyant-Méon station to raise the alarm, while the other trainmen went to the aid of the victims. Six bus passengers were killed more or less outright, and the other 11 were more or less injured. The injured were taken first to three houses in Petit-Saint-Jean before being taken by ambulance to hospital in Saumur. One of the injured succumbed on the journey to Saumur. [1221]

Five of six farmers from Sundridge, Ontario, Canada, lost their lives on Wednesday 27th April 1938 when they set out on a fishing trip, and their car was struck by a Canadian National Railway freight train on a level crossing and burst into flames. Ernest Hornibrook, 50; Gordon Hornibrook, 24; James Jones, 42; and Ray Morrington, 24; were killed more or less instantly. The train was stopped, and Conductor Millar and Engineer Appleton removed Clifford Morrington and Edward Wilton, both 23, from the wreckage. They were treated at the scene by Coroner Dr. M.A. Wittick, loaded into a freight car and

---

[1217] *La Vanguardia*, Barcelona, Spain, 30th March 1938
[1218] *The Times*, London, England, 31st March 1938
[1219] Ibid, 16th April 1938
[1220] *La Charente*, Angoulême, France, 16th April 1938
[1221] *L'Ouest Éclair*, Paris, France, 26th April 1938

the train was run as fast as possible to Bracebridge, covering the 60 miles in 80 minutes. The high-speed run was not enough, however, to save Wilton, who died of his injuries as he was being admitted to hospital. Clifford Morrington was badly hurt, though the exact nature of his injuries was not immediately determined. [1222] The accident was attributed to the driver of the car failing to see the train. [1223]

The 1st of May 1938 was celebrated by a major festival in Viana do Castelo, Portugal, attended by thousands of people, including many from outlying villages and parishes, and to bring them to and from town every available means of transport was employed. Thus it was that 40 people from several parishes in the nearby municipality of Barcelos came to Viana in a lorry, driven by a man named Antonio de Silva Pereira. The journey into Viana passed without incident, and the party presumably enjoyed the festivities, though towards the end of the day at any rate the weather was bad, with heavy rain, but disaster struck on the trip home. The homeward journey was considerably delayed due to traffic congestion, and it was 00:15 on 2nd May when the 'bus' arrived at a level crossing near the town.

The crossing was worked by a gatekeeper, Angelina Brandao, who had the assistance of another gatekeeper, Fernando da Silva. It is not entirely clear from a newspaper report on the accident quite how their duties were divided, though the report suggests that Angelina worked the crossing during the day and Fernando did the night duty, implying that it was a main road which had traffic throughout the 24 hours. Fernando should therefore have been on duty at 00:15, but at midnight he apparently requested that Angelina take over, as in performing his job he had become thoroughly soaked in the torrential rain. Possibly he just wanted his colleague to take over for a few minutes while he changed into some dry clothes. At any rate it was the woman gatekeeper who was on duty when the 'bus' arrived at the crossing. The newspaper report suggests that the crossing keeper was not familiar with the timetable, at least as regards special trains, though her assistant had noted down details of some of them from a visit to the station.

The exact details of the accident are unclear but it would seem that a special train of empty carriages from Porto passed shortly before the accident. When the 'bus' arrived its driver found the gates open, and he started across but having passed over he apparently decided that he had gone the wrong way and turned around. When he reached the crossing again the gates were still open and he began to recross, only to be struck by a passenger train travelling at high speed. One account stated that the crossing gates had been closed until the crossing keeper, harassed by the insistent honking of the driver of a vehicle coming in the opposite direction, had opened them, whereupon the 'bus' had also started across.

The collision was a very violent one. The bus was destroyed and engine No. 2253, in charge of Driver Alberto Ferreira de Silva, came to a stand with two carriages on a metal bridge across the River Lima, with the bodies of five victims, two dead and three badly injured, wrapped around its front. In the darkness and rain Driver de Silva had not seen the vehicle at all and had been surprised to find his train pulling up, the collision having ruptured the brake pipe, causing the brakes to apply automatically. Twenty-one bus passengers died and the other 19 were injured.

[1222] *Winnipeg Evening Tribune*, Winnipeg, Manitoba, Canada, 28th April 1938
[1223] *Winnipeg Free Press*, Winnipeg, Manitoba, Canada, 3rd May 1938

A rescue operation was quickly underway after the accident, though the work was made difficult by the darkness and continued heavy rain. The following morning the police opened an investigation by interviewing both the gatekeeper and the bus driver. The latter had suffered only a slight head injury in the accident but was suffering severely from shock. It was not entirely clear, at least at first, who was to blame for the disaster, and it was thought that a rigorous inquiry would be required.

The train eventually reached Braga three hours late.

The bodies of many of the victims were badly mutilated and could not be immediately identified. As they had come from various parishes they were also unlikely to be known to the survivors. Newspaper reporters who went to the parishes from which the victims had come were questioned as to the names of those, mostly poor peasants, who had died, but were unable to provide any information. [1224]

A train collided with a bus on an unprotected level crossing at Nesselwang, Germany, on the stretch between Kempten and Pfronten, on Sunday 5th June 1938. Seven people died and 21 were injured. [1225]

On Friday 1st July 1938, a bus carrying a circus troupe from Rosita to Torreon, Mexico, was struck by a train on a level crossing near Sabinas. The bus was demolished and thrown 50 yards. Manuel Garcia, the bus driver; Jose Mantecon, his wife and two children; Candelaria Reyes, Natividad Flores and one or two unidentified persons lost their lives. [1226] [1227]

A train of the Ferrocarril Buenos Aires al Pacifico collided with a bus of the Chiquita company on Sunday 10th July 1938 on a level crossing at Coquimbito, Argentina. Eleven bus passengers died as a result of the violent collision and all the remainder were injured, some seriously. The accident occurred at about 21:27 and involved a freight train operated by Driver Juan Angelia, with Fireman Victorio Romero, and bound for Palmira. When approaching the level crossing at 35 kilometres an hour in the darkness the train driver failed to see the bus stopped on the crossing, while there was also a suggestion that the brakes of the train had not worked properly, so that it took a long time for it to pull up. When rumours of an accident became known the offices of the local newspapers were besieged by people wanting to know what had happened. [1228] It was stated that the bus had stalled on the crossing and its driver had been preoccupied with trying to restart it, so that he had failed to notice the approaching train. [1229]

On Thursday 14th July 1938 there was a collision between a bus carrying a party of

---

[1224] *Diario de Lisbòa*, Portugal, 2nd May 1938
[1225] *Katastrophen der deutschen Bahnen Teil II*, by H.J. Ritzau, pg. 239 (in German)
[1226] *Santa Ana Register*, Santa Ana, California, United States, 2nd July 1938
[1227] *Nevada State Journal*, Reno, Nevada, United States, 3rd July 1938
[1228] *El Orden*, Santa Fe, Argentina, 11th July 1938
[1229] *The Railway Gazette*, London, England, 2nd September 1938

women excursionists and a shunting train at Trebsen-Pauschwitz station, Germany. Driving too fast as he approached the level crossing the bus driver realised the danger too late and was unable to pull up in time. Fire broke out following the collision from the shattered fuel tank of the bus and nine people died and 21 were injured. [1230] [1231]

Flashing red warning lights and the fact that two other cars were already stopped at the crossing failed to prevent a bad accident at Hobart, Indiana, United States, early on Monday 1st August 1938. An automobile which was apparently being driven by Werner Boevers, 41, of Cleveland, Ohio was hit by a train. The collision killed Mr Boevers, his wife Clara, 40; their daughter June, 18; Mrs Louise Kuntze, of Chicago, Illinois; and Mrs Alice Wrede, of Cincinnati. The party had apparently been returning to Cleveland after a visit to the Hines Memorial Veterans' Hospital in Maywood, Illinois, where Mrs Wrede's husband Albert was a patient.

The car had been struck by the *Broadway Limited*, crack streamliner train of the Pennsylvania Railroad, which had been travelling at about 70 miles an hour, the crossing being located towards the end of one of the fastest stretches on the route. Wreckage of the vehicle was strewn along 1,000 feet of track, while the impact disabled the engine of the train, and another engine had to be sent to bring the train into Chicago. The train passengers were transferred to a following train, the *Liberty Limited*. The road (Route 6) crossed the line at an acute angle and the view of trains was obscured by lineside buildings. [1232]

At about 18:50 on Friday 5th August 1938 a train collided with a car on Perroquet Crossing, near Adinkerque, Belgium, only 200 metres from the French frontier, and claimed the lives of six people. Mme. Irène Bonithon, her three children, Claude, Jacques, and Michel, aged between 10 and 15; and Mme. Anna Vanrobaey, died at the scene while Mlle. Germain Cappelle, who was probably driving the car, died shortly after admission to Furnes clinic. [1233] The vehicle was carried 750 metres along the line. The Furnes parquet went to the scene of the accident. No protection of any description was provided at the crossing, which had been the scene of a similar, though less serious accident, two months previously. [1234]

A man and his wife, their two young children and their native chauffeur lost their lives on Apex level crossing, between Brakpan and Benoi, South Africa, on about Sunday 14th August 1938, when their car was struck by an electric train which they had failed to give way to. [1235]

Six men from St François de Sales, on their way work in Ile Maligne, were the victims

[1230] *Eisenbahnunfälle in Sachsen* by Claus Schlegel, pg. 168 (in German)
[1231] *The Times*, London, England, 18th July 1938
[1232] *Pharos Tribune*, Logansport, Indiana, United States, 1st August 1938
[1233] *La Charente*, Angoulême, France, 8th August 1938
[1234] *Le Populaire*, Paris, France, 6th August 1938
[1235] *The Times*, London, England, 16th August 1938

of a level crossing accident at Chambord, Quebec, Canada, on Thursday 18th August 1938, when the car in which they were travelling was hit by a Canadian National Railway passenger train which had just left Chambord for Dolbeau. Five of the men: Joseph Boucher, Cetophas Fortin, Louis Lavois, Raoul Martel and Leonce Tremblay were killed, and the sixth man was taken to Roverval Hospital, critically injured. [1236]

A family visit ended in tragedy at Altoona, Kansas, United States, early on Monday 22nd August 1938. A car being driven by a Mrs Kepley, 19, and containing her children Barbara Lee, 2; and Harold Wilbert, 6 months; together with her younger sisters Isabel and Ethel Ware and her younger brother, Billy Ware, was struck by a southbound Missouri Pacific Railroad passenger train. The automobile was carried five blocks through town before the train stopped. All the occupants of the car except Ethel Ware were killed outright. Ethel was critically injured and died several hours later in Neodesha hospital. Mrs Kepley's husband was a Works Progress Administration worker. [1237]

On Thursday 1st September 1938, a car which was returning from an interment at a Jewish cemetery north-east of Riga, Latvia, and carrying five relatives of a deceased industrialist, was struck by a passenger train on a level crossing. The engine and leading coach of the train were derailed, and six people lost their lives, including the widow and daughter of the industrialist and the driver and fireman of the train. Fifteen other people were injured. [1238] [1239]

Five members of a family died when the car in which they were riding was struck by a Boston and Maine Railroad train bound for Concord on a level crossing at Meredith, New Hampshire, United States, on the evening of Sunday 4th September 1938. The victims were Frank Stodd, 45; his wife Sophie, 43; their son Robert, 20; Mrs Stodd's sister, Mrs Thomas Power, 41; and Mrs Power's son John, 14. Miss Marie Stodd, 24, a daughter, had been riding in another car with William Larew, 29. Their car had already passed the crossing and they only learnt of the accident an hour later. [1240]

A passenger train travelling from Barcelona to Bañelos collided with a road tanker carrying 20,000 litres of petrol on a level crossing with the road from Sarriá de Ter to Gerona, Spain, near the former place, at about 21:00 on Wednesday 7th September 1938. The initial collision does not seem to have been that serious and immediately after it had happened the engine driver, the tanker driver and others got out and were examining the damage and deciding (probably arguing) as to who was responsible. People living in the houses adjoining the crossing, alerted by the noise of the collision, had also come out of their front doors or onto their balconies to see what had happened. Then just over three minutes after the collision there was a tremendous explosion and the seven closest

[1236] *The Gladstone Age Press*, Gladstone, Manitoba, Canada, 25th August 1938
[1237] *The Emporia Gazette*, Emporia, Kansas, United States, 22nd August 1938
[1238] *Le Petit Journal*, Paris, France, 2nd September 1938
[1239] *The Times*, London, England, 2nd September 1938
[1240] *The Titusville Herald*, Titusville, Pennsylvania, United States, 5th September 1938

houses, and the six carriages of the train were splashed with flaming petrol and set ablaze.

Most of the train passengers had also alighted from their carriages to find what had happened and according to a local newspaper report the following day they largely escaped from the subsequent explosion and fire. The newspaper wrote that five bodies, yet to be identified, had been taken to the cemetery and there were 30 injured, some it was feared fatally. Francoist sources (the accident had occurred in an area still under Republican control) stated that there had been 19 deaths and 21 injured. Those who died probably included both the lorry and train drivers; the fireman was amongst those taken to Gerona Hospital. [1241] [1242]

A mistake by a crossing keeper led to a serious accident on the electrified line from Villefranche via Vernet-les-Bains to Perpignan at Marquixanes, France, near Prades, on Tuesday 13th September 1938. The barriers of the crossing had been lowered for a train from Prades to Perpignan. This train passed the crossing and with motor vehicles waiting to cross, the keeper raised the barriers. The leading vehicle, a bus carrying 45 people, entered the crossing and was followed by a second vehicle. At that moment another electric railcar, carrying only a motorman and a conductor and travelling at 80 kilometres an hour, appeared. It struck the bus and cut it cleanly in two. The rear of the bus was thrown into a ditch and the front half was carried 60 metres before the train could be pulled up.

Four bodies, those of two men and two women, were found in the remains of the bus, and nine injured persons, among them the bus driver, were transported as a matter of urgency to hospitals in Prades and Perpignan; several of them were seriously injured.

Initial information was that the first train had carried a marker at its rear indicating to the crossing keeper that a second train was following, but that the keeper had failed to notice it.

A fifth victim appears to have died later of his or her injuries. [1243]

A train collided with a postbus at Wiesloch, Germany, on the stretch of line between Heidelberg and Karlsruhe, on Tuesday 4th October 1938. The crossing barriers had been left open. There were 12 deaths and 8 injured. [1244]

A car driver braked as he reached an open crossing between Haidhof and Burglengenfeld, Germany, then continued onto the crossing, perhaps thinking that he or she had time to get across in front of the train. There was not and five people died in the accident on Tuesday 18th October 1938. [1245]

A party on their way to attend a Mission at Stoney Point Roman Catholic Church

---

[1241] *La Vanguardia*, Barcelona, Spain, 8th August 1938
[1242] Webpage: Federación Castellano Manchega de Amigos del Ferrocarril (in Spanish)
[1243] *Le Matin*, Paris, France, 14th September 1938
[1244] *Katastrophen der deutschen Bahnen Teil II*, by H.J. Ritzau, pg. 239 (in German)
[1245] Ibid.

became the victims of a level crossing collision at Stoney Point, Ontario, Canada, at 7 p.m. on Wednesday 26th October 1938. Killed in the accident were Mrs Tourangeau, 48; her son Leo, 22; her daughter Genevieve, 18; Mr Arthur Masse, 83; and Mrs Julia Goudoin, 72. Mrs Tourangeau's husband Peter was critically injured. All were Stoney Point residents, so they must only have been making a short journey when their car was struck by a Canadian National Railway train. Their bodies were carried 1,300 feet and it could not be determined who had been driving the automobile. Although there were no gates or warning signals at the crossing there were unobstructed views along the line. [1246]

An inattentive bus driver collided with a train of the narrow gauge Schönhausen-Genthin Kleinbahn on an unprotected level crossing with the Berlin Chaussee (highway) at Genthin, Germany, on Sunday 30th October 1938. The fuel tank of the bus, damaged in the collision, exploded. Eight people died and 16 were injured. [1247]

A level crossing accident in Italy on about 16th November 1938 had unexpected consequences for some of the more gluttonous local people who witnessed it. A train on the Brescia – Isco line collided with a lorry. There were no reports of anyone being hurt in the collision, but the load of the lorry was spilt all about the place. It appeared to be "chocolate drops" and this confectionary was eagerly consumed by the locals, both adults and children. A short time later there were many calls to doctors and some of the greedier residents found themselves in hospital. The "chocolate drops" had been chocolate-coated laxatives. [1248]

The first and relatively less serious of two level crossing collisions involving school buses in two days in the United States occurred near Frederic, Wisconsin, on Wednesday 30th November 1938. Mrs Covey, 30, a farmer's wife, was employed by the North Star school in Lick Township and was taking the children home from school. The improvised school bus she had been driving was hit by a northbound Minneapolis St Paul and Sault Ste Marie Railroad train from Minneapolis to Duluth at a crossing two miles out of town. Mrs Covey and three of the children – Lenwood and Leroy Lawson, aged 12 and 10 respectively; and Joyce Lind, 6 – were killed. Three of the other five children, two of them Mrs Covey's children, Billy, and Gloria, were critically injured.

There were various theories as to how the accident had happened. It was possible that a frosted windscreen and side windows had prevented Mrs Covey from seeing the train. One of the surviving pupils told the coroner that they had not seen the train until it had been 20 feet away. The coroner himself theorised that the bus might have stalled on the crossing.

Conductor F.E. Burt, in charge of the train, described the bus as having been one made from an old automobile and added that it was the worst wreck that he had ever

[1246] *The Amherstburg Echo*, Amherstburg, Ontario, Canada, 28th October 1938
[1247] *Deutsche Eisenbahn Katastrophen* by Hans Joachim Ritzau, Jorgen Hörstel and Thomas Wolski, pg. 187 (in German)
[1248] *The Havre Daily News*, Havre, Montana, United States, 17th November 1938

seen.[1249] [1250] [1251]

What remains in late 2021 the most serious level crossing collision involving a school bus in the United States and possibly anywhere in the world, occurred at Burgon's Crossing, Utah at 8.43 a.m. on Thursday 1st December 1938. (The Interstate Commerce Commission (ICC) gives the location as 'Near Riverton' and other sources give it as 'Sandy'.) The bus was carrying 38 pupils from their homes to the Jordan High School and was being driven by 29-year-old Farrold Silcox. Twenty-three pupils and driver Silcox were killed outright in the accident and the other 15 pupils were injured to varying degrees; two of them would succumb in the days after the accident.

Approaching the level crossing concerned, the road along which the bus was travelling ran alongside of and south of the single-track of the Denver and Rio Grande Western Railroad for about half a mile. The road then crossed the line at right angles before continuing alongside of the track on its north side. The crossing was protected only by crossbucks but there was a clear view along the line in both directions.

Various laws and regulations governed the operation of school buses over level crossings.

The state road traffic laws stated:

The driver of any motor bus carrying passengers for hire, or of any school bus carrying any child, shall, before crossing at grade any track of a railway stop such vehicle not less than 10 feet or more than 50 feet from the nearest rail of such track, and while so stopped shall listen, and look in both directions along such track for approaching railway trains or cars and for whistles or other warning indicating the approach of a train, and shall not proceed until he can do so safely.

The regulations governing pupil transportation read:

All school buses shall be brought to a full stop at all railroad crossings at least 10 feet and not more than 50 feet from the nearest rail until the driver has ascertained that it is safe to cross. After making the stop the bus shall be kept in low gear until the tracks have been crossed.

Finally, the Jordan School District rules for bus drivers stated:

All busses must be brought to a full stop at least 25 feet from any steam or electric railway crossing. Before proceeding on his way, the driver must be certain that no train is approaching from either side.

The train which struck the bus was westbound second-class freight train No. 31, *The Flying Ute*, consisting of 4-6-6-4 *Challenger* class steam locomotive No. 3708, 12 loaded and 38 empty cars and a caboose, in charge of Conductor J.P. Ryan, with Engineer E.L. Rehmer and Fireman Alfred Elton. It had left Helper, just over 102 miles east of the crossing, at 3.32 a.m., 3 hours, and 48 minutes late; had passed Provo, 27½ miles from the crossing, at 7.54 a.m., now 4 hours and 19 minutes late and was travelling at 52 miles an hour at the time of the collision.

1249 *The Winona Republican Herald*, Winona, Minnesota, United States, 1st December 1938

1250 *The Oshkosh Northwestern*, Oshkosh, Wisconsin, United States, 1st December 1938

1251 *Appleton Post-Crescent*, Appleton, Wisconsin, United States, 1st December 1938

The bus was demolished in the collision, its body was torn from its chassis and thrown to the left and the chassis was carried along the track in front of the engine over 2,000 feet before the train stopped with the front truck (bogie) of the engine derailed.

Engineer Rehmer told the ICC inquiry that he had sounded the whistle approaching the crossing and the whistle had still been blowing when, close to the crossing, his fireman had shouted a warning and he applied the brakes in emergency. The engine bell had been ringing and its headlight burning. It was snowing slightly. Visibility he estimated at about 2,500 feet but he could not see the bus from his side of the engine.

Fireman Elton testified that he had seen the bus some distance ahead as his train was approaching the crossing. He had seen the bus turn to the right and come to a stand about a bus length from the crossing. Then he had seen the bus start off forward again and had called out to his engineer. Visibility had been such that he had been able to see the rear of his train at any time.

Head Brakeman Kelley, who had also been riding on the engine, corroborated the evidence of the fireman.

Conductor Ryan stated that he had been able to see the engine from the caboose all of the time. The emergency brake application had been effective.

Pupil Glenn Kump, aged 15, said that he sat in the rear of the bus. It had been snowing slightly but there was no wind. The bus windows were closed and steamed up so that it was impossible to see through them, but the windscreen had a defrosting window, and the windscreen wiper was operating. He recalled the bus turning to the right and stopping; he did not notice the bus driver open the side door that morning, though he had on other occasions. There had been no difficulty with the operation of the bus. Driver Silcox was very strict and there had been no rowdyism on the bus. Some of the pupils who were members of the school band had had their instruments with them, but these were piled at the front of the bus, and none were being played. One only one occasion during the school year could he (Kump) remember their having previously encountered a train at the crossing. On that occasion the bus driver had given way to the train; then the weather had been better and everyone on the bus could see the train.

Eleven other pupils agreed that though there had been snow on the ground the highway had been clear and the bus had not slipped or skidded at any point. It had been hazy or foggy and impossible to see through the windows. The pupils had been chatting or doing their homework and there had been no abnormal noise, loud talking, laughing, or singing, as the bus driver kept a strict discipline. No one heard a train whistle but a pupil near the front had been heard to shout, "Train," immediately before the collision. [1252]

At least two pupils, one a survivor of the accident and the other a pupil who had been waiting to board the bus, spoke to newspaper reporters:

'One of the survivors was Margie Groves, 16, of south Jordan. "I got on the bus at 8.30 about two miles from the crossing," she said. "I think there must have been about 40 others in the bus with me. When we came to the crossing we stopped, and Mr Silcox, the bus driver, started up again. I thought I heard someone shout 'train' and then I saw something black coming from the right and it hit us. When I woke up I was lying in the

[1252] Interstate Commerce Commission report

snow near the bus. Two students I knew, Mark Bateman and Mabel Smith, were nearby. Mabel was lying near the bus and Mark was lying near a fence. I guess they're all right now. There was a lot of screaming and moaning. They were all just kids – 15 to 18 years of age. And then they came and brought me to a hospital."

'June Wynn, 16, of Sandy, was perhaps the only eye witness. She was standing in the doorway of her home 200 yards from the crossing, she said. "I saw both the bus and the train coming and wondered whether the bus was going to stop. It did stop, and then pulled onto the track directly onto the train. I heard the crash, then screams. When the front of the train got opposite me, I could see the frame of the bus still on the engine. The train started slowing down immediately. My father and I rushed to the track, and train men told us to go back and call ambulances. It was foggy and snowing and I wouldn't have been able to see the train or bus if they both hadn't had their lights on. I did not hear the train whistle." [1253]

Three residents living near the crossing did hear the train whistle.

A traffic check made by the ICC counted 17 trains and 95 motor vehicles, four of them buses, using the crossing in 24 hours, with the busiest period for road vehicles being between 9 and 10 a.m. and 3 and 4 p.m. In a test to determine how audible the train whistle was, it was found that the whistle could not be heard until the train was 300 feet away with the door of a motor vehicle closed; with the door open the whistle could be heard clearly when the train was at the whistle board, 1,430 feet away.

The ICC concluded that if Driver Silcox had taken the proper precautions he would have seen the train. If he had not been able to see through the window he should have opened the side door, when he would have been able to see or hear the train. As he had lost his life in the accident no reason could be given for his failure. Possibly the infrequency of his meeting a train was a factor. It could not be considered sufficient for drivers simply to stop on the approach to crossings and listen for trains with all windows and doors closed and a provision should be included in the regulations and strictly enforced that all bus drivers open the front side doors to listen for trains.

The ICC recommended that

(a) More stringent rules covering the operation of school buses over level crossings should be prescribed and strictly enforced.

(b) That all drivers of school buses should be required to open the front side door when the stop was made at each railroad crossing.

(c) That, whenever practicable, buses should be routed so as to avoid level crossings which were not protected by either watchmen or devices giving visual warning when a train was approaching. [1254]

Mr W.K. Granger, of the Utah Public Services Commission, stated that the accident would cause the commission to push with renewed vigour its campaign to make grade (level) crossings safer. Burgon's Crossing had been included in a recent survey of crossings by the commission but had been considered less hazardous than many others in the state because of the clear view of the tracks and it had thus not been included in a list

---

[1253] *Montana Standard*, Butte, Montana, United States, 2nd December 1938

[1254] Interstate Commerce Commission report

of crossings needing immediate attention. It was, however, on a list of crossings to receive train-operated signals the following year. In view of the accident, crossings used by school buses would be given priority.

Mr Granger added that since the beginning of the crossing campaign the previous spring 14 dangerous crossings had been closed and protective devices added at 80 other crossings in different parts of the state. He said, "I hope this tragedy will arouse the public to the need of greater crossing protection. In every case where we have proposed to close crossings, citizens and local officials have protested, but we have closed many just the same."

The state road commission had reportedly spent several hundred thousand dollars of federal and state funds in the previous few years on grade separation projects and train-actuated crossing devices. Priority had been given to crossings where the traffic was the heaviest and where surveys showed that the hazard was the greatest, though it was realised that any crossing, even where traffic was light, was a potential danger spot. [1255]

Five people died in a level crossing accident a mile and a half from Dresser Junction, Wisconsin, United States, on the evening of Friday 16th December 1938. The automobile had just left the farm of Mr Royal Hazel and his wife, and the collision occurred on a private crossing. Struck by a Minneapolis St Paul and Sault Ste Marie Railroad train en route from Duluth to Minneapolis, the car was tossed 50 feet and the bodies of the victims were strewn over 150 feet. Mrs Hazel, 28; her son James, 4; Charles Bergeson, 26; his wife Bessie, 19; and Ruth Hill, 17, of Taylors Falls, Minnesota, were all killed or fatally injured. Mr Devilah Hazel, 30, the husband of Mrs Hazel, and seven-month-old Carlos Bergeson, son of Mr and Mrs Bergeson, were badly injured. The dead and injured were loaded onto the train and taken to Osceola, Wisconsin, where the injured were admitted to the Ladd Memorial hospital. It was thought that fogged car windows had prevented Mr Hazel from seeing the train at the unobstructed crossing. [1256]

Less than two days after the Dresser Junction accident, early on Sunday 18th December 1938, another five people died in a level crossing accident at Rocky Ridge, Ohio, United States. Mr Arthur Below, 41; his wife Mary, 38; their daughter Orlena, 17, their son Calvin, 14; their niece Alice May Floriana, 9; and a neighbour, Martin Madison, 59, had been returning from a party at Rocky Ridge. The automobile had only travelled a block when it was struck by a westbound New York Central Railroad train. Four of the occupants of the car were killed outright and Calvin Below died in hospital an hour later. Orlena was injured but was able to walk to a restaurant a block away before she collapsed.

There were five tracks at the crossing and a watchman was on duty during the daytime. There were no barriers or warning lights. [1257] [1258]

On Sunday 1st January 1939 a Compañia de Tranvias y Ferrocarriles de Valencia train

[1255] *The Salt Lake Tribune*, Salt Lake City, Utah, United States, 2nd December 1938

[1256] *The Winona Republican-Herald*, Winona, Minnesota, United States, 17th December 1938

[1257] *The Chronicle-Telegram*, Elyria, Ohio, United States, 19th December 1938

[1258] *The Oil City Derrick*, Oil City, Pennsylvania, United States, 19th December 1938 reportedly collided with a Linea de Paterna tram on a level crossing at Paterna, Spain. Seven people died and 16 were injured. [1259]

A Chicago North Shore and Milwaukee Railroad train travelling at a reported 70 miles an hour collided with a car on Nicholson Road Crossing, Oak Creek, Wisconsin, United States, on the afternoon of Sunday 8th January 1939. Mr Herbert Fieldler, 47, Milwaukee department of public works employee; his wife Anna, 35, a high school matron; Mr William Blank and his wife, both 52; and the Blanks' grandson Richard, 7, were the victims. Four of the occupants of the car were killed outright and Mrs Fiedler died later of her injuries. The party had been out on a Sunday afternoon drive.

The road crossed the line at an acute angle at a point where there was good visibility. Motorman Walter Dowse, driving the three-car northbound electric rain, and Conductor Victor Nelson, who was riding in the cab with him, said that the automobile had only been moving slowly and they had assumed that it was going to stop but it kept going and moved onto the track directly in front of their train, the whistle of which had been sounding continuously. Part of the car was trapped under the train and Motorman Dowse said that he had been surprised that his train was not derailed. [1260] It was theorised that Mr Fiedler had been 'confused by the lights,' the reporter presumably meaning the lights of the train. [1261]

Eight people died when a Ferrocarril Central Argentino electric train collided with a car on a level crossing at San Isidro, a suburb of Buenos Aires, Argentina, on the evening of Sunday 5th February 1939. [1262]

Another train/school bus accident happened on the outskirts of Birmingham, Alabama, United States, on the evening of Sunday 26th February 1939. This time no one was killed, though the bus was dragged over 400 metres and demolished. Nineteen children were injured and one of them sustained a fractured skull but was not believed to be critically hurt. One of the pupils, Juliet Williams, 11, leaped from the bus as it was being carried along the line with her two-year-old brother. She escaped with a sprained ankle and he was unhurt. Parents who had rushed to the scene fearing a tragedy found frightened children tangled in the wreckage.

The children had been returning from Sunday School on a bus provided for the parishioners of the Rev. J.A. Bryan, a Presbyterian minister. The bus driver, Lewis Allen, 29, said afterwards that children near the front of the vehicle had blocked his view and that was why he had failed to see the train. He suffered a scalp wound. [1263] [1264]

[1259] Webpage: Federación Castellano Manchega de Amigos del Ferrocarril (in Spanish)

[1260] *The Winona Republican-Herald*, Winona, Minnesota, United States, 9th January 1939

[1261] *The Wisconsin State Journal*, Madison, Wisconsin, United States, 9th January 1939

[1262] *The Times*, London, England, 7th February 1939

[1263] *Thomasville Times-Register*, Thomasville, Georgia, United States, 27th February 1939

[1264] *San Antonio Express*, San Antonio, Texas, United States, 27th February 1939

Only a day after the Birmingham accident, on Monday 27th February 1939, there was another school bus accident involving a train in the United States, and again there was a lucky escape from serious consequences. This time the bus driver, named Harold Higley, was probably not really greatly to blame. When his bus approached the level crossing seven miles south west of Lafayette, Indiana, it was shrouded in a thick fog. Before crossing he had therefore taken the precaution of sending one of the students ahead to check if there was a train anywhere about. There was, but unfortunately the boy failed to hear it, perhaps because the fog had muffled the sound of it. He signalled the driver that it was all right to cross, the bus set off, and its front was struck by a Cleveland Cincinnati Chicago and St Louis Railway (Big Four) train. Only three pupils were injured, only one of whom was detained in hospital. Fourteen other students complained of being shaken up. Another bus took the children on to the Klondike school. [1265] [1266]

Wednesday 26th April 1939 saw yet another school bus colliding with a train in the United States. The location was Commonwealth Avenue Crossing, Concord, Massachusetts. Again, none of the 16 schoolchildren on the was much hurt but the bus driver, James Finigan, suffered a fatal head injury. The bus had just left the school grounds, carrying pupils from the Harvey Wheeler Junior High School and West Concord Grammar Schools, when the accident happened at 3.45 p.m. Arriving at the crossing the bus had stopped to let a passenger train from Boston pass, then according to pupils on the bus the barriers had been raised, and the bus had moved forward only to be struck by a Boston-bound freight train on the level crossing with the Boston and Maine Railroad main line. This, however, was contradicted by the railway company, who stated that the bus had crashed through the gates. The crossing was manned by a crossing tender, James Mitchell, who stated, "I put the gates down for the passenger train and they were down." One of the pupils, 11-year-old James Loring, was credited with quick thinking and quick action after the collision in climbing over the unconscious bus driver and turning off the ignition, thus preventing a likely catastrophe from spilt petrol being set alight. [1267]

A train collided with a bus on an unprotected level crossing at Wulfsen, Germany, between Buchholz and Lüneburg, on Sunday 4th June 1939. Fifteen people died and 23 were injured. [1268] The bus driver reportedly failed to see the train until it was too late and his bus was in the middle of the track when it was struck and carried 20 metres by the train. [1269]

Five people died and one was seriously injured in a level crossing collision between a train and a car at Möckeln, Sweden, between Stockholm and Malmö, on about Saturday 1st July 1939. [1270]

[1265] *The Vidette Messenger*, Valparaiso, Indiana, United States, 27th February 1939
[1266] *Pharos Tribune*, Logansport, Indiana, United States, 27th February 1939
[1267] *Concord Enterprise*, Concord, Massachusetts, United States, 3rd May 1939
[1268] *Katastrophen der deutschen Bahnen Teil II*, by H.J. Ritzau, pg. 240 (in German)
[1269] *The Railway Gazette*, London, England, 9th June 1939
[1270] *Gazette de Lausanne*, Lausanne, Switzerland, 3rd July 1939

An eight-seater automobile which had been converted into a bus was struck and cut in two by the *Alton Limited* passenger express of the Chicago and Alton Railroad on a level crossing near Elwood, Illinois, United States, on Monday 17th July 1939. Mr A.J. Gadberry, aged about 65, the owner and driver of the bus, had been en route from Wilmington to Joliet and had left the main to drive into Elwood. With him in the bus were his son Ray, 35; his daughter-in-law Alice, also 35; his granddaughter Peggy, 5; and Miss Bertha McQueen, about 65. The party were out on a pleasure drive, showers of rain having prevented Mr Gadberry from working on a Works Projects Administration project. (The Works Progress Administration had been renamed the Works Projects Administration in April 1939 by President Roosevelt). All were killed. Travelling at about 45 miles an hour, the train had been travelling from Chicago, Illinois, to St Louis, Missouri. Its engine had whistled for the unguarded crossing and its bell had been ringing. No one on the train was hurt but the cowcatcher of the engine was broken, and engines had to be changed at Bloomington. [1271]

A mistake by the crossing keeper, who opened the crossing barrier prematurely, resulted in a collision between a train and a bus at Potsdam, Germany, on the stretch between Berlin and Wiesenburg, on Saturday 5th August 1939. The bus was carried 300 metres. Twelve people died and 11 were injured. [1272]

Details of an accident on the Palestine Railway on Sunday 17th September 1939 are somewhat meagre and confused, but at least 19 Jews were killed, and others injured when a train collided with an Egged Company bus near Ramleh (now Ramla, Israel), midway between Jerusalem and Tel Aviv. One report speaks of 29 fatalities and the location of the accident is also given as near Sarafend. From a photograph of the accident the bus does not appear to be too badly damaged but there are signs of blackening of its body from a possible post-accident fire. [1273] [1274] [1275]

Two women and seven children died in an accident at Etobicoke, Ontario, Canada, a township just west of Toronto, on the morning of Wednesday 10th October 1939. Mrs Brown, accompanied by her twin sister, Miss Mildred Taylor, was driving the children to the Swamp school at Islington when the car in which they were riding was struck by a train, scattering the bodies of the victims along the track. The children who lost their lives were Mrs Brown's daughter Marion, 6; Marion Tibando, 11; her brother Andrew, 7, and sister Lucy, 6; Gladys Davis, 11, her sister Verna, 6; and Joyce Boyle, 10. A pet dog was also killed but two other Brown children escaped as they had cycled to school. (Other accounts gave the name of the adult accompanying Mrs Brown as Mrs H. Davis.)

Robb Webb, of Toronto, a passenger on the train, described the scene after the accident as a terrible sight. He related, "I saw about four or five bodies strewn around as I

[1271] *The Daily Illini*, Urbana-Champaign, Illinois, United States, 18th July 1939

[1272] *Katastrophen der deutschen Bahnen Teil II*, by H.J. Ritzau, pg. 240 (in German)

[1273] https://www.jta.org/1939/09/18/archive/18-jews-killed-9-wounded-in-train-bus-collision-in-palestine

[1274] https://www.flickr.com/photos/124446949@N06/24752021026

[1275] *Chester Times*, Chester, Pennsylvania, United States, 18th September 1939 got off the train and it sickened me, so I climbed right back into the coach. The train stopped quickly with a terrific grinding sound following the impact."

Visibility at the crossing was good, with an unobstructed view each way for several hundred yards. In what is a fairly common scenario in level crossing accidents Mrs Brown had apparently stopped her car and waited for an eastbound train to pass, then had assumed that the way was clear and pulled onto the track in the path of a westbound Canadian Pacific Railway flyer. [1276] [1277]

A Works Projects Administration employee and almost all his family lost their lives in a level crossing accident on the outskirts of Macomb, Illinois, United States, on Saturday 21st October 1939, when the vehicle was struck by a Chicago Burlington and Quincy Railroad train. Roscoe Johnson, 45; his wife, whose name and age were not stated; and their children Margaret, 12; Duane, 10; Gene, 8; and Josie Lee, 5, were on their way to the home of Roy Lansdowne in Macomb, where the car was going to be repaired. Lansdowne was also a passenger in the car and was the only one who was not killed outright in the accident, but he succumbed from his injuries in hospital the next day. Another Johnson child, Anna Lou, 13, was staying with her uncle, Charles Lansdowne, in Macomb. Whether he was the brother of Roy the newspaper reports do not say, nor is there any indication of how the accident happened, but if the automobile was in need of repair it is conceivable that it might have broken down on the crossing. [1278]

A bad level crossing accident occurred near Moga Thail, India, on the Ludhiana – Ferozepur section of the North Western Railway on Saturday 28th October 1939, when a diesel railcar collided with a motor bus on a class 'C' unmanned level crossing. The bus driver rashly attempted to cross the line in front of the train and stalled his vehicle. Nine people died and 13 were injured. Damage to the railcar amounted to 300 rupees. [1279]

Germany suffered yet another disastrous level crossing collision involving a bus on Friday 1st December 1939 between Dortmund and Witten West. On this occasion the bus crashed through the lowered barriers. Fifteen people died and 17 were injured. [1280]

The failure of a car driver, Dr Charles Hainaut, to notice either an approaching train or the warning gestures of people in another car resulted in the vehicle being struck by a Pennsylvania Railroad freight train on a level crossing with a dirt road on the evening of Thursday 7th December 1939. The party in the car had been returning from a prayer meeting at the home of Howard Holmes, only a few hundred feet from the crossing at McDonald, Pennsylvania, United States. They had been following another vehicle, the occupants of which attempted to warn them of their danger. Mrs Isabel Pletincks, one of those in the front car, told a reporter, "We realised Dr Hainaut might not see the train

---

[1276] *The Winnipeg Tribune*, Winnipeg, Manitoba, Canada, 10th October 1939
[1277] Ibid, 11th October 1939
[1278] *Logansport Press*, Logansport, Indiana, United States, 22nd October 1939
[1279] Report by the Railway Board on Indian Railways for 1939-40
[1280] *Katastrophen der deutschen Bahnen Teil II*, by H.J. Ritzau, pg. 240 (in German)

approaching. It was coming fast and before we could move there was a great crash. We ran down the track to the smashed automobile. We located it by one headlight still burning. Dr Hainaut was still behind the wheel, in a semi-conscious condition."

Dr Hainaut's son, Charles Jr., 3, Mrs Dickinson, Mrs Elizabeth Hookey and Thomas Lewis were killed outright; Mrs Hainaut died from her injuries earlier the next day and Dr Hainaut was seriously hurt. Conductor B.J. Wilson was in charge of the train, with Engineer M.J. Kinary. [1281]

Nineteen workmen of the Lenzing Zellwolle (cellulose) factory were killed and about 20 others injured when a bus in which they were riding was struck by a train on a level crossing at Vöcklabruck, Austria, Germany (near Linz), on Sunday 10th December 1939. An obituary on the victims read: "Sie starben um die Rohstoff-Freiheit Deutschlands" (They died for the freedom of German raw materials.) [1282] [1283]

On the evening of Sunday 10th December 1939 what was described in a French report as a lorry loaded with fertiliser smashed through the crossing barrier at Mesnil-sur-Oger station, France, and was hit by a train. Five people died, 20 were injured and the line was blocked for some time. [1284]

The following day the London *Times* reported a level crossing collision 'in the R.A.F. area' in which a fast goods train collided with a lorry carrying British Army signallers. Five of them were killed outright, a sixth died in hospital and seven were injured, one critically.

Given that Mesnil-sur-Oger is in north-east France, the area where the British Expeditionary Force were located, it seems highly likely that the 'lorry loaded with fertiliser' referred to in a French newspaper was in fact the army lorry detailed in the British report, with the details of the accident being censored in the French press but not the British, other than the location.

The soldiers on the lorry had been returning from a visit to a cinema, while, 'Only a few hours before all the men were cheering the King during his tour of inspection.' They were all office workers from the Hull area and, 'Among those killed were some of the fastest teleprinter workers in Great Britain.'

Those who lost their lives were Lance-Corporal C.W. Smith and Signallers C. Kitching, D.C, Larman, W.M. Sullivan, C.H. Taylor, and N.M. Walker. [1285]

Five people died at Lakewood, Ohio, United States, on the evening of Saturday 16th December 1939, when a car driven by Louis Rupp, 43, was struck by the westbound *Commodore Vanderbilt* passenger train of the New York Central Railroad, bound for Chicago from New York and travelling at 70 miles an hour. Rupp, Mrs Blanche Spance,

---

[1281] *Tyrone Daily Herald*, Tyrone, Pennsylvania, United States, 8th December 1939
[1282] https://www.land-oberoesterreich.gv.at/13783.htm
[1283] *Lubbock Morning Avalanche*, Lubbock, Texas, United States, 12th December 1939
[1284] *Le Petit Parisien*, Paris, France, 11th December 1939
[1285] *The Times*, London, England, 11th December 1939

43, and her three daughters; Mary, 18; Jean, 13; and Donna Louise, six months, were all killed. The automobile was carried 1,000 feet down the line. Engineer Charles Fahlbush said that the bell and whistle of his engine had been sounding with "full force." There were five tracks at the crossing. [1286] Rupp reportedly had overtaken several cars which were stopped waiting for the train at the crossing. [1287]

The evening of Christmas Day, Monday 25th December 1939, saw a tragedy at 95th Street Crossing, Oak Lawn, Illinois, United States, on the outskirts of Chicago. A car carrying George Clark, 33, an oil refinery worker; his wife Elizabeth, also 33; their children Jackie, 10, and George Jr., 8; and Clark's mother Anna, 64, all from Hammond, Indiana, was struck by the *Blue Banner* passenger train of the Wabash Railroad and all were killed. [1288]

The only witness to the accident was his brother-in-law, Lee Moore, whose wife was Mrs Clark's sister, who was following the Clark family in another car and told the inquest on the victims of seeing parts of the Clark car and bodies flying through the air and knowing that they were members of his family. His wife had fainted at the sight. Moore testified, "We were on our way to Glen Ellyn for a Christmas party. Our car was about 50 feet behind Clark's. When we were 75 feet away from the crossing the lights went on. Clark's car was then only 25 feet from the tracks. There wasn't enough time for him to stop. The next thing I knew the accident had happened. I'm sure the lights went on too late. No one could have stopped. It was getting dark, and visibility was poor."

Engineer Prindle stated that the train had been 12 minutes late and had been travelling at 80 miles an hour when it hit the car. He had been blowing the whistle of his locomotive at the time.

The inquest ended with a verdict of 'Accidental Death' accompanied by a scathing attack by the jury on the Wabash company for their having failed to provide adequate protection at the crossing and recommending that either a flagman be provided or that the speed of trains over the crossing be limited to 12 miles an hour. Jury spokesman Charles Ingersoll, an Oak Lawn resident and ex-railwayman, commented, "It's time we stop playing around and do something. The village for years has tried to get this dangerous crossing protected. But what do we find? Nothing is done about it. It seems as though the railroad would rather pay several thousand dollars to the estates of dead people than spend a like amount to protect the crossing. The jury can only find that this was an accidental tragedy, but we strongly urge that the village board, the Illinois Commerce Commission and safety officials start concerted action to get this crossing protected." [1289]

Bad weather with falling snow and slippery roads doubtless contributed to a disaster between Zittau and Hagenwerder, Germany, on Wednesday 17th January 1940. A fully loaded bus skidded through the lowered barriers of a level crossing and was struck by a passenger train. The bus was pushed some distance along the line, the engine of the train

---

[1286] *The Escanaba Daily Press*, Escanaba, Michigan, United States, 17th December 1939
[1287] *The Circleville Herald*, Circleville, Ohio, United States, 18th December 1939
[1288] *Mansfield News Journal*, Mansfield, Ohio, United States, 26th December 1939
[1289] *Blue Island Sun-Standard*, Blue Island, Illinois, United States, 28th December 1939

overturned, and its leading carriage was derailed. Twelve bus passengers died and 20 were injured. [1290] [1291]

A bus was struck by a train on an unprotected level crossing between Cloppenburg and Ocholt, Germany, on Monday 5th February 1940. Eleven people died and 31 were injured. [1292]

A lorry carrying 42 fruit pickers, mostly Mexicans, was struck by a Missouri Pacific Railroad train on Tower Road Crossing, near Alamo, Texas, United States, at 8.06 a.m. on the morning of Thursday 14th March 1940. Twenty-four of the people on the truck, including its driver, Jose Ramon, 40, were killed more or less outright, three were fatally injured and the remainder were more or less injured. Five of the dead were from the Perez family. The truck had been bringing the fruit pickers to work and had approached the crossing along US Highway 83, which paralleled the railroad, before Ramon had turned into Tower Road only 60 feet from the crossing and failing to see the train had driven onto the line in front of it.

The accident was witnessed by Miss Ann Wall, secretary of the Alamo Citrus Association, who said, "I usually look out of my office window, which faces the crossing 200 feet away, nearly every morning when the Missouri Pacific passenger train comes through. I was busy this morning and didn't look up. Then I heard that horrible crash. I glanced out of the window and saw a huge cloud of dust. Then a sheet of flame, and bodies flew out of the flame to lie in a heap along the tracks. The gasoline tank of the truck must have exploded. I tried to scream and couldn't. My boss ran out of the office onto the loading platform, then hollered to me to call every ambulance I could. It was the most horrible thing I have ever seen." [1293]

Witnesses told the inquest that the truck had been travelling at 40 miles an hour along the main highway, slowing to 20 miles an hour as it turned onto Tower Road. No one on the truck apparently saw the train, which was No. 113, en route from Harlingen and Mission. It was in charge of Conductor Tom B. Waite Sr., with Engineer Al Wilson and Fireman J.L. Garrett. Engineer Wilson told the inquest that he had not seen the truck, but his fireman had and had yelled. He had applied the brake a moment before the collision. His train had been travelling at 45 miles an hour. The flames which erupted instantly after the collision had, said the engineer, forced him and his fireman away from the engine windows. His engine had travelled 700 feet before he could pull up, with the truck still wrapped around the front of the engine pilot and on fire. At the conclusion of the inquest Justice of the Peace R.L. Savage returned a verdict of 'Accidental Death'. [1294]

A level crossing accident involving a school bus near New Leipzig, North Dakota, United States, on Monday 6th May 1940, appears to have gone unreported in the press

---

[1290] *Eisenbahnunfälle in Sachsen* by Claus Schlegel, pg. 172 (in German)

[1291] *Gazette de Lausanne*, Lausanne, Switzerland, 19th January 1940

[1292] *Katastrophen der deutschen Bahnen Teil II*, by H.J. Ritzau, pg. 240 (in German)

[1293] *The Brownsville Herald*, Brownsville, Texas, United States, 14th March 1940

[1294] *Valley Morning Star*, Harlingen, Texas, United States, 15th March 1940 until it featured in the review of a court case for damages in July 1941. Mrs Hilda Steiter, the bus driver, had been awarded $1,000 by the Grant county court jury against the Northern Pacific Railway. Her bus had been struck and demolished by a gasoline railcar. No other casualties were mentioned, so the bus had presumably been empty at the time.

Reviewing the evidence, the supreme court decided that the injuries sustained by Mrs Steiter, for which she had claimed the damages, had resulted from her own negligence, and reversed the judgement. The supreme court stated, "In this case the contributory negligence of the plaintiff (Mrs Steiter) is so patent that it bars her recovery as a matter of law. During the last 75 feet of her progress toward the railroad track, Mrs Steiter had an unobstructed view for about 1,500 feet in the direction from which the train approached. The weather was clear. She was familiar with the crossing. She knew the railroad track was there and that trains passed from time to time." [1295]

At 2.15 p.m. on the afternoon of Sunday 30th June 1940 Southern Railway freight train No. 51 collided with a car a mile south of Gretna, Virginia, United States. Engineer Stallings was unable to account for the accident, which killed all five occupants of the car. He said that he had whistled for the crossing and the driver had a clear view in either direction at the crossing, which was on a side road leading from Route No. 29. The automobile had crossed one track when it drove directly in front of his engine. Stallings said that he had been travelling at about 35 miles an hour and his heavy train took half a mile to stop. The occupants of a following car which pulled up for the train witnessed the accident.

Riding in the car had been R.J. Talbott, 55, a Gretna farmer; Mr Bud Anglin, 20, of Danville, the driver of the car; his wife Lucille, also 20; Mr Lester Yeatts, 25, of Gretna; and his wife Ellen Louise, 20. The two women were daughters of Mr Talbott. The Anglins had had dinner at the Tabotts' and the party were out for an afternoon drive. Mr Anglin apparently failed to see the train at all or hear its whistle. [1296]

Rather unusual circumstances led to a level crossing accident which cost the lives of six teenagers – three boys and three girls – in the Chicago suburb of Maywood, Illinois, United States, on the evening of Tuesday 2nd July 1940. The teens had all been students at Oak Park High School and had known one another for years. They left their homes at about 8.30 p.m. to attend a free dance at a party at a tavern, Jack's After Hours Club. Shortly after they arrived at the tavern one of the girls, Lorraine O'Leary, became ill with what was thought to be an attack of appendicitis. The six then set off in a car, heading east, and crossed the tracks of the Minneapolis St Paul and Sault Ste Marie (Soo Line) Railroad. A short distance further on, for some reason, the car was turned around, possibly because someone suggested that Maywood was the closest place to obtain medical attention. When the car reached the Soo Line crossing again the warning lights were flashing, and another automobile was already stopped waiting for the train. The driver of the teen's car, either because he had failed to see the warning lights or because he tried to beat the train in his haste, overtook the waiting car, drove onto the crossing, and the car was struck by a passenger train travelling at 50 miles an hour.

[1295] *The Bismark Tribune*, Bismark, North Dakota, 22nd July 1941
[1296] *The Bee*, Danville, Virginia, United States, 1st July 1940

John Brennan, 18; Catherine Clark, 19; Francis Frehe, 19, the driver of the car; Lorraine Norris, 17; Lorraine O'Leary, 19; and Joseph Santo, 19, were the victims. [1297]

Seven teenagers on their way to a swimming party were all killed outright when the car in which they were riding was struck by a Georgia Railroad fast freight train on Tuesday 30th July 1940. The accident happened on Barton Chapel Road Crossing, Georgia, United States, seven miles from Augusta. George Driggers, who lived near the crossing, said that the collision had "sounded like a small tornado, but no one screamed – just the sound of the car being dragged down the track."

Those who lost their lives were Emma Edith Baker, 16; Julian Baldowski, 19; Corinne Elliott, 16; Helen Curtiss Elliott, 16; R.A. Elliott Jr., 19; Harry Jarrell Sheppard, 19; and William Wheeless, 20.

Engineer C.M. Ray said that he had whistled for the crossing. [1298]

A foreman and five workmen on their way to a rural electrification project in nearby West Springfield lost their lives when the truck in which they were riding, owned by the Melvin F. Burgess Contracting Company of Baltimore, Maryland, was struck by a Boston and Maine Railroad train on Blackwater Road Crossing, New Hampshire, United States on Friday 2nd August 1940. The foreman was Robert Bolick and the men were Waldo Beard, believed to have been driving the truck, Walter Dietz, Raymond Harriman, Willis Logan and Robert Matthews.

Part of the lorry was carried on the cowcatcher of the locomotive to a bridge over the Mascoma River, 1,000 feet along the line, where the wreckage fell into the water. Engineer Bancroft said that he had only seen the lorry when his train had been a short distance from the crossing, when he had jammed on the brakes but had been unable to prevent the accident. Flashing warning lights were provided at the crossing and were working properly when tested afterwards. There was, however, low-lying fog present which might have contributed to the lorry driver failing to see the lights. [1299] [1300]

Between Rheine and Salzbergen, Germany, a train collided with a bus, killing eight persons, and injuring nine, on Saturday 3rd August 1940. The crossing barriers had not been closed. [1301]

A party of six returning from a convention of the Christian and Missionary Alliance at Beulah Beach, Ohio, to their homes in Indiana, were killed in a level crossing collision near Weston, Ohio, United States, on Monday 12th August 1940. The crossing was unprotected, and it appeared that the driver of the car had approached the crossing at too high a speed, as a bystander said he had done all he could to stop when he had seen the

---

[1297] *Oak Leaves*, Oak Park, Illinois, United States, 4th July 1940

[1298] *Thomasville Times-Enterprise*, Thomasville, Georgia, United States, 31st July 1940

[1299] *The Lowell Sun*, Lowell, Massachusetts, United States, 2nd August 1940

[1300] *The Portsmouth Herald*, Portsmouth, New Hampshire, United States, 2nd August 1940

[1301] *Katastrophen der deutschen Bahnen Teil II*, by H.J. Ritzau, pg. 240 (in German)

train but had been unable to do so. Conductor C.W. Day said that the car had been directly across the line when his northbound Baltimore and Ohio Railroad passenger train had struck it. The Rev. Oliver Forester, 57, pastor of the Alliance Gospel Tabernacle at Logansport, Indiana; his wife Bessie, 55; their son Howard, 10; Mrs Ethel Minglin, 50; Miss Lucille Reed, 34; and Donald Woodruff, 18 were the victims. [1302]

A train collided with a lorry on an unguarded level crossing near Pressburg, Slovak Republic (now Bratislava, Slovakia), on Wednesday 21st August 1940. The vehicle was completely demolished, and its six occupants killed. The lorry ended up beneath the locomotive of the train and the removal of the bodies of the victims was difficult. [1303]

An Up light engine collided with a bus on a level crossing between Benapur and Narayangarh, India, on the Bengal – Nagpur Railway on Sunday 15th September 1940. Six bus passengers were killed and seven injured. [1304]

Works Projects Administration (WPA) employees seem to have featured in a number of bad level crossing accidents. Shortly after 7 a.m. on Tuesday 3rd December 1940 a lorry carrying WPA workers was struck by the *Egyptian Zephyr*, a Chicago and Eastern Illinois Railroad gasoline railcar, on a crossing near Benton, Illinois, United States. The truck, en route to a road project several miles further on, had just turned off Route 37 onto the side road leading to the crossing. It was demolished by the impact, while the railcar was derailed. Four workers were killed outright and 12 were injured and taken to Moore Hospital, where another man, George Atkins, died during the afternoon. No one on the train was hurt. Visibility at the crossing was impaired. [1305]

The stalling of a car on a level crossing, as described by a witness, was the probable cause of a level crossing collision at Dalton, Georgia, United States, on Sunday 26th January 1941. Barney Boyd, the eyewitness, told police that it had been raining slightly at the time of the accident. The car was hit by *The Sunbeam* passenger train of the Southern Railway, en route from Chicago to Miami, and carried 100 yards down the track. Ivan Ledford, 27, his 24-year-old wife, their three children aged between two and seven and Mr Ledford's sister, Annie, all died. A watchman was stationed at the crossing from Mondays to Saturdays but there was none on Sundays. [1306]

Another crossing collision with a school bus in the United States on Monday 10th February 1941 was barely recorded in the press. The bus was empty when it stalled on a crossing at Eckerty, Indiana, and was struck by a Southern Railway freight train. The bus

---

[1302] *The Sandusky Register*, Sandusky, Ohio, United States, 15th August 1940
[1303] *Journal de Geneve*, Geneva, Switzerland, 24th August 1940
[1304] Report by the Railway Board on Indian Railways for 1940-41
[1305] *The Daily Independent*, Murphysboro, Illinois, United States, 3rd December 1940
[1306] *The Butler Herald*, Butler, Georgia, United States, 30th January 1941

driver, George Borden, 31, was fatally hurt. [1307]

When a bus arrived at a level crossing at Mariensiel, Germany, on Friday 28th February 1941 the crossing keeper was not at his post. Instead of waiting for him to appear a bus driver asked a passer-by to open the barriers. The latter did so, and a train collided with the bus, killing 21 people, and injuring 28 more. [1308]

Five Works Projects Association men, all from Hamilton, who were returning home from a sewer project, were killed, probably on Saturday 15th March 1941, when their automobile was struck by a Cleveland Cincinnati Chicago and St Louis (Big Four) Railway passenger train near Middletown, Ohio, United States. They were Harry Mills, 44; John Ross, 50; Clarence Russell, 27; D.M. Simmons, 67; and Thomas Webb, 61. [1309]

There were no witnesses other than the engine crew to an accident at Upper Sandusky, Ohio, United States early on Sunday 25th May 1941, which killed eight people. Engineer Dan Shepler of the Pennsylvania Railroad passenger train, *The Trail Blazer*, said that the car had been driven onto the crossing in front of him despite his furious whistling and the flashing warning lights at the crossing. The train was travelling at 70 miles an hour when it struck the automobile at 6.22 a.m. Driving the car had been Billy Elisech, 23, of Saginaw, Michigan, who had only bought the 1936 Pontiac two days earlier. The other victims were the driver's wife Stella, 20; their children Patricia Ann, 3; William, 2; and Ronald, six weeks; Mrs Helen Orsaka, 60, of Saginaw; her daughter Virginia, 14; and Edmund Frank Jurzak, 25, of Detroit.

The train had been westbound from New York to Chicago. It was 13 minutes late. The people in the car were returning from a visit to relatives in Richwood, Ohio. It was initially thought that a ninth person had died as cards bearing both Mrs Elisech's married name and her maiden name were found in the wreckage, causing confusion. [1310] [1311]

As usual 4th July 1941, Independence Day in the United States, was accompanied by a large number of accidents, mostly on the roads. The most serious single accident that Friday occurred at 7.27 p.m. on Sound Avenue Crossing, Mattituck, New York, United States. A crowded 1934 Dodge sedan was struck by an eastbound Long Island Railroad passenger train which was pulling into Mattituck station. The car was pushed almost 600 yards past Westphalia Avenue Crossing to end up within the sight of 100 passengers waiting on the station platform. Mrs Frances Straub, 49; her sons William Straub Jr., 29 (the driver of the car); Alfred, 8, and Richard, 6; Christopher O'Connor, 16; and Carmella Bianco, 14, were all killed. A seventh occupant of the car, Mrs Straub's daughter Rose, 10, was critically injured.

[1307] *Rushville Republican*, Rushville, Indiana, United States, 11th February 1941
[1308] *Katastrophen der deutschen Bahnen Teil II*, by H.J. Ritzau, pg. 240 (in German)
[1309] *The Zanesville Signal*, Zanesville, Ohio, United States, 17th March 1941
[1310] *The Carey Times*, Carey, Ohio, United States, 29th May 1941
[1311] *The Sheboygan Press*, Sheboygan, Wisconsin, United States, 26th May 1941

The train was in charge of Conductor Leo Hantz, with Engineer George Eichhorn. The engineer was said to have broken down and wept when he dismounted from his locomotive and saw the results of the accident. According to various members of the train crew the whistle had been sounded for the crossing. 'R.R.' and 'Stop, Look and Listen' signs were the only protection at the crossing. Visibility there was also not good and on the evening of the accident it was further impaired by drizzling rain. It was assumed that the car driver was either unfamiliar with the crossing (the family were from Brooklyn), was distracted by conversation with the others in the car or had stalled the vehicle on the crossing.

The crossing was regarded as one of the most dangerous in the area and the previous year Mattuituck Chamber of Commerce had campaigned unsuccessfully to have warning lights installed there and at Westphalia Avenue. [1312]

Works Projects Administration workers were the victims in yet another level crossing collision on about Saturday 12th July 1941. Cornelius Caldwell, Edward Cunings, Charles Daniels, Thomas Ellis, and David Ferguson, all of Toledo, were on their way home from a project at Erie Township when their car was struck by a New York Central Railroad train at La Carne station, near Port Clinton, Ohio, United States. All were killed. [1313]

Two West Carrollton families, returning from a family reunion and picnic at Miller Grove, had covered about two thirds of their journey when they reached Gettysburg Pike Crossing, near Dayton, Ohio, United States, on the evening of Sunday 27th July 1941. According to witnesses Otto Kline, 55, the driver of the car, had hesitated momentarily at the unprotected crossing and had then set off again directly in front of a Pennsylvania Railroad train. Conductor H.O. Allen stated that his train, the second portion of a St Louis, Missouri, to New York flyer, was 50 minutes late and had been travelling at 70 miles an hour when it had reached the crossing. Kline, his wife Villa, 53; their son Ralph, 10; Walter Reel, 32; his wife Mildred, 36; and their children Betty, 5; Billy, 3; and Ruth Ann, 2, all lost their lives. [1314] [1315]

Mrs Nelson Pulsifer stalled her car on a level crossing on Route 82 near Aurora, Ohio, United States, on Saturday 2nd August 1941. An Erie Railroad passenger train travelling at an estimated 70 miles an hour struck the vehicle, killing her and her children Patricia, 8; Janice, 4; Wallace, 3; and Elsie, seven weeks. [1316] [1317]

Jew's Creek Crossing, near Molong, New South Wales, Australia, was the scene of a fatal crossing collision at 10 p.m. on Saturday 23rd August 1941. Mr William Herbert Jones, 45, Shire Engineer for the Amaroo Shire Council, was returning with his wife from Orange, where they had gone to collect their daughter and two children of Mr and

---

[1312] *The County Review*, Riverhead, New York, 10th July 1941
[1313] *The Portsmouth Times*, Portsmouth, Ohio, United States, 14th July 1941
[1314] *Washington C.H. Record-Howard*, Washington C.H., Ohio, United States, 28th July 1941
[1315] *The Circleville Herald*, Circleville, Ohio, United States, 28th July 1941
[1316] *Syracuse Herald American*, Syracuse, New York, United States, 3rd August 1941
[1317] *East Liverpool Review*, East Liverpool, Ohio, United States, 4th August 1941

Mrs Errol Blunden for the school holidays. The car was struck by a cattle train. Mrs Dorothy Jones, 37; the Jones' daughter Nell Frances, 10; and the two Blunden children, Betty Flower, 12 and James, 10, were killed outright. Mr Jones suffered multiple injuries and died in Molong Hospital on Sunday afternoon. [1318]

The inquest on the victims held evidence from the crew of the stock train: Guard Oliver Poulter, Driver John Bunt and Fireman Samuel Sutton. Guard Poulter said that the train had given a long whistle a quarter mile from the crossing, and a shorter one close to it.

Fireman Sutton said that the train had been carrying stock from Dubbo to Sydney. Thirty or forty yards from the crossing, as the train had emerged from a cutting he had seen the lights of the car, had sounded the whistle and applied the brakes. It had been a bad night, windy and raining with poor visibility.

Driver Bunt said that the train had left Dubbo at 3.40 p.m. When his fireman had shouted he had applied the air brakes and reversed the engine, then the crash had come. His train had travelled another 250 yards before stopping. Neither he nor his fireman had noticed if the wig-wag signals at the crossing had been working. The leading wheels of his engine were derailed in the collision. The train had been travelling at 15 miles an hour.

Traffic Inspector James White of Orange said that he had arrived at the crossing at 11.15 p.m. The wig-wag signal had not been working then, but it was damaged by the derailed locomotive. He regarded the crossing as dangerous.

Various other witnesses, railwaymen, and users of the crossing, testified as to the wig-wag signals, some stating that they had failed to operate on previous occasions.

At the close of the inquest Coroner J.H. Caldwell returned a verdict of 'Accidental Death', adding a rider that he considered that the train crew had done all that they could to prevent the accident. The evidence showed that there had been driving rain and darkness at the time of the accident, and the engine only had kerosene headlights. The coroner suggested that beacon lights, such as used on passenger trains, should be employed on all trains, also that special attention should be given to the mechanical signals at crossings. [1319]

A brother, sister, and the latter's four young children died as they were returning from a shopping expedition on the evening of Thursday 28th August 1941. William Derrickson, 42, was driving the car, and was accompanied by Mrs Violet Salmons, 40, and children Louise, 10; Grace, 8; Charles, 6; and Stella, 2. Tall standing corn in fields adjoining the crossing with the Pennsylvania Railroad at Harrington, Delaware, United States, probably contributed to the accident. When the car was approaching the crossing a southbound freight train was still some distance away. Derrickson crossed the southbound track in front of the freight but failed to see or realise that a fast northbound freight was close at hand, travelling at about 60 miles an hour. Engineer Alfonso Russell said, "I did not see the car and did not know what the engine had struck until I stopped the train." The engineer, Fireman Oscar Warren and Conductor O.T. Taylor were released on personal recognisances to appear before the coroner at the inquest.

[1318] *Sydney Morning Herald*, Sydney, New South Wales, Australia, 25th August 1941

[1319] *The Leader*, Orange, New South Wales, Australia, 1st September 1941

The coupe was carried several hundred yards and school supplies and clothes bought by Mrs Salmons earlier were scattered along the line. A pet dog was also killed. [1320]

A Grand Trunk Western Railroad passenger train, *The Streamliner*, eastbound from Chicago to Montreal, collided with a car at Highland, Indiana, United States, on Saturday 13th September 1941 and reportedly carried it almost a mile. The driver, Andrew Schafter, 36, of Schererville, Indiana, was killed, as were his wife Anne, 35; their children Harold, 16; Alby, 13; Buster, 12; and Marvin, 6; and Mr Schafter's father-in-law, Mr C.M. Haynes, 60. [1321]

On Monday 29th September 1941, a light sedan car reportedly stalled on Main Street Crossing, Hooker, Oklahoma, United States. It was then struck by a westbound Chicago Rock Island and Pacific Railroad freight train and demolished. The occupants of the car were all killed, and their bodies were badly mangled, so that it took some time to identify them. They were eventually determined to be Mr Florence Cagel, 54; Mr Athol Cline, 21; Mrs Cline, 22; Mr Floyd Fidler, 24; Miss Carol Grimsley, 23; Miss Fredia Hendrix, 24; and Mr Wayne Light, 25. It was not possible to tell who had been driving the vehicle. [1322]

On the afternoon of Friday 10th October 1941 school bus driver Benjamin Priddy picked up a busload of children from the George R. James Elementary School, just outside of Eads, Tennessee, United States. Having dropped off some of them, his route then took him to a level crossing with the Nashville Chattanooga and St Louis Railway. Coming down the hill to the crossing, there was a clear view of the tracks and a passenger train bound for Memphis was nearing the crossing, travelling at 50 miles an hour. One of the children who survived the accident, nine-year-old Nancy Jenkins, related that, "Every child on the bus was screaming for the driver to stop."

A witness on the road said that the bus, "came down the hill, hesitated just a second – though I couldn't say whether it stopped or not – and then rolled out onto the track right in the path of the train." The bus was dragged 75 yards and hurled into a briar patch.

Why Priddy had failed to stop could not be determined as he was killed in the collision, together with seven of his passengers. The other ten pupils were all injured. Engineer Joe Darnell said that his train had been travelling at between 40 and 50 miles an hour and he had not seen the bus until the last moment.

It was reported that Priddy had complained of a headache on the morning of the accident and conjectured that he might have been taken ill. It was also stated that during the 15 years that he had been driving the bus he had never previously encountered a bus at the crossing, though this seems unlikely. The train had been 20 minutes late that day. [1323] [1324]

[1320] *The Salisbury Times*, Salisbury, Maryland, United States, 28th August 1941

[1321] *The Washington Herald*, Washington, Indiana, United States, 13th September 1941

[1322] *The Daily Ardmoreite*, Ardmore, Oklahoma, United States, 29th September 1941

[1323] *The Bee*, Danville, Virginia, United States, 11th October 1941

[1324] https://memphismagazine.com/ask-vance/the-eads-school-bus-crash-of-1941

The Eagle, a streamlined Missouri Pacific Railroad express, killed seven people who ironically were on their way to a funeral, in a level crossing accident at Strasburg, Missouri, United States, on Friday 14th November 1941. The victims had only come five miles and Strasburg was their destination. They were on their way to attend the funeral of Mrs L.M. Sechrist, 68, who had been gored to death by a bull a few days earlier.

Mr Luther Mote, 40, a farmer and the driver of the car, was killed, together with his 32-year-old wife, their children Charles Joe, 5; Donna Lou, 3; and Nancy Agnes, 15 months; Mrs Cochran, 39, Mr Mote's sister; and Mrs Egbert, 23, Mrs Cochran's daughter. Joseph Mote, Luther Mote's father, was standing on the main street of Strasburg talking to friends when the accident happened and was the first person on the scene, to find that his own son and family were involved. The plan had been to leave Nancy Agnes with Mote Senior's family during the funeral.

Engineer L.E. Giffen said that his train had been travelling at 60 miles an hour. He had whistled for the crossing and had seen Mote's sedan apparently stalled on the tracks, with its driver attempting to back it off the crossing. The crossing was unprotected. [1325] [1326]

Seven people, six of them from one family, died in a level crossing accident on a side road seven miles south-east of Paducah, Kentucky, United States, on Thursday 20th November 1941, when their automobile was struck by a Nashville Chattanooga and St Louis Railway train. They were Charles Whalen, 58, a boilermaker's helper at the Illinois Central Railroad shops; his wife Fatie, 51; their children Ira, 16; Evelyn, 15; Juanita, 9; and Wayne, 7; and Randall Petway, 11. Randall's brother Robert witnessed the accident from the Petway home. It was believed that Whalen had expected the train to stop at the crossing, as it usually did. It was not explained why the train stopped at the crossing unless there was a stopping place there. [1327] [1328]

Six of nine occupants of a car died when it was struck by a Canadian Pacific Railway passenger train on a level crossing at Westboro, Ontario, Canada, early on Sunday 4th January 1942. The people in the vehicle had been returning from a family reunion at Westboro. When they reached the crossing, which was on a slight incline and covered with ice, another car was stalled on it. They were able to push the other car, which then made it over the crossing, only to then stall on the tracks themselves. Roger Champagne, 21, the driver of the automobile; Sergeant Yves Nadon of the Royal Canadian Air Force; his wife, Genevieve; his brother, Leading Aircraftman Giles Nadon, Sylvio Boyer, 31; and Celina Bazinet, 27, were all killed or fatally injured. Gabrielle Champagne, 23, the sister of the driver, and Virginia Gallagher, 19, escaped with relatively minor injuries and Maurice Nadon, 22, brother of the dead airmen, was unhurt.

Interviewed after the accident, Maurice Nadon said, "Our windows were coated with

---

[1325] *Macon Chronicle-Herald*, Macon, Missouri, United States, 15th November 1941

[1326] *Daily Capital News*, Jefferson City, Missouri, United States, 15th November 1941

[1327] The *Morning Herald*, Hagerstown, Maryland, United States, 21st November 1941

[1328] *The Daily Independent*, Murphysboro, Illinois, United States, 21st November 1941 frost, and we couldn't see out. When we reached the crossing the other car started and went ahead. Our car stalled and it was then it must have been hit. I didn't see or hear anything. The next thing I knew I was walking into a house carrying a body. I don't know how I got there or when I picked up the body. I must have been going on nerves." [1329]

Taunetaro Konishi, 62, a farmer of Japanese birth who had emigrated to the United States in 1909 along with his wife, was killed together with four of his children at 5.30 p.m. on Monday 9th February 1942 when the pickup truck he was driving was struck by the *Pony Express* passenger train of the Union Pacific Railroad on the Kuner-Empson Crossing in Fort Lupton, Colorado, United States. Mr Konishi had just come from the Twombley ranch, a short distance out of town, to pick up his children, Ruth, 18; Chester, 12; Cynthia Chizuko, 9; and Eugene, 3; from the Japanese school and they were on their way home.

The train, in charge of Conductor Claude Kindig, with Engineer C. J. Miller and Fireman R.H. Angel, was travelling at 70 miles an hour, carried the pickup truck about 90 feet. It took the train half a mile to pull up and the pilot of the engine was damaged so that the train was unable to proceed without a fresh locomotive. The five occupants of the truck were all crowded into its cab, and it was theorised that their breath might have fogged up the windows of the vehicle.

As well as Mrs Konishi an older son, George, another son, James – the latter in the United States Army – and a daughter, Mrs Katie Uchida, survived the accident. [1330]

A detour made necessary by road repairs which involved a driver having to take an unfamiliar route, and early morning fog, were factors in an accident on Thursday 12th February 1942 when a school bus taking pupils to Amite High and Grammar School was struck by an Illinois Central Railroad passenger train bound for New Orleans on a level crossing at Amite, Louisiana, United States. There were about 40 children on the bus, of whom one was killed, 14 seriously injured and 25 slightly hurt. The dead child, Doris Ridgell, 12, was the daughter of farmer Henry Ridgell, the driver of the bus, while his son Lloyd, also 12, was seriously injured and not expected to survive. Ridgell himself was also hurt. The bus was struck in its centre and a large hole was made in its side, but its steel construction was credited with preventing heavier casualties. [1331] [1332]

A bus carrying between 35 and 40 men of the Army Air Corps who had been on leave back to their barracks at the Enid flying school was struck by a Chicago Rock Island and Pacific Railroad train on a level crossing near Enid, Oklahoma, United States, at 12.15 a.m. on Thursday 26th February 1942. A blinding snowstorm was raging at the time of the accident. The bus was being driven by a civilian, Wendell Thomas Smith, 28, of Enid. Seven of the men were killed and 25 more injured, several critically. The bus driver was also hurt.

[1329] *Medicine Hat Daily News*, Medicine Hat, Alberta, Canada, 5th January 1942
[1330] *Greeley Daily Tribune*, Greeley, Colorado. United States, 10th February 1942
[1331] *Hattiesburg American*, Hattiesburg, Mississippi, United States, 12th February 1942
[1332] *The Delta Democrat-Times*, Greenville, Mississippi, United States, 12th February 1942

Explaining the accident, Smith said, "I had shifted down to third gear, and was slowing down to stop at the crossing but the road was so slick from snow I couldn't stop. When I saw the lights of the train bearing down through the snow I speeded up in an attempt to cross but I didn't quite make it. The train hit the back of the bus and turned it round. If I hadn't speeded up I would have landed right in the middle of the track." [1333]

Another American train/school bus collision, also on Thursday 26th February 1942, only received brief attention in the press. A school bus carrying 30 passengers, including members of the Big Sandy High School basketball team, was struck by a Louisville and Nashville Railroad freight train on a crossing at Big Sandy, Tennessee, United States. Pupils said that the bus driver, Lawrence Hinson, had stopped at the crossing and looked both ways before entering. Clearly he had not looked hard enough because he failed to pick out the train. The bus was thrown into a shallow creek but landed upright. Seventeen pupils were injured, none seriously; by the next day all had been released from hospital or were recovering. [1334]

A level crossing collision near Savage, Minnesota, United States, on Saturday 11th April 1942 was followed by a rear collision and is dealt with in the Rear Collisions chapter.

Loud jazz music blaring from a car radio and driver inattention led to the death of five teenagers on a level crossing with County Road 315, three miles from Newark, Ohio, United States, on Wednesday 6th May 1942. The noise of the radio doubtless drowned out the sound of the whistle of the Baltimore and Ohio Railroad train which struck the automobile and the teens also apparently failed to look out for a train, as the view at the crossing was unobstructed. The victims were Jean Bush, 15; Barbara Deutsch, 14; Fred Ehlermann, 18; Kenneth Stockdale, 17; and Harry Sullivan 17. Sullivan had been driving the car.

Engineer O.P. Butcher said that he had not seen the car before his train had hit it. The impact carried the car 150 feet, killing all five occupants outright, though the car radio had still been playing "at full blast" when police officers reached the scene.

D.K. Finney, of the Ohio State Safety Council executive committee, commented that, "A tragedy of this kind is very unfortunate, but people must learn that accidents don't just happen. Railroad accidents usually result from faulty human behaviour." [1335]

A bus which was full of passengers was on its way from Tangier to Larache on Monday 25th May 1942 when it was struck by a train on a level crossing between Tangier and Asilah, Spanish Morocco. Five people were killed and 30 injured, while the train was derailed by the violence of the collision. [1336]

[1333] *The Sedalia Weekly Democrat*, Sedalia, Missouri, United States, 27th February 1942
[1334] *Edwardsville Intelligencer*, Edwardsville, Illinois, United States, 27th February 1942
[1335] *Delphos Daily Herald*, Delphos, Ohio, United States, 6th May 1942
[1336] *Journal de Geneve*, Geneva, Switzerland, 25th May 1942

On Friday 31st July 1942, a Missouri Pacific Railroad freight train collided with a Teche-Greyhound on a level crossing on the outskirts of Franklin, Louisiana, United States. The bus had been en route from Houston, Texas, to New Orleans, Louisiana, and had left Lake Charles with a full load. Eight people died and 18 were injured, mostly only slightly. Seven of those who lost their lives were a Mrs Aldridge, a Mrs Bolner, Postmaster Buren Broadus, Mrs Lee Firmin, 26; her children Mary Lee, 5, and Charles, 3; and Mildred Williams. The eighth victim could not immediately be identified.

Witnesses said that the bus driver, Robert Lupino, had applied the brakes at the crossing, but had been unable to stop the vehicle and the train had torn off the entire rear end of the bus. At the inquest the jury found that defective brakes on the bus had caused the accident and Coroner C.M. Horton exonerated the bus driver, who, he said, "had taken all the precautions he could," to avert the accident. [1337] [1338]

Less than a week after the Franklin accident, a worse accident involving a Greyhound bus and a train happened at Crystal Springs, Mississippi, United States. This time it was not a question of a defective vehicle, but rather of a simple mistake by A.L. Owen, the bus driver, who escaped with slight injuries. On Wednesday 6th August 1942, the bus involved stopped at the crossing and waited for a northbound Illinois Central Railroad freight train to pass. As soon as the train had cleared the crossing the bus had set off again, directly into the path of a southbound troop train.

The bus was reportedly carrying about 65 passengers and was travelling from New Orleans to Chicago. So severe was the collision that the engine at the rear of the bus was torn out, the top of the bus was ripped off and the bus was thrown up into the air and landed on the sidewalk (pavement) where it killed a pedestrian, Mrs W.L. Dodds. Twelve bus passengers also died and over 50 were injured. [1339] The train was not derailed and no one on it was hurt. [1340]

Allegedly because its brakes had failed, a bus smashed through the crossing barrier at Teplitz-Schönau, Germany (now Teplice, Czech Republic) on Monday 10th August 1942. Six people died and 33 were injured in the subsequent collision with a train. [1341]

The recent retiming of a Missouri Pacific Railroad passenger train from Union to Lincoln, such that it ran 20 minutes earlier than before, was a likely factor in a level crossing accident at Weeping Water, Nebraska, United States, just before 10 a.m. on Tuesday 18th August 1942. Farmer William Stubbendick, 42, was taking his pregnant married daughter, Mrs Compton, 17, from the neighbouring farm where she lived, to hospital, accompanied by his wife and their other children, Geraldine, 14; Donald, 13;

[1337] *Moorehead Daily News*, Moorehead, Minnesota, United States, 31st July 1942

[1338] *The Daily Herald*, Biloxi, Mississippi, United States, 1st August 1942

[1339] *Wisconsin Rapids Daily Tribune*, Wisconsin Rapids, Wisconsin, United States, 6th August 1942

[1340] *Dixon Evening Telegraph*, Dixon, Illinois, United States, 7th August 1942

[1341] *Katastrophen der deutschen Bahnen Teil II*, by H.J. Ritzau, pg. 240 (in German)

Ronald, 8; and Elton, 5. They had only covered 30 rods (165 yards) along the lane leading from the farm to the main highway when they reached the railway. Accounts as to the visibility at the crossing were conflicting, it being variously stated that there was a clear view for 300 feet and that the view was blocked by tall weeds. At any rate the train, in charge of Conductor Goodwin and Engineer Simpson, struck the car. The vehicle was dragged 60 feet and demolished. Five of its occupants were killed outright, William Stubbendick, who had been driving, suffered a head injury and a broken back, and his son Donald concussion and a broken hip. Two days after the accident the father died in the Bryan Memorial Hospital, Lincoln. [1342] [1343]

Mrs Susie Szeregnyl, 55, her daughters Mrs Elizabeth Etlinger, 27; Mrs Vera Nemeth, 23; and Mrs Mary Szenas, 23; together with Mrs Szenas's son Fred, 12 months, were killed when a Michigan Central Railroad passenger train struck their car on a crossing at Wayne, Michigan, United States, on Sunday 27th September 1942. Wayne is a suburb of Detroit. Engineer William Black said that he had sounded his whistle several times as he neared the crossing. Other witnesses related that the car had only been travelling at 15 miles an hour. The vehicle was carried 900 feet before the train could stop. The women were all killed outright, the toddler passed away in hospital. [1344] The party were bound for a family reunion.

Detroit, Michigan, United States was again the location of a major level crossing accident on Wednesday 28th October 1942, this time once more involving a bus and in somewhat similar circumstances to the Crystal Springs accident in August. The Detroit Street Railways bus, packed with schoolchildren, office workers and factory employees, had stopped at the Caniff Avenue Crossing for a northbound Grand Trunk Western Railroad freight train. What happened then was described by a surviving passenger, Marie Giles, who had been sat directly behind William Clos, the 25-year-old bus driver. "There was a car ahead of the bus. When the freight train cleared, this car started across. The bus driver asked the people jammed against the front door if everything was clear. He could not see for himself; the bus was too crowded. The people said, 'All clear,' and he started across. All of a sudden I saw the train coming. Then there was a nightmare. I don't know how I got out alive."

Watchman Joseph Levinsky, in the tower at the crossing, stated, "The bus was westbound and had stopped for the red flasher signals. The freight, northbound, cleared the crossing. The red flasher lights were still working. The passenger train was coming fast, southbound, with the engine whistle screaming. I saw the bus start up. I knew it would do no good, but I leaned out of the tower and yelled. I yelled my head off. The bus went right into the path of the train and there was an awful splintering crash, and then screams."

Another witness was a car driver, Arthur Curtis. "I saw the bus start up as the freight train cleared. I caught a glimpse of the approaching passenger train and threw my hands over my eyes and the next thing I heard was an awful crash." Curtis's car was wrecked

---

[1342] *The Telegraph-Herald*, Dubuque, Iowa, United States, 18th August 1942

[1343] *Evening Star Journal*, Lincoln, Nebraska, United States, 20th August 1942

[1344] *The News-Palladium*, Benton Harbor, Michigan United States, 28th September 1942 when the bus was hurled into it but he was unhurt. [1345]

The consequences of the accident were very serious. Sixteen bus passengers were killed or fatally hurt and another 24 were injured.

The passenger train, running from Chicago to Detroit, was in charge of Conductor John Irwin with Engineer W. McDaniels and Fireman S. Babcock. The bus was a regular working on the Plymouth – Southfield route. The collision occurred just before 9 a.m. and school books were littered all over the area. William Clos, who had been operating the bus for two years, was taken into custody by police after the accident; he was unhurt, as the train had struck the bus towards its rear. [1346]

At the end of November the bus driver was questioned by Judge Nicholas Gronkowski at the Hamtramck justice court. He stated that the 25-seat bus had been crowded with 45 passengers, "a normal load for these times." He had stopped at the crossing because the flasher signals were operating, and the freight train was passing. As soon as this train had gone the three cars in front of him had started over the crossing, so he had followed them without noticing if the flasher lights were still working. He was committed for trial on manslaughter charges. [1347] No report of a trial has come to light, however.

Five residents of Hanna, Indiana, who were returning home from work at the Kingsbury ordnance plant on the evening of Friday 6th November 1942, were killed when their car was struck by a Baltimore and Ohio Railroad fast passenger train on a level crossing at Wellsboro, Indiana, United States, 12 miles from La Porte. They were Richard Caesar, Kathleen Gibson, 20, the driver of the car; Rhoda Hull, 19; Kathryn Rosenbaum, 18; and Vivian Rowley, 37. In an all-too-familiar scenario the driver of the car had stopped and waited for a westbound freight train to pass, then thinking all was clear had proceeded into the path of the eastbound flyer. [1348]

Dr G.F. Graham Brown, the Bishop of Jerusalem, lost his life in a level crossing collision between a Palestine Railway train on Monday 23rd November 1942 and the car in which he was returning from a military confirmation service at Aleppo, near the town of Ezzib, Palestine, close to the border with Lebanon. His chauffeur escaped injury in the accident and the third occupant of the car, a soldier who had been given a lift, was only slightly hurt. [1349]

Seven people on their way to a Christmas party lost their lives when their car was struck by a Chicago Burlington and Quincy Railroad passenger train bound for Denver on a level crossing a mile and a half north-east of Princeton, Illinois, United States. The accident happened on Wednesday 23rd December 1942. Five of the victims belonged to the Clan family, being William Clan, aged about 45; his wife, aged about 25; and their children Josephine, 8; Harry Roger, 4; and Virginia Lee, 2. The others were a Mrs

---

[1345] *Joplin Globe*, Joplin, Missouri, United States, 29th October 1942

[1346] *Adrian Daily Telegraph*, Adrian, Michigan, United States, 28th October 1942

[1347] *The Escanaba Daily Press*, Escanaba, Michigan, United States, 28th November 1942

[1348] *The Tipton Daily Tribune*, Tipton, Indiana, United States, 7th November 1942

[1349] *The Times*, London, England, 25th November 1942

Jackson, 54; and her granddaughter, Iola Neff, 6, all of Princeton. The car was dragged almost a mile. There were no witnesses to the accident, which happened only a short distance from the party's destination.

Engineer H.A. Nutter declined to make a statement pending the inquest. No report of the inquest seems to have appeared in the newspapers. [1350]

Marvin Tunstall, 43; his wife Cuba, 35; and their children Ellen, 15; Roy, 8; Ray, 6; Billy, 5; and an unnamed infant were all killed early on the morning of Sunday 17th January 1943 when their car was hit by a southbound Seaboard Air Line Railroad train on a level crossing two miles from Apex, North Carolina, United States and only a mile short of their home. They had been on their way home from Apex. [1351]

A St Louis – San Francisco Railway passenger train in charge of Conductor William Merty, with Engineer A. Phillips, collided with a car carrying five workers to the Douglas bomber plant at Tulsa, Oklahoma, United States, at that place on the morning of Monday 18th January 1943. The workers – Alfreeda Ganz, A.L. Harris, Thomas Hill, Thelma Page and Clarence Webb – were all killed instantly.

Crossing Watchman Andrew Maxwell said that he had stood in the path of the workers' car waving a red lantern, but the car driver apparently failed to see him, and he had to leap aside to avoid being run down. [1352]

Ten people lost their lives and another 16 were injured in a level crossing collision between a train and a motor coach at Montady, France, seven kilometres from Béziers, France, on the evening of Friday 19th March 1943. The crossing was an open one, and visibility was to some degree obscured by the house of the former crossing keeper, while only a few trains used the line. Another factor was that the train involved in the accident, which was in the midst of shunting operations, was being propelled over the crossing, and it was believed that the coach driver, M. Urbain Falliéres, who was one of those who died, had mistakenly supposed that the train had passed over the crossing and was going forward, as he had made no attempt to stop as he approached the crossing. The coach, which was full, was overturned and caught fire, while the leading waggons of the train were derailed and thrown about 15 metres. The surviving bus passengers had been able to get out of the vehicle immediately after the collision but the fire which broke out at once was fanned by a strong wind, making it impossible for them to rescue the driver and passengers who died, trapped in the flames. [1353]

A three-car Illinois Terminal Railroad electric interurban train collided with an intercity bus on a level crossing at Granite City, Illinois, United States, on Saturday 20th March 1943. The bus had almost cleared the last rail when the train 'sent it whirling top-

[1350] *Dixon Evening Telegraph*, Dixon, Illinois, United States, 24th December 1942

[1351] *The Robesonian*, Lumberton, North Carolina, United States, 18th January 1943

[1352] *Miami Daily News-Record*, Miami, Oklahoma, United States, 18th January 1943

[1353] *Le Radical de Marseille*, Marseille, France, 21st March 1943 like, smashed its front on the rebound and rammed its shattered chassis between two rows of telegraph poles,' in the words of a newspaper reporter. The bus was hurled into a ditch by the impact and parts of it were dragged a considerable distance along the track. [1354] Nine bus passengers were killed and 14 injured. Two train passengers were slightly hurt. The train was en route from Peoria to St Louis.

Tollof Stromlund, who lived near the crossing, said, "I've been through three of these things and this is the worst I ever saw." He and his wife were the first on the scene and helped to extricate the injured victims while neighbours called for ambulances. The bodies of five of the victims had been thrown into an adjoining field. Doctors, nurses' aides, and civilian defence workers went to the scene and the injured were taken to Granite City Hospital.

The bus was being driven by Emmett Warden of Granite City, who was only slightly hurt. He stated afterwards that he had stopped the vehicle only 30 feet from the crossing to take up more passengers. When he had started up again the flashing lights at the crossing had not been working. As he pulled up the slight rise to the crossing, his vision was obscured by an oncoming bus. [1355]

Warden was arrested on a charge of manslaughter after the accident and did not testify at the inquest on the victims on the advice of his attorney. In one statement that he made after the accident he related that he had been watching the passengers who had just boarded the bus as he moved towards the crossing. Passengers on the other bus that he had mentioned, said that it had not passed over the crossing before the accident.

Motorman Steve Robinette did testify at the inquest and said that his train had been travelling at about 55 miles an hour as it approached the crossing and at 45 miles an hour at the moment of the collision. His train had dragged the bus 700 feet before it stopped.

The inquest jury was unable to decide whether the crash was an accident or the result of criminal carelessness and left the matter to the grand jury. Representatives of the Illinois Commerce Commission and the Interstate Commerce Commission attended the inquest. [1356]

It does not seem that Warden was indicted by the grand jury.

Argentina suffered a serious level crossing accident on Monday 19th April 1943. A bus carrying workers home from their jobs at the Cordoba Military Powder and Explosives Factory was struck by a train on a level crossing in Villa Maria, Argentina. Seven of the passengers on the bus were killed more or less outright, and nine were seriously injured, two of whom passed away in the course of emergency operations in hospital. Others were slightly hurt.

The accident was the result of a mistake by the crossing keeper, José Díaz, who suffered a severe nervous breakdown in the aftermath of the accident, shouting loudly, "Let me kill myself! Let me kill myself!" He admitted his guilt, saying that he had not looked along the line to see if a train was coming before he raised the barriers to let the bus across, because one usually passed at 12.05, and it being 12.13 when the bus had

---

[1354] *Mason City Globe-Gazette*, Mason City, Iowa, United States, 20th March 1943

[1355] *Joplin Globe*, Joplin, Missouri, United States, 21st March 1943

[1356] *Edwardsville Intelligencer*, Edwardsville, Illinois, United States, 26th March 1943 appeared, he had assumed that the train had gone. Clutching his hair in his hands the crossing keeper repeated that he wanted to kill himself, crying, "Give me a revolver! I will do myself justice!"

Another bus, carrying over 30 more workers, arrived at the crossing soon after the accident.

The provincial governor, Doctor Santiago H. de Castello, when advised of the accident, ordered the local authorities to put themselves at the service of the victims and their families. Numerous private donations flooded in for those victims' families in need and it was arranged that all the local businesses would close on the following Tuesday for the funeral of those who had lost their lives. [1357]

Saturday 8th May 1943 saw a major level crossing disaster in Australia. A bus crowded with 34 military personnel on leave and bound for Albury, New South Wales, for an evening out, was struck by a light engine on a level crossing just across the state border in Wodonga, Victoria. Twenty-three soldiers, one woman – a civilian nurse – and the bus driver were killed and most of the other passengers were injured.

The accident was due to the failure of the bus driver, William Lord, 45, to see the engine, despite the efforts of four soldiers standing by the roadside to attract his attention and warn him of the danger. One of the four was a Sergeant Edward Moloney. He said that he had heard the engine whistle and then saw the bus nearing the crossing. The engine was travelling, he said, at about 15 or 18 miles an hour and he had distinctly heard it whistle several times, the last time when it was about 75 yards from the crossing, this being a prolonged blast. There were lights on the engine, which was running tender first, but they were not very bright.

Another witness, Hilary Campbell Eddy, said, "I saw the bus at the top of a rise some distance from the scene of the accident. It was then travelling fairly slowly but gained momentum as it approached the crossing. When I saw the collision was inevitable my friends and I tried to attract the attention of the bus driver, and when this failed we ran back towards the crossing. We then saw the lights of the bus between the wheels of the engine tender, and then hell seemed to be let loose, and we were showered with glass from the shattered bus. We were only a yard or two from the line when the crash occurred. It was deafening. My first impulse was to run from the horrible scene, but I and my companions gave what assistance we could. Bodies were heaped five feet in the middle of the road. The V.A.D. {Voluntary Aid Detachment – the nurse} was underneath the bodies. The path of the engine was strewn with corpses, and some soldiers were underneath the tender." [1358]

The bus was travelling rather faster than the engine, about 20 or 25 miles an hour. The locomotive reached the crossing first and the bus ran into the side of it. The bus was wrecked, and the engine was derailed. The only fortunate aspect of the accident was that there was no fire, as there were 18 gallons of petrol in the fuel tank of the bus. The bodies of the bus passengers were strewn about the road; that of the bus driver was found in the cattle pit.

[1357] *Ahora Illustrated Periodical*, 23rd April 1943

[1358] *Daily Examiner*, Grafton, New South Wales, Australia, 10th May 1943

Interviewed by reporters Mr Orme Midgley, the manager of Symons' bus service, the owners of the bus, said that the bus driver should have had a clear view of the crossing, as no passengers were permitted to stand at the front right side of the bus. Of the driver he stated that Bill Lord had been with him for eight or nine months and was one of the most steady and obliging chaps that had ever worked for him. "His tragic death has been a terrible blow both to me and to Mrs Midgley. We both had the highest opinion of him." The bus was only licensed to carry 19 passengers, but with the war the position had arisen that it was difficult or impossible to control the volume of passengers, especially on a Saturday evening when all the servicemen on leave were eager to get into Albury on the first possible service. In neighbouring New South Wales official permission had been given for buses with seating for 25 to carry another 15 standing passengers. [1359]

At the inquest upon the victims of the tragedy Mr Midgely stated that the lights and brakes of the bus had been in good order. It had been a good bus to drive, and he had not imposed any limit on the number of passengers carried. In any case it was impossible for the driver to prevent passengers getting on the bus if there was room. The bus was not running to a timetable. Lord was a most careful driver, not prone to speeding.

Sergeant Moloney gave evidence at the inquest that the engine had whistled repeatedly and was carrying fairly distinct lights. Another witness, Mr T.A. Morton, a mechanical engineer, agreed that the whistling of the engine had been loud and continuous. The lights of the engine he described as 'dull and heavy.' [1360]

Driver Robert Cunningham testified that his engine had been going to Bandiana to pick up a train. He had started whistling at the whistle post, about 450 yards from the crossing. Vehicles had been going over the crossing. His fireman, Reginald Rogers, had called out, "Keep it going," and he had closed the regulator. About 100 yards from the crossing he had seen another bus coming from Bandiana. He had whistled again, and his fireman had shouted, "Red light!" He had applied the brakes and kept the whistle blowing. At the crossing, his engine had been running at 18 miles an hour. It stopped 42 yards beyond the crossing. When they left Wodonga there had been three regulation lamps working. They had been brown-out lamps, amber-coloured tissue paper having been placed over the ordinary white lights in accordance with the wartime regulations. Two of the three lights were damaged in the collision. [1361]

At the close of the inquest Acting Deputy Coroner F.W. Whan, J.P., recorded a verdict of 'Death due to Misadventure'. He commented, "It is evident the late W.H. Lord, driver of the motor bus, completely failed to notice the approaching locomotive. What caused his mind to stray from the vigilance necessary in a man who had the safety of so many in his keeping I am unable to say. The evidence discloses that the motor bus was licensed to carry 19 passengers, and that on the night of the collision there were 35 passengers on the vehicle. I am forced to the conclusion that the overcrowding of the motor bus was to some extent at least a contributing factor in the disaster. I say this because the overcrowding of a motor bus causes excessive noise and a stuffy atmosphere due to smoking."

The deputy coroner added, "In view of the heavy traffic on the Wodonga-Tallangatta

---

[1359] Ibid.

[1360] *The Argus*, Melbourne, Victoria, Australia, 29th June 1943

[1361] Ibid, 30th June 1943

highway the railway authorities should take action to provide at this crossing some adequate warning clearly visible to motorists and other road users approaching this intersection at night." [1362]

Express train D 49 collided with a lorry between Saalfeld and Naumberg, Germany, on Sunday 4th July 1943. Ten people lost their lives and three were hurt. The crossing barriers were open. [1363]

A freight train collided with a bus on a level crossing at Aulnois-sous-Laon, France, on Thursday 19th August 1943. Eight bus passengers were killed and 25 passengers and the bus driver, M. Rigaud, were injured, M. Rigaud being seriously hurt. The driver of the train had whistled for the crossing, which was unprotected. The bus was operating a regular service between Guise and Laon and the train was bound for Guise. Visibility at the crossing was reportedly good at the time of the accident. [1364] [1365]

It was thought that a rain-soaked road caused a Canada Coach Lines bus, bound from Waterdown to Hamilton, to skid onto a level crossing at Aldershot, Ontario, Canada, on the evening of Friday 10th September 1943. The bus then stalled on the crossing and before its driver, Fred Smith, could restart it the vehicle was struck by a Canadian National Railway passenger train and sheared in two. Driver Smith and ten bus passengers were killed more or less outright, and an 11th passenger was fatally hurt. The other two people on the bus escaped with relatively minor injuries. It was the opinion of Engineer E. Amory of the train that the accidental skidding of the bus was the cause of the accident. He brought his train, which was also bound for Hamilton, to a stop in a quarter of a mile. Chief Coroner J.H. Mullin was soon on the scene and gave permission for the train to continue its journey, but its engine had been damaged by the impact and a fresh one had to be obtained before it could proceed.

Two intending bus passengers had lucky escapes. Mrs O'Brien of Waterdown set off to catch the bus, then when it began to rain hard she changed her mind and decided to catch a later service. Stanley Nesbitt, 17, who lived near the crossing, ran for the bus, and missed it by a few seconds as it pulled away. He witnessed the crash. "I saw people hurled out of windows. Women and children were strewn along the length of the track, many of them thrown into the ditch." [1366] [1367]

Eight people died on Sunday 19th September 1943 when a crowded Greyhound bus en route from Memphis, Tennessee, to Chicago, Illinois, collided with the motor and baggage portion of a Missouri Pacific Railroad two-car, gasoline-powered train on a level crossing two miles south of Johnston City, Illinois, United States. The bus apparently got

---

[1362] Ibid, 1st July 1943

[1363] *Katastrophen der deutschen Bahnen Teil II*, by H.J. Ritzau, pg. 240 (in German)

[1364] *L' Ouest Éclair*, Nantes, France, 21st August 1943

[1365] *Le Matin*, Paris, France, 21st August 1943

[1366] *The Lethbridge Herald*, Lethbridge, Alberta, Canada, 11th September 1943

[1367] *Winnipeg Free Press*, Winnipeg, Manitoba, Canada, 11th September 1943 out of control on the falling gradient leading to the crossing and rammed the side of the train, overturning it. Fifty-two other people were injured. There had been only three passengers on the train and all the dead and most of the injured were on the bus. Those killed included the bus driver J.R. Hard, 38, while the injured included the Dr Luciana Milla Cisneros, the Honduran minister to Guatemala, and his daughter, Sofia Milla, who had been on their way to Rochester, Minnesota. It was theorised that either the brakes of the bus had failed or that the driver had accidentally let his foot slip off the brake pedal. [1368] [1369]

At the coroner's inquest on the accident the jury returned an open verdict. [1370]

Five minutes after midnight on Friday 24th September 1943, a car containing eight employees of the Moore Enamelling Company, who had just finished work, was struck by an eastbound Pennsylvania Railroad extra passenger train on a level crossing only a short distance from their workplace at West Lafayette, Ohio, United States. Miss Helen Alloway, 17; Mrs Betty Ashman, 22; Charles Barstow Jr., 18; Evolene Carr, 17; Raymond Everhart, 49; and Miss Voneta Reddick, 19, were killed outright and Frank Folger, 52, the driver of the car, died in hospital at 2.30 a.m. the same morning. Mrs Marie D'Ostrophe, 23, was seriously injured. A large number of other shift workers who had been on their way home witnessed the accident.

The crossing was protected by flashing warning lights which were operating at the time. However, a switch engine with several boxcars attached was standing on the siding serving the company plant and it was clear that Mr Folger must have assumed that the switch engine was what was making the warning lights operate. His car was carried along the line and was described as having been 'completely crumpled.' Its occupants were all thrown out of it.

Later in the day Mr D.S. McDill, the county coroner, returned a verdict of 'Accidental Death' on the victims. [1371]

Sunday 26th September 1943 saw the death of another seven people in a level crossing accident at the Arnold Street Crossing, South Bend, Indiana, United States. The automobile involved was struck by *The Pacemaker*, a fast passenger train of the New York Central Railroad running from Chicago to New York. The car was carried six blocks before the train could pull up and the bodies of the victims were badly mangled. Killed in the accident were Mr Joseph Prusinski, his wife, their nine-year-old daughter Christine; Joseph's brother Frank; Genevieve Lagocka, Isadore Michalski, and Valentine Michalski, all of South Bend and all members of the same family.

Crossing Watchman P.C. Myers said that he had been frantically signalling to the car driver to stop but the laughing occupants of the car had not noticed either his signals or the oncoming train. [1372] [1373]

[1368] *The Telegraph-Herald*, Dubuque, Iowa, United States, 20th September 1943
[1369] *Moline Daily Dispatch*, Moline, Illinois, United States, 20th September 1943
[1370] *Carbondale Free Press*, Carbondale, Illinois, United States, 24th September 1943
[1371] *The Coshocton Tribune*, Coshocton, Ohio, United States, 21st September 1943
[1372] *The Daily Mail*, Hagerstown, Maryland, United States, 27th September 1943

A bad accident at Appleford Crossing, England, on the Great Western Railway, on the evening of Friday 1st October 1943, received a brief mention in the national press and became the subject of a question in Parliament, then it appears to have been forgotten, and there was no Ministry of Transport inquiry into it. It involved a motor coach carrying civilian employees of an R.A.O.C. (Royal Army Ordnance Corps) depot which was crossing the railway, presumably a siding, when it was struck by some waggons which were being shunted. The coach was overturned by the impact and one of the waggons fell onto it, crushing it. Of the approximately 30 male and female occupants of the coach, seven, including its driver, were killed and 13 were injured. [1374] In reply to the question in the House of Commons Mr Arthur Henderson, the Secretary of State for War, replied that there would be a Court of Inquiry. [1375]

Six teenage boys died when their car, a 1936 Chevrolet, was driven into the side of the locomotive of a Wabash Railroad freight train on a level crossing at Paris, Missouri, United States, at 10.45 p.m. on Friday 8th October 1943. Carl Blackaby, Lydon Dixon, John Emmons, James Carter Harley, and Roy Thompson, 17, the driver of the car, were killed instantly and Roy's brother Ray, 14, died in hospital a few hours later. Orville Duncan, 16, was seriously injured and still unconscious the next day. In charge of the train was Conductor L.E. Fleming, with Engineer J.H. Cartright. [1376]

Another Greyhound bus, bound from Los Angeles to San Francisco, was involved in a collision with a train at about 8.30 p.m. on Friday 15th October 1943. Six bus passengers lost their lives and about 35 were injured when the bus was struck by the *Exposition Flyer* passenger train of the Western Pacific Railroad on a level crossing at Rhodes, California, United States.

Bus driver Richard Manning, 43, had only been employed by the bus company for five months and had only driven over the route once or twice before. Arrested on a charge of manslaughter after the accident, he said that he had seen the warning lights at the crossing flashing, but the brakes of his bus had failed to act when he had tried to stop. The vehicle had then stalled on the tracks, and he had been unable to get it into reverse gear. At this point the bus passengers had begun to get panicky and he had seen the train about 300 feet away. He had then opened the bus door and alighted from the vehicle together with five or six passengers. The bus had then moved forward on its own accord but failed to clear the crossing and the train struck the rear of it, whereupon it had whirled around and then overturned.

One of those who escaped from the bus with the driver was Mrs C.W. Price, of Alameda, who related, "The driver opened the door and got out with five passengers and myself including a woman with a baby and a child." Just as she alighted, she went on, the bus had rolled forward. "Someone fell down in front of me. I remember hearing the crash

---

[1373] *The Billings Gazette*, Billings, Montana, United States, 27th September 1943
[1374] *The Times*, London, England, 2nd October 1943
[1375] https://api.parliament.uk/historic-hansard/written-answers/1943/oct/13/accident-didcot
[1376] *Moberly Monitor Index*, Moberly, Missouri, United States, 9th October 1943

and something flew over me. Then something struck me. It was strangely quiet aboard the bus after the crash. Then there were a few moans and a man on the road started screaming." [1377] [1378]

An army lorry collided with a train on an open level crossing between Coesfeld and Borken, Germany, on Saturday 16th October 1943. Twenty-one people died and ten were injured. [1379]

Early train (Frühzug) 210 collided with a bus between Munchen Hbf and Ingolstadt, Germany, on Saturday 23rd October 1943. It was foggy and the crossing keeper had failed to lower the barriers in time. Seventeen people died and 25 were injured. [1380]

A shopping trip to purchase presents for the mother of the Rist family ended in tragedy at 9.50 a.m. on the morning of Saturday 18th December 1943, when a car was struck by a Pennsylvania Railroad Pittsburgh – Brownsville passenger train on a level crossing at Courtney, Pennsylvania, United States. Mr Herbert Rist Sr., 34, the driver of the car; his children Herbert Jr., 14; Benjamin, 13; John, 9; and Thomas, 7; his father Benjamin, 65; and Herbert Carden, 9, who had gone along with the family for the ride, were all killed or fatally injured. At the time of the accident Mrs Rist was working with the women's maintenance crew in the Pennsylvania roundhouse at Shire Oaks, two miles away.

The collision happened only a short distance from the Rist home. The automobile had almost cleared the crossing when it was struck by the train and demolished. There was good visibility at the crossing in both directions. Train crew, residents and members of a track maintenance gang gave first-aid to the victims and two hospitals rushed them to hospital, but their efforts were in vain; only Herbert Rist survived long enough to be admitted and he died two hours later.

Mrs Mary Ambrosi, who lived near the crossing, was the only eyewitness to the accident. She said that the car had stalled on the crossing. There had been time, she implied, for its occupants to have got clear, but they were apparently unaware of the approaching train. [1381]

A level crossing accident at Kenton, Ohio, United States, on the afternoon of Tuesday 28th December 1943 killed an elderly woman and four of her grandchildren. The victims were Mrs Maida Blackwood, 67; and her grandchildren Richard Blackwood, 15; Jane, 14; James Jr., 11; and John, 10. The children's mother had died in February, since when Mrs Blackwood had moved in with the children's father.

Richard, the eldest boy, who was almost 16, had a temporary driving permit and was operating the car. There were no witnesses to the accident, but the streets were slippery

[1377] *San Mateo Times*, San Mateo, California, United States, 16th October 1943
[1378] *Tucson Daily Citizen*, Tucson, Arizona, United States, 16th October 1943
[1379] *Katastrophen der deutschen Bahnen Teil II*, by H.J. Ritzau, pg. 240 (in German)
[1380] Ibid.
[1381] *The Charleroi Mail*, Charleroi, Pennsylvania, United States, 18th December 1943

after rain and the car apparently skidded into the path of the Erie Railroad train, a through flyer which was running late and travelling at 65 miles an hour. The party were on their way home after visits to various relatives. The son of Mrs Blackwood and father of the children, James Blackwood Sr., had been ill for some time before the accident. He was placed in the care of a physician after learning of it. [1382]

The year 1944 began badly as far as level crossing accidents were concerned, with a disastrous collision with a bus at Kingman, Arizona, United States, at 9 p.m. on Thursday 6th January. The accident happened at the entrance to Kingman Army Airfield, where the road to the base crossed the Atchison Topeka and Santa Fe Railway outside of the main gate. What was described as a semi-trailer-type bus, returning from night gunnery practice with 34 air force cadets, an officer, and a gunnery instructor, was being driven by Private Fred Smith. As it reached the crossing it was flagged down by the crossing watchman, an army guard, and came to a stand, then, for some reason, it suddenly jerked forward into the path of the Santa Fe freight train. A newspaper reported that, 'The train's cowcatcher caught the vehicle squarely, jackknifing it and dragging the crumpled wreckage several hundred yards down the track. Pieces of the bus were hurled over a wide area.'

Engineer C.L. Hickey said that his train had been travelling at its usual speed of about 45 miles an hour as it had approached the crossing. He said, "I whistled for the crossing and noticed the bus in the glare of my headlight. It appeared to me making a full stop at the crossing. Suddenly it jumped forward and landed squarely in front of us. I didn't even have time to apply the air before we hit it."

A field hospital was set up at the scene and 300 men were ordered to stand by to provide blood transfusions for the injured. Rescuers had to cut through and pry apart the wreckage to release the dead and injured. Doctors, nurses, and men worked under searchlights mounted on trucks to administer morphine and plasma to the badly injured survivors before ambulances took them to the base hospital. Military police roped off the area to keep out unwanted persons.

Twenty-five cadets, the gunnery instructor and the bus driver were killed outright, and eight cadets were critically injured. One officer somehow escaped injury. The subsequent death of one of the injured raised the number of dead to 28.

A military investigation board was set up to look into the cause of the disaster. [1383]

The crossing watchman had a complete nervous breakdown following the accident and had to be admitted to hospital. [1384]

By mid-January plans had been announced to construct an underpass to replace the level crossing. [1385]

Exactly a fortnight after the Kingman disaster a bus/train collision of almost equal

---

[1382] *The Sandusky Register Star-News*, Sandusky, Ohio, United States, 29th December 1943

[1383] *The Wisconsin State Journal*, Madison, Wisconsin, United States, 7th January 1944

[1384] *The Salt Lake Tribune*, Salt Lake City, Utah, United States, 8th January 1944

[1385] *Tucson Daily Citizen*, Tucson, Arizona, United States, 15th January 1944 severity occurred at Hawkesbury River, New South Wales, Australia. This time the victims were all civilians. The bus involved was one providing a local service between Kangaroo Point and Brooklyn. It was carrying 21 passengers, [1386] most of whom had been intending to catch the 9.30 a.m. train from Hawkesbury River into Sydney to go shopping. The bus was caught in the middle by the Kempsey mail train, which did not stop at Hawkesbury River, smashed to pieces and parts of it were carried along the line, two bus wheels being borne along in front of the engine for 400 yards. The train was slightly derailed by the impact with the bus.

Train driver Jack Sponberg said, "I jammed on the brakes, but we were too close, and had no chance of avoiding the bus. The railway runs parallel for some distance along the road, which then turns at right angles to cross the line. I saw the bus coming along the road as I was going down the grade, and to my surprise the bus, which was travelling in the same direction as the train, moved across the track."

Fireman Howard Moller said that as their engine had borne down on the bus he had been able to see the horror on the faces of the women and children on the bus.

Fifteen people, including several children, two nuns and the driver of the bus, Jack Callen (50), were killed outright and five were seriously injured. One of the injured later succumbed. The body of the bus driver had to be cut away from the engine. He had been driving the vehicle in place of his son, who had been called up for military service, and was due home later in the month.

After the accident residents of Brooklyn, the settlement which adjoins Hawkesbury River station, rushed to the scene to give aid and women tore up linen to make bandages. One of the first on the scene, E. Horton, said that the crash had sounded like a bomb explosion. He had rushed to the crossing to find dead and injured lying everywhere. He told a reporter, "The moans of the injured were pitiful. Bus wreckage was scattered all over the line. Navy men and soldiers gave all the help they could, as well as local people. The body of a child was carried 400 yards."

Several sad cases were described in a newspaper report. 'One of the victims, Mrs Mollins, was on her way to buy a lottery ticket. "She came to town to try her luck in the lottery," said her husband after he identified her body. He is left with three children, the eldest of whom is 10.

'The little boy victim, Wally Cox, was unwilling to go on the bus. His uncle, Mr Joe Cox, said that Willie was coming to town to do the shopping, but he must have had a premonition that something would happen, as he did not want to go, which was unusual for him.'

'Little Gloria Sylvia was carried for several hundred yards on the engine after the accident. She was the first to recover consciousness in Hornsby Hospital this afternoon. Her first words were, "I want to go home."

'"We are so happy she is conscious, but she looks so terribly smashed up that we could cry continuously," said Mrs Sylvia. Gloria had been told to take the bus, as her parents thought that it would be safer than riding a bicycle.' [1387]

[1386] *Singleton Argus*, Singleton, New South Wales, Australia, 21st January 1944
[1387] *National Advocate*, Bathurst, New South Wales, Australia, 21st January 1944

The inquest on the victims of the accident did not begin until early April.

Ralpael Italiano, the first witness at the inquest, was carried into the courtroom on a stretcher because of his injuries. He testified that as the bus reached the crossing the gate on the northern side had been open. Near his cabin was the gatekeeper and there were two men on the opposite side of the crossing. He saw no one indicating for the bus to stop. "As the bus went over the crossing the women screamed," he said. "It stopped between the two sets of rails, as the gate on the Brooklyn side was almost closed. I saw the train approaching about five or six carriage lengths away. I heard a whistle at the same time as I saw the train." The next thing he remembered was lying in the gutter. He said that as he had been stood next to the door he could have got out of the bus when it stopped, but he was not a coward and his wife, three children and a grandson had been on the bus. His refusal to abandon his family was to no avail; his wife and children and possibly also his grandson lost their lives.

Evidence was given that from the first point along the line that a train would be visible from the crossing keeper's cabin, it would take a train travelling at 40 miles an hour 29½ seconds to reach the crossing. The crossing was visible from the engine of a train for a quarter mile but because of cottages a bus on the road would only become visible to an engine driver when he was 85 yards from the crossing, a train becoming visible to the bus driver at the same point.

The crossing gates opened outwards. The crossing keeper's hut had small windows giving a view of approaching trains and the whistle of a train in the cutting was audible from the hut. There were no mechanical devices either to operate the gates or to warn of approaching trains. No clock or telephone was provided in the cabin, nor were there any lavatory facilities or water tank. [1388]

The following day of the inquest William Sutton of Brooklyn, who had witnessed the accident, agreed that the gate on the Kangaroo Point side of the crossing had been open for the bus while that on his side had been closed. He could not recall any alterations to the crossing in the 50 years he had lived there, though the road had become much busier.

Edward Marsh, a passenger on the bus, spoke of its stopping on the crossing. He continued, "Directly after the bus stopped a woman behind me called out, 'Look, there's the engine.' I looked round and sighted a green-painted engine. The next second there was a bang, and we were over." The stopping of the bus, the woman's exclamation and the bang had all happened more or less at the same time. He added that the gates at the crossing had not been improved since 1914, when there had been no motor traffic. There had never been any device at the crossing to warn of trains.

Father T.J. O'Brien, priest of St Catherine's Orphanage, Brooklyn, said that he had always considered the crossing dangerous. He considered that a bell should be provided to warn the crossing keeper of trains.

Private Austin Callen, Second Australian Imperial Force, the son of the dead bus driver, said that he had driven the bus for four or five years before joining the Army. It was not possible to see a train approaching the crossing from Sydney or to hear it because of the noise of the bus. He had seen the very light gate on the Brooklyn side of the crossing shake shut after being opened.

[1388] *The Sydney Morning Herald*, Sydney, New South Wales, Australia, 4th April 1944

Warrant Officer John Denness spoke of going over the crossing immediately before the accident and of assisting in the rescue work. He had spoken to the relief crossing keeper, 20-year-old Cecil Peter Tolley, who had appeared very agitated, and had said to him, "You did not have those gates shut."

"No," Tolley had replied. [1389]

When the inquest resumed the next day Detective Martin McMahon testified that Tolley had confessed to him that he (Tolley) had opened the gates half an hour before the accident for a baker's cart and had not closed them again. Tolley had said, "I had not closed the gate, and the bus driver thought the line was clear and drove up on to the line. I heard the noise of a train approaching. I jumped out of my cabin and called out to the bus driver to stop, but the bus driver continued on. The driver looked towards me, and the train struck the bus about the centre of the line. I closed my eyes, and when I opened them again the train was passing me, and I saw wreckage of the bus along the line." Later after being cautioned Tolley had added, "There was a lot of traffic, and it is the most monotonous job on the railway, and I did not close them. You do not realise how serious it is until something like this happens." Tolley said that he was rostered to work a 12-hour shift, from 5 a.m. to 5 p.m., and had no other duties to perform.

Chief Traffic Inspector Albert Alt stated that the gatekeeper would have been given a book of the rules and regulations, plus a working timetable which would show the times of the Kempsey mail. It was not, however, the practice to issue watches to the gatekeepers, or to put clocks in their cabins. The running of trains not shown in the working timetable was frequent. He refused to express an opinion as to whether a watch or clock was necessary for the gatekeeper to perform his duty. [1390]

Driver Sponberg, Fireman Moller and Permanent Gatekeeper Edward O'Brien gave evidence on the final day of the inquest. The proceedings then terminated with the Paramatta Coroner, Mr R.G. Williams, committing Relief Gatekeeper Tolley for trial on a charge of manslaughter. [1391]

At the beginning of June it was announced that the Crown had decided not to proceed with the manslaughter charge against Cecil Tolley. [1392]

The third major level crossing accident involving a bus in January 1944 happened on Sunday 23rd. It was the fault of the very inexperienced driver of the vehicle, Henry Hoerath, who was making his first trip with the Public Service Company bus on the day of the accident. All had gone well until he reached a level crossing with the double-track main line of the Wabash Railroad on the outskirts of St Louis, near the station at Delmar Boulevard, Missouri, United States. A train was approaching, and he stopped his bus to let it pass. When it had done so he moved onto the crossing without considering the possibility that a second train might be coming from the opposite direction on the other line. [1393] The next moment the bus was sheared in two by a Detroit – St Louis flyer. To

---

[1389] Ibid, 5th April 1944
[1390] Ibid, 6th April 1944
[1391] Ibid, 13th April 1944
[1392] *Narandera Argus and Riverina Advertiser*, Narandera, New South Wales, Australia, 2nd June 1944
[1393] *Edwardsville Intelligencer*, Edwardsville, Illinois, United States, 24th January 1944

make matters worse the petrol tank was shattered, and the wreckage of the bus and its passengers were showered with burning petrol.

E.E. Gallagher, one of the first people to reach the scene, said afterwards, "I saw the bus burning from one end to the other and managed to drag five persons through the exit door. Some of them were women. People inside were screaming, 'Save me! Save me!' Their clothing was afire. My wife brought sweaters and coats to wrap around them and smother the flames. Their cries were awful. I saw another woman inside. She was twisting her head, trying to say something. I couldn't reach her. It was too hot. I was too late. Soon she stopped moving her head and I knew then she was dead." [1394]

Another hero of the accident was 11-year-old Guybert Barnes, who although he was injured himself managed to crawl back into the bus through a broken window to save his 16-year-old brother, suffering slight burns in the progress. Barnes related how some of the bus passengers had seen the headlights of the express as the bus had moved onto the crossing and had screamed.

Five passengers died on the bus, trapped and unable to escape. The bodies of three more were found along the right-of-way and seven critically injured patients died in St Louis County Hospital in the following hours, bringing the number of dead up to 15. Nine other bus passengers were less badly hurt. Fifteen pints of blood plasma were transferred from the blood bank at the Barnes hospital to treat the injured. None of the train passengers was injured.

Engineer F.L. Dolson said that he had not seen the bus until immediately before his engine hit it. [1395] During the day, the crossing was manned by a watchman. His hours of duty had ended 45 minutes before the collision.

At about 08:00 on the morning of Thursday 17th February 1944 a local train bound for Leiden passed over the level crossing at Lisse, the Netherlands. The crossing barriers were then raised for road traffic to proceed, and as vehicles began moving over the crossing two lorries came into collision with one another. One of the lorries, coming from Lisse, was carrying about 30 men on their way to work. It was not possible to quickly remove the lorries from the crossing and train No. 1106 from Dordrecht, due in Lisse at 07:49 but 15 minutes late, was nearing the crossing.

The train could not be stopped and collided with the lorry carrying the workmen. The lorry driver had time to save himself but the cabin on the bed of the lorry in which the workmen were housed was almost completely splintered. Six of the workmen were killed outright and several others were seriously injured. Three of the worst injured, who were taken to Haarlem and Leiden, died during the course of the day. A number of the other workmen who were less badly hurt went to their homes. The line was blocked for some time, but normal working resumed later in the day. [1396]

Nineteen children were injured, three seriously, when southbound Chicago Burlington and Quincy Railroad passenger train No. 112 collided with a school bus almost in front of

[1394] *Carbondale Free Press*, Carbondale, Illinois, United States, 24th January 1944

[1395] *The Abilene Reporter-News*, Abilene, Texas, United States, 24th January 1944

[1396] *Dagblad voor Leiden en omstreken*, Leiden, the Netherlands, 19th February 1944 the school building at Bethany, Missouri, United States, at 8.45 a.m. on Thursday 24th February 1944. Engineer E.W. Hautzenroeder of the gasoline-electric train told a reporter that the warning bell at the crossing had been ringing, the wig-wag signal had been operating and he had 'whistled lustily' for the crossing. Despite all this the bus, driven by a man named Floyd Edwards, had tried to cross in front of his train. The bus was carried 30 or 40 feet down the line, then partly overturned, while the train pulled up in 300 feet. Only slight damage was done to the train, which had already been running 30 minutes late. [1397]

Four teenagers, all pupils of Ben Davis High School in Indianapolis, were killed more or less instantly when the car they were in became stalled on a level crossing at Bridgeport, Indiana, United States, and was struck by a Pennsylvania Railroad passenger train on the evening of Sunday 5th March 1944. They were Harley Alsman, 16; Hughey Hughes, 17; Robert Kershaw, 19; and Nancy Sering, 17. The other two occupants of the car, Mildred Schutte, 17; and Private Clarence Alsman Jr., 18, brother of Harley, who was home on a convalescent furlough, were injured. Miss Schutte was very seriously hurt and died in hospital the next day. The car was demolished, bodies and wreckage being scattered 100 feet along the right-of-way. There were warning signs at the crossing but no barriers, light signals, or bells. [1398]

Arnold Street Crossing, South Bend, Indiana, United States, which had been the scene of a bad accident in September 1943, saw another on Sunday 30th April 1944. Enoch Stopczynski, 50, the driver of the car; his wife Lottie, 47; their daughter Elsie, 17; Stanley Babicz, 37; his wife Clara, 36; and Carrie Nowak, 41; were all killed when their automobile was hit by an eastbound New York Central Railroad passenger train. [1399]

On the afternoon of Saturday 6th May 1944 a northbound diesel powered Atlantic Coast Line Railroad freight train collided with a car on Virginia Avenue Crossing, Emporia, Virginia, United States. The car was thrown against the engine of a standing southbound train and then wedged against the waggons of the moving train. Farmer and school bus driver Johnny J. Rook, 49; his wife Lou and their son Johnny Lee, 9; Mrs Pearl Clements and her son William Earl, 7, lost their lives. Another son of Mrs Clements, Clarence, was taken to hospital in Roanoke Rapids, North Carolina. At first he was reported to be critically hurt but later he was reported to be recovering. Conductor R.E. Whitehead was in charge of the train, with Engineer J.R. Cole. [1400] [1401]

Seven teenagers lost their lives on the evening of Thursday 18th May 1944 on what was described as a 'private gravel crossing' less than a mile south of Mt Carmel, Illinois, United States. They had been returning to town from a 4-H meeting (4-H was an

---

[1397] *Tri-County News*, King City, Missouri, United States, 25th February 1944

[1398] *The Vidette-Messenger*, Valparaiso, Indiana, United States, 6th March 1944

[1399] *The News-Palladium*, Benton Harbor, Michigan, United States, 1st May 1944

[1400] *The High Point Enterprise*, High Point, North Carolina, United States, 7th May 1944

[1401] *Kingsport News*, Kingsport, Tennessee, United States, 8th May 1944

American youth organisation). The car in which they were riding was struck by New York Central Railroad passenger train No. 462 from Harrisburg, Illinois, to Mt Carmel. Those who lost their lives were Helen Alka, 15; Patricia Doris Fisher, 15; Dalla Lee Henry, 15; his brother Rolla Jr., 18; Dora Marie Lindstrom, 15; Robert Skelton, 19; and Elizabeth Joanne Veihman, 15. All were killed instantly in the demolished car. [1402]

Six teenagers and young adults on their way from a dance at the Aurora City Park to a swimming pool early on Saturday 17th June 1944 were killed when their light panel truck was struck by a westbound Baltimore and Ohio Railroad passenger train on a level crossing several miles from Aurora, Indiana, United States. They were Thelma Austin, 17; Okel Bowling, 21; Catherine McAlister, 16; her sister Lenamae, 15; George Steele, 17; and Tounger Wishman, 23, the driver of the car. [1403] [1404]

Arthur Martin, 33; his children Kathleen, 6; Beverly, 4; and Carol, 2; Charles Haddock, 3; Virginia Schafer, 14; Roy Schafer, 9; and Hilda Schanda, 10; lost their lives when a light truck was hit by a Western Pacific Railroad near French Camp, California, United States, four miles south of Stockton on the evening of Sunday 3rd September 1944. Dolores Schanda, 18, the other occupant of the truck, was reportedly critically ill in a Stockton hospital. The victims, who were all related, had been returning from a swimming outing. [1405]

Five people lost their lives in a collision between their car and an electric train on an open level crossing near the station at Boronia, Victoria, Australia, on the afternoon of Sunday 29th October 1944. Those who lost their lives were Mr Henry Witchell, 46; his wife Edith, 49; Mrs Emily Witchell, 79, Mr Witchell's mother; Mrs Elizabeth Benwell, 47; and her daughter Betty Esther, 15. The party had been visiting relatives of Mrs Benwell at Boronia and had stopped to buy some soft drinks at a shop near the crossing before setting off again. It was believed that the low sun had interfered with the vision of Mr Witchell and prevented him seeing the 4.30 p.m. passenger train from Ferntree Gully to Melbourne Flinders Street. All except Mrs Benwell were killed outright; she died in the ambulance taking her to hospital. The train was slightly damaged and delayed half an hour. [1406] [1407]

A newspaper editorial complained that residents had been asking the Railway Commissioners for 20 years for warning devices to be installed at the crossing, but their requests had been turned down, the commissioners maintaining that the crossing was not dangerous, despite previous accidents and incidents there. [1408]

[1402] *Carbondale Free Press*, Carbondale, Illinois, United States, 19th May 1944
[1403] *The Coshocton Tribune*, Coshocton, Ohio, United States, 18th June 1944
[1404] *The Telegraph-Herald*, Dubuque, Iowa, United States, 18th June 1944
[1405] *The Lincoln Star*, Lincoln, Nebraska, United States, 4th September 1944
[1406] *The Age*, Melbourne, Victoria, Australia, 30th October 1944
[1407] *The Weekly Times*, Melbourne, Victoria, Australia, 1st November 1944
[1408] *The Herald*, Melbourne, Victoria, Australia, 30th October 1944

A Sunday evening pleasure trip made by the Leonard family ended with the death of five of them on a level crossing at Sioux City, Iowa, United States, on 19th November 1944. The car in which they had been riding was in collision with the caboose of a Chicago Burlington and Quincy Railroad train which was setting back over the crossing. Mr Walter Leonard and his wife, both aged about 30, and their children Larry, 8; Dale, 7; and Doris, 6, were all killed. Another child, Harold, 2, had his right arm amputated but it seemed likely that he would survive. The wreckage of the car was carried almost 50 yards and the victims were trapped for three hours. [1409]

An attorney, Mark Ryan, acting on behalf of the Leonard family, announced that he had discovered that there had been no flagman on duty at the crossing at the time of the accident. However the police stated that they did not know if a flagman had been required at that time and Assistant District Superintendent A.E. Stoll stated that an internal inquiry by the railway company had shown no violation of the operating rules by the train crew. He said that the train had been operated correctly and the lights on the rear of the caboose had been burning brightly. [1410]

A group of women war workers who were sharing a car were involved in a level crossing accident at Pewamo, Michigan, United States, 15 miles from Ionia, on the evening of Wednesday 29th November 1944. They had been returning from their jobs at the General Motors AC spark plug division in Ionia. Six of the women, Miss Mildred Bozung, 23; her sister Marjorie; Mrs Mabel Cunningham, 42; her sister, Miss Ruth Navin, 33, the driver of the car; Miss Margaret O'Connell, 31; and Mrs Stiles were either killed outright or fatally hurt. The seventh occupant of the car, Mrs Schafer, 32, was injured.

One of the first on the scene was Mr Marvin Stiles, a teacher at Pewamo High School and a first-aid instructor who was also the husband of one the victims. He gave what assistance he could to the victims until they were taken away by ambulances. Ironically, he said, this would in any case have been the last time that his wife had shared the car as she had found a more convenient means of transportation.

The Grand Trunk Western Railroad passenger train, in charge of Conductor Claude Owen with Engineer Ed Hunsberger, had killed three people in a level crossing accident at Ionia only nine days earlier. [1411]

Five civilian employees of the Rome Army airfield base were killed when their car collided with a New York Central Railroad milk train No. 64 on the Eastern Rock Corporation Crossing, near Mapledale, New York, United States, on Wednesday 13th December 1944. The train carried the battered automobile for three quarters of a mile before it came to a stop. All five victims were still inside, and state troopers were forced to use crowbars to extricate them from the wreckage. The dead were: Wallace W. Brondstatter, 65, of Cold Brook, carpenter, and driver of the car; Alma V. Wood, 18, of Poland, painter's helper, Jean E. Jones, 18 of Cold Brook, supply clerk; Laura R. Failing,

[1409] *The Burlington Hawk-Eye Gazette*, Burlington, Iowa, United States, 20th November 1944
[1410] *Nebraska State Journal*, Lincoln, Nebraska, United States, 22nd November 1944
[1411] *The News-Palladium*, Benton Harbor, Michigan, United States, 30th November 1944

26. of Cold Brook, storekeeper, and Earl R. Irwin, 70, of Cold Brook, a heating and operating engineer. The accident occurred on the Mapledale-Holland Patent highway, about half a mile west of Mapledale.

Engineer John Caley, said his train had been traveling at about 60 miles an hour as it approached the crossing on the way to Utica. He told the coroner that he saw the automobile approaching at an estimated speed of 26 miles per hour but had assumed that it would stop when it reached the crossing.[1412]

In an ironic accident, Irving Longworth Jr., a village attorney who had spent many years campaigning for the Long Island Railroad to eliminate its level crossings, was killed along with his wife when their car was struck by a Long Island train on a level crossing at Rockville Center, New York, United States on Tuesday 30th January 1945. [1413]

Harvey Johnson, the owner of a sawmill at Berlin, Maryland, had picked up some of his workmen on the morning of Monday 5th February 1945 to take them to the sawmill. He then turned off Route 13 in the panel body truck he was driving at about 7.30 a.m. to pick up three more men at Fruitland, Maryland, United States. Arriving at a level crossing with the Pennsylvania Railroad a freight train was shunting over the crossing. When this train cleared the crossing the warning lights continued to operate but clearly believing that they were only flashing for the shunting train Johnson then set off over the crossing, and his truck was struck by a northbound passenger train. Johnson was killed instantly, as were William Pittman and Harold Townsend. Three other men, Max Creasey, Edward Brewer, and Otho Miles were fatally injured and died on the way to hospital. The remaining five workers on the lorry were all injured, and Julius Pittman died of his injuries the following day. The others were expected to recover. [1414]

William Kerr, 45; his daughter Thelda, 10; Jimmy Custy, 10; Dorothy Custy, 8; Theresa Custy, 6; and Lewis Sunga, 10, were killed when the car which they were riding was struck by an eastbound Union Pacific Railroad hospital train en route from the West Coast to Kansas City on the morning of Monday 5th March 1945. Kerr worked at a dairy a few hundred yards from the crossing and was driving the children to school. The accident happened near Mesa, Colorado, United States, 15 miles east of Denver (the railway location known as Mesa is not the same as the town of Mesa, Colorado, which is some 200 miles west of Denver). Minor repairs had to be made to the engine of the train, which was delayed an hour and 45 minutes. [1415] Thelda Kerr's 10-year-old brother Roger escaped the accident because had not been very well that day and had stayed home from school. [1416]

Seven people died when the car in which they were riding was struck by an eastbound Southern Railway train on a level crossing near Greeneville, Tennessee, United States, on

---

[1412] *Endicott Daily Bulletin*, Endicott, New York, United States, 14th December 1944
[1413] *Daily Journal-Gazette*, Mattoon, Illinois, United States, 30th January 1945
[1414] *The Salisbury Times*, Salisbury, Maryland, United States, 5th and 7th February 1945
[1415] *Albuquerque Journal*, Albuquerque, New Mexico, United States, 6th March 1945
[1416] *The Galveston Daily News*, Galveston, Texas, United States, 6th March 1945

the afternoon of Tuesday 31st July 1945. They were Mrs Stella Metcalf, 29; her children James, 12; George, 11; Mary, 9; Ruth, 4; and Eva, 18 months; and the Metcalf's housekeeper, Mrs Zilla Fillers. The automobile apparently stalled on the crossing. [1417]

Mr Homer Wideman, 31 and his wife, 22, described as 'agricultural workers', were killed together with their children Juanita, 7; twins Marie and Thelma, 4; and Pauline, 2; were all killed when their car was struck by an Atchison Topeka and Santa Fe Railway train near Paige, California, United States, on Wednesday 1st August 1945. The car was thrown 42 feet by the train and ended up in a ditch. The inquest on the victims returned an open verdict. [1418]

A bus crowded with passengers was negotiating Crossing PN 251, near Civens, France, not far from Saint Etienne, on or about Friday 17th August 1945 when it was struck by a train with disastrous results. The force of the impact demolished the vehicle, and its wreckage was carried 130 metres, together with the bodies of some of the victims. Fifteen bus passengers were killed outright, and several were taken to hospital in a critical condition. An investigation by the Société nationale de chemins de fer français placed the responsibility for the accident upon the crossing keeper, who declared that he had not heard the warning bell advising him of the approach of the train. [1419] [1420]

Six people died in a level crossing collision early on Saturday 6th October 1945 when their car was struck by an eastbound Canadian Pacific Railway freight train at Chesterville, Ontario, Canada. The victims were Howard Beckstead, 52, the driver of the car; Herbert Irving, 50; his daughter Gwendolyn, 18; Omer Grouix, 25; Earl Ouderkirk, 26; and Pauline Seguin, 20. Three of the bodies were thrown from the car and the other three were dragged along the track. The wig-wag signal at the crossing had been operating but the car driver had apparently failed to observe it. [1421]

Mrs Ersa Stone, 56; her daughter Mrs Frances Moyes, 21; Mrs Moyes' daughter Sandra, 3; Mrs Stones' daughter-in-law Mrs Vivian Stone, 27; and Mrs Vivian Stone's children Bobby, 10; and Teddy, 7; were killed outright or fatally injured when their car was struck by a New York Central Railroad train on a private farm crossing at St Francisville, Illinois, United States, on Friday 12th October 1945. Another child, Virgil Moyes, 4, was injured. Mrs Moyes, the driver of the car, had stopped the vehicle to open a gate and had then driven onto the crossing without noticing the approaching train, which consisted only of an engine and a caboose. [1422]

[1417] *Kingsport News*, Kingsport, Tennessee, United States, 1st August 1945

[1418] *The Bakersfield Californian*, Bakersfield, California, United States, 9th August 1945

[1419] *La Croix du Nord*, Paris, France, 17th August 1945

[1420] *Gazette de Lausanne*, Lausanne, Switzerland, 18th August 1945

[1421] *Medicine Hat Daily News*, Medicine Hat, Alberta, Canada, 6th October 1945

[1422] The *Daily Journal-Gazette* and Commercial Star, Mattoon, Illinois, United States, 13th October 1945

A crossing collision near Carlisle, Arkansas, United States, on Sunday 21st October 1945 was blamed on the fact that six people had been crowded into the cab of a milk lorry, preventing the driver from getting a clear view of the track. All six of the occupants of the lorry were killed. They were farmer Lonnie Addison, 42; Mrs Hazel Baxter, 24; her daughter Ruth, 3; James Beggs, 29, the driver of the truck; Elizabeth Hamilton, 6; and Virgie Hamilton, 6. Nor were these the only deaths in the accident, for the engine, tender and five baggage cars of the Chicago Rock Island and Pacific Railroad passenger train which collided with the lorry were derailed and Engineer George Kugler, 65; and Fireman A.L. Falkner, 55; also lost their lives in the accident. None of the 122 passengers on the train was hurt. It was believed that the truck had been bound for the Carlisle Cheese plant; its bed was filled with milk cans.

Among the passengers on the train were members of the troupe of the *School for Brides* road show. R.E. Bonbrake, electrician with the troupe, said that they had first become aware of something wrong when the engineer had applied the brakes in emergency. He and several others had been standing in one of the rear cars of the train and when the jar had come he had cautioned them to sit down. Parts of the wrecked lorry had flown past them. Bonbrake said, "After a series of jars, the whole train stopped and then someone looked out of a window and saw the engine and tender in a ditch. Then we knew we had been in a serious train accident." He said that there had been a mild panic among the passengers until they had alighted from the train. Having appeared at Little Rock, Arkansas, on Saturday night the troupe were bound for Birmingham, Alabama, where they were due to perform on Monday night. The baggage car carrying the scenery for the show was smashed but railroad officials said that the scenery itself had not suffered damage and was en route to Birmingham. [1423]

Eight American soldiers of the 84 Infantry Division lost their lives on Monday 22nd October 1945, when their military truck loaded with fuel cans collided with a narrow gauge train on a level crossing near Hesperange, Luxembourg. A fire and explosion followed the collision, and the men were burnt to death. Subsequently a memorial was erected at the site. Their names were Douglas Connell, Roy Davis, Maximo Garcia, Robert Hannabury, Anthony Kunigelis, Charles Taylor, Selby Watkins, and William Weisenmiller. [1424] [1425]

Sixteen German prisoners of war, several of whom had served in Rommel's Afrika Corps, were killed or fatally injured when the lorry in which they were riding was hit by a New York Central Railroad passenger train on a Silberhorn Crossing, near Blissfield, Michigan, United States, at 4.30 p.m. Wednesday 31st October 1945. They had been returning to Blissfield POW camp after finishing work on a beet farm detail. A 17th fatality was Private First Class Edward Loughrin, M.P. who died of his injuries in Adrian hospital. Eight other POWs were injured, as was Pascal Martinez, the owner of the beet farm and the driver of the lorry, who was said to be on the danger list in hospital the next day. Captain Richard Bemont, the commander of the POW camp, said that other POWs

---

[1423] *Blytheville Courier News*, Blytheville, Arkansas, United States, 22nd October 1945

[1424] *The Galveston Daily News*, Galveston, Texas, United States, 23rd October 1945

[1425] https://www.uswarmemorials.org/html/monument_details.php?SiteID=1626&MemID=2138

had been stunned when they had heard what had happened to their colleagues. The conductor of the train was W.B. Sherman, and its engineer was Ford Winbrenner. The war having finished several months earlier, the POWs were awaiting shipment back to Europe at the time of the tragedy. In some of the newspaper reports of the accident they were referred to as Nazis, which was probably not true of all and possibly not of any of them, they would have just been regular soldiers. [1426]

Exactly how the accident happened was not detailed but Martinez, who recovered from his injuries, was exonerated by an Army board of inquiry. [1427]

Crossing Watchman Grover Frame was arrested for manslaughter after the *Pacific Express* passenger train of the Erie Railroad collided with a bus on Avenue O Crossing, Burnham, Illinois, United States, on the outskirts of Chicago, on Sunday 9th December 1945. The bus was the last of three which had followed one another over the crossing. After initially denying that he had done so, the watchman admitted having raising the crossing barriers to let the buses across after a westbound freight train had cleared the crossing, although he could see the headlights of the approaching express. Five bus passengers, Raymond Daniszewski, 18; Josephine Jeglawski, 18; Sophie Elizabeth O'Dobrinak, 22; Paul Petka, 23; and Vincent Zagorski, 50, were killed. The bus driver, Charles Martin, and seven passengers were injured. The three buses had all been en route from Hegewisch to Hammond. [1428]

Six people on their way to a dance on Sunday 16th December 1945 were killed when the car they were riding in was struck by the Chicago, Illinois, to St Louis, Missouri, *Blue Bird* express of the Wabash Railroad on what was variously referred to a Main Street Crossing and Crossing No. 5, Staunton, Illinois, United States. Those who lost their lives were Marine Staff Sergeant Andrew Andrasko, Miss Violet Compton, Private First Class O.V. Fritsche, Seaman George Kasarda, Victor Mueller, and Russell Scheiter, who lived long enough to be taken to hospital. John Majeski Jr, the driver of the car, survived with injuries.

The accident was apparently Majeski's fault. Warning lights protected the crossing and cars driven by Edward Roseman and Emil Beyer were already stopped waiting for the train. Majeski's car overtook the two waiting cars, entered the crossing, and was struck by the train. He blamed a frosted windscreen which had interfered with his view. The car was carried 75 feet by the train. Already 20 minutes late, the express was delayed at least another 40 minutes by the accident. [1429] [1430]

An ice-covered windscreen was likewise blamed for a bad accident on Cole Street Crossing, Lima, Ohio, United States, at 7.04 p.m. on Tuesday 18th December 1945. The

---

[1426] *Monitor-Leader*, Mount Clemens, Michigan, United States, 1st November 1945
[1427] https://www.findagrave.com/memorial/361059/kurt-bernock
[1428] *The Hammond Times*, Hammond, Indiana, United States, 10th December 1945
[1429] *Edwardsville Intelligencer*, Edwardsville, Illinois, United States, 17th December 1945
[1430] The *Daily Journal-Gazette* and Commercial Star, Mattoon, Illinois, United States, 17th December 945

automobile was being driven by Mr Paul Unverferth, 31; and the other victims were his wife Emma Carolyn, 27; Mr John Aldridge, 25; his wife Elizabeth, 28; and their daughter Joyce Ann, 3. Coming into Lima from Delphos, Mr Unverferth had turned from Elida Road into the path of the *Manhattan Express* passenger train of the Pennsylvania Railroad, which was running three hours late. The crossing was protected by warning lights. Engineer L.O. Dennis spoke of 'seeing sparks,' while Fireman C.V. Harter said that he had 'caught a glimpse of something in front of them.' The party in the car had been on their way to a veterans' meeting. [1431]

Freezing rain and poor visibility were blamed for the death of seven persons in a level crossing accident in the centre of Gaffney, South Carolina, United States, shortly before midnight on Christmas Eve, 24th December 1945. The victims were Miss Geraldine Byers, Jack Cothran, Charles Edward Phillips, Henry Phillips, Robert Phillips, G.C. Self and Fred Warren. Their car was struck by a Southern Railway passenger train and carried 50 yards. [1432]

[1431] *The Lima News*, Lima, Ohio, United States, 19th December 1945
[1432] *The High Point Enterprise*, High Point, North Carolina, United States, 26th December 1945

CHAPTER 13

RAIL/RAIL CROSSING COLLISIONS

At locations where railways and roads intersected the idea of having them cross on the level had obvious advantages in terms of economics over the cost of building an over or underbridge, especially if the level crossing were unmanned and the construction of a bridge would require the building of substantial approach embankments or possibly approach cuttings leading to an underbridge. With road users being limited in the early days to pedestrians, cyclists, and horse-drawn vehicles the dangers to a train from a collision on such a level crossing were also likely to minimal. For railways crossing other railways on the level though the situation was very different. A collision with another train on a crossing was likely to be far more serious than a collision with, say, a horse and cart, and furthermore a train driver approaching a level crossing and seeing another train nearing or on it could not pull up in anything like the distance that a road user could. It is no surprise therefore that in most countries, where the railway system was either built directly by the state or under state supervision, that rail/rail crossings were not provided. It was only in countries where the railways were built by competing private companies without state interference that such crossings existed, most especially in the United States, where lines were often built as cheaply as possible and the economy of not needing to construct a bridge was doubtless an important consideration. Thus, the United States had many rail/rail level crossings, Great Britain at one time had about 20 or so and its colonies, such as Ireland and Canada, had a handful. Argentina had a number, while one or two others also existed, such as at Istravoz (later Hilal), Turkey, where the Oriental Railway crossed the Smyrna – Cassaba Railway and at Warthausen, Germany, where a major disaster occurred in 1944 (see following). [1433] As has been pointed out, the American companies also seemed to have a perverse proclivity for building rail/rail crossings where they could easily have been avoided, perhaps because in the early days train speeds were low, trains infrequent and night operation unusual so that the dangers were not fully appreciated. [1434] By 1931 every one of the 15 or so remaining rail/ rail crossings in Great Britain was fully signalled and there were hardly any further collisions on them. In the United States, on the other hand, where many crossings remained with

---

[1433] Wikipedia – Hilal railway

[1434] *A History of Railroad Accidents, Safety Precautions and Operating Practices* by Robert B. Shaw pg. 99

little or no signal protection, collisions on them continued to occur regularly and about 50 of them occurred in the period covered by this volume.

The first accident of this type since the end of 1930 occurred at Trumann, Arkansas, United States, just before 9 p.m. on Thursday 15th January 1931 at the crossing of the St Louis – San Francisco Railway (Frisco) and St Louis Southwestern Railroad (Cotton Belt) lines. Engineer Gus Hon, in charge of the locomotive of the northbound *Florida Special* passenger train of the Frisco, was unable to pull up in time when he found the coal car of a Cotton Belt freight train standing on the crossing. His engine was overturned in the collision and its tender and the leading baggage car were derailed. He was slightly hurt; there were no other casualties. What signalling, if any, there was at the crossing was not stated. [1435]

More serious than the Trumann accident was one at Camp Douglas, Wisconsin, United States, at about 3.47 a.m. on Saturday 28th February 1931. Here the double line of the Chicago Milwaukee St Paul and Pacific Railroad (Milwaukee Road) was crossed diagonally by the single track of the Chicago St Paul Minneapolis and Omaha Railway. The crossing was protected by fixed signals controlled from an electric interlocking plant (tower) at the crossing. On the eastbound Milwaukee line there was a home signal, No. 3-R, 721 feet from the crossing, preceded by a distant signal, No. 141-8, 4,956 feet to its rear. There was also a split-point derail working in conjunction with the signal 546 feet west of the crossing.

Westbound Omaha passenger train No. 515 consisted of engine No. 601, three express cars, two mail cars, two deadhead sleeping cars, a combination baggage car and coach, five sleeping cars and a lounge car, in charge of Conductor Dilzer, with Engineer Coleman and an unnamed fireman.

Eastbound Milwaukee passenger train No. 16, *The Olympian Limited*, consisted of engine No. 6409, two mail cars, an express car, a baggage car, a coach, a tourist sleeping car, a dining car, five sleeping cars and an observation car, in charge of Conductor "Commodore" Griggs, with Engineer James Taylor and Fireman Albert Wood. [1436]

The Omaha train was on the crossing when it was struck by the Milwaukee train, engine No. 6409 striking the tender and leading car of the Omaha train. The colliding engine overturned and was badly damage, the first car of the Milwaukee train climbed onto the leading car of the Omaha train and the second car of the Milwaukee train was also badly damaged. Fireman Wood was killed and Engineer Taylor, Conductor Griggs, six mail clerks, a Pullman employee and 17 passengers, all on the Milwaukee train, were injured. [1437]

Mail Clerk Carl Rezab was interviewed by a newspaper reporter. "I was busy working in the mail car when all at once there was an awful crash, a jolt that threw me, and a general scramble. That's about all I remember because everything was in confusion, and I

[1435] *The Sedalia Capital*, Sedalia, Missouri, United States, 17th January 1931

[1436] *The La Crosse Tribune and Leader Press*, La Crosse, Wisconsin, United States, 28th February 1931

[1437] Interstate Commerce Commission report

was trying to find out how badly I was hurt. We made the 13 miles between Tomah and Camp Douglas in nine minutes, and I estimate we were going about 60 miles an hour when we hit the Omaha train. The explanation I heard given for the wreck was that the air brakes did not work, although I do not know what caused it." [1438]

Engineer Taylor told the Interstate Commerce Commission (ICC) inquiry that he had been able to see signal No. 141-8 at caution for a mile or a mile and a half before reaching it. He had shut off steam, made a 12-pound brake application, the cab signal had shown a caution indication, he had acknowledged the automatic train stop device and made a second brake application. Drifting smoke from his engine had then prevented him from determining his speed until he had been passing the stockyards, about 1,700 feet from the crossing. He had at once applied his brakes in emergency and had told Fireman Wood that they would not be able to stop short of the derail. It transpired from remarks made by Engineer Taylor that he had been considerably concerned about losing time with the train.

Conductor Griggs stated that the brakes of his train had been properly tested and had worked well during the journey. He estimated the speed of his train at 65 miles an hour at the initial brake application, at 40 to 45 miles an hour at the time of the final brake application at the home signal and still about 40 miles an hour at the crossing. He thought that his engineer had been in a normal condition but had noted his concern about losing time and said that he had cautioned the engineer about reducing speed to drop the mail at Tomah and had warned him that they were always held by the Omaha train at Camp Douglas. Engineer Taylor denied this conversation.

Head Brakeman Bartlett, Rear Brakeman McQueeney and Baggageman Waterman of the Milwaukee train could add little of importance, nor could the crew of the Omaha train.

Operator Gilmore at Camp Douglas testified that he had set the route for the Omaha train at 3.40 a.m. This train had arrived at 3.43 a.m. and the Milwaukee train had come on the annunciator circuit at 3.45 a.m. When he had realised that the Milwaukee train was not stopping he had started for the door, but the collision had happened before he had reached it.

Travelling Engineer Little stated that there had been considerable trouble with smoke obscuring the view of enginemen on steam engines of the type involved in the accident due to their long boilers and the blowers not being adequate to take care of the smoke.

Roadmaster McMahon said that he had inspected the derail which had been damaged and had clearly caused the locomotive to leave the track, only for it to rerail itself further on.

Signal Maintainer Stowell stated that the signalling had operated perfectly after the accident.

The ICC found that the collision had resulted from the failure of Engineer Taylor to properly control his train in accordance with the rules and signal indications. The engineer had alleged that the brakes had not held properly but tests had established that the brakes had been in good operating condition. It was clearly apparent that the engineer had greatly underestimated the speed of his train as he had approached the crossing. The fact that his mind had been intent on keeping schedule time and that he had been

---

[1438] *The Winona Republican-Herald*, Winona, Minnesota, United States, 2nd March 1931 concerned about running late on his first trip was doubtless relevant. He had understood the restrictive signal indications. Damaged by the high speed of the train the derail had failed to perform its intended function.

The ICC report concluded with a lengthy discussion on the problem of the rules of the Milwaukee and other companies not requiring the engineer to reduce speed at once upon passing a signal at caution but leaving it to his judgement as to when to apply the brakes. [1439]

A crossing between the Chicago Milwaukee St Paul and Pacific Railroad (Milwaukee Road) and the Great Northern Railway near Wahpeton, North Dakota, United States, was the scene of a collision on Saturday 30th January 1932. Since 1923 the crossing had been protected by an automatic interlocking with home and distant signals on each line, though there were no derails. On the Milwaukee line there was a two position two-arm upper quadrant semaphore signal, showing stop or proceed, 596 feet from the crossing. There was also a fixed distant signal, showing caution, 3636 feet from the crossing. The Great Northern line from the west was protected by home signal No. 3.6, another two position two-arm upper quadrant semaphore signal, 566 feet from the crossing, and signal No. 4.6, 5475 feet from the crossing. The first train to arrive on the approach circuit for its line placed its own signal at clear and maintained the signals for the other line at stop. [1440]

Westbound Milwaukee third-class freight train No. 463 consisted of engine No. 2368, five loaded and six empty freight cars and a caboose, in charge of Conductor Brown, with Engineer Gerald Ostrander and Fireman Timothy O'Connor.

Eastbound Great Northern passenger train No. 2, *The Empire Builder*, consisted of engine No. 2577, a mail car, a baggage car, a smoking car, a coach, a tourist sleeping car, a dining car, four Pullman sleeping cars and a club car, in charge of Conductor Saver, with Engineer English and Fireman Bert Tippings.

The Milwaukee train had left Wahpeton, the last open office, 3¾ miles from the crossing, 3 hours, and 20 minutes late, at 5.30 p.m. and was passing over the crossing at 18 miles an hour.

The Great Northern train passed Kent, the last open office on its line, 8¾ miles from the crossing, at 5.41 p.m., 36 minutes late, and travelling at 35 to 45 miles an hour collided with the Milwaukee train at about 5.51 p.m. The trains met engine to engine, the leading car of the Milwaukee train was destroyed, the second and third cars were derailed as were the front four cars of the Great Northern train. Engineer Ostrander, Fireman O'Connor, and Fireman Tippings were killed and Engineer English, a porter, a newsagent and a waiter on the Great Northern train and a brakeman on the Milwaukee train were injured. [1441]

Engineer English of the Great Northern train told the Interstate Commerce Commission (ICC) inquiry that the brakes of his train had been tested and had worked properly at Fargo. From Moorhead Junction, about 39½ miles from the crossing, his train had run at 55 to 60 miles an hour. Signal No. 4.6 had been yellow, and he applied the

[1439] Interstate Commerce Commission report

[1440] Ibid.

[1441] *The Wisconsin State Journal*, Madison, Wisconsin, United States, 31st January 1932 brakes there. Shortly afterwards he had seen home signal No. 3.6 go from red to yellow and its arm move to the 45 degree position, while his fireman had called, "Clear board." He had released the brakes, the speed of his train being about 35 miles an hour. Although it was still daylight his engine headlight had been burning, the weather was clear, there was no smoke and nothing to obscure his vision. He was positive that the home signal had gone to yellow and its arm to 45 degrees and had not changed afterwards. He could not have seen the Milwaukee train from his side of the engine and had had no warning of the collision.

Conductor Saver put the speed of the train approaching the crossing at 40 or 45 miles an hour. His first warning of anything wrong had been the emergency brake application.

Head Brakeman Murray had likewise noticed nothing wrong until the emergency brake application.

Baggageman Long stated that when he had alighted from the train the engineer had said to him, "I had a clear block, and I can't see what they were doing out there."

Flagman Bentz said that the home signal had been showing stop when he had looked at it after the collision.

Conductor Brown of the Milwaukee train said that his train had approached the crossing at 16 or 20 miles an hour. The gauge in his caboose had shown 70 pounds of air pressure when they had left Ottertail Spur, 2½ miles from the crossing.

Head Brakeman Arnold and Rear Brakeman Fifield of the Milwaukee train agreed with their conductor as to its speed at the crossing. After the accident Fifield saw the Milwaukee signal and Arnold both signals in the stop position. The conductor and both brakeman considered Engineer Ostrander to have been a careful engineer with a very good reputation.

Great Northern Signal Maintainer Sheridan stated that he had examined the automatic interlocking plant after the accident and had found no problems.

Milwaukee Assistant Signal Engineer Alexander testified that given the respective speeds of the two trains the Milwaukee train should have operated the control circuit first.

Great Northern Superintendent of Signals Dunham stated that it was impossible for the signals at the crossing to have been clear for both trains at the same time. According to the speeds of the trains, the Milwaukee train should have operated the control circuit first and had clear signals. However, he accepted the statemen of Engineer English that he had had clear signals. If the contacts on the track relays on the Milwaukee line had become frosted and failed to act the Milwaukee signals would have remained at danger and the signals would have cleared for the Great Northern train. There had been no sudden drop in temperature on the day of the accident though. The Great Northern maintained all the signals at the crossing.

The ICC was unable to determine how the accident had occurred. It was clear that the Milwaukee train should have operated the control circuits first and hence received clear signals and it had made no attempt to stop. Against this was the positive assertion of Engineer English that his signals had been clear and the unsubstantiated suggestion of a failure of the Milwaukee control circuits due to frosting.

Following the accident District Superintendent Gaetz of the Great Northern issued an order limiting the speed of passenger trains over the crossing to 25 miles an hour and of freight trains to 18 miles an hour. In addition smash boards were to be installed at the home signals on both lines and Great Northern semaphore signals Nos. 3.5 and 3.6 replaced by colour lights. [1442]

Trains of the Chicago and Eastern Illinois (C&EI) and Illinois Central (IC) Railroads collided at the crossing of the two lines at Marion, Illinois, United States, on Thursday 3rd August 1933. Both railroads were single track, operated by timetable and train orders with no block signals. The crossing was protected only by stop signals, that for southbound IC trains being 310 feet from the crossing and that for northbound C&EI 221 feet south of the crossing. The signs also governed the crossing of a Missouri Pacific Railroad track. The C&EI sign was black with white letters; the IC sign was unpainted and its letters unreadable. Buildings and oil tanks of the Mid-Continent Oil Company interfered with the views from both lines until trains were near the crossing.

Southbound IC train No. 341 consisted of engine No. 1866, 25 freight cars, a mail compartment car, a combination baggage car, and a coach, in charge of Conductor Brownlee, with Engineer A F "Dick" Gourley and Fireman Reeves.

Northbound C&EI train No. 122 consisted of engine No. 216, a combination mail and baggage car and a combination smoking and chair car, in charge of Conductor Cotton, with Engineer Dan Deffenbaugh and Fireman Harry Hadfield.

The IC train left Carterville, 8¾ miles from Marion, at 7.47 a.m., 48 minutes late, stopped at the stop board, then proceeded at 5 or 10 miles an hour to the crossing and had almost stopped when it was struck by the C&EI train. The latter had left Cypress, 28¾ miles from Marion, at 7.15 a.m., stopped at Marion station, stopped at the stop board, then collided with the IC train at a speed variously estimated as 4 and 17 miles an hour. The engines were derailed and damaged and Engineer Gourley and Conductor Cotton were injured.

Engineer Gourley told the Interstate Commerce Commission (ICC) inquiry that when he had stopped at the stop board his fireman had called, "all clear on the north." He had given two blasts on the whistle and had proceeded at 8 or 10 miles an hour. He had seen the smoke of the C&EI train above the oil tanks, then clearing the tanks had seen the C&EI train, 180 or 200 feet from the crossing, working steam and clearly not stopping. He had applied his brakes in emergency. He had not heard the other engine whistle and was positive that the C&EI train had not been in view when he had started from the stop board. His brakes had been tested and had worked properly.

Fireman Reeves said that he had heard the C&EI train whistle when their (IC) train had been halfway between the stop board and the crossing but had assumed that it had just started from the stop board.

Head Brakeman Miller of the IC train also heard the C&EI train whistle, then saw it approaching at 8 or 9 miles an hour and did not think that it had reduced speed.

Engineer Deffenbaugh of the C&EI train stated that he had stopped at the stop board, whistled twice, and proceeded, his fireman telling him that the way was clear on the left side. Just as he had reached the crossing he had heard one whistle from the IC train and

---

[1442] Interstate Commerce Commission report

his fireman had shouted for him to stop. He had applied his brakes in emergency and the collision had occurred almost at the same moment. He had not heard the IC train whistle as it left the stop board, had not heard it working steam, nor had he been able to see it from his side of the engine. His brakes had been tested before leaving Cypress and had been in order. He was not very familiar with the crossing.

Fireman Hadfield said that after starting from the stop board he had got off his seat box to fire. He was familiar with the crossing and had he remained on the seat box would have been able to see the IC train from his side in time to warn his engineer.

The ICC found Fireman Hadfield mainly responsible for the accident because he had failed to keep a proper lookout. Engineers Deffenbaugh and Gourley were responsible to a lesser extent by failing to keep their trains under full control as they approached the crossing.

Both the state law of Illinois and the rules of the two railroad companies required all trains to stop before negotiating the crossing. Both trains had stopped, and it was obvious that the 27-car IC train must have taken longer to cover the 310 feet from its stop board than the two car C&EI train to cover only 221 feet from its stop board. Had the fireman of the C&EI train kept a proper lookout the accident would have been prevented. However, in view of the lack of signal protection and restricted view a flagman should have been sent ahead of each train, a measure which would not have necessitated extra expense or delay. Ideally an interlocking plant should be provided to protect both this crossing and the adjoining Missouri Pacific crossings. [1443]

Atlantic Coast Line Railroad (ACL) and Seaboard Air Line Railway (SAL) trains collided on Thursday 6th October 1938 on a rail/rail crossing at Archer, Florida, United States. The crossing of the two railroads was protected by gates operated by a crossing watchman; when the gates were open for one line they obstructed the other. Each gate was provided with a standard switch lamp which displayed a red aspect when the gate was closed across the line and a green aspect when the gate was open. A sign beneath each lamp read "STOP" in 8-inch black letters on a white background. All trains were required to approach the crossing under full control, and only proceed at a speed not exceeding 20 miles an hour if the gates were open and a proceed signal was received from the crossing watchman by the display of a green flag or green light. An average of only 2.5 SAL trains and 5 ACL trains used the crossing daily.

Northbound SAL freight train Extra No. 379 consisted of engine No. 379, 40 loaded, 18 empty cars and a caboose, in charge of Conductor Wills, with Engineer Lewis and Fireman J.B. Dormany.

Northbound ACL Extra No. 820, a fruit express, [1444] consisted of engine No. 820, 47 loaded, two empty cars and a caboose in charge of Conductor Livingston, with Engineer Holland and Fireman Howard.

The SAL train had left Gary at 1 a.m., reached Archer at 6.50 a.m., and the gates being open for it, had proceeded over the crossing at a speed of between 10 and 18 miles an hour. The ACL train had left Uceta at 2.26 a.m., approached Archer, just under 126

[1443] Ibid

[1444] *Thomasville Times-Enterprise*, Thomasville, Georgia, United States, 6th October 1938 miles north of there, collided with the closed crossing gate at between 20 and 40 miles an hour and ran into the side of the SAL train. The leading ten cars of the SAL train were derailed, as were the first seven cars of the ACL train, which piled up on one another in a small space. All the crossing gates and the crossing watchman's house were demolished. Fireman Dormany was killed, and Engineer Holland and Fireman Howard were injured.

Engineer Lewis of the SAL train told the Interstate Commerce Commission (ICC) that he had reduced speed to 10 miles an hour as he reached the crossing, whistled for the adjoining State Road No. 5 level crossing, seen that the gates at the rail/rail crossing were open and had called, "Clear crossing," which had been acknowledged by Fireman Dormany and Head Brakeman Simms. The crossing watchman had given him a green flag, he had answered with two blasts of the whistle and begun to work steam. Just as he had reached the crossing he had seen the ACL train and realised that it was not stopping. His first thought had been to brake, then he had thought that it would be better to get his engine and tender over the crossing, so he released the brakes.

Head Brakeman Simms corroborated the engineer's statement. He said that the deceased fireman had jumped from the left side of the engine just before the collision.

Conductor Wills and Flagman Sparkman were unaware of a problem before the collision.

Engineer Holland of the ACL train stated that approaching Archer he had reduced speed from 40 to 20 miles an hour by applying the brakes. About 675 feet from the crossing he had seen the crossing watchman on the embankment. He was positive that the watchman had given him a proceed signal with an unfurled flag, either green or blue in colour. He had replied with two blasts of the whistle and called, "Clear crossing," to the fireman and head brakeman. He had expected the switchman to go to the crossing but did not see him do so and his fireman had then shouted that the SAL train was near the crossing. He had then closed the throttle and applied the brakes in emergency but had not got a full application because of previous service applications. He had not seen the watchman give him a red flag and had only seen the closed gates at the last moment. He admitted that he had not had his train under full control.

Fireman Howard recalled his engineer applying the brakes, then sounding two blasts on the whistle, saying that he had the crossing, releasing the brakes and working steam. Passing the station buildings he had seen the closed gates, the watchman leaving his shanty and running up the bank with a half unfurled red flag and the SAL train. Shouting a warning, he had jumped.

Head Brakeman Crevasse confirmed the statement of Fireman Howard.

Conductor Livingston said that a service brake application had reduced the speed of his train from 45 to 20 miles an hour approaching Archer. Looking through the cupola window of his caboose he had seen the SAL train, between 50 and 100 feet from the crossing, and had realised that a collision was inevitable. As he had called a warning to Flagman Cason the collision had occurred. It was the practice, if the crossing was clear, for the watchman to place himself on the bank where he could be seen by the engineer of an ACL train and reassured by the signal from the watchman, the engineer would approach the crossing at a higher speed than otherwise.

Crossing Watchman Wynn said that the SAL train had arrived first, and he had opened the gate for it and shown it a green flag. He had then heard the ACL train approaching at speed and being alarmed had gone into his shanty, seized a red flag and gone up onto the embankment from where he usually gave proceed signals and waved stop signals with his unfurled flag, which had not been answered.

The accident was witnessed by Agent Cleghorn at the ACL station. He saw the two trains approaching, that the gates were closed against the ACL train and open for the SAL train, and the watchman giving stop signals to the ACL train with a red flag.

The ICC concluded that the accident had been due to the failure of the ACL engineer to operate his train under proper control approaching the crossing. Even if he had misinterpreted the watchman's stop signal as a proceed signal he should have first ensured that the gates were open. Predictably the ICC was critical of the informal practice which had grown out of the watchman positioning himself on the embankment to give a proceed signal to ACL trains to give engineers earlier notice. [1445]

A collision between trains of the Green Bay and Western and Chicago and North Western Railroads at about 7.50 a.m. on the morning of Saturday 8th July 1939 was apparently not the subject of a formal investigation by the Interstate Commerce Commission though several railwaymen were injured. It happened at the crossing of the two railroads at Marshland, Wisconsin, United States. The westbound North Western work train was in charge of Conductor Charles Odikirk, with Engineer Charles Kilmer and Fireman Carl Labenstein and consisted of an engine, two cabooses, several flat cars, one with a crane, and a box car. It had been heading for Bluff Siding, a short distance further on, where track was to be re-laid, and was going over the crossing when it was run into by the Green Bay train. The latter was in charge of Conductor Gus Beyer with Engineer John Ray and Fireman Henry Burmeister and was a combination (mixed) passenger and freight train with about 20 freight cars and a passenger coach. Engineer Ray said that he had seen the signals against him at the crossing but had been unable to stop and his train had collided with the other while still travelling at 15 or 20 miles an hour.

The Green Bay train collided with the second of the two cabooses of the work train, pushing it into the Marshland station buildings, which were partially demolished. Section Foremen William Dettman and Dell Marshall and Roadmaster Walter Daehn, who had been riding in the caboose, plus Brakeman T.H. Bold, who had been standing on the station platform, were badly enough hurt to be taken to hospital. Marshland Station Agent and Operator Frank Gappa was able to jump to safety, though he suffered from shock.

Conductor Odikirk said that he had been on the platform, having just received his orders from Operator Gappa, when the collision occurred. "I was knocked to the ground when the engine hit and as I fell I saw the caboose coming down on top of me," he said. "I did some tall scrambling to get out of the way. I can't understand how Gappa got out of the way of the caboose. I saw him running around the corner of the station as the caboose was struck and learned afterward that he was not hurt." [1446]

Two passenger trains were involved in a side collision at South Denver Colorado, United States, on Saturday 12th August 1939. The crossing concerned was between the

---

[1445] Interstate Commerce Commission report

[1446] *The Winona Republican-Herald*, Winona, Minnesota, United States, 8th July 1939 lines of the Denver and Rio Grande Western Railroad (Rio Grande) and Atchison Topeka and Santa Fe and Colorado and Southern Joint Railway (Santa Fe). Both lines were double track at the point where they crossed one another. The crossing was controlled from a tower with a 26-lever Saxby and Farmer lever frame, controlling semaphore signals. The northbound Santa Fe line was protected by home signal No. 35 and an approach signal, situated 1,515 and 3,021 feet respectively from the crossing. On the southbound Rio Grande line there was home signal No. 20 and its associated approach signal, sited 504 and 1,585 feet respectively from the crossing. Both lines were worked by timetable and train orders.

Northbound Santa Fe passenger train No. 6 consisted of a baggage car, a chair car, a smoking car, a chair car, three standard Pullman sleeping cars, a café lounge car, a standard Pullman sleeping car and a tourist sleeping car, in charge of Conductor A.S. Mock, with Engineer Heath and Fireman Stevens. It had left Pueblo, about 115½ miles south of the crossing, at 6.09 a.m., an hour, and 9 minutes late, passed Englewood, 4 miles from South Denver, at 9.05 a.m., an hour, and seven minutes late, passed home signal No. 35, showing ‘Proceed’, and was passing over the crossing at 10 to 25 miles an hour when it was struck by the Rio Grande train.

Southbound Rio Grande passenger train No. 1 consisted of engines No. 801 and 1800, a baggage car, a mail car, four coaches, two tourist Pullman cars, a dining car, six standard Pullman cars and a lounge car, in charge of Conductor Foster, with Engineers Medae and Williams and Firemen Bovee and Boyle. It had left Denver, 3½ miles north of the crossing, at 9.01 a.m., 16 minutes late; passed Burnham, 1½ miles north of the crossing, at 9.06 a.m., passed signal No. 20, showing *Stop*, and travelling at 18 miles an hour collided with the third car of the Santa Fe train. Engine No. 801 of the Rio Grande train was badly damaged, and the third car of the Santa Fe train was crushed where the engine struck it. Conductor Mock was killed instantly, and 17 passengers and two dining car employees were injured. Passenger Roy Scott succumbed to his injuries a few hours later.

Engineer Heath told the Interstate Commerce Commission (ICC) investigators that signal No. 35 had been clear for him. As he neared the crossing he had seen the Rio Grande train; until the last moment he had thought that it would stop short of the crossing. His testimony was corroborated by Fireman Stevens and by Flagman Haggens and Porter Jones of his train.

Engineer Medae of engine No. 801, the pilot engine, stated that the air brakes of his train had been tested at Denver, though he did not think that there had been time enough to inspect every car. After departure from there he had made a running test, which had been satisfactory. On passing Ellsworth Street, 1,585 feet north of the approach signal, he had called signal No. 20 as red, which his fireman had acknowledged. At the 2,000 railroad crossing sign he had closed the throttle and just north of the distant (approach) signal he had made a 15 lb brake reduction. The speed had not dropped as he had expected, and he had made a further 5 lb reduction. Two car lengths from the home signal it had not appeared that his train was going to stop, and he applied the brakes in emergency; he heard the exhaust, but the speed had not dropped. His train had continued to the crossing and struck the Santa Fe train. He was given a medical examination after the accident and passed.

Fireman Bovee had little to add to his engineer’s testimony.

Engineer Williams on engine No. 1800, the train engine, stated that the speed of the train had been 35 to 40 miles an hour at Ellsworth Street. Seeing the home signal at red he had closed the throttle and opened the sanders. He had felt a brake reduction at the 2,000 railroad crossing sign: none after that. By then smoke from the train engine had obscured his view of the signals. He stated variously that he thought his train was going to stop at signal No. 20 and that the signal must have been cleared. He had not seen the home signal at red again until they were passing it and close to the crossing.

Fireman Boyle corroborated his mate's statement. He had jumped at the last moment.

Conductor Foster testified that he had not realised anything was wrong until the brakes came on in emergency.

Head Brakeman Naples of the Rio Grande train said that when the brakes had been applied he thought that they were to get orders at the tower. Then he had seen the fireman jump and had jumped himself.

Flagman Freeman had also thought that they were to receive orders.

The crew of Colorado and Southern Railroad switch engine No. 605 saw the Rio Grande train approaching at 35 to 40 miles an hour. When they had seen that the train was running past signal No. 20 they whistled a warning.

Colorado and Southern Car Inspector Asbury also witnessed the accident.

South Denver Towerman W.A. Marlett testified that after receiving a report that train No. 6 had left Englewood at 9.05 a.m. he had set the route for that train. The Santa Fe train had passed between 9.09 and 9.10, when he had looked up and seen the Rio Grande train approaching.

The air brakes of the Rio Grande were found to be working normally after the accident. According to the speed recorder fitted to engine 1800, the train had still been travelling at 42 miles an hour 400 feet short of the approach signal, and at 37 miles an hour 400 feet short of the home signal. It was concluded that obviously Engineer Medae had not applied the brakes where he said he had, or the train would have been able to stop at the home signal. Special timetable rules also restricted the speed of trains to 20 miles an hour over seven street crossings just to the north of South Denver.

Engineer Williams was also blamed, the ICC stating that had he taken proper action when he had seen signal No. 20 at red at short range the collision would have been averted. [1447]

There was a lucky escape in a collision on the evening of Wednesday 2nd October 1940 at Shafter, Nevada, United States. Eastbound Western Pacific Railroad fruit train No. 54 struck the gondola car of a local Nevada Northern Railroad freight train at the crossing of the two roads. The gondola car was demolished, two others in the Northern train were derailed, as were the engine and two cars of grapes of the Western Pacific train. No one was hurt, though 135 feet of track had to be replaced. [1448] [1449]

1447 Interstate Commerce Commission report

1448 *Oakland Tribune*, Oakland, California, United States, 3rd October 1940

1449 *The Salt Lake Tribune*, Salt Lake City, Utah, United States, 4th October 1940

Although it involved the death of a railwayman and the injury of 17 passengers and two railwaymen, a collision at Zephyrhills, Florida, United States, on Tuesday 19th December 1940 was apparently all but ignored by the press. It was, however, investigated by the Interstate Commerce Commission (ICC). The accident occurred at the crossing of the lines of the Atlantic Coast Line Railroad (ACL) and the Seaboard Air Line Railway (SAL). The crossing was protected by semaphore home and distant signals and an automatic interlocking. The distant signals were inoperative (fixed). Track circuits were provided in the rear of the home signals on each line. The first train to occupy a track circuit cleared the signals for its line and held the signals for the other line at danger until it had cleared the crossing. Should trains on each line occupy the track circuits simultaneously the signals were all maintained at danger. The maximum speed of ACL trains over the crossing was limited to 20 miles an hour, for SAL trains it was 60 miles an hour.

Southbound SAL first-class passenger train No. 305 consisted of engine No. 824, a passenger baggage car, a coach, a coach diner and two Pullman sleeping cars, in charge of Conductor Williams, with Engineer Ben Green and Fireman McLeod.

Southbound ACL third-class freight train No. 213 consisted of engine No. 1656, 34 loaded and 10 empty cars and a caboose, in charge of Conductor Matthews, with Engineer Boring and Fireman Number.

The SAL train had left Owensboro, 15½ miles north of Zephyrhills and the last open office, at 6.48 a.m., 31 minutes late, passed the home signal at the crossing showing *Proceed* and moving at 30 to 35 miles an hour ran into the side of the ACL train. The ACL train left Vitis, 3¾ miles north of Zephyrhills, at 7.14 a.m., 2 hours, and 39 minutes late, passed the home signal at the crossing showing stop and was passing over the crossing at between 3 and 15 miles an hour when its leading car was struck by the SAL train at about 7.21 a.m. Engineer Green was killed, and Fireman McLeod and a porter were the railwaymen injured. Six cars of the ACL train and the engine and three cars of the SAL train were derailed. The first two cars of the SAL train were badly damaged.

SAL Fireman McLeod said that he had seen the home signal clear for their train and had then tended the fire and his engineer had increased speed. Just as they had passed the home signal his mate had shouted a warning, applied the brakes in emergency and the collision had occurred. It had been daylight and clear but being preoccupied with the fire he had not seen the ACL train.

Conductor Williams stated that the speed of the SAL train had been reduced to 15 miles an hour through Zephyrhills as per the city ordinance, then increased over the crossing.

Engineer Boring testified that his air brakes had been tested and had worked properly en route. He had seen the distant signal at caution, called its indication to his fireman and the brakeman, and had made an 18 or 20 pounds brake application 8 or 10 car lengths beyond the signal, the train's speed then being 40 to 45 miles an hour. Nine hundred feet beyond the distant signal the speed had been 30 to 35 miles an hour and he had seen the home signal showing *Proceed.* He had then released the brakes and called its indication but no one else on the engine had acknowledged it. A second or so later the smoke had cleared from the front of the engine and the fireman had called the home as *Stop*. Seeing himself that it was, he had applied the brakes but had not obtained an emergency application as the brake pipe had not been fully charged. He had also opened the sanders and reversed his engine, but his train had overrun the home signal and just stopped on the crossing when it was struck.

Fireman Number said that his engineer had closed the throttle half a mile north of the distant signal. He had called the distant signal and his engineer had acknowledged it. Shortly afterwards the engineer had called the home signal as clear, but he (the fireman) could not see it so had not replied. As they passed Zephyrhills station he had seen that the home signal was at *Stop* and called a warning. When the engine was 750 feet beyond the station the speed was still 30 miles an hour and his engineer had taken no action, so he had called a second warning. His engineer had then made a service brake application and they passed the home signal at 12 or 15 miles an hour. He believed that if his engineer had braked harder they could have pulled up in time.

Brakeman White was riding on the ACL engine. He stated that he and the fireman had called the home signal at stop at the same time, but the engineer had not braked at that time. The fireman had then shouted a second warning, at which point the engineer had made a service brake application only. He thought that the engineer had been given ample warning to stop the train.

Conductor Matthews agreed that the train brakes had been properly tested. He put the speed of the train at 35 or 40 miles an hour at the distant signal. Smoke had prevented him seeing the home signal at any time. He had only noticed one brake application. He, too, thought that a heavier brake application would have stopped the train.

Post-accident testing of the brakes of the ACL train and of the signalling showed everything to have been working correctly.

The ICC determined that it was clear that the SAL train had entered its approach circuit first and would have received clear signals. Because the ACL train had been moving faster and its approach circuit had been shorter it had reached the crossing first. It concluded that the accident had been due to the failure of the engineer of the ACL train to obey signal indications. It also noted that according to an order issued by the commission in April 1939 the speed of trains over the crossing should have been restricted to 20 miles an hour as it was not protected by operative approach (distant) signals and not the 60 miles an hour authorised by the SAL. [1450]

Kansas City Southern Railway (KCS) and Missouri Pacific Railroad (MP) trains collided near Fuller, Kansas, United States, on Saturday 10th May 1941. The crossing of the two single-track railroads was protected by an automatic interlocking with semaphore home and distant signals on each line. The home signals at the crossing normally displayed *Stop* and the occupation by a train of a track circuit on one of the lines approaching the crossing automatically cleared the home signal for that line provided that there was no conflicting movement on the other line. MP trains were limited to 20 miles an hour and KCS trains to 35 miles an hour over the crossing.

Southbound second-class MP freight train No. 793 consisted on the day of the accident of engine No. 1565, 33 empty cars and a caboose in charge of Conductor Carruthers, with Engineer Priestley and Fireman Lipsey. It passed Fort Scott, nearly 27½

[1450] Interstate Commerce Commission report

miles north of the crossing, at 10.25 p.m. on 9th May, 2 hours and 25 minutes late, passed the approach (distant) signal at 12.30 a.m. on 10th May, passed the home signal showing *Proceed* and was moving over the crossing at 20 miles an hour when it was struck by the KCS train.

Southbound first-class KCS passenger train No. 15 consisted of diesel-electric engine No. 1, three baggage cars, a mail-baggage car, a coach, and a Pullman sleeping car, in charge of Conductor Carter, with Engineer Guinn and Fireman Freeman. It left Kansas City, 120¾ miles north of Fuller, at 10.06 p.m., 11 minutes late, passed Eve, 21¼ miles north of Fuller and the last open office at 12.10 a.m., 14 minutes late, approached the crossing at 60 to 65 miles an hour, passed home signal No. 1229, showing *Stop,* and travelling at 20 to 30 miles an hour ran into the side of the MP train. The engine and leading car of the KCS train and the 29th to 32nd cars of the MP train were derailed and Engineer Guinn and a dining car employee on the KCS train and Conductor Carruthers of the MP train were injured.

Engineer Priestley told the Interstate Commerce Commission (ICC) inquiry that he had seen his home signal change from red to green and called its indication to the fireman. He could also see the KCS home signal showing *Stop*, and the KCS train, apparently some distance away. His train had continued and was struck at 12.33 a.m.

Fireman Lipsey said that he had also seen the home signal clear and the headlight of the KCS train. When he had seen that the latter train was not stopping he shouted a warning to his engineer.

Conductor Carruthers stated that he too had seen their signal clear and that when he had seen the KCS train he could tell that it was not going to stop.

Engineer Guinn of the KCS train was too badly hurt to make a statement, even a month after the collision.

Fireman Freeman testified that the air brakes had been tested at Kansas City and had functioned properly. He had been in the engine room draining the boiler which provided steam heating when he had felt the brakes go on in emergency and the collision had occurred 20 seconds later. He did not remember the engineer whistling for the crossing. He said that on some trips he spent all of his time in the engine room except when passing train order signals.

Conductor Carter could only recall the engineer whistling for some crossings. He said that the first indication that he had had of anything wrong had been the brakes applying in emergency. He confirmed that home signal No. 1229 had been displaying *Stop* immediately after the collision.

Flagman Emerson agreed that the home signal had been at *Stop* when he had gone back to flag after the accident.

Tests of the signals after the accident showed them to be working correctly.

The ICC noted that given the respective speeds of the trains, the MP train must have struck its approach track circuit at least a minute before the KCS activated its track, indicating that the signals would have cleared for the MP train and against the KCS train. On a moonlit night such as on the day of the collision home signal No. 1229 could be seen for 5,000 feet, with an excellent view of it from the approach signal, 2,707 feet away. Engineer Guinn had clearly seen the signal at the last moment when he had braked in emergency. Why he had not done so earlier could not be determined. [1451]

It seems likely that the engineer had dozed off or experienced a microsleep.

An Illinois Central Railroad passenger train, the *Panama Limited*, and a Baltimore and Ohio Railroad freight train collided at the crossing of the two railroads at Odin, Illinois, United States, on Thursday 11th September 1941. The two lines crossed at an angle of 35 degrees. The double-track Illinois Central line was operated by timetable, train order and an automatic cab signal and train stop system, without lineside signals except at interlockings. The single-track Baltimore and Ohio line was operated by timetable, train orders and automatic signals. The crossing itself was controlled from Odin tower, which had 24 levers (18 working) and was operated and maintained by the Illinois Central company. The crossing was protected by home signal No. 24 and approach signal No. 2744, 565 and 5,302 feet respectively from the crossing as far as eastbound Baltimore and Ohio trains were concerned. On the southbound Illinois Central line home signal No. 2, a two-aspect lower quadrant semaphore signal, was 572 feet from the crossing. It was preceded by block indication points 5,889 and 11,568 ft from the crossing. These displayed red and green aspects; a warning whistle rang when an indication changed to red and if the red aspect was not acknowledged within six seconds the brakes were applied automatically.

Southbound Illinois Central first-class passenger train No. 5, the *Panama Limited,* consisted of engine No. 1190, a baggage car, a business car, a club car, two Pullman sleeping cars, a dining car and two more Pullman sleeping cars, in charge of an unnamed conductor, with Engineer Henry Brookbank and Fireman B.B. Boyd. It had passed Champaign, 116½ miles north of Odin, at 3.40 p.m., five minutes late; passed Kinmundy, 15¼ miles north of Odin and the last open office, at 5.29 p.m., two minutes late; passed signal No. 2 showing *Stop,* and was travelling at 20 to 35 miles an hour when it hit the side of the freight train at 5.45 p.m.

Eastbound Baltimore and Ohio freight train Second No. 96 consisted of engine No. 4580, 18 loaded and 50 empty cars and a caboose, in charge of Conductor Ed Schlotter with Engineer Percy Hogan and Fireman Adrian Hunter. It had passed Sandoval, 5½ miles west of Odin and the last open office, at 5.38 p.m., 4 hours and 49 minutes late, passed the automatic stop signal at the east end of Sandoval siding, which was showing *Approach Medium*, passed signal No. 2744, showing *Approach*, passed signal No. 24, showing *Stop,* and stopped with its second car on the crossing, where it was immediately struck by the Illinois Central train. The second car was destroyed, the first car was derailed and the third damaged. More importantly, Fireman Boyd of the passenger train was killed, while Engineer Brookbank was injured. [1452] None of the passengers on the Illinois Central train was hurt. They had included Mr J.L. Beven, the president of the railroad company, riding in his private car second behind the engine, which was derailed along with the baggage car. Riding with Mr Beven had been Divisional Superintendent G.J. Willingham and Divisional Trainmaster A. Gorman. [1453]

Operator Robert Moore at Odin told the Interstate Commerce Commission (ICC)

[1451] Ibid.

[1452] Interstate Commerce Commission report

[1453] *Centralia Sentinel*, Centralia, Illinois, United States, 12th September 1941 inquiry that he had cleared his signals for the Illinois Central passenger train at 5.42 p.m. The signals would thus automatically be set against the Baltimore and Ohio train and the route could not be changed for 3 minutes and 40 seconds because of a time release.

Engineer Hogan of the Baltimore and Ohio train stated that both the approach and the home signal had been displaying *Approach.* He had partially closed the throttle as his train passed signal No. 2744 moving at 40 miles an hour and he and his fireman had kept signal No. 24 under constant observation through 4,400 ft. About 200 ft from the signal, moving at 25 or 30 miles an hour, he had suddenly seen that the signal was showing *Stop* and had applied the brakes in emergency, but the distance too short to stop short of the crossing. He said that he had previously seen both signals showing *Approach* and it was his understanding that they would do so if there was a train in the section beyond the crossing.

Engineer Brookbank of the Illinois Central train testified that his cab signal had shown *Proceed* as he had approached Odin at 70 miles an hour. It was only when he had been 900 ft from the crossing that signal No. 2 had suddenly gone to red and his cab signal to *Stop*. Like Engineer Hogan he had applied the brakes in emergency but too late to be able to stop.

The ICC concluded that Engineer Hogan was at fault inasmuch as he had approached the crossing at too high a speed. According to the regulations he should have reduced speed to not more than 22½ miles an hour after passing signal No. 2744 at *Approach.* Had he done so the accident would have been averted. However, the signalling was also at fault. Tests conducted after the accident showed that signal No. 24 appeared to show a yellow light until the engine of an approaching train was 950 ft from it. This was due to a combination of the red rays from the signal lamp and the sun's rays being reflected by the background of the signal. It transpired that three months earlier a signal maintainer had repainted the signal with black paint, using glossy paint instead of the usual dull paint, and it was this that had caused the problem, described in the ICC report as a 'phantom indication.' [1454]

The rail/rail crossing at Murrow, England, which had already been the scene of an accident in October or November 1925 saw another mishap on Friday 9th October 1941. The London and North Eastern Railway (LNER) March – Spalding line and the London and North Eastern and London Midland and Scottish Joint Railway (LNE & LMS Jt) Wisbech – Peterborough line here intersected one another. The crossing was fully signalled and controlled from Murrow West signal box. On the morning of the accident (which happened at about 3 a.m.) a Down (to Wisbech) LNE & LMS Jt coal train and an Up (to March) LNER fitted goods train were approaching the crossing and Signalman Fred Lawrence decided to give the coal train precedence. The Up outer and inner home signals were therefore necessarily at danger, but the LNER train overran them and then stopped on the crossing to report to the signalman. While it was thus foul of the crossing it was cut through by the coal train and the wreckage from the collision demolished the signal box. Signalman Lawrence just had time to escape by jumping out of a window and a fireman jumped from his engine and landed in a rose bush. No one was hurt, though

[1454] Interstate Commerce Commission report

there was considerable material damage. [1455]

Louisville and Nashville Railroad (L&N) and Seaboard Air Line Railway (SAL) trains collided on the crossing at Wellington, Alabama, United States, on Friday 20th February 1942. Movements over the crossing of the two railroads were controlled by a gate consisting of a triangular arrangement of 1½ inch iron pipes mounted on a post in the south-west corner of the crossing. The gate could be swung across either single track and latched in position. A sign bearing the word 'STOP' in black letters on a white background was mounted on the gate. The gate normally stood closed against SAL trains. On the SAL line 5,288 feet east of the crossing was a sign worded 'RAILROAD CROSSING'. Trains were required to approach the crossing under full control, expecting to find the gates closed against them. The speed of L&N trains was restricted to 10 miles an hour over the crossing and that of SAL trains to 15 miles an hour. An average of 7.1 L&N trains and 12.5 SAL trains used the crossing daily.

Southbound L&N first-class passenger train No. 85 consisted of engine No. 186, a baggage mail car, and two coaches in charge of an unnamed conductor, with Engineer D.L. Harman and Fireman Fred Yarbrough. [1456]

Westbound first-class SAL passenger train No. 9 consisted of engine No. 223, a mail-express car, an express car, a passenger-baggage car, two coaches, a Pullman lounge car and a Pullman sleeping car, in charge of an unnamed conductor, with Engineer P.C. Morgan and an unnamed fireman.

The L&N train had left Gadsden, 16½ miles north of Wellington and the last open office, at 12.03 p.m., 16 minutes late, had stopped at Wellington with the engine tender over the crossing, and was immediately struck by the SAL train. The SAL train had passed Piedmont, 18¼ miles east of the crossing, at 12.06 p.m., 2 hours and 16 minutes late, and travelling at 6 or 8 miles an hour had collided with the L&N train at 12.33 p.m.. The engine of the latter was overturned, and the right side of its cab crushed. Engineer Harman was fatally injured, and Fireman Yarbrough and Engineer Morgan were hurt.

The Interstate Commerce Commission (ICC) determined that a station porter had closed the gate across the SAL line at 10.57 a.m. upon the instructions of the station agent for a southbound L&N freight train. The gate had remained in that position until the accident.

Fireman Yarbrough stated that his engineer had approached the crossing under control. He had seen the gate closed across the SAL line. He had not noticed the SAL train before the collision.

Engineer Morgan testified that his train had been travelling at 55 miles an hour nearing Wellington. Three 10-pound brake applications had reduced the speed of his train to 25 miles an hour at which point he had misread the position of the gate as being open and had called it 'clear' to his fireman, releasing the brakes to stop at the station. When 500 feet from the crossing he had seen that the gate was in fact closed, whereupon he had applied the brakes in emergency, opened the sanders and reversed his engine, but could not stop. The SAL train had not occupied the crossing before the emergency brake application.

1455 Midland and Great Northern Circle Bulletin No. 723

1456 *The Anniston Star*, Anniston, Alabama, United States, 22nd February 1942

The fireman denied hearing his engineman call the gate as clear. When he had seen the closed gate he had been about to shout a warning when his engineer applied the brakes in emergency. He and the engineer had jumped 150 feet from the crossing, the speed had then dropped to 4 or 5 miles an hour.

The conductor and baggageman said that they had assumed that the train was being operated so as to stop short of the crossing.

The ICC concluded simply that the accident had been due to the failure to operate the SAL train in accordance with a crossing signal indication. [1457]

Sunday 29th March 1942 saw a destructive crossing collision at Piqua, Ohio, United States. Engineer Quincy Davis found the signals against his Pennsylvania Railroad train as he had rounded a curve and approached the crossing with the Baltimore and Ohio Railroad line but said that he had been unable to stop. "I threw the engine in reverse and was slowed to a crawl at the time of the crash," he maintained. His engine collided with the middle of a Baltimore and Ohio train of 81 tank cars, overturning some of them, and these immediately caught fire. It took six hours to bring the blaze under control, by which time the small station buildings and the tower at the crossing had been burnt down, while burning oil flowed down a road and into a creek. Railroad and Western Union telegraph lines were also put out of commission. The crew of the Pennsylvania train leapt for safety and the only casualty was an onlooker, Mrs Myrtle Hardacre, whose leg was struck by a fire hose and broken. Twelve tank cars and their contents were lost. [1458] The accident happened at 10.40 a.m. and the lines were blocked for 12 hours.

It emerged after the accident that the Pennsylvania train might have been stopped short of the crossing by a derail had not the device been put out of commission in a minor accident six weeks earlier and not repaired. [1459]

New York Central and Wabash Railroad trains collided at the crossing of the two railroads at Raisin Center, Michigan, United States on Monday 17th August 1942.

The New York Central line was single track and operated by timetable, train orders and a manual block system. The Wabash line was double track, operated by timetable, train orders and an automatic block system. The lines crossed at an angle of 70 degrees and the crossing was protected by an automatic interlocking with colourlight home and approach (distant) signals apart from the Wabash approach signal, which was a three-aspect upper quadrant semaphore. Westbound Wabash trains were governed by home signal No. 546 and approach signal 534, 519 and 7,522 feet respectively from the crossing. Occupation of the approach track circuits by a train on one line automatically placed the signals for that line at clear and held the signals for the other line at danger.

Northbound New York Central freight train Extra 1130 North consisted of engine No. 1130, ten loaded, five empty cars and a caboose, in charge of Conductor August Willnow with Engineer John Clark and Fireman W.J. Kruegher. It had left Lenawee Junction, less than 2¼ miles south of the crossing and the last open office, at 9.30 a.m. Eastern War

---

[1457] Interstate Commerce Commission report

[1458] *The Republican-Courier*, Findlay, Ohio, United States, 30th March 1942

[1459] *The Coshocton Tribune*, Coshocton, Ohio, United States, 30th March 1942

Time, received clear signals on the approach to the crossing, and was moving over it at 20 miles an hour when its ninth car was struck by the Wabash train. [1460]

Westbound New York Central first-class Wabash passenger train No. 1 consisted of engine No. 667, a baggage-mail car, a passenger-baggage car, two chair cars, a club-dining car, a passenger-baggage car, two chair cars, a dining car, and a Pullman car, in charge of Conductor J.C. Madden with Engineer J.A. Hendrick and Fireman Herbert Altman. [1461] This train left Detroit at 7.25 a.m. Central War Time, passed Holloway, about 2½ miles from the crossing and the last open office, at 8.32 a.m., passed signals 534 and 546 and was still moving at 35 miles an hour when it struck the Wabash train. The Wabash engine was badly damaged, the leading two vehicles of the train were damaged, and four cars of the New York Central train were derailed and damaged to a greater or lesser extent. Engineer Hendrick was killed outright, and Fireman Altman was severely burnt and died in hospital three hours later. Conductor Willnow and the front brakeman of the freight train were injured. None of the passengers on the Wabash train was hurt.

The Interstate Commerce Commission (ICC) noted that a track maintenance crew had been welding a frog at the crossing using electric welding equipment. However, tests found no suggestion that this equipment could have interfered with the signalling. Engineer Clark of the New York Central train testified that he had seen the Wabash home signal at danger. None of the surviving crew of the Wabash train had seen the signals. The brakes of the train were functioning normally, and Engineer Hendrick had whistled for a highway crossing close to Raisin Center. Calculations made by the ICC with reference to the relative speeds and distances from the crossing of the two trains showed that the New York Central train must have struck its approach track circuit at least 2¼ minutes before the Wabash train had reached its track circuit and therefore that the signals, which were working correctly, must have been showing 'Stop' to the Wabash train. With the death of the engine crew of that train it could not be determined why it had not obeyed the signals. [1462]

A newspaper account of the accident mentioned large numbers of passengers making telephone calls after the accident, either to relatives and friends to say that they were safe or to businesses to explain that they would be late for meetings and conferences. The poor solitary operator at the Holloway telephone exchange had been completely overwhelmed by the number of calls and eventually some passengers, unable to get through, had walked or hitch-hiked to Adrian to use another exchange. [1463]

An Illinois Central Railroad (IC) passenger train and a Missouri Pacific Railroad (MP) freight train operating over the tracks of the Union Railway collided at the crossing of the two lines at Memphis, Tennessee, United States on Sunday 4th April 1943.

The lines of the two companies crossed one another at right angles. At this point the IC main line was double track and the Union Railway main line single track but various sidings of each company also crossed the lines of the other company so that in total eight

[1460] Interstate Commerce Commission report
[1461] *Adrian Daily Telegram*, Adrian, Michigan, United States, 17th August 1942
[1462] Interstate Commerce Commission report
[1463] *Adrian Daily Telegram*, Adrian, Michigan, United States, 17th August 1942

IC tracks crossed four Union Railway tracks. Stop signs were provided on the IC line 147 feet north of and 202 feet south of the crossing. A single signal was provided at the crossing, which is described in the Interstate Commerce Commission (ICC) report on the accident thus:

Movements over the crossing on either line are governed by a color-light signal, mounted on a pole 20 feet 11 inches high, located in the northwest angle of the crossing. The signal consists of three units arranged vertically to display aspects through 8-inch lenses to the east, west, north, and south. The signal is in the charge of a switchtender and is controlled from a building located in the northwest angle of the crossing. The top and center lenses display red and the bottom lenses display green. The signal is so arranged that when a movement over the crossing on the I. C. is intended the top lenses display red aspects in all directions, the center lenses display red aspects to the east and west and the bottom lenses display green aspects to the north and south.

The IC line was worked by timetable and train order, the Union Railway by yard rules. All trains were required to stop before proceeding over the crossing if the signal was clear for them.

Southbound first-class IC passenger No. 33 train consisted of engine No. 1037, a mail express car and three coaches, all of steel construction. It had left Memphis Grand Central station at 4 p.m., stopped at the stop sign, proceeded past the crossing signal showing 'Proceed', was passing over the crossing at 9 miles an hour when its rear car was struck by the MP train.

Eastbound MP freight train No. 1204 East had left Briant, Arkansas, just under 3 miles west of the crossing, at 3.17 p.m. From there it consisted of engines Nos. 9319 and 1204, an auxiliary water car, 53 loaded cars, an empty car, and a caboose. It stopped 2,000 feet short of the crossing and engine No. 9319 was detached. About 35 minutes later it proceeded and was travelling at about 12 miles an hour when it struck the rear of the passenger train at about 4.06 p.m. The rear car of the passenger train was overturned and badly damaged, the colliding engine was slightly damaged. [1464] One passenger, 15-year-old Catherine Whitsell, was killed and 60 passengers and three railwaymen were injured.

The IC engineer told the ICC inquiry that as he had been passing over the crossing he had seen engine No. 1204 approaching. Its driving wheels had been sliding and it did not look, from its speed, as if it was going to stop short of the crossing, so he had accelerated in an attempt to clear the crossing before the MP train reached it.

The switchtender likewise said that the brakes of engine No. 1204 had been sliding. The first 18 cars of the MP train had passed him and none of their brakes had been applied.

An IC yardmaster who arrived shortly afterwards found that none of the brakes of 37 cars had been applied.

Engineer John McClendon of engine No. 1204 testified that engine No. 9319 had been coupled to his at Briark to assist the train over the Mississippi River bridge. Before leaving Briark he had placed the double-heading cut-out cock in the closed position, the brakes being in charge of the engineer of engine No. 9319. When the second engine had

[1464] Interstate Commerce Commission report

been detached he had placed the double-heading cut-out cock in the open position and attempted to release the brakes but had been unable to do so. Engine No. 9319 had maintained a brake pressure of 90 lb and the feed valve of engine No. 1204 was adjusted to 70 lb. It had taken a considerable time to reduce the brake pressure when members of the train crew had released the brakes on the cars by operating their release valves. He had then proceeded. When his train was 900 feet from the crossing his fireman had shouted that the crossing signal was at *Stop*. He applied the independent brake and opened the sanders, but the speed of the train was not reduced. He had then applied the brakes in emergency, without result. He said that he and other members of the train crew understood that a brake continuity test should have been made after the assisting engine had been detached. It had been omitted to save time.

The ICC concluded that after the auxiliary brake reservoirs had been depleted by the operation of the release valves, the air in them had not been replenished, indicating that despite Engineer McClendon's assertion to the contrary the double-heading cut-out cock on engine No. 1204 had not been opened and the train had proceeded with only its engine brake working. Had the brake test been carried out the error would have been revealed and the accident would have been averted. [1465]

Engineer McClendon was arrested on a manslaughter charge after the accident. [1466]

A Pennsylvania Railroad express, and a Baltimore and Ohio Railroad coal train collided at the crossing of the two lines at Uhrichsville, Ohio, United States, on Sunday 21st November 1943. The freight train apparently reached the crossing first and the express ran into the side of it. The engine and four cars of the express and six cars of the freight train were derailed but casualties were limited to cuts sustained by a Mrs Schwab, presumably a passenger, from broken glass, slight head injuries suffered by Fireman W.E. Eppley of the express and bruises sustained by an express messenger. The Interstate Commerce Commission did not investigate the accident and the brief newspaper accounts give no further details. [1467]

As mentioned in the introduction to this chapter, a serious disaster occurred at Warthausen, Germany, on Thursday 6th January 1944. Here the standard-gauge Friedrichshafen – Ulm main line was crossed on the level by the narrow-gauge Biberach an der Riss – Ochsenhausen line. The narrow-gauge line was protected by trap points on either side of the crossing and a single signal. There were no trap points on the standard-gauge line but *Ausfahr* and *Einfahrsignals* (home and starting signals) at the south end of the station guarded the crossing. A signal box at the crossing controlled the trap points and signals.

On the morning of the accident mixed train GMP 313 from Ochsenhausen to Biberach an der Riss arrived at Warthausen on time at 06:46. It consisted of engine 99.65 and a train of older wooden vehicles. Meanwhile passenger train P 1521 from Friedrichshafen to Ulm, in charge of a class 18.1 steam locomotive, was approaching from the south,

---

[1465] Ibid.

[1466] *The Daily Herald*, Gulfport and Biloxi, Mississippi, United States, 5th April 1943

[1467] *The Salem News*, Salem, Ohio, United States, 22nd November 1943 slightly late. According to the timetable the Ulm train should have passed over the crossing first but as it was running late the signalman decided to give priority to the narrow-gauge train instead and cleared his signal for it. The *Einfahrsignal* applying to the Ulm train was thus automatically held at danger by the interlocking. The morning was foggy; however, visibility was poor, and the driver of the Ulm train failed to see the stop signal and drove into the side of the narrow-gauge train. Four of the wooden vehicles of the smaller train were wrecked and two more damaged; 12 people died, and an unknown number were hurt. An inquiry was held into the accident, but its results are unknown. [1468]

Square Crossing, Dundalk, Eire, which had seen a fatal collision back in September 1874, saw another accident 70 years later, on Wednesday 26th April 1944, when the 12.30 p.m. Great Northern (Ireland) Railway railbus from Dundalk to Drogheda collided with the 12.20 p.m. Dundalk Newry and Greenore Railway (DN&G) passenger train on the crossing. The driver of the railbus was slightly hurt and the railbus itself was badly damaged while the DN&G train was derailed.

In a third accident on Wednesday 18th July 1945 the 5.45 p.m. DN&G passenger train from Greenore to Dundalk collided on the crossing with the 5.15 p.m. interchange goods train from Dundalk Junction station on the GNR (I)) to the Barrack Street goods depot on the DN&G. The goods train, which was being propelled over the crossing, was running 70 minutes late. This time there was damage to rolling stock, but no one was hurt.

The crossing was removed in July 1954 following closure of the DN&G line. [1469]

The first of several rail/rail crossing collisions in 1945 occurred at Brinkley, Arkansas, United States, on Saturday 13th January 1945, at the intersection of the St Louis Southwestern Railway (Cotton Belt) and the Chicago Rock Island and Pacific Railroad (Rock Island). This crossing was busier than some, seeing an average of 33.4 Cotton Belt and 17.3 Rock Island trains daily, but protection against conflicting movements was limited to the provision of stop signs.

Northbound second-class Cotton Belt freight train No. 18 consisted of engine No. 760, 58 cars and a caboose. Starting from a point 200 feet south of the crossing at 5.37 p.m., it was moving over the intersection at 10 miles an hour at about 5.40 p.m. when it was struck by the Rock Island train.

Eastbound first-class Rock Island passenger train No. 50 consisted of engine No. 959, a baggage mail car, two coaches, a chair car, a dining car, a parlor car and a Pullman sleeping car, all of steel construction. The train left Hazen, just over 24 miles west of Brinkley, at 5.14 p.m., passed the stop sign at Brinkley and moving at an estimated 2 miles an hour collided with the Cotton Belt train. The engine of the passenger train and three cars of the Rock Island train were derailed and damaged. [1470] Fireman W.H. Sutton of the passenger train was killed, and Engineer Samuel Pfaff of the same train was injured. [1471]

[1468] Wikipedia – Eisenbahnunfall von Warthausen

[1469] *The Dundalk Newry and Greenore Railway* by D.S.M. Barrie, pg. 53.

[1470] Interstate Commerce Commission report

[1471] *Joplin Globe*, Joplin, Missouri, United States, 14th January 1945

Engineer Pfaff told the Interstate Commerce Commission (ICC) inquiry that his train had been travelling at 70 miles an hour as it approached Brinkley. The brakes had been tested and were working well. A mile from the crossing he had made a 15-pound brake application, then when the speed had dropped to 18 miles an hour he had released the brakes. About 1,600 feet from the crossing he had made another brake application which had been maintained. He said that he had seen another train on the crossing, had thought that he had his train under control and would stop short of the crossing. Then about 500 feet from the crossing he had realised that the speed of his train was not being properly controlled. He had moved the brake valve to the service position in an attempt to pull up but had entered the crossing still moving at 2 miles an hour.

The ICC noted that according to the rules trains were required to stop at the stop signs, then proceed at such a speed as to be able to stop short of the crossing if the way was not clear. The Rock Island superintendent stated that the rule had not been modified but it had been a long-standing practice for eastbound passenger trains not to stop at the stop sign but to continue until they were alongside the station platform and the engine a few feet from the crossing. Had the rules been enforced or an adequate interlocking been provided the accident would have been prevented. The accident was, the ICC found, caused by the failure of the Rock Island Company to enforce its rules governing the movement of trains over the crossing. [1472]

Turkey Lake, Wisconsin, United States, was the location of a collision on Tuesday 13th February 1945 between a Minneapolis St Paul and Sault Ste Marie Railroad (Soo Line) train and one belonging to the Chicago St Paul Minneapolis and Omaha Railway. The crossing of the two single lines was protected by two-position lower quadrant semaphore signals controlled from an adjoining 16-lever tower. Soo line trains were restricted to 35 miles an hour over the crossing, Omaha trains to 25 miles an hour.

Eastbound Omaha freight train No. 423 consisted of engine No. 423, 30 cars and a caboose. It passed signal No. 16 showing *Approach*, signal No. 15 showing *Proceed,* and was passing over the crossing at 20 miles an hour when its second car was struck by the Soo Line train.

Eastbound Soo Line first-class passenger train No. 84 consisted of engine No. 2701, two express refrigerator cars and a passenger-mail-baggage car. It passed signal No. 4 showing *Stop*, and travelling at 20 miles an hour collided with the Omaha train at about 5.30 p.m. The engine and first car of the passenger train and the second to ninth cars of the freight train were derailed. The fireman of the passenger train was killed and its engineer, brakeman and a baggageman were injured.

Train No. 84 had arrived at Turtle Lake at 5.06 p.m., when signals Nos. 3 and 4 had been clear for it, and stopped just short of the crossing. Its crew had then been busy with baggage and express, then the train had set back before moving forward again during shunting operations, but not as far as the crossing. Then at 5.15 p.m. the Turtle Lake operator was advised by the Omaha dispatcher that freight train No. 423 should be at Turtle Lake about 5.25. Deciding that the freight train was likely to reach the crossing before the passenger train was ready to start, he had then replaced signals Nos. 3 and 4 and checked that signal No. 4 was now showing *Stop*. A few minutes later he had seen

[1472] Interstate Commerce Commission report

the freight train approaching and at about 5.27 he had cleared signal No. 15 for it. The passenger train was still engaged in shunting operations at that time. A short time later, as the freight train was on the crossing, the passenger train started forward, passed signal No. 4 at *Stop*, as evidenced by both the operator and the engineer of the freight train. The operator made stop signals with his arms to both trains and the collision occurred. [1473]

The Interstate Commerce Commission stated that tests carried out after the collision had shown that the interlocking was working properly. If signal No. 15 had been replaced to *Stop* the interlocking circuits would have prevented signal No. 4 being cleared for a minute and a half. The accident was due to the engineer of the passenger train failing to operate his train in accordance with signal indications.

An eastbound Erie Railroad express freight train, consisting of an engine, 85 cars loaded with fruit, vegetables, meat, and other perishable traffic had almost cleared the crossing at Erie Junction, Lima, Ohio, United States, at 6.34 a.m. on Tuesday 3rd July 1945, when it was run into by a Baltimore and Ohio Railroad (B&O) freight train. The northbound B&O freight train consisted of an engine, 65 loaded coal cars and a caboose. It struck the 11th car from the rear of the Erie train, upsetting a car of tomatoes and derailing two other cars of the Erie train and the engine and tender of the B&O train. Engineer C.J. Good, Fireman Robert Wildermuth and Brakeman Claude Robinson, who had jumped immediately before the impact, were all unhurt. The only person injured in the accident was Flagman E.L. Steel of the Erie train, who suffered a back injury when he was thrown about in the caboose.

Engineer T.W. Miller continued to Marion with the front of his train after an hour and a half while work began to clear the line. The parallel New York Chicago and St Louis line was unaffected. [1474]

In 1945, at a level crossing between the Baltimore and Ohio Chicago Terminal Railroad (BOCT) and the Atchison Topeka and Santa Fe Railway (Santa Fe) on the outskirts of Chicago at Ash Street Crossing, Illinois, United States (the location is referred to simply as Chicago in the official report), the double-line of the Santa Fe and two other lines intersected the double track of the BOCT and five other lines. Given the multiplicity of tracks the crossing was not surprisingly provided with fixed signals operated from an adjoining tower. The signal at the crossing consisted of a four-arm, two-position semaphore, the two semaphores applying to BOCT trains being placed above those applying to the Santa Fe trains. On the westbound Santa Fe track automatic signal 41, a three-position upper quadrant semaphore was provided 1,302 feet of the crossing, with a stop sign adjoining it. If the stop signal at the crossing was against the Santa Fe train, signal 41 showed *Stop then Proceed.* On the northbound BOCT line automatic signal CC57N, of the colour-position-light type, was situated 180 feet of the crossing. It displayed *Stop then Proceed* if the block ahead of it was occupied, even if the crossing was clear. There was also a stop sign 483 feet south of the crossing. BOCT trains were limited to 30 miles an hour over the crossing, Santa Fe trains to 60 miles an hour.

[1473] Interstate Commerce Commission report

[1474] *The Lima News*, Lima, Ohio, United States, 3rd July 1945

Westbound Santa Fe train No. 13 consisted on Monday 6th August 1945 of gasoline-electric railcar M-185. It left C & WI Junction, 3¼ miles east of the crossing, at 2.20 p.m., an hour, and 8 minutes late, stopped at the stop sign and signal 41, which was showing *Stop then Proceed*, proceeded, passed the crossing signal showing *Stop* and was moving over the crossing at 5 miles an hour when it was struck by the BOCT train at about 2.50 p.m.

Northbound BOCT freight train No. 5570 North consisted of engine No. 5570, three cars and a caboose. It passed Brighton Park, three quarters of a mile south of the crossing, at 2.44 p.m., stopped at the stop sign, proceeded past signal CC57N showing *Stop then Proceed*, passed the crossing signal showing *Proceed*, and was moving over the crossing at 8 miles an hour when it was struck by the Santa Fe train. The impact pushed the railcar 30 feet northwards and ruptured its fuel tank. The escaping gasoline (petrol) ignited, and the interior of the car and engine No. 5570 were severely damaged by the fire. [1475] Brakeman Arthur Trammel of the Santa Fe train was killed and Engineer Harry Trimble, the unnamed conductor, Baggageman Joseph Smolen and three passengers, all also on the Santa Fe train, were injured. [1476]

The Interstate Commerce Commission (ICC) stated that the operator had cleared his signals for the BOCT train at 2.40 p.m.

The engineer of the BOCT train said that he first became aware of the Santa Fe train when his engine was a short distance from the crossing. He had applied the brakes in emergency and the collision had followed. He had not actually stopped at signal CC57N as well as the stop sign as it was not the practice to do so when the crossing signal was clear.

Engineer Trimble told the ICC inquiry that the brakes of his train had been tested and had functioned normally. As his train neared the crossing a switchtender who saw that the crossing signal was at *Stop*, waved a yellow flag at the engineer and shouted to him that the signal was at danger. Engineer Trimble stated that he did not hear the switchtender calling to him and thought that the latter's flag signals were directed at a nearby yard engine. Only when he was 35 feet from the crossing did he finally see the crossing signal at danger, when he applied the brakes in emergency. There had been nothing to distract his attention or obstruct his view.

The ICC found that the accident had been due to the Santa Fe train failing to obey signal indications.

The ICC noted that it previously drawn attention to the danger of fires in accidents involving gasoline motor cars. In six such accidents during the previous five years 54 people had been killed and 94 injured, most of the casualties being due to burning gasoline. In the present accident the railcar had contained 250 gallons of gasoline, though fortunately none of the casualties had been due to the fire (a newspaper account does not quite agree with the ICC report, stating that Engineer Trimble had suffered hip burns). Had, however, anyone been trapped in the railcar, the ICC continued, they would have died in the fire, and it recommended the adoption of a less dangerous fuel than gasoline. [1477]

[1475] Interstate Commerce Commission report

[1476] *Centralia Evening Sentinel*, Centralia, Illinois, United States, 7th August 1945

[1477] Interstate Commerce Commission report

In late 1945 an accident occurred at Danville Junction, Illinois, United States, where the Peoria and Eastern Railway (P&E) crossed the double track of the Wabash Railroad and the single track of the Chicago and Eastern Illinois Railroad (C&EI). The two crossings were protected by what was described as a tilting-target signal mounted on a tall mast. The target was 13½ feet long and 13 inches wide, painted yellow with red ends, and there was a red electric light bulb at each end to provide night-time indications. A leverman controlled the signal from an adjoining building.

The indications of the signal were:

Horizontal – Proceed for P&E trains.

Diagonal – Proceed for Wabash trains.

Vertical – Proceed for C&EI trains

There were stop signs for westbound P&E trains and eastbound Wabash trains situated 457 feet and 718 feet respectively from the crossing.

A bulletin issued in September 1943 regarding the operation of the crossing target read:

Targetman must make sure before changing signal to proceed for an opposing route that no train is closely approaching on the route for which the signal is then set.

He must not take signal away from a train which has received a signal to proceed over the crossing until the rear of the train has passed crossing.

When necessary to set a signal against a train which has received a proceed signal Targetman must have an understanding with the engineer of the train receiving proceed indication that the signal will be taken away from him and given to another train.

Trains using the crossing were required to proceed prepared to stop short of another train or obstruction. An average of 16.6 P&E, 18.2 Wabash and 18.5 C&EI trains used the crossing daily.

On Thursday 6th December 1945 westbound P&E first-class passenger train No. 13 consisted of engine No. 60, a baggage car, a mail car, a coach, and a passenger-baggage car, all of steel construction. It passed Cory, half a mile east of Danville Junction, at 1.20 p.m., 15 minutes late, stopped at the stop sign, then proceeded. The crossing signal was clear for it, and it was passing over the crossing at 10 miles an hour when it was struck by a Wabash train.

Eastbound second-class Wabash freight train Second No. 82 consisted of engine No. 2910, 48 cars and a caboose. It passed Tilton, the last open office, about 4 miles short of the crossing, at 1.10 p.m., four hours late, stopped somewhat short of the stop sign, then proceeded, passed the crossing signal which was showing *Stop* for the Wabash line and moving at 5 miles an hour collided with the P&E train. The engine and leading car of the P&E train and the engine and two cars of the Wabash train were derailed, and the engineer, fireman and brakeman of the P&E train were injured.

The Interstate Commerce Commission (ICC) determined that the crossing signal had been clear for the Wabash train until just before the P&E train had stopped at the stop sign. Seeing the P&E train, the leverman had then cleared the signal for it, thus placing it against the Wabash train, and the P&E train had proceeded. The crew of the P&E train had been unaware of anything wrong before the collision. The engineer testified that it was impossible, because of a lineside building, to see the Wabash train until immediately before the crossing. The fireman of the P&E train was too badly hurt to be questioned as to when he had seen the Wabash train.

The engineer of the Wabash train stated that the crossing signal had been clear for him when he stopped at the stop sign. He had given two short blasts on the whistle and proceeded. After passing the stop signal the crossing signal was not visible from his seat, nor did his fireman or the front brakeman observe the signal. Close to the crossing he had seen the P&E train and applied his brakes in emergency. He had seen the leverman and a section foreman giving stop signals but thought that they were warning him of men working on the track in the vicinity of the crossing.

The leverman said that when he had cleared the crossing signal for the P&E train he had been under the impression that the Wabash train was further away than it actually had been, and that the Wabash engineer would see the crossing signal at *Stop*. The ICC noted that there was nothing to prevent the leverman from changing the route from one line to another at any time regardless of line occupancy. Had a time delay been provided or a signal for each line, properly interlocked, the accident would not have occurred.

The ICC recommended that the three railroad companies should provide an interlocking at the crossing and stated that otherwise a notice would be served upon them requiring them to show why they could not do so. [1478]

[1478] Ibid.

CHAPTER 14

FIRES

Fires often broke out on trains following derailments or collisions. While wooden coaches were still in use in passenger trains, an overturned stove used for heating or spilt engine coals from a steam locomotive might set them ablaze. Later on, the substitution of gas for oil for train lighting proved to be another hazard, and a number of accidents were made much more severe by fires started by leaking or damaged gas cylinders until its dangers led to its replacement by electricity. This chapter will confine itself, however, to what might be termed primary fires, that is, accidents where there was no precipitating derailment or collision but merely a fire. Serious primary fires were considerably rarer than what might be termed secondary fires, but they did occur from time to time from a variety of causes, and more minor fires were fairly common.

Two waggons carrying 37 bales of wool and some empty drums which were attached to the Western mail train caught fire on the morning of Thursday 22nd January 1931 and the waggons and their contents were destroyed before the blaze could be extinguished. The train had left Cunnamulla on time and when it was near Offham Siding, Queensland, Australia, the train crew noticed smoke emanating from one waggon. On investigation two waggons were found to be on fire. They were detached and the train proceeded, having lost an hour and a half. A strong wind fanned the flames and it proved impossible to save either their contents or the waggons themselves, while some damage was caused to the track, a number of sleepers being burnt. The cause of the fire could not be determined but it was attributed to spontaneous ignition of the wool. [1479]

An express train running from Wellington to Auckland, New Zealand, caught fire on the night of Monday 2nd March 1931. It was the mail van of the train which was involved. The fire was discovered when the train was in an isolated rural area in King Country. It pulled up near a stream and the train crew, assisted by some of the passengers, formed a bucket brigade in the hope of extinguishing the fire with water from the stream. Soon, however, the fire became too fierce, the amateur firefighters were driven back and the

---

[1479] *The Daily Mercury*, Mackay, Queensland, Australia, 23rd January 1931 van and most of its contents, including some registered letters, was destroyed. At the start of the blaze a mail clerk had managed to throw some mail bags from the van and so these were saved. [1480]

The passenger train which had left Alexandria for Cairo, Egypt, at 12.30 p.m. on Wednesday 29th April 1931 was very crowded because of the Bairam (Eid al-Adha) Islamic holiday. As it neared the station at Benha a fire broke out in the rear third-class carriage, initially attributed to an overheated axle (hot-box). The fire spread rapidly to the two preceding third-class carriages and a panic ensued amongst the passengers. Meanwhile, unaware of what was happening towards the rear of his train, the driver continued the journey for some minutes before the train was finally brought to a halt. By the time that the train had finally stopped hundreds of passengers had thrown themselves out of the carriages, a number being killed or maimed in the process, some by falling under the wheels of the train.

The Benha station staff saw the burning train approaching and rushed to give help but by the time they reached the burning carriages the fire was too intense to permit any passengers to be extricated. The fire eventually had to be left to burn itself out, leaving the three carriages charred and gutted.

The number of passengers who died was initially put at between 40 and 80 and was later fixed at 48. The number of injured was not stated; they were taken to the Kasrelaini hospital in Cairo. Ten children were believed to have been burnt or trampled to death. The victims were all Egyptians apart from one Frenchwoman. The crowded nature of the train had made it impossible to escape along the corridors of the carriages, which were old, presumably wooden ones, which had been attached to the train only to provide extra capacity for the heavy traffic. After the accident, the train driver was arrested.

An investigation into the disaster determined that the fire had not been due to a 'hot-box' but had been started by a passenger, perhaps one cooking a meal on a stove; there was no suggestion of arson. It was also noted that a regulation was likely to be introduced requiring every carriage to have a communication cord, implying that there had been none in the burning third-class carriages, so that there had been no way to alert the driver. [1481]

A London Midland and Scottish Railway excursion train carrying 250 employees of the Eccles Co-operative Society from Manchester to Barnstaple in Devon ran into trouble as it was passing through the 4-mile Severn Tunnel, England, on the Great Western Railway, at about 5 a.m. on Friday 9th June 1933. A dining car attendant discovered a fire in one of its coaches. Sleeping passengers were rudely awakened and moved to another part of the train, which was stopped at Pilning station, the first one beyond the tunnel. Two carriages and a restaurant car were detached there and shunted into a siding where they were allowed to burn themselves out, the heat being intense enough to melt crockery in the restaurant car and twist the metalwork of the carriages beyond recognition.

After a short delay, the excursionists were taken on to Bristol Temple Meads where they were given breakfast before continuing their journey to Barnstaple. No one was hurt

---

[1480] *The Sun*, Sydney, New South Wales, Australia, 3rd March 1931

[1481] *The Times*, London, England, 30th April and 1st and 8th May 1931 but £15,000 of damage was caused. [1482]

A packet sent by post which "was not in conformity with postal regulations" was blamed for a fire which broke out near Weißenfels, Germany, on Saturday 12th January 1935 in the mail van of a train from Frankfurt-on-Main to Berlin. The post destroyed was letters and parcels from South Germany, mail from Aden, British India, the Portuguese East Indies, the Straits Settlements, Siam, French Indo-China, the Philippines, Bushire (a city and port in Persia, now Bandar Bushehr in Iran) and East Africa carried by the P & O liner *Ranpura*, and mail from Colombo (Ceylon) and Australia carried by the *Orsova*. [1483]

In contravention of the regulations a messenger boy brought onto a train in Japan a tin of petrol wrapped in a cloth on Tuesday 14th January 1936. He then thoughtlessly lit a cigarette and dropped the lighted match onto the cloth, which was soaked with petrol. An explosion and fire followed and before the train could be brought to a stand near Fukui by the pulling of the emergency cord the carriage was ablaze from end to end. There were about 50 schoolchildren in the coach, most of whom were able to make their escape to adjoining carriages but four girls were burnt to death and ten more were seriously injured. The messenger boy responsible himself also caught fire. He jumped from the train, was able to extinguish the flames by rolling in the snow and then decamped. He was later traced and arrested. [1484]

The petrol railmotor which had left Narandera at 6.02 p.m. on Wednesday 22nd January 1936 arrived at Griffith, New South Wales, Australia, at 8.08 p.m. As part of its preparation for the return trip to Narandera, Motorman E. Burke had to refill its petrol tank. Before he had finished doing so the fuel caught alight and the railmotor was set ablaze, while the fire spread to the signal box and the stationmaster's office. People who arrived at the scene reportedly pulled the blazing railmotor away from the buildings and the Griffith fire brigade soon arrived. Stationmaster Hunter said that the fire brigade had done splendid work in putting out the fire in his office and the signal box; unfortunately, they were unable to save the railmotor, which was burnt down to its steel undercarriage.

Motorman Burke suffered some slight burns to his right arm and leg but Guard McLeod and the few passengers who had already been on the train escaped injury. The origin of the fire was unknown, but a lighted cigarette butt thrown away by a passenger alighting from the train was regarded as a likely possibility. [1485]

When the Great Northern (Ireland) Railway mail train from Dublin reached Belfast Great Victoria Street station, Northern Ireland, on the morning of Friday 5th June 1936, smoke and flames were seen to be issuing from the vestibule between the mail van and the first passenger coach. The stationmaster and his staff tackled the fire with an

---

[1482] Ibid, 10th June 1933

[1483] Ibid, 17th January 1935

[1484] *The Railway Gazette*, London, England, 20th March 1936

[1485] *Narandera Argus and Riverina Advertiser*, Narandera, New South Wales, Australia, 24th January 1936 extinguisher and water and were able to put it out, though not before the woodwork and canvas of the vestibule had been considerably damaged. A spark from the engine was thought to have caused the fire. [1486]

Fire broke out on a Southern Railway Channel Islands boat train which had left Southampton Docks for London Waterloo at 4.58 p.m. on Saturday 15th August 1936, between Winchester and Micheldever, England. Only eight passengers were slightly injured, but a number had a lucky escape. Passengers in the second and third carriages had seen smoke and flames while the train was travelling at 45 miles an hour and more than one had pulled the communication cord. Seeing the vacuum gauge fall, the driver had stopped the train and the passengers quickly alighted on both sides of the train. Shortly after the boat train had stopped, a Down train from Waterloo to Bournemouth reached the scene and the passengers who had alighted on the Down side had a lucky escape from being run down, some standing in the six-foot space between the two tracks and others on the cess side of the Down line. One passenger also had to break his way out of the train via a lavatory window.

The fire had originated in the second coach of the train but soon spread to the third coach, the fourth Pullman coach and the leading coach. The four vehicles were detached and drawn forward to Micheldever station, where they were placed in a siding. Winchester Fire Brigade attended and using water from a nearby pond were able to cool gas cylinders in the Pullman car so that they did not explode, there was not enough water, though, to extinguish the fire and the coaches were allowed to burn themselves out. [1487]

The fire was the subject of a full Ministry of Transport report. This noted that neither the signalman at Wallers Ash East box, 2½ miles from the scene of the fire on the up side, nor his colleague at Weston box, 1¼ miles back on the down side, had seen any sign of fire as the train had passed them.

Mr Ellis, a passenger who had been riding with his wife in the second vehicle of the train, told the Ministry of Transport inquiry that he had walked back to see if places were available for tea in the coach next to the Pullman. When he had reached the coach next to his own he had seen "a spurt of flame, as though coming out of a hose," and thick smoke. Calling to another passenger to pull the communication cord, he had returned to the front of the train, warning passengers to close their doors to reduce the draught.

Mr Conner was the passenger who had been in the lavatory. As he had been about to leave it again, he had found smoke and flames outside. He had pulled the communication cord and broken his way out through a window, cutting himself considerably in doing so. By the time he had got outside the gangway had been burning furiously.

Mr Boyce, another passenger, and Pullman Car Attendants Finch and Pickering also described their experiences of the fire. All had pulled the communication cord.

Driver Woof and Fireman Ricketts described taking the burning vehicles to Micheldever.

The inspecting officer, Colonel A.C. Trench, outlined five possible causes of the fire:

[1486] *The Times*, London, England, 6th June 1936
[1487] Ibid, 17th August 1936

(a) Hot-box

(b) Arson – vandalism or sabotage

(c) A discarded cigarette butt or match

(d) A hot cinder from the steam engine

(e) An electrical fault.

The first, he stated, could be dismissed as at variance with the location of the fire and the fact that there was no indication of a problem with any of the axles.

Arson he regarded as inherently improbable and impossible to commit without being observed on a well-filled train with passengers constantly going to and fro.

A discarded cigarette or match was a possibility which was less easily ruled out as many people were careless in discarding cigarette ends and matches, and one could have ignited accumulated fluff or rubbish behind a suitcase or folded table. It seemed unlikely, however, that such a fire could have spread up behind the end panelling of a corridor unobserved until it burst out from under a coach roof, as described by Mr Ellis. Likewise, a fire originating from a cigarette dropped into the underside of the canvas bellows and igniting rubbish there was inconsistent with a fire first observed at the top and not at floor level.

The possibility of a hot cinder having entered a coach through a window or ventilator could be dismissed but it was quite possible that a cinder might have lodged in the loose canvas and bellows of the gangway between the second and third coaches, the fire being fanned by the strong draught. The train had been climbing some 14 miles at 1 in 250 but Driver Woof and Fireman Ricketts said that their engine had been steaming easily with a good fire. On the other hand a passenger, Mr Swaine, referred to having seen burning ash flying past the window at one point after leaving Winchester. A fire which had been still developing might only have produced a little smoke and not been spotted by the Wallers Ash or Weston signalmen.

Regarding an electrical fault as the possible cause of the fire, the only area which could come under suspicion was the branch circuit to the corridor lamp outside of the lavatory door at the trailing end of coach 7180, the third vehicle of the train. A short circuit or low-resistance leak in or adjacent to the corridor lampholder might have sufficiently overloaded a cable without blowing a fuse to initiate a fire in the space between the soft wood roof and the asbestos millboard ceiling, which might have got a good hold on the woodwork before it had burst out as described by Mr Ellis.

In conclusion, Colonel Trench wrote that the probable cause of the fire appeared to be limited to either a spark or cinder from the engine alighting on the gangway or an electrical short circuit. He did not think that it was possible to determine which.

The inspecting officer recommended the replacement of the canvas covers of the gangway buffers with covers composed of a more fireproof material. In addition he recommended various modifications to the electrical installations of the coaches when they became due for overhaul, to reduce the risk of an electrical fire. He did not consider the substitution of all-steel rolling stock could be justified from the point of view of reducing the fire risk. [1488]

A quantity of English and foreign mail was destroyed in a fire on a Bombay – Calcutta mail train shortly after 11 p.m. on Friday 11th December 1936 near Charkhera, India, about 40 miles south of Itarsi Junction. A postal sorter had noticed smoke coming from the leading mail van. The burning vehicle was uncoupled with considerable difficulty and shunted into a siding, after which the train crew, aided by station staff and the crew of a second train, worked to rescue as many mail bags as they could from the blaze. A hole was made in the side of the van with axes and bars and about 100 bags were retrieved before the extent of the fire made further efforts impossible. [1489]

The postmaster general announced that all the letter mail for Calcutta City and for H.M.S. *Enterprise*, all the unregistered letter mail for Burma except Rangoon City, and part of the unregistered letter mail for Bengal and the United Provinces had been lost. [1490]

A disastrous fire broke out on a Hong Kong – Canton express train of the Kowloon – Canton Railway on the evening of Saturday 16th January 1937. The fire occurred near Shihtan station on the Chinese section of the railway. The train was very crowded and according to some reports, of the 150 passengers in the third-class carriage in which the fire broke out 85 died, including 35 women and children, either from burns or from jumping from the moving train, with only 26 escaping with slight injuries. All the casualties apparently occurred in just one coach, No. 92, the second vehicle of the train. The leading four coaches were detached from the remainder of the train and the leading coach was then separated from the second, the second, third and fourth vehicles being left to burn themselves out.

The cause of the fire was not clear, though arson was suspected, theories including explosives set off by smugglers and baskets of celluloid trinkets being set alight by a time bomb. A couple of days after the accident the railway company officially announced that it had been due to incendiarism, though quite why anyone would want to burn to death a large number of ordinary passengers is not clear. [1491] [1492]

The railway company was, in any case, possibly not free from blame in the matter for even at this late date no communication cord was, according to one report, available to the passengers in the third-class coaches for them to stop the train in an emergency, and so it ran on for some distance before its driver realised that there was a problem. [1493]

After the disaster the company announced that improvements were to be made. It was stated that coaches were to be fitted with glass-covered brake cocks instead of the chain type of alarm signal – this implying, of course, that contrary to the earlier report there was some form of communication available, perhaps just a not very efficient one. The previously installed hinged doors of the carriages were also to be replaced by sliding

[1488] Ministry of Transport report
[1489] *The Railway Gazette*, London, England, 8th January 1937
[1490] *The Times*, London, England, 14th and 19th December 1936
[1491] *The Singapore Free Press and Mercantile Advertiser*, Singapore, 18th January 1937
[1492] *Warwick Daily News*, Warwick, Queensland, Australia, 19th January 1937
[1493] *The Times*, London, England, 19th January 1937

doors less likely to cause a jam of passengers in an emergency. [1494]

The final number of dead was given as 112.

Seven people died in a train fire near Moji, Japan, on Monday 27th December 1937, when a discarded cigarette end set fire to a consignment of celluloid toys. Thirty-five other passengers were injured, and two coaches were burnt out. [1495]

The Southern Railway of England suffered what was apparently another fairly significant fire on Thursday 10th March 1938 but this time it was not considered that it justified a full inquiry. The train involved was a steam-hauled passenger service from Dover to London. The passenger who discovered the fire pulled the communication cord, and the train was stopped at a point midway between Swanley Junction and St Mary Cray, with two coaches towards its rear alight. Some of the passengers made their way forward to the front of the train but about 30 alighted onto the track, assisted by a gang of platelayers who chanced to be in the vicinity.

According to a Mr G. James, one of the passengers, firemen had to make their way 100 yards across a field to reach the line. A goods train pulled up alongside to assist but was forced to move away when sparks fell onto it and threatened to set it alight too. Eye-witnesses said that the affected carriages had become a mass of flames in a short time. The vehicles were eventually taken to a siding while the passengers continued their journey on an electric train. The cause of the fire was not stated. [1496]

Typical of the sort of minor mishaps which occur from time to time on any railway was an accident on the Cambrian section of the Great Western Railway at about 7.30 p.m. on Saturday 3rd September 1938, between Abermule and Newtown, Wales. Driver Oliver Gittins was in charge of an engine pulling a dozen waggons of new sleepers. Between these two stations at a place named Penarth he noticed that the second and third waggons from the engine were ablaze. They were uncoupled and he proceeded to Newtown with them, where the fire was allowed to burn itself out. The line was blocked until 7 a.m. on Sunday. Wooden sleepers were treated with creosote as a preservative, but this made them rather flammable, and they were probably set alight by a spark from the engine. [1497]

The explosion of an oil lamp in a wooden car of Baltimore and Ohio Railroad train No. 32 from Chicago to New York was followed by a fire which led to the death of Express Messenger E.D. Owens on the evening of Friday 16th December 1938. Martin Egan, who had been riding in the car with Owens, explained later that the oil lamp had exploded and set fire to the car about five miles from Broad Ford, Pennsylvania, United States. They had attempted to extinguish the fire, but failed, and had pulled the emergency cord to stop the train, but that had not worked either. With no means of

---

[1494] *The Railway Gazette*, London, England, 9th September 1938

[1495] *The Sydney Morning Herald*, Sydney. New South Wales, Australia, 29th December 1937

[1496] *The Times*, London, England, 11th March 1938

[1497] *Mishaps on the Cambrian Railways (1864-1922) Supplement (1922-1947)* by Elwyn V. Jones making their way to adjoining cars, all they had been able to do was cling onto the ladder on the outside of the vehicle. Egan had managed to hold on but near Broad Ford Owens had lost his grip and fallen to his death.

Also near Broad Ford Baggagemaster Leroy Berkey in the car behind had discovered the fire and pulled the emergency cord, which this time did work. The train was stopped, and Conductor Frank Leiberger split the train, the engine and three express cars, the rearmost of which was the one on fire, going forward to Connellsville, where it took firemen 40 minutes to extinguish the blaze. The old wooden car, lacking electric lighting, had been added to the train to cope with the extra Christmas traffic. [1498]

Just before the Bombay (now Mumbai) – Calcutta (now Kolkata) mail train arrived at Jubbulpore, India (now Jabalpur) on Monday 17th April 1939, passengers riding in a third-class carriage became aware of smoke coming from a point between the carriage wall and its roof. The communication cord was pulled and railwaymen, as they thought, extinguished the fire. The train only managed to cover about another mile before the fire broke out again and the train was again stopped. This time more drastic action was taken – the affected area of the carriage was hacked away with an axe, enabling the train to continue to Jubbulpore, where the vehicle was detached. The cause of the fire was not immediately obvious. [1499]

At 08:35 on the morning of Thursday 20th April 1939, railcar No. 372 was two kilometres beyond the station at Daours, France, on the Paris – Lille main line of the Société nationale de chemins de fer français, when Conductor Ronssin discovered a fire in the gangway between the third and fourth carriages. He attempted to warn the motorman, M. Albert Suffys, by pulling the emergency cord, but this failed to work. Instead he made his way through the train to the motorman's cab, during which time smoke and flames began to fill the train. As soon as the train stopped some of the passengers managed to escape via the rear van, but others, in their panic, became wedged in a door and were unable to get out.

The motorman and conductor, assisted by uninjured passengers and some workmen who had been mending an adjoining road at the point where the train stopped, immediately attempted to free the passengers trapped in the fourth car. They tried to smash the windows of the car with stones without success. Finally they managed to force open a door and the terrified passengers were released; by this time some of them had lost consciousness. Two women, Mme Besson and Mme Barge, of Arras, were found to have died of smoke inhalation, and two men, M. Antoine Carlos, a Lille engineer, and M. Armand Collinet, a Belgian national, also lost their lives. Twenty-four other passengers, suffering from smoke inhalation and burns, were taken to hospital in Amiens. It was not thought that any of them was fatally hurt. Among them was Herr Nolda, the German consul at Havre.

The fire having been extinguished, the train was able to continue its journey to

[1498] The *Morning Herald*, Uniontown, Pennsylvania, United States, 17th December 1938
[1499] *The Railway Gazette*, London, England, 12th May 1939

Amiens. An electrical short circuit was given as the cause of the blaze. [1500]

At 9 a.m. on the morning of Wednesday 10th January 1940, fire broke out in the waggon of a goods train running from Bulla Bulling to Coolgardie between Ubini and Calooli, Western Australia. The truck which caught fire was carrying furniture destined for a man named Scott at Norseman. By the time the fire had been discovered the truck was well alight, and all that could be saved of its load was a dressing table and two chairs, while the wooden portions of the waggon were consumed, only the chassis and wheels remaining. The remains of the waggon were eventually put into a siding and the train continued to Coolgardie. Behind it the express from Perth, already 55 minutes late, was delayed another three and a half hours and did not reach Kalgoorlie until 2.15 p.m. The cause of the fire was not known for certain, but it was surmised that a spark from the engine had lodged in the soft packing around the furniture and set it alight. [1501]

The same day as the Western Australia fire, Wednesday 10th January 1940, there was a repetition of the 1938 Broad Ford fire, this time near Durham, Missouri, United States. As a westbound Chicago Burlington and Quincy Railroad train was pulling out of Durham, a kerosene lamp hanging from the ceiling of the mail car began to leak and then exploded, setting the car on fire. Mail Clerk James Whittom was seriously burned trying to put out the blaze. A couple of stations down the line the train stopped at Lewistown, Missouri, where the fire department dealt with the fire, though by this time the roof of the mail van had fallen in, and its contents had been destroyed. The vehicle was towed back to Quincy, Illinois, for possible repair. [1502]

Four soldiers died and 24 were injured when fire broke out accidentally in a carriage of troop train W 162 163 at Rheinhausen, Germany, on the line between Duisburg and Krefeld, on Saturday 18th May 1940. [1503]

A snowshed at Riksgränsen, Sweden, very close to the border with Norway, caught fire at 09:50 on Monday 29th July 1940 as a German army (Wehrmacht) freight train was passing through it. A photograph taken half an hour later shows a German officer staring at the blaze from a safe distance on a nearby hillside. The wooden snowshed is blazing furiously. No one was hurt but both the snowshed and the train within it were completely consumed. [1504]

'Boys leap from Burning Train,' a newspaper headline read on Tuesday 29th April 1941. The train fire from which they had been fleeing had its origin in the antics of some of the boys themselves. Six of them died in the fire, including two sons of M. Pierlot, the

---

[1500] *L'Indépendant des Basses Pyrénées*, Pau, France, 22nd April 1939
[1501] *Kalgoorlie Miner*, Kalgoorlie, Western Australia, 11th January 1940
[1502] *Joplin News Herald*, Herald, Joplin, Missouri, 11th January 1940
[1503] *Katastrophen der deutschen Bahnen Teil II* by H.J. Ritzau, pg. 103 (in German)
[1504] https://postvagnen.com/forum/index.php?id=665840

Belgian Prime Minister, and seven were injured. The accident occurred near Westborough signal box, England, on the East Coast main line of the London and North Eastern Railway. The train involved was the 12.50 p.m. train from London Kings Cross to Newcastle, a relief train to the 12.45 p.m. express, consisting of V2 class 2-6-2 engine No. 4779 and eleven vestibule carriages, in charge of Guard George Dear with Driver Smith and an unnamed fireman. A total of 101 schoolboys from Ampleforth College, a Roman Catholic boarding school in Yorkshire, were travelling in the relief train in two reserved coaches towards its rear, accompanied by a master, the Rev. H. Dunstan Pozzi. The penultimate coach of the train, where the fire began, was occupied by 64 boys. [1505]

Guard Dear told the Ministry of Transport inquiry into the accident that he had walked along the side of the train at Grantham station (about eight miles short of Westborough) and seen nothing wrong. Approaching the scene of the accident he had been writing at his desk in the brake van when he had felt the brakes being applied. Looking out of the window he had seen the Westborough down home signal at danger (it had been replaced by the signalman when he had seen that there was a problem with the train). Looking back he had then seen boys throwing bags from the train and some jumping from it while it was still moving. Flames and smoke had been coming from the roof of the coach. As soon as the train had stopped he had gone back to the affected coach and asked some of the boys who had alighted if they were all out. He had been told that it was thought that they were. Entering the coach he had found it full of flames and black smoke so that it was impossible to walk through it. He said that he had shouted several times, "Is anybody in here?" There had been no reply. He had then made his way along the track to the rear brake van, intending to get his driver to set back so he could uncouple the van. Seeing that the wind was blowing the flames towards the front, he realised that the risk was greater at that end, so had gone forward again and tried to uncouple between the burning coach and the coach ahead. As he had been trying to do so burning timber had begun falling on him and a ganger pulled him clear. Instead, the uncoupling had had to be done at the other end of the eighth vehicle of the train and two coaches and the rear brake van left to the fire. There was fire-fighting equipment on the train, and more was brought from Hougham station, but the fire had been too advanced for the apparatus to be of use.

Driver Smith said that his train had left Grantham 15 minutes late. They had been travelling at 50 to 55 miles an hour when between the Westborough outer and inner distant signals he had felt the engine pulling hard and seen that the vacuum had fallen. Neither he nor his fireman had been able to see anything wrong on looking back and he had been going to continue to Westborough box when the vacuum had fallen further, and he had stopped the train.

The inquiry heard the evidence of 14 boys and the Rev. Pozzi, which was summarised. Most of the seats had been occupied by boys aged between 14 and 18, who had been reading, playing cards and so on. Some of them had been flicking lighted matches at one another, and a burning match had fallen down between a seat and the side of the coach. A boy sitting nearby thought that the match had gone out, but soon afterwards he had smelt burning. He and other boys had removed a seat cushion and tried to beat the fire out, then to put it out with water from the lavatory, carrying it in a hat and a mug. One boy, getting under a table to try and reach the fire, saw that it had spread, and

---

[1505] *The Times*, London, England, 29th April 1941

the fire soon spread very rapidly, fanned by the draught from the sliding-glass ventilators at the top of the windows and the extractor ventilators in the roof. When the boys had become aware of the seriousness of the fire they had gone to tell their master in the next carriage, who had come immediately, told a boy to pull the communication cord and tried to get to the guards' van for a fire extinguisher, but the door had been locked. Thinking that the train was not stopping, the master had then pulled the cord himself.

The boys at the front end of the burning coach had taken their things and moved forward to the next coach while those at the rear had moved to the rear vestibule and some had jumped from the train. Quickly the smoke had become dense and suffocating; at least one of the boys had fainted and others feared that they would never get out. Some had smashed windows and jumped through them. [1506]

Two of the boys described the events leading up to the fire at the inquest. Richard O'Kelly, 15, said that he had seen a lighted match flick past him, and continued: "Ten seconds later I saw a very small flame coming from between my seat and the side of the coach. I tried to put it out with my fingers but, because the seat was close fitting, I could not. Some other boys tried to smother the flames with their coats, but by this time the flames were going up towards the roof."

Asked by the coroner if he was saying that the match had set the carriage ablaze, O'Kelly replied, "I think it was the source of the fire."

Jerry Stewart, 14, who had been sat opposite O'Kelly, said: "I had a box of matches. I lit one and tossed it to the other side of the carriage. I saw it go out, and then I struck another and tossed it in the same direction. I think it went out, but I did not pay particular attention to it. I did not notice any other lighted matches being thrown across, but 50 or 60 seconds later O'Kelly said there was something burning."

Colonel A.C. Trench, the inspecting officer, concluded that it was obvious that the fire had originated by matches being flicked about, a burning match having fallen in an inaccessible position. It was possible that a match might have fallen onto paper or cellophane which had lodged in the recess.

There had been, Colonel Trench remarked, 'among the boys in the coach an unfortunate lack of appreciation of the fact that in a vehicle moving at high speed the draughts which likely exist are likely to cause the rapid spread of any fire, so that immediate action to smother it is essential, however trifling the initial outbreak may appear.'

In the absence of the master in charge some of the more senior boys ought to have put a stop to the dangerous practice of flicking lighted matches, and they should certainly have taken more drastic action to extinguish it when they had noticed it, and alerted Father Pozzi at once. It appeared that the fire had been well under way before he had been alerted.

The inspecting officer considered that Guard Dear had done all that he reasonably could.

Three points, Colonel Trench stated, required consideration:

(a) The time taken for the train to stop after the communication cord had been pulled.

(b) The locking of the gangway doors.

[1506] Ministry of Transport report

(c) The fire risk of the coach fittings.

The definite statement of one of the boys that the communication cord had first been pulled at milepost 112 showed that the stop had taken a mile and a quarter, which given the speed and falling gradient was not unreasonable. Father Pozzi's impression that the train had not slowed after the first pulling of the cord was probably partly due to anxiety and partly to the fact that the speed reduction curve in a brake application was convex and flat in the initial stages. Some discretion must also be permitted to drivers so that they did not stop, for example, on a viaduct or on a set of catch points.

The locking of gangway and corridor doors was prescribed in the regulations of all the British railway companies to separate first and third class carriages and more importantly to prevent thefts of parcels, mail bags and luggage from unoccupied vans and tampering with the brakes. The inspecting officer doubted if the locking of the door to the brake van had any significance in the accident as it was unlikely that an extinguisher would have been effective against the rapidly spreading fire. In general he thought that the locking of doors to unoccupied brake vans was justifiable, but not the locking of doors between carriages of different classes. This point was also raised at the inquest, where Guard Dear had to defend the locking of the doors between the two coaches occupied by the schoolboys and the adjoining coaches as being in accordance with the regulations. The same question would arise again 37 years later, in the aftermath of the 1978 Taunton train fire.

As regards the fire risk of the coach the inspecting officer noted that the design and materials of the seats were substantially identical to those in a modern road motor coach and under normal conditions posed no more fire risk than any ordinary type of seating. On the other hand a burning match falling between the side of the seat and the side of the coach might well have set fire to paper which had lodged there, and to the use of Rexine (artificial leather) which covered plywood there in an inaccessible location. The use of Rexine and wadding in such a location should be omitted, leaving uncovered plywood which was considerably more fire resistant. [1507] Ten years later Rexine was to be implicated in the rapid spread of a fire in a train at Huntingdon.

At the conclusion of the inquest on the six boys who lost their lives Coroner Theodore Norton stated that, "I have come to the conclusion that the fire started as a result of matches used by the boys and also as a result of cigarette ends lying burning and possibly thrown about." He was not prepared to find anything of a criminal nature that could be alleged or supported against any of the boys, and they were all exonerated. A verdict of 'accidental death by burning' was returned. [1508]

When a diesel train from Zarate to Buenos Aires on the Ferrocarril Central Argentino was running between Rio Lujan and Otamendi stations, Argentina, on Wednesday 18th June 1941, flames were seen to be coming from the underside of one of its coaches. It was immediately stopped, and its passengers asked to disembark. The local fire brigade was called but despite their efforts they could not prevent the fire spreading to other vehicles of the train, causing considerable damage, while traffic on both lines through the area was interrupted for some hours. Although the train had been well-loaded, no one was

[1507] Ibid.

[1508] *The Times*, London, England, 29th May 1941

hurt. The cause of the fire was a mystery. [1509]

Longtown, Scotland, on the London and North Eastern Railway main line from Carlisle to Edinburgh, became during the Second World War the site of a Central Ammunition Depot. On the evening of Wednesday 1st April 1942 the Longtown signalman accepted the 9.30 p.m. Carlisle to Longtown freight train, destined for the depot. It was a considerable time before the train arrived. When it finally did appear its guard informed the signalman that the delay was because one of its waggons, carrying explosives, had developed a hot-box, which had begun to set the vehicle on fire. Upon being advised of this the War Department refused to accept the train, which instead was stabled in the branch loop line under a military guard, until, presumably, it was absolutely certain that the fire was out. [1510]

Two carriages of the 4 p.m. Up Great Western Railway express from Cheltenham Spa to London Paddington caught fire between Slough and Iver, England, on Wednesday 16th April 1941. The passengers were evacuated from the burning carriages and moved to the remainder of the train. The affected carriages were detached, and a photograph shows them to be well alight. No one was hurt. [1511]

The 10.15 a.m. London Midland and Scottish Railway (LMS) express from London Euston to Manchester was passing through Leighton Buzzard, England, on Sunday 6th February 1944 when the driver of a ballast train noticed a fire in its third coach. The 'Stop and Examine Train' signal was sent to Chelmscote Bridge box, just north of Leighton Buzzard, and the express was stopped there. The train was separated in front of the third and behind the fourth coach and the fire brigade, ambulances and doctors went to the scene. Six passengers in the crowded train were injured, including Dr Wilkinson, Professor of Dental Surgery at Manchester University, and Mr Hilton, an LMS official, who suffered facial burns. Some of the passengers had jumped from the burning carriage.

Driver J.E. Holt said that the train had been diverted from the fast to the slow line before Leighton Buzzzard. "If it had not been for this the whole train might have caught fire, going at speed through the {Linslade} tunnel, but, as it happened, we were only travelling about 15 m.p.h. at the time."

As might be expected, servicemen were among the rescuers. Private Thomas Gorden, Royal Electrical and Mechanical Engineers, told a reporter, "I got out of the carriage while it was on fire and found three men lying unconscious in a third-class compartment. I got in through the door, but the fumes and smoke nearly choked me when I got inside and threw me back. I tried a second time and dragged an R.A.F. officer out."

The line was blocked for an hour and the express was four hours late reaching Manchester. The cause of the fire was not immediately known. [1512]

[1509] *The Railway Gazette*, 26th September 1941

[1510] *North British Railway Study Group Journal* No. 31, pg. 18.

[1511] *The Railway Gazette*, 30th April and 21st May 1943

[1512] *The Times*, London, England, 7th February 1944

On the evening of Tuesday 23rd May 1944, Crossing Keeper Harry Hesketh at Hibaldstow became aware of a high-pitched whining sound coming from the waggon of a munitions train passing his crossing and realised that it had a 'hot-box'. He got on the telephone to Signalman Walter Ward down the line at Brigg, England, and the latter kept his signals on against the train, then cleared his home signal as the train reached it, stopped its engine at his cabin with a red handlamp, and advised the engine crew of the problem. They could then see that it was the 16th waggon from the engine which had the 'hot-box', which by that point had set the wooden waggon, loaded with 10 tons of explosives, ablaze. Calmly the railwaymen divided the train behind the burning waggon, drew the front portion of it forward, then shunted it back into a siding near the signal box, where there was a plentiful supply of water. Using buckets of water the driver, fireman, guard, and signalman were able to extinguish the blaze. Had the munitions waggon exploded and in turn caused the remaining 49 waggons of the train to explode (500 tons of high explosives) most of the town of Brigg (population nearly 4,000) would probably have been destroyed.

For their "coolness and resource" the four railwaymen were awarded £1 each by the London and North Eastern Railway company – a rather paltry sum, it would seem. [1513]

A few days later a similar situation would arise at Soham but would not end so well (see Chemical Explosions chapter).

A major catastrophe occurred in Japan on Sunday 3rd September 1944. A train bound for Gokurakuybashi (Mount Koya) on the Nankai – Koya line stopped suddenly on an incline in Wakayama Prefecture. Fire then suddenly broke out under the floor of the electric train. Seventy-one people died and 138 were injured. [1514]

When a New York Central Railroad freight train pulled into the Geddes St yards at Syracuse, New York, United States, at about 12.50 a.m. on Thursday 14th June 1945, it was noticed that smoke was coming from one of the boxcars. The fire brigade was called and by the time that they had arrived the car was well ablaze. It was found that it was loaded with hundreds of cartons of Blue Diamond matches made by the American Match Company. The firemen had to chop open the box car to get to the blaze which they were able to extinguish in 23 minutes. Part of the consignment was saved. The cause of the fire was not stated. [1515]

[1513] https://www.grimsbytelegraph.co.uk/news/nostalgia/railway-worker-who-saved-brigg-1780698

[1514] Wikipedia – List of Rail Accidents 1931-45

[1515] *Syracuse Herald Journal*, Syracuse, New York, United States, 14th June 1945

CHAPTER 15

CHEMICAL EXPLOSIONS

Railway accidents involving explosions were of two types: those which might be defined as chemical explosion, and locomotive boiler explosions. Accidental chemical explosions, such as those due to mishaps with blasting powder or munitions, were rare, whereas in the early days in particular, locomotive boiler explosions were much more common, though by the mid-20th century they were becoming much rarer, while chemical explosions were perhaps becoming a little more common. In wartime the latter often involved the accidental explosion of munitions. This chapter will deal with chemical explosions.

As a South African goods train on its way from the Cape to the Rand was less than a mile from the station at Leeudoringstad in the Transvaal on Sunday 17th July 1932, its driver saw a cloud of smoke coming from one of the waggons in the middle of the train. This was a matter of more than usual concern as the train included 31 waggons loaded in total with 10,685 cases of dynamite (about 320 tons) en route to the Rand from the Cape Explosives Company. Stopping his train, the driver was just about to alight from the footplate when there was an explosion. Thanks to the regulation three empty waggons marshalled between the engine and the first waggon loaded with explosives, the driver and fireman escaped death or injury from the blast and they were able to uncouple their engine and take it forward to the station to warn people there.

Less fortunate than the engine crew was Guard Saerl van Wyk, who was blown to pieces as further explosions followed the initial blast. Four occupants of a farmhouse which adjoined the railway also lost their lives, together with most of their livestock, and a trench 300 yards long was reportedly blown in the track. [1516]

An inquiry into the accident blamed a fire originating in an overheated axle ('hot-box') affecting the rear axle of the first waggon loaded with explosives. [1517]

[1516] *The Times*, London, England, 18th July 1932

[1517] https://ruralexploration.co.za/Leeudoringstad.html

A special troop train of the Estrada de Ferro Central do Brasil bound for Juiz de Fóra suffered a disastrous explosion in Entre Rios station, Brazil, on Monday 17th October 1932. The explosion reportedly began in a waggon of ammunition at the front of the train and was followed by a fire which spread to the remainder of the train, accompanied by further exlosions of ammunition. Twelve people died, mostly soldiers but including two railwaymen, and a large number were injured, while the station buildings were damaged. The cause of the explosion was not obvious; it was presumably accidental. [1518] [1519]

The port of La Libertad, El Salvador, was the scene of a disastrous explosion on Wednesday 14th March 1934. Seven tons of dynamite were being unloaded from a ship onto two freight waggons when the explosion occurred. The fire which followed the explosion involved 4,000 cases of gasoline and 15,000 sacks of coffee. It was estimated that 250 people lost their lives and over 1,000 were injured, while all the buildings in the vicinity of the blast were destroyed and not a building in the town escaped damage. Firemen spent all day fighting the blaze while the populace fled into the countryside, doubtless fearing further explosions. The cause of the blast was unknown. Sparks from a locomotive and a rough shunt of the waggons were two theories. [1520] [1521]

In a parallel with the South African explosion four years earlier, it was dynamite destined for gold and silver mines in nearby El Oro for use in blasting out the ore, which caused a disastrous explosion in the Ferrocarril Nacional de México station at Tultenango, Mexico, on Thursday 26th March 1936. A fire reportedly broke out in a waggon loaded with explosives at the junction station about 100 miles from Mexico City. The explosion killed 20 people, injured 60 and demolished the station buildings, a restaurant, several warehouses and in all, half of the buildings of the village. A number of those injured were not expected to survive. Among those who lost their lives were the stationmaster, a ticket agent, and several permanent way labourers. Several hundred yards of track were ripped up by the blast, disrupting train services in the area. [1522] [1523] [1524]

Presumably because it was a location which would be safe from air raids, a decision was made to store a large quantity of munitions in an underground station in the Ventas district of Madrid, Spain. This decision had disastrous consequences on Monday 10th January 1938 when at about 07:00 the munitions accidentally exploded as a train was entering the station. First reports were that 100 people had died but later reports put the number of dead at between 600 and 700, adding that a large part of the surrounding area had been devastated. Strict censorship of news of the accident was imposed by the Republican government, doubtless for morale purposes, but the story leaked out to

---

[1518] *O Estado de São Paulo*, São Paulo, Brazil, 18th, 19th and 21st October 1932
[1519] *The Railway Gazette*, London, England, 9th December 1932
[1520] *The Evening News*, Rockhampton, Queensland, Australia, 16th March 1934
[1521] *Nevada State Journal*, Reno, Nevada, United States, 15th March 1934
[1522] *Chester Times*, Chester, Pennsylvania, United States, 26th March 1936
[1523] *The Piqua Daily Call*, Piqua, Ohio, United States, 26th March 1936
[1524] *The Charleston Gazette*, Charleston, West Virginia, 27th March 1936

foreign sources. [1525]

Another catastrophic explosion occurred in Spain at 11:19 on Sunday 9th July 1939. Stabled in a dock at Peñaranda de Bracamonte station, 40 kilometres from Salamanca, were three waggons containing 23,000 kg of trinitrotoluene (TNT), which had been left there two days earlier. Nearby, four quartermaster soldiers were loading straw into two trucks, while on another line freight train No. 352, consisting of engine No. 873, a boxcar and ten waggons, had arrived from Salamanca at 10:40. A fire broke out in one or another of these adjoining waggons, started perhaps by an overheated axle box or possibly by a discarded match or cigarette end carelessly thrown into the straw. At any rate the fire spread to the three waggons loaded with TNT and they exploded. Worse still, close by was an Air Force store of 300 tonnes of bombs which had been there since the beginning of the civil war, containing the equivalent of another 107 tonnes of TNT, and these exploded in turn. Tremendous damage was caused, half the buildings of the town of 4,500 inhabitants being destroyed. Fifteen railwaymen comprising the station staff and the crew of train No. 352, the stationmaster's family, the four soldiers who had been loading the waggons, two Civil Guards looking after the Air Force store, and at least 80 citizens of the town lost their lives and about 1,500 people were injured. [1526]

Eight years after the Leeudoringstad explosion (see above) another train carrying explosives from Cape Town to Rhodesia was involved in a similar accident as it was crossing Foley Bridge, in the Bechuanaland Protectorate (now Botswana), on the evening of Saturday 15th June 1940. Two hundred tons of dynamite reportedly exploded, shattering 14 explosives trucks, and wrecking the trainsmen's caboose, a van, and other waggons. The bridge itself was also destroyed, a crater was blown in the riverbed, the ground was churned up for a considerable distance, trees nearby were uprooted and smashed and those further away were stripped of leaves, bark, and branches. Goods from trucks of general merchandise were scattered over a radius of 200 yards and portions of waggons were picked up from a much greater distance away. A 'caboose native' and three 'natives' guarding the bridge were killed. Two European railwaymen were seriously injured, six others escaped with slight injuries and shock. The concussion brought down the ceiling of a ganger's cottage half a mile away but the engine of the train was little damaged and was able to proceed under its own steam. The sound of the explosion was heard in Bulawayo, 158 miles away.

The line was blocked until a temporary bridge could be completed on 21st June. An inquiry into the cause of the explosion was opened. [1527]

Sweden found itself in an awkward position in the Second World War. It was neutral but had land borders with German-occupied Norway and Germany's ally Finland, while German-occupied Denmark was only a short distance away across the Øresund strait. As a concession to Germany it agreed to allow German troops and munitions to pass by rail

---

[1525] *The Daily News*, Huntingdon, Pennsylvania, United States, 13th January 1938

[1526] Webpage: Federación Castellano Manchega de Amigos del Ferrocarril (in Spanish)

[1527] *The Railway Gazette*, London, England, 23rd August 1940 across its territory between the Norwegian and Finnish borders. Thus at 01:45 on Saturday 19th July 1941 when *fjärrgodståg* (long-distance goods train) No. 4532 started from Hallsberg in Sweden it included 13 'foreign' waggons, seven of which were loaded with ammunition which had been sent by ship from Germany to Norway and were destined for Finland. The train arrived at Krylbo, Sweden, with Driver B. Nyman and Fireman Karlsson at 04:58, made up of an engine, 59 waggons and a van. As the train was coming into Krylbo Driver Jansson, in charge of an engine standing in the station, noticed that one of the 'foreign' waggons, about the middle of the train, had a 'hot-box'. Smoke and flames coming from the waggon were also noticed by Waggon Inspectors Melin and Dammström and Station Foreman Lind.

The fire in the burning waggon, which was one of those loaded with ammunition, intensified and at 05:13 the waggon exploded. Two effects of the explosion were damage to the overhead line equipment (OLE) in the station area, and the spread of burning material, so that other waggons of the train, waggons in the station yard and the overnight train from Jämtland to Stockholm, which was standing in the station, were all set ablaze. Attempts to pull unaffected waggons out of danger using an electric engine were unsuccessful due to the damage to the OLE but a steam locomotive was used to save some vehicles including a mail van. The fire service were also quickly on the scene and were able to prevent the fire spreading further, though not to prevent further explosions which continued until 11:00 or 12:00. Perhaps surprisingly no onc was killed by the explosions or fire, though about 30 people were injured by 'shrapnel', 24 of whom went to hospital, though only six of them were kept in. About 130 passenger and goods vehicles were damaged and there was extensive damage to the station buildings. [1528]

Several theories as to the cause of the fire and explosions were advanced after the accident, mostly that the fire had been accidentally started either by an overheated axle ('hot-box'), by sparks arising from braking or spontaneous combustion, though there was also a suggestion that it had been sabotage carried out by the British Special Operations Executive (SOE) and the SOE representative in Stockholm was expelled the next day. [1529] The 1941 police investigation, which included statements from the railwaymen involved, and which indicated that there had been a 'hot-box' in Norges Statsbaner (Norwegian State Railway) waggon 40256, was not made public until 2011. According to a German lieutenant interviewed by the police the waggons of the train coming from Norway had been examined on their arrival in Sweden at Charlottenberg and the axle boxes had been found to be lacking oil, so they had been refilled there. [1530]

An overheated cannon stove in the crew van of a military train caused an explosion on Tuesday 3rd February 1942 at Liégnitz, Germany, on the line between Kohlfurt and Arnsdorf. Twenty soldiers died and 15 more were injured. [1531]

On the evening of Wednesday 14th April 1943 the Tua to Braganca mail train left

---

[1528] Malin Ekenberg (maj 2007). "Krylbosmällen - D-uppsats Historia". Luleå tekniska universitet, Institutionen för Industriell ekonomi och samhällsvetenskap. Läst 21 Oktober 2019.
[1529] Wikipedia – Krylbosmällen
[1530] Krylbosmällen 70 år. Utredningens PM här
[1531] *Eisenbahn-Katastrophen in Deutschland* by Hans Joachim Ritzau, pg. 130

Mirandela, Portugal, at 19:25. It had travelled about a kilometre when there was a powerful explosion in a compartment of a third-class carriage which caused a panic among the passengers, some of whom tried to escape by throwing themselves from the train. The carriage caught fire following the explosion and was wrecked. The initial report on the accident spoke of seven dead and a number of injured. Because the bodies of those who died were badly charred they were not readily identifiable. There had been a fair in Mirandela on the day of the accident and the train was crowded with people from neighbouring towns and villages who were returning home from it.

It emerged that the explosion had not been a bomb (although the Second World War was raging Portugal was, of course, neutral) but rather a firecracker carried by a passenger from Macedo de Cavaleiros which had been accidentally set off. At least two of the injured later succumbed.

The line was not damaged, and trains were running normally the next day. [1532] [1533] [1534]

As a Denver and Rio Grande Western Railroad freight train pulled into Grand Junction, Colorado, United States, early on Monday 28th June 1943, its crew discovered that one of its cars was on fire. They then detached two cars and shunted them into a siding about half a mile west of the station. One or both of the cars contained munitions and from about 1.30 a.m. a series of explosions rocked the town. Joseph Lowe, a newspaper man, commented that, "It scared the daylights out of people," while Fire Captain Joe Brady declared, "It was a glorified fireworks with lots of noise and pyrotechnics, except that it was a week early for the Fourth of July."

A newspaper wrote that, 'Hundreds of townsmen, some clad in pajamas, hurried down Main street as close as they dared to the tracks on the city's western outskirts. The ammunition-fed fire glowed brightly in the Denver and Rio Grande railroad yards. Shells burst in increasing tempo, dying down finally about 5 a.m.'

Clarence Kurtz, hardware, and lumber dealer, commented, "The freak damage done by the explosion is comparable to cyclonic damage."

Pieces of shells were hurled as much as a mile and at least four people were injured, though no one was killed. Worst hurt was Fire Chief Charles Downing, whose right arm, torn by shell fragments, had to be amputated. Bob Walraven, a railroad workman, who had been standing on a caboose half a mile away, suffered a possible fractured jaw, while Fireman Henry Tebo and Mrs Virginia Buckley were also injured by shrapnel.

The area was subsequently placed under military guard while demolition crews removed unexploded munitions.

An investigation into the origin of the fire was begun by Army ordnance officers and railroad officials. One possibility was that a hot-box had caused the fire. [1535]

On Friday 2nd June 1944 a reporter for the London Times wrote from 'a

---

[1532] *Diario de Lisbòa*, Lisbòa, Portugal, 15th April 1943
[1533] Diario de Zamora de Falange Española de las J.O.N.S.
[1534] https://www.cm-mirandela.pt/pages/313
[1535] *The Iola Register*, Iola, Kansas, United States, 28th June 1943

Cambridgeshire town' that, 'Widespread devastation was caused was caused in this small Cambridgeshire town about 1.45 a.m. when a railway ammunition wagon containing bombs exploded in the station. The station buildings were wrecked, and houses nearby were so severely damaged that the occupants were made homeless.' [1536]

The town in question, whose name the paper had to omit due to wartime censorship, was Soham, England, on the London and North Eastern Railway secondary line between Ely and Newmarket. The train involved in the accident was the 11.40 p.m. freight train from Whitemoor Yard, March, to Ipswich, consisting of 2-8-0 *Austerity* class engine W.D.7337, 51 open waggons and a guards van, in charge of Guard H.G. Clarke with Driver Benjamin Gimbert and Fireman James Nightall. Forty-four of the waggons were loaded with United States aircraft bombs and seven with bomb components. Thanks to the heroism of the engine crew the waggon which exploded had been detached from the remainder of the train, thus preventing the whole trainload of bombs from exploding with the likely total destruction of the town. Fireman Nightall was killed outright by the explosion; Signalman Frank Bridges was fatally hurt and Driver Gimbert seriously injured. The accident formed the subject of a comprehensive Ministry of Transport inquiry.

The bombs carried by the train formed part of a consignment which had arrived by ship at Immingham Docks and was bound for White Colne, close to the wartime air base at Earl's Colne. A train of 61 waggons had left Immingham at 2.55 a.m. on 1st June, arriving at March at 9.45 a.m. Ten waggons were detached there and the train left at 12.15 a.m. on 2nd June as the 11.40 p.m. service to Ipswich. At 1.05 a.m. the train passed Ely North Junction; it arrived at Ely Dock Junction, at the far end of Ely station, at 1.15 a.m. and was held there until 1.24 a.m. for an opposing train to clear the single line from Barway Siding and Soham, before proceeding. Signalman G. Francis at Ely Dock Junction was confident that there had been no indication of a fire or an overheated axle box when the train had passed him. The signalmen at Ely North Junction and Ely Station North also denied seeing anything amiss with the train.

Signalman C.D. King at Barway told the inquiry into the accident that the train had passed him at 1.31 a.m., exchanging tokens via the lineside apparatus. He had noticed no sign of fire or smoke in any of the waggons, nor any unusual smell. However, when the train had been about a third of the way to Soham he had seen a "pinkish glow" on the steam or smoke from the engine. Twelve minutes after the train passed him he had heard the noise of the explosion and found that he could not contact his colleague at Soham either via the telephone or the block bell.

Due to his injuries it was almost seven weeks before Driver Gimbert was able to give evidence. He said that he had approached Soham under clear signals at 15 or 20 miles an hour. Just after passing the Soham distant signal he had noticed flames coming from the leading waggon. Describing them, he said, "The flames appeared to be getting all over the bottom of the waggon and seemed to be spreading very rapidly, which seemed to suggest that something very inflammable was alight." Whistling to warn his guard he had braked gently to stop the train, aware that it was carrying bombs. He had then instructed Fireman Nightall to uncouple the leading burning waggon, and this being done he had moved forward slowly through the station. Opposite the signal box he had stopped to warn Signalman Bridges to stop any opposing trains, knowing that the mail train was

[1536] *The Times*, London, England, 3rd June 1944

nearly due. The latter had come out onto the platform with a fire bucket of water. At that moment the explosion had occurred, and he remembered no more.

Guard Clarke stated that he had first seen that the leading waggon was on fire when his train was rounding the right-hand curve before the distant signal. He had assisted his driver by lightly applying his hand brake. By the time that the train had come to a stand there was a "tremendous blaze." He had heard the sound of uncoupling, had seen the engine and waggon moving forward, then the explosion had thrown him down. When he had recovered himself he had walked back to Barway box, 2¼ miles, laying detonators on the way. Signalman King said Guard Clarke had reached him at about 3.30 a.m., "in a very dazed condition."

Sub-ganger W. Fuller, whose house adjoined the station, said that he had heard shouts for assistance from Signalman Bridges, who had clearly seen that the train was on fire. He had been watching the train from his window as he got dressed and described the appearance of the fire. The explosion buried him, his wife and daughter in the debris of their house.

Having dealt with the immediate events leading up to the explosion and its results, the Ministry of Transport report went on to consider in detail the loading of the waggons at Immingham Docks and previous journeys made by the waggon which had exploded, Southern Railway 12-ton open goods waggon No. 29416.

Checker H.W. Frow, 'who appeared to be a reliable witness,' testified that he had considerable experience in loading aircraft bombs. He was emphatic that he had personally inspected each waggon before it was loaded, and that none had been left containing straw or other inflammable material, though suchlike often needed removing from incoming waggons. He thought that the load in the waggon concerned would have been 'a nice fit' which would not have needed wooden scotches.

As regards the previous loads of waggon No. 29416, the records showed that on 4th May it had travelled from Victoria Docks, London, to Luton with a load of bulk sulphur (used to produce sulphuric acid). Even after unloading and sweeping of the waggon some small amount of sulphur had probably remained.

Concluding his report, Major G.R.S. Wilson, the inspecting officer, declared that it was unlikely that the fire had been caused by either an act of sabotage or an overheated axle (hot-box). The fire must therefore have been accidental and have originated inside the waggon. It was unlikely that on the level route approaching Soham the engine would have produced a glowing cinder large enough to ignite directly either the waggon sheeting or the woodwork, though this had happened the following day at Gomshall on the Southern Railway, on a line with severe gradients, where a cinder had set alight the woodwork of a waggon carrying ammunition. It must be assumed, therefore, that there had been some substance present in the waggon which was especially sensitive to ignition from a trifling spark. The remains of the previous load of sulphur were the likely suspect, and that the fire had begun at an earlier point in the journey but had not spread sufficiently to be noticed by one of the signalmen.

Regardless of the exact cause of the fire, the inspecting officer stated that he was pleased to record that Driver Gimbert and Fireman Nightall had been awarded the George Cross for preventing an incomparably greater disaster. It seems a pity, though, that no award was made to Signalman Bridges.

Major Wilson noted that despite the countless loads of ammunition which had been carried on the railway since the beginning of the war, this was the first instance of an explosion in a running train. Since the accident a strong reminder had been issued to all railway and dock companies regarding the importance of waggons being "clean" before they were loaded. Waggons loaded with sulphur should be cleaned especially thoroughly.

Properly secured sheeting was a protection against fires, but sheeting was in short supply due to the war, and to avoid unnecessary delay as a wartime measure the regulations had been relaxed to permit the carriage of certain categories of bombs in open unsheeted waggons. [1537]

An American train of 40 waggons of munitions jumped the track 600 metres from Saint-Egrève station, near Grenoble, France, on the evening of Thursday 2nd November 1944. Several violent explosions followed, and railwaymen hurried from Grenoble to the scene. They found that there was great material damage and there were an unspecified number of victims. The Grenoble – Valence and Grenoble – Lyon lines were blocked. [1538]

One of the first reports of an explosion on an Atchison Topeka and Santa Fe Railway train at Tolar, New Mexico, United States, on Thursday 30th November 1944, came from a Mrs J.D. Leftwich, of Lubbock, Texas, who had been driving her car along an adjoining highway at the time. Her automobile was one of five 'demolished' by the blast, but she was only slightly hurt.

As a carload of bombs were involved and it was wartime, the explosion was inquired into by the Federal Bureau of Investigation (FBI). It was soon determined, however, that the explosion was accidental in cause, and not the result of enemy sabotage. The 81-car freight train was in charge of Conductor Darry Winn with Engineer Byron Clark and Fireman J. Worley. As it was passing Tolar the journal of the oil car which was 7th in the train broke, and 39 cars were derailed. Fire then broke out in the derailed vehicles. Engineer Clark managed to draw away the front few cars and a following freight train was able to remove the rear cars, but all the derailed vehicles were eventually consumed by the fire. While Conductor Winn was on the telephone to Clovis advising them of the accident, the carload of bombs which was marshalled 21st in the train blew up, damaging every building in Tolar and demolishing several. No one on the train was hurt but Jesse Brown, a Tolar resident who had been standing in front of his house, was fatally injured and others were less seriously hurt. A newspaper commented that the accident had very much brought home to the residents of Tolar the realities of the distant war. [1539]

A train which was carrying munitions, weapons, fuel, and many passengers, almost all of them soldiers, exploded at Haebaru, Japan, on the Itoman line, on Monday 11th December 1944. About 210 soldiers, eight students and three railwaymen were killed in the worst railway accident, in terms of the number of fatalities, in Japanese history. The

---

[1537] Ministry of Transport report

[1538] *Gazette de Lausanne*, Lausanne, Switzerland, 4th November 1944

[1539] *Clovis News-Journal*, Clovis, New Mexico, United States, 30th November and 1st December 1944 cause of the explosion was unknown but the mixings of munitions, fuel, and passengers on one train was clearly a recipe for disaster, made necessary, no doubt, by the pressures of the war. [1540]

The Second World War in Europe was almost over when at 7.10 p.m. on the evening of Thursday 22nd March 1945 a through freight train left Workington, England for Bankfield goods depot, Liverpool, both on the London Midland and Scottish Railway. The train was in charge of Guard T. Wilkinson with Driver H. Goodall and Fireman H.N. Stubbs. On arrival at Corkickle Yard, Whitehaven, shunting was performed and from there the train consisted of Class 3F 0-6-0 engine No. 3579, seven waggons of depth charges from the Admiralty Depot at Camerton, on the branch to Cockermouth; two waggons of non-explosive Admiralty stores, seven waggons loaded with ordinary goods, an empty waggon for repair, two empty coaches for repair, 38 empty waggons and a guards' van.

Corkickle Yard was left at 8.50 p.m., 47 minutes late, and the train proceeded southwards. The first ten miles to Sellafield were single track, and 15 minutes were lost at Nethertown waiting for a northbound freight train from Barrow to clear the single line. The signalman at Drigg, 4 miles beyond Sellafield, watched the train as usual as it passed him and saw nothing wrong with it. The next box at Eskmeals was out of switch and the station closed, then came the signal box and station at Bootle, 6½ miles from Drigg. Here, as the train neared his box, Signalman J. Southward saw a white glow in the sixth waggon of the train. His exact description of what he saw was reproduced in the Ministry of Transport report on the accident: -

"I saw flames rising over the top of the waggon and they would be about 10 or 12 feet above the waggon side, as much as that; it seemed to be ablaze all inside, or in other words one level, big, flame inside the waggon, but there was no flame outside the waggon, and I could see the darkness of the waggon side."

Signalman Southward immediately replaced his up starting signal, sent the 'Stop and Examine Train' signal to the next box at Silecroft and telephoned there. As a result a Down passenger train was stopped there.

At about the same time that Signalman Southward saw the fire it was also observed by Porter G. Bradshaw as he was about to enter Bootle box, and by Fireman Stubbs, who looked back from the engine as the train was passing the Bootle station platform and saw the reflection of the fire on the station buildings. Driver Goodall stopped the train just under half a mile south of the station, the fireman uncoupled behind the burning waggon, the first six waggons were drawn forward about 80 yards, the fireman uncoupled in front of the burning waggon, and the first five waggons were drawn forward another 80 yards. Fireman Stubbs then made his way up the line towards Silecroft and had put down one detonator to protect the Down line when he was thrown across the track by a terrific explosion. Quickly recovering himself, he continued up the line and put down three more detonators.

Fireman Stubbs had been 200 yards from the burning waggon when the explosion occurred. Driver Goodall was a good deal closer; he was killed. The only other casualty was Guard Wilkinson, who suffered from shock. Major G.R.S. Wilson, the Ministry of

[1540] Wikipedia – Okinawa Prefectural Railways

Transport inspecting officer, commented that, 'There is little doubt that the prompt action of the enginemen in isolating the burning waggon, without regard to personal risk, prevented a far more serious explosion.'

Little trace was left of the waggon which had exploded; two of the leading five waggons and 11 of the rear waggons were damaged. The 52 depth charges which exploded (9¼ tons of high explosive) formed an oval crater 105 ft long, 45-60 ft wide and 50 ft deep. Bulldozers were used to fill the crater; single-line working was opened up at 7 p.m. on 24th and the line was fully reopened at 5.40 p.m. on 25th after 68 hours.

Mr. J.F. Burgess, Assistant Foreman of Stores at Camerton Depot, and Mr W.B. Kirkby, who had been in charge of the contractors' gang who had loaded the waggons, told the inquiry into the accident that strict discipline had been observed in cleaning of the waggons and the loading of the depth charges. Porter J. Simon at Camerton was certain that waggon No. 9386, that which had exploded, had been properly sheeted. Stationmaster Fraser at Camerton stated that he had never had to complain about the unsatisfactory sheeting of waggons from the depot during the whole of the war. No problems, in particular with overheating axles, had been observed during examinations of the train at Workington and Corkickle.

Major Wilson concluded that the probable cause of the waggon fire had been a spark from the engine chimney lodging in and igniting the sheet covering the waggon, the fabric of the sheet having become exposed as a result of ordinary wear and tear. It was unfortunate that the waggons loaded with depth charges had been marshalled next to the engine. Prior to the war Rule 240 (clause 10) had required that waggons loaded with explosives must be marshalled in the middle of trains and had limited the number of waggons carrying explosives to a maximum of five per train. Early in the war the regulations were amended to permit trains of 60 waggons of explosives to be operated and Rule 240 (clause 10) had become inoperative.

Since the accident, the inspecting officer continued, the Companies had issued instructions that where possible when less than full train loads were involved, the waggons should be placed in the middle of the train.

The sheeting had been of the red war-time pattern which was less fire resistant than the pre-war pattern. It was therefore suggested after the accident that to reduce the risk of sheeting catching fire, open waggons carrying munitions should travel unsheeted, with special attention being paid to the waggons being "clean." The suggestion was rejected, the fire risk being considered less important than the advantages of sheeting for weather protection and security. It was hoped that improved sheeting would become available with the end of the war.

Another wartime change was that the labels of the munitions waggons, which before the conflict had read 'Explosives' had for security reasons been altered to read 'Shunt with great care'. It seemed that Driver Goodall and Fireman Stubbs had been unaware that there were any explosives on the train. The rules were to be amended so that the guards were required to tell their engine crews if explosives were being carried on their trains. [1541]

[1541] Ministry of Transport report

Slightly less than a month after the Bootle explosion, on Wednesday 18th April 1945, a London and North Eastern Railway freight train consisting of an engine, 12 waggons loaded with unfused 500 lb aircraft bombs, 37 empty waggons and a guards van was entering the yard at Gascoigne Wood, England, near Selby, via a facing connection, when its driver and a signalman noticed that the 8th waggon from the locomotive was on fire. In contrast to his counterpart at Bootle the driver here was aware that the waggons were carrying bombs. He uncoupled the front seven waggons and drew them forward. A few minutes later there was a violent explosion which killed Fireman George Crowe on the engine of another train and injured several other people. A crater with a diameter of 90 ft and 50 ft deep was formed. It was found that 98 of the 132 bombs in the 8th, 9th and 10th waggons had detonated, though none of those in the 11th and 12th had exploded.

The waggons had been loaded at Cardiff Docks. According to the guard, the three waggons concerned had been tied over on one side in the ordinary way and tucked inside, over the bombs, on the other side. The sheeting had not been carried out in accordance with the regulations despite the issuing of a special reminder to staff after the Soham explosion (see above). The weather at the time had been warm and dry and the fire which had preceded the explosion was blamed on a spark from the engine igniting the sheeting.

The accident was not the subject of a separate Ministry of Transport inquiry but was referred to in the report on the Bootle explosion. [1542] [1543]

Eleven people died when part of a munitions train exploded in the marshalling yard at Berg am Laim, Munich, Germany, on Tuesday 8th May 1945. [1544]

Shortly after the end of the Second World War a munitions train caught fire and some of its contents then exploded near Peine-Horst, Germany, on Wednesday 16th May 1945. Fortunately the accident happened in open country, though portions of the munitions landed in Peine itself. No casualties were reported. [1545]

A Finnish munitions train blew up on about Sunday 22nd July 1945 at Tornio, Finland, on the Gulf of Bothnia close to the border with Sweden, as the munitions were being charged, Fifteen soldiers lost their lives. [1546]

The explosion of a munitions train about 5 kilometres south of the Adriatic port of Ancona, Italy, on Friday 31st August 1945, reportedly cost 100 lives and resulted in severe property damage. Smokers aboard the train were blamed for the catastrophe. [1547]

Fire broke out in the open waggon destined for the Red Army barracks at Stern

---

[1542] Ibid

[1543] *The Times*, London, England, 19th April 1945

[1544] Wikipedia – Liste von Eisenbahnunfällen in Deutschland

[1545] Ibid.

[1546] *Journal de Geneve*, Geneva, Switzerland, 24th July 1945

[1547] *Joplin Globe*, Joplin, Missouri, United States, 1st September 1945

(Buchholz), Germany, on Monday 24th September 1945. Seeing the blaze, the crossing keeper at km 9.2 of the Schwerin-Görries-Parchim branch had sounded the alarm and the train was stopped. At that moment there was a terrific explosion. Ten covered waggons, the station buildings, the crossing keeper's house, and associated buildings were wrecked, and 15 acres of beech forest were burnt down. The stationmaster and his wife escaped with slight injuries.

The circumstances of the accident were thoroughly investigated. Train No. 19120 should have started from Ganzlin on 15th September, but the loading of its munitions had not been completed. The next day the line was blocked by an accident and the train did not start until Saturday 22nd. It consisted of a locomotive, two empty waggons, a covered barrier waggon, 43 waggons of munitions, a third-class coach for the accompanying crew and a waggon of wood (fuel for the engine). According to the regulations there should have been six barrier waggons between the engine and the first munitions waggon as a protection against sparks. For some reason the train did not take the direct route and the route it did take meant a reversal at Wittstock, from where only the wood waggon and the coach acted as barrier waggons. Then at Wittenberge there was another reversal and here the train was remarshalled, so that all four non-munitions waggons and the coach were placed directly behind the engine.

The yard at Schwerin, the intended destination of the train, was full, and at 01:50 on 24th September it was left at Holthusen. Its engine was detached and continued with its Soviet crew. The same day another engine arrived at Holthusen with a fresh crew. The supervisors, but probably not the new train crew, were aware that it was a munitions train. From Holthusen the train was pushed to the junction at Krebsförden, then it was pulled from there, now without any barrier waggons behind the engine. On a gradient the 1,134-tonne train stalled, and it had to be divided. The engine took the front portion on to its destination, returned for the second part and was on its way with that when not only the crossing keeper but also the engine crew observed that the sixth or seventh waggon was on fire. The crew uncoupled the engine and brought it to safety.

It was concluded that a spark from the engine had started the fire. The report added that with the use of open waggons for the carriage of explosives there had been a risk of a fire throughout the whole journey, irrespective of the lack of a barrier waggon on the final stretch of the trip. [1548]

[1548] *Eisenbahnunfälle bei der Deutschen Bahn* by Erich Preuß, pg. 169. (in German)

CHAPTER 16

BOILER EXPLOSIONS

The railways of the early 20th century were still powered largely using steam engines, though electric traction was making inroads, mostly on suburban and underground railway systems and by the end of the 1920s gasoline or diesel powered locomotives were also coming into use to some degree. Locomotive boiler explosions were, however, more common than chemical explosions. Although the explosion of a locomotive boiler could often prove fatal to the engine crew it was rare for train passengers to suffer and probably because of this information, regarding boiler explosions in countries outside of Britain and the United States has been hard to obtain. By the period dealt with in this book there were no full boiler explosions in Great Britain and the majority of those dealt with below occurred in the United States, though towards the end of the Second World War several boiler explosions also occurred in Germany and Japan, doubtless due to poor wartime maintenance and/or inexperienced staff, while there were one or two crown sheet failures in Great Britain involving imported American locomotives.

A serious boiler explosion in Germany on 19th May 1893 has just come to light. It occurred at Ilmenau in what was then the duchy of Sachsen-Weimar (now part of Thuringia). According to brief reports in several American newspapers the explosion on the morning of that day killed five people outright and another eight were so seriously hurt that it was thought that they were dying. [1549]

Another 29th-century boiler explosion which was not found in time to feature in the 1881-1900 volume of this series occurred at Cardenas, Mexico, on the Ferrocarril Nacional de México on Wednesday 16th August 1899. Seven men were killed, four of whom were American engineers named Fitzgerald, Gibson, Hussey, and Simon who had been in the cab of the locomotive, and three others were probably fatally injured, among them another American engineer named Lockhart who was standing near the engine at the time and was thrown 100 feet. The names of the three Mexican fireman and wood packers who were killed outright were not stated. Described as being of a special pattern and great size, the engine was used to haul trains over the mountains. It was wrecked and

---

[1549] *Evening Journal*, Wilmington, Delaware, United States, 19th May 1893 'heavy pieces of iron were scattered over the mountain.' [1550] [1551]

Missing from the 1901-15 volume of this series are two boiler explosions. One was on the Central of Georgia Railroad at Macon, Georgia, United States, on Thursday 2nd January 1902. At 07:30 on that day the boiler of engine No. 1475 gave way in the roundhouse of the company, killing Engineer Tesh Hodges, Machinist John McDonald, Helper Uruah Cornelius, James O'Neill and an unidentified fifth man. Several other men were injured. The engine had been in the roundhouse for repair for several days and it was noticed that the safety valve of the engine was not working properly. It was while the men were working to fix the problem that the explosion occurred. [1552]

In the other missing accident five railwaymen died and one was injured in a boiler explosion on the Coal and Coke Railway at Yankee Dam, West Virginia, United States, at 2 p.m. on Saturday 30th October 1909. Conductor J.T. Riddle, Engineer J.R. Rogers, Fireman W.E. Carruthers, Brakeman Roy Thomas, and Student Fireman F.J. Findley all lost their lives. Brakeman Charles Patten was injured and Flagman R.A. Brake, riding at the end of the freight train being pulled by the engine, escaped injury and was able to go and raise the alarm. It was supposed that the other six had all been riding on the locomotive to save time during switching operations. Flagman Brake telephoned news of the accident from a nearby sawmill and the engine and baggage van of a passenger train were detached at Clay Court House and went to the scene with surgeons, while a second relief train was despatched from Gassaway. The boiler of the engine was blown into the adjoining Elk River. No cause was stated for the explosion. [1553]

The engine of a short Erie Railroad work train was slowly backing towards Middletown, New York, United States, at 9.35 a.m. on Saturday 18th August 1931 when its boiler exploded with a noise likened to 'several boxes of dynamite detonating simultaneously.' The engine cab was found in a field 200 feet from the scene of the explosion, near Red Onion Switch, and the massive boiler was itself thrown 20 feet. Engineer Robert Kerr was killed instantly, and Fireman R.H. O'Neil died shortly after reaching hospital. The accident happened at an inaccessible location half a mile from the nearest road and a mile from a telephone. No damage was caused to the vehicles which had been on either side of the locomotive or to the track, all the force of the explosion having been directed upwards.

Coroner Sheldon was soon at the scene and after examining the wreckage returned a verdict of 'accidental death' without attempting to determine the technical cause of the accident. Railroad officials, however, were sure that it had been due to low water in the boiler. [1554]

Early on Wednesday 16th September 1931, the boiler of an Erie Railroad pusher (banking) engine which was assisting a 100-car freight train up an incline at Graham, New York, United States, exploded. Parts of the boiler were thrown several hundred feet.

[1550] *Lawrence Chieftain*, Mt Vernon, Missouri, United States, 24th August 1899
[1551] *The Commercial Tribune*, Cincinnati, Ohio, United States, 18th August 1899
[1552] *The Montgomery Monitor*, Mount Vernon, Georgia, United States, 9th January 1902
[1553] *The Evening Times*, Cumberland, Maryland, United States, 1st November 1909
[1554] *Middletown Times Herald*, Middletown, New York, United States, 20th April 1931

Engineer William Mallet and Brakeman Edward Brewster were killed instantly, and Brakeman George Hamilton died as he was being taken to hospital. Three other men, Conductor Henry Smith, Fireman Edward Crotty, and Flagman L.Z. Dodge were hurt. Fireman Crotty, despite his injuries, had staggered down the line to protect the obstruction and flagged the train Second No. 78 before collapsing from his injuries. He died in hospital the following morning. The conductor and flagman were only slightly hurt by flying wreckage. The pusher engine, caboose and several trucks were derailed, and the line was blocked for a couple of hours. [1555] [1556] [1557]

On Christmas Day, Friday 25th December 1931, the Southern Pacific Railroad's southbound *West Coast Limited* passenger train from Portland, Oregon, to Los Angeles, California, was two miles west of Richvale, California, United States, and travelling at 25 miles an hour when the boiler of its engine, a GS-1 class 4-8-4, exploded, killing Engineer Joseph Sturm and Fireman Walter Stone. None of the passengers was hurt. The explosion separated the brake pipe behind the engine so that the brakes came on automatically, stopping the train. Other trainmen then ran forward to the engine and found that the boiler had been torn from its frame, while the engine and tender had been thrown from the track.

When the explosion occurred the train had been passing the federal rice experimental station at Richvale. Mr L.L. Davis and his wife, in charge of the station, were the first outsiders to reach the wreck and were able to telephone Sacramento for help.[1558]

Damage caused by the accident was estimated at $25,632.

A board of inquiry laid the blame for the explosion on the engine crew. It concluded, "Responsibility rests with Engineer Joseph Sturm and Fireman W.T. Stone for allowing water in boiler to get at least five inches below the highest point of crown sheet, which caused the sheet to become overheated and tear away from stays." [1559]

A 52-car Wabash Railroad extra freight train was being hurried to Truesdale, Missouri, United States, 'running in the face of' passenger train No. 14, when the boiler of its engine, No. 2418, gave way about two miles short of there at 6.40 a.m. on Sunday 10th January 1932. The train was travelling at 20 to 25 miles an hour at the time. After the explosion Conductor C.L. (Buck) Merritt went forward to flag the passenger train and warn it of the accident, and the passenger train was then backed to Truesdale, the dispatcher was notified, and an ambulance arranged to be sent to the scene.

Meanwhile the other uninjured railwayman, Rear Brakeman J.E. Bassett, stayed at the scene to give what help he could to the men who had been on the engine, all of whom were badly scalded and otherwise injured. Fireman Frank Foster, who had been firing at the time, was thrown 50 feet and was the worst hurt. He died in the ambulance half an hour after the accident. Engineer Louis Zeller and Front Brakeman J.C. Christy were slightly less badly hurt. [1560] Initially hopes were entertained for their recovery but the brakeman died a few days later from his injuries. Engineer Zeller did survive but was in

---

[1555] *The Cattaraugus Republican*, Salamanca, New York, United States, 16th September 1931

[1556] *Olean Times*, Olean, New York, United States, 16th September 1931

[1557] *Middletown Times Herald*, Middletown, New York, United States, 17th September 1931

[1558] *The Daily Inter Lake*, Kalispel, Montana, United States, 26th December 1931

[1559] *Woodland Daily Democrat*, Woodland, California, United States, 30th December 1931

[1560] *Moberly Monitor Index*, Moberly, Missouri, United States, 11th January 1932 hospital for eight months. [1561]

A spectacular boiler explosion in Belgium on Friday 19th February 1932 was fortunately without serious consequences other than the damage to the locomotive itself. Société nationale des chemins de fer belges (SNCB) engine No. 905 was being prepared to work a passenger train from Weelde, Belgium, near Antwerp. Its fire had been lit and steam had been got up to the point where it was able to move under its own power to the water column, where it was to take a further supply of water and steam pressure could be brought up to the working level. At this point its boiler exploded, greatly damaging the engine but without killing or injuring anyone or causing much other material loss. It was theorised that the boiler had been weakened by corrosion at the point where repairs by riveting had been made during the wartime German occupation. [1562] (Another source gives the date of the accident as Monday 22nd February.) [1563]

The explosion of the boiler of the engine of a stationary train at Nag Hamadi, Upper Egypt, on about Saturday 4th June 1932, killed eight workmen and seriously injured another 11. [1564]

The boiler of a locomotive exploded at Oviedo, Spain, on Friday 5th January 1934, and was reportedly thrown 80 metres. The train driver was killed, the fireman and another railwayman were seriously injured, and ten passengers were slightly hurt. [1565]

As noted in the introduction, it was unusual for a boiler explosion to claim much in the way of victims other than the engine crew and anyone else who chanced to be on the engine, but the explosion of a locomotive of the Elkhorn Piney Coal Company at McDunn (Powellton), West Virginia, United States on Thursday 27th December 1934, was a notable exception. The boiler of the engine gave way as the train, which consisted of the engine and four coaches filled with miners on their way to work, was half a mile short of the mine. When the crown sheet of the locomotive gave way its boiler shot up into the air and then descended again upon the first coach of the train, in which were 53 of the 350 miners on the train. The boiler smashed the roof of the coach, and it was then filled with jets of searing steam.

Seven ambulances and many private automobiles took the injured to the little hospital at Montgomery, where the injured were housed four to a room. Seven of them had sustained fractured skulls.

Miner Charles Kitchen had been riding in the second coach and was treated for steam burns to his wrists. He spoke to a reporter. "Everyone was laughing and joking in the cars and the special had stopped to take on some men. It was 6.30 a.m. and just getting light. I heard a roar up front that shook the train, then another roar followed by groans and screams. The boiler had been thrown into the air and fell on the first car, causing another noise which sounded like a second explosion. The boiler burst and steam rushed in all

---

[1561] Ibid, 15th September 1932

[1562] http://biblio.naturalsciences.be/rbins-publications/annales-des-mines-de-belgiques/1934%20ADM/stat-3.pdf

[1563] https://www.mariusbroos.nl/Uit%20Belgie/Het%20Bels%20Lijntje.html

[1564] *The Times*, London, England, 6th June 1932

[1565] *Journal de Geneve*, Geneva, Switzerland, 6th January 1934 directions, making thick, scalding clouds as it hit the cold air.

"The first explosion ripped the front end of the first car apart, then the boiler came down in the middle of the car and crashed through it. The steam fairly whistled for a few minutes, then the noise died down and we could hear only the screams of the hurt and the cries of excitement as the men unloaded and ran forward to try to help the victims. There wasn't much to do for a lot of them. They were mangled, some beheaded. My wrists were stinging from the steam and my ankles were burned."

Early reports stated that Engineer William Blankenship, Fireman Delmar Oxley and 11 passengers (miners) had been killed and 13 miners injured. [1566] The day after the explosion the number of dead had risen to 16 and eight of the remaining injured were critically ill. [1567]

Miner Okey Litton died in Montgomery Hospital on 3rd January 1935 and was probably the last to succumb. His death brought the total number of fatalities in the disaster up to 18. [1568] Meanwhile the Interstate Commerce Commission inspector L.D. Allision had stated that "the explosion was caused by an overheated crown sheet which became too hot because of low water." [1569] The boiler had been regularly inspected and there appeared to have been no problem there. What was strange was that Engineer Blankenship was known as a "high water man," who was always anxious to keep a good amount of water in his boiler. [1570]

The boiler of Chemins de fer de Paris à Orléans et du Midi engine 141 C-623, pulling train No. 658 from Culoz to Ambérieu, exploded just beyond the station at Tenay-Hauteville, France, on 2nd August 1935. The line climbed for 24 kilometres from Culoz to La Burbanche with a maximum gradient of 12 ‰, then fell for 5 kilometres to Tenay-Hauteville, partly also at 12 ‰. According to the Flaman track recorder the train speed had fallen to 50 kilometres an hour at La Burbanche and had risen to 80 kilometres an hour at the point of the accident. The train weighed 619 tonnes.

Both driver and fireman were killed in the explosion. The boiler was carried over the telegraph wires at an estimated height of 18 metres and landed 83 metres from the point of the explosion.

The investigation into the accident blamed it on low water in the boiler and stated that about nine minutes must have elapsed between the melting of the first fusible plug and the explosion. [1571] Because both the engine crew had both lost their lives it could not be determined why they had not drawn the fire during that time and thus prevented the explosion. [1572]

Early on the morning of Friday 20th March 1936 the boiler of a Western Pacific Railroad helper locomotive exploded at Silver Zone, Nevada, United States. Brakemen C.E. Dockendorf and G.F. Howell, who had been riding in the caboose directly in front of the helper engine, were killed, and Conductor Fred Black and Special Agent O.C. Cherry,

---

[1566] *The Chronicle-Telegram*, Elyria, Ohio, United States, 27th December 1934
[1567] *The Raleigh Register*, Beckley, West Virginia, 28th December 1934
[1568] *Montgomery News*, Montgomery, West Virginia, United States, 3rd January 1935
[1569] *The Daily News*, Huntingdon, Pennsylvania, United States, 31st December 1934
[1570] *Beckley Post Herald*, Beckley, West Virginia, United States, 29th December 1934
[1571] Revue Générale des chemins de fer: Note sur une explosion de Chaudière by M. Chan
[1572] *The Railway Gazette*, London, England, 23rd August 1935

in the caboose, and Engineer J.P. Knowles and Fireman Reuben Weaver were all injured. The caboose of eastbound freight train No. 62 was derailed but the remainder of the train was not damaged. Silver Zone lies at the summit of the divide over the Toana Mountains. [1573]

About mid-morning on Tuesday 7th May 1936 a boiler explosion occurred a mile east of the station at North Alexander, New York, United States. Farmer James Belbin, who had been working in his fields beside the track, had just waved to the fireman of a Delaware Lackawanna and Western Railroad freight train when, in his words, "the locomotive blew apart." He continued, "The train was going slow when the explosion occurred. The cab and the whole thing flew up into the air over a hundred feet. I saw something white fall into the creek beside the track and I think it was the body of the fireman."

The locomotive and about ten cars of the train, carrying cattle, were thrown down an embankment. The train had been travelling at 10 or 15 miles an hour. Some of the cattle, released from the broken cars, ran about the countryside while others, which were injured, were shot by state police. The bodies of Engineer Stephen Hasfurter, Fireman Oscar Arthur, and Head Brakeman E.P. Arwine were found lying close together alongside the track. All were dead. Conductor J.P. Reidy, in the caboose at the rear of the train, was unhurt. [1574]

On Saturday 15th August 1936 the locomotive of a southbound Chicago and Eastern Illinois Railroad freight train suffered a boiler explosion at Pimento, Indiana, United States. The engine was thrown onto its side, several cars at the head of the train were derailed and 200 feet of track were torn up. Engineer Gene Vierling and Fireman Harry Irwin were killed. Conductor J. Stakeman and Brakeman J.C. Shaw were unhurt. The conductor said that the train had been moving at 45 miles an hour at the time of the explosion. The cause of the accident was not immediately clear. While the line was blocked trains were diverted over the rails of the Chicago Milwaukee St Paul and Pacific. [1575]

Low water was blamed by railroad officials for the explosion of an engine on a bridge near the Chesapeake and Ohio Railway's White Stick station, West Virginia, United States, on its Piney branch, at 4.32 a.m. on Thursday 20th August 1936. The large *Mallet* engine was running tender first in front of a second engine and a train of 97 empty cars. Conductor Blaine Simmons, who was riding on the engine, was killed instantly; Engineer Brown Nunnaly died either before or immediately after his admission to hospital and Fireman Ray Singer died a few hours after admission to hospital. [1576]

Much of the force of the blast was directed downwards to the White Stick Creek, 25 ft below, but the engine was derailed and steam and hot water which was sprayed around killed all the vegetation within a 200 ft radius. The explosion occurred a long way from the nearest telephone, and help was delayed until the crew of the other engine could propel the train back to Quinnimont.

The single track branch was stated to be one of the busiest in the area. Wrecking crews cleared the single line during the afternoon.

[1573] *Reno Evening Gazette*, Reno, Nevada, United States, 20th March 1936
[1574] *Syracuse Herald*, Syracuse, New York, United States, 7th May 1936
[1575] *Linton Daily Citizen*, Linton, Indiana, United States, 15th August 1936
[1576] *The Raleigh Register*, Beckley, West Virginia, United States, 20th August 1936

General Superintendent E.L. Bock of Huntington stated, "As far as know the explosion was caused by the engineer permitting water to get low in his boiler." [1577]

A Wabash Railroad freight train from Detroit to Chicago had just left the depot at Adrian, Michigan, United States, on the night of Sunday 3rd January 1937 when there was a terrific explosion. Brakeman M.J. Madden had just left the engine cab and was reportedly walking on a freight car halfway between the engine and the caboose at the time. He said that he ran along the tops of the freight cars to the caboose and asked Conductor J.V. Gray if the train had been carrying dynamite. No, he had been told, and the two men had hurried to the engine, where they found that Engineer Ray Mead, Fireman M.R. Johnson, and Brakeman C.E. Hudson were all dead.

The force of the explosion had torn up 250 ft of the eastbound track and a small fragment of the boiler had gone through the door of the residence of Hubert Estes, 300 yards north of the tracks, while every window of his home had been broken. Neither the locomotive nor any of the cars of the train was derailed though.

Water had just been taken at Adrian. The explosion was blamed, though, on low water in the boiler. [1578]

The boiler of the engine of the Sevillano express exploded about a kilometre beyond Tortosa Junction, Spain, on the main line from Barcelona to Valencia, on Sunday 7th February 1937. The engine was blown to pieces and two coaches were derailed and smashed. The driver, fireman and a third railwayman were killed, and 16 passengers were injured. The accident was unofficially attributed to the generally poor condition of engines in Republican Spain due to wartime conditions. [1579]

Fireman Roy Carr and Brakeman Charles Dickinson were killed when the boiler of a Baltimore and Ohio Railroad freight train exploded at about 3.14 a.m. on Tuesday 22nd June 1937 rather more than a mile east of Osgood, Indiana, United States. Engineer George Gottberg was critically injured. Eight cars at the front of the train were derailed, but Conductor W.P. Clause and Flagman G.T. Prall in the caboose at the rear of the train were unaware of anything being wrong other than that the train had stopped suddenly. Flagman Prall went back to protect the rear of the train while the conductor went forward to investigate. Two tank cars of oil and two of gasoline were involved and there was some leakage of their contents, but no fire. [1580] Engineer Gottberg died of his injuries a day or so later. [1581]

Chicago Great Western Railway engine No. 887, pulling freight train No. 65 from Oelwein to Des Moines, suffered a boiler explosion just outside of Melbourne, Iowa, United States, at about 4.30 a.m. on Friday 28th January 1938. Engineer F.A. Holmes and Fireman W.F. Krautz were killed outright, and Brakeman Tillman Hartzer was critically injured. The train had consisted of 53 cars and the engine of the 800 class, was said to have been one of the largest in use in the mid-west. Riding in the caboose were

---

[1577] *The Charleston Gazette*, Charleston, West Virginia, United States, 21st August 1936
[1578] *Defiance Crescent News*, Defiance, Ohio, United States, 4th January 1937
[1579] *The Railway Gazette*, London, England, 12th February 1937
[1580] *Seymour Daily Tribune*, Seymour, Indiana, United States, 22nd June 1937
[1581] *The Hammond Times*, Hammond, Indiana, United States, 24th June 1937

Conductor G.H. Matson and Rear Brakeman H.L. Schaler. All that they knew of the accident was the noise of the blast and the sudden stoppage of the train.

A reporter described the scene after the accident. 'The bodies of the engineer and fireman were thrown clear of the wreckage and were found on opposite sides of the track. Neither was mangled but both had been badly scalded with steam. Parts of the locomotive were strewn for a quarter of a mile over the prairie. There were no farmhouses nearby. Some of the pieces of metal were the size of a man's hand. Others were two feet square. The cab of the locomotive was demolished completely. The rear part of the boiler was left intact and came to rest on the side of the tracks, ahead of the locomotive trucks. The flues were blasted through the front end of the engine and the fore part of the boiler was reduced to small bits. After the explosion the momentum of the cars pushed the wreckage of the engine for almost a quarter of a mile. Thirteen cars were piled in one huge heap of twisted wood and metal. The right of way was torn up for a city block.' [1582]

According to the Interstate Commerce Commission investigation of the accident, 17 empty refrigerator cars were in fact derailed as a result of the explosion. The commission found that the explosion had resulted from insufficient water getting to the boiler and that a sudden inflow of cold water had then apparently caused the overheated crown sheet to "let go." Engineer H.A. Franklin stated, "It is not possible to explain why the water could have reached the low level that it did, without the notice of either the engineer or fireman, whose business it is to know at all times the level of the water in the boiler." He added that all parts and appurtenances had been found in satisfactory condition with no evidence of stoppages or failures and that there had been no neglect or mismanagement on the part of the railway company. [1583]

Four men and a teenager lost their lives as the result of the explosion of the boiler of a Northern Pacific Railway Z-6 class locomotive at Willis, Montana, United States, at about 11.30 p.m. on Thursday 7th July 1938. They were Engineer Ernest Westin, Fireman Tiny Dunlap, Brakeman Ernest Bedillion, and two transients – Ernest Nissen, of Missoula, and a 15-year-old boy, known only as Raymond and believed to be from Norfolk, Nebraska. The engine and nine freight cars were derailed by the force of the blast and the roadbed was badly damaged. The force of the explosion blew the engine boiler to the bank of the Missoula River and the engine cab ended up in the river. It was considered that the damage to the engine was such that it was beyond repair. Conductor Raymond Abbey and Brakeman H.P. Van Pelt in the caboose at the rear were not hurt and were not apparently immediately aware of what had happened, though the conductor thought that it had been an unusually hard stop.

The train was eastbound fast freight No. 602, which had left Missoula just after 7.30 p.m. There were an estimated 70 transients aboard it, mostly riding in the fifth car from the engine. Apart from the two who were killed several others of them were hurt, the worst being Charles Zuremba, of Plainfield, New Jersey, whom it was thought had broken his back. He had already been wearing a steel brace following a previous back injury. He was taken to hospital.

One of the transients, Taumo Rinne, of Virginia, Minnesota, spoke to a newspaper reporter. With his brother, Arve, and a friend, William Takala, of Angora, he had been

---

[1582] *The Oelwein Daily Register*, Oelwein, Iowa, United States, 28th January 1938

[1583] Ibid, 18th February 1938

sitting at the open door of a freight train when the blast happened. He related, "One more car and we would not be here to tell about it. After the first blast there was a series of crackling sounds as the first four cars piled up. There were 50 men sprawled over the floor of our car. They all slid forward and ended up in a great pile at the front of the car. A few of us, sitting in the doorway, just slid out and were unhurt. Some of the others were badly shaken up and bruised. But I don't think anyone in the car was badly hurt.

"When our car came to a halt, it was sidewise of the track and was in bad shape with cracks all over the side and top. We ran up forward to see what had occurred and to see if we could give any help. The engine, a twisted mass of metal, was lying half in, half out of the river. The first four cars were badly wrecked. They were all closed cars, and it was impossible for anybody to be riding in them. We saw one man lying across the rails, a big gash in his head. He must have been dead."

After a few minutes, Rinne continued, "Someone had taken charge of the situation and warned all survivors to stay clear of the front part of the train because the wreck had snapped off a couple of service poles and there was a danger of live wires."

In the confusion after the accident Rinne and his companions lost their sense of direction, boarded a westbound Chicago Milwaukee St Paul and Pacific passenger train, and ended up back in Missoula instead of Deer Lodge, their intended destination (the Milwaukee track paralleled the Northern Pacific one for some miles). "We're going to get to Deer Lodge yet," Rinne added. [1584]

At 4.30 a.m. on Thursday 13th October 1938, the boiler of a westbound Chicago Milwaukee St Paul and Pacific Railroad freight train of 99 cars travelling from Chicago to Savanna, Illinois, exploded three miles north west of Elgin, Illinois, United States. Engineer George Watson, Fireman James Mangan and Brakeman George Hachtel were killed and Brakemen Ambros Grady and Rear Brakeman Walter Darr, who had been riding in the caboose, were injured. The wreckage of the engine was strewn along the right-of-way for several hundred feet but none of the cars of the train left the track.

In a preliminary examination of the remains of the locomotive, Coroner Victor Peterson noted that the crown sheet of the engine was scored and concluded that there had been an insufficient supply of water to the boiler. He stated, "The grade between Elgin and Almora is one of the steepest in this section and the engineer may have tried to make the grade with a low supply of water, fearing that addition of cold water would lower the head of steam and result in loss of power." [1585]

An Interstate Commerce Commission inspector agreed with the coroner. He stated that the explosion had been caused by, "overheating of the crown sheet due to low water." Since all the men on the engine had been killed, "no information could be obtained with respect to difficulties that may have been encountered in the operation of the locomotive." [1586]

The boiler of engine 03 174, pulling express train D 17, exploded between Angermünde and Pasewalk, Germany, on Monday 20th March 1939, due to lack of water. The engine was derailed, the driver and fireman were killed, and two other people were

---

[1584] *The Daily Inter Lake*, Kalispel, Montana, United States, 8th July 1938

[1585] *Freeport Journal-Standard*, Freeport, Illinois, United States, 14th October 1938

[1586] *Dixon Evening Telegraph*, Dixon, Illinois, United States, 14th November 1938 injured. [1587]

Only a fortnight after the Angermünde accident the boiler of another engine pulling an express train, 02 101, gave way at Rothenstadt, Germany, in the Upper Palatinate, on Monday 3rd April 1939. Low water was again given as the cause. [1588]

A Texas and Pacific Railway football special carrying 151 fans was on its way to Sweetwater to pick up the Texas Christian University football team when it was involved in an accident on Tuesday 27th September 1939. The train was then bound for Los Angeles, California, where the team were due to play on the Friday. All went without incident until the train reached Millsap, Texas, United States, where a massive blast derailed the engine and four cars of the train. The baggage car was thrown into a field and overturned, and the sleeping and dining cars were thrown from the rails but did not overturn. Engineer Leonard Perry and Conductor Newt McNeely were both killed; the body of the engineer was found 300 feet away in a field and that of the conductor was in the baggage car. The fireman, J.B. Hammer, was badly scalded and half a dozen others were less badly hurt. [1589]

Fireman Hammer said later, "There was no warning of anything wrong when the locomotive was rocked by an explosion. I remember being covered with hot water before I lost consciousness." [1590]

The passengers were transferred to another train and continued their journey early Wednesday. [1591]

Fires from the engine set fire to the grass alongside the right-of-way and fire engines had to be sent from Mineral Wells to put out the fire. The accident also attracted a crowd of 2,000 sightseers and this led to a third death when a car containing a party of rubberneckers overturned, killing Miss Vera Smith, 17, of Fort Worth, and injuring two others in the car. [1592]

The engine of a Chicago Rock Island and Pacific Railroad local train consisting of only three cars and a caboose blew up on Christmas Day, 25th December 1939, two miles east of Choctaw, Oklahoma, United States. It had just delivered coal to the Oklahoma Gas and Electric Company's plant at nearby Harrah. Engineer Ollie Henderson and Fireman Walter Crowe were killed; Conductor J.W. Lewellyn and Brakemen Roy Eckert and Victor Godwin dived to the floor of the caboose when they heard the sound of the explosion and narrowly escaped serious injury from pieces of flying metal. Only the driving wheels and part of the frame of the engine remained on the track, but the tender and the cars behind it were not derailed. [1593]

The boiler of the locomotive of a northbound Chicago Milwaukee St Paul and Pacific Railroad freight train exploded shortly before midnight on Sunday 4th February 1940 at

[1587] *Katastrophen der deutschen Bahnen Teil II* by H.J. Ritzau, pg. 95 (in German)
[1588] Ibid.
[1589] *Denton Record-Chronicle*, Denton, Texas, United States, 27th September 1939
[1590] *Oakland Tribune*, Oakland, California, United States, 27th September 1939
[1591] *The Brownsville Herald*, Brownsville, Texas, United States, 27th September 1939
[1592] *The Daily Texan*, Austin, Texas, United States, 28th September 1939
[1593] *The Daily Ardmoreite*, Ardmore, Oklahoma, United States, 25th December 1939

New Lisbon, Wisconsin, United States. Engineer Richard Goerick and Head Brakeman Otto Winters were killed, and Fireman Alvin Ganz was seriously injured. The brakeman had been riding on the engine. The conductor and rear brakeman in the caboose were unhurt. The boiler was hurled 300ft, but neither the wheels of the engine nor any of the cars was derailed. [1594]

On Sunday 25th February 1940 the boiler of engine 56 2753 exploded at Neubeckum, Germany, in North Rhine – Westphalia. This time the cause was not low water but worn studs in the copper firebox. [1595]

The first failure of a boiler for 19 years on a British locomotive occurred on Tuesday 10th September 1940, near Carstairs, Scotland, on the London Midland and Scottish Railway. The 10 a.m. express from Glasgow Central to London Euston consisted of 16 coaches hauled by 4-6-2 *Coronation* class engine No. 6224. The driver and fireman took over at short notice, they had not worked together before, and the fireman had never fired on that class of engine. On the climb from Glasgow to Craigenhill summit the fireman failed to feed the fire sufficiently to prevent the steam pressure falling. The train managed to breast the summit but on the succeeding descent to Carstairs the engine firebox gave way. A massive escape of high-pressure steam seriously scalded the driver and fireman and the latter died of his injuries later in the day.

Employment Inspector J.L.M. Moore found that due to lack of water the crown sheet had become red hot until a small portion of it had given way, though the whole of the crown sheet was bulged. In their efforts to avoid delaying the train the enginemen had failed to use the injectors, or to keep an eye on the water gauges. [1596]

The boiler of a Southern Pacific Railroad locomotive pulling a train of 94 cars en route to the army replacement centre at Naciemiento exploded at Cooper, California, United States, on the afternoon of Saturday 3rd May 1941. Engineer Philip Jolly, Fireman Floyd McKean, Brakeman Herb Horan, and Student Brakeman Alvin Schoendahler were all killed. Nineteen cars loaded with coal, oil, lumber, and tractors were derailed and two lineside houses were damaged. A Miss Patrick, in one of the houses, was injured, as was an unidentified man found in a lettuce field 400 yards away, who was struck by flying debris and badly hurt.

The accident was witnessed by a soldier travelling in a truck on the nearby Castroville highway, who said, "The engine appeared literally to go up in smoke. The boiler simply disappeared, and parts of it flew in every direction. A piece of iron sheeting shot across the highway in front of the truck and landed in a lettuce field." [1597]

The explosion was provisionally blamed on the boiler having been low in water. [1598]

A Delaware and Hudson Railroad freight train from Mechanicville to Binghampton left the former place at 2.08 a.m. on Tuesday 15th July 1941. It stopped at the Mohawk Yard in Schenectady, where more cars were taken on and water was taken, then

[1594] *The Capital Times*, Madison, Wisconsin, United States, 5th February 1940
[1595] Wikipedia – Kesserzerknall
[1596] *Locomotive Boiler Explosions* by C.H. Hewison, pgs. 121-2
[1597] *Nevada State Journal*, Reno, Nevada, United States, 4th May 1941
[1598] *Reno Evening Gazette*, Reno, Nevada, United States, 5th May 1941

proceeded and was climbing the grade between Cobleskill and Richmondville at about 30 miles an hour. Conductor Albert Griffin and Brakemen Howard Doolittle and Rowland Keeby, riding in the caboose 45 cars back from the engine, noticed the train stop but did not know why until they went forward. They then found that the boiler of the engine had exploded and that Engineer Charles Smith, Fireman Elbert Price and Brakeman James Cleary had all been killed. The boiler of the locomotive had been blown 430 feet and other parts as much as 800 feet, with only the chassis and driving wheels remaining.

The Interstate Commerce Commission (ICC) report on the accident appeared at the end of August and blamed it on low water. Perhaps significantly, it had been reported a week before the explosion that the right-hand injector was not feeding water into the boiler. The problem was supposedly corrected.

The ICC reported, "The accident which is the subject of this report is the sixth boiler explosion to be caused by overheating of the crown sheet due to low water that has occurred in the last four months. These six accidents caused the death of 11 persons and the injury to six persons. Ten of the deaths occurred in connection with three of the six accidents in which especially large boilers, which were blown from running gears, were involved.

"The increasing number and violence of crown sheet failures should be given more thoughtful concern, instruction and discussion by those whose duties involve design, maintenance and operation of steam locomotives, in order that unnecessary loss of life, injury to persons and destruction of property may be reduced to a minimum."[1599]

On Saturday 8th March 1941 the boiler of engine 50 123 failed at Groß Strehlitz, Germany, in Upper Silesia (since 1945 Strzelce Opolskie in Poland). Low water was given as the cause. [1600]

Engine 50 846, pulling a freight train, suffered a failure of its boiler on Wednesday 23rd July 1941 at Kenzingen, Germany. A failed weld in the boiler had caused it. [1601]

New York Chicago and St Louis Railroad (Nickel Plate) manifest freight train Second No. 49 from Sandusky, Ohio, to St Louis, Missouri, was a mile east of Red Key, Indiana, United States, early on Tuesday 31st March 1942, when its engine exploded. Fireman Walter Hershberger and Brakeman J.E. Schletter were killed instantly and Engineer E.L. Benner, who was blown 100 ft, died shortly after he was found by rescuers. Parts of the engine were blown over a wide area. Three empty tank cars were blown sideways across the track and a stock car also ended up at right angles across the right of way.

The train had taken water at Celina, Ohio, about 30 miles short of Red Key, but the coroner's report blamed the explosion on low water. [1602]

The leading (pilot) engine of a double-headed Atlantic Coast Line Railroad westbound troop train, consisting of 20 cars carrying 1,500 men, exploded rather more than two miles out of Bainbridge, Georgia, United States, at about 8.30 p.m. on Thursday 7th May 1942. Engineer J.W. Withington, Fireman Gordon Mills and Brakeman C.L. Nicholls

---

[1599] *The Cobleskill Index*, Cobleskill, New York, United States, 28th August 1941

[1600] Wikipedia – Kesserzerknall

[1601] Ibid.

[1602] *The Lima News*, Lima, Ohio, United States, 31st March 1942 were killed. Both locomotives, two baggage cars and a Pullman car were derailed. Engineer E.D. Watts and Fireman R.H. Pickett on the train engine were pinned beneath wreckage but escaped with slight injuries. Two passengers were detained in hospital and several others treated and released. The telegraph wires were blown down and commercial telephone lines had to be resorted to. [1603] [1604]

Engineer J.M. Carmichael and Fireman C.B. Butts were severely injured when the boiler of their Baltimore and Ohio Railroad engine, which was running light, exploded three miles east of Ravenna, Ohio, United States on Monday 14th September 1942. Their condition was later reported to be 'fair' in Ravenna Hospital. The engine had been en route from New Castle to Akron. [1605]

Another Baltimore and Ohio Railroad engine blew up at 1.20 a.m. on Saturday 19th September 1942 at Listie, Pennsylvania, United States, on the company's Somerset and Cambria branch. Engine No. 4240 was the second of two locomotives pulling a string of coal cars. Its engineer, John Berkley, was killed outright and his fireman, Clarence Ohler, was fatally injured and died in hospital. Also injured were Engineer Jacob Cook and Fireman Burdine Wilson on the pilot engine. A resident told a reporter that he thought that the world was coming to an end when the boiler crashed near his house. Another said that he thought that it was an air raid. [1606]

The inquest on the driver and fireman at the end of September returned a verdict of 'accidental death'. The chief witness at the inquest was Baltimore and Ohio General Foreman Thomas Cunningham, who stated that the boiler was examined monthly, and more thoroughly annually, and had been inspected on 1st September. At the time of the accident the two locomotives had been pulling a train of 5,875 tons, with two further engines at the rear of the train. The weight of the train was well within the capacity of four engines. He had examined the boiler again after the accident and had found no fault with it. He attributed the accident to low water. [1607]

Northbound Seaboard Air Line Railroad passenger train No. 4 from Atlanta, Georgia, to Raleigh, North Carolina, was between Hoffman and Marston, North Carolina, United States, on Tuesday 13th October 1942 when the boiler of its engine exploded at 11.47 a.m. Engineer J.A. Robertson and Fireman J.L. Stafford (or Trafford) were killed, and it was feared that the body of a third railwayman might be buried in the wreckage. Baggagemen J.A. Kennedy and E.A. Mason and a soldier, R. D. Spearn were slightly hurt. Five express and one mail car were described as 'telescoped like an accordion' and three hundred feet of track was torn up.

A former newspaper reporter who chanced to be a passenger on the train ran half a mile to a telephone to report the accident to the Associated Press. He said, "I was riding about seven cars back, when the train, which was running about 45 miles an hour, I suppose, ground to a sudden stop. Hastily I got off with a number of passengers, thinking that there was some minor trouble ahead. The side of the track was shrouded in steam

---

[1603] *The Decatur Daily*, Decatur, Alabama, United States, 8th May 1942
[1604] *The Lee County Journal*, Leesburg, Georgia, United States, 8th May 1942
[1605] *The Evening Independent*, Massillon, Ohio, United States, 15th September 1945
[1606] *Somerset Daily American*, Somerset, Pennsylvania, United States, 19th September 1942
[1607] Ibid, 29th September 1942

from the engine, so heavy that it looked as if a huge flame thrower had been at work.

"I made my way up the tracks slowly as it was hard to see because of the density of the steam vapour. I found the engine lying in a field on its side, the coal car was crumpled, and three or four cars were crushed in like an accordion. I heard a trainman say that at least three of the train crew were killed, and two of them were the engineer and the fireman. Without checking any further I dashed for a telephone to call the Associated Press." [1608]

On Saturday 14th November 1942, the boiler of the engine of a westbound Pennsylvania Railroad westbound oil special gave way just east of Cresson, Pennsylvania, United States, at 8.15 a.m. Engineer A.H. Schroddell and Brakeman Joseph Ward were killed, Fireman John Hogan was critically injured, Conductor Edgar Baird was less badly hurt, and Flagman W.R. Bressler escaped injury. A number of the 86 empty tank cars of the train were thrown from the track. The explosion occurred near OM tower, two fifths of a mile east of Cresson, while the train was travelling at speed. The force of the explosion demolished and set fire to a dwelling at the top of the cutting and slightly injured a man and a girl. The train was bound from Altoona to Conway. [1609]

The side wall of the firebox of engine No. 92.576 was forced inwards between Danzig (now Gdansk, Poland) and Saspe (now Zaspe, Poland) Germany, on Saturday 13th March 1943. The failure was blamed on worn side studs. Then on Wednesday 14th July 1943 the firebox of engine No. 5043 (a French engine from Alsace-Lorraine) was ripped open at Falk-Hargarten, Germany. Again, worn side studs were blamed for the accident. [1610]

Four different investigations were opened into a boiler explosion which wrecked the crack *Twentieth Century Limited* express of the New York Central Railroad at 4.34 a.m. on Tuesday 7th September 1943 at Canastota, New York, United States. In addition to the internal inquiry by the railway company the State police, the Interstate Commerce Commission (ICC) and the Federal Bureau of Investigation were involved. Engineer George Pierce, Fireman John Larsen and Travelling Fireman Charles Wriker were all killed, and 29 people, railwaymen, mail clerks and passengers, were injured. The eastbound (Chicago – New York) train was travelling at speed when the explosion happened, and it continued down the line for half a mile before coming to a stand. The remains of the engine and tender and the leading mail car then toppled over, the leading passenger car fell down the embankment and the nine following cars were derailed. The five rear vehicles stayed on the track. Some of the passengers were thrown from their berths but none was hurt, an escape attributed by railwaymen to the all-steel rolling stock of the train. As might be expected most of the passengers, many of them returning home after the Labour Day holiday, were asleep.

According to newspaper reports the passengers were warm in their praise for the railwaymen. Thus, 'Miss Diana Perry, a singer returning from San Francisco, said, "There was no panic, despite the wreckage. All the passengers were quiet. Too much praise can't be given the train crew. They did an excellent job."

'David J. Long, Navy civil consultant, of the Bronx, said, "The train crew handled the

---

[1608] *Gastonia Daily Gazette*, Gastonia, North Carolina, United States, 13th October 1942
[1609] *The Altoona Mirror*, Altoona, Pennsylvania, United States, 14th November 1942
[1610] Wikipedia – Kesselzerknall

situation – and it was a bad one – wonderfully."

'L.C. Strite, Chicago machinery salesman, told how the car he occupied rolled down an embankment into a marshy field, saying, "and it was lucky for us that marsh was there," to reduce the impact and to hold the car upright.'

The line was blocked for eight hours, reopening shortly after noon. The undamaged cars of the train arrived at New York Grand Central Terminal at 4.38 p.m. [1611]

The crew which had worked the locomotive from Buffalo to Syracuse testified to the ICC that they had had some difficulty with a water pump connection but had fed water to the boiler with the injector. The ICC concluded that the explosion had resulted from overheating, due to low water, and recommended that the New York Central Company provide crown sheet protection against low water by the use of 'suitable devices.' [1612]

On Wednesday 17th November 1943 the boiler of engine 50 3158, in charge of a transfer freight train, failed at Crange, Germany, near Herne. [1613]

Three boiler explosions occurred in England in nine months in 1943 and 1944, all to 2-8-0 tender goods locomotives imported from the United States for use in Europe post invasion but which up until that time were used on English railways. They all arose from the same cause, the unfamiliarity of English engine crews with the differently designed American water gauges. The first happened on the Great Western Railway at Honeybourne on the same day as the Crange explosion in Germany, Wednesday 17th November 1943. The train involved was the 2.35 p.m. working from Banbury to Margam in South Wales, via Leamington and Stratford, with engine No. 2403. As the train was approaching Honeybourne East Junction at 11.55 p.m. the engine firebox crown collapsed. Driver Arthur Hill was lucky to escape with only slight scalds, but Fireman John Sirrell was badly scalded by the escape of steam. He was treated on the spot by Signalman C. Taylor, who had nine years' ambulance training, but died a short time later in Evesham Hospital. Inspection of the engine afterwards found that the steam valve spindle was slightly bent. This meant that the steam valve had not been opened fully when the driver believed that it had been, due to its being held by the bent spindle, and as a result a false reading of the level of water in the boiler had been given. [1614] [1615]

Exactly a week after the Crange and Honeybourne accidents, on Wednesday 24th November 1943, the boiler of engine 17 269, pulling special leave train for front line troops SF 2181, failed at Torgau, Germany. Lack of water was blamed. [1616]

The third boiler explosion in Germany in quick succession occurred at Heydebreck, when the boiler of engine 44 1639, hauling freight train G 672, failed on Monday 6th December 1943. Low water was again found to be responsible. [1617]

1611 *The Troy Record*, Troy, New York, 8th September 1943
1612 Ibid, 6th October 1943
1613 Wikipedia – Kesselzerknall
1614 *Locomotive Boiler Explosions* by C.H. Hewison, pgs. 123-6
1615 Ministry of War Transport report
1616 Wikipedia – Kesselzerknall
1617 Ibid.

A Louisville and Nashville Railroad freight train was just taking water at Letohatchie, Alabama, United States, at 8 a.m. on Monday 3rd January 1944 when the boiler of its engine exploded with terrific force, killing Engineer Morgan Collier. Fireman W.W. Burch was seriously injured, and Station Agent A.W. Davis and Porter Henry Farrier were slightly hurt. A witness remarked that the explosion had sounded "like a whole lot of dynamite going off." It was estimated that the boiler was lifted 75 ft into the air, and it landed on a concrete road overbridge 150 ft away, causing it to cave in. In addition, the station depot was set on fire and badly damaged and power lines were brought down. Trainmen attributed the blast to the running of cold water into an almost empty boiler. The tubes of the wrecked boiler were reportedly twisted like pieces of boiled spaghetti. [1618]

At 12.40 a.m. on 12th January 1944 a London and North Eastern Railway freight train from Ipswich to Whitemoor was nearing the station at Thurston, England, when the crown sheet of American engine No. 2363 gave way. The explosion was more severe than it had been at Honeybourne the previous November, though the consequences were slightly less serious. Driver W. Nicholls suffered a broken thigh from the blowing off of the firebox door, while Fireman H.W. Leek was blown off the engine altogether, escaping with burns and bad bruises. The train ran for about a mile without anyone in charge of the engine, till a rising gradient brought it to a halt.

Employment Inspector J.L.M. Moore found that the water gauge steam valve had been extremely stiff so that it needed the full force of two hands to fully open it, and the engine crew had clearly believed it to be fully open when it was not. He blamed the driver for failing to check that the steam valve was fully open or to use the test cocks to check the water levels. [1619]

The first section of *The Texan* passenger train of the Texas and Pacific Railway was wrecked at 8.05 a.m. on Monday 31st January 1944, a mile east of Forney, Texas, United States. This was a long train, consisting of an engine, a wooden baggage car, three Pullman cars, a kitchen car, six Pullman cars, a kitchen car, three Pullman cars and a caboose, in charge of Conductor A.R. Hayes with Engineer B.H. Williford and Fireman Gerald Puckett. Engineer Williford, Fireman Puckett and Brakeman W.V. Trumble, who was riding on the engine, were seriously injured and Fireman Puckett died in hospital two days after the accident. [1620] When the explosion occurred the train was on an 8 ft embankment and the boiler was thrown down it. The baggage car, loaded with barracks bags and officers rolls, was thrown down the embankment on the other side, the leading Pullman car went down the embankment in front of the engine and the next two Pullman cars fell onto the baggage car, crushing it. Only eight soldiers suffered bruises, the remainder escaping with a shaking, due it was said to the strength of the modern rolling stock. [1621]

On about 20th March 1944 the engine of a workman's train reportedly carrying hundreds of passengers and running between Suez and Port Teflik, Egypt, exploded. The train was derailed, and 21 persons were killed, 28 seriously injured and over 100 slightly

---

1618 *The Decatur Daily*, Decatur, Alabama, United States, 3rd January 1944
1619 *Locomotive Boiler Explosions* by C.H. Hewison, pgs. 126-8
1620 *The Brownsville Herald*, Brownsville, Texas, United States, 3rd February 1944
1621 *The Kauffman Herald*, Kauffman, Texas, United States, 3rd February 1944

hurt. The initial investigation into the accident attributed it to sabotage but a boiler explosion seems a more likely possibility. [1622]

On Wednesday 30th August 1944, the 2.45 a.m. London and North Eastern Railway freight train from Neasden to Woodford consisted of American engine No. 1707, 19 waggons and a brake van. It was passing through South Harrow Tunnel at 3.45 a.m. when the firebox crown collapsed. Driver S. Dyer and Fireman J. Horton were both severely scalded; the fireman died in a few hours and the driver four days later. The train came to a stand just beyond the tunnel. The guard thought that the noise of the explosion was the report of a bomb falling nearby and did not realise what had happened until the train came to a stand and he went forward to the engine.

It was clear that the accident had resulted from low water, but it was not clear how this had happened. The water gauge steam valve had been fully open. The most likely cause was thought to be that in the darkness the driver and fireman had read the water gauge incorrectly, thinking it to be full when it was empty. That the enginemen had had no idea that anything was wrong was indicated by an ante-mortem remark by Driver Dyer about a burst water tube. [1623] [1624]

The boiler of a Baltimore and Ohio Railroad yard engine exploded at Streets Road Crossing, in Baldwin, Pennsylvania, United States, on Sunday 10th December 1944, while the engine was pushing several cars up an incline. The explosion reportedly shot live steam from the boiler and flames from the firebox 100 ft, shattering the windows of nearby buildings. Engineer Thomas Quail and Flagman William Wall were critically injured and Fireman Charles Scrofani, Conductor Aaron Owen, and Mrs Mary Zabarchak, who had been sat in a car at the level crossing, were also hurt. The conductor and flagman were injured when the caboose was damaged in the explosion. Mrs Zabarchak was scalded by escaping steam. Fire broke out in the wreckage and three fire brigades attended. [1625]

The same day three railwaymen died in the explosion of a Southern Railway locomotive hauling a freight train from Greenville to Atlanta near Lula, Georgia, United States. Engineer J.G. Elliott, Fireman J.R. McCombs and Head Brakeman F.M. Rooks all lost their lives. Eighteen cars of the train were derailed and ten of them burned. Traffic was diverted over Seaboard Air Line tracks while the mess was being cleared up. [1626]

The boiler of the engine pulling the eastbound *Pacemaker,* the premier passenger train of the New York Central Railroad, which competed with the Pennsylvania Railroad's *Trail Blazer*, exploded at Elkhart, Indiana, United States, on Wednesday 27th December 1944. Engineer R.F. Donnell was killed outright, and Fireman John Wongan died in hospital the next day. None of the train passengers was hurt. [1627] [1628]

[1622] *L'Italie Nouvelle*, Paris, France, 22nd March 1944

[1623] Ministry of War Transport report

[1624] *Locomotive Boiler Explosions* by C.H. Hewison, pg. 128

[1625] *Huntingdon Daily News*, Huntingdon, Pennsylvania, United States, 11th December 1944

[1626] The *Gastonia Daily Gazette*, Gastonia, North Carolina, United States, 12th December 1944

[1627] *The Vidette Messenger*, Valparaiso, Indiana, United States, 27th December 1944

[1628] The *Linton Daily Citizen*, Linton, Indiana, United States, 29th December 1944

As was usually the case, the engine crew were the only victims of a boiler explosion near the station at Barnesville, Maryland, United States, early on Wednesday 10th January 1945. The Baltimore and Ohio Railroad engine which failed was hauling a passenger train of 13 cars, carrying 'largely military personnel,' which was apparently a troop train bound for Fort Meade, a recruit training centre. Only the leading car was derailed and none of the passengers was more than shaken up. The locomotive, however, was reportedly 'entirely destroyed' and Engineer Lloyd Frederick and Fireman H.P. Brown lost their lives. Their bodies were found lying beside the tracks. The passengers remained on the train after the accident and its coaches were eventually taken forward by a replacement engine. [1629]

On Monday 23rd April 1945 Southern Railway *Lord Nelson* class No. 854 set off from Bournemouth Central with the 14 coaches of the 11.02 a.m. express train to London Waterloo. It had only covered 12 miles and was between Hinton Admiral and New Milton, England, when its firebox gave way. The violent burst of steam hurled the fireman into the tender with scalds to which he succumbed three days later. The driver scrambled onto the running plate, from where he was able to apply the brakes and stop the train. He escaped serious injury.

It transpired upon investigation that another driver and fireman had earlier taken the engine light to Wimborne, from where it had returned with coaches for the Waterloo train. That the top of the water could not be seen in either of the two water gauges was observed by the first driver, but he had accepted the word of his fireman that the water must be out of sight at the top. When the fresh crew had taken over at 11 a.m. they had assumed the same thing and had not bothered to check the water level by testing one of the gauges. It was clear that in fact the water level had been below the bottom of the gauge glasses all morning, leading to the accident. [1630]

On Saturday 8th September 1945 the boiler of engine 41 309, hauling through fast freight train De 5043, exploded between Garßen and Eschede, Germany, on the line between Hamburg and Hannover. Three people lost their lives, probably the driver, fireman and a third person on the footplate of the engine. [1631]

Engine No. D52-209 suffered a boiler explosion as it was passing through Samegai station on the Tokaido main line in Shiga Prefecture, Japan, at an unstated date in October 1945. [1632]

The boiler of engine 50 2764, pulling train 93 249, exploded at Königsmoor, Germany, on the line between Bremen and Hamburg, on Wednesday 5th December 1945. Three people, presumably the engine crew, died. [1633]

[1629] *The Frederick Post*, Frederick, Maryland, United States, 11th January 1945

[1630] *Locomotive Boiler Explosions* by C.H. Hewison, pgs. 128-30

[1631] *Die Katastrophenszene der Gegenwart* by Hans Joachim Ritzau and Jürgen Hörstel, pg. 12 (in German)

[1632] Webpage – Japanese Railway Safety and the Technology of the Day by Masao Saito

[1633] *Die Katastrophenszene der Gegenwart* by Hans Joachim Ritzau and Jürgen Hörstel, pg. 12 (in German)

On the evening of Thursday 20th December 1945 a northbound St Louis Southwestern Railway (Cotton Belt) train was in trouble three miles from Weiner, Arkansas, United States. The boiler of the engine exploded, and ten of the 53 cars of the train were telescoped. For once there were no deaths but Engineer J.R. Holt, Fireman R.L. Hortenber and Brakeman Harvey Hamm were all injured. The brakeman was pinned under the engine for almost ten hours before he could be released. [1634]

Engine No. D52-371 suffered a boiler explosion between Yoshinaga and Mitsuishi on the San'yo main line in Okayama Prefecture, Japan, at an unstated date in December 1945. Like the Samegai accident in October the explosion was attributed to bad workmanship, using poor quality metal and welding rods. A third explosion from the same cause affected wartime austerity class engine No. D52-82 as it was passing through Mantomi station on the San'yo main line in Okayama Prefecture at an unstated date in 1945. [1635]

[1634] *Moberly Monitor-Index*, Moberly, Missouri, United States, 21st December 1945

[1635] Webpage – Japanese Railway Safety and the Technology of the Day by Masao Saito

CHAPTER 17

VANDALISM

'Accidents' due to vandalism and sabotage have always been a problem for the railways. They are obviously very closely related, inasmuch as both represent deliberate attempts to cause trains to crash, as opposed to accidents caused through non-deliberate carelessness, negligence or purely accidentally. Here they will, however, be treated separately, as they are slightly different. This separation is perhaps a little artificial, and if the perpetrator of a given act is not caught or his or her motives not sufficiently obvious, it may be difficult to decide which it was, but it has nevertheless been felt that the distinction is worth making.

Vandalism is usually, though not invariably, carried out by either children or young adults and its results are most commonly, though not always, less serious than those following from acts of sabotage. The motives for train accidents resulting from vandalism are usually either simple curiosity, in the case of younger children – 'I wanted to see what would happen' – or in the case of older children and teenagers, a reflection of general anti-social behaviour; so that such individuals who attempt to derail trains also carry out other acts of 'mindless vandalism', and perhaps also commit burglaries or carry out arson attacks.

Sabotage on the other hand is usually undertaken by adults, who employ methods more likely to cause death and injury, so that while a vandal might for example just place an object upon the track or jam a stone into a set of points, a saboteur would be more likely to employ methods such as blowing up a section of track using explosives.

A Southern Railway passenger train from Goldsboro to Asheville struck three large rocks which had been placed on the track near the station at Drexel, North Carolina, United States, on Friday 17th April 1931. The engine, tender, and three cars of the train left the track and Engineer P.I. McAllster, Fireman B.W. Chaney, express messenger C.R. McLean and news butcher J.B. Honeycutt were somewhat injured. Passengers in the third car, which was only slightly derailed, were jolted. The vehicles in front of that car overturned in a cutting and completely blocked the line. The passengers were transferred to another train and trains were diverted until the line could be cleared.

Six-year-old Charles Whitener Jr. was found to have placed the rocks on the line. He said afterwards, "Yes, I wanted to wreck the train, and I didn't care if some people were killed." [1636]

Representatives of the railway stated that they had no intention of prosecuting the child and at a hearing the following Monday Judge J.F. Bowers placed the boy in the custody of his grandfather on condition that he be taken to hospital to have his adenoids and tonsils removed. The boy was described as being very small and under-developed. [1637]

Football fans returning home from the 1931 F.A. Cup Final (in which West Bromwich Albion beat Birmingham City 2-1) were somewhat delayed early on Sunday 26th April 1931 at Rugby after the driver of an up London Midland and Scottish Railway goods train from Liverpool to London reported at 1.25 a.m. that he had run into an obstruction. A light engine was despatched to examine the line, accompanied by an inspector and a shunter. A mile south of the station at Brinklow, England, they found three iron chairs and some wooden posts placed on the up fast line. The objects were removed without difficulty and traffic was resumed. Forty minutes before the goods train had struck the items the line had been clear. 'The police are investigating the matter.' [1638]

An act of vandalism at Lockland, Ohio, United States, on Friday 15th May 1931 resulted in the death of a woman resident of a lineside house and was the subject of a full Interstate Commerce Commission (ICC) inquiry.

Cleveland Cincinnati Chicago and St Louis Railway (Big Four) eastbound passenger train No. 124 from Cincinnati to Cleveland consisted of engine No. 6521, a baggage car, two coaches, five Pullman sleeping cars and a dining car in charge of Conductor Thomas, with Engineer Clarence Lee and Fireman Fred Heard. It had left Cincinnati on time at 5.30 p.m., passed Ivorydale Junction, about 4¾ miles west of Lockland, at 5.56 p.m., six minutes late, and was travelling at 40 to 45 miles an hour when it was derailed at about 6.01 p.m. The engine, tender and leading three vehicles of the train were wholly or partly derailed, and the engine struck a two-storey frame house, completely demolishing its summer kitchen. [1639] Mrs Rose Earhart, 76, was buried in the debris of her kitchen and grievously injured. She was extricated and rushed by car to Cleveland General Hospital, where she died a short time later. Her husband Frank, Engineer Lee, Fireman Heard, Baggagemaster James Salrey and newsagent Leon Leonard were all injured. [1640]

Engineer Lee told the inquiry that he thought it was the trailer truck (bogie) of his engine which had derailed first, otherwise it would not have travelled so far as it had before overturning, though he could not be sure.

Fireman Heard at first agreed with his engineer, then later changed his mind, and said that he thought that the front truck had derailed first.

Conductor Thomas said that the brakes of his train had been tested and were satisfactory before the accident. Afterwards he could find no defects in the train which might have caused the accident.

Baggageman Salrey, Brakeman Herrigel and Flagman Lyons put the speed of the train at 40 miles an hour. The flagman thought that the train had struck something placed on

---

[1636] *Burlington Daily Times*, Burlington, North Carolina, United States, 18th April 1931
[1637] Ibid, 20th April 1931
[1638] *The Times*, London, England, 27th April 1931
[1639] Interstate Commerce Commission report
[1640] *The Hamilton Daily News*, Hamilton, Ohio, United States, 16th May 1931

the rail.

Riding in the second car of the train were Master Mechanic Karibo and Assistant Master Mechanic Mellon. They said that they had examined the track after the derailment and had found a crushed piece of bone, which they did not think, however, had caused the derailment. A boy named Sharpe had then joined them and had pointed out a spike lying beside the right of way fence, which was crushed and damaged. They were in no doubt that the bone and spike had been placed on the rail and that the spike had caused the derailment. About an hour before the accident Assistant Mechanic Mellon had seen three boys near the track, and he said that he had immediately connected their presence with the derailment.

The three boys were brothers Walter and Alfred Humphrey, aged six and 12, and Voyd Hasty, aged seven. Voyd said that the younger Humphreys boy had twice placed the objects on the rails, to see if a train would flatten them. Each time he had kicked them off. He had then gone off to play with another boy and had not seen them put on a line a third time. [1641]

In the racist language of the time, the derailment of a goods train on Sunday 20th September 1931 five miles from Warmbaths, South Africa, was ascribed to the action of 'three piccanins' aged 10, 11 and 12, who had tampered with a set of points, diverting a goods train into a dead-end siding. It ran off the end of the line and there was considerable damage to rolling stock. The engine overturned but the driver and fireman had miraculous escapes, while the guard suffered a severe shaking. Police dogs were able to track down the offenders. [1642] In an article on such acts of vandalism in South Africa written earlier in the year, probably by the same person, a correspondent to *The Railway Gazette* had commented, 'In the large majority of cases the culprits are native piccaninnies and Indian youths, and their excuse is invariably that they were curious to see what would happen. Cuts with the cane are administered as punishment but it does not deter others from committing like offences, notwithstanding the prominence given to the matter in the press.' [1643]

A New York Central Railroad freight train was derailed at Mapes, Pennsylvania, United States, on Tuesday 29th September 1931, fatally injuring its fireman and a brakeman.

Westbound freight train Extra 5108 consisted of engine No. 5108, 30 empty cars and a caboose, in charge of Conductor White with Engineer William Hudson and Fireman Arthur Kinkald. It had left Mill Hall, just over 10 miles east of Mapes, at 4.38 p.m., and moving at 30 or 35 miles an hour left the track at the switch points at the east end of the passing track at Mapes at about 5.15 p.m. It was partly derailed at the switch, collided with a flat car loaded with rails on the passing track, then with some stored cars which were derailed and demolished, while the engine was badly damaged.

Engineer Hudson told the Interstate Commerce Commission (ICC) inquiry that approaching Mapes he had seen from the target that there was something wrong with the switch. He had at once applied the air brakes in emergency, but his engine had then been only two or three lengths from the switch. He was reportedly blown through the back of

[1641] Interstate Commerce Commission report

[1642] *The Railway Gazette*, London, England, 6th November 1931

[1643] Ibid, 5th June 1931

the locomotive and into an empty car. Engineer Hudson said that afterwards he had examined the switch and found it set for the siding with the lock lying on the ground. No one had been near it. He said that had he been looking ahead as he approached the switch, he might have been able to pull up, but he had been looking back to check the running of his freshly reconditioned engine.

Conductor White stated that he had also seen the open switch and discarded lock after the derailment, as did Brakeman Miller.

Brakeman Gottschall stated that he had gone back to flag immediately after the accident and had not seen the switch for several hours.

Brakeman Johnson, of eastbound Extra 3888, testified that his train had deposited a car on the passing track at about 11 a.m. He had locked the switch and pulled it to check, as he had believed, that it was secure, although he admitted that it might not have been. He had seen no one in the vicinity.

Conductor Fries of train 3888 said that he had looked at the switch from his caboose as his train departed and saw the lock lying in the staple, but he could not say definitely that it had been locked.

The next morning two boys, Leonard Hahn, 11, and his brother Arthur, 9, who lived close to the scene, confessed that they had been sent to the section house for water because the pump at the Hahn home was broken. Proceeding to the section house they had placed two stones on the track to see if a train would smash them. On their way back they had found that the switch was not locked. Arthur, wanting to see how it worked, had taken the lock out of the staple, and the two of them had raised the lever but had not been able to throw the switch completely over. Hearing a train coming they had tried to close the switch but had not been able to do so and had continued home, reaching there just as the train was wrecked. Leonard added that he did not possess a key to the switch lock.

The ICC concluded that Brakeman Johnson clearly had not properly locked the switch or else the boys could not have opened it. [1644]

A boy aged 12 appeared at East Grinstead Magistrates' Court on Monday 4th April 1932 charged with having placed large stones on the metals at East Grinstead, England, with a tree branch added to keep them in position. A Southern Railway passenger train passed over them and although not derailed, it rocked. The culprit told police that he had committed the act 'because he had read about it.' He was sentenced to three stokes of the birch and six months' probation. [1645]

The derailment of a San Antonio bound freight and sheep train at Sabinal, Texas, United States, at about 11.35 a.m. on Sunday 24th April 1932 bore similarities to that at Loveland, Ohio, the year before. The Southern Pacific Company train was 'travelling at a fair speed' when it derailed close to Sabinal station, the engine and seven cars leading the rails. Three of the derailed vehicles were refrigerator cars, two were empty and two were stock cars. None of the train crew was hurt but 60 sheep, which had been loaded at Spofforth, were killed. The line was blocked by the accident and east and westbound afternoon passenger trains exchanged passengers and mails. [1646]

After the derailment Deputy Sheriff Ware found an iron spike lying on the track

---

[1644] Interstate Commerce Commission report

[1645] *The Times*, London, England, 5th April 1932

[1646] *Del Rio Evening News*, Del Rio, Texas, United States, 25th April 1932 which had been flattened by the locomotive wheels. Several boys had been seen playing on the track a short time before. [1647]

An Interstate Railroad freight train was derailed near Josephine, Virginia, United States, on the evening of Friday 1st December 1932. One railwayman lost his life and three were injured.

Eastbound freight train Extra No. 20 consisted of 2-8-8-2 engine No. 20, 29 loaded coal cars, 2-8-0 pusher engine No. 6 and a caboose, in charge of Conductor Collier, with Engineer John Jervis and Fireman Frank Moneyman on the front engine and Engineer Stout and Fireman Smith on the pusher engine. The train was derailed at a set of facing points. The front engine and tender were derailed, the leading four coal cars rammed the tender, the seventh car was derailed and the eleventh and twelfth cars telescoped one another. Fireman Moneyman was killed.

Engineer Jervis told the Interstate Commerce Commission inquiry that as his train had approached the switch at about 20 miles an hour he had been able to see the switch lamp showing white, indicating that the switch was closed. The engine brakes had been tested and were working properly and the engine headlight had been burning brightly. As his engine had reached the switch there had been an unusual metallic noise, he had seen fire flying from the front of his engine and had closed the throttle and applied the brakes in emergency. He had already passed the accident point four times earlier that day, twice in each direction.

Conductor Collier and Engineer Smith testified that they had examined the switch immediately after the derailment. The points were partly open, the target was showing red, the switch lamp was not burning, the switch lock was unlocked and damaged as if it had been struck by a rock, and the switch lever was partly thrown. No train had used the switch that day. The conductor added that two or three weeks earlier he had found indications of tampering with a switch at Dorchester Junction, less than half a mile from Josephine.

Superintendent McKenzie, who was at the scene soon afterwards, agreed that the switch had been tampered with. He said that the switch lamp had not been burning but had been still warm. There was no indication of dragging equipment. It was possible that the switch had been cocked open just enough for a white to still have been shown to the engineer, and then been extinguished by the derailment.

Master Mechanic Smith said he had found no defect with the engine.

Roadmaster Wright agreed that the switch had been tampered with.

Section Foreman McCracken stated that four years previously the switch had been interfered with. More recently the lenses of the switch lamp had been broken and oil had been stolen from it.

Eight boys were initially arrested and three of them, all under 10 years of age, were temporarily taken into custody. One of them admitted under interrogation that some of the others had interfered with the switch, cocking it and placing a piece of iron between the switch rail and the stock rail. No reason for doing so was given. The identity of the culprit was not immediately determined. [1648] [1649]

[1647] *The San Antonio Light*, San Antonio, Texas, United States, 25th April 1932

[1648] Interstate Commerce Commission report

[1649] *Bluefield Daily Telegraph*, Bluefield, West Virginia, United States, 2nd and 3rd December 1932

It quickly became clear that the derailment of a Wabash Railroad passenger train at Jacksonville, Illinois, United States on the evening of Wednesday 28th December 1932 was the result of vandalism, when it was found that the lock of a switch had been sawn through, the points opened, and the red warning light of the target removed.

Eastbound passenger train No. 2 consisted of engine No. 692, a combination mail and baggage car, a dining car, a chair car, and a Pullman car, in charge of Conductor Redman, with Engineer John Rapp and Fireman H.L. Meyer. It had passed Chapin, 10 miles from Jacksonville, at 7.19 p.m., six minutes late, and was travelling at 45 to 50 miles an hour when it was diverted onto the spur track at about 7.28 p.m. Standing on the spur track was an oil tank car with which the engine collided, overturning both the tank car and the engine and tender, while the three leading vehicles of the train were derailed. The contents of the tank car were splashed over the engine and mail and baggage car and ignited. Engineer Rapp and Fireman Meyer were killed and a brakeman, an express messenger, a cook and a waiter were injured. The passengers were severely shaken. [1650]

Conductor Redman told the Interstate Commerce Commission investigators that the brakes of his train had been tested and were working properly. Approaching Jacksonville the whistle had been sounded for a highway crossing, the brakes had been applied in emergency and he had felt the train swerve as it entered the spur track, immediately followed by the severe jolts of the derailment and collision. Having attended to his passengers after the accident he had then examined the switch and found it set for the spur track, with the hasp of the lock sawn off, and the switch lamp extinguished.

Flagman Fleming stated that he had gone past the switch when he had gone back to flag. He confirmed the observations of his conductor.

Section Foreman Gregory described seeing saw marks on the switch lock. He said that he normally cleaned and filled the switch lamp every five days and had done so on 24th December. When he had checked it after the derailment, it had been half full of oil, and he had been able to relight it without adjusting the wick. There had been sand on the rails for 200 feet west of the switch. There was nothing in the condition of the switch or track which could have caused the derailment.

About three hours after the derailment an iron bar and a nut were found on the rails about 1,200 feet west of the switch. [1651] About three weeks after the derailment 16-year-old Charles Willis, an orphan, was arrested following a tip-off that he might have been responsible. He confessed to having sawn through the lock and turned the switch "to see what would happen." He said that he had then been unable to restore it and had left the scene. He was charged with murder. It had earlier been theorised that an act of sabotage with the idea of robbing the train was involved. [1652]

Willis pleaded guilty to the vandalism in March 1933. It emerged that previously he had committed acts of arson which he said he had carried out "for the thrill of seeing the fire." He had been judged mentally deficient and committed to the Lincoln State School and Colony, then later transferred to the Decatur Boys Home. [1653]

In mid-April Circuit Court Judge Walter Wright announced that the sanity of Willis "was a borderline case," but nevertheless sentenced him to life imprisonment. [1654]

[1650] *The Hutchinson News*, Hutchinson, Kansas, United States, 29th December 1932
[1651] Interstate Commerce Commission report
[1652] *The Daily Independent*, Murphysboro, Illinois, 23rd January 1933
[1653] *Alton Evening Telegraph*, Alton, Illinois, 16th March 1933
[1654] *The Urbana Daily Courier*, Urbana, Illinois, United States, 15th April 1933

Southern Railway passenger train No. 40 from Birmingham to Alabama was derailed on the morning of Tuesday 14th February 1933, near Bremen, Georgia, United States. Its engineer was killed and its fireman, baggageman and two passengers were injured.

Train No. 40 consisted of a 4-8-2 steam locomotive No. 1483, a combination baggage and mail car, a combination baggage car and coach, a coach, a Pullman car, a dining car, and a coach, in charge of Conductor Davis with Engineer James Cox and Fireman W.A. Goree. It had left Tallapoosa, 10¼ miles from Bremen, at 9.58 a.m., five minutes late, and approaching Bremen at 40 to 45 miles an hour was derailed at about 10.15 a.m. The engine, tender and leading four cars left the rails and the engine and tender overturned.

Fireman Goree told the Interstate Commerce Commission inquiry that nothing out of the ordinary had happened until they had neared Bremen. He had just been returning from the tender when he had felt something unusual and heard a peculiar noise. Before he had had time to alert his engineer the engine had been turning over. He did not know if the engineer had closed the throttle or applied the brakes.

Baggageman Dowdle said that he had had no warning before the derailment occurred.

Conductor Davis likewise said that he had had no indication of a problem before the train began to leave the track. He had tried to reach the emergency cord, but the severe jolting of the car had prevented him from doing so.

Flagman McLaughlin said that going back to protect the train he had observed marks on the ties indicating that the initial derailment had occurred a quarter to half a mile back from where the engine had finally left the rails. Half a rail length further back there were the remains of crushed rock on the south or outside rail and pieces of broken rock lying beside the line. Small footprints in the wet soil suggested that a child had placed the rocks on the track.

Superintendent Lawler said that the track had been in good condition. He believed that the crushed rock had caused the initial derailment of the leading wheels of the engine, with the final derailment occurring at west switch of the passing track.

Engineer Maintenance of Way Tobien stated that the track had been in a first-class condition and that he had come to the same conclusion as the superintendent as to the cause of the derailment.

Master Mechanic Goff said that the engine had been fitted with a "Little Watchman" safety device which in case of the engine truck becoming derailed was intended to open the air brakes in emergency. Unfortunately it appeared that the safety chains had held the truck high enough to prevent the device from working. [1655]

Inquiries led to the arrest of a negro boy named Henry Lowe, aged 9 or 10 according to different accounts. Lowe was 'quoted by the officer as saying he placed the rock on the track "jes to see de train wreck."' Sheriff White of Bremen announced that the boy would be charged with manslaughter. [1656] What happened to Lowe is unknown.

On 3rd May 1933 a Southern Railway freight train was derailed near Stacey, North Carolina, United States. Initially it was reported that the engineer and fireman were recovering in hospital from scalds, but this proved too optimistic, for the engineer succumbed to toxaemia (sepsis) two days later.

Northbound second class freight train No. 52 consisted of engine No. 4893, 48 cars and

[1655] Interstate Commerce Commission report

[1656] *The Butler Herald*, Butler, Georgia, United States, 16th February 1933 a caboose, in charge of Conductor Miller with Engineer George Allen and Fireman L.O. Woodson. It left Pomona at 3.35 p.m., passed Reidsville, 7 miles from Stacey and the last open office, at 4.18 p.m., an hour and two minutes late, and was travelling at 45 to 50 miles an hour when it left the rails at about 4.35 p.m. The engine, tender and leading 28 cars were derailed, the engine overturned and 27 cars piled up in a heap. The conductor and four poultry caretakers were also injured, as was a hobo riding on top one of the cars, while later a spectator was slightly injured by a bull from one of the stock cars.[1657]

Conductor Miller told the Interstate Commerce Commission (ICC) inquiry that he had been riding on the engine. As they had neared Stacey some dust had blown up into his face and looking over the side he had seen ballast and rock flying up. As he had shouted to the engineer to apply the brakes the latter had been doing so anyway. The brakes had been working properly. They had not been applied at Stacey before his engineer operated them.

Engineer Allen told visitors before he died that it had all happened so quickly that he had not realised what was happening.

Fireman Woodson was briefly interviewed in hospital by ICC investigators and said that the engine had been running smoothly until, like the conductor, he had seen ballast flying up. He had shouted to the engineer that something was dragging, and the latter had applied the brakes as the derailment had occurred.

Head Brakeman Gray stated that he had also seen some dust flying up, though not enough to cause him alarm. He added that the brakes had been tested after four cars had been set out at Pomona.

Patrolman Lowder said that he had gone back some time after the derailment and had found marks on the ties a quarter mile back where the derailment had begun.

Flagman Whitler stated that he had been watching the air gauge in the caboose because trespassers often turned the angle cocks. He had seen no application of the brakes before the accident.

Roadmaster Anderson and Signal Supervisor Snead stated that they had been on a following train, southbound No. 135, which had been flagged at Ruffin, two miles north of Stacey. They had proceeded to the scene of the accident together with Section Foreman Parris. They had traced wheel marks due to the derailment from a tool house and found a spike which appeared to have been run over.

As at Bremen (see above) engine No. 4893 had been equipped with a Wright "Little Watchman" device and as at Bremen the device had failed to prevent the accident. Superintendent Woodall found that it had tripped, he thought this had probably happened at the time of the final derailment.[1658]

A clue as to the perpetrator of the vandalism came from the engineer of a previous freight train who said that he had seen a group of boys near the line. Southern Railway Special Agent W.G. Marley elicited a confession from seven-year-old Junior Caldwell, who said that he and two other boys had got off the Ruffin school bus at the crossing near the derailment scene. Caldwell said that he had picked up the spike from a pile of scrap and had placed it on the outside rail, watched by Eldridge Robertson and Thurman Golden, both aged 12. He added, "I didn't mean to wreck the train. I just wanted to see the engine flatten out the spike." Special Agent Marley stated that no warrants had been issued against the boys because of their age; they had been turned over to their parents for

---

[1657] *The High Point Enterprise*, High Point, North Carolina, United States, 4th May 1933

[1658] Interstate Commerce Commission report

correction. He understood that they had been whipped. [1659]

The driver of the engine of a Southern Railway express train from Cardiff to Portsmouth noticed that it was "dancing about" after it had passed through the station at Bitterne, England, on Tuesday 3rd April 1934. At first he thought that it was slipping on greasy rails. Pulling up to investigate, he found that the leading bogie of the locomotive was off the track, and it was discovered that the permanent way had been damaged over half a mile. In what seems a remarkably short length of time the passengers were reportedly transferred to another train to continue their journey to Portsmouth in only a quarter of an hour. Children had recently been seen placing stones on the rails in the vicinity and a police watch had been established in the hope of catching the offenders. [1660]

'Irresponsible children' were blamed for the wrecking of a goods train at Darling Harbor, New South Wales, Australia, on the evening of Wednesday 2nd January 1935. The train of more than 60 trucks containing about 720 tons of goods and perishables had just pulled out of Darling Harbor for Bourke at dusk and was travelling at about 20 miles an hour when it was derailed. The engine, tender and four leading waggons left the rails, as did six trucks and vans in the middle of the train, which were telescoped and piled up 30 feet into the air. The contents of the waggons were scattered about the track, over 50 yards of which was torn up. One of the train crew was knocked unconscious and all of them were said to have had narrow escapes from being killed.

Guard Payne was the railwayman who was knocked senseless when his van struck the truck in front of it with great force, throwing him onto his head. Driver W. Kelly was dazed through being flung about the engine cab, but immediately began to draw the fire of his engine, which with the tender had been thrown against the side of a cutting.

A large piece of planking which had been laid across the rails was found by police. It had wheel marks from the engine upon it and was clearly the cause of the derailment. The children who had placed the obstruction on the line reportedly watched the crash and then took to their heels. [1661]

A Missouri Pacific Railroad freight train was wrecked as it was approaching Little Rock, Arkansas, United States, on Sunday 23rd June 1935. Three trespassers died in the accident, and a fireman and three trespassers were injured.

Second-class northbound freight train First No. 76 consisted of 2-8-2 steam engine No. 1535, 67 cars and a caboose, in charge of Conductor Murray with Engineer Young and Fireman J.R. Gaskett. It had passed QQ tower, 28 miles south of Little Rock, at 11.44 a.m., 2 hours and 9 minutes late, and was travelling at 40 to 45 miles an hour when it became derailed at 12.30 p.m. The engine, tender and first 28 cars left the track. Twenty-four of the cars piled up. [1662] Among the derailed vehicles were some oil cars, damage to one of which, containing gasoline, caused a fire to break out immediately after the derailment which spread quickly through the wreckage. [1663] In addition to the other casualties four city firemen (firefighters) were burned fighting the fire.

[1659] *The Bee*, Danville, Virginia, United States, 6th May 1933
[1660] *The Times*, London, England, 4th April 1934
[1661] *The Northern Herald*, Cairns, Queensland, Australia, 5th January 1935
[1662] Interstate Commerce Commission report
[1663] *San Antonio Express*, San Antonio, Texas, United States, 24th June 1935

Engineer Young told the Interstate Commerce Commission (ICC) inquiry that the first that he had known of the accident was the front of the engine jumping or being lifted upward. The air brakes of the train had been tested and were working properly.

Fireman Haskett likened the derailment of the engine to it running upon a coal pick to repair a stuck wedge. He had not noticed any obstruction on the line.

Head Brakeman Dillon said he had been unaware of anything wrong before the accident.

Conductor Murray said that he had been in the caboose with Swing Brakeman Holloway and Flagman Thompson. He said that after the accident, assisted by some bystanders, he had found three spikes, two of which appeared to have been run over. These, he believed, had been the cause of the derailment.

Section Foreman Jones agreed with the conductor as to the cause of the derailment. He said that he had been performing general repair work on the track and the day before the derailment had delivered an unopened keg of new spikes to the extra gang foreman.

Extra Gang Foreman Smith said that the track had been in good condition at the end of the day's work on Saturday. The keg delivered to him had been left in the vicinity of the work with some spikes remaining in it.

Roadmaster Ward said that he had inspected the accident site at 1 p.m. New indentations on the rails showed where spikes had been placed on them. Two of the spikes which had been run over had been handed to him.

Master Mechanic Hanna said that he had found no defects in the engine which could have contributed to the derailment.

Assistant Chief Special Agent Monroe said that several boys had been seen running away from the scene five minutes before the derailment. Four youths had been apprehended and one of them, David Barnes, 16, at first admitted and then denied having caused the derailment. He apparently said variously that he had just wanted to see what would happen and that he had hoped to get freight or valuables from the wreck. The other three, Cisco Barnes, 18; Lloyd Johnson, 20; and Joe Johnson, 25, were released. [1664] [1665]

The ICC blamed the track maintenance staff, noting that, 'The rules and instructions for the maintenance of way and structures provide that foremen and others will be held strictly responsible for all tools and materials left in their charge, and that all tools and supplies, when not in use, must, as far as possible, be kept locked in the tool box or tool house provided for the purpose, or where they can be carefully stored. The keg of spikes from which the spikes which caused this accident were thought to have been obtained, should have been disposed of according to rule.' [1666]

The front axle of the locomotive of a passenger train was derailed near Großhartmannsdorf, Saxony, Germany, on Thursday 24th September 1936. Two seven-year-old children had placed three fist-sized stones on the rails. [1667]

Two nine-year-old boys appeared at Birmingham Magistrates Court on 29th September 1936 charged with having placed a 20 lb piece of metal on the track at Aston, England, on the London Midland and Scottish Railway. As a result a goods engine had

---

[1664] *The Kingsport Times*, Kingsport, Tennessee, United States, 25th June 1935
[1665] *Biloxi Daily Herald*, Biloxi, Mississippi, United States, 26th June 1935
[1666] Interstate Commerce Commission report
[1667] *Eisenbahnunfälle in Sachsen* by Claus Schlegel, pg. 163 (in German)

been derailed, 60 yards of track was displaced, the line was blocked for three hours and £50 damage was caused. A passenger train had passed a short time previously. The magistrates were unable to agree on a suitable punishment and the case was adjourned, the boys being sent to a remand home.

At the resumed hearing the boys were each fined 40s, Mr J.H. Howard, the chairman of the magistrates, said that the boys' parents would have to pay the fines, which was appropriate, for if they had exercised more supervision their children would not have got into trouble. [1668]

Stones placed on the rails derailed passenger train P 1571 between Nienhagen and Jerxheim, Germany, on Thursday 28th October 1937. Two people died and 48 were injured. [1669]

On Friday 23rd September 1938 a New York Central Railroad passenger train was derailed 2½ miles west of Adams, New York, United States, killing the engineer and fireman and injuring the flagman and an express messenger. None of the 18 passengers on the train was hurt.

Westbound passenger train No. 8 consisted of a milk car, two baggage cars, a baggage and mail car, a baggage car, a coach, and a Pullman sleeping car, in charge of Conductor U.S. Grant, with Engineer Jay Slover and Fireman Owen McGinnis. [1670] It had left Watertown, 13¾ miles from Adams, at 5.58 p.m., eight minutes late; had left Adams, the last open block station, at 6.28 p.m., seven minutes late, and had become derailed on the open switch leading to the Valley Filling Station, at about 6.34 p.m. Proceeding along the spur track the engine had struck a derail without being derailed, and at the end of the track the engine, tender and three leading vehicles had gone down a 25 ft embankment.

Conductor Grant told the Interstate Commerce Commission inquiry that the brakes of the train had been tested and had worked properly. Approaching the switch he had heard the engine whistle, then the brakes had been applied in emergency. His first thought had been that an automobile had stalled on a nearby crossing but looking out had seen no vehicle. The train, which he estimated had been travelling at 30 miles an hour, had then derailed. After the accident he had examined the switch and found it set for the spur track; the switch lock had been smashed, the switch lamp had been out, and its red lens broken.

Head Brakeman Peterson and Flagman Hall put the speed of the train at 50 to 55 miles an hour approaching the switch, reduced to 20 to 25 miles an hour at the moment of derailment.

Section Foreman Tanner, in charge of the stretch, stated that after the derailment he had found the switch open and latched and two switch lamp lenses broken.

Section Foreman Hughes, in charge of a nearby section, said he had found a battered tie-plate at the scene.

Section Foreman Palloni stated that the switch had been properly locked when he had passed at about 1.30 p.m.

Divisional Engineer Barker and Trainmaster Walrath agreed that the switch lamp, if lit, could be seen for three quarters of a mile.

[1668] *The Times*, London, England, 30th September and 4th October 1936

[1669] *Katastrophen der deutschen Bahnen Teil II* by H.J. Ritzau, pg. 92 (in German)

[1670] *The Titusville Herald*, Titusville, Pennsylvania, United States, 24th September 1938

Superintendent Hamler testified that the speed recorder fitted to the engine showed a speed of 57½ miles an hour immediately before the derailment, thus considerably higher than estimated by the surviving members of the train crew.

Brothers Kenneth Beach, 13, and William Beach, 10, who lived close to the line, admitted that they had been delivering newspapers on their bicycles on the day of the accident. When they had reached the level crossing adjoining the switch Kenneth had thrown stones and broken the red lens of the switch lamp, whereupon William had thrown more stones and broken the green lens, upon which the light had gone out. Delivering a paper to a nearby residence they had then returned, and William had pounded the switch lock with a tie-plate, without effect. His older brother had taken over and after several more blows the lock had sprung open. Kenneth had then thrown the switch. When they had tried to restore the switch to the main line they had found that they could not do so, and had gone home, hearing the crash of the derailment five or ten minutes later. They said that they had realised that they might have caused a serious accident but had been frightened to tell their parents what they had done. They had tampered with the switch because they wanted to see if they could operate it. They had never done so before. [1671]

On Wednesday 12th April 1939 an engine pulling ten waggons of coal destined for a ship was derailed on a set of points at Alloa Harbour, Scotland, on the London and North Eastern Railway, and toppled over in a dangerous position. As a result the loading of the ship was delayed for a day, involving the shipping company in a loss of possibly hundreds of pounds. Two boys, aged nine and eleven, subsequently admitted having fitted an iron bolt in the points. They appeared at Alloa Juvenile Court on 24th April. The presiding magistrate pointed out that in addition to the probable financial loss caused by the vandalism the lives of the engine crew had been endangered. He deferred his decision as to the punishment of the offenders for six months. [1672]

The former Caledonian Railway branch to Ardrossan duplicated that of the rival Glasgow and South Western Railway and after both had become part of the London Midland and Scottish Railway in 1923, the ex-Caledonian line lost its ordinary passenger services in 1932. It remained in use, however, for goods and excursion trains and it was one of the latter, the 12.30 p.m. express from Glasgow Central to Ardrossan, which was involved in a serious accident near the end of its journey, at Saltcoats, Scotland, at about 1.21 p.m. on Saturday 5th August 1939.

Miners having their dinner at the Lanarkshire Miners' Convalescent Home in Saltcoats saw the engine of the train rise into the air and then fall down a 20ft embankment of the back garden of the home. The leading coach followed the engine, 0-6-0 No. 4315, and its tender down the embankment and the second and third coaches went partly down the embankment, while the fourth and fifth were derailed. The last two coaches remained on the track. Fireman Patrick Brown was buried beneath his engine and killed while Driver Joseph Birell was thrown clear but severely scalded and died later in Kilmarnock Infirmary. Two passengers – Miss A. Wilson, a Shawlands clerk, and James McKillop, a Glasgow iron driller – lost their lives, six passengers were seriously injured and 53 slightly hurt.

[1671] Interstate Commerce Commission report

[1672] *The Times*, London, England, 25th April 1939

The miners rushed to the scene and using timber from a nearby hoarding were able to prop up the coaches to prevent them falling further. Unemployed miner David Usher, who had been reading a paper in the garden, bravely clambered onto the engine, and shut off the steam which was pouring from it. The Miners' Home was turned into a casualty clearing station and ambulances and doctors attended from Saltcoats, Ardrossan, Stevenston and Irvine, together with Ardrossan Fire Brigade.

Superintendent James Halliday, of the Miners' Home, related, "I was sitting at dinner at the time. I heard the noise of the engine approaching, and then saw it leap into our garden. All the inmates at once left their dinner and rushed out to help in the work of rescue. The driver of the train was either thrown clear of the wreckage of the engine as it fell, or jumped clear, but he was badly scalded by the escaping steam, and was taken into the home and given first-aid treatment by the nurse. The worst feature of the disaster was the screams of the injured passengers. We could see the hand of the fireman protruding from the wreckage, but he was so completely buried in the debris that we could not hope to extricate him." [1673]

The Ministry of Transport inquiry into the accident was told by Guard Hall that the train had been travelling quite smoothly until there had been a jar, the van had risen into the air slightly and the train had stopped abruptly. There had been no application of the brakes.

Stevenston Signalman William Clark said that there was a lot of trespassing on the line. He had taken some boys to the stationmaster's office on the Tuesday before the accident.

District Permanent Way Inspector C. Ramage said that the track at the accident site had given no trouble with maintenance. He agreed that there was much trespassing. About 3 p.m. he had arrived at Saltcoats and had found pieces of stone on the line. He believed that the stone, which had been at exactly first point of the derailment, had been the cause of it.

Ganger John Edgar said that his men had had a good deal of trouble with trespassers, especially during the school holidays. [1674]

Colonel A.C. Trench, the inspecting officer, said that although the derailment of an engine by a crushable stone on the rail was a rare occurrence, the evidence was so well defined and convincing that he felt no doubt that the presence of one or more stones on the left-hand rail was the primary cause of derailment. Pieces of slag further back on the right-hand rail further back might have been a contributory factor. The curvature of the track and a worn rail were other possible factors.

Concluding his report the inspecting officer said that he had no recommendations, adding that, 'I can only express a hope that the death of four persons which lies at the door of those placing the obstruction on the rails, may be a warning to others of the deplorable results which may follow such michevious interference.' [1675]

An Atchison Topeka and Santa Fe Railway freight train was derailed between Plaines and Joliet, Illinois, United States, on Sunday 1st October 1939, killing its fireman and front brakeman and injuring the engineer.

Westbound freight train Extra 3264 consisted of engine No. 3264, 48 loaded and 18 empty cars and a caboose, in charge of Conductor Bennett, with Engineer George Breault

[1673] Ibid, 7th August 1939

[1674] Ibid, 10th August 1938

[1675] Ministry of Transport report

and Fireman E.M. Carroll. [1676] It had left Joliet Yard at 11.54 a.m. and was moving at between 20 and 35 miles an hour when it was derailed at 12.05 p.m.. The engine and leading 11 cars left the track, the cab of the engine was crushed, and seven cars were destroyed.

The Interstate Commerce Commission inquiry was told by Engineer Breault that the brakes of the train had been tested before it had left Corwith Yard and had been working properly. Approaching the point of derailment his engine had given a slight jar, which he had attributed to a flat wheel or broken tyre. He had made a service brake application, had felt his engine leave the rails and had applied the brakes in emergency.

The evidence of Conductor Bennett and Brakeman Jurkovitch did not add anything of importance.

Divisional Engineer McKibben stated that he had examined the site of the accident. He had found nothing wrong with the engine which could have caused the derailment. He described various marks on the track and the finding of a spike which had clearly been run over by the train.

Trainmaster West likewise described the marks on the track and the spike which had been found.

Master Mechanic Goodrich stated that he had found nothing wrong with the engine and that he believed that the spike had been the cause of the derailment.

Three girls, aged 9, 11 and 16, had been walking along the right-of-way just before the derailment. The two younger girls stated that the oldest girl had placed a spike on a rail. The youngest girl had removed it, then the two younger girls had gone away before the oldest girl and therefore did not know if she had replaced the spike. The oldest girl admitted placing a spike on the rail, saying that the youngest girl had removed it. All three girls said that two boys had chased them away. A schoolboy, Dominic Vironda, spoke of seeing both the three girls and the two boys on the tracks. [1677]

Eastbound Baltimore and Ohio Railroad passenger train No. 32 from Chicago to New York, travelling at 35 miles an hour, was derailed at Ursina, Pennsylvania, United States, on the afternoon of 5th October 1941. Fortunately only the leading truck (bogie) of the double-headed train left the rails. No one was hurt. The train was delayed about an hour. Officials found limestone ballast piled up on both rails. [1678] At the rear of the train was a car carrying Western gasoline company officials who were en route to a Washington conference with government authorities regarding the motor fuel shortage. [1679]

The following Thursday 17-year-old Irvin Morrison was arrested. He had been seen at the site of the derailment immediately after it had happened and showed great interest in the event. When a trainman had asked his name he had run off. Morrison confessed that he had placed about 30 pieces of ballast on the rail on the creek side of the track and about the same number on the other rail. Then he had gone about 100 yards towards Ursina and placed 25 or 30 stones on the rail closest to the hillside on the sharpest part of the curve. This had been done about 4.30 p.m. after two light engines had passed. His work accomplished, he had sat down in the adjoining woods to wait for the express to see if it would be derailed. He said that when he had been questioned by the railwayman he

---

[1676] The *Morning Herald*, Hagerstown, Maryland, United States, 2nd October 1939

[1677] Interstate Commerce Commission report

[1678] *The Daily Courier*, Connellsville, Pennsylvania, United States, 6th October 1941

[1679] The *Morning Herald*, Uniontown, Pennsylvania, United States, 6th October 1941 had become frightened and had run off. [1680] Morrison apparently added that, "I thought that I might be able to collect some valuables if the train was wrecked," [1681] implying that, also given his age, his act might come closer to an act of sabotage than of vandalism.

On 10th November he pleaded guilty to charges of malicious injury to railroad property and carrying a concealed gun and was awarded an indeterminate sentence in the Pennsylvania Industrial School at White Hill, with a recommendation that he be held until he was 25. It was the most severe sentence passed by the court that day. [1682]

The *Wolverine* passenger train of the New York Central Railroad, en route from Chicago to New York, was derailed near Hughart, Michigan, United States, on Thursday 19th August 1943. Its fireman lost his life and a baggageman was injured.

Eastbound first-class passenger train No. 108, the *Wolverine,* consisted of 4-6-2 steam locomotive No. 4626, a baggage car, a mail baggage car, a passenger-baggage car, a coach, a Pullman sleeping car and a passenger café car, in charge of an unnamed conductor, with Engineer E.D. Underwood with Fireman John McCarty and Student Fireman C.W. Jackson. It had left Grand Rapids, 1¾ miles from Hughart, at 12.35 p.m., passed Hughart at 12.41 p.m., two minutes late, and was travelling at 30 miles an hour when the engine and leading four cars were derailed at 12.42 p.m. The cab of the engine was almost demolished, and the first four cars were badly damaged, the next two slightly so.

Engineer Underwood told the Interstate Commerce Commission investigation that the train brakes had been tested at Grand Rapids and had been working properly. As he had neared the point of derailment he had heard an unusual noise and the engine had lurched and swayed. He had applied the brakes in emergency at the moment that his engine had left the track. [1683]

Student Fireman Jackson told a newspaper reporter, "We were pulling out of town, going about 30 miles an hour, when I felt the locomotive begin to rock and jump. Standing near the rear of the cab, I shouted to McCarty, he had better jump, but he didn't hear me I guess. I jumped clear of the locomotive just as it left the tracks." The engineer stayed on the engine and was unhurt, but the fireman was scalded to death by escaping steam, trapped by the jack-knifed tender. [1684]

Investigation had revealed no defect with the engine.

It was found that the derailment had been caused by three small boys, two aged five and one aged six, who had laid stones on the rails, "to see what would happen." Having placed the stones they had waited for a train to pass but when none had appeared they had got bored and had gone home before the derailment had occurred. They told a police inspector that they had had no intention of causing any trouble. The police said that no action would be taken against them. [1685]

A three-car St Louis – San Francisco Railway passenger train en route from Springfield was derailed on the southern outskirts of Kansas City, Missouri, United

---

[1680] *Somerset Daily American*, Somerset, Pennsylvania, United States, 11th October 1941

[1681] *The Morning Herald*, Uniontown, Pennsylvania, United States, 14th October 1941

[1682] *Somerset Daily American*, Somerset, Pennsylvania, United States, 11th November 1941

[1683] Interstate Commerce Commission report

[1684] *The Escanaba Daily Press*, Escanaba, Michigan, United States, 20th August 1943

[1685] *The News Palladium*, Benton Harbor, Michigan, United States, 20th August 1943

States, late on Wednesday 20th October 1943 while travelling at about 45 miles an hour and four people were slightly hurt. There had been about 100 passengers on board. Engineer J.W. Balke said that the train had struck an open switch. [1686]

Two seven-year-olds, a boy and a girl named Norman Cochran and Patricia Friedh, were soon apprehended as the culprits. They said that they had been walking home from school when they had seen the bright brass lock of the switch, so they pulled it until it opened and then tugged the switch into another position. Probation officer Andrew Steele remarked that the two were just infants [1687] and the police decided to take no further action in the matter other than taking them before the juvenile judge to show them the seriousness of their actions. [1688] This was duly done and Judge Ray Cowan, in dismissing the case, said, like Mr Howard at Birmingham, England, back in 1936, that it was, "more the fault of the parents than anyone." [1689]

Twenty-five minutes out of Boston the *Southwestern Limited* of the Boston and Albany Railroad, bound for St Louis, struck a tie (sleeper) which had been placed upon the track on the outskirts of Natick, Massachusetts, United States, on Monday 28th February 1944. The cowcatcher of the engine cut the tie in two, throwing one piece aside while the other caught under the cowcatcher and was pushed along until the train was flagged to a stop by Car Inspector Arthur Pearson.

Federal Bureau of Investigation agent A.E. Soucy announced that two 14-year-old boys had confessed to putting the tie on the tracks "to see what would happen."

United States Attorney Edmund Brandon said that because of the extreme youth of the boys they would not be prosecuted under the wartime sabotage statutes.

Natick Police Chief Thomas Evans assured him that the culprits would be brought before the juvenile court in that town. [1690]

A local newspaper took a very dim view of the matter, writing, 'Making every allowance for the youth of these juvenile vandals and with full knowledge that they did not comprehend the enormity of grief which might result from their work, they cannot be excused on the theory that "boys will be boys." They must be penalized to a point that will make them fully aware of their misdemeanours and fully decided that they will not duplicate their stunts in the future.' [1691]

[1686] *Joplin Globe*, Joplin, Missouri, United States, 21st October 1943
[1687] *Ames Daily Tribune*, Ames, Iowa, United States, 21st October 1943
[1688] *Washington Evening Star*, Washington, D.C., United States, 22nd October 1943
[1689] *The Taylor Daily Press*, Taylor, Texas, United States, 22nd October 1943
[1690] *Nebraska Star Journal*, Lincoln, Nebraska, United States, 2nd March 1944
[1691] *The Lowell Sun*, Lowell, Massachusetts, United States, 4th March 1944

CHAPTER 18

SABOTAGE

Four distinct motives for sabotaging trains may be distinguished (after Robert B. Shaw).

(a) As an aid to intended robbery, either of the train passengers or to steal, for example bullion being carried by the train.
(b) To satisfy a personal grudge against the railway company, such as by an ex-employee who feels that he has been wrongly dismissed
(c) To hinder the operation of the railway, as for example during a strike
(d) As an act of guerrilla warfare or terrorism, perhaps against an occupying enemy power. Sabotage against the railways has been a feature of many conflicts from the American Civil War onwards.

The period covered by this volume saw numerous instances of sabotage in various countries in the 1930s, notably in India, China, and Manchukuo, and in the later part of the decade in Palestine, while in the early 1940s there was a massive increase of sabotage in those parts of Europe occupied by the Nazis, especially in Eastern Europe and in France, with fewer instances in countries such as Belgium, Denmark, and Norway. In order to underline their extent, the acts of sabotage carried out in Western Europe during the Second World War have been described in considerable detail.

An announcement of drastic cuts in railway staff from 1st March was believed to lie behind three concerted attacks on different railways in Buenos Aires, Argentina, on the morning of Tuesday 20th January 1931.

The first attack involved the explosion of a bomb on an incoming electric train at the Once terminus of the Buenos Aires Western Railway. Three coaches of the train were wrecked, three persons were killed and several injured.

The second explosion of a bomb was at the Constitución terminus of the Buenos Aires Great Southern Railway. Several people were injured in this attack.

The final attack was on a train of the Central Argentine Railway as it was stopped at Palermo station. A coach was wrecked and six people injured. [1692]

The railways of India and Burma were the subject of a number of acts of sabotage in

---

[1692] *Janesville Daily Gazette*, Janesville, Wisconsin, United States, 20th January 1931 the first half of 1931, though many minor ones probably went unreported in the press, for the annual report to the railways board stated that there had in total been no less than 145 accidents due to train wrecking and attempted train wrecking in the financial year 1930-31, rising to 154 in 1931-32. [1693]

On Friday 6th March 1931 the Great Indian Peninsula Railway Nagpur express was near the Saki Bridge, between Manmad and Nandgaon, India, when it experienced a sudden lurch and there was a crashing noise. It was at once brought to a standstill and an investigation found that fishplates, bolts, nuts, and key pins had been removed over a considerable length of the track. The engine was slightly damaged, and the leading bogie of the train was derailed. [1694]

A special goods train was derailed between Ambala and Ghaziabad, India, on the North Western Railway, on Saturday 28th March 1931, fishplates having been removed from the track. The contents of the Down Calcutta Mail had to be transhipped and other trains were diverted.

The express which had left Delhi for Simla just after 21:00 on 5th April, carrying many families of government officials who were making their annual exodus to the hills to escape the summer heat, was derailed near Ambala, India, early on Monday 6th April 1931. No. 83 Up passenger train was thrown off the track at mile 135/7, between Lalru and Ghaggar, India on the Ambala – Kalka section of the North Western Railway line. A length of rail had been removed on the inside of a curve. The engine and four coaches left the track and the engine crew, and one or two passengers were slightly hurt. Had the train left the rails on the other side it would have fallen over a *khud* (a ravine) about 15 feet deep. [1695] [1696]

The Railway Gazette commented, 'Malicious actions of this nature have not been rare in recent years, and the perpetrators have generally eluded detection. When on their nefarious acts they are possibly mistaken for railway employees, although it is not usual for platelayers to work late in the evening. If an attempt were made to rouse public indignation against miscreants who would not hesitate to jeopardise the lives of hundreds of people, it might be possible to obtain information of suspicious persons working on the railway lines at unusual hours.' [1697]

A passenger train from Calcutta was derailed on Tuesday 12th May 1931 between Saratnagar and Dilpassar, India. The engine fell down an embankment and six carriages were badly damaged. The driver and fireman were slightly hurt but all the passengers escaped. Several fishplates had been removed from the track. [1698]

An act of sabotage was carried out in neighbouring Burma on Monday 1st June 1931. Armed disturbances had broken out in the country the previous December and they lasted into 1932. They were part of a nationalist movement to drive out the British. The Rangoon – Mandalay mail train was derailed near Penwegon, Burma, the engine and three third-class bogie carriages overturning. Six passengers were injured. The rebels had first blown up the first pier of a bridge and then removed two rails. [1699]

[1693] Report by the Railway Board on Indian Railways for 1931-32
[1694] *The Railway Gazette*, London, England, 3rd April 1931
[1695] Report by the Railway Board on Indian Railways for 1931-32
[1696] *The Times*, London, England, 8th April 1931
[1697] *The Railway Gazette*, London, England, 8th May 1931
[1698] *The Times*, London, England, 14th May 1931
[1699] Ibid, 4th June 1931

In mid-1931 Macedonian Revolutionaries (*komitajis*) were reportedly anxious to prevent a rapprochement between the governments of Bulgaria and Yugoslavia. In order to further their aims they carried out several terrorist attacks, beginning on 20th or 21st July with the sabotage of a Yugoslav troop train crossing a bridge over the River Morava near Vranya, Yugoslavia (now Vranje, Serbia), as a result of which the engine and several carriages fell into the river. Two Macedonians (the engine crew?) were said to have lost their lives. The casualties to the troops were not stated, all news of the affair being supressed by the Yugoslav government. Of the three *komitajis* who carried out the attack one was killed a few days later, one was captured and the other escaped. Two further attacks were carried out on 24th July. [1700]

A bomb exploded on a Paris – Munich – Zagreb – Belgrade express between Zemun and Zemun-Novigrad, Yugoslavia, on Monday 3rd August 1931. Three people were killed and 16 gravely injured. Macedonian *komitajis* or Bulgarian Communists were believed responsible. [1701] The bomb exploded in the compartment occupied by Professor Brunetti, of Belgrade University, and his 11-year-old son was killed outright, and his wife fatally wounded. A third, unidentified man was also killed and 18 people were injured, comprising 17 Yugoslavs and a Belgian. The same day at Rosenbach, Austria, five bombs were discovered on a train bound for Yugoslavia. A sixth was overlooked and exploded. [1702]

The Basel – Frankfurt – Berlin express was derailed at about 21:45 on Saturday 8th August 1931 by an explosive charge detonated electrically from some undergrowth about 200 metres from the line near Jüterbog, Germany. The engine had already passed the spot before the charge was detonated and the engine, tender and leading mail van remained on the line, while the remainder of the train was derailed. The derailed coaches ploughed their way through soft sand and the effects of the derailment were thus mitigated. Between 60 and 80 passengers were slightly hurt, mostly by cuts from broken glass.

The government and the railway company separately offered substantial rewards for the apprehension of the culprits. The owner of a Berlin electrical supply shop was able to describe a man who had apparently bought the wire used for the detonation of the bomb and suspicion early fell on a naturalised Irishman named Carnell, but this proved to be a false lead. [1703] The real identity of the saboteur did not become apparent until a further, much more serious attack (see Biatorbagy, below).

A Great Northern (Ireland) Railway goods train from Belfast proceeding towards Armagh was stopped by means of a red light near Rice Hill, Northern Ireland, on the morning of Saturday 15th August 1931. Driver Henry Bell and his fireman were then ordered to disconnect the engine from the waggons and run it forward over a section of track which the armed gang had ripped up, using crowbars, pickaxes and similar tools taken from a permanent way hut the gang had broken into. Driver Bell protested that it was unfair to ask him to deliberately derail his engine but was told, "If you don't do it,

---

[1700] *The Times*, London, England, 4th August 1931

[1701] *The Star Journal*, Sandusky, Ohio, United States, 3rd August 1931

[1702] *The Times*, London, England, 4th August 1931

[1703] Ibid, 10th, 12th and 13th August 1931

you will get a bullet," and a shot was fired into the air to show that the gang meant business. The engine was duly sent forward and jumped the rails but did not overturn as the gang had hoped.

The saboteurs were Orangemen (Protestants) who wished to block the railway to prevent trains taking people to an Ancient Order of Hibernians (Catholic) demonstration in Armagh as a reprisal for the suppression of Orange celebrations at Cootehill. [1704] [1705]

At about 00:15 or 00:20 on Sunday 13th September 1931 the international express which had left Budapest at 23:30 the previous day for Vienna, Genoa, Ostend, and Paris was blown up as it was crossing the viaduct near Biatorbagy, Hungary. As a result the engine and all but the rear two coaches out of seven, the train fell about 25 metres from the viaduct. It was perhaps surprising, therefore, that only 25 people were killed. Another 14 were seriously wounded and about 100 slightly hurt. As might be expected on an international train the victims included two Austrians, a Frenchman, a Belgian and an American, while an Englishwoman, Miss Hilda Fowlds, was killed outright and another Englishman, Mr Harry Clements, passed away in hospital.

It was reported that the outrage had been caused by a handbag filled with dynamite, connected by a wire to an electric pocket lamp battery 'of Bohemian make.' Nearby was a scrap of paper secured by a stone, reading, "Brother Proletarians! If the capitalist State cannot provide work we shall get it otherwise. We have plenty of petrol and explosives."

The sabotage was at first blamed on a gang of international terrorists, probably the same ones who had derailed the train at Jüterbog. [1706] The first assumption proved to be wrong, the second was correct.

A man named Szilvestre Matushka soon came to the attention of the authorities. Rather than fleeing the scene of destruction he had stayed to assist in the rescue of the people whose train he had just wrecked, then claimed that he himself had been a passenger on the train, and that he was therefore entitled to compensation. It soon became apparent that it had been Matushka, acting alone, rather than an international terrorist gang, who had been responsible for the disaster and that he had also been the perpetrator of the Jüterbog derailment, as well as two failed attempts to wreck trains in Austria. It also became apparent that he was mentally deeply disturbed and psychotic. In court he claimed to be possessed by a demon called Leo the Ghost and, though he had had a career as a businessman, to be a radical revolutionary socialist. He would also weep uncontrollably, shout at no one in particular, get down on his hands and knees in court and so on.

Matushka first stood trial in Austria, where he was sentenced to six years' imprisonment. Meanwhile he was sentenced to death in Hungary. When his death sentence was commuted to life imprisonment, probably because of his mental state, he was transferred to Hungary, where he remained in prison at least until 1944. It was rumoured that at the end of that year he was released by the Germans because of his knowledge of explosives. His ultimate fate is unknown. [1707]

A Soviet troop train was apparently blown up 'during manoeuvres' south-west of Leningrad, USSR, on about Saturday 19th September 1931. The locomotive and 12

[1704] Ibid, 17th August 1931

[1705] *The Railway Gazette*, London, England, 21st August 1931

[1706] *The Times*, London, England, 14th and 15th September 1931

[1707] https://europebetweeneastandwest.wordpress.com/tag/szilvestre-matuschka-mass-murder/

carriages were totally destroyed and there was heavy loss of life. [1708]

The derailment of a Peiping – Mukden passenger train at Jaoyengho, Manchuria, on Saturday 26th September 1931, was initially blamed by the Chinese on Japanese airmen (the Japanese had invaded Manchuria on 18th September) but later bandits were held responsible. The driver, fireman, and five passengers died, and a large number were injured. The dead passengers included an Englishman, a Russian and a 'Hindu' (presumably an Indian). A number of foreigners were injured. Robbery was apparently the motive for the attack as 30 passengers had possessions taken by the bandits.

The Chinese government complained that shortly after the attack a detachment of 20 Japanese soldiers had arrived at the scene on another train. After a cursory inspection they had left again without offering assistance. [1709]

Chinese bandits armed with machine guns derailed an international train from Yunnanfu, the capital of Yunnan Province, in the hills near Lielang, French Indo-China, on Wednesday 30th September 1931. A French railway official was murdered, and the bandits made off, without seizing any booty. Troops were sent in their pursuit 'with the co-operation of the Chinese authorities.' (It has not proved possible to identify the location of Lielang on a modern map, but it is assumed that the sabotage occurred in French Indo-China, given that a French railway official was involved and that it was stated that Chinese officials were co-operating in chasing the saboteurs.) [1710]

The most serious of several attempts to derail trains on the Bulawayo – Mafeking occurred near Francistown, Bechuanaland, on the night of Tuesday 24th November 1931. Five sleepers, four steel and one wooden, were laid across the track before the arrival of the mail train from Bulawayo. The driver of the latter observed the obstructions and was able to reduce speed sufficiently to prevent the train from being derailed and the only damage was to the engine cowcatcher. Local 'natives' were blamed. No motive was suggested, and the incidents possibly came closer to an acts of vandalism. The police were investigating. [1711]

An attempt to derail a train on the Grand Chord of the East Indian Railway failed when a rail was discovered to be out of alignment and an express was stopped just in time on Wednesday 16th December 1931. [1712] The next day, Thursday 17th December, there was an attempt to derail Dehra Dun express near Sitarampur, India. A railwayman on a trolley was on his way to repair a signal wire when his trolley was derailed, revealing the fact that fishplates had been removed, the rail shifted and wedged in its new position with the fishplates to prevent it slipping back into its proper position. [1713]

On Sunday 20th December 1931 the Up Dinapore express from Howrah was derailed just beyond Oyaria station, India, No one was hurt. Several fishplates had been removed,

---

[1708] Ibid, 2nd October 1931
[1709] *Lima Sunday News*, Lima, Ohio, United States, 27th September 1931
[1710] *The Evening Journal*, Washington, Iowa, United States, 30th September 1931
[1711] *The Railway Gazette*, London, England, 22nd January 1932
[1712] Ibid.
[1713] *The Times*, London, England, 19th December 1931

leaving a rail loose. [1714]

During fighting between Chinese irregulars and Japanese in Manchuria on Monday 11th January 1932, an armoured train sent to relieve a Japanese company was derailed by removal of the rails at Hsinmintun, Manchuria. [1715]

Japanese troop trains approaching Harbin, Manchuria, on the morning of Sunday 31st January 1932 were attacked by 2,000 anti-Kirin troops who were routed after four hours' fighting, leaving 400 casualties, while the Japanese lost 21 dead and 40 wounded. Whether any of the trains were derailed was not stated. [1716]

The Calcutta – Benares express was derailed near Dinapore, India, (now Danapur) at 03:00 on Monday 25th January 1932, 'owing, it is believed, to the rails having been maliciously torn up.' The driver and a passenger were injured. The same train was derailed again possibly the next day. 'Investigating officers attribute the interference with the rails to discharged railway servants.' [1717]

The express from Laredo to Mexico City which connected with the Missouri Pacific Railroad Sunshine Special was wrecked by bandits who tore up the track of the Ferrocarril Nacional de México in Queretaro State, between Mariscala and Jacoqui stations, Mexico on Saturday 19th March 1932. The engine, two express cars, a second-class coach and a first-class day coach overturned. The wreck occurred at 11.30 p.m. After 30 minutes' fighting the soldiers who formed the armed escort of the train drove off the bandits. An express messenger and one bandit were killed, according to one account the fireman also died. One soldier was wounded. The train had been due in Mexico City at 6.35 a.m. the following morning. [1718] [1719]

Another international train from Laredo was derailed by sabotage at Obregon, in Guanajuato State, Mexico, on the Ferrocarril Nacional de México at 10 p.m. on Sunday 10th April 1932. The engine went into a ditch, overturned and oil spilling from it started a fire which destroyed a baggage car and a second-class coach. An engine rushed from Chirimoya pulled the other coaches away from the fire, including sleeping cars from St Louis, Missouri, and San Antonio, Texas. Trainman Jose Perez Vertis was crushed in the wreckage and killed and another railwayman was badly burned. In contrast to the Mariscala incident the military escort on the train did not have chance to engage the saboteurs.

When news of the attack reached Mexico City the war department issued orders to the military commanders of Guanajuato and other states to apprehend if possible, and deal summarily with those responsible for the derailment. Meanwhile three bandits implicated in the earlier attack had already been executed and 30 more were under arrest awaiting execution. [1720]

[1714] *The Railway Gazette*, London, England, 22nd January 1932
[1715] *The Sedalia Capital*, Sedalia, Missouri, United States, 12th January 1932
[1716] *The Times*, London, England, 2nd February 1932
[1717] Ibid, 25th and 28th January 1932
[1718] *Nevada State Journal*, Reno, Nevada, United States, 21st March 1932
[1719] *Gallup Independent and Evening Herald*, Gallup, New Mexico, United States, 22nd March 1932
[1720] *San Antonio Express*, San Antonio, Texas, United States, 12th April 1932

A train of Japanese troops returning from a campaign against Chinese rebels at Fangcheng was dynamited and derailed 17 kilometres east of Harbin, Manchuria, on the Chinese Eastern Railway, on Tuesday 12th April 1932. Two officers and 12 men belonging to motor transport units of Major General Murai's brigade were killed and over 50 wounded. A relief train was sent from Harbin together with an investigation commission. A report from Tokyo stated, 'Some press reports laid the blame for the wreck on "Communists" of unknown nationality. Others said it was caused by elements interested in causing friction between Russia and Japan.'

Two hours before the attack a train carrying Maj. Gen. Hasebe and his staff from Fangcheng had escaped a wreck when railway guards discovered dynamite of German manufacture three kilometres outside of Harbin. It was found only half an hour before the train was due to pass. [1721]

Two attempts to derail trains at Dehra Dun were followed on Monday 11th July 1932 by the derailment of a mixed train on the Shahdara – Saharanpur Light Railway near Shahdara, India. Three carriages and nine goods waggons left the rails and two coaches went over an embankment but only one man was injured. There had been 84 passengers on the train. Two sets of fishplates had been removed from the track. [1722] [1723]

An accident occurred with a cattle train of the Ferrocarril Provincial de Santa Fe proceeding from Santa Fe to Rosario midway between Arocena and Coronda stations, Argentina, on Tuesday 9th August 1932. Several waggons were destroyed when the train was derailed and a large number of beasts were killed. An examination of the track showed that several fishplates had been removed and fish bolts loosened. It was believed that the intended target of the attack had been a passenger train carrying a member of the provincial government, which had been following the cattle train. [1724]

Four passengers were shot and wounded when bandits derailed the Shanhaikwan express on Saturday 20th August 1932 at an unstated location in the Mukden area. In a separate incident, an engineer and four Japanese policemen were killed in an attack on a train between Mukden and Hialung, Manchukuo, while there had been a previous attack on a train a few days earlier. [1725]

In the first of a series of attacks in Manchuria on about Wednesday 7th September 1932, bandits derailed a passenger train on the South Manchurian Railway 400 kilometres east of Harbin, Manchukuo, by piling stones on the track. The passengers were robbed of personal belongings and nearly 100 Russians and Manchurians were taken hostage. [1726]

In another attack on the South Manchurian Railway on Friday 9th September 1932 a Japanese troop train was derailed between Taschihchiao and Haicheng, bandits having

[1721] *The Evening Star*, Washington, D.C., United States, 13th April 1932
[1722] *The Times*, London, England, 13th July 1932
[1723] *The Railway Gazette*, London, England, 5th August 1932
[1724] Ibid, 23rd September 1932
[1725] *Indiana Evening Gazette*, Indiana, Pennsylvania, United States, 20th August 1932
[1726] *Berkeley Daily Gazette*, Berkeley, California, United States, 9th September 1932

loosened the tracks. Its coaches were overturned, and a number of soldiers were injured. [1727]

Early on Sunday 11th September a southbound passenger train from Harbin to Changhung, carrying about 600 Chinese and Japanese passengers, was derailed near Suangcheng, about 50 kilometres south of Harbin, Manchukuo. Bandits had loosened the rails in the darkness. Some of its coaches reportedly fell over a 30 ft cliff. Twelve or 15 passengers were killed according to different accounts, 36 were wounded and 36 kidnapped, while the others were robbed. [1728] Doctors and nurses were sent to the scene from Harbin at 4 a.m. [1729]

In another attack on the same Sunday, a northbound express was derailed near Chengkaotze, Manchukuo. One report stated that six passengers had been killed and 100 kidnapped; another, presumably relating to the same attack, stated that four Japanese soldiers had died plus an unknown number of Chinese while two Japanese, five Russian, one Korean and an unknown number of Chinese civilians had been wounded. [1730] [1731]

In a third attack a Japanese hospital train was derailed and wrecked close to the scene of the previous attack near Suangcheng. Six Japanese were reportedly killed, several wounded and others were missing. [1732]

Bandits derailed a train at Penjamo, Mexico, on about Thursday 6th October 1932. Federal troops were said to be scouring the country looking for them. No casualties were reported. [1733]

A train from Beisan to Haifa, Palestine, was derailed between Beisan and Jisr el Mejami on about Friday 21st October 1932 and four Jews were slightly injured. The train was known as Rutenberg's Special as most of its passengers were workers at the Rutenberg electrification works, and it was believed that they had been the target. [1734]

On about Friday 25th November 1932 a Japanese armoured train was derailed at Kirin, in northern Manchukuo, and 11 soldiers died. Sabotage was suspected. [1735]

Twenty-seven Chinese and Japanese were wounded on Wednesday 18th January 1933 when a train on the Southern Manchurian Railway was derailed between Mukden and Dairen, Manchukuo, the tracks having been dynamited. [1736]

At the beginning of 1933 a strike was in progress on the Great Northern (Ireland) Railway. On Wednesday 1st February 1933 a passenger train left Dublin Amiens Street at 9 a.m. for Belfast, manned by 'volunteer labour,' in other words scabs, and nine passengers. It got as far as Dromiskin, Republic of Ireland, when it was derailed, rails

[1727] *The Daily Courier*, Connellsville, Pennsylvania, United States, 10th September 1932
[1728] *The Times*, London, England, 13th and 14th September 1932
[1729] *The Monroe News-Star*, Monroe, Louisiana, United States, 12th September 1932
[1730] *The Times*, London, England, 14th September 1932
[1731] *The Times*, London, England, 13th and 14th September 1932
[1732] *Medicine Hat News*, Medicine Hat, Alberta, Canada, 12th September 1932
[1733] *Del Rio Evening News*, Del Rio, Texas, United States, 13th October 1932
[1734] *The Sentinel*, Chicago, Illinois, United States, 4th November 1932
[1735] *Volkskrant*, 'S-Hertogenbosch, the Netherlands, 28th November 1932
[1736] *The Winona Republican-Herald*, Winona, Minnesota, United States, 18th January 1933

having been lifted on both the Up and Down lines. Most of the train went down a 14 ft embankment. A mail clerk named John Paterson was killed outright and John MacWilliams, a train lighting foreman, was fatally injured. Guard Duffy, the engine driver and a dining car attendant were injured.

It was stated that the affair had caused the greatest indignation throughout the country and that sabotage had always been rare, even during the worst periods of labour disputes. The engine drivers' union disclaimed all responsibility for the attack. All train services between Dublin and Belfast were abandoned after the derailment. [1737]

In the early 1930s the island of Cuba, though nominally independent, was still the subject of United States military intervention and economic dominance and in 1933 it was under the control of President Machado, who refused to step down at the end of his term of office. This led to a series of uprisings [1738] and it was against this background that several attacks on trains of the American-owned Consolidated Railways of Cuba and the British-owned United Railways of Habana occurred in February and March 1933.

On the night of 23rd February damage to a switch and a culvert resulted in the derailment of a train at Colorado, Cuba, in Camaguey Province. The same night, a train was derailed between Puerto Tarafa and Ciego de Avila in the same province. Then on 25th February a bomb derailed a train in Matanzas province. [1739] The next day a 17-car United Railways of Habana train was derailed at Tumbala Berra Crossing, between Esperanza and Santa Clara, by a bomb (or more correctly, a mine) placed on the tracks. No deaths or injuries had resulted from the first three attacks, but this time Engineer José Oviel and Conductor Estanislao Rodriguez were killed. [1740] [1741] Two brakemen were said to be missing.

Finally on Saturday 4th March a train hauling sugar cane was derailed on an open switch in northern Oriente Province, near Vasquez Bello, and Engineer Camilo Cobas and Conductor Adolfo Flores were killed. [1742]

A passenger train from Vladivostok to Harbin was derailed by guerrillas in Mataoshi Tunnel, Manchuria, on Friday 31st March 1933. Manchu-Japanese guards on the train repelled the guerrillas after the derailment but 'scores of casualties were reported.' Two of the guards were also killed and five were injured. [1743]

A passenger train on the Kirin – Tunhua line was wrecked 30 kilometres west of Tunhua, Manchukuo, on Wednesday 19th April 1933, in what was clearly an act of sabotage as communication lines had also been severed. Eleven Japanese and four Manchurians lost their lives. [1744]

No. 2 Down Punjab Mail of the East Indian Railway, bound for Calcutta, was derailed just before midnight on Tuesday 2nd May 1933 at Mile 273 19/20, between Dumra and

[1737] *The Times*, London, England, 2nd, 3rd and 4th February 1933
[1738] Wikipedia – History of Cuba
[1739] https://history.state.gov/historicaldocuments/frus1933v05/d233
[1740] Lebanon Daily News, Lebanon, Pennsylvania, United States, 28th February 1933
[1741] *The Emporia Gazette*, Emporia, Kansas, United States, 28th February 1933
[1742] *Sunday Journal and Star*, Lincoln, Nebraska, United States, 5th March 1933
[1743] *The Wisconsin State Journal*, Madison, Wisconsin, United States, 31st March 1933
[1744] *The Bakersfield Californian*, Bakersfield, California, United States, 19th April 1933

Birhee stations, India. The engine and six coaches were derailed, five of the coaches overturned and five passengers were killed and 13 wounded, all Indians. Damage to the sum of 48,000 rupees was caused. The derailment was the result of malicious interference with the track, a rail having been removed. [1745] [1746]

A Buenos Aires Western Railway mixed train, consisting of an engine, three goods waggons, a postal van and seven carriages, en route from Buenos Aires to Colonia Alvear, was partly derailed about 120 kilometres from the Argentinian capital on Monday 22nd May 1933. The engine and leading vehicles kept to the rails but most of the coaches were derailed and a second-class coach overturned. Fortunately, no one was hurt. It was found that the rails had been almost completely sawn through in two places. The front portion of the train had traversed the damaged track safely and it was theorised that the weight of the train had completed the fracture of the rails, derailing the rear vehicles. [1747]

Railwaymen of the United Railways of Habana were on strike in Matanzas Province, Cuba, in late 1933, in protest of a rule preventing railwaymen transferring from one division to another, which they claimed prevented veteran workers from taking employment in other divisions on the basis of superiority should they be laid off in their own due to lack of business. In pursuance of the strike, acts of sabotage were committed to prevent the continued running of trains. Two passenger trains from Santiago de Cuba to Habana were targeted on Saturday 21st October 1933. In the first attack in the early morning nearly 400 feet of track was blown up near Empalme, derailing the engine and two baggage cars of an express. In the second attack an open switch derailed another train at Jovellanos. No one was hurt in either derailment. [1748]

What was described as a Trans-Siberian express on the Chinese Eastern Railway was derailed by bandits who had unbolted the rails while it was travelling at 50 miles an hour, about 40 miles east of Tsitsihar, Manchukuo, on Sunday 26th November 1933. The bandits then opened fire upon the train, which was carrying about 600 passengers. Eight Japanese soldiers on the train returned the bandits' fire and reportedly routed them. Four soldiers were killed and eight wounded, while ten passengers were also badly wounded. Some of the passengers were captured by the bandits but were later released. [1749]

Another express was derailed by bandits, reportedly 45 miles west of Tsitsihar, just after midnight on Thursday 14th December 1933, again by loosening the rails. Three hundred bandits then swooped on the wrecked train from their hiding places, looting the mail car and stripping the passengers of valuables. Some of the wealthier-looking passengers were kidnapped for ransom. According to a Reuters report six Manchus, three Japanese civilians, three Japanese soldiers, a Russian and a European, stated in some reports to have been a German but probably M. Bauer, a Czech employee of the Skoda armaments company, were killed and 11 Manchu civilians, three Manchu soldiers and a Russian were injured. The survivors suffered terribly from the cold, the temperature being 50 degrees below zero Fahrenheit (about -28°C). It is unclear whether those killed

---

[1745] *The Times*, London, England, 4th May 1933
[1746] Report by the Railways Board on Indian Railways for 1933-34
[1747] *The Railway Gazette*, London, England, 7th July 1933
[1748] *The Brownsville Herald*, Brownsville, Texas, United States, 22nd October 1933
[1749] *The Times*, London, England, 28th November 1933

lost their lives in the initial derailment or in the short gunfight which took place between the bandits and the small military escort of the train. Several hours after the derailment a Japanese armoured train arrived, drove off the bandits and administered aid to the survivors. [1750] [1751] [1752] [1753]

Following the election of a centre-right party in Spain a short-lived anarchist insurrection began in December 1933 characterised by violent clashes with police, the burning of churches, the destruction of property and railway sabotage. [1754] The latter included the blowing up of a bridge between Puig and Puzol, outside of Valencia, by what was described as a contact bomb, as train No. 702 was passing was passing at 02:30 on Saturday 9th December 1933. The tender engine No. 4532 was blown to pieces and four coaches were hurled into the Puig gorge. Others remained suspended in mid-air and the restaurant car lay across the line, with the rear coaches of the train piled up on top of it. The bridge itself was damaged and distorted but did not collapse.

The casualties were initially stated as five dead, 11 seriously wounded and 27 less badly hurt. Those hurt were taken to Valencia. Local villagers gave help, carrying lanterns and torches. The train passengers were mostly third-class.

A later report stated that the number of dead had risen to 30. [1755]

In Cuba in September 1933, a short-lived Provisional Revolutionary Government had been formed which amongst other measures abrogated the 1901 Platt Amendment, which gave the United States the right to interfere in Cuban affairs. [1756] On Thursday 21st December pro-government supporters of its demonstration against the Platt Amendment travelled by train to Habana. Two United Railways of Habana trains carrying them, together with a relief train, were derailed by unknown persons. Train No. 1 was derailed between Guareiras and Manguito in Matanzas Province, causing the death of a fireman. Train No. 2 was derailed near Aguica, also causing the death of a trainman. A relief train sent from Sagua to the second derailment was derailed near Cascajal. Finally, a powerful bomb exploded in the station at Santa Clara, as an excursion train was leaving for Habana. [1757] [1758]

A westbound international express was wrecked 20 miles west of Harbin and between there and Manchuli, Manchukuo, on Thursday 18th January 1934. Having left the rails part or all of the train plunged over a cliff and fire broke out, consuming seven Pullman cars and a mail van. Nine or ten passengers died according to different reports, including four Russians, a Belgian and two Catholic missionaries. The injured included seven Chinese, six Japanese, several Russians, a German and M. Laurens, a correspondent for the *Paris Midi*, who lost his right eye.

Initial reports stated confidently that the derailment was the work of bandits who had

---

[1750] *Defiance Crescent-News*, Defiance, Ohio, United States, 15th December 1933
[1751] *Biloxi Daily Herald*, Biloxi, Mississippi, United States, 15th December 1933
[1752] *The Daily Mail*, Hagerstown, Maryland, United States, 15th December 1933
[1753] *The Times*, London, England, 16th December 1933
[1754] Wikipedia – Anarchist insurrection of December 1933
[1755] *The Times*, London, England, 11th and 12th December 1933
[1756] Wikipedia – History of Cuba
[1757] *The Daily Times-News*, Burlington, North Carolina, United States, 21st December 1933
[1758] *Mason City Globe-Gazette*, Mason City, Iowa, United States, 21st December 1933

torn up the track, but some doubt was cast on this by the fact that no trace of any bandits could be found. There were, however, unsuccessful attempts made to blow up a Japanese armoured train carrying Major-General Kawahara and a Harbin – Vladivovstock express at locations on the eastern branch of the Chinese Eastern Railway. [1759] [1760] [1761] A special train carrying over 100 survivors from the Harbin – Manchuli accident was itself attacked by bandits 50 miles west of Tsitsihar. 'The frequency and daring of these attacks lead the Japanese and Manchukuo authorities to attribute them to a group of terrorists seeking to embroil Soviet Russia and Japan in war.' [1762]

A decree by the right-wing government of António Salazar banning collective action, trade unions and strikes led to an uncoordinated general strike in Portugal at the beginning of 1934. Strikes took place in Lisbon and various other towns and on Thursday 18th January the actions included the derailing of a train. [1763] Caminhos de Ferro Portugueses train No. 2105, due to leave Lisboa Santa Apolónia at 01:10 for Entroncamento, consisted of an engine, 50 mostly empty waggons and a guards' van, in charge of Brakesman Luiz dos Santos Novais with Driver Agostinho Pires and Fireman Cipriano Sanches Capinha. It started somewhat late and did not reach Sacavém station until after 03:00, where goods train No. 2007 from Entroncamento to Lisboa passed it, going in the opposite direction. A little further on, at Kilometre 15,620, near Santa Iria station, where the line was sandwiched between a road and a salt marsh, the train ran into a heap of ballast which had been piled onto the track, and it was badly derailed at 03:35.

Driver Pires told a newspaper reporter later that his engine had suddenly jumped up in the air. He had realised that it was derailed and gently applied steam in reverse, then there had been a terrific bang and the waggons had piled up behind his engine. As soon as his engine had come to a stand he checked that his fireman was unhurt and then they had then gone looking for the brakesman. They found him screaming with pain and had only been able to free him from the wreckage with considerable difficulty. The driver was sure that the saboteurs had been aware of the timings of the two trains and had intended that train No. 2105 should become derailed immediately in front of train No. 2007, which would then have ploughed into the wreckage. It was only the late running of train No. 2105 which had frustrated their plan.

Two rescue trains were sent to the accident site from Campolide, with a squad of workmen to clear the line. Photographs of the accident show a large pile of wrecked waggons, and it was reported that the clearance operation was a lengthy and difficult one. [1764]

A bomb exploded about 20 miles north of Zagreb, Yugoslavia, behind a seat in a second-class compartment of the through carriage from Berlin to Sushak of a Wien-Zagreb express, on Monday 22nd January 1934. Three of the 13 passengers in the coach were killed and their bodies so badly burnt as to be unrecognisable. An Austrian from Graz was wounded. It was believed that the bomb had been planted by Croat émigrés

---

[1759] *The Hammond Times*, Hammond, Indiana, United States, 19th January 1934
[1760] *Edwardsville Intelligencer*, Edwardsville, Illinois, United States, 19th January 1934
[1761] *The Times*, London, England, 19th January 1934
[1762] Ibid, 20th January 1934
[1763] Wikipedia – Portuguese general strike of 1934
[1764] *Diario de Lisboa*, Lisboa, Portugal, 18th and 19th January 1934

residing in Austria or Germany, keen to spoil the local effect of a conference of the Little Entente Ministers. Following the outrage, orders were given that all through carriages from Austria were to be stopped at the Yugoslav frontier and their passengers requested to change to Yugoslav coaches. [1765]

Cuba was in political turmoil in March 1934, with workers threatening a general strike. On the evening of Wednesday 14th March 1934 a mixed train from Santiago de Cuba was dynamited five miles from Camaguey. With the railwaymen on strike the train was being driven by Lieut. Palmer. He and the privates acting as fireman and brakeman were badly scalded by escaping steam when the locomotive and four cars were wrecked. Fourteen other cars were derailed. The only passengers on the train were 25 soldiers acting as its escort. They were unhurt. [1766]

The night express from Wien to Köln and Paris, running at high speed, collided with some obstructions on the line and was derailed near Wels, Austria, early on Tuesday 10th April 1934. The engine, three luggage and postal staff and two through coaches for Paris left the track. Buried beneath a pile of coal the fireman was suffocated and a mail clerk was also killed, while 17 people, mostly travelling post office workers, were injured, and were removed to Wels and Linz. Several lengths of rail had been unscrewed and thrown across the line.

A reward of 5,000 schillings was offered for the apprehension of those responsible. Two previous derailments due to sabotage had occurred at the same location in the previous three years, for which men had been convicted and sentenced. [1767]

The suggested motive for the sabotage was an attempt by left-wing radicals to assassinate Austrian Vice Chancellor Emil Fey, a right-wing politician who had commanded the Heimwehr paramilitary group during the brief Austrian Civil War in February of the same year. He was a passenger on the train. [1768] [1769]

Repeated attacks on the eastern section of the Chinese Eastern Railway (now renamed the North Manchurian Railway) culminated on Saturday 26th May 1934 in an attack on a Japanese military train in which 12 soldiers were killed and 26 wounded. Six feet of track had been removed at Santunghotze Bridge, near Hsiachengtze, Manchukuo, and an engine and four carriages were derailed. Five hundred bandits then poured a withering fire into the train from all sides. The soldiers scrambled out of the carriages and drove off the attackers, who left 40 dead behind them and carried off a score more. [1770]

An estimated 100 bandits wearing Manchurian Army uniforms derailed a passenger train on the North Manchurian Railway near Jablory, Manchukuo, probably on Thursday 31st May 1934. One passenger was killed and two were wounded. [1771]

1765 *The Times*, London, England, 23rd January 1934

1766 *Thomasville Times-Enterprise*, Thomasville, Georgia, United States, 15th March 1934

1767 *The Times*, London, England, 11th April 1934

1768 *Biloxi Daily Herald*, Biloxi, Mississippi, United States, 14th April 1934

1769 Wikipedia – Emil Fey

1770 *The Times*, London, England, 29th May 1934

1771 Butte *Montana Standard*, Butte, Montana, United States, 2nd June 1934

The resumption of through trains after three years between Peking and Mukden was marred by the explosion of a bomb in a third-class carriage of a crowded Mukden-bound express at a location 70 miles beyond Tientsin, China, on Tuesday 31st July 1934. A Chinese boy was killed, and eight Chinese were seriously wounded. [1772]

Chicago Burlington and Quincy Railroad passenger, mail, and freight train No. 72, in charge of Conductor J.E. Elliott with Engineer Edward Robertson and Fireman C.J. Jones, was derailed at a dynamited bridge four miles north of Waltonville, Illinois, United States at 4 a.m. on Tuesday 7th August 1934. The engineer and fireman were slightly hurt. The train was bound for Paducah, Kentucky.

Farmers living in the vicinity said that they had heard two or three explosions just after 1 a.m., while the blast were heard and felt in Mt Vernon, eight miles away, where windows and doors were shaken, and people thought it was an earthquake. Both tracks of the railroad were torn up and the bridge over the Little Muddy River was damaged.

The line involved was one of the main feeders of the Southern Illinois coalfield and the sabotage was linked to a bitter feud which was ongoing between the Progressive Workers of America union, the more right-wing United Mine Workers of America union and the Peabody Coal Company, which had cut miner's wages. [1773]

In a further act of sabotage connected to the Southern Illinois mining conflict just after midnight on Wednesday 22nd August 1934 what was thought to be a time bomb exploded under a Chicago and Illinois Midland Railroad train at Springfield, Illinois, United States after the engine and first few waggons had passed the spot, derailing 24 cars, most of them loaded. No one was hurt. [1774]

The locomotive and four vehicles of a Japanese armoured train were derailed by bandits on the North Manchurian Railway between Hsiaochiu and Mifeng stations, Manchukuo on Thursday 23rd August 1934. 'The casualties are feared to be heavy, and a rescue train has been sent out.' [1775]

The express from Harbin to Hsinking on the North Manchurian Railway was derailed by bandits shortly before 10 p.m. on Thursday 30th August 1934 between Wuchia and Schangcheng, Manchukuo. As the engine and three coaches tumbled down an embankment the attackers opened what was described as a withering fire on the wreckage. Japanese soldiers riding on the train scrambled out and returned fire and after an hour's fighting the bandits withdrew with six hostages. Ten Japanese were reportedly killed, seven severely and two slightly wounded, while three Russians and a Chinese were also wounded.

The hostages included a Dane, M.E.F. Johansen, manager of the Far Eastern division of the Metro-Goldwyn-Mayer Film Company and Mr R. Lury, an American who was the Manchukuo representative of the same company. Lucker was another American, Mr. J.J. Russell, a sales manager of the company, who was in a room the locked door of which the bandits were unable to smash with their rifles. The bandits reportedly seized

---

[1772] *The Times*, London, England, 2nd July 1934
[1773] *Centralia Evening Sentinel*, Centralia, Illinois, United States, 7th August 1934
[1774] *Dixon Evening Telegraph*, Dixon, Illinois, United States, 22nd August 1934
[1775] *The Times*, London, England, 24th August 1934

possessions from the living and removed bracelets, earrings, and rings from the dead. [1776]

What was described as a disastrous wreck occurred on the southern section of the North Manchurian Railway on about Saturday 6th October 1934 when bandits derailed a train at an unstated location in Manchukuo. Two foreigners and nine Japanese were kidnapped, and many Japanese were killed or wounded. A correspondent for the London *Times* was informed by a foreign passenger that he had seen the bandits drag out passengers, enquire as to their nationality, and if they said that they were Japanese they were immediately shot. The Japanese taken prisoner had only survived because they had claimed to be Mexicans, Filipinos, Hawaiians, or Koreans. [1777]

A freight train was derailed near Maltrata, Mexico, on the line between Vera Cruz and Mexico City, on Thursday 22nd November 1934, the track having been destroyed by 25 Cristeros (Catholic) rebels, led by a local sexton named Corona. According to one account the derailment occurred in a tunnel; another unconfirmed account stated that the engineer and fireman had been shot by the rebels. The Catholic Church had been in conflict with the government since the passing of an anti-clerical constitution in 1917. [1778] Government forces under General Lucas Gonzales and Lt. Col. Francisco Padillo reportedly easily defeated the rebels and repaired the track. [1779] [1780]

On Saturday 15th December 1934 No. 3 Up Mail was derailed at Bridge No. 173 between Penwegon and Tangywein, Burma, on the Rangoon – Mandalay main line of the Burma Railway. A woman third-class passenger was fatally injured, the driver and fireman were seriously injured, and 12 passengers were hurt. Damage to engine and rolling stock amounted to 36,800 rupees. Both pairs of fishplates had been removed from the joints immediately preceding the bridge by unknown persons. [1781] There had been a previous attempt at sabotage at the same spot. [1782]

A log placed upon the track derailed the engine and several cars of a passenger train between El Salto and Durango City, Mexico, on Thursday 24th January 1935. A half-hour gun battle then ensued between the soldiers escorting the train and the saboteurs, who were described as outlaws or rebels, before the attackers withdrew. Engineer José de los Rios and Fireman A. Garcia were killed in the first exchange of fire. [1783]

On Sunday 24th February 1935 a 32-car Chicago and Illinois Midland Railroad freight train was derailed by sabotage at Andrew station, Illinois, United States. Damage to a gasoline tank car in the train caused a fire which destroyed 10 cars, but no one was hurt. [1784]

Late on Thursday 2nd May 1935 Chinese bandits derailed a train on a new Japanese-

---

[1776] Ibid, 1st September 1934
[1777] Ibid, 10th October 1934
[1778] Wikipedia – Cristero War
[1779] *Valley Morning Star*, Harlingen, Texas, United States, 22nd November 1934
[1780] *The North Adams Transcript*, North Adams, Massachusetts, United States, 22nd November 1934
[1781] Report by the Railway Board on Indian Railways for 1934-35
[1782] *The Times*, London, 17th December 1934
[1783] *San Antonio Express*, San Antonio, Texas, United States, 25th January 1935
[1784] *Freeport Journal Standard*, Freeport, Illinois, United States, 10th August 1935

built line between Hsinking and North Korean ports. The attack occurred near Tashihtoutzu, Manchukuo, east of Tunhua. After the derailment guards escorting the train battled the attackers, reportedly numbering 100, for half an hour before Japanese troops came to their aid and the attackers fled, leaving 13 dead. Five of the train guards were slain, 14 passengers were injured, and a dozen were taken hostage. Most of the passengers were Japanese. [1785]

A train which had left Hsinking on the evening of Monday 29th July 1935 for Korea was derailed between Yingchengtze and Tumenling, Manchukuo, 40 miles from Hsinking, about 100 bandits having removed four rails from the track. The locomotive, luggage van and dining car left the track and the attackers, who had concealed themselves nearby, then poured repeated volleys into the train, killing two of its guards and wounding and overpowering the other 10 in the darkness.

A railwayman ran to a telephone and gave the alarm, whereby two armoured trains, which were held constantly in readiness, rushed to the scene. Before they had arrived, the bandits had fled in a north-westerly direction towards their mountain hideout, taking hostages with them. 'Troops, police and railway guards are hot on their trail and are now throwing a cordon around the entire mountain.'

Ten Japanese and two Chinese were killed, five Japanese and three Chinese were wounded, and five Japanese and 20 Chinese were kidnapped.

'The capital {Hsinking} was deeply shocked by the news of this outrage, the first which has yet occurred in the district midway between Hsinking and Kirin, which since the founding of the new State of Manchukuo, has been considered among the most peaceful parts of the country.' [1786]

The Southern Illinois mine war dragged on and early on Wednesday 5th June 1935 a Chicago and Illinois Midland Railroad train was derailed near Athens, Illinois, United States, by dynamite laid under the track. A hole several feet deep was formed by the explosion. The locomotive pitched forward and then onto its side, taking several cars with it. Railwayman W. Sapp was slightly injured. The train had been carrying flour and general goods. The usual precaution of having a handcar carrying special agents preceding the train had been employed; presumably the handcar was not heavy enough to set off the mine or it was remotely detonated after the handcar had passed.

On Saturday 10th August 1935 a train of the Illinois Central Railroad was the target of another act of sabotage near Springfield, Illinois, United States. Shortly after midnight a dynamite blast derailed the engine and ten empty freight cars two miles south of Springfield. Engineer Tony Hoff escaped injury, but Fireman Ira Fitzgerald and Conductor W.W. Westerbrook were badly enough hurt to be taken to hospital. A mine set off from a vantage point along the line was apparently used. [1787]

Eleven people died and nine were wounded when bandits derailed, wrecked, and looted and eastbound express on the Hsinking – Tumen line near Weihuling, Manchukuo, 35 miles west of Tunhua on Monday 23rd September 1935. A relief train sent to the scene was also derailed at an unstated location with 19 casualties – whether these were dead, or

[1785] *The Evening Sun*, Hanover, Pennsylvania, United States, 3rd May 1935
[1786] *The Times*, London, England, 1st August 1935
[1787] *Freeport Journal Standard*, Freeport, Illinois, United States, 10th August 1935

wounded was not stated. The victims were Japanese, Manchoukuans and Koreans. [1788]

Bolts removed from the rails caused the derailment of the engine and six waggons of a goods train near Aligarh, northern India, and the death of a driver on about Monday 16th December 1935. [1789]

Robbery was apparently the motive for an attack upon a Vera Cruz – Mexico City express late on Monday 6th April 1936. The train had left Vera Cruz at 7.30 p.m. with an estimated 150 passengers. It arrived at Paso del Macho, Mexico, just before 9.10 p.m. and here an oil-burning locomotive was substituted for its electric locomotive for the steep gradients of the next stage of the journey. A few minutes after leaving there it was crossing a bridge over a 33 ft ravine at Kilometre 354 when it exploded a dynamite bomb placed there by bandits. The engine, express car, two first-class and two second-class cars had crossed the bridge when the bomb went off under three wooden Pullman cars. Two of the cars fell into the ravine and the third was left hanging from the bridge. Oil from the locomotive then sat fire to the wreckage.

The fireman of the train escaped injury and ran back to Paso del Macho station to report the accident. Relief trains were sent to the scene from there, from Orizaba and from Vera Cruz, the latter train being preceded by a troop train ordered out in pursuit of the bandits by General Heriberto Jara, the commander of the Vera Cruz military district.

Passengers on the train included three candidates for the post of Vera Cruz state governor, and the superintendent of the terminal company of Vera Cruz, who was carrying a large sum of money. There was also more money being carried in the express car.

Initial estimates were that at least 90 people had died in the accident. [1790] This proved to be a considerable exaggeration; the number of dead was later corrected to 18. Four days later bandits derailed another train near Chalco, this time without causing casualties. [1791]

A bomb exploded beneath a Japanese troop train east of Tientsin, China, on Friday 29th May 1936. Three horses in the box car beneath which the explosion had occurred were injured but the blast was not powerful enough to derail the train. Armed with machine guns and cameras, men were sent to the scene to obtain evidence of the attack. [1792]

Until the First World War, Palestine had been part of the Ottoman Empire. Following the defeat of Turkey, Palestine was ruled by the British, and the Balfour Declaration of 1917 declared it a national home for the Jewish people. Increased Jewish immigration following the war led in 1936 to an Arab revolt which lasted until 1939. The Palestinian railways suffered a number of acts of sabotage during this period, the first of which occurred on Monday 1st June 1936, when a light engine was derailed between Jerusalem and Lydda, several fishplates having been removed. About 100 yards of the permanent way was ploughed up. [1793]

[1788] *San Antonio Express*, San Antonio, Texas, United States, 23rd September 1935
[1789] *The Times*, London, England, 18th December 1935
[1790] *The Racine Journal-Times*, Racine, Wisconsin, United States, 7th April 1936
[1791] *San Antonio Sunday Light*, San Antonio, Texas, United States, 12th April 1936
[1792] *The Times*, London, England, 1st June 1936
[1793] Ibid, 3rd June 1936

A patrol train was derailed near Haifa, Palestine, on Wednesday 10th June 1936, a rail having been sawn through. No one was hurt but traffic was interrupted. [1794]

On Friday 12th June 1936, as the morning train from Haifa was leaving the station at Qalqiliya, Palestine, for Jerusalem, Jaffa and Tel Aviv, an Arab, threw a bomb into a third-class carriage at the rear of the train. The coach was crowded with mostly Jewish passengers, 14 of whom were wounded, four seriously. The escort of the train then fired at the bomb thrower, but his retreat was covered by return fire from accomplices and all the attackers made their escape. [1795]

Two attacks on Palestinian railways were made on the night of Thursday 18th June 1936. A goods train was derailed by the removal of a rail between Lydda and Jaffa, some damage was done, and its guard was fired at, though not hit. Meanwhile near Qaqun on the Lydda – Haifa line a sawn through rail derailed a military patrol train. [1796]

A patrol trolley was fired upon near Gaza, Palestine, late on Monday 22nd June 1936, then derailed by the explosion of an old Turkish bomb, slightly wounding a British officer in the foot. [1797]

A more serious attack occurred near Kafr Jinnis, five miles north of Lydda, on Friday 26th June 1936. An empty train with an escort from the Cheshire Regiment was proceeding towards Lydda. The Arabs had removed two rails above a culvert, causing the engine to drop 10ft and the derailment of two coaches. The driver and fireman were scalded to death in the overturned engine, while Lance Corporal A. Houtram was also killed and Private J. Hooley was wounded. The attackers then opened fire on the wrecked train, wounding Lieutenant J.L. Gardner Brown, and Gunner A. Murray of the Royal Horse Artillery. When help arrived from Lydda the Arabs withdrew, leaving a wounded man behind. [1798]

Italy had invaded Abyssinia in October 1935 and the war which followed lasted until February 1937. The Italians did not have things all their own way and in early July 1936 the Ethiopians repossessed the town of Mojjo, blocking the railway line between Addis Ababa and Djibouti in French Somaliland. The Italians hastily sent two armoured trains from Addis Ababa on about Monday 6th July 1936. These were intercepted by the Ethiopians. One was derailed and the other was forced to withdraw back to Addis Ababa. Many Italian soldiers were reportedly killed. [1799]

A bomb exploded under the seat of a carriage in the station at Acre, Palestine, on the morning of Tuesday 14th July 1936. Some damage was caused but no one was hurt. [1800]

The Spanish Civil War broke out on Friday 17th July 1936. A few days later, on about

[1794] Ibid, 12th June 1936
[1795] Ibid, 13th June 1936
[1796] Ibid, 20th June 1936
[1797] Ibid, 23rd June 1936
[1798] Ibid, 27th June 1936
[1799] *Oakland Tribune*, Oakland, California, United States, 10th July 1936
[1800] *The Times*, London, England, 15th July 1936

Wednesday 22nd July, the rebels (Nationalists) derailed a train by sabotage in the Bilbao region of Spain, killing four people and injuring 27. [1801]

The Haifa – Lydda line was blocked by the derailment of a goods train near Qalqiliya, Palestine, on the evening of Friday 24th July 1936, the rails having been tampered with. The engine and two waggons were derailed. It was thought that the evening passenger train had been the target. Collective fines were imposed on the villages of Qalqiliya and Jiljiliya, between which the sabotage had occurred. [1802]

A military patrol train was derailed through sabotage while crossing a bridge near Athlit, Palestine, on the Haifa – Lydda line, then fired upon on Wednesday 12th August 1936. There were no casualties. [1803]

A double-headed goods train was derailed by sabotage near Qalqiliya, Palestine, on the afternoon of Friday 4th September 1936. An unnamed 'native' driver and a British soldier on the pilot engine were killed, and Driver Sidney Monk, Fireman Petty Officer H. Shorter, R.N., Driver T. Oates of The Cheshire Regiment and Fireman W. Brown, R.N., were wounded. The blocking of the line meant that all evening trains were cancelled. [1804]

A Japanese troop train carrying Manchurian and Japanese soldiers was derailed and ambushed by Chinese bandits during the night of Monday 14th September 1936 near Taimakou, Manchukuo, 60 miles west of Suifeno. Immediately after it had been derailed a hail of bullets was poured into the train. Those on the train fought back and the bandits eventually fled, reportedly leaving many dead and dying. Twenty-five Japanese were killed and 65 wounded. [1805]

A strike on the Louisiana and Arkansas Railway, precipitated by the railroad company making a 20 per cent pay cut, turned violent on Monday 28th September 1936 when a train was derailed at Moore, Louisiana, United States. [1806]

Southbound passenger train No. 3 consisted of engine No. 392, three baggage cars, a combination mail and baggage car, a chair car, a Pullman sleeping car and a chair car, in charge of a 'replacement', that is, scab, train crew consisting of Conductor J.W. Barnage with Engineer Shelton Beauchamp and Fireman Walter Delco. The train left Winnfield, 6 miles north of Moore, at 11.40 p.m., one minute late, and travelling at between 30 and 40 miles an hour was derailed at about 11.50 p.m. Engineer V.N. Vaught and Special Agent J.T. Dempsey were killed and Engineer Beauchamp and Fireman Delco were injured. [1807]

Engineer Beauchamp told the Interstate Commerce Commission (ICC) inquiry that when his train had been 400 or 500 feet from the switch where the train was derailed he had seen the switch lamp was green and had called it to his fireman. He had then seen an object which he at first thought was an animal before he realised that it was a man. He had then seen the man jerking the switch lever before running away to the nearby woods,

---

1801 *The Daily News*, Huntingdon, Pennsylvania, United States, 28th July 1936
1802 *The Times*, London, England, 27th July 1936
1803 Ibid, 14th August 1936
1804 Ibid, 5th September 1936
1805 Ibid, 16th September 1936
1806 *Monroe Morning Star*, Monroe, Louisiana, United States, 4th October 1936
1807 *Blytheville Courier-News*, Blytheville, Arkansas, United States, 29th September 1936

hiding his face.

Fireman Delco said he had seen the switch lamp at green but not the man.

Conductor Barnage spoke of having found the switch points open an hour after the derailment.

Section Foreman Carrier said that the switch had been in good condition when last examined on 25th September. He had found the lock to the switch in the grass about 4½ feet from it.

Assistant Engineer Danforth described in detail the conditions found after the derailment.

Sheriff Sholars stated that tracker dogs had found a trail leading to the switch stand and back through the woods to a gravel road, from where the saboteur had apparently left by automobile.

The ICC concluded that the switch had been cocked or partly opened with malicious intent directly in front of the approaching train. [1808]

In another derailment on the Louisiana and Arkansas Railway near Jamestown, Louisiana, United States, early on Thursday 12th November 1936, Engineer W.M. Johnson, Fireman T.H. Bost and Brakeman Joe Jacobs were slightly injured. A switch had been tampered with. The same day a train was derailed three miles south of Springhill, Louisiana, United States, spikes and angle bars having been pulled out. No one was hurt in this derailment. [1809] [1810]

A train carrying a thousand of Generalissimo Franco's foreign rebel troops was derailed on about Tuesday 9th February 1937 by government agents who had made their way behind rebel lines. The sabotaged train fell into a ravine near Cordoba, Spain. The casualties were not stated. [1811]

Two acts of sabotage, aimed at disrupting the movement of Japanese troop trains, were carried out in Manchukuo in late July 1937. On Thursday 22nd a train was derailed six miles south of Hsinking, killing eight Japanese and wounding 42 Japanese and seven foreigners. Exactly a week later, on Thursday 29th another train was derailed between Chungliancheng and Tientsin, 12 miles further east. There were five deaths, including one or more foreigners, and 60 were wounded. [1812]

On about Thursday 16th December 1937 a Japanese troop train was derailed, it was presumed by sabotage, about midway between Tientsin and Tangku, China, and 50 soldiers, riding in the leading car, were killed. After the accident soldiers were deployed in the fields adjoining the line in the expectation of an attack.

Also on the train were about a dozen foreigners, trying to reach Tientsin from Tangku, the ocean terminus of the line, whom the Japanese had permitted to ride on it.

A reporter described the events after the derailment: 'While deploying his men the Japanese officer in charge, speaking English, had told the foreigners to leave hurriedly

---

[1808] Interstate Commerce Commission report

[1809] *Hattiesburg American*, Hattiesburg, Mississippi, United States, 12th November 1936

[1810] *The Mobile Times*, Mobile, Alabama, United States, 12th November 1936

[1811] *Lawrence Daily Journal*, Lawrence, Kansas, United States, 9th February 1937

[1812] *Moorhead Daily News*, Moorhead, Minnesota, United States, 31st July 1937 because he expected an attack. There was great confusion. Soldiers milled about, there were sounds of escaping steam, the crackling of fire as it spread along the train, the cries of those pinned in the wreckage. A Russian mother could not find one of her sons. Frantic, she appealed to other foreigners. A.P. Tucker, a Chinese representative of a tobacco company, said he would investigate. At first a Japanese officer blocked the way, but he was told about the boy. "Go back, but hurry," he said. Tucker found the boy, crushed to death.

'The expected attack did not occur, but the foreigners, loaded down with luggage, had to face a 13-mile walk back to Tangku. The Japanese would not permit them to go further towards Tientsin because they thought marauders were in the neighbourhood. An American woman, struggling along the railroad ties with a heavy suitcase banging against her knees, suddenly put down the bag and cried, "They can't do this!"

'"What do you mean, they can't do this?" asked Tucker, arms full and perspiring.

'"They can't wreck this train. The Boxer treaty says the railroad must be kept open between Peiping and the sea."

'"Lady," said Tucker, lapsing for a moment into his boyhood Virginia talk, "They've already done it." The party finally got to Tangku and came up to Tientsin on a river barge a few days later.' [1813]

Having occupied Manchuria (which by now had become the puppet state of Manchukuo) the Japanese had declared war on China in July 1937 and soon captured the cities of Peking, Shanghai and Nanking, plus much territory, but they were unable to defeat Chinese Communist forces, who retreated into the hinterland and waged a guerrilla war with much sabotage. [1814] A correspondent to the London *Times* wrote in early March 1938, 'These guerrillas are causing the Japanese much anxiety. Railway services are interrupted almost every day by night attacks in which the guerrillas blow up sections of line. Five days ago {Friday 25th February 1938} they unpinned several rails on a gradient south of Shihchiachwang, wrecking a Japanese supply train and killing a number of Japanese soldiers.' [1815] Another supply train was reportedly wrecked on Sunday 28th February at an unstated location and again a number of Japanese soldiers were killed. [1816]

A Kiev – Moscow express was derailed near Konotop, in the Ukrainian Soviet Socialist Republic of the USSR, on the night of 6th to 7th March 1938, the track being damaged. A senior Soviet official, who was not identified in the report, was a passenger on the train. The authorities became aware of the intended sabotage and extensive unspecified precautions were said to have been taken, presumably involving the train passing through the area where the attack was expected at much reduced speed, so that although it was derailed the accident was not serious. [1817]

After an apparent break of a year and a half Arab attacks on railways in Palestine resumed with the derailment of a goods train between Haifa and Lydda, on Tuesday 15th March 1938. [1818]

[1813] *Daily Capital News*, Jefferson City, Missouri, United States, 18th December 1937
[1814] Wikipedia – Second Sino-Japanese War
[1815] *The Times*, London, England, 4th March 1938
[1816] *The Evening Star*, Washington, D.C., United States, 3rd March 1938
[1817] *Haagsche Courant*, Den Haag, the Netherlands, 11th March 1938
[1818] *The Times*, London, England, 16th March 1938

Late on Thursday 24th March 1938 a bomb exploded under a trolley carrying a military patrol near Khan Yunis, Palestine, close to the border with Egypt. The trolley was wrecked, Drummer Thompson, of the Royal Scots, was killed and Private Young was slightly wounded. Traffic was disrupted for several hours. [1819]

A goods train was derailed, and oil waggons were set on fire at Ras-el-Ain, Palestine, on the Lydda – Haifa line, on Thursday 7th April 1938. [1820]

An express from Algiers to Oran was derailed near Orléansville, Algeria, on the morning of Sunday 17th April 1938. The locomotive and four carriages, one of them a sleeping car, left the rails and five Europeans were killed and about 10 injured. Sabotage of a set of points by a recently dismissed railwayman was suspected and he was arrested. [1821]

A Japanese troop train was blown up between Chochow and Liuliho by Chinese irregulars on about Thursday 28th April 1938. [1822]

Removal of rails derailed a military trolley near Ramleh, Palestine, on about Saturday 30th April 1938. [1823]

In 1938 a conflict had arisen in Mexico between the government of President Lázaro Cárdenas and General Saturnino Cedillo, who had his own private army. [1824] On about Tuesday 24th May 1938, 40 of Cedillo's men dynamited a repair train between El Gato and El Borrego, on the San Luis Potosi – Tampico line, killing two trainmen and wounding four more. [1825]

On Thursday 26th May 1938 Cedillo forces derailed the *Sunshine Special* international train between Mexico City and Laredo near Jaral de Berrios, Mexico, in Guanjuanto state at 6.30 a.m. [1826] The engine and five coaches overturned but no one was hurt. [1827]

A freight train of ten cars and a caboose was derailed near Las Tablas, Mexico, in San Luis Potosi state, by Cedillo rebels on Saturday 4th June 1938. One car was blown up and nine overturned. A trainman was wounded. [1828]

Rebels derailed a Southern Pacific of Mexico freight train near Ruiz Mayarit, Mexico, on about Friday 10th June 1938, killing two trainmen. Spikes had been removed from the track. Ten box cars were derailed, and the line was blocked for 12 hours. [1829]

[1819] Ibid, 26th March 1938
[1820] Ibid, 9th April 1938
[1821] Ibid, 18th and 19th April 1938
[1822] Ibid, 3rd May 1938
[1823] Ibid, 2nd May 1938
[1824] Wikipedia – Saturnino Cedillo
[1825] *The Charleston Daily Mail*, Charleston, West Virginia, United States, 25th May 1938
[1826] *San Mateo Times*, San Mateo, California, United States, 26th May 1938
[1827] *The Times*, London, 28th May 1938
[1828] *El Paso Herald Post*, El Paso, Texas, United States, 4th June 1938
[1829] *Arizona Republic*, Phoenix, Arizona, United States, 12th June 1938

An international express which had left Mexico City the previous evening was derailed on Monday 27th June 1938, 21 miles north of San Luis Potosi, Mexico, injuring the engineer, fireman and other members of the train crew. A spike had been removed from a rail, causing the engine and express cars to overturn. A relief train forwarded the passengers and baggage to Laredo. [1830]

Cedillo rebels dynamited a passenger train between El Gato and El Borrego, Mexico, within a mile of the location of the May attack, on Friday 29th July 1938, having first captured a party of track workers to prevent their inference. A second-class coach of the Ferrocarril Nacional de México train was wrecked and two passengers, an army lieutenant and an infant, were killed and others slightly hurt. The rebels were driven off by the military escort of the train and their leader, Leonarda Tapia, a Cedillista who had previously been granted an amnesty by President Cárdenas, was killed. [1831] [1832]

A freight train was derailed between Monterrey and Tampico, Mexico, on Monday 7th November 1938, killing the engineer and a brakeman. Several freight cars left the track. Tampering with the track by bandits, presumably Cedillo rebels, was blamed. [1833] This would be the last act of sabotage committed by the rebels. The rebellion was supressed, and Cedillo was killed in January 1939. [1834]

The Punjab mail train from Calcutta on the East Indian Railway was derailed near Madhupur, Bihar, India, on Tuesday 7th June 1938. The engine, tender, and five leading coaches were derailed, capsized, and rolled down an embankment. The sixth coach was also derailed; the remaining two remained on the track. The Indian driver and a postal sorter were killed and 33 passengers, one a European, and a sorter were wounded. [1835]

The Senior Government Inspector announced at the end of the month that the derailment had been due to the deliberate removal of rails and other interference with the track. The East India Company offered a reward of 5,000 rupees for information leading to the arrest of the culprits. [1836]

An afternoon goods train between Haifa and Lydda, Palestine, was derailed on Friday 10th June 1938, presumably by sabotage. Traffic was interrupted for several hours. [1837]

A goods train was derailed due to sabotage at an unstated location in Palestine on about Saturday 9th July 1938. [1838]

A band of eight Arabs captured a platelayer at 9 a.m. on Thursday 28th July 1938, presumably to prevent him interfering, and having let a goods train pass unmolested removed the dog spikes from the track over a length of 40 yards north of Lydda,

---

[1830] *The Corpus Christi Times*, Corpus Christi, Texas, United States, 28th June 1938
[1831] *Nebraska State Journal*, Lincoln, Nebraska, United States, 30th July 1938
[1832] *The Times*, London, 2nd August 1938
[1833] *San Antonio Express*, San Antonio, Texas, United States, 8th November 1938
[1834] Wikipedia – Saturnino Cedillo
[1835] *The Times*, London, England, 9th June 1938
[1836] Ibid, 29th June 1938
[1837] Ibid, 13th June 1938
[1838] Ibid, 11th July 1938

Palestine, so that a passenger train bound for Kantara was derailed at 11 a.m. The engine and an empty van overturned, and an empty sleeping car and a first-class carriage were derailed. No one was apparently hurt, and the passengers were transferred to a relief train. [1839]

The engine and one passenger coach of an afternoon train were derailed on Saturday 6th August 1938 five miles outside of Jaffa, Palestine. [1840]

Early on Sunday 14th August 1938 a goods train was derailed by a mine near Tulkarm, Palestine. No one was hurt but the line was blocked all day. [1841]

A land mine wrecked a southbound goods train near Tulkarm, Palestine, early on Tuesday 30th August 1938. The engine tender and 13 trucks left the rails, and the driver was slightly hurt. [1842]

Having failed in an attempt to derail a passenger train the previous day, saboteurs derailed a goods train near Ras-el-Ain, Palestine, on Sunday 4th September 1938. [1843]

On Monday 5th September a goods train was derailed between Lydda and Wadisrar, together with an engine between Battir and Deirsheikh, Palestine, on the Jerusalem line. [1844]

A Haifa – Lydda goods train was derailed by sabotage on Tuesday 13th September 1938. The engine and 13 waggons were thrown over an embankment, the driver was killed and the fireman and guard wounded. [1845]

Two land mines exploded under a military trolley near Tulkarm, Palestine, on the Haifa – Lydda line, on Sunday 23rd October 1938. A British private was slightly injured. [1846]

A land mine exploded beneath a trolley between Haifa and Affuleh, Palestine, on Sunday 4th December 1938, injurung two of its Arab crew. [1847]

On the morning of Friday 20th January 1939 a landmine exploded under a military trolley piloting a train to Jerusalem near Ramleh station, Palestine. Private W. McKay, of Belfast, of the 1st Battalion, The Buffs, was killed and Lance-Corporal R.H. Carson, of London; Private J.K. Mullens, of Folkestone; and Sapper H. Hinchcliffe were wounded. [1848]

[1839] Ibid, 29th July 1938
[1840] Ibid, 8th August 1938
[1841] Ibid, 15th August 1938
[1842] Ibid, 31st August 1938
[1843] Ibid, 5th September 1938
[1844] Ibid, 6th September 1938
[1845] Ibid, 14th September 1938
[1846] Ibid, 24th October 1938
[1847] Ibid, 5th December 1938
[1848] Ibid, 21st January 1939

A land mine explosion derailed the engine and nine trucks of a goods train near Qalqiliya, Palestine, on Wednesday 15th February 1939, injuring its driver and fireman. As had happened before, a pilot trolley which had preceded the train had not been heavy enough to set off the mine. [1849] The Arab driver of the train died of his wounds the next day.[1850]

The Jaffa – Jerusalem train was derailed outside an Arab village in Palestine on Sunday 26th February 1939, bolts having been removed from the track. The engine and one coach overturned and the Arab driver, fireman and brakeman, and two Arab passengers were injured. [1851]

A land mine exploded beneath a train on the Jerusalem – Lydda line on Monday 27th March 1939, derailing three vans and slightly injuring six Arab labourers and a British soldier. [1852]

Five British soldiers died and one was wounded when a military trolley preceding a train between Tel Aviv and Lydda, Palestine, was derailed by a land mine on Monday 10th April 1939. Killed were Lance Corporal Robert Sydney Gould; Privates Henry Charles Watting, George Herbert Barlow, and Freeman Cecil George Jackson, all of the 1st Battalion, the Bedfordshire and Hertfordshire Regiment; and Sapper Robert Henderson Blair, 8th Railway Company, R.E. Private P.W. Shipley, 1st Battalion, The Sherwood Foresters. The following day two Arabs caught trying to lay another land mine were killed by a railway patrol. [1853]

Two land mines exploded beneath a train at 6 a.m. on Thursday 29th June 1939 between Acre and Haifa, Palestine, derailing two carriages but causing no casualties. A third unexploded mine was found further along the line. This was the last act of sabotage in Palestine until after the Second World War. [1854]

The Reformed Government of China was a puppet state which existed in the Japanese-occupied part of China between 1938 and 1940. [1855] On about Saturday 1st October 1938 a train in which its officials were travelling from Shanghai to Nanking was attacked by guerrillas who had destroyed the tracks and cut the telegraph and telephone wires near Wushi, China. The train was duly derailed, and the guerrillas subjected it to machine-gun fire, driving its passengers into the fields. Japanese troops then arrived and according to Chinese sources 80 Japanese were killed, while six guerrillas were killed and 12 wounded in the fighting which followed. [1856]

[1849] Ibid, 16th February 1939
[1850] Ibid, 17th February 1939
[1851] Ibid, 27th February 1939
[1852] Ibid, 28th March 1939
[1853] Ibid, 11th and 12th April 1939
[1854] Ibid, 30th June 1939
[1855] Wikipedia – Reformed Government of the Republic of China
[1856] *The Times*, London, England, 6th October 1938

No. 18 Down Punjab Express on the East Indian Railway was derailed as it was approaching Bhadaura station, India, early on Sunday 16th October 1938. This was the same train as was involved in the Bihta disaster the year before. The consequences this time were much less serious; one passenger was killed and 40 were injured. Damage to rolling stock and permanent way was estimated at 4,743 and 1,718 rupees respectively. The accident was due to sabotage, a rail having been removed. [1857] [1858]

Chinese irregulars derailed a goods train near Langfang, China, on Wednesday 11th January 1939, wrecking it. They then fired on its cargo, which consisted of cotton bound for Japan. Had an express from Mukden preceded instead of followed the goods train, that would have been wrecked. A reporter wrote, 'Last night a trainload of Japanese troops was sent from Peking to the scene of the wreck to engage the guerrillas, but the fact that artillery fire was clearly audible here {Peking} today indicates the possibility that large guerrilla forces are in the neighbourhood of Peking. Passengers who arrived this afternoon after walking past the wrecked train from the Tientsin side to the Peking side, said that hundreds of coolies were busily working to clear the wreckage; Japanese soldiers swarmed over the countryside and firing was audible. Ten cars were completely wrecked and many damaged, and the track was torn up for a considerable distance.'

On Tuesday 7th February 1939 guerrillas blew up a culvert or bridge near the scene of the January attack. The Japanese considered building a loop line around the damaged area as they had done in January. [1859] Meanwhile in a separate attack on the Tientsin – Pukow line the Chinese dynamited a Japanese troop train near Kwantien, China, killing or wounding an estimated 300 soldiers. [1860]

Chinese guerrillas blew up a Japanese military train near Samsui, China, on about Tuesday 11th April 1939, causing a reported 200 casualties. [1861]

Travellers arriving by the Peking – Hankow railway said that they had seen a long Japanese military train completely wrecked a few miles north of Paoting, China, on about Tuesday 18th April 1939. [1862]

No. 9 Up Dehra Dun express from Howrah was derailed on the Grand Chord of the East Indian Railway between Chichaki and Hazaribagh Road, Bihar, India, several rails having been removed, on Thursday 12th January 1939. The middle coaches of the train, two of which were third class, overturned, caught fire, and were gutted. Initially it was reported that seven persons had been killed and 49 wounded and the final toll was 21 dead and 77 wounded. [1863] Guard R. Macdonald was one of those injured. [1864] The fire which followed the derailment lasted for 36 hours. There was some doubt initially as to whether the derailment had been caused by sabotage but that it had was the view of both

---

[1857] *The Railway Gazette*, London, England, 21st October 1938
[1858] Report by the Railway Board on Indian Railways for 1938-39
[1859] Ibid, 8th February 1939
[1860] *The Evening Star*, Washington, D.C., United States, 7th February 1939
[1861] *The Times*, London, England, 12th April 1939
[1862] Ibid, 20th April 1939
[1863] Report by the Railway Board on Indian Railways for 1938-39
[1864] *The Times*, London, England, 13th January 1939

the Senior Government Inspector of Railways and the Bihta police. [1865] A tribunal was subsequently set up to investigate the Hazaribagh Road accident and four other recent derailments and attempted derailments on the East Indian Railway and concluded that, "We have not the slightest hesitation in finding that the cause of the derailment in each of the three cases was malicious tampering with the track by disconnecting a length of rail." [1866]

Chinese guerrilla bands wrecked two Japanese trains on about Friday 30th June 1939, one 25 miles from Peking and the other at Weihwei, China, in northern Honan Province, on the new loop line to Kaifeng which the Japanese had built across the Yellow River. [1867]

About 400 guerrillas dynamited the track of the Shanghai – Nanking railway near Wushi, China, on Saturday 29th July 1939, derailing the engine and eight coaches of a local train crowded with farmers. The guerrillas then opened fire with machine-guns. Both the passengers and the Japanese guards on the train reportedly suffered heavy losses. [1868]

Travellers arriving by the Peking – Hankow railway reported that a Japanese express had been blown up north of Chengtingfu, China, by Chinese guerrillas on Friday 17th November 1939. [1869]

What was described at the time as the worst train wreck in the history of the state happened at 9.33 p.m. on the evening of Saturday 12th August 1939 near Harney, Nevada, United States. The train involved was the westbound streamlined *City of San Francisco* express from Chicago, Illinois, to Oakland, California. Nine passengers and 15 dining car employees died and 99 passengers, 11 dining car employees, four railwaymen and a stewardess were injured. The railroad was double track at the site of the derailment, the eastbound track being operated by the Western Pacific Railroad and the westbound, on which the accident happened, by the Southern Pacific Railroad.

The City of San Francisco, train No. 101, consisted of three diesel engines (power cars), an auxiliary power and dormitory car, two chair cars, two kitchen dining cars, a dormitory club car, seven Pullman sleeping cars, and a lounge car, in charge of Conductor Robert Edwards, with Engineer E.F. Hecox and Fireman James Kelly. It had left Carlin, 16 miles east of the point of derailment, at 9.15 p.m., 29 minutes late, and was derailed at 9.33 p.m., 170 feet short of Bridge 518.54, while travelling at 60 miles an hour. All but the three rear vehicles of the train were derailed, the front kitchen dining car struck the bridge truss and caused the bridge to collapse and some of the cars fell from the bridge. [1870]

Several of the train passengers spoke to reporters after the accident. 'Nick Saurlan, Chicago rug dealer en route to the Fair {the Golden Gate International Exposition} said: "I was in a coach, the first car behind the power unit, the car most seriously damaged. Everyone had settled down, when there was a sudden jerk, and the car turned over, slithered on its side, and then turned over again. We all tried to get out of the top when

---

[1865] *The Railway Gazette*, London, England, 10th February 1939
[1866] Ibid, 16th June 1939
[1867] *The Times*, London, England, 3rd July 1939
[1868] Ibid, 31st July 1939
[1869] Ibid, 20th November 1939
[1870] Interstate Commerce Commission report

the car landed on its side but were unable to do so. We then crawled along the bottom row of windows or over the back door. It was dark, but several men had flashlights. We walked back up the hill and just stood there and looked at the gruesome sight. No one paid attention to us because they were too engrossed in the wreck. There was remarkably little panic, and everyone seemed calm." [1871]

'E.A. Bells, research professor at Penn State, was in the club car three minutes before the tragedy. "I was watching three of my friends playing cards. Suddenly I remembered I wanted something at the rear of the train and left them. There were fifty-three persons in that car. All of a sudden the lights went off, then there was a terrific crash. I missed death by three minutes."'

'Robert Schmidgull, of Chicago, auto electric worker, was in the chair car where none were killed. "The train hit the ties before we reached the bridge. Sparks began to fly, and I thought there was another train bumping into us. Ours was the last car to get across. I groped my way out and went back to the bridge where I heard Negroes groaning. One had both his legs broken, but he had plenty of pluck."' [1872]

Engineer Hecox told the Interstate Commerce Commission inquiry into the derailment that the train had been running smoothly as he had approached the site of the derailment and the automatic signals had shown *Proceed.* As his train had rounded a curve he had seen an object on the track, which turned out to be a green tumbleweed. As he reached the spot his engine left the track and he thought at first that he had struck a rock. His power car did not overturn. He had walked to Harney for relief and returned about 11 p.m., when he had thrown the tumbleweed down the embankment, and he had seen a track clawbar which had been thrown down there previously. He had then found that a rail had been disconnected and the track had been otherwise interfered with.

Fireman Kelly said after the derailment he had been busy with the rescue work, and had not examined the track until 3 a.m. He corroborated his engineer's statement and said that in his opinion the displaced rail had been arranged so as to form a derailer.

Numerous other witnesses gave evidence, all of which pointed to the same conclusion, that the derailment had been the result of sabotage. The person or persons responsible had been careful not to disturb the bond wires, so that the automatic signals remained at clear, while the tumbleweed had been carefully placed in position so as to hide the disturbed track from the view of the engineer of the oncoming train. [1873]

There was of course a massive search for the culprit or culprits responsible and a number of suspicious-looking characters were arrested and questioned by local police and the Federal Bureau of Investigation. Thus, 'A slight, twenty-eight-year-old partially bald man was under arrest here {Reno, Nevada} today and five more men were being held in Salt Lake City for questioning in connection with the alleged deliberate wrecking of the Southern Pacific's crack streamliner, "The City of San Francisco." ... The man in custody here gave his name as Bob La Duceur, formerly of Lewiston, Mont., who, officers said, "is sore at the railroads." La Duceur's right ear is partly missing, and half of his left foot was severed in a train accident in Montana two years ago. Reno police said J.T. McLaughlin, Reno agent for the federal bureau of investigation, had gone to Pyramid lake to investigate La Duceur's story that he had eaten breakfast with members of an S.P. section crew yesterday morning. Pictures of La Duceur were taken to Pyramid by

[1871] *Nevada State Journal*, Reno, Nevada, United States, 13th August 1939

[1872] *Reno Evening Gazette*, Reno, Nevada, United States, 13th August 1939

[1873] Interstate Commerce Commission report

McLaughlin. The man was arrested in the Sparks railroad yards today by William Sked of Sacramento, Calif., a special investigator for the Southern Pacific. Sked and Ralph Sterling of Sparks, another S.P. investigator, questioned La Duceur today. He has denied any knowledge of the streamliner wreck.

'Chief of Police Andy M. Welliver broadcast a state-wide alarm yesterday for a man whose description fits that of the man under arrest. It was said that the man sought had "acted in a suspicious manner" at Fernely, Nev. Yesterday and had asked, "Has anything happened?" Bob Ralchlen, of 842 Lander street, Reno, who has visited the wreck and was en route home, noted that the man was extraordinarily curious about the wreck. Ralchlen furnished Reno police with the description of the man.

'The five men held in Salt Lake City were taken by officers from an eastbound freight train, at the request of Wells, Nev. police. Authorities in western Nevada began a round-up of all transients yesterday morning after word was received that the wreck of the streamlined train had been caused by sabotage. Hobo camps at the outskirts of Reno and Sparks were raided by officers, who questioned the transients camped there.'

Neither La Duceur nor any of the transients turned out to have been the saboteurs. All were released after questioning and the real criminals were never found. [1874]

A Japanese military train bound for Nanking was reportedly derailed halfway between Shanghai and Nanking, China on Sunday 1st October 1939 by Chinese guerrillas. Thirty soldiers were allegedly killed and others wounded. [1875]

Russia invaded Finland on 30th November 1939 and the Winter War between Finland and the USSR continued until mid-March 1940. At an unknown date in late December 1939 or early January 1940 irregular Finnish forces operating on skis derailed a Russian train on the railway between Leningrad and Murmansk, killing a number of soldiers. [1876]

The Central Chinese News Agency claimed that about Japanese had been killed or injured in the collision of two trains at Suchow, China, on Friday 22nd March 1940. One of the trains, they said, had been attacked by Chinese guerrillas on the Lunghai line east of Suchow and in an attempt to escape at high speed it had collided with the other, telescoping several coaches. [1877]

In two other attacks on the same day Chinese guerrillas were claimed to have completely wrecked a Japanese troop train on the Tientsin – Pukow line, with 500 soldiers killed or wounded, and mined a train near Wenshi, killing 200. The *Railway Gazette* commented, 'The actual numbers of casualties may be exaggerated, but there can be little doubt that much damage and loss of life was and is frequently being caused by the Chinese. [1878]

Train No. 8 Down, the Dacca – Sealdah (Calcutta) mail of the Eastern Bengal Railway was derailed between Chaudanga and Jairampur, India, at 02:50 on Monday 5th August 1940. The train had left Goaland at 22:45 the previous evening and was due in

---

[1874] *Reno Evening Gazette*, Reno, Nevada, United States, 14th August 1939
[1875] *Morning Tribune*, Singapore, 2nd October 1939
[1876] *Journal de Geneve*, Geneva, Switzerland, 13th January 1940
[1877] *Reno Evening Gazette*, Reno, Nevada, United States, 25th March 1940
[1878] *The Railway Gazette*, London, England, 31st May 1940

Calcutta at 05:32. The engine, tender and three coaches left the track and went down an embankment and three more coaches were derailed. The Darjeeling Mail was quickly on the scene of the accident and three relief trains were sent from Calcutta. The first relief train reached the scene at 05:50. An early report stated that there were 34 dead, including a member of the Bengal Legislative Council. The Darjeeling Mail reached Calcutta at 14:05 with 55 injured and the bodies of three of the dead; three other injured passengers had been removed at Kanchrapara to the hospital there. Fireman Gwyther died after admission to Calcutta Hospital. Another railwayman, N.C. Mookherji and Parcels Clerk Dwijen Chakraverty were also among those killed and railwayman S.S. Ghose was injured. The derailment was clearly the result of sabotage, a rail having been removed.

Mr Fazlul Rahman, a Member of the Legislative Assembly, had a lucky escape. He was interviewed by a reporter, "Khan Saheb, Dr Maizuddin Ahmed and myself were travelling in a second-class compartment, being the fifth from the engine. We were all asleep. Suddenly the train lurched and a shrieking noise woke us up. We tried to get out of the compartment, but found the doors jammed. It was dark. I came out through the window, lighted by a torchlight which one of my co-passengers produced. Dr Maizuddin was helped out by another passenger, a Sub-Divisional Collector. We found that the bogie to which our bogie was attached was fixed between the two ends of the culvert which was damaged. The bogie behind ours was the mail van which was intact. The one before us, which was the combined first and second class coach, was found in an inclined position off the rails. One Mr Paul and his family were in this bogie. Mr Paul's wife and son were helped out of the train with difficulty. The two third-class bogies was {sic} completely smashed as well as the brake van and a small third-class compartment attached to the engine. The engine was off the rails and lying on the slope." [1879]

The railway company offered a reward of 5,000 rupees for the apprehension of the culprits. [1880]

Early on Sunday 22nd September 1940 an attempt was made to derail a train carrying President Lázaro Cárdenas to Tampico. Rebels tore up the rails a mile east of Zacape, Mexico, in San Luis Potosi Province, but their efforts were frustrated by the fact that a freight train was preceding the President's train and was derailed instead. [1881] [1882]

A train carrying prominent Chinese and Japanese officials to Nanking for the formal recognition by Japan of the pro-Japanese regime of Wang-Ching-Wei on 30th November 1940, was dynamited the day before, Friday 29th November 1940, just outside Suchow, China. Five coaches were derailed, and 100 persons were reportedly killed or wounded. Most of the prominent Chinese and Japanese officials escaped because they had waited for another train. The majority of those who died were Chinese coolies who had been riding in the vehicle immediately behind the engine. [1883] The entire railway service between Shanghai and Nanking was disrupted by the accident, it was reported. [1884] The Japanese later announced that 73 Chinese and one Japanese had been killed and 70 others

[1879] *The Bombay Chronicle*, Bombay, India, 6th August 1940
[1880] Ibid, 21st August 1940
[1881] *The Circleville Herald*, Circleville, Ohio, United States, 23rd September 1940
[1882] *The Charleston Gazette*, Charleston, West Virginia, United States, 23rd September 1940
[1883] McAllen Daily Press, Mc Allen, Texas, United States, 29th November 1940
[1884] *The Railway Gazette*, London, England, 6th December 1940

were seriously wounded. [1885] Forces under Chiang-Kai-Shek were said to have been responsible. The Japanese promised compensation to the families of the victims who were killed while villages near the scene of the attack were burnt down because it was alleged that the villagers must have assisted the demolition party. [1886]

Chinese guerrillas derailed a Japanese troop train with a mine at Yangtzi, west of Tangshan, China, on the Lunghai line, on Tuesday, 3rd December 1940, causing heavy casualties. Guerrilla forces were said to be very active in the area. [1887]

At the end of December *The Railway Gazette* wrote, 'Land mines and other forms of demolition are now causing the Japanese-managed railways so much trouble that traffic on the Shanghai – Hangchow line is again interrupted almost as soon as the track has been restored and traffic resumed. On the Tientsin – Pukow Railway all trains are at present being run at 8 mph to prevent serious accident if they are wrecked by guerrillas or by the Chinese 8th Route Army, which is systematically operating against the entire Japanese-controlled railway system, despite its being heavily guarded and patrolled.' [1888]

A goods train was derailed about 364 miles from Howrah, India, on the night of Sunday 9th February 1941, fishplates having been removed from the track. Nine waggons were derailed but no one was hurt. [1889]

A train bound for Germany which included 40 oil tank cars was derailed at midnight on Monday 10th February 1941 near Dragoman, Bulgaria, close to the border with Yugoslavia. Fire broke out in the wreckage and both train and track were severely damaged. Eight people died in the accident and many were injured. The derailment was variously attributed to the train running away down a gradient and to sabotage. [1890]

The United States suffered a serious act of railway sabotage on the late evening of Sunday 16th March 1941, on the Pennsylvania Railroad, near Baden, Pennsylvania. The train involved was eastbound passenger train No. 316 from Cleveland, Ohio, to Pittsburgh, Pennsylvania, known as *The Buckeye*. The train consisted of 4-6-2 steam locomotive No. 3663, a baggage car, two coaches, a café coach, and a Pullman car, in charge of Conductor White with Engineer A.R. Weigel and Fireman C.H. Millburn. It had left Cleveland, Ohio, 117¼ miles from Baden, at 6.17 p.m., two minutes late, left Rochester, 5¼ miles from Baden, at 8.56 p.m., on time, passed the tower at East Conway, a quarter mile short of Baden, at 9.02 p.m., and travelling at between 60 and 70 miles an hour was derailed half a mile beyond Baden at 9.03 p.m. The whole train was derailed, the cab of the engine was crushed, the leading two coaches went down an embankment and partly into the Ohio River and the second was penetrated by a pair of rails. Engineer Weigel, two passengers and two off-duty railwaymen were killed and 102 passengers, 10 railwaymen off duty, five railwaymen on duty, three dining car workers and a Pullman

---

[1885] *The Daily News* Record, Harrisonburg, Virginia, United States, 2nd December 1940
[1886] *The Railway Gazette*, London, England, 7th March 1941
[1887] *Journal de Geneve*, Geneva, Switzerland, 14th December 1940
[1888] *The Railway Gazette*, London, England, 27th December 1940
[1889] Ibid, 25th April 1941
[1890] *The Times*, London, England, 12th February 1941

porter were injured. [1891]

A newspaper reporter wrote: 'Screams of the injured – men, women and children – rang out in an atmosphere made eerie by the wind and snow and orange-red glare of steel furnaces on the opposite side of the river. Authorities learned of the tragedy when an injured passenger, his clothes sopping wet and blood dripping from his head and arms, ran to a Baden residence and had a woman telephone police. The casualty list grew rapidly as the night wore on, but many persons were only slightly hurt and were released after hospital treatment.' [1892]

Fireman Millburn told the Interstate Commerce Commission (ICC) inquiry that the brakes of the train had been tested before leaving Cleveland and had worked properly en route. The East Conway signals had been clear, and his engineer had then increased the speed to 65 or 70 miles an hour. Beyond Baden station he (the fireman) had called the indication of the next signal and before the engineer could reply, the engine had begun to rock, had thrust hard to the right, and left and then overturned to the right down the embankment. He had not noticed anything wrong with the track before the derailment and did not think that his engineer had.

Conductor White agreed that a terminal air brake test had been made before leaving Cleveland. He put the speed of his train at 60 miles an hour when the brakes had been applied in emergency. Soon after the accident he had seen a displaced rail which he did not consider had been caused by the engine but must have been displaced prior to the arrival of the train.

Superintendent Rex stated that he had arrived at the scene at 9.59 p.m. He had found nothing wrong with the engine or vehicles of the train. One hundred feet in the rear of the last vehicle of the train he had found a displaced rail, from which the spikes, angle bars, angle bar bolts, nuts and washers were missing. The conditions were not such as would have existed had the rail been displaced by the train. Track spikes, angle bars, tieplates, bolts and so on had later been found scattered down the embankment, also a wrench and a claw bar.

Engineer of Maintenance of Way Crew said that it was apparent that the angle bars had been removed before the passing of the train which had preceded No. 316 but the movement of the rail beneath the train would probably not have been noticeable to those on it. He considered that the track had been interfered with by someone experienced in work on the permanent way. A test done on 18th March indicated that one man using a standard wrench and a claw bar took just under 27 minutes to remove the rail in the same way as had happened on the day of the derailment.

Section Foreman Liposcack said that no member of the permanent way staff had been discharged for several years.

Supervisor of Telegraph and Signals Hendricks stated that a rail could be moved 18¼ inches from its normal position without disconnecting or breaking the bond wires and thus putting the signals to danger.

Various other witnesses gave evidence.

The ICC concluded that the derailment had been caused by malicious tampering with the track. [1893]

[1891] Interstate Commerce Commission report

[1892] *Lock Haven Express*, Lock Haven, Pennsylvania, 17th March 1941

[1893] Interstate Commerce Commission report

The saboteur or saboteurs responsible were never caught. [1894]

Several hundred Japanese soldiers were allegedly killed when a troop train detonated a large land mine near Chienshien, in Shansi Province, China on Friday 28th March 1941. Chinese guerrillas attacked the survivors of the wreck. [1895]

Several people were killed and wounded when a Shanghai – Hankow express was dynamited by Chinese guerrillas at an unstated location on Thursday 15th May 1941. The engine and one carriage overturned. [1896]

A passenger train detonated a mine at an unstated location on the Peking – Hankow railway on 3rd June 1941. Many passengers were killed or wounded. [1897]

Chinese guerrillas claimed to have killed or wounded over 200 Japanese soldiers when they dynamited a Japanese troop train in northern Ahnwei Province on about 30th August 1941. [1898]

Romania had joined the Axis powers in November 1940 and participated in the invasion of the Soviet Union in June 1941. On about Monday 7th July 1941 Soviet parachutists landed in Romania and derailed two military trains between Bucharest and Jassy before making their escape. Rail traffic between the two cities was paralysed for some time. [1899]

On Sunday 20th July 1941 Czech saboteurs blew up a German munitions train, destroying 21 trucks of artillery shells, near Presov, Czechoslovakia. Forty-four soldiers – three officers and 41 other ranks – who were travelling with the munitions, were killed. [1900] [1901]

On about Tuesday 19th August 1941, four cars of a German train were derailed by sabotage at Vanves-Malakoff, France, not far from Paris Montparnasse station, by the removal of bolts from a double length of track. This was despite the empowerment of emergency courts to impose death sentences for sabotage, even at this relatively early stage of the war. [1902] In another derailment of a French train at Poissy, France, perhaps a couple of days later, the line was blocked for 20 hours. The Germans arrested the drivers and firemen of both trains. [1903] In another derailment at the entrance to Valdonne Tunnel, east of Marseilles in Vichy France, probably on Friday 22nd August 1941, 30 miners were

---

[1894] https://www.post-gazette.com/local/community-eyewitness/2011/04/03/Eyewitness-1941-Saboteurs-never-found-in-deadly-train-disaster/stories/201104030144

[1895] *Mansfield News-Journal*, Mansfield, Ohio, United States, 28th March 1941

[1896] *The Times*, London, England, 17th May 1941

[1897] *The Railway Gazette*, London, England, 25th July 1941

[1898] *The Times*, London, England, 3rd September 1941

[1899] Ibid, 10th July 1941

[1900] *The Circleville Herald*, Circleville, Ohio, United States, 21st July 1941

[1901] *Tucson Daily Citizen*, Tucson, Arizona, United States, 21st July 1941

[1902] *Edwardsville Intelligencer*, Edwardsville, Illinois, United States, 26th August 1941

[1903] *Tucson Daily Citizen*, Tucson, Arizona, United States, 25th August 1941 injured. [1904]

A report from Sofia, Bulgaria, stated that five people had died and 40 had been injured in the wreck (presumably a derailment) of a passenger train between Niš and Lescovac, Serbia, Yugoslavia, late on Wednesday 15th October 1941. [1905] A Reuters report stated that ten people had died and that the derailment had resulted from an act of sabotage. [1906]

Chinese guerrillas dynamited a Japanese troop train near Hangchow, Chekiang Province, China, on about Thursday 23rd October 1941, 'causing heavy casualties.' [1907]

An announcement from Shanghai, China, on 30th October 1941 stated that four Japanese trains had been derailed by sabotage a few days earlier in Honan Province, one of them a munitions train, and that significant casualties and material damage had resulted. [1908]

On about Monday 3rd November 1941 Chetnik guerrillas derailed a train in Bosnia, Yugoslavia. Seven passengers were killed, the survivors were made to undress, and the saboteurs made off with their clothing. [1909]

'Criminals,' as they were called in the German-controlled French press, derailed a train at Abbeville, France, in Normandy, jamming points and damaging the track. The train driver saw the problem in time to reduce speed, but the engine of his train was derailed in the attack in late October or early November 1941. [1910]

Probably on about Wednesday 24th December 1941 a British troop train was sabotaged on the main line between Baghdad and Mosul in Iraq. A train carrying Indian troops was attacked as it was crossing a bridge. Three *wagon-lits* coaches carrying British officers fell from the gap in the bridge and ten other coaches were wrecked. There were an unknown number of dead and wounded. It was estimated that it would take five days' work to repair the damage. [1911]

There had been a brief war between Britain and Iraq in May 1941 after a *coup d'état* by the anti-British Rashid Ali. Since the end of May and the defeat of Rashid Ali the British had occupied the country in order to safeguard their access to the oil resources. [1912]The sabotage was probably the work of Iraqi nationalists anxious to secure their country's independence.

The daily passenger train to Laredo was derailed 75 miles north of Mexico City, Mexico, on Monday 26th January 1942 by a large rock placed on the rails. No one was

---

1904 *The Capital Times*, Madison, Wisconsin, United States, 24th August 1941

1905 *Canandaigua Daily Messenger*, Canandaigua, New York, United States, 18th October 1941

1906 *The Railway Gazette*, London, England, 24th October 1941

1907 *The Times*, London, England, 27th October 1941

1908 *Journal de Geneve*, Geneva, Switzerland, 31st October 1941

1909 *Joplin Globe*, Joplin, Missouri, United States, 5th November 1941

1910 *L'Œuvre*, Paris, France, 27th November 1941

1911 *Gazette de Lausanne*, Lausanne, Switzerland, 29th December 1941

1912 Wikipedia – Anglo-Iraqi War

hurt. [1913]

Two German troop trains were derailed by the French Resistance at Abbeville, France, in January 1942. [1914]

In an action by partisans along the coast of the Sea of Azov near Melitopol, USSR, a German train laden with tanks, trucks, and munitions was wrecked. At least five waggons loaded with shells exploded and the tanks and trucks were wrecked, according to the Soviet information bureau. The location of the accident was 200 miles west of Rostov-on-Don, on the main route from the Crimea. [1915]

A German troop and supply train was derailed by sabotage between Belgrade and Niš, Yugoslavia, on about Tuesday 3rd February 1942, reportedly by Bulgars. Two hundred and sixty officers and men were stated to have been killed. [1916]

A Japanese troop train was blown up by Chinese guerrillas south of Tangying, China, causing allegedly 400 casualties, on about Friday 20th February 1942. An engine and four coaches were demolished. [1917]

Russian sources reported that on about Saturday 21st March their guerrillas had slipped through the German lines surrounding Leningrad, USSR, and mined the railway, blowing up a troop train and killing about 200 men. [1918]

Three very serious derailments due to sabotage, in three different countries, reportedly occurred about the end of March 1942. In Romania 72 German officers and men died in a derailment near Craiova. The government arranged for police to supervise peasants planting spring crops to prevent them from committing acts of sabotage. In Greece over 100 soldiers were said to have died in the derailment of a German troop train between Herrai and Drama, while much ammunition was destroyed. Guerrillas machine-gunned the train after it had left the rails. In Italy 135 German soldiers were reportedly killed in the wrecking of a troop train at an unstated location. [1919]

Chinese forces in Chungking stated on 7th April 1942 that their guerrilla fighters had derailed two Japanese supply trains in the vicinity of Hankow, China, during the previous fortnight. As well as a high number of casualties there was much material damage. [1920]

The French Resistance derailed a German leave train on Thursday 16th April 1942 at Airan, France, in Normandy. The Maastricht – Cherbourg working was carrying sailors of the Kriegsmarine, 28 of whom were killed and a further 19 wounded according to

[1913] *San Antonio Express*, San Antonio, Texas, United States, 28th January 1942
[1914] *The Helena Independent*, Helena, Montana, 19th January 1942
[1915] Ibid
[1916] *The Helena Independent*, Helena, Montana, United States, 9th February 1942
[1917] *Kingsport Times*, Kingsport, Tennessee, United States, 22nd February 1942
[1918] *The Sheboygan Press*, Sheboygan, Wisconsin, United States, 23rd March 1942
[1919] *Syracuse Herald-Journal*, Syracuse, New York, United States, 7th April 1942
[1920] *Journal de Geneve*, Geneva, Switzerland, 7th April 1942

German sources. Several metres of track had been removed. [1921]

Savage reprisals were carried out in the wake of this sabotage. On 20th April it was announced that 30 French hostages would be shot in retaliation, and that 80 more would suffer the same fate if those responsible were not turned over by Thursday 23rd. As a further measure French citizens would in future be forced to ride on all-German troop trains to ensure their safety. [1922]

An announcement from Moscow on 30th April 1942 stated that guerrillas had derailed an armoured train in Ukraine and that with the return of Spring guerrilla activity in Ukraine had intensified, with the derailing of six trains carrying German troops or supplies during April. [1923]

The guerrilla activity would greatly increase in the following months. At the end of the year Radio Moscow would announce that 148 trains composed of 2,656 waggons had been derailed. [1924]

The Japanese had invaded Burma in January 1942 and Rangoon, in the south of the country, fell to them on 8th March. The speed of the Japanese advance then increased and there was a chaotic retreat of the British forces plus large numbers of refugees northwards and westwards towards India. The railways were used as a means of transport during the retreat but were the subject not only of Japanese air attacks but also of sabotage by members of the Burma Independence Army, who by collaborating with the Japanese hoped to win independence for their country from British rule.

At 22:00 on Saturday 7th March 1942 a train carrying members of the Burmese Military Police was approaching a bridge near Wanetchaung when the pilot engine and then the train engine fell from a bridge which had been sabotaged. The train then came under fire from what was presumed to have been a mixed force of Burma Independence Army and Japanese, but the attackers were driven off, leaving about 20 dead. Major Lall Khan and other members of the police were killed or wounded; the major was killed.

At about 10:00 on Saturday 18th April 1942 a train carrying Burmese Military Police from Ywataung was derailed at Kinu, north of Schwebo, the track having been sabotaged. Subedar Moli Ram Thapa of the 2nd Rangoon Battalion of the police was killed, and five other ranks were injured. There were also an unstated number of civilian casualties. [1925]

In May 1942 an Australian war correspondent wrote: 'In the early days of the retreat the Burmese Fifth Columnists caused the greatest difficulties. They deliberately delayed trains by dispatching fake telegrams describing mythical trains, which, they said, were blocking the line, and by emptying water tanks and thus immobilising locomotives. North of Sho Webo the line was blocked by the wreckage of a troop train dive bombed by the Japanese. North of Naba the rear carriages of another train became mysteriously uncoupled, running downhill over an embankment. About 25 appeared to be alive in the tangled wreckage but there appeared to be no means of rescuing them.' [1926]

[1921] https://www.dday-overlord.com/en/battle-of-normandy/resistance
[1922] *The Bakersfield Californian*, Bakersfield, California, United States, 20th April 1942
[1923] *Journal de Geneve*, Geneva, Switzerland, 30th April 1942
[1924] Ibid, 31st December 1942
[1925] http://www.rothwell.force9.co.uk/burmaweb/RangoonBattalionsBMP.htm
[1926] *The Barrier Miner*, Broken Hill, New South Wales, Australia, 27th May 1942

Despite the German reprisals which had followed the Airan sabotage, another attack was carried out on about Sunday 3rd May 1942 when a German munitions train was derailed at Vire, France, near Caen. Fifteen German soldiers were reportedly killed in the derailment but so were 21 French hostages who had been forced to accompany the train. In addition 30 hostages were shot at Caen and probably 20 more at Romorantin. A house-to-house search was meanwhile being conducted to find three parachutists who had reportedly landed just before the Vire attack, and a curfew was imposed in western Normandy from 19:30.[1927]

In early May 1942 a Dutch workman's train from Zaandam to Amsterdam was derailed by sabotage with many casualties, according to a report reaching London, England.[1928]

A couple of American newspapers reported on 10th May 1942 that free Belgian quarters in London had received news that 56 German soldiers on leave had been killed in a collision near Amiens, Belgium, in an accident which it was presumed was due to sabotage. Amiens is of course in France, though not far from the Belgian border. Collisions, as opposed to derailments, due to sabotage are also rare.[1929]

On about Sunday 10th May 1942 a German freight train was derailed by the Resistance at Parthenay, in the occupied zone of France.[1930]

On Thursday 14th May 1942 a detachment of guerrillas from White Russia reportedly destroyed a railway bridge and derailed two troop trains. 'More than 600 Germans were killed.'[1931]

An act of sabotage apparently not related to the Second World War occurred in India on Saturday 16th May 1942 when the Punjab Mail, en route from Karachi to Lahore, was derailed about midnight between Oderlal and Tando Adam stations, India. Armed dacoits, members of the Sind criminal sect known as Hurs, were blamed for the derailment. Twenty-four passengers were killed, including Munwar Hindayatullah, the son of Sir Ghulam Hussain Hingayatullah, the Home Minister of Sind. Mr Nichaldas Vazirani, the Revenue Minister, had a narrow escape, while a number of Europeans were among the injured. The motive was apparently robbery; after the derailment the dacoits held up the surviving passengers at gunpoint and stole money and valuables from them.

Militia from Karachi pursued the Hurs and captured four of them.

A further act of sabotage on Wednesday 20th May was prevented when a driver noticed damaged track in time to pull up. It was then decided to provide trains with armed escorts, plus police protection at stations, while as a further measure martial law was declared at the beginning of June. Despite this there was a further attack on 20th June when six people were killed after dacoits had stopped a train by placing boulders on the

---

[1927] *The Times*, London, England, 4th and 7th May 1942

[1928] *Wichita Daily Times*, Wichita Falls, Texas, United States, 12th May 1942

[1929] *The Cedar Rapids Gazette*, Cedar Rapids, Iowa, United States, 10th May 1942

[1930] *El Paso Herald-Post*, El Paso, Texas, United States, 12th May 1942

[1931] *The Times*, London, England 16th May 1942

line. [1932]

Towards the end of September four Hurs were sentenced to death for the May attack. Twelve others were acquitted for want of evidence, released, then re-arrested under the defence regulations. [1933] Pir Pigaro, the Hurs leader, was hanged at Hyderabad in March 1943. [1934]

Probably in late April 1942 a band of 2,000 guerrillas reportedly ambushed a troop train and killed 250 Germans and Bulgarians in Greece. Later they attacked and killed half of a German punitive force. [1935]

An act of sabotage in Yugoslavia on about Saturday 16th May 1942 was said to have killed 150. [1936]

A report from London on 22nd May 1942 stated that Greek guerrillas had attacked the Salonika – Edirene line in at least five places and blown up a troop train at Drama, killing or wounding 200 Germans. (This could be the same attack as the one between Herral and Drama in March.) Another troop train was attacked near Larissa where the rails were torn up and at least 25 Germans killed in a three-hour fight. [1937] A separate report from Jerusalem on 26th May, very possibly relating to the Larissa attack, spoke of an Axis train being derailed with 40 dead and 100 wounded. [1938]

A German troop train was derailed by the Resistance in the occupied zone of France and caught fire on about Wednesday 10th June 1942, while attempts were made to wreck two other trains. [1939]

A Polish correspondent reported in mid-July 1942 that the occupying German authorities were making desperate and ruthless attempts to stamp out sabotage attempts on trains going towards the Russian front. At Plaszow, near Krakow, seven railwaymen had been hanged at the end of June for their alleged involvement in the derailment of a train. [1940] Some months later further details emerged. A bridge had been mined using explosives which had been kept hidden since September 1939 and a German military transport train was derailed, killing or wounding about 30 officers and men. In reprisal the Germans had rounded up not just seven railwaymen, as originally reported, but 150 Poles, who were executed on the spot and their bodies left hanging along the railway line. [1941]

A freight train was derailed by sabotage between Amiens and Sotteville, France, on

---

1932 Ibid, 20th, 21st, 22nd, and 29th May and 3rd, 22nd, 23rd, and 24th June 1942
1933 Ibid, 22nd September 1942
1934 Ibid, 22nd March 1943
1935 *The Tweed Daily*, Murwillumbah, New South Wales, Australia, 4th May 1942
1936 Author's personal notes – source mislaid.
1937 *The Vidette-Messenger*, Valparaiso, Indiana, United States, 22nd May 1942
1938 *The Times*, London, England, 27th May 1942
1939 *The Daily Ardmoreite*, Ardmore, Oklahoma, United States, 18th June 1942
1940 *The Times*, London, England, 18th July 1942
1941 Ibid, 12th October 1942

about Saturday 11th July 1942. [1942]

BBC Radio reported on Tuesday 28th July 1942 that saboteurs had derailed a German leave train travelling between Cherbourg and Paris, France, killing 40 soldiers. At about the same time the Germans had executed 28 Frenchmen, mostly miners, at Lille, for sabotage, for possessing weapons and for 'Communist activities.' [1943]

A Japanese train was attacked and derailed at Bitaogan, Panay Island, in the Philippines, on Monday 24th August 1942.

In a further attack on Panay Island Saturday 29th August 1942, guerrillas placed two boxes of dynamite on the line at Man-it and a diesel-hauled train carrying Japanese soldiers was derailed. First Lieutenant Matao Asahi, Second Lieutenant Nisibe and 20 other officers (*shôko*) and men were killed. Colonel Senô escaped death because he had been seated in a coach at the rear of the train. [1944]

Belgian partisans derailed a train near Lessines on the line to Grammont on Thursday 17th September 1942. This was the first attack on a train in Belgium during the war. [1945]

Information reaching Yugoslav circles in London indicated that during September 1942 patriots had derailed 46 trains in Croatia and three more in Banat Province, Yugoslavia. [1946]

The Peshawar Express of the Great Indian Peninsula Railway was derailed on Saturday 3rd October 1942 near Chalisgaon, India. Twelve passengers, a fireman and a special duty constable were killed, and 28 passengers and the guard were wounded. Most of the victims were third-class passengers. The derailment was due to sabotage, a rail having been removed. The Bombay Government offered a reward of 10,000 rupees for the apprehension of the saboteurs. [1947]

Partisans blew up a military train at Divaca, Yugoslavia, near Trieste, on about Monday 5th October 1942, killing 25 men and wounding over 100. [1948]

Partisans wrecked four German trains near Orel, USSR, in the Vitebsk – Smolensk – Velikiye Luki triangle on about Monday 12th October 1942. On about the same date Minsk guerrillas wrecked a train transporting lorries to the front and 56 lorries were destroyed. [1949] On Friday 16th October Kalinin guerrillas reportedly derailed three German military trains, smashing 23 waggons, and derailing 61, while 'many of the enemy were killed.'

1942 *The Daily Herald*, Biloxi, Mississippi, United States, 13th July 1942
1943 *The La Crosse Tribune*, La Crosse, Wisconsin, United States, 28th July 1942
1944 https://www.thenewstoday.info/2010/11/19/the.blood.and.mud.in.the.philippines.html
1945 http://humanite-socialiste.over-blog.com/article-25575220.html
1946 *The Times*, London, England, 15th October 1942
1947 *The Bombay Chronicle*, Bombay, India, 7th October 1942
1948 *The Times*, London, England 19th October 1942
1949 Ibid, 13th and 16th October 1942

On Saturday 10th October 1942 an Up goods train bound for Viramgam was derailed between Wadhwan Junction and Bala Road, India, due to the removal of fish bolts and dog spikes at about 00:15. The leading twelve waggons, six loaded and six empty, were derailed, and overturned but no one was hurt. [1950] Two days on, late on Monday 12th October 1942, the Bombay mail train was derailed 80 miles from Calcutta, India, killing 5 persons and injuring 31. A railway spokesman confirmed that the derailment was again due to sabotage, saying that eight fishplates and bolts had been removed from a length of rail at the accident site. [1951]

On about Wednesday 14th October 1942 Chinese guerrillas derailed with a mine a Japanese troop train travelling from Nanchang to Kiukiang (described as an important Yangtze River port between Nanking and Hankow) near Shahsiatu, China. Six coaches were blown up, a river bridge was damaged and about 500 soldiers were reportedly killed or wounded. [1952]

Soviet partisans derailed four German military trains in the Leningrad region on about Tuesday 20th October 1942. [1953]

Rails were blown up by dynamite simultaneously at four points around Warsaw on about Saturday 24th October 1942. In one of these attacks, near the important junction at Nasielska, Poland, an ammunition train was derailed and exploded. Nine trucks were totally destroyed, and 17 damaged. In retaliation 55 Poles were taken hostage and were executed a few days later. [1954]

The Russian High Command reported on Friday 20th November 1942 that between 13th and 18th November Soviet guerrillas had derailed 66 enemy trains. 'According to incomplete data they wrecked two armoured trains, 16 engines, 200 trucks and open trucks, and blew up 10 railway bridges. Many Germans have been killed as a result of these operations.' [1955]

A German train, loaded with supplies for the Eastern front, was wrecked in an explosion due to saboteurs at Charleroi, Belgium, on about Friday 20th November 1942, according to an announcement by *Inbel*, the Belgian News Agency. [1956]

In Greece on about Wednesday 25th November 1942 a military train carrying ammunition and 700 Italian soldiers was derailed was derailed by sabotage. Many of its coaches fell into a ravine, the ammunition exploded, and many soldiers were killed. [1957]

A Wehrmacht train, W 17 774, was derailed by Polish partisans between Bialystok

---

[1950] *The Bombay Chronicle*, Bombay, India, 15th October 1942
[1951] *Jefferson City Post-Tribune*, Jefferson City, Missouri, United States, 13th October 1942
[1952] *The Times*, London, England, 19th October 1942
[1953] Ibid, 23rd October 1942
[1954] Ibid, 2nd November 1942
[1955] Ibid, 21st November 1942
[1956] Ibid, 25th November 1942
[1957] Ibid, 10th December 1942

and Wolkowysk, Poland on Saturday 12th December 1942. The engine and 15 coaches were derailed and caught fire. Twenty-one soldiers died and 46 were wounded. [1958]

Three derailments of trains in the Lublin area of Poland on about Friday 1st January 1943 were reported. One involved a train carrying German colonists on their way to take over farms in what had become the General Government area of the country; some were killed and wounded. The other two trains attacked were military trains carrying troops and war material. [1959]

A mine exploded beneath a military train between Louvain and Aerschot, Belgium, on Saturday 2nd January 1943

A German leave train was mined and derailed at Wilsele on Friday 8th January 1943. Another was derailed between Corbeek and Loo on Saturday 16th January. A length of rail was removed, derailing a freight train, at Binche, Belgium on Wednesday 3rd February 1943. [1960]

About 30 passengers were injured when a bomb exploded on a train near Botad, India, on about Monday 8th February 1943. The alarm chain was pulled, but as the train did not stop at once some of the passengers jumped from it, injuring themselves in the process. Only slight damage was done to the carriages. [1961]

In Belgium on 8th February 1943 a Vienna – Bruxelles express was derailed by a mine between Saventhem and Bruxelles. Many coaches were wrecked and over 100 German officers and men were killed or wounded. The next day a freight train was attacked at Velthem, Belgium, on the Louvain – Bruxelles main line, and an engine and 18 waggons were derailed. The line was blocked for 24 hours. On Wednesday 10th February a train was derailed near Louvain on the line to Tirlemont; three waggons left the track, and the line was blocked for 12 hours. On Thursday 11th February a freight train was derailed near Courtrai on the line to Gand (Ghent).

A German leave train was derailed near Louvain on the line to Tirlemont on Tuesday 16th February 1943. Traffic was interrupted for 12 hours. A freight train was derailed at Velthem, Belgium, on Saturday 20th February; 16 waggons were involved, and the line was blocked for 36 hours.

Dakar radio announced on 21st February 1943 that a German troop train had been derailed, presumably by sabotage, near Chalon sur Saône, France, and that 50 officers and men had been killed. [1962]

The Malines – Louvain line was sabotaged at Muysen, Belgium, on Thursday 4th

---

[1958] *Katastrophen der deutschen Bahnen Teil II Chronik 1845 – 1992* by H.J. Ritzau, pg. 105 (in German)
[1959] Ibid, 12th January 1943
[1960] http://humanite-socialiste.over-blog.com/article-25575220.html
[1961] *Bombay Chronicle*, Bombay, India, 9th February 1943
[1962] *The Hammond Times*, Hammond, Indiana, United States, 21st February 1943

March 1943. A Wehrmacht train was derailed and 'dozens of enemy soldiers were killed.'

The removal of a length of rail in a tunnel at Gedinne, Belgium, on the Dinant – Bertrix line on Thursday 11th March 1943 caused the derailment of 17 waggons of a freight train and the blockage of the line for 5 days. [1963]

General Charles de Gaulle broadcast in London on Friday 12th March 1943 that the French Resistance had sabotaged a German troop train 'bound for the East' (presumably meaning the Russian front) at Chagny, France. The location of the attack had, it was said, been well chosen, as the train had been travelling at full speed. Over 250 Germans were said to have been killed and 100 injured. Two further trains had been derailed in the Côte d'Or department. One of them, which had been carrying munitions, was completely destroyed. The other was carrying coal destined for Italy; 22 of its waggons were wrecked. [1964]

A train carrying German war supplies was derailed by sabotage in the Cote d'Or department, France, on about Friday 5th March 1943. The locomotive and 22 trucks were smashed. [1965]

The Bruxelles – Liege main line was sabotaged at Vertryck, Belgium, on Sunday 21st March 1943, a Wehrmacht train was derailed, and the line was blocked for six hours. [1966]

A communique from Moscow on Wednesday 14th April 1943 stated that on Saturday 20th March guerrillas in the Smolensk region of the USSR had derailed a German troop train. The engine and five carriages had been destroyed and 50 Germans killed or wounded. Two days later guerrillas in another sector had mined the track, derailing, and wrecking an engine and 13 carriages and killing 300 soldiers. [1967]

On Friday 26th March 1943 Russian partisans derailed a German troop train in the Smolensk region. The same day the partisans derailed another troop train 60 kilometres to the south of the first attack. Several hundred troops were said to have died in the two actions. [1968]

In Belgium the newly refurbished Louvain – Liège line was sabotaged between Corbeek and Loo and a freight train derailed, with many of its waggons overturned, on Monday 29th March 1943.

A freight train was derailed at Malines, Belgium, on Wednesday 31st March 1943.

A German freight train was derailed at Braine-le-Comte, Belgium, on Thursday 8th

[1963] http://humanite-socialiste.over-blog.com/article-25575220.html
[1964] *The Times*, London, England, 12th March 1943
[1965] *The Corsicana Daily Sun*, Corsicana, Texas, United States, 6th March 1943
[1966] http://humanite-socialiste.over-blog.com/article-25575220.html
[1967] *Gazette de Lausanne*, Lausanne, Switzerland, 15th April 1943
[1968] *The Times*, London, England, 14th April 1943

April. The engine and seven waggons were destroyed, and 18 waggons were badly damaged. The line was blocked for many days. The same day another train on the Visé – Tongres line was derailed at Tongres. [1969]

On the night of 9th to 10th April 1943 a train was wrecked between Elboeuf and Rouen. France. Ten waggons were derailed, and the line was blocked for three days. [1970]

Partisans derailed an engine at Aux Hayettes, Belgium, on Monday 12th April 1943. A following troop train was stopped for 12 hours. [1971]

A freight train was derailed at Soignies, Belgium, on Thursday 15th April 1943, and an engine and a number of waggons badly damaged. [1972]

On 16th April 1943 the Secretary of State for India was asked in the British Parliament whether there were still cases of sabotage on the Indian railways and if so, how many had occurred in the previous three months. He replied that there had been 61 reported cases – 28 in January, 16 in February and 17 in March. Only four cases had been serious or had involved loss of life. These were two derailments in Assam, one in January and one in March, a derailment in Bihar in February and a bomb explosion at Cawnpore station (not involving a train) in February. Only one attempted derailment and four minor attempts at sabotage had been reported since the beginning of April. [1973]

A Polish correspondent to the London *Times* reported that 17 trains had been derailed in Eastern Poland in late March and early April, damaging nine locomotives, destroying 58 trucks, and killing 424 Germans. A petrol train had been completely destroyed. [1974]

Damage to the Ninove – Grammont line at Schendelbeke, Belgium, on Friday 23rd April 1943 caused the derailment of a German SS train; the engine and a vehicle were demolished. [1975]

On about Wednesday 28th April 1943 Polish partisans derailed a troop train and a coal train in the Lublin area. [1976]

A leave train was derailed near Louvain, Belgium, on the line to Tirlemont on Friday 30th April 1943.

On Saturday 1st May 1943 a freight train was derailed at Erbisoeul by the *Partisans Armés* (Belgian Resistance) an engine and two waggons being wrecked. The same day they derailed an engine and five waggons at Godarville and dynamited a train of oil tank

---

[1969] http://humanite-socialiste.over-blog.com/article-25575220.html
[1970] France d'Abord, Paris, 1st May 1943
[1971] http://humanite-socialiste.over-blog.com/article-25575220.html
[1972] Ibid.
[1973] *The Railway Gazette*, London, England, 30th April 1943
[1974] *The Times*, London, England, 20th April 1943
[1975] http://humanite-socialiste.over-blog.com/article-25575220.html
[1976] *Nebraska State Journal*, Lincoln, Nebraska, United States, 3rd May 1943

waggons on the Courtrai – Gand line near Courtrai. [1977]

An announcement from London on 3rd May 1943 stated that the *Francs-Tireurs et Partisans* (a resistance group organised by leaders of the French Communist Party) had recently carried out two attacks. In one, a German leave train was derailed near Beard, France, between Nevers and Chagny. Fifty-two German soldiers were killed and 150 seriously injured. In the other attack, a German troop train had been derailed at Rebecourt, France, on the outskirts of Paris. One hundred and thirty German soldiers were killed or injured, and the line was blocked for three days. [1978]

A German troop train was derailed at Harelbeke, Belgium, on the Courtrai – Gand line on Monday 3rd May 1943. The train was badly damaged, and an unknown number of Germans were killed. On Wednesday 5th May the *Partisans Armés* removed a rail from a bridge at Hayettes, near Hayne St Pierre. Six waggons of a freight train fell 15 metres to the stream below and the engine and nine other waggons were damaged.

What was allegedly the first successful act of railway sabotage in the Netherlands was carried out on the night of Wednesday 5th to Thursday 6th May 1943 at Wierden by a resistance group led by Henk Heerdink. (It would appear, in fact, to have been the second, the first being the Zantem – Amsterdam attack a year earlier.) The bolts securing the track were undone and the rails bent, with the result that the engine and 13 waggons of a freight train were derailed, blocking the line for some days. There were no more attacks on trains in the Netherlands until a general order for sabotage was issued in September of the following year. [1979] The lack of more previous acts of railway sabotage might be related to the German reaction to an unsuccessful attack on a German military train in Rotterdam in mid-1942, when at least five Dutch hostages were shot in retaliation. [1980]

The removal of a rail derailed a train at Erbissoeul, Belgium, on Saturday 15th May 1943. The following day, Sunday 16th May, a military train was derailed at Groenendael, on the Bruxelles – Namur line. Most of the vehicles of the train were destroyed and there were many dead and wounded. The same day a freight train was derailed at Rhodes St Genese on the Bruxelles – Charleroi line. [1981]

Partisans wrecked a fully laden goods train near the station at Ralye, Serbia, Yugoslavia, on Monday 17th May 1943. [1982]

Moscow announced on 22nd May 1943 that a German munitions train had been blown up by a mine in the Chitomir (now Zhytomyr) region of Ukraine and a locomotive and 17 waggons destroyed, while in a second attack partisans had derailed a train loaded with food for Germany. [1983]

[1977] http://humanite-socialiste.over-blog.com/article-25575220.html
[1978] *Journal de Geneve*, Geneva, Switzerland, 3rd May 1943
[1979] https://johannesterhorst.nl/spoorwegsabotage
[1980] *The Times*, London, England, 17th August 1942
[1981] http://humanite-socialiste.over-blog.com/article-25575220.html
[1982] *The Railway Gazette*, London, England, 3rd September 1943
[1983] *Journal de Geneve*, Geneva, Switzerland 22nd May 1943

On Sunday 23rd May 1943 a military train was blown up at Kušatk, Yugoslavia and 140 Germans were reportedly killed, on Tuesday 25th May a goods train was blown up near Barič, Yugoslavia, and about Saturday 29th May another train was wrecked near Niš, Yugoslavia. 'This sabotage is a great worry for the Germans, and their reprisals increase against the population. The Nazis recently transferred to Serbia a Railway Battalion which previously was busy in the Caucasus.' [1984]

A report from Moscow on 7th June 1943 announced that guerrillas had derailed two German military trains in the Zhirardovo region of Poland, and that 20 officers and a large number of men had perished. [1985]

On about Friday 18th June 1943 patriots chloroformed guards at Barentin, France, reversed a set of points and engineered a collision between two trains, according to a broadcast from French underground radio. [1986] In other attacks trains were derailed at Bethune and Orleans. [1987]

Soviet partisans mined a German troop train near Velikte Luki, in the Zhitomir region of the USSR, on about Monday 28th June 1943, wrecking an engine and eight coaches, and killing or wounding an officer and 60 men. The same report stated that in the Leningrad area guerrillas had derailed nine German trains in 10 days, destroying large quantities of war material and killing large numbers of German troops. [1988]

A freight train was derailed by the Resistance at Villers Bretonneux, France, on Thursday 1st July 1943. Twenty-two tank waggons were derailed, two caught fire and the line was blocked for 96 hours. The same day a German train was derailed near Mauberge, France, on the Paris – Charleroi line. Thirty Germans were killed and 11 wounded. [1989]

In late June or early July 1943 Soviet partisans mined a German troop train in the Kamenets-Poldolsk area of south-west Ukraine, USSR, wrecking a locomotive and six coaches. It took the Germans three days to reopen the line. When they had done so the partisans wrecked in turn the first train to use the restored line, wrecking a locomotive and 17 vehicles. "Scouts saw the Germans carrying away large numbers of dead and wounded in trucks," the Soviet communique added. [1990]

On Tuesday 5th July 1943 a freight train was derailed by sabotage at Grand Couronne, France, blocking the line for 24 hours. The same day a leave train was derailed between Amiens and Montdidier, France. [1991]

[1984] *The Railway Gazette*, London, England, 3rd September 1943
[1985] *Journal de Geneve*, Geneva, Switzerland 7th June 1943
[1986] *The Helena Independent*, Helena, Montana, United States, 20th June 1943
[1987] *Arizona Republic*, Phoenix, Arizona, United States, 21st June 1943
[1988] *Tucson Daily Citizen*, Tucson, Arizona, United States, 29th June 1943
[1989] *France d'Abord*, Paris, France, 15th July 1943
[1990] *The Delta Democrat-Times*, Greenville, Mississippi, United States, 6th July 1943
[1991] *France d'Abord*, Paris, France, 15th July 1943

Partisans wrecked a German troop train near Latgale, Latvia, on about Friday 9th July 1943, reportedly killing "hundreds" of soldiers. Some of those who escaped death were captured and then recruited by the partisans, according to the account in a Swedish newspaper. [1992]

A German leave train was derailed by the Resistance between Rennes and Noyal-Acigné, France, on Saturday 10th July 1943, the track having been interfered with. The wreckage was then run into by a German materials train coming from the opposite direction with disastrous results. There were reportedly 200 dead and 300 wounded, 500 metres of track was destroyed, and the line was blocked for 60 hours. Between Ligugé and Isteuil, France, on the Paris – Bordeaux line, the same day, a freight train was derailed, the engine fell down a five metre embankment, 30 waggons were wrecked, 300 metres of track was destroyed, and the line was blocked for four days.

On the night of Sunday 11th to Monday 12th July 1943 at Blanzy-Trouville, France, on the Amiens – Chaulne line a locomotive and 12 mineral (coal?) waggons were derailed and the line was blocked for 72 hours.

On about Monday 12th July 1943 an engine and 18 waggons were wrecked at Ancy-le-Franc, France. A leave train was derailed at Sommery, France, on the Rouen-Amiens line and five vehicles were 'pulverised.' Another leave train was derailed between St Mesmin and Savièvres, France, on the Paris – Belfort line. The engine overturned and 17 vehicles were wrecked. A cattle train bound for Germany was targeted at Rouilly-Sacey, France, on the Troyes – Brienne line. Fourteen waggons were wrecked, one German was killed, some of the cattle escaped and the line was blocked for 40 hours. Finally, a German materials train was sabotaged between Vannes and Auray, France. Thirty waggons were wrecked, 400 metres of track were ripped up and the line was blocked for 30 hours. [1993]

The *Francs Tireurs et Partisans* reported the following attacks on trains in France between Saturday 17th and Friday 30th July 1943: Between Igny and Palaiseau-Villaine a double-headed freight train was derailed. Near Courville on the Paris – Le Mans line a German materials train was targeted and the engine and 12 waggons derailed. The line was blocked for 24 hours. On the Paris – Cherbourg line the *Tunisie* detachment derailed a light engine. A leave train was derailed at Venables. Five coaches were wrecked, 13 passengers seriously wounded, and the line was blocked for 12 hours. Fifteen waggons of a train loaded with pit props were wrecked at Thézy on the Amiens – Montdidier line. Near Pont de l'Arche on the Rouen – Paris line seven vehicles of a German train were derailed, and the line was blocked for 43 hours. [1994]

On Sunday 18th July 1943 an empty waggon train was derailed on the Châlon – Dôle line by the loosening of the rails by No. 1 group of the 1789 detachment of the *Francs Tireurs et Partisans*. [1995]

At 01:30 on Saturday 24th July 1943 a mine exploded under a German leave train near

[1992] *The Salt Lake Tribune*, Salt Lake City, Utah, United States, 12th July 1943
[1993] *France d'Abord*, Paris, France, late July 1943
[1994] Ibid, 1st August 1943
[1995] *France d'Abord*, Paris, France, 5th September 1943

Miribel station, France, on the Ambérieu line some kilometres from Lyon. The train was derailed, and a number of carriages were damaged. Many soldiers were wounded but none was killed. Until the line could be cleared trains were diverted via Bourg or via Valence. [1996] The attack was carried out by No. 2 group of the *La Convention* detachment of the *Francs Tireurs et Partisans.* [1997]

A train was derailed on the Lourdes – Pau line, in France, on Wednesday 28th July 1943, six waggons of iron ore leaving the rails.

On Thursday 29th July 1943 the *Jean Catelas* group of the *Francs Tireurs et Partisans* derailed a train carrying German troops and material near Grancourt, France, on the Amiens – Arras line. A leave train travelling towards Arras collided with the wreckage and 45 vehicles were wrecked, 130 Germans were reportedly killed, and many were wounded, 100 seriously. On the same day the *Gabriel Peri* group derailed a train at Vacherie-Plessis, France. Also on about the same day a leave train travelling at 75 kilometres an hour was derailed between Chagny and Meursault and eight coaches were wrecked, near Fontaine a troop train was derailed, and 14 vehicles telescoped, while the derailment of a light engine on the Paris – Belfort line blocked it for 24 hours.

A German troop train was dynamited at Vieux Heverlee, Belgium, on the Louvain – Ottignies line, on Friday 30th July 1943. In what was the most serious act of sabotage carried out in Belgium during the war, no less than 285 Germans were killed, including a senior officer. The engine and 10 vehicles were destroyed. The same day a leave train was derailed at Velthem by the removal of a rail. [1998]

A German materials train was derailed six kilometres from Les Arcs, France, on Friday 30th July 1943, and the engine and five waggons were wrecked.

Eight kilometres from Paris on the Paris – Chateau-Thiéry line a train was derailed on Saturday 31st July 1943. [1999]

Partisans in Serbia, Yugoslavia, were said to have derailed a German military train at a point about 300 kilometres south of Belgrade, killing over 500 officers and men and wiping out an entire Bulgarian cavalry squadron. The sabotage took place at an unstated date, probably in late July 1943 [2000] The casualties claimed in some of these acts of sabotage doubtless need to be taken with a large pinch of salt as they are probably frequently exaggerated for propaganda purposes. This is presumably the same attack reported by another source as having taken place near Zaječar. [2001]

Paris Radio announced on 2nd August 1943 that 11 "terrorists" had been sentenced to death by an Amiens tribunal for acts of sabotage which included the wrecking of a train.

[1996] *Journal de Geneve*, Geneva, Switzerland, 29th July 1943
[1997] *France d'Abord*, Paris, France, 5th September 1943
[1998] http://humanite-socialiste.over-blog.com/article-25575220.html
[1999] *France d'Abord*, Paris, France, 20th August 1943
[2000] *Gazette de Lausanne*, Lausanne, Switzerland, 3rd August 1943
[2001] *Moberly Monitor Index*, Mobberly, Missouri, United States, 31st July 1943

2002

An announcement from Greek sources in Cairo on 7th August 1943 stated that a party of guerrillas had recently blown up, at night, a railway bridge at Dereli, between Athens and Salonika, as a train of 45 oil waggons was about to cross it. Ten or twelve waggons fell into the ravine before the train could be pulled up and the guerrillas then fired on the engine boiler. The bridge, which consisted of a single metal span, was rendered unusable for some time. [2003]

A freight train was derailed early on Saturday morning, 7th August 1943, in Paradis Tunnel, France, between Bellegarde and Culoz, the engine and a dozen waggons leaving the rails. The track had been interfered with. All traffic to and from Switzerland via Cornavin was halted. It was expected that it would take two days to clear the line. [2004]

On Sunday 8th August 1943 the Marcel Cachin detachment of the *Francs-Tireurs et Partisans* derailed an oil train near La Varenne on the Paris – Marseille line. Some minutes later a freight train coming in the opposite direction collided with the wreckage, whereupon 14 tank waggons caught fire. The blaze consumed an engine and 23 waggons and lasted 18 hours. [2005]

An engine and two waggons were derailed by sabotage on Sunday 15th August 1943 at Wezemael, Belgium and another engine and pair of waggons at Zolder on Wednesday 18th August. [2006]

The clandestine Radio France broadcast on 23rd August 1943 that amongst other operations carried out by members of the *Francs-Tireurs et Partisans* (FTP) a German leave train had been derailed in the Ille-et-Vilaine department. A short time later a freight train on the opposite running line had crashed into the wreckage. It was said that there were 200 dead and 300 injured, while the line was blocked for 60 hours.

In a separate action carried out by the FTP on the Paris – Bordeaux line a long freight train was derailed, and the engine and 30 waggons fell into a ravine and were wrecked. The track was torn up for 300 metres and both lines were blocked for four days. [2007]

The *Francs Tireurs et Partisans Patrie* group derailed a Todt organisation materials train between Bordeaux and Pessac on Wednesday 11th August 1943. An engine and five waggons were wrecked, 15 waggons were damaged, and the line was blocked for 48 hours.

On Saturday 14th August 1943 the *14 Julliet* group derailed a materials train between Formerie and Gaille-Fontaine on the Rouen – Amiens line. The engine and 30 waggons were wrecked, and the line was blocked for 5 days. Between Nuits-sous-Ravières and

---

[2002] *The Times*, London, England, 3rd August 1943

[2003] *Journal de Geneve*, Geneva, Switzerland, 7th August 1943

[2004] Ibid.

[2005] *France d'Abord*, Paris, 1st September 1943

[2006] http://humanite-socialiste.over-blog.com/article-25575220.html

[2007] *Radio France* No. 29, 23rd August 1943

Aisy-sur-Armançon the *Vaubon* detachment derailed a German messenger train, wrecking 17 vehicles and blocking the line for 40 hours. At Butancourt, between Amiens and Montdidier, the *Bir Hackeim* detachment derailed a materials train, destroying 15 waggons and blocking the line for 72 hours. The *Marceau* group derailed an electric locomotive and 20 waggons at Drouet sur Drouette, destroying 300 metres of track and blocking the line for 31 hours.

A train was derailed at La Charité, on the Paris – Nevers line; on Sunday 15th August 1943, also German materials train at d'Imphy on the Nevers – Chagny line and a leave train on the Amiens – Paris line.

Tuesday 17th August 1943 saw the derailment of the engine and 19 waggons between St Just and Clermont, blocking the line for 30 hours and the wrecking of an engine and five vehicles at Harnincourt (name illegible) on the Amiens – Montdidier line.

Near Ste Maxence, between Creil and Compiègne, on Wednesday 18th August 1943, 12 waggons loaded with asbestos and coal were destroyed. Numerous deaths and injuries reportedly resulted from the derailment of a leave train the same day on the Longeville – Montereau line.

Saturday 21st August 1943 saw the derailment of a freight train at kilometre 63 of the Rouen – Elboeuf line by the *Marseillaise* group. An engine and 35 waggons were wrecked, and the line was blocked for 35 hours. The same day a freight train was derailed at Chatelet on the Paris – Reims line, while on the Paris – Le Mans line a train was derailed between Epernon and Rambouillet. An electric engine and 21 waggons were wrecked, 300 metres of track were ripped up and the line was blocked for 36 hours. [2008]

Forty waggons of a freight train were derailed between Dol and St Malo on Sunday 22nd August 1943; 12 were completely destroyed and the line was blocked for 48 hours. The same day a freight train was derailed between Chateaubourg and Vitré, with major damage to and engine and 17 waggons. [2009]

At 04:37 on Monday 23rd August 1943 a rail was loosened near the station at Bengy, France, on the Bourges – Saincaize line in Cher department. Freight train No. 6387 was derailed and passenger train No. 1016 coming from the opposite direction ran into the wreckage almost immediately. The leading vehicles of the passenger train were wrecked and overturned. Eleven people died and 22 were wounded according to one report; another spoke of only seven dead and three gravely injured. Trains were diverted via the Bourges and Cosne line, but then a train was derailed in turn on this line at Veaugnes, delaying a German mail train and two leave trains. [2010] [2011]

On Friday 27th August 1943 a German freight train was derailed on the Paris – Troyes line, the engine and 14 waggons being derailed. Near Clermont de l'Oise on the Paris –

[2008] *France d'Abord*, Paris, France, 15th September 1943

[2009] Ibid, 1st September 1943

[2010] Wikipedia – Liste des accidents ferroviaires den France au XX siècle

[2011] *Paris D'Abord*, Bulletin des Chemins de fer, November 1943

Amiens line a troop train was derailed. 10 coaches were wrecked, there were numerous dead and injured and the line was blocked for 38 hours. Near Fontenay-Eleury on the Paris – Granville line a train loaded with war material was derailed, the engine and 17 waggons were wrecked, and the line was blocked four days.

The locomotive and 28 40-tonne waggons of a coal train were put out of action in a derailment on the Arras – Amiens line near Achiet-Miraumont on Saturday 28th August 1943. The line was blocked for 72 hours. A bomb derailed one waggon of a train at Vendeuvre, on the Paris – Belfort line, blocking it for six hours. The *Maurice Romagon* group derailed a train of livestock bound for Germany at Piney on the Troyes – Vitry; 17 waggons were wrecked, while a German train was derailed at Uchizy on the Chalons – Lyon line.

Thirty waggons and the engine of a coal train were wrecked in a derailment at Bestancourt on the Amiens – Montdidier line on Monday 30th August 1943. The line was blocked for 48 hours. Seven vehicles of a military train were wrecked at Paray-le-Minial and the line was blocked for 12 hours. Numerous injured resulted from the derailment of a leave train at Santenay, near Chagny; eight coaches were derailed, and the line was blocked for 36 hours. [2012]

In January 1944 the French Commissariat for the Interior reported that during August 1943, 45 trains had been derailed by the Resistance, including 15 troop trains, 58 locomotives were sabotaged, and 45 other acts of sabotage had been carried out against rolling stock or track. In the single week between 16th and 22nd August 1,138 trains had been temporarily immobilised, 471 of them for more than six hours. [2013]

Sabotage by the *Osvald* group wrecked a German military train at the Kværner Bru, Oslo, Norway, on Tuesday 7th September 1943, causing much damage. [2014]

The Bruxelles – Mons line was dynamited at Nimy, Belgium, on Wednesday 8th September 1943. The engine and three coaches of a leave train were wrecked, 60 metres of track were damaged, there were numerous wounded, and the line was blocked for 17 hours.

On Saturday 11th September the Louvain – Malines line was dynamited between Corbeek and Loo and the engine and 11 waggons of a freight train were derailed. The line was blocked for 14 hours. [2015]

Resistance forces tampered with the track at Mars, France, on the Paris – Vichy line, on Friday 10th September 1943 and derailed a train. A train coming from the opposite direction then ran into the wreckage and the line was blocked for at least 24 hours. [2016]

Resistance forces in Belgium derailed a German military train in the Louvain area on

---

[2012] Ibid, 15th September 1943
[2013] *The Times*, London, England, 14th January 1944
[2014] http://larsborgersrud.no/artikler/til_minne_om_de_falne.pdf
[2015] http://humanite-socialiste.over-blog.com/article-25575220.html
[2016] *Paris D'Abord*, Paris, France, Bulletin des Chemins de fer, November 1943

about Wednesday 15th September 1943. Thirty Germans were stated to have been killed and about 100 injured. [2017]

On Friday 17th September 1943 a freight train carrying tanks and aircraft parts was dynamited near Hasselt, Belgium, on the Hasselt – Masseik line, which was blocked for 10 hours. The same day another freight train, which included oil tank waggons, was derailed at Blaregnies by the removal of a rail. The engine and 15 waggons left the rails and three tank waggons overturned and lost their contents. The line was blocked for 48 hours. [2018]

Wehrmacht express Dm W 31 was derailed by Polish partisans between Bialystock and Prostken, Poland, on Sunday 19th September 1943. Twenty-three soldiers died and 33 were injured. [2019]

Italy changed sides in September 1943, signing an armistice on 3rd of that month (made public on 8th) upon which the Germans immediately attacked their former allies. On Sunday 19th September Italian troops who had been occupying part of France withdrew to their own country by train via the Fréjus or Mont Cenis Tunnel with German forces in pursuit. Having negotiated the tunnel, the retreating Italians deliberately blew up a portion of it about 1,800 metres from the Bardonèche (Italian) portal to stop the following Germans. The pursuing German train, travelling at high speed, crashed into the obstruction, and was derailed. Thirty Germans were killed and 150 injured. A little later a light engine sent to examine the line ran inro a new obstruction at a different point. The French driver and fireman managed to jump to safety but ten of the escorting Germans were killed and two injured. It was clear that the repair of the tunnel was going to be a long and difficult operation as it was damaged over 800 metres; meanwhile all rail traffic between the French frontier and Turin was interrupted.

In other attacks in France at about the same date an armoured train was derailed by sabotage at Chignan, 19 kilometres from Chambery, a train was partly derailed by sabotage near St Amour and a German fuel train was machine-gunned by a Royal Air Force fighter aircraft between St Amour and Loubrans, and five of its escorting soldiers were killed. [2020]

A mineral (coal?) train was derailed at Fourmies, France, on Sunday 19th September 1943 with much material damage. [2021]

Radio Free Yugoslavia announced that on about Monday 20th September 1943 their partisans had derailed a train near Jakovac, in Bosnia. [2022]

A number of derailments were carried out in France on Monday 20th September 1943 on the 150th anniversary of the French victory against Prussia at the Battle of Valmy. A

[2017] *Gazette de Lausanne*, Lausanne, Switzerland, 20th September 1943
[2018] http://humanite-socialiste.over-blog.com/article-25575220.html
[2019] *Katastrophen der deutschen Bahnen Teil II Chronik 1845 – 1992* by H.J. Ritzau, pg. 110 (in German)
[2020] *Gazette de Lausanne*, Lausanne, Switzerland, 21st September 1943
[2021] *France d'Abord*, Paris, France, 1st October 1943
[2022] *Journal de Geneve*, Geneva, Switzerland, 25th September 1943

coal train was derailed at Compans on the Paris – Laon line. The engine was 39 waggons left the track, 400 metres of line were damaged, and the line was blocked for 4 days. A German freight train was derailed on the Valenciennes – Orachies line and the locomotive and eight waggons wrecked. Two engines were derailed on the Valenciennes – Douai line. A German freight train was derailed near Nemours on the Moret – Montargis line; the engine and nine waggons were wrecked. Seventeen waggons of a freight train were wrecked at Crouy-sur-Orucq on the Paris – Reims line. On the Paris – Orléans line a train was wrecked at Etampes, the engine and 30 waggons were derailed. The locomotive and six waggons of a freight train were derailed at Epluches on the Paris-Persant-Beaumont line. The engine and numerous waggons were derailed at Cintray on the Chartres – Mans line. Finally, a brand new light engine was derailed at Clayes and the line was blocked for 24 hours.

Another light engine was derailed between Vaincourt and Eux on Tuesday 21st September 1943. A German train was derailed between Albert and Heilly and another between Crépy and Valemont Bethisy, a freight train between Creil and Persant-Beaumont, and a German train between Rennes and La Brohinière[2023].

The French railway system in Savoy and the adjoining area to the east and south east of Lyons was of importance in view of its proximity to the Italian frontier, which meant that it was the only district through which the Germans could send reinforcements to their armies in northern Italy, as Switzerland would not permit the transport of military material through its country.

The *Railway Gazette* carried an article on sabotage in the area, 'The 24-mile Bellegarde – Annemasse line, to the south of Geneva, has been the object of special attention after having been reopened to through traffic only on August 1…Towards the end of August the line was blown up near Viry station, 12 miles to the east of Bellegarde, and a three day suspension of traffic resulted. In the first fortnight of September the rails were torn up by a bomb near Bossey-Veyrier station, 20 miles to the east of Bellegarde, but the line was reopened to traffic after a delay of a few hours. On September 21 it was reported that near the same station an attempt had been made to blow up a goods train; three waggons were derailed. Two days later the line was again blown up, this time near Valleiry station, 8 miles to the east of Bellegarde. About the same date an attempt was made against the main line railway linking Aix-les-Bains with Annecy, La-Roche-sur-Foron, and Annemeasse; this was near Evires, between Annecy and La-Roche, 14 miles to the north of Annecy. The Bayonne-Toulouse-Annemasse fast train was derailed, and two carriages fell down the embankment. It took a week to restore the line to working order.

'Damage to the line, and a derailment were caused through an explosion near St Amour station on the Lyons-Besançon main line in the second half of September. St Amour is 78 miles to the north east of Lyons and is an important junction station. Traffic was disorganised on both the Dijon-Bourg-Chambéry-Modane and the Lyons- Besançon main lines. Almost simultaneously it was reported that a German armoured train had been derailed through a bomb explosion near Chignin-les-Marches station on the Chambéry-Modane main line, 6 miles to the south east of Chambéry, disorganising the traffic also on the connecting lines from Chambéry to Lyons, and from St Pierre d'Albigny (to the east of Chambéry) to Albertville and Bourg-St Maurice, as well as on the Montmélian-

---

[2023] *France d'Abord*, Paris, France, 1st October 1943

Grenoble main line. In the same region during the night of September 12-13, the rails were unfastened for a considerable length near Chamousset station, 19 miles to the east of Chambéry, on the main line to Modane. A week later a viaduct on the Toulon-Nice main line, near Anthéor-Cap-Roux station, 12 miles to the west of Cannes, was blown up.' [2024]

An engine was derailed by sabotage at La Hulpe, Belgium, on the Bruxelles – Namur line, on Thursday 23rd September 1943. [2025]

On about Friday 24th September 1943 a German freight train was derailed at Brie-Comte-Robert, France; the engine left the rails, and seven waggons were wrecked.

Monday 27th September 1943, a train carrying food and livestock bound for Germany was wrecked at La Charité, France.

A German train carrying troops and material was derailed on the Paris – Cherbourg line on Tuesday 28th September 1943. The locomotive and 17 vehicles left the track, there were numerous dead and injured and the line was blocked for three days. The same day a train was derailed at Villers Bretonneux, France, on the line between Amiens and Chaulnes. The line was blocked for four days. [2026]

The engine and 10 waggons of a freight train were derailed at Mignault, Belgium, on Wednesday 29th September 1943. The line was blocked for 24 hours. [2027]

A German goods train on the Warsaw – Radom line was derailed by Polish partisans at Zalesie, Poland, at midnight on Monday 27th September 1943. Twenty-seven people took part, using a kilogram of plastic explosive and a detonator. The wrecked train was machine gunned, petrol bombs were used to set waggons and vehicles on fire, and a mine was laid for the engine of the breakdown train to set off. [2028]

On 30th September 1943 Yugoslav partisan forces stated that they had derailed a train between Brod and Zagreb, killing 60 soldiers. [2029]

Free Yugoslav radio announced at the end of September 1943 that two goods trains and an armoured train had been derailed on the line between Bosansko and Novi – Prijedor, Bosnia, Yugoslavia and 63 German soldiers had been killed. [2030]

A train was derailed by sabotage at Soignies, Belgium, on Thursday 30th September 1943. Two engines were badly damaged, and the line was blocked for 12 hours. [2031]

2024 *The Railway Gazette*, London, England, 29th October 1943
2025 http://humanite-socialiste.over-blog.com/article-25575220.html
2026 *France d'Abord*, Paris, France, 20th October 1943
2027 http://humanite-socialiste.over-blog.com/article-25575220.html
2028 http://www.sosabowski.com/stancv.html
2029 *Journal de Geneve*, Geneva, Switzerland, 1st October 1943
2030 *The Times*, London, England, 30th September 1943
2031 http://humanite-socialiste.over-blog.com/article-25575220.html

About Sunday 3rd October 1943 a German train was derailed on the Zagreb – Belgrade main line between Ruma and Indjija, Yugoslavia. An engine and 23 vehicles were destroyed. At about the same time an armoured train was ambushed in north-eastern Croatia, between Osijek and Nasice, Yugoslavia. [2032]

A German train was derailed at Genly, Belgium, on Monday 4th October 1943. It was presumably a troop or leave train as many Germans were said to have been killed. The line was blocked for 72 hours. The same day the engine and many waggons were damaged in the derailment of a freight train at Velthem on the Louvain – Bruxelles line and there was what was described as a 'double derailment' at Naveau on the Arlon – Bruxelles line in which four locomotives were 'definitely wrecked' and there was enormous disruption to traffic for 48 hours. [2033]

Near Sarrou, France, on the Creil – Compiègne line a freight train was attacked, and the engine and ten waggons were derailed on Monday 4th October 1943. [2034]

Two mineral trains were derailed at Florenville, Belgium, on Tuesday 5th October 1943 and the same day a train was derailed in a tunnel at Hachy by the removal of a rail. The next day, Wednesday 6th October, a freight train was stopped at Ciply by the *Partisans Armés* and its engine crew made to alight. It was then set off at full speed towards Mons and the engine and 17 waggons were derailed. [2035]

'Des terroristes saboteurs font dérailler un train,' a French newspaper wrote, reporting on an act of sabotage at Chalon sur Saône, France, on Wednesday 6th October 1943. A freight train was derailed and a Paris – Marseille express crashed into the wreckage. Two of the coaches of the express then caught fire. An initial report stated that there were 32 dead and a large number of wounded. The freight train was reportedly carrying packages and medicines dispatched by the Red Cross for French prisoners of war in Germany. The newspaper report went on to write of the peaceful passengers added to the hunting table of the criminals as a result of the abominable attack. [2036]

On 12th October it was announced that the number of dead had risen to 39 with the discovery of eight more charred and unrecognisable corpses in the wreckage. [2037]

A major act of sabotage was carried out in Norway on the evening of Thursday 7th October 1943 at Ryggkollen, on the Sørlandsbanen between Daler and Mjøndalen. The train targeted was the Norges Statsbaner night express, train No. 509, which consisted of an electric locomotive, six bogie carriages, a four-wheeled goods waggon and type 26c steam locomotive No. 380 at the rear. The attack took place at a point where the line ran on an embankment alongside the River Drammen. The saboteurs, members of the *Osvald* group, had planted three explosive charges under the track. Only one exploded, but this

---

[2032] *The Daily Times*, New Philadelphia, Ohio, United States, 13th October 1943

[2033] http://humanite-socialiste.over-blog.com/article-25575220.html

[2034] *France d'Abord*, Paris, France, 20th October 1943

[2035] http://humanite-socialiste.over-blog.com/article-25575220.html

[2036] *L'Œuvre*, Paris, France, 8th October 1943

[2037] *Gazette de Lausanne*, Lausanne, Switzerland, 13th October 1943 was sufficient to derail the train, and one of its carriages, No. 632, was thrown from the embankment into the river and wrecked. How many died is unclear. The train was carrying a large number of German soldiers and figures between 8 and 70 of them having been killed have been quoted. Twenty-five Germans and three Norwegians were admitted to Drammen hospital.

The Nazis carried out reprisals in the wake of the derailment. Sixty-three or 64 Norwegians from Eiker, Mjøndalen and Drammen were arrested. A week later five of them were shot and six others were sent to a concentration camp in Germany. None of the saboteurs was arrested but one of those who was shot had possibly supplied the explosives. [2038]

An engine was derailed by sabotage and badly damaged near Gand, Belgium, on the Gand – Bruxelles line on Friday 8th October 1943. [2039]

On Friday 8th October 1943 a train was derailed by the Resistance near Varonne-le-Grand, France. The same day a munitions train was derailed and exploded at St Florentin, France. The derailment of a leave train on the Amiens – Abbeville line near Fontaine-sur-Somme, France, reportedly cost 55 lives as well as leaving many wounded. The same day a light engine was derailed at Landevant, France. [2040]

The *Francs Tireurs et Partisans* announced that 20 derailments had been carried out between Friday 8th October and Wednesday 20th October 1943. These included one of significance, an attack on a leave train between Landevant and Hennebont in which four coaches were wrecked; there were reportedly 200 dead and wounded and the line was blocked for 50 hours. [2041] [2042]

Radio Free Yugoslavia stated on 12th October 1943 that partisans had derailed a train between Rouma and Indija on the Zagreb – Belgrade main line, destroying a locomotive and 23 waggons. In another attack they had ambushed an armoured train on the Osijek – Nasice line in Slovenia. [2043]

The *Partisans Armés* caused the derailment of the engine and 10 waggons at the entrance to the station at Tamines, Belgium, on Sunday 17th October 1943. [2044]

A train of tanks and lorries was derailed at Chagny, France, on Friday 22nd October 1943, and 30 Germans appear to have died in the attack. [2045]

The Wolkowysk – Warsaw – Berlin overnight express carrying Wehrmacht soldiers home on leave from the Eastern front was derailed by 30 Polish partisans on Sunday 24th

---

2038 *På Sporet* (magazine of Norsk Jernbaneklubb) No. 169 pgs. 38 (in Norwegian)
2039 http://humanite-socialiste.over-blog.com/article-25575220.html
2040 *France d'Abord*, France, 20th October 1943
2041 Ibid, 1st November 1943
2042 Ibid, 31st December 1943
2043 *Gazette de Lausanne*, Lausanne, Switzerland, 13th October 1943
2044 http://humanite-socialiste.over-blog.com/article-25575220.html
2045 *France d'Abord*, France, 20th November 1943

October 1943 at Plochocina, Poland. The train was then strafed with machine-gun fire and grenades were thrown at it. Heavy fire was returned and after eight minutes the partisans withdrew. [2046]

Three trains in succession were derailed at Florenville, Belgium, on Monday 25th October 1943. Four engines and 46 waggons were wrecked, and the line was blocked for 96 hours. The same day the track was dynamited at Ressaix on the Erquelinnes – Haine St Pierre line, a locomotive was derailed, and the line was blocked for 12 hours. [2047]

A German troop train bound for the Eastern front was derailed by *Francs Tireurs et Partisans* at Le Quesnoy on the Le Tréport – Abbeville line on Wednesday 27th October 1943. It was travelling at 80 kilometres an hour at the time and 27 of its vehicles were said to have been 'pulverised' with 180 Germans killed and 200 seriously injured. [2048]

Early on Sunday 31st October 1943, Marseille – Lyon – Paris express No. 110, which was packed with passengers on the eve of the All Saint's Day holiday (1st November is a public holiday in France) was derailed near the station at Tournus, France. At 01:10 the express had passed through the little station at Uchizy and travelling at 90 kilometres an hour was approaching Tournus when it was derailed at a culvert at kilometre 414. The engine fell into a ditch, the leading six coaches passed over it and the seventh ended up in a lineside wood and was wrecked. Fourteen metres of rail had been removed by the saboteurs on the approach to the culvert. At first it was reported that nine passengers had died, 17 had been gravely wounded and 30 slightly so. Two of the badly wounded soon died, raising the number of dead to 11. As in the report on the Chalon sur Saône derailment, the saboteurs were referred to as terrorists by the German controlled press. [2049]

In October 1943 the armoured train of Hans Frank, the Nazi Governor General of occupied Poland, was derailed by partisans between Cracow and Warsaw. A gun battle then ensued between the partisans and the soldiers guarding the train before an engine was attached to the rear of it and the vehicles which remained on the track were drawn back to Cracow. Frank subsequently travelled by aircraft to Warsaw. [2050]

The track was dynamited at Nossegem, Belgium, on Tuesday 2nd November 1943, and an engine derailed. The next day, Wednesday 3rd November, a similar attack was carried out at Neufville, a train was again derailed, 100 metres of track was damaged and the line blocked 14 hours. [2051]

A German leave train was derailed two and a half kilometres north of Magny, on the Rouen – Amiens line, on Wednesday 3rd November 1943. Four hundred of the enemy were reportedly killed or injured in this attack. On the same day a coal train was derailed at Heilly on the Albert – Amiens line and a second train loaded with potatoes destined for

---

[2046] http://www.sosabowski.com/stancv.html
[2047] http://humanite-socialiste.over-blog.com/article-25575220.html
[2048] *France d'Abord*, Paris, France 20th November 1943
[2049] *Le Journal*, Paris, France, 2nd November 1943
[2050] *The Lowell Sun*, Lowell, Massachusetts, United States, 7th December 1943
[2051] http://humanite-socialiste.over-blog.com/article-25575238.html

Germany crashed into it.

On Thursday 4th November 1943 a materials train was derailed between St Loup and Domigny, France, another was attacked on the Persan – Pontoise line on Friday 5th November and a third at Champagne-sur-Oise on Saturday 6th November 1943. [2052] In what was presumably a different attack by the *Vodli* detachment at an unstated date in early November 1943, a troop train was derailed at St Loup, leaving 180 dead or wounded. [2053]

On about Friday 5th November 1943 guerrillas derailed a military train loaded with tanks near the village of Verniski, Bulgaria, on the Sofia – Plovdiv line, killing five German soldiers. In retaliation the Germans arrested three Bulgarian stationmasters and a number of subordinates. [2054] [2055]

The Warsaw – Berlin express, carrying a large number of German passengers, was derailed by partisans in the suburbs of Warsaw at Celestynow, Poland, also on about Friday 5th November 1943. Hundreds were reportedly killed. A telegram to London stated that it was in the reprisal for the German execution of Polish hostages. [2056]

A German train carrying troops and material was derailed between Chagny and Fontaine on Saturday 6th November 1943. Seventeen waggons were wrecked, including eight loaded with bombs and three carrying oil. Fifty-three Germans were said to have died. The line was blocked for 60 hours. [2057]

The *Francs-Tireurs et Partisans* derailed 23 trains on Armistice Day, Thursday 11th November 1943. The measures taken by the Germans to prevent such attacks notwithstanding, a troop train carrying 500 officers and men from France to the Eastern front was derailed at an unstated location in northern France. Fifteen coaches were said to have been completely destroyed and a large number of Germans were killed and wounded. Meanwhile between Chourette and Ribécourt on the Creil – Tergnier line a German leave train was derailed, and 100 officers and men were allegedly killed. Two Resistance fighters lost their lives during the withdrawal from the scene. Amongst other actions an engine and eight waggons were destroyed and 460 tonnes of mineral (coal?) were spilt at Donchery on the Charleville – Sedan line; 48 other trains were held up for 48 hours. [2058]

A Francs-Tireurs et Partisans detachment derailed a German leave train at Bourg les Valence, France, on Friday 12th November 1943. Eight carriages were derailed and seriously damaged and a number of soldiers were injured. [2059]

[2052] *France d'Abord*, Paris, France, 15th December 1943
[2053] Ibid, 31st December 1943
[2054] *The Amarillo Daily News*, Amarillo, Texas, United States, 11th November 1943
[2055] *Gazette de Lausanne*, Lausanne, Switzerland, 12th November 1943
[2056] *Sunday Journal and Star*, Lincoln, Nebraska, United States, 7th November 1943
[2057] Ibid, 15th December 1943
[2058] *France d'Abord*, Paris, France, 15th December 1943
[2059] *Forces Françaises Interieuses*, No. 28 15th December 1943

On Friday 12th November a freight train was derailed near Montataire on the Beauvais – Creil line.

At Cramoisy, also on the Beauvais – Creil line 22 waggons of a train of livestock destined for Germany were derailed on Saturday 13th November 1943.

The engine and 26 waggons of a mineral train were wrecked at Cauffy upon the Creil – Amiens line on Sunday 14th November 1943; one member of the Resistance lost his life during this attack.

A train of wine bound for Germany was derailed at Neufchâteau on Monday 15th November 1943 and a freight train was attacked near Frouard on Tuesday 16th November 1943.

Things did not go exactly according to plan on Wednesday 17th November 1943. The track had been ripped up inside a tunnel between Luxeuil and Fontaine, in the expectation that a German train would be derailed in the tunnel. This train chanced to be late and was preceded by an omnibus train carrying local French passengers which had to be stopped by exploding a charge beneath its engine at the entrance to the tunnel.

On Sunday 21st November 1943 a troop train was derailed between Beaucourt and Albert on the Paris – Lille line. Seven coaches were wrecked and there were dead and wounded. Freight trains were derailed between Artres and Le Poirier (Aulnoye – Valenciennes line) on Monday 22nd November and between St Didier and St Arnoux (Nevers – Chagny line) on Tuesday 23rd November 1943. [2060]

On 23rd November 1943, the Belgian government in London announced that patriots had derailed the Paris – Cologne express near Nimy. No casualties were mentioned but the line was said to have been blocked for days. [2061]

On Thursday 25th November 1943 *Francs-Tireurs et Partisans* guerrillas derailed a train in a tunnel at Veynes, France, destroying a locomotive and 12 waggons and blocking the line for 40 hours. [2062]

A German armoured train (Panzerzug 1) was badly damaged when it was derailed due to Soviet partisans blowing up the track near Viyasma, Poland, on Sunday 28th November 1943, the engine and eight waggons leaving the track. Two recovery teams worked around the clock under enemy fire to rerail the train, but as it withdrew it was then struck by a bomb and further damaged, so it had to go to Konigsberg to be completely rebuilt. [2063]

On Sunday 28th November 1943 Resistance forces derailed a train on the Grenoble – Valence line. They loosened the rails over a distance of about a dozen metres at

[2060] *France d'Abord*, Paris, France, 15th December 1943
[2061] *Gazette de Lausanne*, Lausanne, Switzerland, 24th November 1943
[2062] *Forces Françaises Interieuses,* No. 27, 30th November 1943
[2063] https://www.feldgrau.com/ww2-german-railway-armored-trains/

Goncelin, France, about 27 kilometres from Grenoble. An express arrived. Its engine left the rails and overturned, and its leading van was completely telescoped by the following vehicles. The momentum of the train then carried the remainder of it past the locomotive and destroyed van. The driver, fireman, conductor and two agents who had been riding in the van were all killed. Both lines were completely blocked, and it was anticipated that trains would have to be diverted for some days. [2064]

Yugoslav partisan forces announced on Monday 29th November 1943 that two German military trains had been derailed in eastern Bosnia. One hundred and fifty Germans had been taken prisoner and the same number killed. [2065]

A communique from Moscow on 12th December 1943 announced that partisans had derailed five German troop trains in the Leningrad region between 24th November and 8th December, destroying 100 coaches and killing 200 Germans. [2066]

A materials train was derailed by sabotage on Wednesday 1st December 1943 at Jemeppes, Belgium.

On Friday 3rd December 1943 a train was derailed at Masny St Pierre on the line between Mons and Braine-le-Comte and 11 waggons were wrecked. [2067]

At 02:00 on Thursday 2nd December 1943 a passenger train from Ternier to Laon was derailed between Versigny and La Fère, France. Four coaches piled up one on top of the other. There were ten dead and 45 wounded were taken to hospital in La Fère. The derailment was not categorically put down to sabotage though the implication seemed to be that it was.

On the morning of Friday 3rd December 1943 mixed train No. 9380 was derailed due to sabotage at Km 8.700 of the Chagny – Dòle line at Demigny. [2068] The engine, tender and four coaches left the track. There were two dead and many wounded. The same day an engine was derailed by sabotage a few hundred metres from the station at Tencin-Goncelin, France, 40 kilometres from Grenoble. A short time later an express for Nice, which included carriages from Toulouse, collided with the derailed engine before it could be stopped. The locomotive of the express, which was travelling at 70 kilometres an hour, ploughed into the ballast, and four carriages left the rails with a violent shock which threw their occupants one upon the other. A rescue operation was quickly organised, and ambulances were called from Grenoble. It was found that four passengers, one of them an infant, had lost their lives, and 20 were more or less injured, some being unconscious. It was believed that about a dozen individuals, operating in broad daylight, were responsible. [2069]

2064 *Journal de Geneve*, Geneva, Switzerland, 30th November 1943
2065 *Gazette de Lausanne*, Lausanne, Switzerland, 30th November 1943
2066 Ibid, 13th December 1943
2067 http://humanite-socialiste.over-blog.com/article-25575238.html
2068 *Le Petit Parisien*, Paris, France, 4th December 1943
2069 *Gazette de Lausanne*, Lausanne, Switzerland, 6th December 1943

The German night express troop train on the Warsaw – Siedlice line, carrying Wehrmacht soldiers back to the Eastern front, was derailed by Polish partisans at Skruda, Poland, on Saturday 4th December 1943 but remained upright. The coaches were then machine gunned and grenades were thrown at them. When the Germans began to return fire a whistle was blown, and the partisans withdrew. Twenty-five partisans took part; one was killed. [2070]

Several attacks were carried out in different parts of France by *Francs-Tireurs et Partisans* guerrillas on Saturday 4th December 1943. At Vienne (Isère department) a German freight train was derailed; the engine and many waggons were badly damaged, and the line was blocked for 27 hours. At Villers Rotin (Côte D'Or department) on the Dijon – Dole line detachment X derailed the engine and eight waggons of a freight train. The line was blocked for 30 hours. Finally, at Arc Senas (Jura department) the engine and eight waggons of a freight train were derailed, blocking the line for 32 hours. [2071]

Fonchercans, France, was the location of the derailment of a German leave train by *Francs-Tireurs et Partisans* guerrillas on Tuesday 7th December 1943. The locomotive and two carriages were wrecked, 12 Germans were killed or seriously injured, and traffic was interrupted for 48 hours. [2072]

On about Wednesday 8th December 1943 five German military trains bound for the Eastern front were blown up by partisans near Mogilev, USSR. 'Four engines, 33 trucks carrying troops, and several dozen trucks laden with tanks were smashed.' [2073]

Another train was derailed by sabotage at Jemeppes, Belgium, on Wednesday 8th December 1943 and the same day a leave train was derailed at Liège on the line to Visé. The line was blocked for 24 hours. [2074]

Francs-Tireurs et Partisans guerrillas derailed a German leave train on the night of Thursday 9th December 1943 at Montelimar, France. Seven carriages were thrown from the track and badly damaged; two soldiers were killed and 34 wounded. [2075]

Partisans from the Ramenetspodolsk region of the USSR reported in early January 1944 that on 8th December 1943 they had blown up a train carrying guns and searchlights, on 10th December they derailed a train carrying lorries, the next day one carrying tanks was derailed, and in a week they had accounted for seven enemy trains. [2076]

In a major action on Friday 10th December 1943 the *Partisans Armés* mined a leave train at Voroux Goreux, Belgium. The engine and many coaches were wrecked, and 85 Germans were reportedly killed and 130 were injured. The line was blocked for 24 hours.

[2070] http://www.sosabowski.com/stancv.html
[2071] *Forces Françaises Interieuses,* No. 29, 31st December 1943
[2072] Ibid.
[2073] *The Times*, London, England, 10th December 1943
[2074] http://humanite-socialiste.over-blog.com/article-25575238.html
[2075] *Forces Françaises Interieuses,* No. 28, 15th December 1943
[2076] *The Times*, London, England, 3rd January 1944

2077

Detachment X of the *Francs-Tireurs et Partisans* (FTP) guerrillas derailed and attacked a German leave train at Porte-les-Valence, France, on Friday 10th December 1943. The locomotive and eight carriages of the train were wrecked, and the guerrillas then attacked the remainder of the train with grenades and machine guns. Twenty-one soldiers were killed and about 200 wounded. Two resistance fighters were killed. The FTP report of the action ended '*Glorie immortelle à nos héros*' [2078]

Detachment X of the *Francs-Tireurs et Partisans* guerrillas were in action again on the night of Sunday 12th to Monday 13th December 1943 when they derailed two trains at Grezat *poste sémaphoriqe* (signal box) on the Cahors – Brive line in France. An electric locomotive, a steam locomotive and 45 waggons were 'pulverized' and traffic was interrupted for 48 hours. [2079] Also on Sunday 12th a troop train was derailed between Laon and Reims. An engine and 30 waggons were demolished, there were numerous dead and artillery pieces were destroyed, while a train of iron ore was wrecked at Baroncourt. [2080]

On 15th December 1943 Yugoslav partisan forces announced that their partisans had derailed a train between Doboi and Derventa, Bosnia. Thirty Germans were killed and 70 taken prisoner. [2081]

The Vintimille – Paris express which had left Nice at 15:40 on Thursday 14th December 1943 was much delayed and well over two hours late leaving Lyon Perrache station. Running at full speed it had just passed the station at Villevert-Neuville, France, 17 kilometres from Lyon, when it reached the spot where a rail had been removed. This was a point especially suitable with the line running on an embankment and twisting as it followed the course of the River Sâone, between which and the railway was the national highway, while there were broad meadows on the other side of the embankment.

Following the derailment the leading *Wagon-Lit* coach was forced into the engine and caught fire from hot coals from the firebox; Driver Léon Achalé and Fireman Jean Pasquiet were amongst those killed, the driver still clutching his rag, according to a reporter. '"I heard the train coming," a young girl says, "I heard the rhythm of the axles on the curve. Then there was a terrible noise…then silence…I got up. You see, we live in this little white house – barely 300 metres from the accident – I cried, 'Mama, get dressed, a train is derailed!' We saw enormous jets of white steam coming from the foot of the embankment. I took with me everything I could, water, cotton wool. In the pasture there was a veritable stampede. Half-dressed men were jumping over the fence. The flames rose and lit the scene. A woman in a chemise near the third carriage was half screaming, half moaning, in a monotone tone, like a wounded animal. 'My husband is dead.' I will never forget her cries. [2082]

Another source stated that there had been eight fatalities in all. The exact time of the

---

2077 http://humanite-socialiste.over-blog.com/article-25575238.html

2078 *Forces Françaises Interieuses,* No. 28, 15th December 1943

2079 Ibid. No. 30, 10th January 1944

2080 *France d'Abord*, Paris, France, 15th January 1944

2081 *Journal de Geneve*, Geneva, Switzerland, 15th December 1943

2082 *Le Journal*, Paris, France, 16th December 1943

derailment was not stated, only that it had occurred early on Friday 16th. The conductor and postal employees escaped with injuries. [2083] One of the passengers who died was Paul Carbone, a leading figure of the Marseille underworld.

As usual the German-controlled French newspapers referred to the saboteurs as 'terrorists', 'criminals' and 'bandits'. The Resistance had apparently targeted the train because it was likely to be used by many German soldiers travelling on leave.

Four passengers died in a derailment due to sabotage between Jeugny and Saint-Phal-Chamoy, France, on Thursday 16th December 1943. [2084]

On the night of Thursday 16th to Friday 17th December 1943 a troop train was derailed on the Lens – Douai line between Beaumont and Quéry la-Motte, France, the engine and ten vehicles leaving the track.

A train was derailed by sabotage at Lesterny, Belgium, on Monday 20th December 1943, with the destruction of an engine and six waggons. [2085]

A troop train was derailed by sabotage at Liège, Belgium, on Wednesday 22nd December 1943 on the line to Namur. [2086]

Francs-Tireurs et Partisans guerrillas derailed a German military train at Die (department of Drôme), France, on Wednesday 22nd December 1943. Two waggons of munitions exploded, an oil tank caught fire and the train was totally wrecked. Forty Germans were reportedly killed and 300 wounded. [2087]

A freight train was derailed between Brou and Courtelain on Christmas Eve, Friday 24th December 1943.

A troop train was derailed at Bortmeerbeek, Belgium, on the Louvain – Malines line, on Thursday 23rd December 1943, the engine and a number of vehicles being wrecked. [2088]

The derailment of a freight train due to sabotage near Pons, France, north of Bordeaux, on Monday 27th December 1943, was followed by the collision of a Paris – Bordeaux express with the wreckage. One person died and 12 were injured. The same day a German troop train was derailed between Valenciennes and Le Quesnoy, France, on, and 25 soldiers were killed or wounded.

On Tuesday 28th December 1943 an engine and 24 waggons of DCA (?) were derailed between Crépy and Compiègne, six Germans were killed, 300 metres of track were torn up and the line was blocked for a week. The same day freight trains were derailed

---

[2083] *Le Petit Parisien*, Paris, France, 17th December 1943

[2084] Author's personal notes. Source mislaid.

[2085] *Le Patriote du Pas-de-Calais*, 1st January 1944

[2086] http://humanite-socialiste.over-blog.com/article-25575238.html

[2087] *Forces Françaises Interieuses,* No. 30, 10th January 1944

[2088] http://humanite-socialiste.over-blog.com/article-25575238.html

between Somain and Péruwelz (over the border in Belgium) and between Valenciennes and Onning. [2089]

On New Year's Day, Saturday 1st January 1944, a train of freight and coal was derailed at Presney, on the line between Busigny and Tergnier, France, and a coal train was derailed near Donchery, on the Mézières – Sedan line.

On Monday 3rd January a materials train was derailed at Aremberg, while on Tuesday 4th January 1944 a train of freight and oil was derailed between Cernay and Franconville. [2090]

In a more significant attack on 4th January a leave train was derailed between Monteray and Paray-le-Monial. The engine and four vehicles fell into a ravine, 30 other vehicles were damaged, and an unknown number of Germans lost their lives. [2091]

Probably in early January 1944 Greek partisans derailed a German supply train and destroyed a bridge between Katerini and Aylos Demetrios, Greece. [2092] In Yugoslavia on about the same date Slovenian partisans in Yugoslavia severed the Zagreb – Fiume railway at several points and derailed a train which fell into the Drave river, 'killing several Nazi troopers.' [2093]

A report from London on 6th January 1944 stated that Croatian partisans had derailed a German armoured train on the Zagreb – Belgrade main line near Brod, Yugoslavia. An engine and seven waggons were wrecked. [2094]

A freight train was derailed between Valenciennes and Lille, France, on Thursday 6th January 1944 and a coal train between Douai and Arras. [2095]

The *Partisans Armés* derailed a freight train near Ans, Belgium, on the Ans Guillemins line, on Friday 7th January 1944. [2096]

On Saturday 8th January 1944 at Givet, France, on the Charleville – Mézières line, a troop train was derailed, three vehicles being destroyed and 18 much damaged. Ten Germans were reportedly killed and 50 seriously injured. Two hundred metres of track were torn up, and the line was blocked for 48 hours. Near Perigieux, France, on the Perigieux – Mussidan line in the Dordogne, the same day, the Nième detachment of the *Francs-Tireurs et Partisans* guerrillas derailed 17 waggons of a German freight train, the load of which included 15 cannons. A number of German soldiers accompanying the train were killed and injured. [2097]

[2089] *France d'Abord*, Paris, France, 15th January 1944
[2090] Ibid.
[2091] *La Voix de Paris*, 10th February 1944
[2092] *The Charleston Gazette*, Charleston, West Virginia, United States, 7th January 1944
[2093] *The Lowell Sun*, Lowell, Yugoslavia, United States, 7th January 1944
[2094] *Gazette de Lausanne*, Lausanne, Switzerland, 7th January 1944
[2095] *La Voix de Paris*, Paris, France, 10th February 1944
[2096] http://humanite-socialiste.over-blog.com/article-25575238.html
[2097] *Forces Françaises Interieuses,* No. 33, 10th February 1944

A coal train was derailed between Blanzy and Montchanin, France, on Monday 10th January 1944. [2098]

Francs-Tireurs et Partisans guerrillas derailed a German train at Porte-les-Valence, France, on Tuesday, 11th January 1944, and a locomotive and seven waggons were seriously damaged. [2099]

A coal train was derailed between Denain and Aniche, France, on Wednesday 12th January 1944. [2100]

A series of actions were carried out by *Francs-Tireurs et Partisans* guerrillas on Friday 14th January 1944 in different parts of France. At Theimellers on the Paris – Belfort line a freight train was derailed, blocking the line for 96 hours. At Chamelet (Rhône department) a freight train bound for Germany was derailed. An engine was seriously damaged, and 19 waggons were broken up. The line was blocked for 4 days. Between Aulnoye and Mauberge a materials train was derailed. At Lus la Croix Haute on the Grenoble – Veynes line a locomotive was seriously damaged and five waggons were broken up in a derailment. The line was blocked for 47 hours. After a dozen guards had been apprehended and 'neutralised' a train was derailed in Teche Tunnel, near Vinay on the Grenoble – Valence line. An engine was seriously damaged, and traffic was interrupted for 36 hours. Finally, a train was derailed at La Tour du Pin on the Grenoble – Lyon line. [2101] Also, on about Saturday 15th January in an unusual act of sabotage a waggon loaded with concrete pylons was parked on a running line at Saint-Pierre-du-Rumilly. A freight train collided with it and was seriously damaged. [2102]

On Saturday 15th January 1944 *Francs-Tireurs et Partisans* guerrillas derailed an engine and 13 waggons at La Pouzir, a freight train was derailed at Templeuve on the Lille – Valenciennes line and on the night of Saturday 15th to Sunday 16th a train of bauxite was derailed at Nyons, France. [2103]

A light engine was derailed between Laon and Hirson on Sunday 16th January 1944 and a mineral train between Sedan and Mézières.

A mineral train was derailed between Hirson and Aulnoye on Tuesday 18th January 1944 and the following day a coal train between Hirson and Liart. [2104]

Passenger train No. 3513, running as an express between Pau and Toulouse, was derailed at 06:10 on Wednesday 19th January 1944, the track having been damaged between Ossun and Juillan, France, a few kilometres from Troyes. Two carriages and a

---

[2098] *La Voix de Paris*, Paris, France, 10th February 1944
[2099] *Forces Françaises Interieuses,* No. 32, 31st January 1944
[2100] *La Voix de Paris*, Paris, France, 10th February 1944
[2101] *Forces Françaises Interieuses,* No. 32, 31st January 1944
[2102] *Journal de Geneve*, Geneva, Switzerland, 17th January 1944
[2103] Ibid.
[2104] *La Voix de Paris*, Paris, France, 10th February 1944

van were wrecked and the bodies of 25 passengers were taken from the wreckage. Another 50 were injured. The victims were said to have all been French civilians, most of whom were going to work in factories and offices. The final number of dead was probably 27.

In a second 'criminal derailment,' as a newspaper referred to it, on the same day or possibly the following day, a passenger train was derailed near Le Buissonet, France (location also given as Attichy, a nearby settlement.) An initial report spoke of six dead and 20 injured. Another report spoke of 12 dead. Both acts of sabotage were blamed on *'Des communistes.'* The victims of this derailment were also stated to have all been French. The *compiègne parquet* (public prosecutor) went to the scene.[2105]

On Thursday 20th January 1944 *Francs-Tireurs et Partisans* guerrillas derailed a train on the Toulouse – Sète line in France. Seven waggons were damaged along with three aircraft being carried on them. The line was blocked for seven hours. [2106] Between Hirson and Aulnoye another coal train was derailed. There was also another very serious derailment of a leave train at Saint-Lié, on the Paris – Belfort line. The engine and 20 vehicles were reportedly completely wrecked, about 100 Germans were killed and about 200 seriously wounded.

A leave train was derailed on Friday 21st January 1944 at Sous-le-Bois, France, between Aulnoye and Mauberge. The engine and seven vehicles were badly damaged, and a number of Germans were killed and wounded. The same day a coal train was derailed at Liévin. A light engine was thrown off the line between Hirson and Busigny the next day, Saturday 22nd, while two died and seven were wounded in the derailment of a leave train at Valentigny, between Troyes and St Dizier. Another leave train was derailed the same day between Montceau and Blanzy. Eight coaches were wrecked and 60 were said to have been killed.

The engine and six coaches of a troop train were wrecked on Sunday 23rd January 1944 at Lembersat, France, between Lille and Armentières. An unknown number of soldiers were killed and wounded. A coal train was thrown off the rails near Blanc Misseron.

A freight train was derailed near Varenne on Monday 24th January 1944 and another between Dourges and Ostricourt. [2107]

Group No. 2 of the Nième detachment of the *Francs-Tireurs et Partisans* guerrillas derailed a German leave train at Porte-les-Valence, France, on Friday 28th January 1944. The locomotive and six carriages were precipitated down an embankment. About 150 soldiers were killed and injured. [2108]

On about Sunday 30th January 1944 partisans derailed two German trains on the Belgrade – Zagreb line near Novska, Yugoslavia and a third train was derailed by a

---

[2105] *L' Ouest Éclair*, Nantes, France, 21st January 1944

[2106] *Forces Françaises Interieuses* No. 39, 10th March 1944

[2107] *La Voix de Paris*, 10th February 1944

[2108] *Forces Françaises Interieuses,* No. 33, 10th February 1944 second group of partisans near Barova Jaruga. [2109]

On Monday 31st January 1944 a locomotive was thrown down an embankment on the connection between the Lens – Arras and Lens – Lille lines. [2110]

Partisans derailed a German fuel train on the Trieste – Pola line in Yugoslavia on about Tuesday 1st February 1944. [2111]

Tuesday 1st February 1944 saw *Francs-Tireurs et Partisans* guerrillas derail a freight train on the Aine-Ussel line.[2112] The same day a coal train was thrown into a ravine on the Mézières – Givet line.

A freight train was derailed at Sévérac-le-Château on Friday 4th February 1944 and many waggons fell into a ravine. [2113]

A freight train was derailed by *Francs-Tireurs et Partisans* (FTP) forces at La Mothe, France, on Saturday 5th February 1944. An engine and 18 waggons were damaged, and the line was blocked for some time.

The following day a German military train bound for the eastern front was intercepted in La Planche Tunnel and a locomotive and 40 waggons were wrecked. In the engagement which followed, a German captain and 14 soldiers were reportedly killed without any losses on the FTP side. [2114]

A freight train was derailed by *Francs-Tireurs et Partisans* guerrillas at Saint Laurent d'Oingt, France on Monday 7th February 1944. The engine and 10 waggons were wrecked, and merchandise destined for the occupying forces was destroyed. The line was blocked for five days. [2115] Another freight train was derailed at Négrondes in the Dordogne. Two days later, Wednesday 9th February, a freight train on the Thiviers – Limoges line was derailed near Thiviers with the destruction of a locomotive and 22 waggons and the loss of many drums of benzine. [2116]

The staff at Gimel station, France, were 'neutralised' by *Francs-Tireurs et Partisans* guerrillas on Friday 11th February 1944; the track was sabotaged, and a train derailed. An engine and seven waggons were damaged. In the fight which followed two German officers were killed and a number injured with the resistance suffering any casualties. [2117]

The same day a train was derailed at Tournon (Ardèche department) with the wrecking of an engine and 17 waggons and the interruption of traffic for 72 hours. [2118]

On Saturday 12th February 1944 a tank train was derailed at Ligueux on the Limoges – Bordeaux line and on Sunday 13th February a coal train on the Montchanin – Chagny

---

2109 *The Daily News*, Huntingdon, Pennsylvania, United States, 31st January 1944
2110 *La Voix de Paris*, Paris, France, 10th February 1944
2111 *Salt Lake Tribune*, Salt Lake City, Utah, United States, 4th February 1944
2112 Ibid, No. 39, 10th March 1944
2113 *France d'Abord*, Paris, France, 28th March 1944
2114 *Forces Françaises Interieuses,* No. 33, 10th March 1944
2115 Ibid, No. 35, 6th March 1944
2116 Ibid, No. 41, 20th March 1944
2117 Ibid, No. 39, 10th March 1944
2118 Ibid, No. 41, 20th March 1944

line. A freight train was derailed at Forteron on the La Guerche – Beffes line on Wednesday 16th February. [2119]

Several attacks on trains in France were undertaken by *Francs-Tireurs et Partisans* guerrillas on the night of Wednesday 16th to Thursday 17th February 1944. Having 'neutralised' the guard, detachments Y and Z damaged the track in a tunnel at Champvans, on the Dole – Dijon line. A freight train was derailed. Six waggons were destroyed, and the two tracks were blocked for 72 hours. Detachment X damaged the track of the Dole – Chalons-sur-Saône line between Beauvoisin and Neutland (name unclear). A freight train bound for Germany was derailed, 20 waggons were damaged, and the line was blocked for some hours. Group No. 1 of detachments X and Y damaged the track between Neuville-sur-Saône and Saint-Romain-au-Mont-d'Or. The engine and six waggons of a freight train were wrecked, and Group No. 2 derailed a train in the Lyon area. A locomotive and three waggons were destroyed, and the line was blocked for 16 hours. [2120] Also on 17th February a freight train was derailed at Tain L'Hermitage (Ardèche department). [2121]

On Friday 18th February 1944 (date illegible) the track was damaged in a tunnel at La Chapelle, France, (Savoie department) by *Francs-Tireurs et Partisans* guerrillas, derailing a train. The locomotive and 14 waggons of a train were derailed, and the line was blocked for two days. [2122] In the Ardèche department at Saint Peray on 18th February another train was derailed.

Two days later, on the morning of Sunday 20th February, they made a much more serious attack. The target was a German leave train, which was derailed at Châteauneuf-du-Pape. The locomotive and many carriages were derailed, and the train was then attacked with grenades and machine guns. The Germans reportedly suffered 29 dead and 80 wounded and the resistance no losses. [2123] The same day the *Nème* detachment derailed a coal train in a tunnel at Alès (Gard department) blocking the line for 32 hours. [2124]

Four tank waggons containing 100,000 litres of oil were derailed and wrecked at Sprincoutre (name illegible) on the Conflans – Longuyon line on Saturday 19th February 1944. [2125]

On about 21st February 1944 partisans in Yugoslavia derailed a train near Rakek, blocking the Trieste – Ljubljana main line for two days. [2126]

Partisans in Greece, accompanied by a British officer, carried out a major act of sabotage on Tuesday 22nd February 1944 on the Athens – Salonika – Berlin main line. A German military express was derailed during a violent snowstorm as it was crossing a

---

[2119] *France d'Abord*, Paris, France, 28th March 1944
[2120] *Forces Françaises Interieuses*, No. 35, 6th March 1944
[2121] Ibid, No. 41, 20th March 1944
[2122] Ibid, No. 35, 6th March 1944
[2123] Ibid, No. 39, 10th March 1944
[2124] Ibid, No. 41, 20th March 1944
[2125] *France d'Abord*, Paris, France, 28th March 1944
[2126] *The Evening Star*, Washington, D.C., United States, 24th February 1944

bridge over the River Pinion near Mount Olympus. The engine, the leading armoured car and the ten following carriages fell into the river. The occupants of the armoured car survived the initial attack and engaged the guerrillas in combat. The Germans suffered heavy losses – 400 were reportedly killed. Those travelling on the train included a general and his staff, returning from an important conference. The attack had been carried out on orders from Cairo. [2127]

On Tuesday 22nd February 1944 five coaches of a leave train were wrecked at Montleranet on the Paris – Granville line. Three hundred metres of track were destroyed, and an unknown number of Germans were killed and wounded. [2128]

Three days of attacks by *Francs-Tireurs et Partisans* guerrillas began on Wednesday 23rd February with the derailment of a leave train near Mouchard, France, on the line to Arc Sensus. The engine and two carriages were 'completely pulverised' and the line was blocked for four days. German casualties were heavy but could not be estimated.

Two kilometres from Égletons a freight train was derailed, and the locomotive and 12 waggons damaged on Thursday 24th February. The same day at nearby Brive-Capdenac another freight train was derailed, [2129] and the locomotive and 18 waggons damaged on the Thiviers – Saint Jean line. A freight train was derailed at Nontron [2130] and a light engine was derailed between Varenne and Senecey-le-Grand.

On Friday 25th February there was another major attack at La Voulte (Ardèche department). A German military train was derailed, 22 Germans were killed and many wounded, and the engine and 22 waggons damaged. [2131]

Also on 25th February a freight train was derailed on the Longuyon – Sedan line, a materials train was wrecked between Laventy and Sailly, and a train of material for fortifications was derailed between Saint Brieue and Cesson.

A light engine was derailed between Lepinay and Monchaux on the Abbeville – Le Tréport line and a freight train Morlaix and Pluigneau on Saturday 26th February. [2132] On the night of Saturday 26th to Sunday 27th February 1944 the French Resistance derailed a train at Mervans, France. The locomotive and 23 waggons overturned. The work of clearing the track was made difficult by the workmen suffering continual harassment from the resistance. [2133]

A report from Chiasso on 28th February 1944 stated that partisans had derailed a German military train near Cadorago, Italy, on the Milan – Como line. [2134]

2127 *Journal de Geneve*, Geneva, Switzerland, 2nd March 1944
2128 *France d'Abord*, Paris, France, 28th March 1944
2129 *Forces Françaises Interieuses*, No. 39, 10th March 1944
2130 Ibid, No. 41, 20th March 1944
2131 Ibid, No. 39, 10th March 1944
2132 *France d'Abord*, Paris, France, 28th March 1944
2133 *L' Ouest Éclair*, Nantes, France 4th March 1944
2134 *Gazette de Lausanne*, Lausanne, Switzerland, 29th February 1944

A freight train was derailed near Précy, France, on the Creil – Beaumont line on Monday 28th February 1944. [2135]

A communique from Moscow on 29th February 1944 stated that guerrillas had derailed a train in the Pinsk region of the USSR.[2136]

A German leave train was derailed by *Francs-Tireurs et Partisans* guerrillas between Dole and Rochefort (Jura department), France, on Wednesday 1st March 1944, after they had 'neutralised' four *gardes-voies*. The Germans suffered heavy but undetermined casualties and the line was blocked for 48 hours. The same day a freight train was derailed near Varennes-le-Grand between Chalon and Mâcon.

In other actions on 1st March a train was derailed in a tunnel near La Souterraine (line blocked 48 hours), a freight train was derailed at Selles-sur-Cher on the Vierzon – Tours line, [2137] and a train bound for Germany was derailed and fell into a ravine at Louhans on the Dijon – Bourg line. [2138]

The next day a train was derailed between Mouchard and Surans, the locomotive and four vehicles being wrecked. The Germans left a large number of dead and injured 'on the ground.' [2139]

In a misguided act of sabotage the French Resistance (who were described as armed and masked bandits in a German-controlled newspaper) damaged the track and derailed a train between Amiens and Abbeville, France, at 03:00 on Thursday 2nd March 1944, derailing the locomotive and 14 carriages of the train. The train which they had targeted was not carrying German soldiers or munitions but evacuees from the Pas-de-Calais bound for Nièvre. Fortunately, the speed of the train had been considerably reduced, and no one was hurt, though there was considerable material damage. [2140]

The same day a mineral train was derailed between Aubenton and Matigny in the Ardennes. [2141]

On about Sunday 5th March 1944 a German troop train carrying 800 men was wrecked by Polish partisans near Lipa, Poland, on the Lublin – Rozwadow line, with reportedly 350 casualties. Handbills were left amid the wreckage, inscribed, "Revenge for German cruelties committed in Poland."

Following previous sabotage attempts on the line the train was preceded by a pilot engine, while the troop train itself was equipped with guards armed with machine guns and anti-tank weapons. The partisans let the pilot engine pass before exploding the mine beneath the troop train and opened fire with machine guns on the wreckage. [2142]

On 5th March 1944 *Francs-Tireurs et Partisans* guerrillas derailed a train near

---

[2135] *France d'Abord*, Paris, France, 28th March 1944
[2136] *Journal de Geneve*, Geneva, Switzerland, 29th February 1944
[2137] *Forces Françaises Interieuses,* No. 41, 20th March 1944
[2138] Ibid, No. 42, 31st March 1944
[2139] Ibid. No. 39, 10th March 1944
[2140] *L' Ouest Éclair*, Paris, France, 9th March 1944
[2141] *France d'Abord*, Paris, France, 1st June 1944
[2142] *The Times*, London, England, 4th April 1944

Grenoble on the Grenoble – Chambéry line, wrecking two engines and 18 waggons and blocking the line for 36 hours. [2143] The same day a train of material for fortifications was derailed in Quimereck Tunnel between Quimereck and Landerneau, and a materials train was derailed at La Galocherie. [2144]

Francs-Tireurs et Partisans guerrillas derailed a train near Correze, France, on Tuesday 7th March 1944, wrecking the engine and seven waggons,[2145] and another train was derailed near Ugine on the Ugine – Albertville line. [2146] In the Correze area the next day a train coming Aix was derailed near Ussel on the Ussel – Clermont line. [2147]

Near Plouaret, between Vieux and Marchém on Wednesday 8th March 1944 a freight train was derailed, and a munitions train was derailed and fell into a ravine near Bannalec.

A report from Chiasso, Italy, on the 10th of March 1944 announced that partisans in Tuscany had derailed a train carrying petrol at Vernio on the Bologna – Florence line. [2148]

On Friday 10th March 1944 *Francs-Tireurs et Partisans* guerrillas derailed a train at Saint-Amour (Jura department), France. [2149]

On Saturday 11th March a freight train was derailed on the Poitiers – Bordeaux line at Ligugé. [2150]

Three trains were derailed by partisans on about Tuesday 12th March 1944. Thirty German soldiers were reportedly killed and 17 wounded in the derailment of a troop train between Bjelovar and Garesnica in Slovenia. The second derailment was between Bjelovar and Krizeviei and the third between Kranj and Skofja Loka in Slovenia. [2151]

A freight train on the Lyon-Paray-le-Monial line was derailed by *Francs-Tireurs et Partisans* guerrillas at Chamelet, France, on Tuesday 14th March 1944 and a second train ran into the wreckage. The line was blocked for four days. [2152] The same day, a materials train was derailed at Saucelle-Monelin on the Paris – Charleville line. [2153]

Vichy Radio announced late on Saturday 18th March 1944 that the Paris – Bordeaux express had been derailed between Anger and Ingrandes-sur-Vienne, France, early that morning by malicious action, causing at least 50 deaths. Other sources state that a German leave train was targeted. [2154] [2155]

[2143] *Forces Françaises Interieuses,* No. 42, 31st March 1944
[2144] *France d'Abord*, Paris, France, 1st June 1944
[2145] *Forces Françaises Interieuses,* No. 39, 10th March 1944
[2146] Ibid. No. 42, 31st March 1944
[2147] Ibid, No. 41, 20th March 1944
[2148] *Gazette de Lausanne*, Lausanne, Switzerland, 11th March 1944
[2149] *Forces Françaises Interieuses,* No. 42, 31st March 1944
[2150] *France d'Abord*, Paris, France, 28th March 1944
[2151] *The Salt Lake Tribune*, Salt Lake City, Utah, United States, 15th March 1944
[2152] *Forces Françaises Interieuses,* No. 42, 31st March 1944
[2153] *France d'Abord*, Paris, France, 1st June 1944
[2154] *Syracuse Herald American*, Syracuse, New York, United States, 19th March 1944

The *Nème* detachment of the *Francs-Tireurs et Partisans* guerrillas derailed a freight train near Lyon, France, on the Lyon – Ambérieu line on Friday 24th March 1944, wrecking a locomotive and 20 waggons. [2156]

Four kilometres from Sévérac-le-Château, *Francs-Tireurs et Partisans* guerrillas derailed a train on Sunday 26th March 1944. [2157] The train, carrying artillery pieces, ran at full speed into an obstruction, overturned and destroyed a bridge over National Road No. 9.

On Monday 27th March an engine was wrecked on the Amiens – Poix line.

On about Tuesday 28th March 1944 an express which had been diverted via Bourg was derailed two kilometres from Pont d'Ain, France. About 20 passengers were injured, most of them seriously; they were taken to Bourg Hospital. The derailment was attributed to sabotage. [2158]

The same day a freight train was derailed at Baisieux on the Lille – Bruxelles line. [2159]

The *Nème* detachment of the *Francs-Tireurs et Partisans* guerrillas derailed a freight train at Chaufailles, France, on Wednesday 29th March 1944, wrecking two engines and 34 waggons.

The following day the same company derailed a train at Creches-sur-Saône, on the Paris – Lyon line, wrecking a locomotive and 18 waggons. The line was blocked for 72 hours, and three cranes were needed to clear up the mess.[2160]

Also early on Thursday 30th March 1944 a freight train bound for Switzerland was derailed due to an act of sabotage between Torcieu and Saint Rambert stations, France. Five of the 40 waggons of the train went forward and arrived in Geneva, Switzerland at 11:30. [2161]

As seen in some of the earlier accounts in this chapter regarding acts of sabotage in France, Poland and elsewhere, one of the problems which resistance forces faced when planning acts of sabotage against an occupying power was that a ruthless enemy could carry out reprisals against innocent civilians. The most obvious examples during the Second World War are probably the massacres at Oradour-sur-Glane in France and Lidice in Czechoslovakia. Another mass killing followed the derailment of a train carrying the 12th SS Panzer Division Hitler Youth near Ascq in Northern France on Saturday 1st April 1944. The troops were returning to Normandy from the Russian front. The direct consequences of the derailment were minor; two soldiers were slightly wounded. However, the commander of the train, SS Obersturmführer Walter Hauck, was furious at the attack and ordered reprisals. The Ascq stationmaster was the first person to be shot dead, followed by the remainder of the local railwaymen, and it was then

---

[2155] https://fusilles-40-44.maitron.fr/spip.php?article150633
[2156] *Forces Françaises Interieuses,* No. 40 bis, 20th April 1944
[2157] Ibid. No. 42, 31st March 1944
[2158] *Gazette de Lausanne*, Lausanne, Switzerland, 29th March 1944
[2159] *France d'Abord*, Paris, France, 1st June 1944
[2160] *Forces Françaises Interieuses,* No. 40 bis, 20th April 1944
[2161] *Gazette de Lausanne*, Lausanne, Switzerland, 31st March 1944

intended to kill all the male inhabitants of Ascq (population 600) over the age of 15. The mayor of Ascq, himself wounded, telephoned the local German police, who intervened and managed to halt the killings. In all, 85 men were shot. The massacre led to a strike of 60,000 workers in nearby Lille. After an investigation by the Gestapo six men were charged with having derailed the train and executed. [2162] [2163]

On about Sunday 2nd April 1944 (date unclear) *Francs-Tireurs et Partisans* guerrillas derailed a train at Argenton, France, wrecking a locomotive and 28 waggons. [2164]

The *Nème* detachment of the *Francs-Tireurs et Partisans* guerrillas derailed a train at Miribel, France, on Monday 3rd April 1944 on the Lyon – Ambérieu line. On the same day another group derailed a train at Fayzin (Rhône department) wrecking an engine and 10 waggons, [2165] and a mineral train was derailed at Braux in the Ardennes. [2166]

On the night of Tuesday 4th to Wednesday 5th April the *Nème* detachment of the *Francs-Tireurs et Partisans* guerrillas derailed a German troop train at Saint-André-le-Gaz, France, on the Lyon – Grenoble line. A locomotive and 16 carriages were damaged and 'significant losses' caused. The line was blocked for 47 hours.

The same night a freight train was derailed at Voreppe (Isère department) on the Grenoble – Moirans line. A train of caterpillar vehicles and German troops was passing at that moment and was also involved in the attack. Two engines and about 30 waggons were wrecked, an unknown number of troops were killed and injured, and the line was blocked for 53 hours.

The following day the *Nème* detachment, after 'neutralising' seven *gardes-voies* derailed another train at Creches-sur-Saône. [2167]

Later on 5th April a freight train was derailed in Bouziès Tunnel on the Capdenac – Cahors line. A petrol tank waggon caught fire and an ammonia tank waggon exploded. [2168]

On Thursday 6th April a materials train was derailed between Voincourt and Eu and on Friday 7th April a freight train between Pleumeur and Lézardieux. [2169]

On Wednesday 12th April 1944 *Francs-Tireurs et Partisans* guerrillas derailed a train at Chaufailles, France, wrecking a locomotive and 22 waggons and blocking the line for 72 hours. On the night of 12th to 13th another group derailed a train at La Tour du Pin on the Grenoble – Lyon line. Then on Thursday 13th on the Grenoble – Valence line a train was derailed at Voreppe (Isère department). Also on 13th a train was derailed near Capdenac on the Cahors – Capdenac line. Thirty waggons of fuel were destroyed. [2170]

[2162] Wikipedia – Ascq Massacre
[2163] La Nouvelle République, 10th April 1944
[2164] *Forces Françaises Interieuses,* No. 41 bis, 30th April 1944
[2165] Ibid. No. 40 bis, 20th April 1944
[2166] *France d'Abord,* Paris, France, 1st June 1944
[2167] *Forces Françaises Interieuses,* No. 40 bis, 20th April 1944
[2168] Ibid. No. 41 bis, 30th April 1944
[2169] *France d'Abord,* Paris, France, 1st June 1944
[2170] *Forces Françaises Interieuses,* No. 41 bis, 30th April 1944

A report from Cairo on 14th April 1944 stated that partisans had recently derailed a German troop train near Messorrakhi station, causing heavy losses. [2171]

A double-headed materials train was derailed between Pommeret and Lamballe, France, by *Francs-Tireurs et Partisans* guerrillas on Tuesday 18th April 1944 and parcels (?) train between Saint-Just-en-Chausée and Breteuil on Wednesday 19th. [2172] Also on the latter day a freight train on the Chambéry – Modane line was derailed near Chambéry. The locomotive and 16 waggons were wrecked, and the line was blocked for three days. [2173]

On the night of Sunday 23rd to Monday 24th April *Francs-Tireurs et Partisans* guerrillas derailed a train at Goncelin, France, (Isère department). An engine and 13 waggons were destroyed, and the line was blocked for 48 hours. [2174]

A report from London, England, on 4th May 1944 stated that in Yugoslavia Slovenian partisans had derailed two trains between Ljubljana and Trieste. The trains had been carrying 35,000 gallons of petrol. Meanwhile, in Croatia, partisans had derailed three trains in three days of fighting and the Germans had suffered heavy losses in terms of dead and wounded. [2175]

Having cleared the line at Annecy, France, a breakdown train was returning to its depot at Bellegarde when it was stopped by resistance forces on the afternoon of Sunday 14th May 1944 at Evires. Its crew was ordered to disembark, and the train was then run into the tunnel there, where the track had been sabotaged, and was derailed. [2176]

On about Monday 22nd May 1944 in the Lyon area, a train carrying mostly flour was derailed by the French Resistance and most of the 'precious commodity' was lost. [2177]

A German troop train was reportedly derailed at Menetrol, France, by sabotage at an unknown date, probably at the end of May or beginning of June 1944. [2178]

Following the Allied invasion the activities of the French Resistance increased even more and became bolder. Trains were derailed on D-Day (6th June 1944) between Vesoul and Gray, between Paris and Bâle, between Paris and Vitry-le-François and in the Nancy area, while other lines were ripped up or obstructed. [2179]

On Thursday 8th June 1944 a freight train was derailed between Auvers-sur-Oise and Pontoise, France. The next day, Friday 9th June, a train was derailed in Maison-Rouge

[2171] *Gazette de Lausanne*, Lausanne, Switzerland, 15th April 1944
[2172] *France d'Abord*, Paris, France, 1st June 1944
[2173] *Forces Françaises Interieuses,* No. 41 bis, 30th April 1944
[2174] Ibid.
[2175] *Gazette de Lausanne*, Lausanne, Switzerland, 5th May 1944
[2176] *Journal de Geneve*, Geneva, Switzerland, 15th May 1944
[2177] *Gazette de Lausanne*, Lausanne, Switzerland, 23rd May 1944
[2178] *The Amarillo Daily News*, Amarillo, Texas, United States, 5th June 1944
[2179] *France d'Abord*, Paris, France, 30th July 1944

Tunnel, near Coulommiers and between Baroncourt and Audin-le-Roman. There were at least four derailments on or about Saturday 10th June. A German train carrying troops and material was derailed at Fours with numerous dead and wounded and much material damage. A freight train was derailed between Marsla and Tour and a German train between Flamboin and Montereau, blocking the line for 48 hours. Between Lussiaux and Vitray a German materials train was derailed with numerous deaths. The line was blocked for four days.

On Sunday 11th June 1944 a train was derailed at Coudray on the Orleans – Chartres line, the track having been ripped up. Two following trains carrying tanks and lorries were held up for eight hours.

A German troop train was derailed at an unstated location between Paris and Strasbourg on Wednesday 14th June 1944. Twelve vehicles were thrown off the track, 20 soldiers were killed, 15 were wounded and the line was blocked for 10 hours.

A train carrying tanks was derailed on the Bordeaux – La Rochelle line at Mosnac-sur-Seugne on Thursday 15th June. The line was blocked for two days.

At Boulay on Friday 16th June a train of troops and war material was derailed, the engine, two coaches and five waggons being wrecked. No casualties were mentioned.

A train carrying torpedoes was derailed between Champagne-sur-Oise and L'Isle Adam on Saturday 17th June 1944, causing great congestion on the line. [2180]

A freight train was derailed at Le Grand-Lemps on Sunday 18th June 1944 and three freight trains between Rives and Voiron on Tuesday 20th. [2181]

Two trains of munitions were derailed between Vierzon and Orleans on Thursday 22nd June with great material loss. [2182]

On Saturday 24th June 1944 a materials train was derailed between La Ferte-Gaucher and Meilleray, France, the rails having been loosened. A number of tanks, including five Tigers, were thrown into the ditch, 400 metres of track were destroyed, nine German soldiers were killed, many were injured, and the line was blocked for six days. [2183]

The same day the French resistance derailed a train loaded with flying bombs at Boncourt, France, on the Merouville – Merz line. [2184]

Also on Saturday 24th June in Haute-Savoie department the Maquis derailed two locomotives in Evires Tunnel.

On Sunday 25th June a German military train was derailed on the Verdun – Sainte Menehould line, which was blocked for three days.

On Monday 26th June the Resistance stopped a train at Croisy-le-Plot, its crew were

---

[2180] Ibid, 15th July 1944

[2181] *Le Dauphiné Libre*, July 1944

[2182] *France d'Abord*, Paris, France, 30th July 1944

[2183] Ibid, 24th July 1944

[2184] *Journal de Geneve*, Geneva, Switzerland, 13th July 1944 ordered to get off, and the empty train was again set in motion, upon which, the track having been interfered with, it was derailed. [2185] The same day a freight train was derailed between Comière and Forges on the Verdun – Sedan line. A train was derailed between Le Grand-Lemps and Rives on Tuesday 27th June.

A locomotive was derailed between Vierzon and Orleans on Friday 30th June 1944.

On Sunday 2nd July 1944 a train was derailed by a mine between Epinal and Illvilers, France, and there presumably being no alternative route, a number of munitions trains were stopped behind it.

On Thursday 6th July an engine was derailed at Caudan and on Saturday 8th July a German military train was derailed in a tunnel at La Basse with much material damage. [2186]

A Todt organisation train was derailed between La Chapelle and Vendemoise on Monday 10th July 1944.

The derailment of a train between Pont-sur-Yonne and Champigny on Friday 14th July 1944 caused a considerable backup of other trains which were then attacked by Allied aircraft. [2187] The same day a munitions train was derailed at Caudos, between Bordeaux and Irun. The line was blocked for 24 hours, and a German general was killed. A freight train was derailed at Lantenay, on the Paris – Dijon line and a train of lorries was wrecked at Lamotte-Brebière on the Amiens – Lille line.

The next day a train was derailed between Hirson and Vervin (Aisne department). [2188]

On about 16th July 1944 Italian partisans ripped up about 20 metres on track in a tunnel between Lesa and Arona on the Domodossola – Arona line. A train was derailed, and the line was obstructed.[2189]

In France on 16th July 1944 a munitions train was derailed between Villers and Monghams on the line between Saint Michel and Verdun. The line was blocked for three days. Another munitions train was derailed at Monthecourt on the Lerouville – Verdun line. On the Amiens – Tergnier line a train of flying bombs was derailed, the engine and 10 waggons being damaged. Many Germans were killed, and the line was blocked for 36 hours. At Lantenay a freight train was derailed; the engine was put out of action and 14 waggons of goods for the German army were seized by the Resistance.

A German troop train was derailed between Hirson and Vervin on Monday 17th July 1944. Ten soldiers were killed, and the line was blocked for 24 hours. The same day a train of lorries was derailed between Verdun and Sainte Menehould, with great destruction of material and a 48-hour blockage of the line, and on the same line between

---

[2185] Ibid, 27th June 1944
[2186] *France d'Abord*, France, Paris, 30th July 1944
[2187] Ibid, 5th August 1944
[2188] Ibid, 12th August 1944
[2189] *Gazette de Lausanne*, Lausanne, Switzerland, 19th July 1944

Balencourt and Dombastes a train of light artillery and DCA was wrecked.

A train was again derailed between Verdun and Sainte Menehould on Thursday 20th July 1944. The next day a train was derailed between Hirson and Busigny and a light engine between Béthesy-Saint Pierre and Saintines on the Crépy – Verberie line.

Two troop trains were derailed at Moncornet on Saturday 22nd July 1944. Another troop train was derailed near Albert on the Amiens – Lille line. Seven coaches left the track, 25 soldiers were killed, and a number injured. On the Troyes – Saint Florentin line at Saint Léger four waggons loaded with tanks were derailed the same day. One German was wounded, and the line was blocked for 20 hours. [2190]

Evires Tunnel, France, in Haute-Savoie department, seems to have been a favourite target for the resistance forces, and again on Sunday 23rd July 1944 an engine was derailed in it, blocking the line. [2191]

In the Oise department a coal train was derailed on Monday 24th July 1944, blocking the line for 48 hours. A munitions train was derailed between Verdun and Sainte Menehould and near Corbie, on the Amiens – Lille line, a German troop train was derailed; 70 Germans were killed, and the line was blocked for three days. [2192]

A German munitions train was derailed on Friday 28th July 1944 at Uzsok, Hungary, cutting an important line of communication for several days, while a number of German soldiers were killed. The following day two trains were derailed on the Vienna – Budapest main line near Herceghalom, Hungary, and another important transport route was blocked for 36 hours. [2193]

The German forces were in retreat from Yugoslavia at the beginning of September 1944 when one of their troop trains was derailed between Skopje and Zeles. About 1,000 soldiers were reportedly killed or wounded. [2194]

Following a general order to proceed with railway sabotage in the Netherlands issued at the beginning of September 1944 by the Dutch government in exile in London, there were attacks on the railway at various places. On Sunday 5th September 1944 a train was derailed on the Delden – Goor line. A bridge was blown up and the engine and two waggons of a train ended up in a ditch. [2195]

On the night of Sunday 5th to Monday 6th the track was interfered with near Ede, using tools stolen from a platelayer's hut. At 04:30 on Monday a German military train of 30 vehicles, hauled by a French locomotive and carrying men and horses from France, was derailed and its vehicles telescoped one another. Fifteen men died and 38 were injured, though a waggon of brandy escaped damage. The line was badly blocked, but a bypass

[2190] *France d'Abord*, Paris, France, 12th August 1944
[2191] *Gazette de Lausanne*, Lausanne, Switzerland, 9th August 1944
[2192] *France d'Abord*, Paris, France, 12th August 1944
[2193] *Gazette de Lausanne*, Lausanne, Switzerland, 16th August 1944
[2194] *The Twin Falls Telegram*, Twin Falls, Idaho, United States, 8th September 1944
[2195] https://johannesterhorst.nl/spoorwegsabotage

line was brought into use on Wednesday 8th with a 5 kilometre an hour speed restriction and the line was finally reopened after another nine days, trains meanwhile being diverted via Amersfoort and Oldenzaal. The wounded were accommodated in local homes before being taken by train to Arnhem. This train was twice attacked by Allied aircraft, wounding 12 more. The Germans did not carry out any reprisals against the local population, possibly because they believed the train had suffered from an air attack. [2196]

Then on Tuesday 7th September two engines were derailed on the Enschede – Hengelo line. Driver Anton Vorlog was injured, possibly fatally. [2197]

The Dutch resistance derailed an engine at the Karnemelksloot near Naarden, the Netherlands, on Tuesday 26th September 1944. In retaliation the Germans destroyed the nearby villa of the Hamers family. [2198]

Few trains ran on the Dutch railways in the latter part of the war, railwaymen having been instructed to commence a general strike on 17th September 1944 which lasted until May 1945.

A mine exploded under the engine of a Danish goods train on about Saturday 30th September 1944. Half a dozen waggons were derailed and some overturned. A German train should have preceded the Danish one. On about Thursday 5th October 1944 the Danish Resistance set off a mine beneath a northbound German military train between Hjordkær and Rødekro, Denmark. The German locomotive and three vehicles were derailed, and 12 vehicles were tipped over. Five persons died and 40 were injured. Relief trains were sent from several locations. [2199]

In October 1944, in an effort to counter growing Danish resistance to the German occupation, such as the Hjordkær sabotage, the Petergruppe, led by Henning Brøndrum, a Danish Nazi, carried out a terrorist attack by planting a time bomb on a train. The bomb exploded at 19:45 on Sunday 8th October 1944 at Tobberup, Denmark, just north of Hobro, on passenger train No. 25, killing ten passengers, seriously wounding seven and slightly wounding many more. Henning and six other members of the Petergruppe were executed in May 1947. [2200]

Greek partisans derailed a German train of 60 waggons near Florina on Thursday 2nd November 1944. [2201]

A mine exploded beneath a German train at 19:00 on Sunday 5th November 1944 between Kilbæk and Troldhede, Denmark, breaking one axle of the engine and damaging the track. The line was reopened at 23:55.

Late on Monday 6th or early on Tuesday 7th November 1944 the track was damaged

---

[2196] Wikipedia – Aanslag op trein bij Ede
[2197] https://johannesterhorst.nl/spoorwegsabotage
[2198] https://gooiland.50plusser.nl/weblog/print.php?warticle_id=51787
[2199] https://www.kobenhavnshistorie.dk/index.php/13-bog/dbbtb/180-dbbtb?start=5
[2200] https://nordjyske.dk/nyheder/togterror-og-raedsel-i-tobberup/d7c6c636-0604-47a4-993c-bded4bcd3317
[2201] *Gazette de Lausanne*, Lausanne, Switzerland, 4th November 1944

between Sæby and Frederikshavn, Denmark and a goods train with German waggons was derailed. The engine and first three waggons overturned and there was considerable material damage, but no one was hurt. The line was blocked for a considerable time. [2202]

Train No. 6 of the East Indian Railway, the Down Punjab Mail, was derailed at mile 375 between Behea and Karisath, India, at 00:40 on Wednesday 22nd November 1944. The engine and six bogies left the line and crashed into the side of a cutting and a postal van was derailed. It was initially reported that six people had been killed and 60 injured. A complete Red Cross military train left Dinapur at 6 a.m. for the scene of the accident and acted as a base hospital, and further help followed.

An Indian newspaper reported that, 'There were many miraculous escapes and praiseworthy actions on the part of passengers as could be gathered here. Some of the coaches which overturned were lying in shallow water and there was a danger that those lying inside would be drowned. Mrs Hunt, wife of Major E. Hunt, now in Iraq, saw the legs of an American officer sticking out from the debris. She pulled aside the luggage rack and released the injured man who was lying in water.

'A Lt.-Colonel who left Dinapur with the ambulance train stated that a large proportion of the praise for the early work immediately after the accident was due to an Indian Major in the I.A.M.C. {Indian Army Medical Corps} and to Lt-Col. Turner, of the East Yorks. Regiment, who was on his way from the Staff College in Quetta. These officers, although shaken, quickly organised what help they could for the injured and those trapped. They also located the dead and got the bodies free from the wreckage.' [2203]

A statement by the Bihar Government indicated that the derailment had resulted from an act of sabotage, a rail having been removed on the approach to a bridge. The casualties were later reported as eight dead, 16 seriously and 86 slightly wounded. [2204]

The tank waggon of a goods train exploded due to sabotage between Hol and Geilo, Norway, on Friday 24th November 1944. [2205]

On Sunday 10th or early on Monday 11th December 1944 a mine exploded under a train entering Gredstedbro station, Denmark. Two carriages were completely derailed and the bottom of one was demolished, the other was less badly damaged. No one was hurt.

At 18:00 on Monday 11th December 1944 there was an explosion beneath a goods train between Hovslund and Over Jersdal, Denmark, whereby four empty German waggons were derailed with all wheels and the line was blocked until further notice.

A bomb exploded beneath a goods train on Monday night, 11th December 1944, on the main line between Ryeskov and Borup, Denmark, derailing 11 waggons and a coach. The regular Danish train was involved, it having preceded rather than followed a German train. No one was hurt. One line was cleared between 07:00 and 08:00, it was anticipated that the other line would be cleared during the course of the morning.

Probably late on Friday 15th December 1944 there was an explosion beneath a train between Hovslund and Rødekro, Denmark, whereby two brake vans, two second-class

[2202] https://www.kobenhavnshistorie.dk/index.php/13-bog/dbbtb/180-dbbtb?start=5
[2203] *The Bombay Chronicle*, Bombay, India, 24th November 1944
[2204] *The Railway Gazette*, London, England, 1st January 1945
[2205] *På Sporet* (magazine of Norsk Jernbaneklubb) No. 140, pg. 25 (in Norwegian)

coaches and two closed goods waggons overturned, and a second-class coach and a fish waggon telescoped one another. The engine crew and a German soldier were slightly hurt. Relief trains were sent from Padborg and Fredericia, including a breakdown crane from the latter place. Passengers and luggage had to be transhipped past the site of the derailment, at least until the afternoon. [2206]

Goods train No. 5511 was again sabotaged between Hol and Geilo, Norway, at almost the same spot as the November action, on Saturday 16th December 1944. This time there were several explosions, five tank waggons were wrecked and one leaked its oil, without it catching fire. Afterwards the Germans found a further explosive charge on one of the waggons which had not gone off. [2207]

Saturday 6th January 1945 there was an explosion beneath a special train carrying workers from Oksbøl to Esbjerg, Denmark. Two coaches were derailed but no one was hurt.

On about Wednesday 10th January 1945 a band of Filipino guerrillas, led by an American, Capt. Robert Lapham, in anticipation of the American invasion of the Philippines, derailed a Japanese train between Munoz and Guimba in Nueva Ecija province. "They took up the rails on a curve and sent a troop train of three cars carrying supplies and two with troops off into the rice swamps." [2208]

In June 1944 the Norwegian government-in-exile in London had given the green light for acts of sabotage to be carried out in Norway, despite the possibility of Nazi reprisals, especially with a view to interfering with the movement of German troops from Norway (where 400,000 of them were stationed) to deal with the Allied invasion of France. In late December 1944, in an effort to disrupt the movement of troops from north Norway, an unsuccessful attempt had been made to destroy a railway bridge at Grana. On Saturday 13th January 1945, a repeat attempt was made, this time on the bridge at Jørstad. This succeeded, and it led to what was the worst 'accident' in Norwegian railway history.

The five saboteurs were: Robert Andersson, Ludvig Kruksve, Harald Larsen, John Moan and Oddvar Østgård. They left their base, named Woodlark, just inside Sweden, early on Thursday morning and set off across country on skis towards their target. On Thursday evening they collected a stash of explosives which had been left hidden by the group which had attempted the Grana sabotage and spent the night at Kjenstadsetra. On Friday they continued for the remainder of the distance to the bridge. Friday evening, after darkness, a reconnaissance of the bridge was made, and it was found that it was unguarded – the operation was on. One man was posted as a lookout and the remaining four worked on attaching the dynamite to the bridge. After two failed attempts the explosives were secured in position and three of the group withdrew to an observation position above the bridge, leaving Larsen and Østgård. Larsen held a three-minute fuse, Østgård lit it, and they sprinted away to their friends. A powerful explosion shortly followed, and the 20 kilograms of dynamite damaged the north end of the bridge and caused it to collapse; the Norlandsbanen was cut. German troops would be forced to

---

2206 https://www.kobenhavnshistorie.dk/index.php/13-bog/dbbtb/180-dbbtb?start=5

2207 *På Sporet* (magazine of Norsk Jernbaneklubb) No. 140, pg. 25 (in Norwegian)

2208 *The Telegraph-Herald*, Dubuque, Iowa, United States, 16th January 1945 come southwards by sea, at the risk of attacks from the British navy. The saboteurs then retreated to a ski *hytta* at Rathaugfjellet, where they rested for an hour. Before they continued back to Sweden they left 'visiting cards' at the *hytta*, implying that British commandos had carried out the sabotage, in the hope of thus forestalling German reprisals against their countrymen.

On Sunday, back in Sweden, they heard on the radio that their sabotage action had resulted in the wrecking of a German troop train, with heavy loss of life.

Southbound troop train No. 7762 consisted of engine No. 57.1261 and 31 vehicles, in charge of Conductor Gunnar Jørgensen with Assistant Conductor Hilmar Tyvold, Driver O. Dahlen and Fireman Rolf Pedersen. It was loaded with German troops, their equipment and 200 horses.

Jørstad Stationmaster Johan Skotvoll was off duty at the time of the accident but from his rooms on the first floor he and his wife Helga heard the troop train go by, as normal, but a short time later it was clear that something was wrong and the steam from the engine of the train had suddenly disappeared. There had obviously been an accident.

Conductor Jørgensen wrote in his report on the accident that his train had left Grong at 06:10, 80 minutes late. It had passed Jørstad station at about 08:20 and a short time afterwards there had been a violent shock, and everything in his van, at the rear of the train, had been thrown about. By the time that he had recovered himself sufficiently to apply the brake the train was already at a stand. He had immediately jumped out of his van and with Tyvold had gone forward to meet a terrible sight. The hissing noise from the engine was mixed with moans and shouts from the wounded soldiers and whinnying from the horses. Only 14 of the vehicles remained on the track, the remainder, more or less destroyed, lay in a smoking heap in the river with the remains of the bridge. He had gone to the engine to see what had happened to his driver and fireman. It seemed unlikely that they could have survived, and they had not, though the fireman lived for a short time before succumbing to his wounds. Conductor Jørgensen said that he had not lingered long by the engine for fear that it might explode. He had then made his way to the nearest farm, only 400 or 500 metres away, from where he had telephoned Jørstad station, where Telegraphist Toldnes had said that he already knew of the accident.

Rescue trains were sent to the site from Valøy and Mosjøen. The work of clearing the line went on night and day and German pioneer troops managed to erect a temporary bridge after a week's work. There were also repeated interrogations of local railwaymen, not only regarding the sabotage itself but as to why the troop train had not been stopped.

On Monday 15th the German senior commander arrived at Jørstad at 05:00 from Trondheim. It appeared that the local railwaymen were going to be arrested but in the end nothing happened, it was supposed because the Germans blamed themselves for the sabotage for having withdrawn the soldiers guarding the bridge. From then until the end of the war guards were posted at all stations.

No official figures were forthcoming for the number killed in the accident, but sources agree that about 100 German soldiers lost their lives, together with Driver Dahlen and Fireman Pedersen. In addition about 200 horses were killed or had to be put down and much military materiel was destroyed. [2209]

Four explosions occurred beneath a train between Hampen and Ejstrupholm, Denmark, on Sunday 21st January 1945 and derailed a tender.

[2209] *På Sporet* (magazine of Norsk Jernbaneklubb) No. 70, pgs. 23-26 (in Norwegian)

On Monday 22nd January 1945 a mine derailed two passenger coaches and a luggage van of a train between Silkeborg and Funder, Denmark.

On Thursday 25th January 1945 an explosion damaged both tracks between Marslev and Odense, Denmark as a train was passing. One vehicle was derailed but no one was hurt. [2210]

An explosion derailed a goods train consisting of a locomotive and six waggons about seven miles on the Buenos Aires side of Cordoba, Argentina, on the British-owned Central Argentine Railway on Thursday 25th January 1945. Sixty-five yards of track was destroyed. There was also apparently an obstruction on the line which the engine driver saw and was able to pull up short of before the explosion. No lives were lost. Sabotage was presumed. [2211]

There was an explosion beneath a train between Nyborg and Hjulby, Denmark, on Friday 26th January 1945 which derailed a waggon, but no one was hurt.

An explosion beneath a train on Saturday 27th January 1945 derailed the wheel of a locomotive between Odense and Holmstrup, Denmark. The same day an explosion under a train between Hjulby and Ullerslev derailed a coach and a goods waggon overturned, blocking both lines. No one was hurt.

There was an explosion beneath a train from København to Nykøbling F. at Vordingborg, Denmark, on Thursday 1st February 1945. The engine remained on the line, but the coaches were overturned and blocked the opposite track. A German soldier was wounded, and a woman passenger suffered shock. The line was blocked until the afternoon. The passengers eventually reached Nykøbling F. on another train.

A mine exploded beneath a train between Lunderskov and Vamdrup, Denmark, at 22:05 on Friday 2nd February 1945. The engine and conductor's van were derailed, and three goods waggons telescoped one another. No one was hurt.

On Saturday afternoon, 3rd February 1945, there were three explosions beneath a special train between Ellishøj and Støvring. The engine and leading coach were derailed but no one was hurt. In another attack the same day a transport train was derailed between Vammen and Møldrup. Two vehicles overturned, eight were derailed and 110 metres of track was damaged. Some people were slightly hurt. On Saturday night there was an explosion beneath a transport train between Brøup and Vejen. A vehicle was derailed but no one was hurt.

On the night of Tuesday 5th February 1945 there was an explosion beneath a German leave train between Tim and Ulfsborg, Denmark. A coach was damaged, but no one was hurt.

[2210] https://www.kobenhavnshistorie.dk/index.php/13-bog/dbbtb/180-dbbtb?start=5
[2211] *The Times*, London, England, 26th January 1945

An explosion beneath a double-headed German leave train between Struer and Humlum on Wednesday 6th February 1945 derailed one of the Danish engines and two coaches whilst five open waggons loaded with cars overturned. A breakdown train was sent to the spot by the State Railway.

Earlier, at 01:00, there had been an explosion beneath a German leave train between Farris and Sommersted. Two coaches were derailed and five tipped over; no one was hurt.

An explosion beneath a goods train entering Herlev station, Denmark, at 04:00 on Sunday 17th February 1945, overturned four waggons and blocked the line. No one was hurt.

A mine planted by the Danish Resistance exploded under a German supply train between Horning and Gødstrup, Denmark, at 23:00 on Monday 19th February 1945. The engine was derailed and ended up across the rails, five waggons telescoped one another, and several derailed. Eight people were killed and 32 wounded, including three women and three children. All were taken to Horning Hospital. Conductor Thaarup was slightly hurt. [2212]

At 07:00 on Thursday 22nd February 1945 there was an explosion beneath a German military train between Tølløse and Hvalsø, Denmark. The last 14 coaches were derailed or overturned, 100 metres of track was torn up and three soldiers were injured. In another attack the same day a German transport train was derailed between Ulleslev and Langeskov. The engine was damaged, and a coach derailed.

Late on Friday 23rd or early on Saturday 24th February 1945, a German military train was derailed at Hæstrup, Denmark, six kilometres from Hjørring, by a resistance group led by Johannes Kragh. Two lengths of rail had been blown up before the train arrived. The locomotive ended up sideways across the track with its tender down an embankment, and 11 following vehicles overturned on either side of the track. Four of them immediately caught fire, probably because overturned stoves had set fire to straw lining the bottom of them. Included in the train were several waggons of munitions and the fire spread to them, setting off explosions. At least 16 German and Hungarian soldiers died, resistance sources put the number at 25 to 30. Ambulances took 25 of the wounded to Hjørring and relief trains were sent from Aarlborg and Frederikshavn. It was expected that the line would be blocked for several days. [2213] [2214]

Early on Saturday 24th February 1945 a special train carrying Georg Lindemann, the recently appointed *Wehrmachtsbefehlshaber* (Commander-in-Chief) of the German troops in Denmark, from north-west Jutland to Silkeborg, was sabotaged by the Danish Resistance between Struer and Hjerm, Denmark. Two waggons of the train were wrecked, and a *Wagons-Lit* coach was derailed but there is no record of anyone being

---

2212 http://www.illegalpresse.dk/images/information/Information%20XI_29%20januar%201945%20-%2021%20februar%201945_merged.pdf

2213 http://www.vendsysselunderbesaettelsen.dk/128730238

2214 https://www.kobenhavnshistorie.dk/index.php/13-bog/dbbtb/180-dbbtb?start=5

hurt. Nevertheless, General Lindemann was said to have become 'white with rage' at the attack and ordered reprisals, though these were far less harsh than those carried out by the SS at Ascq in France the previous year. No one was put to death, but the farmhouse of Immanuel and Hulda Espensen, only a few metres from site of the derailment, and nearby gatehouse No. 113, were burnt down on the suspicion that their residents might have been involved in the sabotage. Lindemann was not prosecuted after the war for the arson by the Danes. [2215] [2216] [2217]

The October 1944 terrorist attack at Tobberup, Denmark, was followed by a similar attack in the same location at 22:50 on Saturday 24th February 1945, again carried out by the pro-Nazi Petergruppe, this time led by Henning Bothilsen Nielsen, probably in retaliation for the attacks on German trains in the previous 24 hours. Phosphorus time bombs were placed in a toilet of train No. 25 from Aarlborg to København and again 10 people died in the explosion and subsequent fire. Eleven other passengers were seriously wounded, and many others were slightly hurt. Nielsen was another of the seven members of the Petergruppe executed in 1947. [2218]

About 17:00 on Wednesday 28th February 1945 there was an explosion beneath a train from Roskilde to Kalunborg at Darupvejen, Denmark. The leading coach was derailed. Five people were seriously wounded and admitted to hospital and a number were slightly hurt.

On about Monday 5th March 1945 there was an explosion beneath a train between Farris and Sommersted, Denmark. A mail and luggage van was derailed all wheels, but no one was hurt. A breakdown train reached the scene at 04:00.

A mine exploded under a German refugee train from København to Frederikshavn (carrying refugees from East Prussia fleeing the advancing Russians) between Hjørring and Sønderskov, Denmark, on Thursday 15th March 1945. The train consisted of a German locomotive and 11 vans. The engine and tender were derailed, five vehicles overturned and three telescoped one another. Four passengers were killed and four seriously wounded; the latter were brought to Hjørring. An ambulance train was sent from Aarlborg and a breakdown train from Frederikshavn. The line was expected to be blocked for 24 hours. [2219]

On Thursday 22nd March 1945 there was an explosion beneath a German hospital train bound for Odense between Viborg and Rinsholm, Denmark. A four-axled coach was derailed but no one was hurt. A relief train was sent to the scene. Later the same afternoon there was an explosion under the engine of a Danish goods train south of Braband. Driver K.K. Jensen was injured by glass splinters and suffered shock. The locomotive was badly damaged and must be replaced by another.

[2215] *Gazette de Lausanne*, Lausanne, Switzerland, 26th February 1945

[2216] Wikipedia – Georg Linemann

[2217] www.folkebladetlemvig.dk.artikel – Ung DSB-mand var ombord på generalens tog

[2218] https://nordjyske.dk/nyheder/togterror-og-raedsel-i-tobberup/d7c6c636-0604-47a4-993c-bded4bcd3317

[2219] https://www.kobenhavnshistorie.dk/index.php/13-bog/dbbtb/180-dbbtb?start=5

On Sunday 25th March 1945 there was an explosion beneath a German transport train between Ryomgaard and Skjerm. The engine was overturned, the tender derailed and the nine leading waggons wrecked. Fireman Poulson and nine other people were slightly hurt. Later in the afternoon there was an explosion between another German transport train between Aarhus and Lystrup. Eleven coaches were derailed, eight of which overturned. Nine people, including members of the train crew, were wounded.

On Tuesday 27th March 1945 there was an explosion under a German transport train between Trudstrup and Kolind. The engine overturned and ten vehicles were derailed and some wrecked. No one was hurt. The same day there was an explosion beneath a train at 22:00 at Darupbroen, 400 metres in the rear of the Roskilde *indkørselsignal* (home signal). The track was ripped up and 10 waggons were derailed. No one was hurt.

On Tuesday 3rd April 1945 a train bound for Tirstrup works was derailed, a rail having been loosened on the Randers – Ryomgaard line in Denmark. Some of its coaches were derailed, and 45 workmen were injured, 17 of them badly enough to be taken to hospital. Two workmen, Rasmus Rasmussen, 60, and N.P. Nielson were killed. The workmen had been given veiled warnings the previous day that something was going to happen. An ambulance train and road ambulances were sent from Randers. The injured had to be freed from the wreckage.

A mine exploded under a German leave train on the evening of Sunday 8th April 1945 between Holstebro and Navr, Denmark. The three leading coaches were derailed and the other two tipped over. One person was slightly hurt. The same day there was an explosion beneath a hospital train between Kavslunde and Nørre Aaby. Four coaches were derailed; no one was hurt.

On Wednesday 11th April 1945 there were explosions beneath hospital trains at Holmstrup and between Farris and Sommersted. No one was hurt but the engines were wrecked.

On Saturday 14th April 1945 there was an explosion under an empty hospital train between Dørstrup and Skærbæk. The engine tender was derailed, and three coaches overturned but no one was hurt. The same day a train was derailed between Lunderskov and Elstrup. Ten vehicles were derailed. No one was hurt. On Saturday afternoon one person died and eight were wounded in an explosion under a German transport train between Lysbro and Øster Bording. In another attack between Aarhus and Braband the engine crew were slightly wounded, and a mail van was derailed.

An explosion occurred beneath a German transport train late on Saturday 14th or early on Sunday 15th April 1945 between Farris and Sommersted, Denmark. Four Germans were killed, eight seriously and six slightly injured. The train crew escaped unhurt. [2220]

On Tuesday 17th April 1945 there was an explosion under a German leave train between Hjortkær and Bolderslev, Denmark. Seven German coaches were derailed, three of which overturned. Three German soldiers were seriously and ten slightly wounded. A

[2220] Ibid.

rescue train was sent from Flensborg.

On the evening of Friday 20th April 1945 there was an explosion under a goods train between Herlev and Ballerup and an engine and ten waggons were derailed.

There was an explosion beneath a workman's train bound for Vojens on Tuesday 24th April 1945. Six vehicles were derailed but no one was hurt.

A goods train which was also carrying passengers was derailed by an explosion between Odense and Svendborg on Thursday 26th April 1945 and a coach, a luggage van and five waggons were derailed. One passenger suffered shock.

On about Saturday 28th April 1945 a German refugee train en route to Skive was derailed between Ulstrup and Bjerringbro, Denmark, on track which had been damaged by an explosion the night before. Twenty-one vehicles were derailed, four refugees were killed, 36 were badly and 20 slightly wounded. [2221]

Late on Tuesday 1st May or early on Wednesday 2nd May 1945 a train carrying 1,000 German refugees en route to Brænderup was derailed near Langesø. No one was hurt.

On the evening of Thursday 3rd May 1945 there was an explosion beneath a German hospital train between Langaa and Ulstrup. A coach was derailed but no one was hurt. These were the last two acts of railway sabotage in Denmark. [2222]

At the end of the Second World War the Nazis formed several schools of sabotage to conduct operations after their final defeat in various parts of Czechoslovakia, especially the Sudetenland. One of these terrorist groups derailed a train carrying guards near Choumoutov, Czechoslovakia on about Tuesday 15th May 1945. [2223]

At the end of May 1945 there was an influx of French troops into Syria despite French assurances that the Syrians would be granted full independence. This caused a crisis with fighting and numerous deaths while on Monday 28th May 1945 a train was derailed north of Hamas, Syria. No one was apparently killed in the derailment, but a number were injured, including three Syrian deputies. [2224] The Syrian Republic was recognised by the United Nations in October 1945 and became a *de facto* sovereign state with the final withdrawal of French troops in April 1946. [2225]

A Trinidad Government Railways train was derailed at Savonetta, Trinidad, just north of San Fernando, on Monday 9th July 1945. A railway official said that pins had been removed from a switch, causing the engine to overturn, and wrecking two passenger coaches and two freight waggons, wounding 102 people. No motive for this apparent act

[2221] Ibid.
[2222] Ibid.
[2223] The *Montana Standard*, Butte, Montana, United States, 3rd August 1945
[2224] *Norwalk Reflector-Herald*, Norwalk, Ohio, United States, 29th May 1945
[2225] Wikipedia – First Syrian Republic

of sabotage was suggested. [2226]

In contrast to the numerous acts of railway sabotage committed in occupied Europe during the Second World War, very few such actions are recorded in the war against Japan, mostly odd attacks in the Philippines, as noted above. However, right at the very end of the war, on Monday 23rd July 1945, a commando force from the USS *Barb* submarine succeeded in derailing a train with explosives planted on the track near Natoma Town in Karafuto Prefecture (in the then Japanese part of Sakhalin Island, now part of Russia). Unaware of the sabotage, the Japanese attributed the derailing of the train to an air attack. [2227]

Two young Navajo Indians attempted to derail a passenger train on Sunday 28th October 1945. Their plan was foiled when a light engine preceding an Atchison Topeka and Santa Fe Railway passenger train was derailed instead. Clah Benally, 23, and Paul Franklin, 18, removed 52 spikes and angle irons from the main line about six miles west at Williams, Arizona, United States. Railroad officials declared that had the engine not been in front of *The Scout* there might have been heavy loss of life. The two were railroad section hands and were disgruntled by loss of Sunday overtime pay. They were soon arrested and charged with violation of the Federal wreck statute and faced a maximum penalty of 20 years' imprisonment or a $10,000 fine or both. [2228]

[2226] *Del Rio News-Herald*, Del Rio, Texas, United States, 10th July 1945
[2227] *The Stars and Stripes*, Honolulu, Hawaii, 21st August 1945
[2228] *Tucson Daily Citizen*, Tucson, Arizona, United States, 2nd November 1945

CHAPTER 19

MISCELLANEOUS ACCIDENTS

The vast majority of railway accidents fall into one of a small number of categories such as derailments, collisions, bridge failures and so on. A few, though, do not easily fall into any obvious group, hence this chapter on miscellaneous accidents. As this volume covers the Spanish Civil War, the Second Sino-Japanese War and in particular the Second World War a good proportion of it is unavoidably devoted to air attacks on trains.

Trains are of course heavy things, but every so often the power of the wind has proved too much for them and is sufficient to throw them from the rails. On Saturday 21st February 1931 the island of Fiji was struck by a cyclone, which with its accompanying flooding caused perhaps 200 deaths. A train from Ba to Lautoka was also a victim of the storm and was blown over, causing many injuries amongst its passengers. [2229]

Three months after the Fiji accident the eastbound Great Northern Railway *Empire Builder* passenger train was struck by a tornado near Fargo, North Dakota, though the accident happened over the state line in Minnesota. The train was moving at 50 miles an hour across the prairie when it was struck by a tornado and the whole of its 12 coaches were lifted from the track and deposited on the ground. Only the engine and tender were left on the track. Engineer B.E. McKee and Fireman Al Klinfehn said that they had noticed the menacing cloud of the tornado but thought no more of it until the coaches had suddenly been jerked away from their engine at 4.22 p.m. The coupling between the tender and the leading coach had snapped and the engine had trembled but had stayed on the track.

The passengers had little warning before the windows began breaking and the cars started pitching wildly. Many saved themselves from serious injury by throwing themselves to the floor. One passenger, Andrew Anderson, was thrown through a window and killed outright and 83 of the 175 passengers were more or less injured, mostly slightly. Seven passengers were trapped in the club car for half an hour because its doors had jammed. Conductor Edward Brandt was somewhat hurt. The fact that the train

---

[2229] *The Herald*, Melbourne, Victoria, Australia, 27th February 1931 was composed of all-steel coaches probably prevented a worse accident.

Passing motorists took news of the accident to nearby Moorhead and transported the injured to hospitals in Moorhead and Fargo. [2230]

A second passenger, Mrs Havens, of Conrad, Montana, died of her injuries in hospital at Moorhead on 8th June. [2231]

Japan had gained control over Manchuria after the September 1931 Mukden incident but wished to increase its influence in the Chinese port of Shanghai. The Japanese deliberately stirred up trouble at the end of January 1932 and the situation quickly deteriorated into an armed conflict between Japan and China, which included the aerial bombing of Shanghai by Japanese aircraft operating from an aircraft carrier on 28th January. On Friday 5th February 1932 Japanese airmen spotted a troop train at Chapei (now Zhabei), a Shanghai suburb, bringing up reinforcements of the 19th Chinese Army, and bombed it. A newspaper reporter wrote, 'They swooped low to cut loose with their bombs and one of them struck a closed car, splitting it from end to end and scattering the bodies of Chinese over a wide area. The *Associated Press* correspondent counted 16 bodies strewn along the railroad tracks. The bodies of mangled horses also lay among the wreckage.' [2232]

The conflict ended with a ceasefire on 5th May. [2233]

A brief report on an accident to a passenger at Tosna, near Leningrad, USSR, stated that it had killed six people, while 13 others were injured. The railwaymen involved had been tried by a railway court and the provisional stationmaster sentenced to six years' forced labour and others to two years' forced labour. The accident presumably occurred in either late January or early February 1932. Whether it was a collision, or a derailment was not stated. [2234]

India was affected by a severe cyclone in mid-December 1933 and on Saturday 16th a passenger train was reportedly blown over by the wind near Chidambaram, on the South Indian Railway. It was initially reported that 28 lives had been lost. [2235] [2236] A follow-up report stated that the casualties had been greatly exaggerated and that in fact only one passenger had been slightly bruised. [2237]

A violent thunderstorm was raging over much of Belgium on the night of Tuesday 17th April 1934. While it was in progress a passenger train was struck by lightning near

[2230] *The Bismark Tribune*, Bismark, North Dakota, United States, 28th May 1931
[2231] Ibid, 8th June 1931
[2232] *Ogden Standard Examiner*, Ogden, Utah, United States, 5th February 1932
[2233] Wikipedia – January 28th incident.
[2234] *The Times*, London, England, 22nd February 1932
[2235] *The Advertiser*, Adelaide, South Australia, 22nd December 1933
[2236] *The Railway Gazette*, London, England, 26th January 1934
[2237] 16th March 1934

Namur. The glasses of the carriage windows broke, and a passenger was blinded. [2238]

The small Central American country of El Salvador was devastated by a severe hurricane in early June 1934, with very excessive rainfall which caused widespread flooding, great damage, and much loss of life. Among the victims of the hurricane was a train of the International Railways of Central America. A report from the pilot of an aircraft indicated that the track had been washed away over a long distance and that there was a passenger train lying upside down in a valley several hundred feet below where the line had formerly run. It was assumed that all its passengers and crew had been killed. The disaster probably occurred on Thursday 7th June 1934. [2239]

It was the turn of Japan to suffer a railway disaster due to strong winds on Friday 21st September 1934. On the morning of that day the western part of the country was struck by a typhoon, the most severe for 17 years, with winds whose velocity was up to 60 metres per second (120 miles an hour). No less than ten trains were reportedly derailed or overturned by the wind, including one, probably the Tokyo – Shimonoseki express with 250 passengers, which was struck on the Setagawa bridge, in Shiga Prefecture between Kusatsu and Ishiyama, by a wind with a velocity of 41 metres per second. Sixteen people died and 216 were injured in this accident, which was blamed on the train being operated when the wind speed had exceeded the permitted maximum, due to a mistake in judging the weather conditions, a delay in transmitting information, and unclear operating rules. The rules were revised after the disaster and anemometers were installed at high-risk locations. [2240] [2241] [2242]

Triple collisions, that is three separate collisions involving four trains, are very rare, but one occurred on Wednesday 13th March 1935 at Kings Langley, England, on the London Euston – Rugby main line of the London Midland and Scottish Railway. It was not a major disaster, but a driver lost his life, and a fireman and a guard were injured.

The events which led up to the accident began with the breaking down on the Up fast line between Nash Mills signal box and Kings Langley station of the 4.55 p.m. Up express meat train from Alexandra Dock, Liverpool, to London Broad Street. This was due to the breakage of the vacuum brake pipe on the engine, whereby the brakes had come on automatically. It only took Driver J.A. Mawdsley and Fireman W.P. Donoghue six minutes to remedy the defect, and they were just setting off again when their train was run into from behind by the 5.50 p.m. Up milk train from Stafford to London Euston. The wreckage from this collision fouled the Down Fast and Down Slow lines and Driver William Buckley of the milk train immediately sent his fireman, Edward Newns, forward to stop the 10.30 p.m. Down express from London Euston to Edinburgh and Aberdeen, which he knew must be due. Meanwhile Driver Buckley himself went back to try and stop the 12.25 p.m. Up coal train, Toton to Willesden. As he was going back he was

---

2238 *The Times*, London, England, 19th April 1934
2239 Ibid, 12th June 1934
2240 Ibid, 22nd September 1934
2241 *The Railway Gazette*, London, England, 28th September 1934
2242 Webpage: Japanese Railway Safety and the Technology of the Day by Masao Saito

struck and killed by the 10.30 p.m. Down partially fitted freight train from Camden to Holyhead in the second collision. A short time later the Toton coal train ploughed into the wreckage in the third collision. Guard G.W. Bond of the meat train and Fireman R.A.R. Mayes of the coal train were the two railwaymen injured.

Fireman Donoghue told the Ministry of Transport inquiry that following the first two collisions he had said to his driver that he thought their train must have become divided, derailed, and that some of its waggons had fouled the adjoining running line, to be run into by the Down freight. He had then met a guard who had told him that all the lines were blocked, so he had run forward towards Kings Langley, putting down five detonators on the Down Fast line. It was only when he had reached Kings Langley box at 11.40 p.m. that he had realised that his train had been run into. He had understood from Signalman W. Gibbs there that "with us not showing up, he thought he had made a mistake on the bells regarding the 5, and he rubbed us off the instrument and accepted the milk."

Guard Bond, who was unconscious for ten hours after the accident, said, when he had recovered enough to be interviewed, that everything had been going well, though his train had been somewhat late. He remembered passing Nash Mills box, after which his train had first slowed up and then stopped. He remembered seeing the fireman coming back and thought that he had got down from his van but could remember little more.

Fireman Newns of the milk train said that they had been practically stopped at Nash Mills. He had been firing when his driver saw the tail lamps of the meat train, shouted and applied the brakes. His driver had then sent him forward to stop the express and his driver's last words had been, "I will get off and attempt to stop the coal train." Unaware of the freight train on the Down Slow he had concentrated on protecting the Down Fast line and had been putting down detonators on it when the Camden freight had passed him on the Slow line.

Guard O.W. Tew of the milk train confirmed that it had almost been stopped at Nash Mills. He estimated the interval between the first two collisions at four minutes.

Driver P.J. Tyler of the Camden freight stated that he had had clear signals at Kings Langley. As he had passed the front of the meat train he had seen that it appeared to be divided, then his train had collided with the wreckage. The Up coal train had then run into other trains almost at once.

Guard Melleney of the Camden freight thought that there had been only a few seconds between the second and third collisions and photographs indicated that they must have been almost simultaneous.

Driver W. Claridge of the Up coal train indicated that he had had no warning of any obstruction before the third collision.

Fireman Mayes of the coal train confirmed his driver's account and spoke of seeing, "Black objects which seemed to be spreading on to the Down Slow line."

Signalman C.E. Caffall at Boxmoor (the box on the down or country side of Nash Mills) spoke of hearing a telephone conversation at 11.12 between Signalman Gibbs and Control. It was regarding the stopping of the Down express at Kings Langley because Signalman Gibbs was unsure what trains were on the way from Nash Mills. He (Caffall) had interposed that, "a 5 and a 1-1-3 had cleared Nash Mills." Nothing more had happened until he at Boxmoor had received the 'Obstruction Danger' signal from Nash

Mills at 11.21 p.m. Signalman Caffall added that the 6.12 p.m. Up passenger train from Stafford to London Euston was always known as "the Stafford" – there were no particular names for the meat and milk trains.

Signalman B.A. Branscombe at Nash Mills said that he had only begun working the box on Monday 11th March and had not worked a main-line box before. He said that the 6.12 p.m. passenger train from Stafford, for which the bell code was four beats, had been accepted by Kings Langley at 10.44 p.m., had passed him at 10.53 p.m. and Kings Langley had cleared it at 10.58 p.m. He knew that the train came from Stafford but not its number, what time it left there, or when it was due at Euston.

The meat train had followed on the Up Fast, had been accepted by Kings Langley at 10.58 p.m. and had passed him at 11 p.m. He had replaced his signals behind it, cleared it back to Boxmoor and had been offered and had accepted the milk train. Boxmoor had sent on the milk train at 11.04 p.m.

At 11.05 or 11.06 p.m. Signalman Gibbs had telephoned him and asked, "What have you got coming up fast?"

He replied, "Got the milk approaching my home signal."

Gibbs had asked, "Haven't I cleared for the 5?"

He answered, "No, that is just what I am waiting for."

Continuing with his statement, Signalman Branscombe said that Gibbs had at once cleared, "my block instrument in respect of the meat train and I then offered the milk train to him and he accepted at once and placed the instrument again to 'Line Clear', which permitted me to lower my home signal, which I did, the milk train not having quite come to a stand at it. The time which I have booked for the acceptance of the milk train is 11.6. The milk train passed my box at 11.8 when I transmitted the 'Entering Section' signal to Kings Langley, and I returned my signals to danger behind the train. When the train had proceeded ¼ mile I gave the 'Train out of Section' signal to Boxmoor and it is entered in my register at 11.8 p.m. After that, the next thing that happened was that the guard of the coal train arrived at my box at about 11.23 p.m., when I got to know of the accident for the first time, and when I immediately sent the 'Obstruction Danger' signal."

Signalman Branscombe said that the enquiry from Gibbs as to whether he had cleared back the meat train was not something which had happened before, but he had no doubt that the train had gone through the section. He was definite that Gibbs had referred to "the 5" and not to "the Stafford", which was a "4" and not a "5" and had cleared Kings Langley eight minutes previously. He was emphatic that he did not refer to any train by its title or book name.

With Signalman Gibbs in Kings Langley box on the night of the accident was Relief Signalman S. Chownes, who was there to take the marks in connection with an appeal from the three signalmen that the box be upgraded from a Class 3 to a Class 2. The clock at Kings Langley was apparently at least a minute slow. According to his train register Signalman Gibbs had accepted the Stafford passenger train at 10.42 p.m., he had received the 'Entering Section' signal for it from Nash Mills at 10.53 p.m., had cleared it back to Nash Mills at 10.55 p.m. and it had been cleared back from Watford North End box at 10.58 p.m.

Gibbs continued, "At 10.55 p.m., after giving 'Out of Section' for the Stafford, I was offered a "5" at 10.56 p.m., and it is recorded in my register as approaching on the Up Fast line. This train is recorded as 'entering section' at 10.58 p.m. at which time I offered it to North End box and got it accepted from North End and it is recorded in my book accordingly. This train did not arrive, and at 11.3 p.m. I mentioned it to the local relief signalman (Chownes) that it was time this train put in an appearance; he, however, passed no comment. I then called up the signalman at Nash Mills and I asked him where the '5' was, and he replied that I had not cleared the 'Stafford,' and that he had a 1-1-3 close to his box. I then came to the conclusion that I had booked a train in error on the up fast, and, as nothing had shown up, the signalman at Nash Mills must have been correct that I had not cleared the 'Stafford.' I therefore looked towards Nash Mills, and, as I could see nothing coming and heard nothing, I definitely concluded that the signalman at Nash Mills was correct, and I then cleared my block instrument to Nash Mills and turned it to normal, whereupon he offered me a 1-1-3 by transmitting a bell signal which I accepted and placed my block instrument to the 'Line Clear' position. The time of this is not recorded in the register, but I have entered against my entry for this train the words, 'Ctd to 1-1-3.'"

Signalman Gibbs had had to replace his signals before accepting the milk train (the block controls meant that the Up Fast home signal must be replaced to normal before the instrument could be placed to 'Line Clear') and he had cancelled the milk train to Watford North End.

As to what had happened then, Gibbs said, "At 11.12 p.m. I again called the signalman on the telephone and asked him where the 1-1-3 was. I had the 10.50 p.m. from Euston offered on the Down Fast (recorded as accepted at 11.12 p.m.) which I passed on to Nash Mills and at the same time the signalman at Boxmoor spoke on the telephone and told me the 1-1-3 had passed Nash Mills and also a '5' was in advance. It was at this time that I came to the conclusion that I was quite correct that I had taken the '5' Up Fast, and that the signalman at Nash Mills had misinformed me regarding the 'Stafford,' and I commented to Signalman Chownes that I was not allowing anything else to go to Nash Mills box until I found out the position of affairs in the section. Then, of course, there was a lull for a few minutes and the express arrived at my Down Fast home signal. The first I knew of anything serious happening was at 11.20 p.m., when I received the 'Obstruction Danger' signal from Nash Mills, also receiving information that a guard had arrived there and informed him that an accident had happened, and all lines were blocked."

The Ministry of Transport inspecting officer, Lieut. Colonel A.H.L. Mount, pointed out that the meat train, from the times in the Kings Langley train register, should have passed there at 11.01 p.m., given that it had passed Nash Mills at 10.59 p.m. (Kings Langley time) and that Gibbs had therefore had six minutes to stop and caution the Down Camden freight train, the meat train having been an unusually long time in section. Gibbs said in regard to this, "I called him {Branscombe} up at 11.3 p.m. at which time the freight train was not out of the tunnel; that was when I had the information that I had not cleared for the 'Stafford.' Had I been told that I had cleared for the 'Stafford' I should have known that the meat train had stopped in the section, and I am quite sure and positive that I should have been able to stop the freight train."

Gibbs said that he could not remember accepting the meat train. He admitted, "To be frank, I was concerned with counting up the trains and discussing the matter with

Chownes. As the 'Stafford' left we counted up the trains, and I suggest that is how I came to overlook this {meat} train. I am quite sure I put my signals to danger behind the passenger train. I was very busy at the time and had a train on each road with the exception of the Down Fast; in fact I was dealing with three trains, viz. the train on the Up Fast {meat} which I cancelled, the train approaching on the Down Slow {freight} and a coal train on the Up Slow." The Aberdeen express was accepted by Gibbs at 11.12 p.m., entered the section at Watford North End at 11.14 p.m., and was stopped at the Kings Langley down fast home signal at 11.16 p.m.

District Relief Signalman Chownes stated that at 11 p.m. he had been counting up the marks for the first hour when Signalman Gibbs had looked around and said, "Nash Mills says I haven't cleared for the 'Stafford.'

He (Chownes) had replied, "You told me you had taken a '5.'" He had then gone on counting the marks.

A couple of minutes later Gibbs had remarked, "There is something amiss, I should have two trains coming up, one on the Slow and one on the Fast."

Chownes said that he knew Gibbs had accepted a 5 because while he had been counting up the marks Gibbs had been calling out every acceptance to him.

As regards Gibbs' actions Chownes said, "Block Telegraph Regulation No. 11 states that in the case of a train being a long time in section, trains in the opposite direction should be cautioned. I should say a long time in the case of this particular meat train would have been about five minutes, and I should not have expected Gibbs to act in this respect before this, and I should not have done so myself, but I should have called up Nash Mills and asked him where the meat was, and if I then confirmed the train had passed Nash Mills I should have out in the caution arrangements."

Signalman A.E. Hadaway at Watford North End box stated that he had accepted the meat train from Kings Langley at 10.58 p.m. Watford No. 2 box had cleared out the 'Stafford' at 11.04 p.m. and had accepted the meat train from him but he had not received the 'Train Entering Section' signal for it from Kings Langley. Instead at 11.05 p.m. Gibbs had telephoned him and said, "Nash Mills had informed him that he had not cleared the 'Stafford.'"

With this, Signalman Hadaway said that he had replaced his Up Fast signals so that he could give another 'Line Clear' release to Kings Langley. "I then called up Watford No. 2 on the block instrument," Hadaway continued, "and when I received his acknowledgement I cancelled the (meat) train to him, at the same time receiving the cancelling signal from Signalman Gibbs. This was at 11.5 p.m. and after he had cancelled the signal to me he immediately offered me the 'Is Line Clear' by giving me 1-1-3 on the bell for a milk train, which I then accepted, and I passed the same on to Watford No. 2 and got that also accepted. And I pulled off my signals again a second time. This train, however, did not arrive, and I subsequently heard of the accident, but in the meantime when talking on the telephone to Boxmoor, the signalman there told me he had clearance from Nash Mills for both the meat train and the milk train when I realised than an accident had happened."

As regards the use of the telephone and Block Regulation 11 Hadaway said, "We do occasionally ring up the box in the rear to ascertain what trains are in the rear, but I do not ring up the box in the rear as to whether I have given 'Out of Section' and if the book is kept correctly you have the book for reference and there is no need to ask such a question. The block instrument works in conjunction with the book, but I always rely on my block instruments first and foremost. In the case of a train being an unusually long time in section, say, in the case of milk or meat train, unless I could see the train approaching I should carry out Rule 11 by immediately placing my signals to danger on the Down road and on the trains arriving caution them through; that applies in clear weather. I should certainly not ring up the signalman at Kings Langley to ask if I had given 'Out of Section' for the train."

At the beginning of the conclusions to his report Lt. Col. Mount stated that there could be no doubt that the accident had been caused by the negligence of Signalman Gibbs.

Although the accounts of Signalman Branscombe and Gibbs as to the telephone conversation between them at 11.05 p.m. varied it was clear that the intention of the call as far as Gibbs was concerned was to ascertain whether or no he had given the 'Out of Section' signal for the meat train. Branscombe had replied that that was what he was waiting for, as the milk train was close at hand. Gibbs had then immediately cleared out the meat train without having seen it, under the plea that Branscombe had referred to the 'Stafford', which had passed Kings Langley ten minutes earlier.

The evidence of Guard Bond was clearly affected by his severe concussion but it seemed that he had been a little slow in reacting when his train had stopped. However, he had given his evidence honestly and had an excellent record.

Driver Buckley had lost his life in a commendable endeavour to prevent a further collision. The enginemen of the freight and coal trains had clearly been on the alert.

It was most unlikely that Signalman Branscombe had used the word 'Stafford' in his conversation with Signalman Gibbs. He had no reason to do so and it was difficult to understand why Signalman Gibbs should think he was referring to a train which was a '4' rather than a '5' and had passed ten minutes earlier. It was regrettable that Signalman Branscombe's suspicions had not been aroused by his colleague's inquiry but, the inspecting officer stated, 'I consider, however, that no blame attaches to signalman Branscombe, although a great deal of Gibbs' evidence and explanation was spent in a regettable endeavour to implicate him. It was only his third day on duty in this main line box, but he had been in training during the previous week, and he felt confident and fully capable of taking over his duties. I have no reason for thinking that he was not fit for the responsibilities attaching to his post.'

As regards Signalman Gibbs, his own words as to his method of working condemned him. "I have not done such a thing as this before, but I have many a time rung up the box in rear to ask if I have not cleared a train and I have always had the correct answer... I should say this occurs in my case perhaps twice a week. I will explain how it occurs; there are occasions when we have possibnly three or four trains passing together and after getting the signals back for the various trains, one is apt to overlook clearing a block instrument and only when I look round to see if the instruments are correct I sometimes find an instument remaining at 'Train on Line'. It is at that moment that I call up the box in the rear concerning the instrument to verify whether I have cleared or not."

Gibbs pleaded, without justification, Lt-Col Mount considered, that he had been under undue pressure at the time of the accident, and admitted that he had been preoccupied with the counting up of the marks. The inspecting officer was not impressed: 'Indeed, there is no redeeming feature in this case, and I find nothing to relieve signalman W. Gibbs of responsibility for this accident, which fortunately had no worse consequences.'

Kings Langley was on one of the most heavily worked quadruple-track main lines in the country and the latest interlocking block equipment was provided. Kings Langley box was new and left nothing to be desired in respect of comfort or ease of operation. Nothing short of continuous track circuiting, which could hardly be justified, would ensure safety in the hands of a signalman who had so little regard for, or interest in, the significance of his block working as to be content to rely on the telephonic advice of the signalmen on either side of him to verify that he had carried out his own fundamental duty. 'In 99 cases out of 100, this may have saved Gibbs trouble, and he was enabled to operate by rote, avoiding the individual conscientious care which is essential, if such duty is to be performed as the public expects. In the 100th case, however, the resulting lack of confidence, and the verbal misunderstanding which the block instument is there to avoid, brought about the breakdown of the most carefully designed and efficient equipment.'

Signalman Gibbs was intelligent and fully cognisant of the rules and regulations. With plenty of time, had he been concentrating on his work, he could have stopped and cautioned the Down Camden freight train in accordance with the provisions of Block Regulation 11, as he had subsequently stopped the down Aberdeen express. 'In the circumstances, however, his failure to obey this Regulation, which materially worsened the results of his initial mistake, and caused the second and third collisions, came about through lack of alertness, consequent upon his preparedness to perform his work indolently, depending upon every train passing normally without hitch and upon the men on either side of him to safeguard his errors or omissions. In fact, Gibbs' failure and subsequent effort to place the blame on his colleague for failing to advise him that the meat train had not passed him, confirms the view that the high sense of discipline, which is so essential among signalmen, and I am glad to say generally exists, was lacking in his case, rendering him unsuitable to retain the resonsible position he held.'

As for Relief Signalman Chownes, he was not acting in a responsible position, but he was a man of long experience, and it was surprising that although he knew that the meat train had been accepted he had been content to allow it being cancelled without protest or even remark. It seemed that both signalmen were more interested in the mark taking than with the operation of the traffic.

Lt-Col Mount had little to say in the way of recommendations other than stressing that telephone messages should not be used routinely to determine the state of the block section and to stress that if the 'Out of Section' signal had not been sent or received it should always be assumed that the train was still in the section, rather than that it had been forgotten to clear it out.

He rejected a suggestion that engines should be provided with searchlights, which would be of no use on a curved line and had other, inherent, disadvantages. [2243]

Four people died and 14 were injured when a hurricane struck and derailed a passenger train on the outskirts of Merida, the capital of Yucatan Province, Mexico, on

[2243] Ministry of Transport report

Monday 8th April 1935. [2244]

By August 1936 Spain was in the throes of a civil war. On 3rd August it was reported that an aircraft flown by transatlantic flyer Pablo Rada, who had volunteered to fight with the Republican government, had bombed two rebel (Nationalist) provision trains en route from Valladolid to the besieged Zaragoza garrison. One train had been set ablaze and there were numerous casualties, the second had been derailed. [2245]

Fascist (Nationalist) aircraft bombed Aranjuez, Spain, on Thursday 8th October 1936, killing an estimated 300 people. Several bombs fell on a loaded troop train and a station and section of track were destroyed, blocking communication between Madrid and Alicante. [2246]

Republican aircraft bombed a train of 30 vehicles carrying insurgent troops as it was crossing a bridge between Montoro and Pedro Abad, Spain, on about Thursday 25th March 1937. After the train had been derailed the aircraft continued to bomb and fire upon its occupants.

Accidents involving trains running into trees which have been blown down across the line are quite commonplace during stormy weather with high winds but are rarely very serious. That which occurred on Saturday 13th March 1937 at Corquoy, near Châteauneuf-sur-Cher, France, was an unfortunate exception.

At about 12:50 a violent gust of wind uprooted an old elm tree, about 20 metres high, alongside the line, across which it fell. M. Lemoine, the owner of the land on which the elm had stood, heard it fall, as did a ganger, M. Guillot, from his hut about 100 metres away. Aware that express train No. 83 from Paris to Le Mont-Dore was shortly due, the ganger ran down the line with some detonators and managed to cover 400 metres before the express reached him.

M. Lemoine, meanwhile, had gone and told the Châteauneuf-sur-Cher stationmaster but the latter was unable to do anything effective to prevent the accident. The wind having brought the telegraph wires down, he was unable to contact the next station at Lunery to warn them of the danger and have the express cautioned. All that the two men could do was to attempt to saw up the tree, though this was a hopeless task in the short time available. They had barely begun their work when the express appeared at 13:08. Because there was a sharp curve on the approach to the point where the tree had fallen Driver Picot of the express only saw it when he was about 50 metres away and although he at once applied the brakes the train was travelling at 80 kilometres an hour, and the distance was too short for any appreciable reduction in speed. It appeared that Driver Picot had failed to hear the explosion of the detonators laid by the ganger above the noise of the gale, and his catching sight of the tree had been his first warning of danger.

[2244] *The Telegraph Herald*, Dubuque, Iowa, United States, 8th April 1935
[2245] *Middletown Times Herald*, Middletown, New York, 3rd August 1936
[2246] *Joplin Globe*, Joplin, Missouri, United States, 9th October 1936

The train was badly derailed when it struck the tree and the wooden third-class carriage which was the second vehicle of the train was telescoped and wrecked by the leading van. Thirteen passengers died in the accident and three more were injured. Driver Picot sustained a broken arm, while Fireman Dumat suffered slight concussion. District Chief Dumas, who was also riding on the engine, escaped with slight injuries, while Chief Conductor Pierre Mangeat and Conductor Henri Gilbert, who had both been riding in the leading van, somehow escaped injury. The train, which had left Paris at 08:58, had been mostly carrying skiers bound for Mont-Dore, though two of those who died were railwaymen, Driver Fougeres and Apprentice Driver André Sabourin, who had been 'travelling on the cushions' en route to or from a job elsewhere.

At 14:00 the injured railwaymen were taken to the *Hôtel-Dieu* (hospital) in nearby Bourges, but it was impossible to extricate the passengers in the wrecked third-class carriage, who were entangled in broken wood and twisted metal, until the arrival of the breakdown crane from Vierzon, which finally reached the scene at 19:00. Meanwhile the 13:18 passenger train from Montluçon to Paris had arrived from the opposite direction, and with the aid of its locomotive it was possible to move the fateful tree from the line. This accomplished, the train resumed its journey to the capital. Newspaper accounts spoke of the rescuers being able to hear the heart-breaking cries of the injured who were trapped in the wreckage amongst the badly mutilated bodies of the dead and whom it was impossible to help until the 50 tonne breakdown crane had lifted the derailed van. This was finally done about 21:00.

M. Baticle, of the Ministry of Public Works, went to the scene of the accident to open an inquiry into its cause and M. Marx Dormoy, the Minister of the Interior, also visited the place.

It was stated that the railway company, the Chemins de fer de Paris à Orléans et du Midi, had previously requested that M. Lemoine cut down the elm because its great size and its proximity to the line had posed a danger to trains, but he had refused to do so. [2247] [2248]

The Second Sino-Japanese War had begun in early July 1937. At the beginning of the conflict Japanese aircraft machine-gunned Chinese troop and supply trains in southern Hopei Province on Sunday 18th July 1937, leading the Chinese government to make a vigorous protest. In response the Japanese admitted that their aircraft had fired on trains at Yuanshih, 275 kilometres south of Peiping, but declared that the troops on the trains had first fired on their aircraft. [2249]

On about Friday 30th July seven Japanese aircraft dropped 12 bombs on Chinese troop concentrations at Paotingfu, damaging the station, blowing up a munitions train and killing ten soldiers. Near Paotingfu a passenger train was bombed, killing 20 Chinese civilians.

[2247] *L'Homme Libre*, Paris, France, 14th March 1937

[2248] *L'Œuvre*, Paris, France, 14th March 1937

[2249] *Stevens Point Daily Journal*, Stevens Point, Wisconsin, United States, 19th July 1937

On about Tuesday 3rd August 1937 Japanese aircraft bombed Chinese troop trains near Yulin and at Hsiahuayuen and an armoured train at Fentaokow, China. [2250] Chinese troops detraining at Hsinpaoan had also been bombed, the Japanese stated, "with satisfactory results." [2251]

Japanese naval planes bombed Suchow, China, at the junction of the Tientsin – Pukow and Lunghai railways, on the afternoon of Monday 30th August 1937, and a munitions train blew up. [2252]

Defying British, French, and American demands that she remove her warships and troops from the immediate vicinity of Shanghai, on Wednesday 8th September 1937 a Japanese aircraft made what was described in contemporary American newspapers as a ferocious bombing attack on a Shanghai – Hangchow passenger train at Sungkiang, near Shanghai. The train was packed with Chinese refugees from Nanking, Three hundred refugees were said to have been killed and 400 wounded in the five coaches of the train. According to the Chinese there had been no troops either on the train or in its vicinity. [2253]

Japanese aircraft reportedly destroyed two Chinese troop trains, probably in the Chochow area, on about Wednesday 15th September 1937. [2254]

On Sunday 10th October 1937 the Japanese captured the important railway junction at Shihchiachwang and the Chinese retreated at speed down the railway to Shunteh, about 110 kilometres to the south. 'Overloaded troop trains carrying the retreating armies are furnishing targets for Japanese airmen.' One troop train was reportedly 'destroyed beyond recognition.'

A couple of days later troop trains were attacked at Neikiu, and on 14th October it was reported that, 'On all the railways in Hopei, Shansi, and Suiyon, Japanese airmen are raiding Chinese troop trains, stations and strategic points, in advance of the pursuing Japanese troops.' [2255]

On the morning of Sunday 24th October 1937 Japanese aircraft attacked a slow passenger train which had left Canton at 6.45 a.m. at Namsa. Six bombs were dropped, three of which reduced the station to debris and the other three of which struck two passenger carriages. The train stopped and its passengers fled to the adjoining rice fields. The aircraft then returned and machine-gunned the train. Two people died and four were wounded. [2256]

2250 *The Hammond Times*, Hammond, Indiana, United States, 3rd August 1937
2251 *The Evening Democrat*, Fort Madison, Iowa, United States, 3rd August 1937
2252 *The Times*, London, England, 1st September 1937
2253 *Berkeley Daily Gazette*, Berkeley, California, United States, 8th September 1937
2254 *New Castle News*, New Castle, Pennsylvania, United States, 16th September 1937
2255 *The Times*, London, England, 12th, 14th and 15th October 1937
2256 Ibid, 25th October 1937

The attacks on Chinese trains continued. On Wednesday 3rd November an express from Canton on the Hankow line was bombed between Pakshektu and Taipingli. It had come to a stand and its passengers were taking refuge in the countryside when a bomb demolished a third-class carriage from which passengers were still alighting. Other coaches were machine-gunned. The Japanese, it was thought, believed that General Chan-Chaitong, the Canton leader, was on the train. Twenty-five passengers were killed and 30 injured. [2257]

More than 200 passengers were reported killed or wounded when six Japanese aircraft bombed the southbound *Blue Express* at Sishipho station, China, near Suchow, on the Tientsin – Pukow railway, on Monday 8th November 1937. The train was one much used in normal times by foreign tourists, but none were on it that day. Nor, according to the Chinese, were there any troops aboard it. Seven of the ten coaches of the train were wrecked. [2258]

Missionary reports stated that Japanese bombers had destroyed a Chinese train carrying troops and civilians near Suancheng, China, 50 kilometres south east of Wuhu, on Wednesday 24th November 1937. 'Casualties were said to run into hundreds.' [2259]

A Chinese troop train was attacked south of Tsinan, China, on Tuesday 30th November 1937. [2260]

On Christmas Day, Saturday 25th December 1937, it was reported that the Japanese advance into China was continuing, and that having crossed the Yellow River their aircraft had carried out 'vigorous' attacks on concentrations of Chinese troops, in one of which a military train had been destroyed seven miles east of Tsinan. [2261]

A single Chinese aircraft bombed a Japanese troop train near Pengpu, China, on about Thursday 10th March 1938. [2262]

The Japanese reported that they had bombed a Chinese troop train on the Lunghai line near Yunho, China, and set it alight on about Thursday 31st March 1938. [2263]

On about Wednesday 27th April 1938 the Japanese attacked Chinese troop trains on the Lunghai Railway in the Suchow area. [2264]

[2257] Ibid, 6th November 1937
[2258] Ibid, 9th and 10th November 1937
[2259] *Nebraska State Journal*, Lincoln, Nebraska, United States, 29th November 1937
[2260] *The Times*, London, England, 2nd December 1937
[2261] Ibid, 27th December 1937
[2262] Ibid, 12th March 1938
[2263] *The Evening Star*, Washington, D.C., United States, 1st April 1938
[2264] *The Times*, London, England, 29th April 1938

In a morning raid on Canton on Wednesday 8th June 1938 Japanese aircraft scored a direct hit on a 16-coach train which had just arrived at Wongha station from Hankow, setting it on fire and destroying it. It was not stated if the passengers had already left it; presumably they had. [2265]

Japanese aircraft bombed a mixed train at Sheklung station, China, on Thursday 9th June 1938. An engine and a coach were damaged but there were no casualties. [2266]

On Tuesday 5th April 1938 Government hospital train No. 20, stopped in the station at San Guim de Freixenet, Spain, was bombed by Nationalist (Fascist) aircraft. Six people were killed, and one was possibly wounded. [2267]

A hospital train stopped in the station at Sant Vicenç de Calders, Spain, was badly damaged when the place was bombed by a squadron of Nationalist aircraft on Wednesday 31st August 1938. None of the five bombs fell directly onto the train but flying glass and bomb fragments killed six wounded men and wounded 12 more. [2268]

A second air attack on a train at Sant Vicenç de Calders, Spain, was a much more serious affair than the previous one. On Saturday 8th October 1938 a Nationalist seaplane bombed a passenger train near the coastal village. Sixty passengers were reportedly killed and over 100 injured. Most of the victims were women returning to Tarragona after a day in the country. [2269]

The Japanese claimed that their aircraft had attacked 50 Chinese trains on the Peiping – Hankow railway south of Sinyang, China, on Saturday 15th October 1938. [2270]

At the end of October 1938 the Nationalists in Spain were carrying out daily air raids along the Mediterranean coast to hinder the arrival of food and other supplies. In one of these a passenger train was bombed as it was leaving Algemesi station. Because of the limited train service and the lack of other means of communication the train was overcrowded. Twenty-one people were killed and almost 80 more wounded. [2271] [2272]

In the first of three air raids by Nationalist aircraft on Tarragona, Spain, on

[2265] Ibid, 9th June 1938
[2266] Ibid, 10th June 1938
[2267] Webpage: Federación Castellano Manchega de Amigos del Ferrocarril (in Spanish)
[2268] Ibid, 2nd September 1938
[2269] Ibid, 10th October 1938
[2270] *Nebraska State Journal*, Lincoln, Nebraska, United States, 17th October 1938
[2271] *The Times*, London, England, 1st November 1938
[2272] *The Railway Gazette*, London, England, 4th November 1938

Wednesday 4th January 1939 two passenger trains in the station were bombed. As at Algemesi they were packed, and early estimates were that there were 30 dead and 40 injured. [2273]

In the closing days of the civil war a train carrying members of the 49th Mixed Brigade of the Republican Army was bombed by five Nationalist aircraft as it stood in the station at Xàtivia, Spain at about 10:30 on Sunday 12th February 1939. One hundred and nine people were killed outright, 20 were fatally wounded and over 200 were less badly hurt. The majority of the victims were soldiers, but some were civilians hoping to greet relatives or friends, including 14 women and three children. [2274]

On Tuesday 21st February 1939 Japanese aircraft attacked a mixed train a mile and a half on the Hong Kong side of the border with China, near Lowu station. Bombs fell on either side of the train and its passengers, 'including two Europeans,' fled into the open. One Chinese was slightly hurt. Sir Stafford Northcote, the Governor of Hong Kong, sent a strong protest to the Japanese at the bombing of British territory. [2275] The Japanese occupation of Hong Kong was less than three years in the future.

Although the Second World War had not quite begun it was clearly imminent by August 1939 and as a precautionary defensive measure against a possible German invasion the Belgian authorities had ordered the mining of various strategic bridges, one of which was the metal railway bridge over the Meuse River at Val-Benoit, just outside of Liège. This precaution led by chance to an unfortunate accident on the evening of Thursday 31st August 1939, the day before war broke out. As a Luxembourg – Liège train was approaching the bridge at 18:45 a violent thunderstorm was raging. A stroke of lightning struck the bridge, set off the dynamite charges and the bridge collapsed, immediately in front of the train and a light engine approaching on the other running line. The distortion of the rails caused by the explosion stopped the trains and prevented them falling into the river, but the engine and coaches of the passenger train were overturned. A few minutes later the old parallel Ougree road bridge was struck by lightning in turn. At 22:00 it was stated that there were 10 dead, five of whom were soldiers who had been guarding the bridge. The others were the driver of the passenger train, the driver and fireman of the light engine, which had left Liège Guillemins station at 18:39, and a man and a woman who had been on the new road bridge at the moment of the explosion. According to a French newspaper the train passengers were not much hurt, though the carriage windows were shattered. [2276]

Germany invaded Poland on Friday 1st September 1939 and the war was only a few hours old when Germain aircraft machine-gunned a passenger train which had left Gdynia for Warsaw at 14:40 on Thursday afternoon. The train had 35 or 40 coaches plus

---

[2273] *The Times*, London, England, 5th January 1939
[2274] Wikipedia – Bombing of Xàtivia
[2275] *The Times*, London, England, 22nd February 1939
[2276] *L'Intransigeant*, Paris, France, 2nd September 1939

a few waggons. There were some soldiers on it but most of the passengers were the wives and children of civil servants, railwaymen and officers who were being evacuated to the interior. At about 06:00 the next morning the passengers saw the German bombers as the train was near Kutno, Poland, but did not realise what they were until a shower of bullets hit the carriages. The train stopped and the passengers jumped out into a meadow and a ploughed field, only to be machine-gunned again. About a dozen were killed. [2277]

Fifty German aircraft bombed and partially destroyed a Red Cross train near Szydlowicz, Poland, at 9 a.m. on Tuesday 5th September 1939, according to the Polish Telegraph agency. [2278] The next day a hospital train carrying a party of wounded soldiers and clearly bearing Red Cross markers on the roofs of its carriages was bombed as it was drawn up at Warsaw East station. [2279]

As noted previously, the Winter War or First Soviet-Finnish War began on 30th November 1939 and on Thursday 21st December 1939 Russian aircraft machine-gunned a passenger train travelling from Turku to Helsinki, Finland, then a second train travelling in the opposite direction between the two places. Two or three passengers died in the first attack. [2280]

On Saturday 20th January 1940 Russian aircraft machine-gunned several passenger trains between Helsinki and Viipuri, Finland, [2281] and on Sunday 28th January an ambulance train was reportedly machine-gunned near Kuhmoniemi, Finland, despite it bearing Red Cross markings. [2282]

A month later the boot was on the other foot, as it was reported that on about Wednesday 28th February 1940 the Finns had bombed several Russian troop trains. [2283]

Twenty-seven Japanese aircraft attacked a lofty steel bridge on the Kumming – Haiphong railway, 60 kilometres from the border with French Indo-China on Friday 2nd February 1940. A northbound passenger train was involved in the attack, in which over 100 persons, 'including 10 foreigners,' were killed or wounded. After the accident, 'British officers, naval doctors, and sailors from two Yangtze gunboats, who arrived on the scene in a southbound train, on their way from Chungking to Haiphong, assisted in rescue work.' A tunnel adjoining the bridge was blocked by the wreckage of the train, stopping traffic on the line. [2284] A follow-up report stated that over 110 people had been killed and many injured.

2277 *The Times*, London, England, 2nd September 1939
2278 *Altoona Mirror*, Altoona, Pennsylvania, United States, 5th September 1939
2279 *The Times*, London, England, 7th September 1939
2280 *Corona Daily Independent*, Corona, California, United States, 21st December 1939
2281 *The Times*, London, England, 22nd January 1940
2282 Ibid, 30th January 1940
2283 Ibid, 1st March 1940
2284 Ibid, 3rd February 1940

The "lace" bridge was attacked again on 13th February. [2285]

Germany invaded Norway on 3rd April 1940 and rapidly overran the southernmost part of the country, despite British support for the Norwegians. On Wednesday 24th April the last train northwards with retreating Norwegian soldiers arrived at Telneset station, where two carriages carrying local railwaymen and civilians were attached. Before the train could set off again, however, it was attacked by German aircraft with both machine-gun fire and bombs. The engine was wrecked and the two leading coaches badly damaged. Six soldiers were killed or fatally wounded. The engine of a goods train in the yard continued northwards with the undamaged coaches. [2286]

The German advance continued and on Tuesday 30th April the last train from Dombås to Åndalsnes met with an accident just west of Lesjaverk station, when at 01:15 it ran into a bomb crater. The train was double-headed and there were two more locomotives further back along it. The two leading locomotives, type 45a No. 5 and type 26c No. 435, were overturned and four Norwegians and seven British soldiers in the leading carriage lost their lives. The engine crew escaped unhurt except for Driver B. Skreden, who suffered a broken arm. Driver Skreden's fireman, Ingvald Hessen, had his skis with him on the engine and after the accident, realising that he could do nothing more at the scene, he made his way homewards across country on his skis. [2287] The remaining troops continued by road to Verma and thence by train to Åndalsnes, from where 1,500 of them escaped by boat shortly before the Germans arrived. [2288]

The Germans invaded the Netherlands, Belgium, Luxembourg, and France on Friday 10th May 1940 and in the first reported Allied air attack on trains of the war, four German armoured trains were reportedly put out of action. One of them was blown up with the railway line near Venlo in the Netherlands. The next day German supply trains were attacked in the area between the Rhine and the Dutch frontier. [2289] There were further attacks on German trains in the following days and weeks, though on nothing like the scale that such attacks were to reach towards the end of the war. Supply and munitions trains were the usual targets, though a troop train was bombed at North Geldern, Germany, on Friday 24th May 1940. [2290] Air attacks were also carried out by the Germans, including an attack on a hospital train plainly marked by Red Crosses at an unstated location in France in late May. [2291]

No. 7 passenger train of the South Indian Railway from Madras Egmore to Dhanushkodi was struck by a cyclone and overturned on Tuesday 28th May 1940 between Panchalam and Tindiwanam, India. Five passengers died and six were injured. [2292]

[2285] Ibid, 15th February 1940
[2286] https://museainordosterdalen.no/24-april
[2287] *Norges Statsbaners Krigshistorie 1940-1945 Hamar district* by Arvid Östby, pgs. 59-60
[2288] *På Sporet* (magazine of Norsk Jernbaneklubb) No. 82, pg. 25 (in Norwegian)
[2289] *The Times*, London, England, 11th and 13th May 1940
[2290] Ibid, 25th May 1940
[2291] Ibid, 28th May 1940
[2292] Report by the Railways Board on Indian Railways for 1940-41

Dhanushkodi was a port from which there was a ferry service to Ceylon, served by the metre gauge line from Madras.

On Wednesday 3rd July 1940 a German aircraft made an attack on a train in England, the first such of the war. The censored press reported that, 'A train travelling between two South Coast towns was bombed last night. The driver was killed. The only passengers, two women and two children, had a lucky escape while they crouched on the floor. One of the women had the toe of her right shoe blown off. The windows were smashed, and the carriage was filled with fragments of glass. The train stopped, but later proceeded to its destination, where the driver died from his injuries. The aeroplane dropped a number of bombs near the railway line.[2293] The train had been attacked near Bishopstone Beach Halt, on the Seaford branch of the Southern Railway.

Spontaneous German air attacks on British trains continued. Those which involved isolated aircraft attacking trains tended to be minor but on Friday 6th December 1940 a passenger train became involved in a bombing raid on the city of Bristol and 14 Great Western Railway train passengers died at Bristol Temple Meads station. One hundred people in total were estimated to have died that night. [2294]

An indirect result of the German bombing campaign against Great Britain was accidents resulting from trains running into bomb craters, of which there seem to have been a good number, though none, apparently, had very serious consequences. Thus there were, for example, accidents on the Southern Railway at St Denys on Wednesday 14th August 1940, between Littleton and Badsey, and Evesham on the Great Western Railway on 30th August 1940, at Spon Lane, Birmingham, on the London Midland and Scottish Railway, on 19th November 1940, and between Ampleforth and Coxwold on the London and North Eastern Railway on 16th April 1941. [2295]

Italy had joined the Second World War in June 1940 on the Axis side and threatened British colonies in East Africa. Thus it was that in March 1941 aircraft of the South African Air Force (SAAF) began to attack Italian trains in Abyssinia (now Ethiopia) which had been conquered by Italy in 1935. In the first such attack goods trains on the line between Keren and Asmara, in Eritrea, were machine-gunned and damaged on Tuesday 11th March 1941. [2296] Further attacks on Italian trains occurred throughout March and into early April, after which time the Italians were in full retreat and were no longer in possession of the railway.

Five passengers died and 11 were injured when train P 752 was struck by an aircraft

---

[2293] Ibid, 4th July 1940

[2294] http://humanities.uwe.ac.uk/bhr/Main/ww2/1_8.htm

[2295] Author's personal notes taken from various railway magazines and society journals.

[2296] *The Times*, London, England, 12th March 1941

and caught fire at Fellheim, Germany, on Saturday 14th June 1941.[2297]

On 22nd June 1941 the Germans invaded the Soviet Union (Operation Barbarossa) and in the following months there were sporadic reports in the British press of German air attacks on Russian trains and vice versa. Meanwhile the British were carrying out air attacks on trains in France, Germany, Italy, and North Africa.

On the evening of Sunday 20th July 1941 an accident with unusual features occurred in Italy on the Milano – Chiasso main line between Como San Giovanni and Albate Camerlata. Southbound freight train No. 5571 included a crane waggon, the crane on which was apparently not properly secured, for in the course of the journey its boom came loose, struck the parapet of a bridge which the train was crossing, and then swung around and fouled the opposite running line. At that moment passenger train No. 3588 from Milano was passing. This consisted of eight carriages filled with Italian workers bound for jobs in Germany. Five of the carriages of the passenger train were struck by the crane and three of them were badly damaged. Thirty workers were killed outright and 50 were injured and taken to Como Hospital. Signor Janelli, the Undersecretary of the State Railway, went to the accident scene to begin an investigation with railway officials.[2298]

A similar accident was to happen at Beighton in England the following year (see below).

An accident like that which had occurred at the Val Benoit Bridge in Belgium in 1939 (see above) happened at Chillon in Switzerland on 25th September 1941. In the late 1930s, with the growing prospects of a European war, encouraged by the French building of the Maginot line, and, partly as a measure to relieve unemployment, the Swiss began to improve their border fortifications in an extension of an 1880s plan for a National Redoubt against foreign invasion.[2299] One of these was a fort at Chillon, on the edge of Lac Léman, protecting the road and railway lines running along the lake shore, work on which was begun in 1941.

At 16:30 on the afternoon of the day of the accident, shortly after the passage of a passenger train for Villeneuve, and as a freight train bound for Lausanne was running past Veytaux station, mines laid in the track as part of the fortifications exploded. The noise of the blast was heard a great distance away and the effects of the explosion were severe. The driver and conductor of the train, a sentry stood nearby, and three people who chanced to be going along the cantonal road at the time were all killed outright. A mass of rocks fell onto the train, derailing the engine and a dozen waggons. Thirty metres of cantonal road 780 were destroyed and the gas, water, telephone, and electrical installations were all wrecked. Two neighbouring houses were damaged and the occupants of four floors of one had to find accommodation elsewhere. The other residence suffered less. An immediate rescue operation was launched and the injured, who were probably mostly the occupants of the two houses, were taken to hospital in

[2297] *Eisenbahnunfälle in Europa* by Erich Preuß, pg. 122 (in German)
[2298] *La Stampa*, Torino, Italy, 21st July 1941
[2299] Wikipedia – National Redoubt (Switzerland)

Montreux.[2300] A seventh person appears to have died later.[2301]

It was hoped to reopen one line on 1st October and both on 3rd October.[2302] In the meantime passengers were transhipped past the site by boat along the lake while freight trains were diverted. A tramway running along the road was not back in use until April 1942.

A northbound troop train, which had probably started its journey at Chatham, on the Southern Railway, and was bound for Scotland, was involved in an unusual accident near the station at Beighton, England, on the London and North Eastern Railway main line between Nottingham and Sheffield, at 9.56 p.m. on Wednesday 11th February 1942. As it was passing Holbeck colliery sidings its carriages came into contact with a heavy steel plate which was projecting over the side of a plate waggon stabled on No. 1 track of the colliery sidings. Displaced during shunting operations, the plate penetrated deeply into the sides of the carriages, especially the sixth from the engine and to a lesser degree the seventh. Damage to the train also severed the vacuum brake, bringing it to a stand 280 yards beyond Beighton signal box. Within a few seconds an Up mineral train passed on the other running line and struck the protruding wreckage of the sixth carriage, but it was not thought that it caused any further casualties.

There had been nearly 400 soldiers and sailors travelling in different portions of the train. The sailors escaped but the soldiers were in the affected part of the train. Fourteen of them were killed or fatally injured, another 35 were badly enough injured to be detained in hospital and a number of others were slightly hurt.

The accident happened at time that the railwaymen were changing shifts so there were plenty of staff on hand to conduct the rescue operation, which was promptly organised by Stationmaster Edgar Allan, while miners and villagers also lent assistance. Twenty-three ambulance attended the scene, together with doctors and police, and all the seriously injured had been taken to hospital by midnight.

The train consisted of a class B 7 4-6-0 mixed traffic steam engine, 13 corridor bogie coaches and two four-wheeled vans. The coaches were electrically lit, the vans had no lighting.

The plate which had caused the accident was one of a batch of seven en route from Frodingham to West Tinsley. Each measured 9 ft 6 in x 8 ft 7 in x ⅞ in and weighed 25 cwts. The plates being wider than the inside width of the 20-ton plate waggon in which they were carried, they had been loaded with one edge on the floor and the far end resting on the top of the opposite side. Packing pieces were used to prevent the plates moving. When the waggon was examined after the accident it was seen that the packing pieces had become displaced. The four plates at the south end of the waggon had fallen onto the floor, two of the plates at the north end had moved towards the middle and were slightly overlapping the waggon, and the top plate, the one which had struck the troop train, was missing.

[2300] *Gazette de Lausanne*, Lausanne, Switzerland, 26th September 1941

[2301] https://blogs.letemps.ch/christophe-vuilleumier/2015/06/07/les-catastrophes-ferroviaires-suisses-du-xixeme-au-xxieme-siecle/

[2302] *Gazette de Lausanne*, Lausanne, Switzerland, 30th September 1941

The waggon had been loaded by an experienced worker at the Appleby-Frodingham Steel Company in good light on the morning of Sunday 8th February. An examiner had passed the loaded waggon, which had not left the yard until 3.30 a.m. on the day of the accident. It had arrived at Broughton Lane Sidings, Sheffield, at midday. From there the waggon had continued at 6.05 p.m. in a train in charge of Guard Walter Helliwell. Seventeen waggons had been left at High Hazels and the remainder of the train had then continued to Holbrook, arriving there at 8.15 p.m. Here the train had been shunted from the down line to No. 3 colliery siding and Guard Leonard Calladine, acting as shunter, had taken charge of the movements, leaving Guard Halliwell to control the vehicles in the sidings.

Several shunting movements were made until only the plate waggon and a common owner waggon were attached to the engine. The road was then set for track No. 1, Guard Halliwell gave the engine a green light to set back so that he could uncouple the waggon, then a red light as he wished to reverse the points to let the common owner waggon run into track 2. Guard Halliwell said that he had had to break into a gentle run to get to the plate waggon to pin down its brake. He had then dealt with the common owner waggon, and with that the shunting operations were complete. Neither guard had noticed anything wrong when the plate waggon had come into contact with other vehicles in the siding.

The Ministry of War Transport inspecting officer. Mr J.L.M. Moore, concluded that the movement of the plate had occurred during its final loose shunting into No. 1 siding. There was evidence that Guard Calladine had been careful with the shunting but that the brake levers on the plate waggon had not been applied. Tests made with a similar waggon showed that if the brakes were applied as stated by Guard Halliwell the plate waggon came to a halt well short of the vehicles already stabled in the siding. It was clear, the inspecting officer said, that either the guard did not apply the brake at all, or else in the darkness and conditions underfoot he had not applied it properly.

The inspecting officer stated, perhaps not very charitably, that Guard Halliwell, who was over 60 years of age, 'was apparently not over alert, either in mind or body.' In addition the guard had been on duty for eight hours and had not expected to have to play any part in the shunting at Holbrook. At any rate the responsibility for the accident lay with him but the extent to which he was to blame must not, the inspector continued, be judged by the results. The person doing the shunting had to think and act quickly. In the few moments during which he could see the vehicle with his hand lamp he must consider its speed, what type of brake it had and what must be done to stop it at the desired point. Should there be a problem with the brake he also had to consider his own safety and whether he would in such circumstances be justified in abandoning the brake and letting the waggon run. It was unlikely that the guard had anticipated any serious results in doing so on the night in question. Mr Moore said that he hesitated, therefore, to criticise Guard Halliwell unduly for his failure, though he felt that a younger, more active man, would have checked the vehicle.

The inspecting officer pointed out that as well as establishing the responsibility for the accident the tests had shown how just a slight impact could be enough to move the plates in the waggon. As there was no practicable means of better securing loads carried as the plates in the manner in which those involved in the accident had been, there was no alternative but to immediately discontinue this method of loading them, a proposal to which the London and North Eastern and other railway companies had readily agreed to.

Finally he pointed out that the clearance between No. 1 siding and the down main line was less than what was normal in modern practice. [2303]

German bombing raids on Great Britain (the Blitz) ended in mid-1941 with Hitler concentrating his resources on Russia, but sporadic attacks by isolated aircraft continued into the following year, thus in March 1942 a Great Western Railway train was machine-gunned at an unstated location between Westbury and Weymouth and on Thursday 16th April 1942 another train of the same company was shot at near Dawlish Warren. The very last attacks by the Luftwaffe were as late as Saturday 3rd March 1945, when London and North Eastern Railway trains were machine-gunned between Gilling and Pilmoor, at Northallerton, and at Temple Hirst. [2304]

During their rapid advance through Burma Japanese aircraft dive bombed a Chinese troop train and a Red Cross train near Shwebo on about Saturday 30th April 1942. [2305]

The frequency of Allied attacks on trains in the occupied countries of Western Europe and Germany increased as 1942 wore on, making the life of a railwayman in say France a dangerous one. On Friday 28th August 1942 Royal Air Force (British) aircraft bombed and machine-gunned a French passenger train at Menars, France, near Blois. Twenty-six people died and another 50 were injured. This prompted the French Vichy government to send a protest to London via Madrid, pointing out that quite apart from the Blois attack 17 French railwaymen had been killed and over 50 injured in air attacks on trains between 1st May and 26th August. [2306] [2307]

Allied air attacks on Germany further increased as the war went on and towards the end of April 1943 an aeronautical correspondent wrote in the London *Times*, 'Germany's communications, which have for some time been hard pressed to meet the demands of war transport between the Reich and the scattered occupied territories in Europe, are becoming an even more difficult problem for the enemy as a result of the Anglo-American attacks from this country. It is known, for instance, that locomotives stolen from France have had to be sent back because dozens of trains were immobilized, and in Germany itself the enemy is being forced to transfer engines and goods trains to the areas where their target is most urgent.' [2308]

Meanwhile in Eastern Europe, in mid-May, the Russians announced that they had made numerous attacks on German trains on the Eastern Front at various points. [2309]

The London *Times* of 17th December 1942 carried a censored report of a Luftwaffe attack on a British passenger train. 'A train was attacked in one area of the Home Counties and a number of passengers were taken to hospital with bullet wounds. Two of

---

[2303] Ministry of War Transport report

[2304] Author's personal notes taken from various railway magazines and society journals.

[2305] *The Courier Mail*, Brisbane, Queensland, Australia, 25th May 1942

[2306] *Manitowoc Herald-Times*, Manitowoc, Wisconsin, United States, 29th August 1942

[2307] *Le Journal*, Paris, France, 31st August 1942

[2308] *The Times*, London, England, 24th April 1943

[2309] Ibid, 11th, 12th and 13th May 1943

the passengers taken to hospital died last night. Operations were performed on some of the serious cases. Not all the passengers admitted to one hospital had been identified last night.'[2310]

The attack took place just after the train had left the Southern Railway station at Bramley and Wonersh, England, on the branch line from Guildford to Horsham. Seven people died. A photograph shows the train to have been severely damaged by the machine-gun attack, though the fact that it had a side corridor running along the side which came under fire was said to have prevented more deaths.[2311]

A passenger train was attacked by Allied aircraft between Waldbröl and Boxberg, Germany, on Friday 10th September 1943. Four adults and two children died.[2312]

What was perhaps an especially tragic incident of soldiers being killed by friendly fire occurred on Friday 28th January 1944. The Ponte Paglia bridge at Allerona, Italy, near Orvieto, was attacked by B26 bombers of the United States Army Air Force as the pilots thought a German troop train was crossing it. In fact the train was carrying 800 Allied prisoners of war who were being transferred from PG Camp 54 at Fara-in-Sabina, near Rome, to Germany in cattle cars, in anticipation of the Allied advance. When the bridge came under attack the driver stopped it on the bridge, leaving the prisoners, who were padlocked in the cars, to their fate. It would seem in fact, according to witness statements, that the train had already crossed the bridge and had stopped at the station beyond when the Allied aircraft had appeared and had then been backed onto the bridge, perhaps in the vain hope that the presence of the train on the structure might save it from destruction. Two spans and about 100 metres of the bridge were destroyed, together with ten cars of the train, others being damaged and derailed. About 400 American, British, and South African prisoners died according to local records and witness testimony.[2313]

A number of the surviving prisoners later described what happened. Private Russel Eugene Kurzhal 168 Regiment 34 US Infantry Division said:

'And it so happens, they always put the Americans up front for some reason or another. We were in the front car behind the engine. And then they put the British and then the South African negroes. There were a bunch of them. They had a direct hit on the ones in the centre of the train. We figured over 300 were dead out of the 800. Killed them. Dumped them right in the river.

'We were on the move again when the bombing started. By the time the first wave had dropped their bombs, the train was stopped. There was five waves of planes that flew over. The first wave dropped theirs, the second wave dropped theirs, and by the third wave, I was out of the car by that time... The car I was in... some of the shrapnel hit it 'cause it knocked the slider door half way off – a door that slides back and forth...The guys took that toilet and used it as a battering ram and knocked the rest of that door off the hinges. Then they went out that door. And it also knocked those boards – some right

[2310] Ibid, 17th December 1942

[2311] *Branch Lines to Horsham* by Vic Mitchell and Keith Smith, plate 12.

[2312] http://www.geschichtspolitik.de/Alliierte_Tieffliegerangriffe.pdf

[2313] Wikipedia – List of friendly fire incidents

under that window… the one with the wires crossed over it. The shrapnel hit there… I went out the window. What happened is this. I looked at the window and the guys lined up to go out the door. I was going to go whichever way was open. I looked back at the window and the guys were all gone already, so I grabbed those two rails above the window and swung my feet through that window and down to the ground. About a 15 foot drop to the ground. And hit the ground running… we lost four Americans there. How many of them were our bunch, I don't know. Four more got killed on that train… It couldn't be helped, I guess. They didn't know it was a prison train and they bombed it. It seemed funny, though. I thought our intelligence should have been smarter than that, cause we'd been on the road now for two days. They should have known there was a POW train going up through there. It wasn't marked, though, as a prison train, but there at each end of each boxcar, there was a little hallway built and that's where the guards sat. That's the way it was on all those prison trains.' [2314]

The British and American air attacks on Germany, Italy, and occupied Europe continued, up to and beyond the invasion of France in June 1944, as did Russian attacks on trains in Eastern Europe. On a smaller scale there were also air attacks on Japanese trains in Burma.

A military train loaded with tanks was bombed between Lingen and Ems, Germany, on Monday 21st February 1944, killing its driver and fireman. [2315]

Berlin reported that on Tuesday 7th March 1944 low-flying Anglo-American aircraft had bombed and strafed a train carrying Italian refugees at an unstated location north of Roma, Italy, killing 40 and wounding many more. [2316]

A passenger train was attacked by aircraft while it was in the station at Coesfeld, Germany, on Friday 21st April 1944. An unspecified number of railway staff and three passengers were killed outright, and seven passengers were fatally wounded. [2317]

Two, presumably stationary trains and a military train which was passing through were attacked by Allied aircraft in Berga-Kelbra station, Germany, on Friday 5th May 1944. Twenty-eight people died and many were wounded. [2318]

On Tuesday 1st August 1944 Allied aircraft attacked a passenger train between Lohr and Wertheim, Germany. Three died and 20 were wounded. [2319]

[2314] https://bombedpowtrain.weebly.com/
[2315] http://www.geschichtspolitik.de/Alliierte_Tieffliegerangriffe.pdf
[2316] *Le Matin*, Paris, France, 18th March 1944
[2317] http://www.geschichtspolitik.de/Alliierte_Tieffliegerangriffe.pdf
[2318] Ibid.
[2319] Ibid.

A passenger train was attacked between Neustadt and Rothenfels, Germany, on Tuesday 8th August 1944. Again three passengers were killed and 20 wounded. [2320]

A train accident in Kent, England, on Wednesday 16th August 1944, in which nine people died, was reported two days later in a national newspaper in which it was stated that the train, which had been full to capacity, had left the track on a curve between Gillingham and Sittingbourne. The engine and front part of the train had been derailed and wrecked but the rear portion had escaped damage, and its passengers, which had included a number of doctors and nurses as well as holidaymakers and members of the Services, had hurried forward to assist their less fortunate fellow travellers up ahead. The injured driver had also helped in the rescue work until taken to hospital, while his fireman, dazed and bleeding, had set off down the track to the next station to protect the line. In addition to those who had lost their lives 35 injured were detained in hospital and 100 others were given first-aid for minor injuries. [2321]

The cause of the derailment was not mentioned in the newspaper report, and it was not until almost a month later that it was revealed that it had been due to a V1 flying bomb which had destroyed a bridge over a country lane right in front of the train. [2322]

Because the derailment had been the direct result of enemy action it was not the subject of a Ministry of Transport investigation, but research by local historians has uncovered more details of the accident. The train involved was the 3.35 p.m. Southern Railway service from London Victoria to Ramsgate. Driver Charles Barnett saw the smoke from the damaged bridge when his train was close at hand but too late to pull up. His engine overturned, the leading two carriages were crushed and mounted by the third carriage, while the fourth carriage was left overhanging the gap where the bridge had been. The bridge crossed Oak Lane at Upchurch, near Rainham.

The VI had been spotted and chased by RAF Flight Lieutenant Charles Malloy, who after failing to shoot it down had manoeuvred his aircraft alongside the missile to tip it over and cause it to crash short of its destination, a common procedure employed by pilots. He succeeded in upsetting the delicate gyro mechanism of the V1 and forcing it down, it was just unlucky that it had landed where it did.

Doctors, ambulances, passers-by, and members of the local Women's Voluntary Services were soon on the scene after the accident and the injured were taken to a transport café on the A2 Dover road, *The Rest Tea Rooms*, which was converted into a temporary hospital. Most were later transferred to St Bart's Hospital in Rochester. [2323] Eight of those who died were train passengers, the ninth was a railwayman, Alfred Naylor, who had been sheltering under the bridge at the time. [2324]

[2320] Ibid.

[2321] *The Times*, London, England, 18th August 1944

[2322] Ibid, 14th September 1944

[2323] Webpage: History of Rainham, Kent – The Upchurch Railway Train Crash Disaster of August 1944

[2324] Webpage: Historical Research Group of Sittingbourne – The Train that Jumped the Gap – 16 August 1944

The magnitude of Allied attacks on German trains by late 1944 may be gauged by a single day's report on air operations: 'Long-range fighters of the United States 8th Air Force continued their offensive against communications behind the German lines in Belgium and Holland {on Friday 1st September 1944} strafing freight cars and half-tracked vehicles. They also disabled 69 locomotives, attacked a canal lock at Eindhoven, in Holland, with 500lb bombs, and bombed a number of marshalling yards.

'Air Defence of Great Britain fighters on offensive and bomber escort missions in Flanders and the Pas de Calais strafed enemy transport of all kinds. By early evening they had attacked at least 45 barges, 60 road vehicles and 20 trains (many of them packed with troops), a dozen locomotives, as well as over 100 parked railway trucks, in addition to many isolated enemy armoured fighting vehicles, motor cyclists, staff cars and horse transport.'

Passenger train P 526 from Osnabrück to Rheine was attacked by Allied aircraft near the railway bridge at Elterstraße, Germany, on Tuesday 5th September 1944. Fourteen people died and many were wounded. [2325]

A passenger train was attacked by Allied aircraft at Odenhausen, Germany, on the Lumdatalbahn on Saturday 9th September 1944. Sixteen died and 40 were wounded. [2326]

Allied aircraft attacked a Teutoburger Wald Eisenbahn train at Marienfeld, Germany, on Sunday 10th September 1944. Four passengers, two of them children, died and 22 were wounded. [2327] The same day a passenger train of the Hohenzollerischen Landesbahn was attacked between Gauselfingen and Burladingen, Germany. Eight schoolchildren and two young women lost their lives. [2328]

Fifteen died when a train was attacked by Allied aircraft between Losheim and Ahrdorf, Germany, on Monday 11th September 1944. [2329]

A train carrying refugees from that part of Germany close to the Belgian border, and which had just left Eschweiler, was attacked by Allied aircraft at Hücheln, Germany on Friday 15th September 1944. An unknown number of passengers, mostly women and children from nearby Stolberg, were killed and wounded.

The midday train on the Uplandbahn was attacked by Allied aircraft between Korbach and Briton Wald, Germany, on Monday 25th September 1944. The driver and a number of

---

2325 http://www.geschichtspolitik.de/Alliierte_Tieffliegerangriffe.pdf
2326 Ibid.
2327 Ibid.
2328 Ibid.
2329 Ibid.

passengers were injured. [2330]

Passenger train P 2041 from Frankfurt to Usingen was attacked by a low-flying Allied aircraft just before it reached Köppern, Germany, on Wednesday 4th October 1944. Thirty-one people died. [2331] The wounded passengers had to be taken to hospital on a flat waggon with straw as bedding. [2332]

Allied aircraft attacked a passenger train at Allendorf, Germany, on Friday 6th October 1944. Nine people died. Eyewitness Christine Born said, "There stood the train and wounded lay about screaming in pain." [2333]

A freight train was attacked by Allied aircraft at Riedenberg, Germany, on Monday 9th October 1944. [2334]

A passenger train carrying German soldiers and including a munitions waggon was attacked by Allied aircraft between Wels and Passau, Germany, on Monday 16th October 1944. Seven soldiers were killed, and others were wounded. The munitions waggon also exploded, setting a farm on fire, and damaging other houses.

Fireman Franz Anton Wollman was fatally wounded when a train was attacked by Allied aircraft at Coesfeld, Germany, on Tuesday 24th October 1944. The same day 18 people died, and many were wounded in an another more serious attack on a train on the Uplandbahn between Korbach and Briton Wald, Germany. [2335]

A passenger train from St Goarshausen was machine-gunned by Allied aircraft at Friedhof, Germany, leaving dead and wounded, on Thursday 26th October 1944. [2336]

About 60 people were wounded, 25 of them seriously, when a passenger train was attacked by Allied aircraft in the station at Lotte, Germany, on Monday 30th October 1944. [2337]

The train which had left Silkeborg at 10:55 was attacked by Allied aircraft just outside of the station at Hampen, Denmark, on Tuesday 31st October 1944. Three passengers were killed, and Driver S.A. Knudsen was fatally wounded and died after admission to hospital. Seven passengers and Fireman Niels Borch were wounded, one of the

---

[2330] Ibid.
[2331] Wikipedia – Liste von Eisenbahnunfällen in Deutschland
[2332] http://www.geschichtspolitik.de/Alliierte_Tieffliegerangriffe.pdf
[2333] Ibid.
[2334] Ibid.
[2335] Ibid.
[2336] Ibid.
[2337] Ibid.

passengers seriously. [2338]

A passenger train was attacked by Allied aircraft at Bachlet, Germany, near Friedrichshafen, at an unstated date in October 1944. Ten died and a number were wounded. [2339]

A military train stopped on a road overbridge at Staelkottenweg, Germany, was attacked by Allied aircraft on Wednesday 8th November 1944. Soldiers from the train attempted to shelter beneath the bridge but it received a couple of direct hits and over 70 soldiers and others taking refuge there were killed. [2340]

The same day a passenger train was attacked at Krombach, Germany. Eleven died and 30 were wounded in this attack. [2341]

A train was attacked by Allied aircraft at Dersdorf, Germany, on the Vorgebirgsbahn, on Monday 13th November 1944. Seven people died and between 30 and 40 were wounded. [2342]

A stationary train containing horses which had been requisitioned by the Germans was attacked by Allied aircraft near Baarn, the Netherlands, on Saturday 25th November 1944. The attack caused the boiler of the Belgian G8 class locomotive to explode, and the driver and fireman were killed. It was said that local people feasted on horsemeat for a long time afterwards. [2343]

Five women and two 16-year-old apprentices were killed, and 20 others were wounded when an Allied aircraft attacked a passenger train at Herbern, Germany, on Monday 27th November 1944. [2344]

Two passenger trains were attacked in the station at Lauda, Germany, by Allied aircraft on Thursday 30th November 1944. Fifteen died. [2345]

A German munitions train was attacked from the air at Naarden, the Netherlands, by Allied aircraft on 30th November 1944 until it exploded, causing much damage to the

---

[2338] https://www.kobenhavnshistorie.dk/index.php/13-bog/dbbtb/180-dbbtb?start=5
[2339] http://www.geschichtspolitik.de/Alliierte_Tieffliegerangriffe.pdf
[2340] Ibid.
[2341] Ibid.
[2342] Ibid.
[2343] https://japan-forward.com/75-years-more-than-100-deaths-later-attack-on-fukuoka-trains-still-raises-questions/
[2344] http://www.geschichtspolitik.de/Alliierte_Tieffliegerangriffe.pdf
[2345] Ibid.

town.[2346]

A passenger train was attacked by Allied aircraft at Trennfeld, Germany, between Lohr and Wertheim on the Maintalbahn, at an unstated date in November 1944. There were many killed and wounded. Also in November 1944 there were a number of attacks on trains in the Dieringhausen area which left numerous dead.[2347]

A passenger train was attacked by Allied aircraft at Groß Reken, Germany, on Sunday 3rd December 1944. Thirty-nine people died and many others were seriously wounded.[2348]

Sixty-four people reportedly died and 120 were seriously wounded in an attack on a passenger train by Allied aircraft at Bornhein, Germany, on Monday 4th December 1944.[2349]

On Saturday 9th December 1944 an express train from Tübingen to Stuttgart was attacked by Allied aircraft in Bempfingen station, Germany. Three people died.[2350]

A low-level attack was made by three Allied fighter aircraft on two express trains on the Kaisersburg – Krobach line near Leoben, Austria, Germany, at 13:15 on Monday 11th December 1944. Ten passengers were killed outright in the attack, 20 were seriously wounded and the same number slightly wounded. Two of those badly hurt later died of their wounds.[2351]

A passenger train on the Nördlingen – Dinkelsbühl line was attacked by Allied aircraft at Marktoffingen station, Germany, on Thursday 14th December 1944. Eighteen died and 15 were seriously wounded.[2352]

Allied aircraft attacked a passenger train at Eisweiler, Germany, on the Köllertalbahn on Saturday 16th December 1944. This time 16 people died, and many were wounded.[2353]

A broadcast from the Japanese-controlled Malayan radio in early January 1945 reported a major catastrophe at Padang Panjang, in west Sumatra, Dutch East Indies, on

---

2346 https://gooiland.50plusser.nl/weblog/print.php?warticle_id=51787
2347 http://www.geschichtspolitik.de/Alliierte_Tieffliegerangriffe.pdf
2348 Ibid.
2349 Ibid.
2350 Ibid.
2351 https://www.nicospilt.com/index_leger.htm
https://www.landesarchiv.steiermark.at/cms/dokumente/11683563/3ccacc24/69%20bis%20158%20aus%20Mitteilungen%2038-Der%20Luftkrieg%20in%20der%20Steiermark%201941%20-%201945.pdf
2352 http://www.geschichtspolitik.de/Alliierte_Tieffliegerangriffe.pdf
2353 Ibid

Christmas Day, Monday 25th December 1944. According to the broadcast 224 people had died in the accident and 245 more were injured. No further details were given as to the catastrophe, not even whether it was a derailment or a collision or even, perhaps, a bridge failure. The broadcast did add, however, that a collection of 28,000 florins had been made from the inhabitants to aid the injured and the relatives of those who had lost their lives. [2354]

Niederlahnstein station, Germany, was bombed and totally destroyed in an attack which began at 12:30 on Boxing Day, Tuesday 26th December 1944, while an ambulance train was stopped there. One hundred and sixty people died in total, it is not clear how many of them were on the train. [2355]

A passenger train on the Tecklenburger Nordbahn was attacked by Allied aircraft at Osnabrück Germany, on Thursday 28th December 1944. Twenty people died and a large number were injured, though many quickly left the train and sought cover when they realised that they were under attack. [2356]

A passenger train was attacked by Allied aircraft between Kundl and Wögl, Germany, on Thursday 4th January 1945. Five people died.

A train carrying munitions and fuel was bombed by Allied aircraft at Edenkoben, Germany, on Friday 5th January 1945 and a massive explosion destroyed the station area. [2357]

While it was travelling between Walldürn and Hardheim, Germany, a train was attacked by Allied aircraft on Saturday 6th January 1945. Two young women passengers and the conductor were killed. [2358]

Allied aircraft attacked a passenger train at Winsen an der Aller, Germany, on Monday 22nd January 1945. Four young girls lost their lives.

An ambulance train coming from Köln was attacked on Saturday 27th January 1945 at Meschede, Germany. Fifteen people were killed, and 50 others were wounded. [2359]

Four people died and one was wounded when an engineering train was attacked by

---

[2354] *The Lethbridge Herald*, Lethbridge, Alberta, Canada, 11th January 1945
[2355] http://www.geschichtspolitik.de/Alliierte_Tieffliegerangriffe.pdf
[2356] Ibid.
[2357] https://de.wikipedia.org/wiki/Liste_von_Anschl%C3%A4gen_im_Schienenverkehr#1931_bis_1940
[2358] Ibid.
[2359] Ibid.

Allied aircraft in Reckenfeld station, Germany, on Monday 29th January 1945. [2360]

Many people were killed and wounded when trains were attacked by Allied aircraft at unknown dates in January 1945 at Natrup-Hagen, Germany. [2361]

A narrow-gauge train was attacked by Allied aircraft at Nedderaverbergen station, Germany, on Wednesday 14th February 1945. Two women who had been evacuated from Hamburg, Irma Henni Hein, née Lottman, and Anna Martha Stender, née Leimann, were killed and many others were wounded, including children on their way home from school in Verden. [2362] The same day 30 people died when a passenger train was attacked at Königsbronn station Germany. They were buried in a common grave. [2363]

A munitions train was attacked by an Allied aircraft in Garßen station, Germany on Thursday 15th February 1945 and exploded, killing a German soldier and the pilot of the attacking aircraft.[2364]

A passenger train bound for Attendorn was attacked by Allied aircraft in that station at Heggen, Germany, on Friday 16th February 1945. Ten people were killed and 44 wounded, four fatally. On the same day many died in an attack on a passenger train at Ahauser Stausse, Germany.

On Saturday 17th February 1945 Allied aircraft attacked a passenger train at Bad Camberg, Germany. Twenty-two people were killed and many wounded. In another attack that day a passenger named Johan Ebert was killed and four other passengers were wounded in an attack on a train of the Halle-Hettstedter Eisenbahn between Halle Klaustor and Hettstedt, Germany. [2365]

A goods train hauling tank waggons was attacked by Allied aircraft on Sunday 18th February 1945 and some of its waggons were set on fire at Edikofen, Germany. [2366]

Seven died in an air attack on a passenger train at Dischingen, Germany, on Monday 19th February 1945. Another train was attacked that day at Wienhausen, Germany, between Celle and Gifhorn, killing passengers Margarethe Homann and Robert Riedel and Conductor Karl Niemann. [2367]

[2360] Ibid.
[2361] Ibid.
[2362] Ibid.
[2363] Ibid.
[2364] Ibid.
[2365] Ibid.
[2366] Ibid.
[2367] Ibid.

On Tuesday 20th February 1945 a passenger train was attacked by Allied aircraft on the line between Klachau and Bad Mitterndorf, Germany. Passenger Frieda Kienberger was killed, one person was seriously wounded and five slightly hurt. The same day at Unterriedenberg, Germany, a transport train was attacked. Two men (the driver and fireman?) were seriously injured, and the engine was badly damaged. In a third attack at Rangendingen, Germany, on the Hohenzollerischen Landesbahn, two people died and three were badly wounded when the train was machine-gunned. [2368]

Crowded express train D 90 from Nordhausen to Sangerhausen was attacked by Allied aircraft between Berga-Kulbra and Auleben at Kurieth, Germany, on Wednesday 21st February 1945. Forty-one died and 80 were injured. In another attack that day a troop train carrying French members of the Waffen-SS was attacked in Visselhövede station, Germany, killing six and injuring six more. Another troop train was attacked at Wilmersdorf, Germany. [2369]

At least 12 died and 24 were seriously wounded when a passenger train was attacked by Allied aircraft between Künzing and Pleinting, Germany, on Thursday 22nd February 1945. The same day a refugee train was attacked between Kirchlinteln and Salzwedel, Germany, killing 20 and wounding over 50 more. [2370]

A munitions train was attacked by Allied aircraft at Titisee station, Germany, and exploded on Saturday 24th February 1945. Twenty-seven people died. [2371]

At 16:00 on Monday 26th February 1945. two low-flying British aircraft attacked the station at Hohenbrugg an der Raab, Austria, Germany, and a *Schanzarbeiterzug* (field workers' train?) approaching the place from Jennersdorf. The engine of the train was destroyed and its carriages badly damaged. Sixteen people, all from the local Feldbach area and all between the ages of 16 and 20, were killed in the attack and 20 badly wounded. [2372]

Another train on the Hohenzollerischen Landesbahn was attacked by Allied aircraft at Jungnau, Germany on Tuesday 27th February 1945. Thirty-two people died and eight were injured. Among those who lost their lives were a farmer from Neufra who was on his way with his children to Sigmaringen Hospital where his wife was giving birth. His children survived.

[2368] Ibid.
[2369] Ibid.
[2370] Ibid.
[2371] Ibid.
[2372] https://www.landesarchiv.steiermark.at/cms/dokumente/11683563/3ccacc24/69%20bis%20158%20aus%20Mitteilungen%2038-Der%20Luftkrieg%20in%20der%20Steiermark%201941%20-%201945.pdf

A passenger train entering Spalt station, Germany, was attacked by Allied aircraft on Wednesday 28th February 1945, wounding Chief Engine Driver Karl Amnon, a conductor, and a tax inspector (?). The same day a passenger train from Lübeck to Dassow was attacked at Herrenberg, Germany. Two were killed and three wounded. [2373]

On Thursday 1st March 1945 a refugee train was attacked by Allied aircraft at Altenschwand, Germany. Twenty-eight people died and over 70 were wounded. [2374]

A passenger train was attacked by Allied aircraft at Niedenstein, Germany, on Tuesday 13th March 1945. Seven people died and many were wounded. [2375]

Two troop trains were attacked by Allied aircraft in Niederselters station, Germany on Sunday 18th March 1945. One soldier from Konstanz was killed. Also on 18th March, a work train using forced labour was attacked in the station at Großen Buseck, Germany, and four of the labourers were killed.

A narrow-gauge train was attacked between Engelskirchen and Blumenau, Germany, by Allied aircraft on Monday 19th March 1945. Thirty-one people died and others were wounded. [2376]

A fully loaded munitions train of 28 waggons was attacked on Tuesday 20th March 1945 in Willingen station, Germany. [2377]

On Wednesday 21st March 1945 a train carrying 600 children from Recklinghausen was attacked by Allied aircraft at Winterberg, Germany. Four schoolchildren and two teachers were killed and 16 were seriously wounded. [2378]

A passenger train of the Oberrheinischen Eisenbahngesellschaft was attacked by Allied aircraft at Dossenheim, Germany, on Friday 23rd March 1945. Many passengers were killed and wounded. [2379]

On Saturday 24th March 1945 a train carrying Russian prisoners of war was attacked by an Allied aircraft in the station at Lauda, Germany. Forty-three were killed. [2380]

A passenger train was attacked by Allied aircraft at Plettenberg-Eiringhausen station,

---

2373 http://www.geschichtspolitik.de/Alliierte_Tieffliegerangriffe.pdf
2374 Ibid.
2375 Ibid.
2376 Ibid.
2377 Ibid
2378 Ibid.
2379 Ibid.
2380 Ibid.

Germany, on Sunday 25th March 1945. Eleven died and 60 were seriously injured. Also on Sunday 25th, an express train going towards Tübingen was attacked at Bempfingen, Germany. Six died and many were injured in this attack. [2381]

A midday train was attacked by Allied aircraft at Winterberg, Germany, on Friday 30th March 1945. An unknown number of people were killed and wounded. [2382]

A number of trains were attacked by Allied aircraft at different points in Germany on Saturday 31st March 1945. At Görsdorf on the Werratalbahn a military train was targeted, and munitions waggons exploded. There were numerous dead and wounded. Twenty-six died at Wilhelmsglück when a passenger train was attacked. [2383]

A munitions train was attacked at Waldkappel. It exploded and 17 people died. [2384] Parts of the train were thrown between four and five kilometres and 138 houses, barns and stables were destroyed.

At Mühlfeld a passenger train carrying a good number of soldiers was attacked. Fifteen soldiers on the train were killed, as were another 12 who were fired on as they fled to the station area and neighbouring meadows. The wife of a teacher, riding on the train with 31 children, was also killed. [2385]

Many people died in Allied air attacks on trains at St Peter station, Germany, in March 1945.

Eight people died and ten were seriously injured when Allied aircraft attacked a freight train at Oberkochen station, Germany on Sunday 1st April 1945. A munitions train was attacked the same day at Zapfendorf, Germany, and exploded. The community of Zapfendorf was reduced to 'rubble and ashes.'

A hospital train was attacked by Allied aircraft at Biberach, Germany, on Monday 2nd April 1945. Fourteen patients died. [2386]

Allied aircraft made at least three attacks on trains in Germany on Sunday 8th April 1945. The two locomotives of a refugee train were shot up at Kalwang, Austria, Germany. A passenger train was attacked between Leißling and Goseck, killing two civilians and injuring possibly seven (some servicemen were also victims?). In the third attack 13 people died on a passenger train at Moorwinkelsdamm. [2387]

Two passengers were fatally wounded when a train was attacked by Allied aircraft at

---

[2381] Ibid.
[2382] Ibid.
[2383] Ibid.
[2384] Wikipedia – Liste von Eisenbahnunfällen in Deutschland
[2385] http://www.geschichtspolitik.de/Alliierte_Tieffliegerangriffe.pdf
[2386] Ibid.
[2387] Ibid.

Worphausen, Germany on Tuesday 10th April 1945. This was the least serious of three attacks on trains in the country that day. Thirty-eight people died in an attack on a local train on the Weldenbahn at Aystetter, in the Augsburg district. The third attack occurred between Schwandorf and Irrenlohe. Forty-one were killed outright and 28 of the seriously injured succumbed later in hospital. [2388]

A passenger train from Halle was attacked by Allied aircraft at Aschersleben, Germany, on Wednesday 11th April 1945, and 'completely destroyed.' Casualties unknown. [2389]

Allied aircraft attacked a passenger train in Kühren station, Germany, on Saturday 14th April 1945. Twenty-three people died. [2390]

Several people died in an attack on a train at Neumarkt, Germany, on Sunday 15th April 1945, and seven in an attack at Freiburg in Sachsen (Saxony), Germany.

A number of trains were attacked by Allied aircraft in Germany on Monday 16th April 1945. One of them was entering the station at Zöblitz-Pobershau in Sachsen; eight people died and 72 were wounded. Other trains were attacked in Sachsen at Lößnitzal (two died), Dahlen and Frauenstein. Fourteen soldiers and civilians died in an attack on a train between Eger and Wiesau. The most serious attack was at Weiden station in Bayern (Bavaria) where what described as a goods train carrying explosives (so basically a munitions train) blew up. Nine houses nearby were wrecked, 60 people were killed and 225 wounded. [2391]

Soviet aircraft attacked two hospital train carrying wounded soldiers from the Oder-Neisse front while they were in Senftenberg station on Wednesday 18th April 1945. Eighteen soldiers were killed.

An attack on a passenger train by Allied aircraft left at least nine dead and many wounded between Wommen and Herleshausen, Germany on Thursday 19th April 1945. According to some accounts at least 35 died. [2392]

By the beginning of 1944 Allied bombing raids on aircraft factories had seriously affected the output of fighter aircraft. In response the decision was made to construct three massive concrete bunkers, safe from air attack, near Landsberg am Lech in Bayern (Bavaria). In the absence of other sources of labour Jews from Dachau concentration camp were forced to build the bunkers and new subcamps were opened at Kaufering, where the prisoners were kept in extremely harsh conditions, with very little food, in huts

[2388] Ibid.
[2389] Ibid.
[2390] Ibid.
[2391] Ibid.
[2392] Ibid.

infested with vermin. When the area was about to be overrun by advancing Allied troops the Nazis decided to move the remaining prisoners further south, either on foot or by train. [2393] Ironically during this evacuation there were at least two instances of trains being attacked by Allied aircraft, with great loss of life amongst the prisoners.

On Wednesday 25th April 1945 the United States Army was nearing Kaufering Camp IV (there were 11 sub-camps of Dachau in all) at Herlach, Germany. That evening the Jewish prisoners were ordered to assemble on an embankment 600 metres from the camp to await a train to take them forward. They waited on the grass in the freezing cold of the night from 17:00 until 09:00 the next morning, Thursday 26th, when finally, an engine pulling four goods waggons and a flak waggon arrived from Lagerlechfeld. About a quarter hour later, while the sick prisoners were still boarding the train, it was attacked by two hunter-bombers which first fired tracer bullets at it and then dropped bombs. Only about 30 prisoners appear to have escaped the burning train and fled to adjoining woods, while about 200 others are stated to have died. A local farmer related years later that he saw a second burning train (of prisoners?) further north, but this information could not be confirmed. [2394]

The following day, Thursday 27th April 1945, the Herlach tragedy was repeated at Schwabhausen, Germany. At 08:00 on the morning of that day a Luftwaffe anti-aircraft and catering train was stabled on a track there. This train was spotted by an American observation aircraft at 09:30. Aware of this, the Nazis stabled a train containing about 3,500 prisoners, 95 per cent of them Jews, alongside it as protection. Informed of the location of the Luftwaffe train American fighter bombers appeared at about 10:00. Obersturmführer Müller, in charge of the train, refused to allow any of the prisoners to leave the train, stating, “Anyone who tries to leave the train will be shot, even during an air raid.” At about 10:15 the aircraft began firing at the locomotive and leading vehicles of the train. At this the SS guards fled into the neighbouring forest and the prisoners, the ones who were fit enough to do so, followed suit. The attack was over in ten minutes, leaving, according to an eyewitness, 136 dead and 80 seriously wounded (another source gives about 170 dead – possibly this includes the wounded who died later). Dr Grinberg (or Grünberg), a prisoner who survived the attack, said that it had been impossible to treat those wounded in the attack because of lack of resources, while when they had been sure that the attack had ended, the SS had reappeared. The sick and wounded had been left to suffer the rest of the day without bread or water and in the evening the train had resumed its journey. [2395] [2396]

A transport train was attacked by Allied aircraft at Poing, Germany, on Saturday 28th April 1945, killing or injuring 200 prisoners of war. The same day the midday train from

[2393] Wikipedia – Kaufering concentration camp complex
[2394] *Katastrophen der deutschen Bahnen 1945-1992 Teil 1* by Hans-Joachim Ritzau, pg. 13 (in German)
[2395] Ibid.
[2396] https://www-sotavent-de.translate.goog/judengr%C3%A4ber-in-schwabhausen/beschie%C3%9Fung-des-transportzuges-mit-schwerkranken-juden/?_x_tr_sl=de&_x_tr_tl=en&_x_tr_hl=en&_x_tr_pto=nui,sc

Straubing to Neufahrn was attacked at Sallach, Germany. Five people were killed, six seriously and seven slightly wounded. [2397]

United States 14th Air Force Thunderbolt aircraft attacked a Japanese troop train on Tuesday 10th July 1945 between Peiping and Hankow, China[2398], killing an estimated 300 soldiers.

Just a week before the end of the war in Asia, and between the dropping of the atom bombs on Hiroshima and Nagasaki, American aircraft machine-gunned two passenger trains on the Nishitetsu – Omuta line near Nishitetsu Chikushi station on Wednesday 8th August 1945. At about noon the first Down train was arriving from Fukuoka with about 200 passengers. It was considerably delayed, ironically due to air-raid warnings, when it was attacked. A surviving passenger who tumbled out onto the station platform said, "Bodies were piled on top of bodies in carriages that were awash with blood." He counted over 100 bullet holes in the walls of the carriages. Fifty-six people were reportedly killed and over 100 wounded in an attack which only lasted about five minutes.

In the opposite direction an Up passenger train from Kurume had left Tsuko station and was between there and Nishitetsu Chikushi when it too was attacked. Eight people were reportedly killed, and others wounded.

The number of victims of the attacks is disputed. Officially there were only 20 passengers on the Up train but a passenger on it stated that every seat had been taken and there had been possibly between 120 and 130 passengers in its two cars. This witness said that the plane had been so low that it was possible to see the pilot's face and they could not believe that only eight passengers had died. It has been concluded that the figure of 64 is possibly the number of people who were killed instantly and that the total number of deaths might have been 100.

The attacking aircraft were apparently P-51 Mustangs. [2399]

Many unexploded bombs remained in much of Western Europe after the end of the Second World War and some are still being discovered in 2021. On Thursday 29th November 1945, one which had been dropped by the Allies on Germany exploded beneath a passenger train at Burkau, on the line between Kamenz and Bischofswerda. As a result the engine was damaged and the leading van was wrecked, six railwaymen riding in it being killed. [2400] [2401]

[2397] http://www.geschichtspolitik.de/Alliierte_Tieffliegerangriffe.pdf

[2398] *The Cedar Rapids Gazette*, Cedar Rapids, Iowa, United States, 10th July 1945

[2399] https://japan-forward.com/75-years-more-than-100-deaths-later-attack-on-fukuoka-trains-still-raises-questions/

[2400] *Katastrophen der deutschen Bahnen 1945-1992 Teil 1* by Hans-Joachim Ritzau, pg. 284 (in German)

[2401] https://de.wikipedia.org/wiki/Bahnstrecke_Kamenz%E2%80%93Bischofswerda

INDEX

TABLE OF MAJOR RAILWAY ACCIDENTS 1931-1945

Date	Location	Railway Company	Description	Passenger Fatalities
12 Jan 1931	Kochi, Japan		Derailment, excessive speed through crossover, train fell from bridge.	7
19 Jan 1931	Gdynia, Poland		Collision	5
6 Mar 1931	Apasco, Mexico		Derailment, track defects	7
19 Mar 1931	Etampes, France		Derailment Points moved under train?	10
20 Apr 1931	Near Taipo, Hong-Kong		Derailment, track subsidence	12
29 Apr 1931	Benha, Egypt		Fire	48
9 May 1931	Near Pendennis, KS, USA	Missouri Pacific	Derailment, hot-box, broken journal?	6
27 May 1931	Leushkovskaya, USSR		Derailment, inadequate protection of PW work	6
20 Jul 1931	Near Vranya, Yugoslavia		Explosion, sabotage, part of train fell from bridge	?
7 Aug 1931	Near Bialystock, Poland		Rear collision	5
16 Aug 1931	Henterberg-Goss, Austria		Rear collision	14
c19 Aug 1931	Near Leningrad, USSR		Collision	7
12 Sep 1931	Near Govan, SK, Canada	Canadian Pacific	Derailment, broken rail	5
13 Sep 1931	Near Biatorbagy, Hungary		Derailment, sabotage, part of train fell from bridge	23
19 Sep 1931	'South-west of Leningrad,' USSR		Derailment, sabotage	?

26 Sep 1931	Joayangho, Manchuria	Peking and Mukden	Derailment, sabotage	5
2 Jan 1932	Kosino, USSR		Double collision	68
17 Jan 1932	St-Just-en-Chaussée, France		Derailment, broken axle? Track defects?	11
5 Feb 1932	Near Chapei, China		Train bombed by Japanese aircraft	?
Feb 1932?	Tosna, USSR		?	6
18 Mar 1932	Napoli, Italy		Rear collision in tunnel	6
12 Apr 1932	Near Harbin, Manchuria		Derailment, sabotage	14
15 May 1932	Zelonaya, USSR		Collision	?
23 May 1932	212 Km east of Harbin, Manchukuo	Chinese Eastern	Derailment, rear collision	40
30 Jun 1932	Elk Creek Siding, OH, USA	Cincinnati and Lake Erie	Head-on collision, single line	7
2 Jul 1932	Near Vlašim, Czechoslovakia		Head-on collision, single line	7
9 Jul 1932	Between Polati and Beylik Keupru, Turkey		Derailment	19
11 Sep 1932	Near Suangcheng, Manchukuo	Chinese Eastern	Derailment, sabotage	12
11 Sep 1932	Near Chengkeotze, Manchukuo	Chinese Eastern	Derailment, sabotage	?
12 Sep 1932	Near Suangcheng, Manchukuo	Chinese Eastern	Derailment, sabotage	6
14 Sep 1932	Turenne, Algeria		Derailment, track subsidence	56
16 Oct 1932	Lublino, USSR		Collision	36
16 Oct 1932	Cérences, France		Collision with shunting goods	8
17 Oct 1932	Near Temishoara, Romania		Derailment, defective points?	24
c25 Nov 1932	Kirin, Manchukuo		Derailment, sabotage?	11
10 Jan 1933	Near Bucharest Nord, Romania		Rear collision in blizzard	8

13 Feb 1933	Between Moscow and Kazan, USSR		Collision	?
16 Mar 1933	Between Chengchiatun and Ssupingkam, Manchukuo		Rear collision	50
31 Mar 1933	Mataoshi Tunnel, Manchukuo		Derailment in tunnel, sabotage	?
10 Apr 1933	Near Ventaquemada, Columbia	Ferrocarril de Nordeste	Derailment, traction shock? Track defects?	21
19 Apr 1933	Between Tunhua and Kirw, Manchukuo		Derailment, sabotage	11
29 Apr 1933	Sir John's Run, WV, USA	Baltimore and Ohio	Derailment, defective vehicle	5
2 May 1933	Between Dumra and Birhee, Bihar, India	East Indian	Derailment, sabotage	5
25 May 1933	Between Wimbledon and Raynes Park, England	Southern	Derailment, inadequate protection of PW work, collision	5
4 Jun 1933	Les Blottereaux, France	Chemin de fer de Paris à Orléans	Derailment, excessive speed through crossover, traction shock	14
9 Jun 1933	Near Sazak, Turkey		Derailment, track washed away	7
21 Jun 1933	Near Dupnitza, Bulgarie		Collision in tunnel	15
11 Jul 1933	Between Sivapur and Tarikere, India	Mysore	Runaway vehicles, collision	17
1 Aug 1933	Between Leningrad and Moscow, USSR		Collision	15
21 Aug 1933	Between Kiukiang and Nanchang, China		Derailment, train fell from bridge	200
29 Aug 1933	Near Hargis, NM, USA	Southern Pacific	Bridge collapsed under train	8
5 Sep 1933	Binghampton, NY, USA	Erie	Rear collision	14
24 Oct 1933	Between Conches-en-Ouche and La Bonneville-sur-Iton, France	Chemins de fer de l'Etat	Derailment, track defects, high speed with unstable engine, part of train fell from	37

			bridge	
9 Dec 1933	Between Puig and Puzol, Spain		Explosion, train fell from bridge, sabotage	28
10 Dec 1933	Trujillo, Peru		Derailment	8
14 Dec 1933	Near Tsitsehar, Manchukuo		Derailment, sabotage	14
15 Dec 1933	Poznan, Poland		Rear collision in fog	10
23 Dec 1933	Pomponne, near Lagny, France	Chemins de fer de l'Est,	Rear collision in fog	201
7 Jan 1934	Near Donnybrook, South Africa		Derailment	5
c9 Jan 1934	Near Stalingrad, USSR		Rear collision?	?
18 Jan 1934	Between Harlin and Manchuli, Manchukuo		Derailment, fire, sabotage?	9
18 Feb 1934	Near Populonia, Italy		Head-on collision, single line, fire	13
20 Feb 1934	Near Villaneuva de la Reina, Spain		Head-on collision, single line, fire	9
26 Feb 1934	Pittsburgh, PA, USA	Pennsylvania	Derailment, excessive speed	9
4 Mar 1934	Near Moscow, USSR		Rear collision	19
12 Mar 1934	Tavatui, USSR		Collision with shunting goods	33
17 May 1934	Pforzheim, Germany		Sidelong collision	7
21 May 1934	Near San Baudilio de Llobregat, Spain		Head-on collision, single line	9
26 May 1934	Santunghotze Bridge, Manchukuo	North Manchurian	Train derailed and fired on by bandits.	12
7 Jun 1934	El Salvador		Derailment, track washed away?	?
9 Jul 1934	La Vergne, TN, USA	Nashville Chattanooga and St Louis	Derailment, broken wheel	8
10 Jul 1934	Near Morrisville, NC, USA	Southern	Derailment, fire, explosion	10
9 Aug 1934	Verba, Bulgaria		Derailment	5

12 Aug 1934	Avignon, France	Chemins de fer de Paris à Lyon et à la Méditeranée	Derailment, excessive speed? Traction shock?	8
23 Aug 1934	Between Hsiaochiu and Mifeng, Manchukuo	North Manchurian	Derailment, sabotage	?
30 Aug 1934	Between Wuchia and Shwangcheng, Manchukuo	North Manchurian	Train derailed and fired on by bandits	10
6 Sep 1934	Port Eglington Junction, Scotland	London Midland and Scottish	Head-on collision	6
20 Sep 1934	Between Kesinga and Rupra Road, India	Bengal-Nagpur	Derailment, track washed away	13
21 Sep 1934	Between Kusatsu and Ishiyama, Japan		Train blown over by the wind	16
28 Sep 1934	Winwick Junction, England	London Midland and Scottish	Rear collision	10
2 Oct 1934	Krzeszowice, Poland		Rear collision	10
6 Oct 1934	Manchukuo		Derailment, sabotage	?
28 Nov 1934	Near Eremo, Italy	Vesuvius	Runaway train, derailment (rack railway)	7
22 Dec 1934	Schlemisweiler, Germany		Head-on collision, single line	7
25 Dec 1934	Dundas, ON, Canada	Canadian National	Train diverted onto wrong line, collision	15
27 Dec 1934	Powellton, WV, USA	Elkhorn Piney Coal Co	Boiler explosion	14
6 Jan 1935	Near Torbino, USSR		Rear collision, fire	23
8 Jan 1935	Between Koyshug and Kayal, USSR		Double collision?	6
31 Jan 1935	Abu-Zabal, Egypt	Egyptian State	Train diverted onto wrong line, collision	6
8 Feb 1935	Karian-Stroganovo, USSR		Head-on collision, fire	18

6 Mar 1935	Saint-Loubès, France	Chemins de fer de Paris à Orléans et du Midi	Derailment, broken tyre	5
2 Apr 1935	Near Chengching, Manchukuo		Head-on collision, single line?	7
16 Apr 1935	Near Pogronichnaya, Manchukuo	North Manchurian	Collision	?
2 May 1935	Near Tashihtoutzu, Manchukuo		Derailment, sabotage	5
c12 Jun 1935	Near Blagovetchens, USSR		Collision	5
15 Jun 1935	Welwyn Garden City, England	London and North Eastern	Rear collision	13
21 Jun 1935	Near Deodoro, Brazil		Collision in fog	6
c30 Jun 1935	Near Omsk, USSR		Rear collision, fire	6
29 Jul 1935	Between Yingchengtze and Tumenling, Manchukuo		Train derailed and fired on by bandits	10
10 Aug 1935	Near Springfield, OH, USA	Cincinnati and Lake Erie	Head-on collision, single line	6
9 Sep 1935	Noventana, Italy		Head-on collision	5
23 Sep 1935	Near Weihuling, Manchukuo		Derailment, sabotage	11
16 Oct 1935	São Francisco Xavier, Brazil		Rear collision	9
27 Sep 1935	Near Kawamaye, Japan		Derailment, train ran into landslide	9
24 Dec 1935	Großheringen, Germany	Deutsche Reichsbahn	Sidelong collision	33
28 Jan 1936	'East of Yamagata,' Japan		Derailment, train ran into or struck by avalanche	23
11 Feb 1936	Near Asparuchovo, Bulgaria		Collision	18
6 Apr 1936	Near Paso del Macho, Mexico		Explosion, sabotage, fire, part of train fell from bridge	18
10 Jun 1936	Near Bufola, Italy		Derailment	15
22 Jun 1936	Karymskaya, Russia		Rear collision	51

23 Jun 1936	Tunnel No. 27, near Fragas, Spain		Head-on collision, single line, in tunnel	12
1 Jul 1936	Near Posen, Poland		Derailment	10
3 Aug 1936	Between Valladolid and Zaragoza, Spain		Fascist train attacked by government aircraft, fire	?
10 Sep 1936	Near Klondike, MO, USA	Missouri Kansas Texas	Derailment, broken wheel? Defective vehicle? Fire	5
14 Sep 1936	Near Taimakou, Manchukuo		Train attacked and fired on by bandits	25
23 Sep 1936	Near Lourdes, France	Chemins de fer de Paris à Orléans et du Midi	Rear collision	16
1 Oct 1936	Lamberg, Poland		Collision	20
3 Oct 1936	Between Rieti and Contigliano, Italy		Head-on collision, single line	13
8 Oct 1936	Aranjuez, Spain		Government troop train bombed by Fascist aircraft	?
10 Oct 1936	Near Alban, Columbia	Cundinamarca	Runaway coaches, derailment, sabotage?	40
22 Oct 1936	Between Hengyang and Shaoguan, China		Runaway vehicles, collision	60
26 Oct 1936	Foscani, Romania		Derailment	5
23 Nov 1936	Trnovce, Czechoslovakia		Derailment?	8
24 Nov 1936	Granville Avenue, Chicago, IL, USA	Chicago Rapid Transit	Rear collision	11
21 Dec 1936	Vlakteplaats, South Africa		Derailment	13
10 Jan 1937	La Volcana, Colombia	Ferrocarril de Antioquia	Rear collision	14
15 Jan 1937	Myslowice, Poland		Head-on collision	8
16 Jan 1937	Near Shihtan, China		Fire, arson?	112
9 Feb 1937	Near Cordoba, Spain		Derailment, sabotage	?
13 Mar 1937	Corquoy, near Châteauneuf-sur-	Chemins de fer de Paris à Orléans et	Collision with tree, derailment	13

	Cher, France,	du Midi,		
19 Mar 1937	Siberia, USSR		Derailment	14
25 Mar 1937	Montoro and Pedro Abad, Spain		Fascist troop train bombed by Government aircraft	?
2 Apr 1937	Battersea Park, England	Southern	Rear collision	9
12 Apr 1937	Dzerzinsky, USSR		Head-on collision	6
13 Apr 1937	Sestao, Spain		Derailment	6
11 Jun 1937	Between Hirschberg and Schmiedeberg, Germany		Runaway waggons, collision	5
28 Jun 1937	Barón, Valparaiso, Chile	Ferrocarriles del Estado	Derailment on points, collision	6
10 Jul 1937	Near Le Mans, France	Chemins de fer de l'Etat	Rear collision	9
17 Jul 1937	Near Bihta, Bihar, India	East Indian	Derailment, high speed with unstable engine	106
22 Jul 1937	Near Hsinking, Manchukuo		Derailment, sabotage	8
29 Jul 1937	Okayama, Japan		Rear collision	6
29 Jul 1937	Between Chungliangcheng and Tientsin, Manchukuo		Derailment, sabotage	5
29 Jul 1937	Near Villeneuve St Georges, France	Chemins de fer de Paris à Lyon et à la Méditeranée	Derailment, points moved under train, fire	29
30 Jul 1937	Near Paotingfu, China		Train bombed by Japanese aircraft	20
31 Jul 1937	Near Makwa, ON, Canada	Canadian National	Derailment, broken rail?	8
3 Aug 1937	Near Yulin, China		Chinese troop train bombed by Japanese aircraft	?
3 Aug 1937	Hsiahuayuen, China		Chinese troop train bombed by Japanese aircraft	?
3 Aug 1937	Near Hsinpaoan, China		Chinese troop train bombed by Japanese aircraft	?
5 Sep 1937	Holzheim, Germany		Derailment, track defects	17

8 Sep 1937	Sungkiang, China		Chinese train bombed by Japanese aircraft	300
15 Sep 1937	Near Chochow, China		Chinese troop trains bombed by Japanese aircraft	?
20 Sep 1937	Vars, France	Chemins de fer de Paris à Orléans et du Midi	Rear collision	8
22 Sep 1937	Ciucea, Romania		Collision	12
10 Oct 1937	Between Shihchiachwang and Shunteh, China		Chinese troop trains attacked by Japanese aircraft	?
12 Oct 1937	Neikiu, China		Chinese troop trains attacked by Japanese aircraft	?
14 Oct 1937	China		Chinese troop trains attacked by Japanese aircraft	?
25 Oct 1937	Near Tsinghuayuan, China		Head-on collision	?
31 Oct 1937	Barra Mansa, Brazil	Estrada de Ferro Central do Brazil	Derailment, broken axle, part of train fell from bridge	5
2 Nov 1937	Mesquita, Brazil	Estrada de Ferro Central do Brazil	Rear collision	7
3 Nov 1937	Between Pakshektu and Taipingli, China		Chinese train bombed and machine-gunned by Japanese aircraft	25
8 Nov 1937	Sishipho, China		Chinese train bombed by Japanese aircraft	?
19 Nov 1937	Alanis de la Sierra, Spain		Runaway train, collision	68
23 Nov 1937	Kerman, CA, USA	Southern Pacific	Derailment, defective vehicle	6
24 Nov 1937	Near Suancheng, China		Chinese train attacked by Japanese aircraft	?
30 Nov 1937	'South of Tsinan,' China		Chinese train bombed by Japanese aircraft	?
4 Dec 1937	Freginals and Planes de Montsiá, Spain		Collision with waggon, part of train fell from bridge	12
4 Dec 1937	Valencia, Spain		Sidelong collision	7

10 Dec 1937	Castlecary, Scotland	London and North Eastern	Rear collision	35
c16 Dec 1937	Between Tientsin and Tangku, China		Derailment, sabotage? fire	50
25 Dec 1937	Near Tsinan, China		Chinese train attacked by Japanese aircraft	?
27 Dec 1937	Near Moji, Japan		Fire	7
2 Jan 1938	Between Canton and Shaokwan, China		Derailment, track subsidence	?
3 Jan 1938	Between Canton and Wuchung, China		Collision with debris which had fallen from tunnel roof following Japanese air raid	42
10 Jan 1938	Madrid, Spain		Explosion in tunnel	?
30 Jan 1938	Between Zabalata and Cabañas, Columbia	Ferrocarril de Antioquia	Derailment, defective vehicle? Excess speed? Track defects?	12
6 Feb 1938	Near Randonnai, France		Runaway train, head-on collision	5
25 Feb 1938	South of Shihchiachwang, Manchukuo		Derailment, sabotage	?
26 Feb 1938	West Burlington, IA, USA		Rear collision	5
c5 Mar 1938	Near Reus, Spain		Sidelong collision	19
10 Mar 1938	Near Pengpu, China		Japanese troop train bombed by Chinese aeroplane	?
13 Mar 1938	Between Bicocca and Passo Martino, Italy		Head-on collision, single line	9
26 Mar 1938	Near Ratana, New Zealand		Derailment, excessive speed	6
29 Mar 1938	Meliana, Spain		Collision with petrol tanker on open? Level crossing, fire, explosion	21
31 Mar 1938	Near Yunho, China		Chinese troop train bombed by Japanese aircraft	?
3 Apr 1938	Near Sargan, Yugoslavia		Derailment, train ran into rockfall	6
4 Apr 1938	Between Vakaranga and Ramaquebane,		Head-on collision, single line	26

	Southern Rhodesia			
5 Apr 1938	San Guim de Freixenet, Spain		Hospital train bombed by Fascist aircraft	6
15 Apr 1938	Hardwar, India	East Indian	Buffer stop collision	8
17 Apr 1938	Near Orléansville, Algeria		Derailment, sabotage?	5
c27 Apr 1938	Near Suchow, China		Chinese troop trains bombed by Japanese aircraft	?
28 Apr 1938	Between Chuchow and Liuliho, China		Derailment, sabotage	?
13 May 1938	Between Bellevue and K'beu, French Indo China		Runaway train, derailment	15
17 May 1938	Milokang, China		Head-on collision	?
17 May 1938	Between Charing Cross and Temple, England	London Passenger Transport Board	Rear collision in tunnel	6
15 Jun 1938	Near Kumayama, Japan		Derailment, track subsidence, collision	25
19 Jun 1938	Custer Creek Bridge, Saugus, MT, USA	Chicago Milwaukee St Paul and Pacific	Bridge collapsed under train	43
24 Jul 1938	St Trond, Belgium		Derailment	5
30 Jul 1938	Near Balaclava, Jamaica	Jamaica Government	Derailment, excessive speed	27
21 Aug 1938	Between Ayyalur and Vadamadura, India	South India	Derailment, track washed away	31
31 Aug 1938	Sant Vincenç de Calders, Spain		Hospital train bombed by Fascist aircraft	6
8 Sep 1938	Reguiny, France		Derailment	5
10 Sep 1938	Figaredo, Spain		Derailment, track subsidence	10
20 Sep 1938	Tortuga, CA, USA	Southern Pacific	Train diverted onto wrong line, collision	8
25 Sep 1938	Near Martorell, Spain		Head-on collision, single line	65
26 Sep 1938	Borken, Germany		Collision with light engine	13
1 Oct 1938	Near Wushi, China		Train derailed and fired on by guerrillas	?

6 Oct 1938	Between Rauha and Tainionkowski, Finland		Collision in fog, leakage of sulphur dioxide	11
8 Oct 1938	Sant Vincenç de Calders, Spain		Train bombed by Fascist aeroplane	60
15 Oct 1938	'South of Sinyang,' China		Chinese trains bombed by Japanese aircraft	?
15 Oct 1938	Ciudad Real, Spain		Collision	12
31 Oct 1938	Algemesi, Spain		Train bombed by Fascist aeroplane	21
22 Nov 1938	Near San Gottardo, Italy	Società Veneta	Bridge collapsed under train	19
16 Dec 1938	Between Alpera and Bonette, Spain		Head-on collision, single line?	7
19 Dec 1938	Between Joâo Ayres and Sitio, Brazil	Estrada de Ferro Central do Brazil	Head-on collision, single line	59
20 Dec 1938	Xala, Mexico		Derailment	5
24 Dec 1938	Near Etulia, Romania		Head-on collision, single line	150
1 Jan 1939	Castro Urdiales, Spain		Derailment	7
4 Jan 1939	Tarragona, Spain		Trains bombed by Fascist aeroplane	30
14 Jan 1939	Bélmez and Villanueva del Rey, Spain		Collision	11
16 Jan 1939	Albigasta River Bridge, near Frias, Argentina		Bridge collapsed under train	15
7 Feb 1939	Near Kwantien, China		Derailment, sabotage	?
11 Feb 1939	Between Plaza Molino and Gracia, Spain		Runaway train, rear collision, runaway train, collision	32
12 Feb 1939	Xàtiva, Spain		Troop train bombed by Fascist aircraft.	129
12 Feb 1939	León, Nicaragua		Head-on collision	27
13 Feb 1939	Daljosephat, South Africa		Derailment	5
13 Mar 1939	Near Châteauroux, France		Collision with cow, derailment, collision	18

10 Apr 1939	Between Tel Aviv and Lydda, Palestine	Palestine	Explosion, sabotage	5
11 Apr 1939	Near Samshui, China		Derailment, sabotage	?
13 Apr 1938	Near La Llave, Mexico		Rear collision	32
17 Apr 1939	Majdia, Bengal, India	Eastern Bengal	Rear collision	34
18 Apr 1939	'North of Paoting,' China		Derailment, sabotage	?
c19 Apr 1939	Near Vinnycja, Ukraine, USSR		Head-on collision, single line	29
22 May 1939	Campanillas, Spain		Head-on collision	7
5 Jun 1939	Kochi, Japan		Derailment	13
7 Jun 1939	Göschwitz, Germany		Derailment, track distorted by the heat	6
15 Jun 1939	Mittelgrund, Czechoslovakia, Germany		Derailment, excessive speed through crossover	10
28 Jun 1939	Chanpur Siau and Haldaur, United Provinces, India	East Indian	Derailment, culvert washed away	10
12 Jul 1939	Clodul Marc, Romania		Collision	22
16 Jul 1939	Powsin, Poland		Head-on collision, fire	9
29 Jul 1939	Near Wushi, China		Train derailed and fired upon by guerrillas	?
12 Jul 1939	Near Harney, NV, USA	Southern Pacific	Derailment, sabotage	24
8 Aug 1939	Between Puebla and Matamoros, Mexico		Derailment	7
13 Aug 1939	Petroaïa, Romania		Derailment, points moved under train?	15
1 Sep 1939	Near Kutno, Poland		Train machine-gunned by German aircraft	?
2 Sep 1939	Les Aubrais, France		Rear collision	35
5 Sep 1939	Near Szydlowicz, Poland		Red Cross train bombed by German aircraft	?
6 Sep 1939	Warsaw East, Poland		Hospital train bombed by German	?

			aircraft	
8 Sep 1939	Between Metzdorf and Wintersdorf, Germany		Collision	12
11 Sep 1939	Brancourt, France		Rear collision	11
1 Oct 1939	Between Shanghai and Nanking, China		Derailment, sabotage	30
1 Oct 1939	Between El Tio and La Francia, Argentina	Ferrocarriles del Estrada	Rear collision	12
8 Oct 1939	Gesundbrunnen, Germany	Deutsche Reichsbahn	Rear collision, fire	23
12 Oct 1939	Nikko, Japan		Derailment	17
21 Oct 1939	Tolosa, Mexico		Derailment, track defects, fire, explosion	40
26 Oct 1939	St Valentin, Austria, Germany		Derailment, excessive speed	20
30 Oct 1939	Lambrate, Italy	Fellovie dello Stato	Head-on collision	14
30 Oct 1939	Karwitz, Germany		Rear collision?	9
31 Oct 1939	Between Barent and Sonnenwalde, Germany		Head-on collision, single line	5
5 Nov 1939	Between Eger and Pilsen, Czechoslovakia, Germany		Derailment	5
12 Nov 1939	Between Rosengrund and Langlieben, Germany		Head-on collision, single line	43
17 Nov 1939	'North of Chengtingfu,' China		Derailment, sabotage	?
19 Nov 1939	Spandau West, Germany		Collision, driver intoxicated?	9
26 Nov 1939	Niederwöllstadt, Germany		Collision	11
1 Dec 1939	Near Sibiu, Romania		Runaway train, derailment	20
12 Dec 1939	Between Hagen-Vorhalle and Wetter, Germany		Train diverted onto wrong line, head-on collision	15
15 Dec 1939	Vennebeck, Germany		Collision	6
22 Dec 1939	Genthin, Germany	Deutsche Reichsbahn	Rear collision	186
22 Dec 1939	Between Markdorf and Kluftern, Germany	Deutsche Reichsbahn	Head-on collision, single line	97

30 Dec 1939	Torre Annunziata, Italy		Rear collision	29
2 Jan 1940	Between Leningrad and Murmansk, USSR		Derailment, sabotage	?
3 Jan 1940	Matsui, Japan		Head-on collision	10
6 Jan 1940	Near Orly, France		Double collision in fog	7
18 Jan 1940	Near Krakow, Poland		Rear collision?	5
21 Jan 1940	Saint Julien, near Troyes, France		Sidelong collision	7
29 Jan 1940	Ajikawaguchi, Japan		Derailment, points moved under train	181
2 Feb 1940	''About 65 km from the Indo-Chinese border, China		Train bombed by Japanese aircraft	25
12 Feb 1940	Near Pisa, Italy		Derailment, excessive speed	5
25 Feb 1940	Las Adjuntas, Mexico	Ferrocarril Nacional de México	Head-on collision, single line, fire	12
4 Mar 1940	Near Mount Yokone Tunnel, Yamagati, Japan	Japanese Government	Train fell from bridge destroyed by avalanche, fire	107
5 Mar 1940	Iitala, Finland		Head-on collision, fire	21
9 Mar 1940	Near Santa Caterine, Mexico	Ferrocarril Nacional de México	Head-on collision	5
12 Mar 1940	Between Harviala and Turenki, Finland		Head-on collision, single line, fire, explosions	39
17 Mar 1940	Between Zaluka and Ozalj, Yugoslavia		Derailment, train ran into and struck by landslide	12
22 Mar 1940	Suchow, China		Train attacked by guerrillas, collision	?
22 Mar 1940	Between Tientsin and Pukow, China		Derailment, sabotage	?
22 Mar 1940	Near Wenshi, China		Derailment, sabotage	?
22 Mar 1940	Near Magé, Brazil	Estrada de Ferro Central do Brazil	Sidelong collision?	16
19 Apr 1940	Little Falls, NY, USA	New York Central	Derailment, excessive speed, traction shock	28
24 Apr 1940	Telneset, Norway	Norges Statsbaner	Troop train bombed by German aeroplane	6
1 May 1940	Near Lesjaverk, Norway	Norges Statsbaner	Derailment, train fell into bomb crater	11
3 May 1940	Tangen, Norway	Norges Statsbaner	Head-on collision, single line	8
4 May 1940	Between Vallon and Urçay, France		Derailment, track subsidence	33
15 May 1940	Jekot, India	Bombay Baroda and	Collision with	21

		Central India	shunting goods	
18 May 1940	Morgny, France		Rear collision	53
28 May 1940	Between Panchalam and Tindivalam, India	South Indian	Train blown over by the wind	5
29 Jul 1940	Between Beugen and Oeffelt, The Netherlands		Head-on collision, single line	6
31 Jul 1940	Cuyahoga Falls, OH, USA	Pennsylvania	Head-on collision, single line, fire	42
5 Aug 1940	Between Chaudanga and Jairampur, India	Eastern Bengal	Derailment, sabotage	34
4 Sep 1940	Near Sambre, Belgium		Collision	15
17 Sep 1940	Gubbero, Sweden	Statens Järnvägar	Head-on collision, single line	9
12 Oct 1940	Wembley, England	London Midland and Scottish	Derailment, obstruction on line	9
29 Oct 1940	Rheinhausen Ost, Germany		Rear collision?	6
4 Nov 1940	Norton Fitzwarren, England	Great Western	Derailment on trap points	26
14 Nov 1940	Diegem, Belgium		Collision	21
16 Nov 1940	Between Eberswalde and Frankfurt (Oder), Germany		Head-on collision, single line	9
19 Nov 1940	Near Hommelvik, Norway	Norges Statsbaner	Head-on collision, single line	21
29 Nov 1940	Near Suchow, China		Derailment, sabotage	74
3 Dec 1940	Velilla de Ebro, Spain		Head-on collision, single line	53
6 Dec 1940	Bristol Temple Meads, England	Great Western	Train bombed by the Luftwaffe	14
14 Dec 1940	Iguala, Mexico		Runaway train, derailment	25
27 Dec 1940	Near Kyujo, Formosa		Head-on collision, single line?	?
27 Dec 1940	Near Arad, Romania		Collision	6
29 Dec 1940	Kodersdorf, Germany		Double collision	5
30 Dec 1940	Between Nennhausen and Rathenow, Germany		Collision	6

3 Jan 1941	Villers-La-Tour, Belgium		Train diverted onto wrong line, collision	7
8 Jan 1941	Near Nerettyo Ujfalu, Hungary		Collision, explosion	20
22 Jan 1941	Isenbüttel, Germany		Rear collision	103
27 Jan 1941	Between Dreileben and Drackenstedt, Germany		Rear collision?	8
c4 Feb 1941	Near Smyrna, Turkey		Derailment, track washed away	12
10 Feb 1941	Near Dragoman, Bulgaria		Derailment, runaway train? Sabotage?	8
10 Feb 1941	Near Brentwood, England	London and North Eastern	Rear collision, fire	7
c15 Feb 1941	Warsaw, Poland		Head-on collision	10
15 Feb 1941	Near Zumaya, Spain		Train blown off bridge by the wind	26
8 Mar 1941	Near Palos Quemados Tunnel, Chile	Chilean Longitudinal	Derailment, excessive speed?	8
16 Mar 1941	Between Ars and Metz, Germany		Rear collision?	23
c20 Mar 1941	Near Oporiise, Bulgaria		Train fell from bridge	10
27 Mar 1941	Gardelegen, Germany		Sidelong collision in blizzard	7
28 Mar 1941	Near Chienshien, China		Derailment, sabotage	?
6 Apr 1941	Rengershausen, Germany		Double collision	6
23 Apr 1941	Near Kampala, Uganda		Derailment	20
28 Apr 1941	Westborough, England	London and North Eastern	Fire	6
3 May 1941	Güsow, Germany		Sidelong collision	16
3 Jun 1941	Between Peking and Hankow, China		Derailment, sabotage	?
14 Jun 1941	Fellheim, Germany		Train struck by aircraft, fire	5
2 Jul 1941	Dolphin Junction, England	Great Western	Head-on collision	5
20 Jul 1941	Near Presov, Czechoslovakia		Derailment, sabotage	44
24 Jul 1941	Between La Pola de Gordón and La Robla, Spain		Rear collision	12
c25 Jul 1941	Near Warsaw, Poland		Collision	?

c30 Aug 1941	Northern Anhwei, China		Japanese troop train dynamited by Chinese guerrillas	?
30 Aug 1941	Forbach-Gausbach, Germany		Derailment, points moved under train	8
1 Sep 1941	Tablada, Spain		Derailment	8
14 Sep 1941	Holmes Chapel, England	London Midland and Scottish	Rear collision	9
16 Sep 1941	Aboshi, Japan		Rear collision	63
21 Sep 1941	Rabsztyn, Germany		Rear collision	7
23 Sep 1941	Near Keisen, Switzerland	Chemins de fer fédéraux suisses	Head-on collision in fog during temporary single line working	5
1 Oct 1941	Between Kumamoto and Oita, Japan		Bridge collapsed under train	100
8 Oct 1941	Tejares, Spain		Derailment	11
c11 Oct 1941	Cseska Trehova, Czechoslovakia		Collision	15
15 Oct 1941	Niš and Lescovac, Serbia, Yugoslavia		Derailment, sabotage?	5
23 Oct 1941	Near Hangkow, China		Japanese troop train dynamited by Chinese guerrillas	?
27 Oct 1941	Honan, China		Japanese troop train dynamited by Chinese guerrillas	?
29 Oct 1941	Between Berlin-Ost-Kreuz and Grünau, Germany		Rear collision	6
2 Nov 1941	Paris Austerlitz, France	Société national de chemins de fer Français	Sidelong collision	21
3 Nov 1941	Bosnia, Yugoslavia		Derailment, sabotage	7
9 Nov 1941	Near Dunkirk, OH, USA	Pennsylvania	Derailment, obstruction on line (engine defect), fire	11
17 Nov 1941	Eichkogel, Austria, Germany		Collision	5
19 Nov 1941	Near Kami-Nagatoto, Japan		Collision	20
3 Dec 1941	Unterhofen-Solgen, Germany		Sidelong collision	6
17 Dec 1941	Suchsdorf, Germany		Collision	7
20 Dec 1941	Cercola, Italy	Circumvesuviana	Derailment on points	25
c24 Dec 1941	Between Baghdad and Mosul, Iraq		Derailment, sabotage, part of train fell from bridge	?
27 Dec 1941	Leichholz, Germany	Deutsche Reichsbahn	Rear collision in blizzard, explosion, fire	41

28 Dec 1941	Between Lavente and La Gorgue, France		Head-on collision, single line	55
29 Dec 1941	Langenhagen, Germany		Train diverted onto wrong line, collision	27
30 Dec 1941	Eccles, England	London Midland and Scottish	Sidelong collision in fog	23
3 Jan 1942	Banjaluka, Bosnia, Yugoslavia		Train diverted onto wrong line, head-on collision	7
5 Jan 1942	Ortelsburg, Germany		Collision	8
9 Jan 1942	Saint-Denis-Bovesse, Belgium		Rear collision	7
17 Jan 1942	Near Kausala, Finland		Collision	19
30 Jan 1942	Cowlairs East Junction, Scotland	London and North Eastern	Collision with light engine	12
3 Feb 1942	Belgrade and Niš, Yugoslavia		Derailment, sabotage	260
3 Feb 1942	Liégnitz, Germany		Fire, explosion	20
6 Feb 1942	Near Lawang, Java, Dutch East Indies	Nederlandsch Indische Spoorweg Maatschappij	Head-on collision, single line	26
11 Feb 1942	Near Beighton	London and North Eastern	Train struck by load projecting from waggon on adjoining siding	14
12 Feb 1942	Tyringe, Sweden		Train diverted onto wrong line, collision	5
14 Feb 1942	Between Gaswitz and Meuselwitz, Germany		Derailment, train ran into a snowdrift, collision	6
17 Feb 1942	Between Neudietendorf and Ritschenhausen, Germany		Collision?	8
27 Feb 1942	Södertälje Central, Sweden		Collision, fire	5
28 Feb 1942	Between May Pen and Sevens Siding, Jamaica		Head-on collision, single line	14
c21 Mar 1942	Near Leningrad, USSR		Derailment sabotage	200
29 Mar 1942	Between Herral and Drama, Greece		German troop train derailed and machine-gunned by Greek partisans	100
March 1942	Near Craiova, Romania		Derailment, sabotage?	72
March 1942	Italy		Derailment,	135

			sabotage?	
16 Apr 1942	Airan, France		German troop train dynamited by the French resistance	288
26 Apr 1942	Exchange Place, NJ, USA	Hudson and Manhattan	Derailment, excessive speed, motorman intoxicated?	5
Apr 1942	Near Sho Webo, Burma	Burma	Troop train bombed by the Japanese	?
Apr 1942	Near Naba Junction, Burma	Burma	Runaway coaches, Derailment	?
Apr 1942	Greece		Derailment, sabotage	250
3 May 1942	Near Vire, France		Derailment, sabotage	36
c5 May 1942	Near Amiens, France		Collision, sabotage?	56
c10 May 1942	Between Zaandem and Amsterdam, The Netherlands		Derailment, sabotage	?
c14 May 1942	USSR		German troop train attacked by Russian partisans	300
c14 May 1942	USSR		German troop train attacked by Russian partisans	320
16 May 1942	Between Oderolal and Tando Adam, India		Derailment, sabotage	24
c16 May 1942	Yugoslavia		Derailment, sabotage	150
18 May 1942	Between Kehl and Straßburg, Germany		Head-on collision during temporary single line working?	11
May 1942	Drama, Greece		Derailment, sabotage	?
May 1942	Near Larissa, Greece		Train derailed and fired on by partisans	?
2 Jun 1942	'North of Hsuchowfu,' China		Derailment	300
14 Jun 1942	Kittrell, NC, USA	Seaboard Air Line	Rear collision in fog	8
c20 Jun 1942	Plaszow, Poland		Derailment, sabotage	?
24 Jun 1942	Werbig, Germany		Runaway engine, Germany	10
8 Jul 1942	Near Norristown, PA, USA	Lehigh Valley Transit	Head-on collision, single line	11
c24 Jul 1942	Between Paris and Cherbourg, France	Société nationale des chemins de fer français	Derailment, sabotage	40
29 Jul 1942	Near Llay-Llay, Chile		Rear collision	9
13 Aug 1942	La Verdure, Algeria		Train diverted onto wrong line, collision	5
28 Aug 1942	Menars, France	Société nationale des chemins de fer français	Train bombed and machine-gunned by the RAF	26

29 Aug 1942	Man-it, The Philippines		Derailment, sabotage	22
11 Sep 1942	Herstal Belgium		Collision	14
24 Sep 1942	Near Dickerson, MD, USA	Baltimore and Ohio	Double collision in fog, fire	11
28 Sep 1942	Nikolsdorf, Austria, Germany		Derailment, track washed away	19
2 Oct 1942	Daucher, Switzerland		Head-on collision, single line	10
3 Oct 1942	Near Chaligaon, India	Great Indian Peninsular	Derailment, sabotage	13
5 Oct 1942	Divaca, Yugoslavia		Derailment, sabotage	25
8 Oct 1942	Neustadtl, Germany		Collision	11
12 Oct 1942	'80 miles from Calcutta,' India		Derailment, sabotage	5
14 Oct 1942	Near Shahsiatu, China		Derailment, sabotage	?
16 Oct 1942	Stuttgart Bad Cannstatt, Germany		Rear collision	12
3 Nov 1942	Bor, Turkey		Collision	30
16 Nov 1942	Betweem Falconara Albanese and Paola, Italy		Runaway train, derailment (rack railway)	18
17 Nov 1942	Near Hanoi, Vietnam, French Indo-China		Head-on collision	25
c25 Nov 1942	Greece		Derailment, sabotage	?
27 Nov 1942	Nocera Inferiore, Italy		Collision	28
27 Nov 1942	Sliedrecht, The Netherlands		Rear collision	18
12 Dec 1942	Between Bialystock and Wolkowysk, Poland		Derailment, sabotage, fire	21
12 Dec 1942	Between Tito and Baragiano-Ruoti, Italy		Runaway train, derailment	29
16 Dec 1942	Near Bramley and Wonersh, England	Southern	Train machine-gunned by the Luftwaffe	7
17 Dec 1942	Lüneberg, Germany		Rear collision	10
27 Dec 1942	South Belgium		Collision	9
27 Dec 1942	Almonte, ON, Canada	Canadian Pacific	Rear collision	36
4 Jan 1943	Between Hanover and Wunstorf, Germany		Rear collision in blizzard	25
31 Jan 1943	Tantow, Germany		Collision	38

8 Feb 1943	Between Saventhem and Bruxelles, Belgium	Société nationale des chemins de fer belges	Derailment, sabotage	?
20 Feb 1943	Near Chalon sur Saône, France	Société nationale des chemins de fer français	Derailment, sabotage	50
28 Feb 1943	Galkau, Germany		Collision	25
2 Mar 1943	Ashausen, Germany		Collision	6
4 Mar 1943	Muysen, Belgium	Société nationale des chemins de fer belges	Derailment, sabotage	?
c10 Mar 1943	Chagny, France	Société nationale des chemins de fer français	Derailment, sabotage	250
20 Mar 1943	Near Smolensk, USSR		Derailment, sabotage	?
21 Mar 1943	Caltanisseta Xirbi, Italy		Runaway train, collision, explosions	137
22 Mar 1943	Near Smolensk, USSR		Derailment, sabotage	?
26 Mar 1943	Near Smolensk, USSR		Derailment, sabotage	?
26 Mar 1943	Near Smolensk, USSR		Derailment, sabotage	?
14 Apr 1943	Near Mirandela, Portugal	Caminhos de Ferro Portugueses	Explosion	9
c25 Apr 1943	Near Beard, France		Derailment, sabotage	52
c25 Apr 1943	Near Rebecourt, France	Société nationale des chemins de fer français	Derailment, sabotage	?
3 May 1943	Harelbeke, Belgium	Société nationale des chemins de fer belges	Derailment, sabotage	?
16 May 1943	Groenendael, Belgium	Société nationale des chemins de fer belges	Derailment, sabotage	?
18 May 1943	Jaleswar, India	Bengal-Nagpur	Rear collision	14
23 May 1943	Kušatk, Yugoslavia		Derailment, sabotage	140
23 May 1943	Near Minson, NJ, USA	Pennsylvania	Derailment, excessive speed	12
4 Jun 1943	Near Hyde, New Zealand		Derailment, excessive speed, driver intoxicated?	21
4 Jun 1943	Near Akola, India	Great Indian Peninsular	Rear collision	83
c4 Jun 1943	Zhiradovo, Poland		Derailment, sabotage	?
c4 Jun 1943	Zhiradovo, Poland		Derailment, sabotage	?
28 Jun 1943	Near Velikte Luki, USSR		Derailment, sabotage	?
30 Jun 1943	Between Orvieto and Chiusi, Italy		Collision	
30 Jun 1943	Dortmund Hbf,		Collision	28

	Germany			
c1 July 1943	Kamenets-Podolsk, USSR		Derailment, sabotage	?
1 Jul 1943	Near Mauberge, France	Société nationale des chemins de fer français	Derailment, sabotage	30
6 Jul 1943	Between Bezwada and Guntakal, India	Madras and Southern Mahratta	Derailment	20
c9 Jul 1943	Near Latgale, Latvia		Derailment, sabotage	?
10 Jul 1943	Between Rennes and Noyal-Acigné, France	Société nationale des chemins de fer français	Derailment, sabotage	200
12 Jul 1943	Sommery, France	Société nationale des chemins de fer français	Derailment, sabotage	?
12 Jul 1943	Between St Mesmin and Savières, France	Société nationale des chemins de fer français	Derailment, sabotage	?
29 Jul 1943	Near Grancourt, France	Société nationale des chemins de fer français	Derailment, sabotage, collision	130
30 Jul 1943	Vieux Heverlee, Belgium	Société nationale des chemins de fer belges	Derailment, sabotage	285
July 1943	Near Zaječar, Serbia, Yugoslavia		Derailment, sabotage	500
4 Aug 1943	Warden, WA, USA	Chicago Milwaukee St Paul and Pacific	Sidelong collision	10
c14 Aug 1943	Ille-et-Vilaine, France	Société nationale des chemins de fer français	Derailment, sabotage	200
18 Aug 1943	Between Longeville and Montereau, France	Société nationale des chemins de fer français	Derailment, sabotage	?
23 Aug 1943	Near Bengy, France	Société nationale des chemins de fer français	Derailment, sabotage, collision	11
27 Aug 1943	Near Clermont de L'Oise, France	Société nationale des chemins de fer français	Derailment, sabotage	?
30 Aug 1943	Wayland, NY, USA	Delaware Lackawanna and Western	Sidelong collision	26
6 Sep 1943	Shore, PA, USA	Pennsylvania	Derailment, 'hot-box,' broken journal	80
10 Sep 1943	Between Waldbröl and Boxberg, Germany		Train attacked by allied aircraft	6
c15 Sep 1943	Near Louvain, Belgium	Société nationale des chemins de fer belges	Derailment, sabotage	30
15 Sep 1943	Falkenberg (Pom), Germany		Sidelong collision in fog	18

19 Sep 1943	Between Bialystock and Prostken, Poland		Derailment, sabotage	23
19 Sep 1943	Frejus Tunnel, Italy		Collision with portion of tunnel roof destroyed by sabotage	30
19 Sep 1943	Frejus Tunnel, Italy		Collision with portion of tunnel roof destroyed by sabotage	10
19 Sep 1943	Between St Amour and Loubrans, France	Société nationale des chemins de fer français	Train machine-gunned by the RAF	5
c26 Sep 1943	Between Brod and Zagreb, Yugoslavia		Derailment, sabotage	60
28 Sep 1943	Between Paris and Cherbourg, France	Société nationale des chemins de fer français	Derailment, sabotage	?
Sep 1943	Between Bosanico and Novi Prijedor, Yugoslavia		Derailment, sabotage	63
4 Oct 1943	Genly, Belgium	Société nationale des chemins de fer belges	Derailment, sabotage	?
6 Oct 1943	Near Chalon sur Saône, France	Société nationale des chemins de fer français	Derailment, sabotage	39
7 Oct 1943	Ryggkollen, Norway	Norges Statsbaner	Derailment, sabotage	?
9 Oct 1943	Fontaine-sur-Somme, France	Société nationale des chemins de fer français	Derailment, sabotage	55
c15 Oct 1943	Between Landevant and Hennebont, France	Société nationale des chemins de fer français	Derailment, sabotage	20
15 Oct 1943	Between Pedro Martinez and Moreda, Spain		Derailment, broken rail	20
17 Oct 1943	Schüpheim, Switzerland		Derailment, points moved under train	5
22 Oct 1943	Chagny, France	Société nationale des chemins de fer français	Derailment, sabotage	30
27 Oct 1943	Quesnoy, France	Société nationale des chemins de fer français	Derailment, sabotage	180
29 Oct 1943	Tsuchiura, Japan		Train diverted onto wrong line, double collision, part of train fell from bridge	57
31 Oct 1943	Near Tournus, France	Société nationale des chemins de fer français	Derailment, sabotage	11

3 Nov 1943	Magny, France	Société nationale des chemins de fer français	Derailment, sabotage	?
c3 Nov 1943	St Loup, France	Société nationale des chemins de fer français	Derailment, sabotage	?
c5 Nov 1943	Celestynow, Poland		Derailment, sabotage	?
5 Nov 1943	Near Verniski, Bulgaria		Derailment, sabotage	5
6 Nov 1943	Between Chagny and Fontaine, France	Société nationale des chemins de fer français	Derailment, sabotage	53
11 Nov 1943	Northern France	Société nationale des chemins de fer français	Derailment, sabotage	?
11 Nov 1943	Between Chourette and Ribécourt, France	Société nationale des chemins de fer français	Derailment, sabotage	100
14 Nov 1943	Near Sendanur, India	South Indian	Derailment	39
15 Nov 1943	Near Rignano Flamino, Italy	Roma Nord	Head-on collision, single line	100
21 Nov 1943	Between Beaucourt and Albert, France	Société nationale des chemins de fer français	Derailment, sabotage	?
25 Nov 1943	Eastern Bosnia, Yugoslavia		Derailment, sabotage	?
25 Nov 1943	Eastern Bosnia, Yugoslavia		Derailment, sabotage	?
2 Dec 1943	Naumberg Hbf, Germany		Collision	35
2 Dec 1943	Versigny and La Fère, France	Société nationale des chemins de fer français	Derailment, sabotage?	10
3 Dec 1943	Near Lisp, Belgium	Société nationale des chemins de fer belges	Rear collision	24
7 Dec 1943	Fonchercans, France	Société nationale des chemins de fer français	Derailment, sabotage	?
10 Dec 1943	Voreux Goreux, Belgium	Société nationale des chemins de fer belges	Derailment, sabotage	85
10 Dec 1943	Portes les Valence, France	Société nationale des chemins de fer français	Derailment, sabotage	21
12 Dec 1943	Between Laon and Reims, France	Société nationale des chemins de fer français	Derailment, sabotage	?
c13 Dec 1943	Between Doroi and Derventa, Bosnia, Yugoslavia		Derailment, sabotage	30
15 Dec 1943	Near Villevert-Neuville, France	Société nationale des chemins de fer	Derailment, sabotage, fire	6

		français		
16 Dec 1943	Near Rennert, NC, USA	Atlantic Coast Line	Derailment, broken rail, collision	72
18 Dec 1943	Neuhaus, Germany		Sidelong collision	23
20 Dec 1943	Between Praha-Kbely and Praha-Satalice, Czechoslovakia		Head-on collision, single line	42
22 Dec 1943	Die, France	Société nationale des chemins de fer français	Derailment, sabotage, explosion, fire	40
25 Dec 1943	Between Korschen and Lötzen, Germany		Rear collision	15
27 Dec 1943	Between Valenciennes and Quesnoy, France	Société nationale des chemins de fer français	Derailment, sabotage	?
28 Dec 1943	Between Crépy and Compiègne, France	Société nationale des chemins de fer français	Derailment, sabotage	6
3 Jan 1944	Tunnel No. 20, Torre del Bierzo, Spain	Red Nacional de los Ferrocarriles Españoles	Runaway train, double collision in tunnel, fire	500
4 Jan 1944	Between Monteray and Paray-le-Monial, France	Société nationale des chemins de fer français	Derailment, sabotage	?
6 Jan 1944	Warthausen, Germany		Side collision on rail/rail level crossing	12
8 Jan 1944	Givet, France	Société nationale des chemins de fer français	Derailment, sabotage	10
11 Jan 1944	Arévalo, Spain		Rear collision	41
13 Jan 1944	Novice, Texas	Gulf Colorado and Santa Fe	Rear collision	7
16 Jan 1944	Ilford, England	London and North Eastern	Rear collision in fog	9
18 Jan 1944	Müncheberg, Germany		Rear collision	54
19 Jan 1944	Between Ossun and Juillan, France	Société nationale des chemins de fer français	Derailment, sabotage	27
19 Jan 1944	Le Buissonet, France	Société nationale des chemins de fer français	Derailment, sabotage	12
20 Jan 1944	Porta Westfalicia, Germany	Deutsche Reichsbahn	Rear collision, fire	79
20 Jan 1944	Saint-Lié, France	Société nationale des chemins de fer français	Derailment, sabotage	100
21 Jan 1944	Sous-le-Bois, France	Société nationale des chemins de fer français	Derailment, sabotage	?
22 Jan 1944	Between Montceau	Société nationale	Derailment, sabotage	60

	and Blanzy, France	des chemins de fer français		
23 Jan 1944	Lembersat, France	Société nationale des chemins de fer français	Derailment, sabotage	?
28 Jan 1944	Ponte Paglia Bridge, Allerona, Italy		Train crossing bridge bombed by Allied aircraft	400
28 Jan 1944	Porte-les-Valence, France	Société nationale des chemins de fer français	Derailment, sabotage	?
c14 Feb 1944	Near Skopje, Yugoslavia		Collision	100
20 Feb 1944	Châteauneuf-du-Pape, France	Société nationale des chemins de fer français	Derailment, sabotage	29
22 Feb 1944	River Pinion Bridge, Greece		Train wrecked by partisans	400
22 Feb 1944	Montleranet, France	Société nationale des chemins de fer français	Derailment, sabotage	?
23 Feb 1944	Near Mouchard, France	Société nationale des chemins de fer français	Derailment, sabotage	?
25 Feb 1944	La Voulte, France	Société nationale des chemins de fer français	Derailment, sabotage	22
28 Feb 1944	Breifoss, between Geilo and Hol, Norway	Norges Statsbaner	Runaway train, collision, fire	?
1 Mar 1944	Between Dole and Roquefort, France	Société nationale des chemins de fer français	Derailment, sabotage	?
2 Mar 1944	Between Mouchard and Surans, France	Société nationale des chemins de fer français	Derailment, sabotage	?
5 Mar 1944	Near Lipa, Poland		Train derailed and machine-gunned by partisans	?
7 Mar 1944	'North of Roma,' Italy		Train bombed by Allied aircraft	40
11 Mar 1944	Between Bjelovar and Garesnica, Yugoslavia		Derailment, sabotage	30
18 Mar 1944	Between Anger and Ingrandes-sur-Vienne, France	Société nationale des chemins de fer français	Derailment, sabotage	50
20 Mar 1944	Between Suez and Port Teufik, Egypt		Boiler explosion? Derailment	21
25 Mar 1944	Near Tucuruvi, Brazil		Derailment	31
8 Apr 1944	Pontecagnano, Italy		Collision with shunting train	115
c10 Apr 1944	Near Messorrakhi,		Derailment, sabotage	?

	Greece			
21 Apr 1944	Coesfeld, Germany		Train attacked by Allied aircraft	10
5 May 1944	Berga-Kelba, Germany		Trains attacked by Allied aircraft	28
10 Jun 1944	Fours, France	Société nationale des chemins de fer français	Derailment, sabotage	?
10 Jun 1944	Between Lussiaux and Vitray, France	Société nationale des chemins de fer français	Derailment, sabotage	?
14 Jun 1944	Between Paris and Strasbourg, France	Société nationale des chemins de fer français	Derailment, sabotage	20
24 Jun 1944	Between La Ferte Gaucher and Meilleray, France	Société nationale des chemins de fer français	Derailment, sabotage	9
26 Jun 1944	Between Bessamorel and Rosières, France	Société nationale des chemins de fer français	Runaway train, derailment	12
6 Jul 1944	Near High Cliff, TN, USA	Louisville and Nashville	Derailment, excessive speed and track defects, fire	33
16 Jul 1944	Between Amiens and Tergnier, France	Société nationale des chemins de fer français	Derailment, sabotage	?
17 Jul 1944	Between Herson and Vervin, France	Société nationale des chemins de fer français	Derailment, sabotage	10
22 Jul 1944	Albert, France	Société nationale des chemins de fer français	Derailment, sabotage	25
24 Jul 1944	Corbie, France	Société nationale des chemins de fer français	Derailment, sabotage	70
28 Jul 1944	Uzsok, Hungary		Derailment, sabotage	?
4 Aug 1944	Stockton, GA, USA	Atlantic Coast Line	Derailment, broken rail	47
16 Aug 1944	Near Rainham, England	Southern	Derailment, underbridge damaged by flying bomb	8
1 Sep 1944	Between Skopje and Zeles, Yugoslavia		Derailment, sabotage	?
3 Sep 1944	Wakayama, Japan		Fire	71
5 Sep 1944	Elterstraße, Germany		Train attacked by Allied aircraft	14
6 Sep 1944	Near Ede, The Netherlands		Derailment, sabotage	15
9 Sep 1944	Odenhausen, Germany		Train attacked by Allied aircraft	16

10 Sep 1944	Between Gauselfingen and Burladingen, Germany		Train attacked by Allied aircraft	10
11 Sep 1944	Between Losheim and Ahrdorf, Germany		Train attacked by Allied aircraft	15
14 Sep 1944	Near Dewey, IN, USA	Chicago and Eastern Illinois	Head-on collision, single line in fog	27
15 Sep 1944	Hücheln, Germany		Train attacked by Allied aircraft	?
28 Sep 1944	Missouri Valley, IA, USA	Chicago and North Western	Sidelong collision	9
4 Oct 1944	Köppern, Germany		Train attacked by Allied aircraft	31
5 Oct 1944	Between Hjordkær and Rødekro, Denmark		Derailment, sabotage	5
6 Oct 1944	Allendorf, Germany		Train attacked by Allied aircraft	9
8 Oct 1944	Tobberup, Denmark	Danske Statsbaner	Bomb exploded on train	10
16 Oct 1944	Between Wels and Passau, Germany		Train attacked by Allied aircraft, explosion, fire	7
24 Oct 1944	Between Korbach and Briton Wald, Germany		Train attacked by Allied aircraft	18
26 Oct 1944	Friedhof, Germany		Train attacked by Allied aircraft	?
Oct 1944	Bachelt, Germany		Train attacked by Allied aircraft	10
2 Nov 1944	Craiova, Romania		Collision	60
7 Nov 1944	Aguadilla, Puerto Rico		Runaway train, derailment, boiler explosion	14
8 Nov 1944	Near Chanyunis, Palestine		Derailment, track washed away	7
8 Nov 1944	Staelkottonweg, Germany		Train attacked by Allied aircraft	70
8 Nov 1944	Krombach, Germany		Train attacked by Allied aircraft	11
8 Nov 1944	Near Colfax, CA, USA	Southern Pacific	Derailment, excessive speed	8
13 Nov 1944	Dersdorf, Germany		Train attacked by Allied aircraft	7
20 Nov 1944	Bad Zwischenahn, Germany		Rear collision	29
22 Nov 1944	Near Karisath, India		Derailment, train fell from bridge, sabotage	8
23 Nov 1944	Near Wiesau, Germany		Collision, fire	56
24 Nov 1944	Barwald Średni, Poland		Head-on collision, single line	130

27 Nov 1944	Herbern, Germany		Train attacked by Allied aircraft	7
30 Nov 1944	Lauda, Germany		Train attacked by Allied aircraft	15
Nov 1944	Trennfeld, Germany		Train attacked by Allied aircraft	?
3 Dec 1944	Groß Reken, Germany		Train attacked by Allied aircraft	39
4 Dec 1944	Bornheim, Germany		Train attacked by Allied aircraft	64
11 Dec 1944	Haebaru, Japan		Explosion	218
11 Dec 1944	Leoben, Austria, Germany		Trains attacked by Allied aircraft	12
14 Dec 1944	Marktoffingen, Germany		Train attacked by Allied aircraft	18
16 Dec 1944	Eisweiler, Germany		Train attacked by Allied aircraft	16
25 Dec 1944	Padang Panjang, Dutch East Indies		Derailment? Collision?	224
26 Dec 1944	Niederlahnstein, Germany		Train bombed by Allied aircraft	?
27 Dec 1944	Stará Kremnička, First Slovak Republic		Runaway train, buffer stop collision, fire, explosions	370
28 Dec 1944	Osnabrück, Germany		Train attacked by Allied aircraft	20
31 Dec 1944	Bagley, UT, USA	Southern Pacific	Rear collision in fog	48
4 Jan 1945	Between Kundl and Wögl, Germany		Train attacked by Allied aircraft	5
10 Jan 1945	Between Munoz and Guimba, The Philippines		Derailment, sabotage	?
10 Jan 1945	Ballymaccarrett Junction, Northern Ireland	Belfast and County Down	Rear collision in fog	22
11 Jan 1945	Rozières-sur-Mouzon, France	Société nationale des chemins de fer français	Collision	21
13 Jan 1945	Jørstad Bridge, Norway	Norges Statsbaner	Bridge destroyed by sabotage; train fell into river	100
15 Jan 1945	Herzebrock, Germany		Rear collision	28
17 Jan 1945	St Valéry-en-Caux, France	Société nationale des chemins de fer français	Runaway train, buffer stop collision	89
27 Jan 1945	Meschede, Germany		Train attacked by Allied aircraft	15
Jan 1945	Natrup-Hagen, Germany		Train attacked by Allied aircraft	?
1 Feb 1945	Cazadero, Mexico		Rear collision, fire	127
9 Feb 1945	Near Lyons, France	Société nationale des chemins de fer français	Collision with waggon	5
14 Feb 1945	Königsbronn,		Train attacked by	30

	Germany		Allied aircraft	
16 Feb 1945	Heggen, Germany		Train attacked by Allied aircraft	14
16 Feb 1945	Ahauser Stausee, Germany		Train attacked by Allied aircraft	?
17 Feb 1945	Between Mikawa-Makihara and Mikawa-Kawai, Japan		Derailment, train ran into rockfall	20
17 Feb 1945	Bad Camberg, Germany		Train attacked by Allied aircraft	22
19 Feb 1945	Between Horning and Goodstrup, Denmark		Derailment, sabotage	8
19 Feb 1945	Dischingen, Germany		Train attacked by Allied aircraft	7
21 Feb 1945	Kurieth, Germany		Train attacked by Allied aircraft	41
21 Feb 1945	Visselhövede, Germany		Train attacked by the RAF	6
21 Feb 1945	Wilmersdorf, Germany		Train attacked by the RAF	?
22 Feb 1945	Between Künzing and Pleinting, Germany		Train attacked by Allied aircraft	12
22 Feb 1945	Between Kirchlinteln and Salzwedel, Germany		Train attacked by Allied aircraft	20
23 Feb 1945	Hæstrup, Denmark	Danske Statsbaner	Derailment, sabotage, fire, explosions	16
24 Feb 1945	Tobberup, near Hobro, Denmark	Danske Statsbaner	Bomb exploded on train, fire	10
26 Feb 1945	Hohenbrugg an der Raab, Austria, Germany		Train attacked by British aircraft	16
27 Feb 1945	Jungnau, Germany	Hohenzollerischen Landesbahn	Train attacked by Allied aircraft	32
1 Mar 1945	Altenschwand, Germany		Train attacked by Allied aircraft	28
13 Mar 1945	Niedenstein, Germany		Train attacked by Allied aircraft	7
19 Mar 1945	Between Engelskirchen and Blumenau, Germany		Train attacked by Allied aircraft	31
21 Mar 1945	Winterberg, Germany		Train attacked by Allied aircraft	6
22 Mar 1945	Jungshahi, India		Rear collision	24
23 Mar 1945	Dossenheim, Germany	Oberrheinischen Eisenbahngesellsch aft	Train attacked by Allied aircraft	?
24 Mar 1945	Lauda, Germany		Train attacked by Allied aircraft	43

25 Mar 1945	Plettenberg-Eiringhausen, Germany		Train attacked by Allied aircraft	11
25 Mar 1945	Bempfingen, Germany		Train attacked by Allied aircraft	6
30 Mar 1945	Winterberg, Germany		Train attacked by Allied aircraft	?
31 Mar 1945	Verne, Germany		Rear collision	5
31 Mar 1945	Görsdorf, Germany		Train attacked by Allied aircraft, explosion	?
31 Mar 1945	Wilhelmsglück, Germany		Train attacked by Allied aircraft	26
31 Mar 1945	Mühlfeld, Germany		Train attacked by Allied aircraft	16
Mar 1945	St Peter, Germany		Trains attacked by Allied aircraft	?
1 Apr 1945	Oberkochen, Germany		Train attacked by Allied aircraft	8
2 Apr 1945	Biberach, Germany		Hospital train attacked by Allied aircraft	14
8 Apr 1945	Moorwinkelsdamm, Germany		Train attacked by Allied aircraft	13
10 Apr 1945	Aystetter, Germany		Train attacked by Allied aircraft	38
10 Apr 1945	Between Schwandorf and Irrenlohe, Germany		Train attacked by Allied aircraft	69
11 Apr 1945	Aschersleben, Germany		Train attacked by Allied aircraft	?
14 Apr 1945	Kühren, Germany		Train attacked by Allied aircraft	23
15 Apr 1945	Freiberg, Germany		Train attacked by Allied aircraft	7
16 Apr 1945	Between Eiger and Wiesau, Germany		Train attacked by Allied aircraft	14
16 Apr 1945	Zöblits-Pobershau,		Train attacked by Allied aircraft	8
18 Apr 1945	Senftenberg, Germany		Trains attacked by Soviet aircraft	18
19 Apr 1945	Between Wommen and Herleshausen		Train attacked by Allied aircraft	9
26 Apr 1945	Hurlach, Germany		Train attacked by Allied aircraft, fire	200
27 Apr 1945	Schwabhausen, Germany		Train attacked by Allied aircraft, fire	170
28 Apr 1945	Sallach, Germany		Train attacked by Allied aircraft	5
28 Apr 1945	Poing, Germany		Train attacked by Allied aircraft	?
21 May 1945	Near Manirampur, India	East Indian	Rear collision	14
13 Jun 1945	Between Baschi		Head-on collision,	70

	and Attigliano, Italy		single line, fire	
15 Jun 1945	Povrly, Czechoslovakia		Head-on collision during temporary single line working	46
15 Jun 1945	Near Milton, PA, USA	Pennsylvania	Derailment, 'hot-box,' broken journal, collision	16
24 Jun 1945	Near Ouarzirah, French Morocco		Runaway train, derailment, fire	160
25 Jun 1945	Near Andimeshk. Iran		Derailment	50
1 Jul 1945	Near Champigneulles, France	Société nationale des chemins de fer français	Collision	12
10 Jul 1945	Between Peiping and Hankow, China		Japanese troop train attacked by American aircraft	300
12 Jul 1945	Near Victoria de las Tunas, Cuba		Derailment	19
13 Jul 1945	Near Weingarten, Germany		Collision	6
16 Jul 1945	Near Aßling, Germany		Rear collision	102
28 Jul 1945	Near Moghalsarai, India	East Indian	Sidelong collision	17
28 Jul 1945	Saint-Fons, France	Société nationale des chemins de fer français	Rear collision, fire, explosion	7
8 Aug 1945	Near Nishitetsu Chikushi, Japan		Train machine-gunned by American aircraft	56
8 Aug 1945	Between Tsuko and Nishitetsu Chikushi, Japan		Train machine-gunned by American aircraft	8
9 Aug 1945	Michigan, ND, USA	Great Northern	Rear collision	34
13 Aug 1945	Near Goch, Germany		Head-on collision, single line	45
20 Aug 1945	'On the Omuta line,' Japan		Collision	40
27 Aug 1945	Tamagawa Bridge, Between Komily and Haijima, Japan		Head-on collision single line, part of train fell from bridge	105
31 Aug 1945	Between Diosek and Galanta, Czechoslovakia		Collision	31
4 Sep 1945	Kédange, France	Société nationale des chemins de fer français	Train diverted onto wrong line, collision, fire	40
6 Sep 1945	Sassago, Japan		Buffer stop collision	60
9 Sep 1945	Between Iturbe and Tres Cruces, Argentina	Ferrocarriles del Estrado	Derailment, excessive speed, part of train fell from bridge	34
19 Sep 1945	Köpenick,		Collision with	5

	Germany		shunting train	
21 Sep 1945	Near Valence, France	Société nationale des chemins de fer français	Sidelong collision? fire	30
24 Sep 1945	Between Port-de-Piles and Saint Maure, France	Société nationale des chemins de fer français	Rear collision	6
30 Sep 1945	Bourne End, England	London Midland and Scottish	Derailment, excessive speed	41
8 Oct 1945	Near Erzurum, Turkey		Collision	30
10 Oct 1945	Nauen, Germany		Rear collision, fire	16
12 Oct 1945	Between Berlin and Hamburg, Germany		Derailment	?
6 Nov 1945	Atraighat, India	Assam-Bengal	Head-on collision, single line	9
7 Nov 1945	Tronzano Vercellese, Italy		Train diverted onto wrong line; buffer stop collision	6
15 Nov 1945	Between Gérgal and Fuente Santa, Spain		Head-on collision, single line	10
18 Nov 1945	'On the Sinyu line,' Japan	Kobe-Dontetsu	Derailment	45
25 Nov 1945	Mala, India		Head-on collision?	10
26 Nov 1945	'On the Ibaragi line,' Japan	Ibaragi Kotsu	Derailment	10
29 Nov 1945	Near Burkau, Germany		Bomb exploded under train, derailment	6
4 Dec 1945	Near Neddemin, Germany		Head-on collision, single line	38
6 Dec 1945	'On the Kouya line,' Japan	Nankam	Derailment	27
16 Dec 1945	Berlin, Germany		Collision	8
20 Dec 1945	Near Matillas, Spain		Collision with light engine	7

ABOUT THE AUTHOR

The author has been interested in railways for many years, ever since he was invited up to a signal box at the age of 14 back in 1966. Not long afterwards he 'cabbed' the engine of a goods train that was shunted out of the way of a faster train and its driver showed him a copy of *Red for Danger*, the late L. T. C. Rolt's book about British railway accidents. This was the first railway book that he ever bought, several months' saved up pocket money going towards a paperback copy. Later he came across a copy of *A History of Railroad Accidents, Safety Precautions and Operating Practices* by Robert B. Shaw, dealing with American railroad accidents, and his interest in the subject grew.

When the local signal boxes were all done away with in a modernisation scheme in the mid-1980s he began a more detailed study of world railway accidents. There was no internet back then, of course, and many hours were spent in the local reference library slogging through bound volumes of *The Times*, writing out longhand in large notebooks details of every accident found. Later it was possible to view the paper on microfilm and print pages, which made things a little easier, and with the internet the amount of information available vastly increased.

Some years ago, the author contributed to *Railway Disasters of the World* by P. W. B. Semmens (which deals only with 20th-century accidents) and more recently the idea came of writing his own book or books on world accidents, starting at the beginning. This book is the fifth in the series, with previous volumes covering the years 1811-1880, 1881-1900, 1901-1915 and 1916-30. Further volumes will hopefully follow.

Made in the USA
Middletown, DE
05 November 2026

36589026R00457